RODD'S CHEMISTRY OF CARBON COMPOUNDS

ELSEVIER SCIENTIFIC PUBLISHING COMPANY
335 JAN VAN GALENSTRAAT
P.O.BOX 211, AMSTERDAM, THE NETHERLANDS

DISTRIBUTORS FOR THE U.S.A. AND CANADA:

ELSEVIER NORTH-HOLLAND INC.
52 VANDERBILT AVENUE, NEW YORK, N.Y. 10017

WITH 14 TABLES

ISBN: 0-444-40664-6 (series)
ISBN: 0-444-41363-4 (vol. IVE)

LIBRARY OF CONGRESS CARD CATALOG NUMBER 64-4605

PRINTED IN THE NETHERLANDS

RODD'S CHEMISTRY OF CARBON COMPOUNDS

ADVISORS

The late Professor Sir ROBERT ROBINSON, O.M., M.A. (Oxon.), D.SC. (Manc.), HON.D.SC. (Lond., Liv., Wales, Dunelm, Sheff., Belfast, Bris., Oxon., Nott., Strath., Delhi, Sydney, Zagreb), HON. SC.D. (Cantab.), HON. LL.D. (Manc., Edin., Birm., St. Andrews, Glas., Liv.), HON. D. PHARM. (Madrid and Paris), HON. F.R.S.E., F.R.S., *London*

Professor A. R. BATTERSBY, M.SC. (Manc.), PH.D. (St. Andrews), D.SC. (Bris.), M.A , SC.D. (Cantab.) F.R.S., *Cambridge*

Professor R. N. HASZELDINE, M.A., PH.D., SC.D. (Cantab.), PH.D., D.SC. (Birm.), F.R.I.C., F.R.S., *Manchester*

Professor R. D. HAWORTH, D.SC., PH.D. (Manc.), B.SC. (Oxon.), F.R.I.C., F.R.S., *Sheffield*

The late Professor Sir EDMUND HIRST, M.A., PH.D. (St. Andrews), D.SC. (Birm.), HON.LL.D. (St. Andrews, Aberdeen, Birm., Strath.), HON. D.SC. (Dublin), F.R.I.C., F.R.S., *Edinburgh*

Professor Lord TODD, M.A. (Cantab.), D.SC. (Glas.), D.PHIL. (Oxon.), DR.PHIL. NAT. (Frankfurt), HON. LL.D. (Glas., Edin., Melb., Calif.), HON. DR. RER. NAT. (Kiel), HON. D. MET. (Sheff.), HON. D.SC. (Oxon., Dunelm, Lond., Exe., Leic., Liv., Adel., Alig., Madrid, Stras., Wales, Strath.), F.R.I.C., F.R.S., *Cambridge*

RODD'S CHEMISTRY OF CARBON COMPOUNDS

VOLUME I

GENERAL INTRODUCTION

ALIPHATIC COMPOUNDS

*

VOLUME II

ALICYCLIC COMPOUNDS

*

VOLUME III

AROMATIC COMPOUNDS

*

VOLUME IV

HETEROCYCLIC COMPOUNDS

*

VOLUME V

MISCELLANEOUS

GENERAL INDEX

*

RODD'S CHEMISTRY OF CARBON COMPOUNDS

A modern comprehensive treatise

SECOND EDITION

Edited by
S. COFFEY
M.Sc. (London), D.Sc. (Leyden), F.R.I.C.
formerly of
I.C.I. Dyestuffs Division, Blackley, Manchester

VOLUME IV PART E

HETEROCYCLIC COMPOUNDS

Six-membered monoheterocyclic compounds containing oxygen, sulphur, selenium, tellurium, silicon, germanium, tin, lead or iodine as the hetero-atom

ELSEVIER SCIENTIFIC PUBLISHING COMPANY
AMSTERDAM OXFORD NEW YORK
1977

CONTRIBUTORS TO THIS VOLUME

R. LIVINGSTONE, B.SC., PH.D., C.CHEM., F.R.I.C.
Department of Chemical Sciences, The Polytechnic, Huddersfield HD1 3DH

The late Sir ROBERT ROBINSON, O.M., M.A., D.SC., F.R.S.

R. E. FAIRBAIRN, B.SC., PH.D., F.R.I.C.
formerly of Research Department, Dyestuffs Division, I.C.I. Ltd., Manchester 9 *(Index)*

PREFACE TO VOLUME IV E

The first four sub-volumes of this edition of *Rodd's* Chemistry of Carbon Compounds are concerned with the chemistry of those organic compounds in which a three-, four- or five-membered ring containing a single hetero-atom is present. The three chapters in the present volume, IV E, deal with six-membered ring compounds in which the ring contains a single hetero-atom from three Groups of the Periodic Table as follows: oxygen, sulphur, selenium or tellurium from Group VI; silicon, germanium, tin or lead from Group IV; iodine from Group VII. The first two chapters dealing with the systematic chemistry of the various types of parent compounds and their derivatives, as well as the very extensive range of important and extremely interesting, naturally occurring substances containing these same heterocyclic systems, are concerned with the following main classes of compounds: Chapter 20, (1) pyran and its derivatives; (2) hydropyrans; (3) benzo[*b*]pyrans (chromenes); (4) chromans; (5) isobenzopyrans; (6) isochromans; (7) xanthenes; (8) 6*H*-dibenzo[*bd*]pyrans; (9) naphthopyrans; (10) benzo- and dibenzo-xanthenes and their various derivatives: Chapter 21, (1) thiopyran derivatives; (2) benzo[*b*]thiopyrans (thiochromenes); (3) isothiochromene and its derivatives; (4) thioxanthenes; (5) bridged-ring sulphur compounds (thiabicycloalkanes, thiaphenalenes, naphthothioxanthenes); (6) selenopyrans; (7) tellurane and related compounds; (8) compounds containing a Group IV element, silicon, germanium, tin or lead; (9) compounds containing a heterocyclic iodine atom.

A truly prodigious amount of literature has been published on this subject since the first edition of *Rodd* appeared and the author of these two main chapters, Dr. R. LIVINGSTONE, is indeed to be congratulated on the way he has presented a truly fascinating subject of interest not only to organic chemists but also to readers interested in botany, biology, pharmacology and medicine.

The third chapter is a reproduction of Chapter IX of Volume IV B of the original edition on brazilin and haematoxylin, vegetable products related to catechin, contributed by (the late) Sir ROBERT ROBINSON. As chromanol derivatives, a description of these compounds could well have found a place in Chapter 20, but, as no significant new material on the chemistry of the two substances has been published since 1960, Chapter IX has been reproduced here as a tribute to Sir ROBERT's memory and his long connection with *Rodd*'s Chemistry of Carbon Compounds as Chairman of the Advisory Board from its inception over 25 years ago, particularly as it describes, in his own words, one of his many elegant contributions to the development of structural organic chemistry.

The editor wishes again to thank Mr. E. B. ROBINSON for his help in preparing the manuscripts for the press.

March 1977 S. COFFEY

CONTENTS

VOLUME IV E

PREFACE . . . VII
OFFICIAL PUBLICATIONS; SCIENTIFIC JOURNALS AND PERIODICALS . . . XV
LIST OF COMMON ABBREVIATIONS AND SYMBOLS USED . . . XVII

Chapter 20. Six-membered Ring Compounds with One Hetero Atom: Oxygen
by R. LIVINGSTONE

1. Pyran and its derivatives . . . 2
   a. Pyrans, pyranols and pyrylium salts . . . 2
      (*i*) 2- and 4-Pyrans (2*H*- and 4*H*-pyrans, α- and γ-pyrans), 2 – (*ii*) Pyranols, 3 – (*iii*) Pyrylium salts, 4 –
   b. 2*H*-Pyran-2-ones (α- or 2-pyrones) and 4*H*-pyran-4-ones (γ- or 4-pyrones) 9
      (*i*) 2-Pyrones, 10 – (*ii*) 4-Pyrones, 20 – (*iii*) Hydroxy-4-pyrones (hydroxy-γ-pyrones), 30 –
   c. 4-Pyronecarboxylic acids . . . 36
2. Hydropyrans . . . 39
   a. Dihydropyrans and derivatives . . . 39
      (*i*) 2,3-Dihydro-4-pyrans, 39 – (*ii*) 5,6-Dihydro-2-pyran, 44 – (*iii*) Naturally occurring furo- and pyrano-dihydro derivatives, 45 –
   b. Tetrahydropyrans and their derivatives . . . 48
      (*i*) Tetrahydropyrans, 48 – (*ii*) Hydrogenated 4-pyrones, 51 –
3. Benzo[*b*]pyrans (5,6-benzopyrans, chromenes) and their derivatives . . . 51
   a. Chromenes and their oxidation products . . . 52
      (*i*) Chromenes, alkyl- and aryl-chromenes and flavenes, 52 – (*ii*) Naturally occurring derivatives of 2,2-dimethyl-2*H*-chromene, 63 – (*iii*) Chromenols (5,6-benzopyranols, benzo[*b*]pyranols), 67 – (*iv*) Benzopyrylium or chromylium salts, 68 – (*v*) The anthocyanins and anthocyanidins, 81 – (*vi*) Coumarin (2*H*-benzo-[*b*]pyran-2-one, 5,6-benzo-2-pyrone, 5,6-benzo-α-pyrone) and its derivatives, 96 – (*vii*) Hydroxycoumarins, 111 – (*viii*) Dihydroxycoumarins, 120 – (*ix*) Trihydroxycoumarins and derivatives, 125 – (*x*) Furocoumarins, 126 – (*xi*) Pyranocoumarins, 134 – (*xii*) Chromones (4*H*-benzo[*b*]pyran-4-one, 5,6-benzo-4-pyrone, 5,6-benzo-γ-pyrone), 138 – (*xiii*) Chromone and alkyl-chromones, 149 – (*xiv*) Hydroxychromones, 151 – (*xv*) Furochromones, 155 – (*xvi*) Pyranochromones, 159 – (*xvii*) Flavones (2-phenylchromones, 2-phenyl-4*H*-benzo[*b*]-pyran-4-ones), 162 – (*xviii*) Flavonols (3-hydroxyflavones, 3-hydroxy-2-phenyl-4*H*-benzo[*b*]pyran-4-ones), 170 – (*xix*) Flavone and flavonol pigments, 174 – (*xx*) Biflavonyls, 201 – (*xxi*) Furoflavones, 204 – (*xxii*) Pyranoflavones, 206 – (*xxiii*) Isoflavones, 3-phenylchromones, 4-oxo-3-phenyl-4*H*-benzo[*b*]pyrans, 207 – (*xxiv*) Naturally occurring isoflavones, 214 – (*xxv*) Naturally occurring homoisoflavones, 218 –
4. Chroman, dihydrochromene, 3,4-dihydro-2*H*-benzo[*b*]pyran and derivatives . . . 222
   a. Chromans . . . 222
      (*i*) Syntheses, 222 – (*ii*) Chemical properties, 225 –

b. Naturally occurring chromans 226

(*i*) Tocopherols, 226 – (*ii*) Fuscin and cannabicyclol, 231 –

c. Phenylchromans (flavans and isoflavans), 3,4-dihydrophenyl-2*H*-benzo-[*b*]pyrans 231

(*i*) Synthesis, 231 –

d. Chromanols 234

e. Flavanols: the catechins and related condensed tannins 237

(*i*) Catechins, 238 – (*ii*) Condensed tannins, 244 – (*iii*) Flavan-4-ols and flavan-3,4-diols, 247 – (*iv*) Isoflavan-4-ols, 250 –

f. Chromanones, dihydrobenzo[*b*]pyranones 251

(*i*) Chromanones, 251 – (*ii*) Chromanochromanones, rotenone and related substances, 257 –

g. Flavanones, 2-phenylchroman-4-ones, 2,3-dihydro-2-phenyl-4*H*-benzo-[*b*]pyran-4-ones 268

(*i*) Synthesis, 268 – (*ii*) Properties, 271 –

h. Hydroxyflavanones 273

(*i*) Naturally occurring hydroxyflavanones, 277 –

i. Biflavanones 285

j. Isoflavanones, 3-phenylchroman-4-ones, 2,3-dihydro-3-phenyl-4*H*-benzo[*b*]pyran-4-ones 286

5. 1*H*-Benzo[*c*]pyran, isobenzopyran, 3,4-benzopyran and its derivatives . . 288

a. Isocoumarins, 1*H*-benzo[*c*]pyran-1-ones 290

(*i*) Synthesis, 290 – (*ii*) Properties and reactions, 293 –

6. Isochroman, 3,4-dihydro-1*H*-benzo[*c*]pyran and derivatives 297

a. Isochromans 297

(*i*) Synthesis, 297 – (*ii*) Reactions, 299 –

b. Isochromanones 300

7. Xanthene, 2,3-5,6-dibenzopyran, dibenzo[*a*,*e*]pyran and its derivatives . . . 304

a. Xanthene, alkyl- and aryl-xanthenes 304

(*i*) Xanthene, 304 – (*ii*) Alkyl- and aryl-xanthenes, 307 –

b. Xanthydrols and xanthene colouring matters 310

(*i*) Fluorones, 3*H*-xanthen-3-one and fluorimes, 311 – (*ii*) Xanthene dyes, 312 –

c. Xanthones 316

(*i*) Halogenoxanthones, 319 – (*ii*) Nitro- and amino-xanthones, 320 – (*iii*) Hydroxyxanthones, 321 – (*iv*) Naturally occurring hydroxyxanthones, 324 –

8. 6*H*-Dibenzo[*b*,*d*]pyran, 3,4-benzochromene and its derivatives 334

9. Naphthopyrans, benzochromenes and some of their derivatives 336

a. Some derivatives of 2*H*- and 4*H*-naphtho[1,2-*b*]pyran, 7,8-benzochrom-3-ene and 7,8-benzochrom-2-ene 337

b. Some derivatives of 1*H*- and 3*H*-naphtho[2,1-*b*]pyran, 5,6-benzochrom-2-ene and 5,6-benzochrom-3-ene 339

c. Some derivatives of 2*H*- and 4*H*-naphtho[2,3-*b*]pyran, 6,7-benzochrom-3-ene and 6,7-benzochrom-2-ene 341

d. Some derivatives of naphtho[1,8-*bc*]pyran and 1*H*,3*H*-naphtho[1,8-*cd*]-pyran, perinaphthopyran 343

10. Benzo- and dibenzo-xanthene derivatives 345

Chapter 21. Six-membered Ring Compounds with one Hetero Atom: Sulphur, Selenium, Tellurium, Silicon, Germanium, Tin, Lead or Iodine
by R. LIVINGSTONE

1. Thiopyran derivatives 347
a. Thiopyrans, thiopyrylium salts and thiopyrones 347
(*i*) Thiopyrans, 347 – (*ii*) Thiopyrylium salts and thiopyranyl 1,1-dioxide anions, 350 – (*iii*) Thiopyranones, 354 –
b. Di- and tetra-hydro-thiopyrans and -thiopyranones 359
(*i*) Dihydrothiopyrans, dihydrothiins, dihydrothiapyrans, 359 – (*ii*) Tetrahydrothiopyrans, thianes, tetrahydrothiapyrans, 362 –
2. Benzo[*b*]thiopyrans, benzo[*b*]thiins, thiochromenes, 5,6-benzothiapyrans and derivatives, and some naphthothiopyrans and derivatives 367
a. Benzo[*b*]thiopyrans and benzo[*b*]thiopyranones and related naphtho-derivatives 367
(*i*) Benzo[*b*]thiopyrans, benzo[*b*]thiins, thiochromenes, 367 – (*ii*) 2*H*-Benzothiopyran-2-ones, 2*H*-benzothiin-2-ones, thiocoumarins, 370 – (*iii*) 4*H*-Benzo-[*b*]thiopyran-4-ones, 4*H*-benzo[*b*]thiin-4-ones, thiochromones and 2-phenyl-4*H*-benzo[*b*]thiopyran-4-ones, thioflavones, 372 –
b. 2,3-Dihydro-4*H*-benzo[*b*]thiopyrans, 2,3-dihydro-4*H*-benzo[*b*]thiins, thiochromans and 2,3-dihydro-4*H*-benzo[*b*]thiopyran-4-ones, 2,3-dihydro-4*H*-benzo[*b*]thiin-4-ones, thiochroman-4-ones 376
(*i*) 2,3-Dihydro-4*H*-benzo[*b*]thiopyrans, 376 – (*ii*) 2,3-Dihydro-4*H*-benzo[*b*]-thiopyran-4-ones, thiochroman-4-ones, 379 –
3. 1*H*-Benzo[*c*]thiopyran, isothiochromene, 3,4-benzothiopyran and its derivatives 384
a. 1*H*-Benzo[*c*]thiopyran, isothiochromene and 1*H*-benzo[*c*]thiopyran-1-ones, isothiocoumarins 384
b. 3,4-Dihydro-1*H*-benzo[*c*]thiopyran, isothiochroman and its derivatives 385
4. Dibenzothiopyran, thioxanthene and derivatives 388
5. Bridged ring sulphur compounds and related compounds 396
a. Thiabicycloalkanes 396
(*i*) Thiabicyclo[3.2.1]octanes, 396 – (*ii*) Thiabicyclo[3.3.0]nonane, 397 – (*iii*) Thiabicyclotridecenes, 398 – (*iv*) Thia-adamantane, 398 –
b. Perinaphthothiopyrans, thiaphenalenes 399
c. Cerothiene derivatives 401
6. Selenopyrans and related compounds 403
a. Selenopyrans and derivatives 403
b. Selenopyrans containing fused rings 405
(*i*) Benzoselenopyrans and derivatives, 405 – (*ii*) Dibenzoselenopyran, selenoxanthene and derivatives, 409 – (*iii*) Naphthoselenopyrans, 411 –
7. Tellurane, tetrahydrotelluropyran derivatives and related compounds . . . 412
8. Compounds containing an element from Group 4 413
a. Silicon compounds 413
(*i*) Silacyclohexanes and their derivatives, 413 – (*ii*) Silacyclohexenes and silacyclohexadienes, 417 – (*iii*) Spirocyclosilanes, 419 – (*iv*) Polycyclic silanes with six-membered rings, 419 –

b. Germanium, tin and lead compounds 420
(*i*) Germanium compounds, 420 – (*ii*) Tin compounds, 422 – (*iii*) Lead compounds, 424 –
9. Compounds containing iodine 425

Chapter 22. Brazilin and Haematoxylin
by SIR ROBERT ROBINSON

1. The brazilin group 427
2. The haematoxylin group 443

Index 447

Titles of other parts of Volume IV

HETEROCYCLIC COMPOUNDS

Vol. IV A: Three-, four- and five-membered heterocyclic compounds with a single hetero-atom in the ring

Vol. IV B: Five-membered heterocyclic compounds with a single hetero-atom in the ring: alkaloids, dyes and pigments

Vol. IV C: Five-membered heterocyclic compounds with two hetero-atoms in the ring from Groups V and/or VI of the Periodic Table

Vol. IV D: Five-membered heterocyclic compounds with more than two hetero-atoms in the ring

Vol. IV F: Six-membered heterocyclic compounds with a single nitrogen atom in the ring: pyridine, polymethylenepyridine, quinoline, isoquinoline and their derivatives

Vol. IV G: Six-membered heterocyclic compounds with a single hetero-atom from Group V of the Periodic Table: polycyclic derivatives related to pyridine; monoheterocyclic compounds containing phosphorus, arsenic, antimony or bismuth; alkaloids of the pyridine, quinoline, isoquinoline, acridine and related series

OFFICIAL PUBLICATIONS

B.P.	British (United Kingdom) Patent
F.P.	French Patent
G.P.	German Patent
Ger. Offen.	German Patent Application, open for inspection
Sw.P.	Swiss Patent
U.S.P.	United States Patent
U.S.S.R.P.	Russian Patent
B.I.O.S.	British Intelligence Objectives Sub-Committee Reports, H.M. Stationery Office, London.
C.I.O.S.	Combined Intelligence Objectives Sub-Committee Reports
F.I.A.T.	Field Information Agency, Technical Reports of U.S. Group Control Council for Germany
B.S.	British Standards Specification
A.S.T.M.	American Society for Testing and Materials
A.P.I.	American Petroleum Institute Projects
C.I.	Colour Index Number of Dyestuffs and Pigments

SCIENTIFIC JOURNALS AND PERIODICALS

With few obvious and self-explanatory modifications the abbreviations used in references to journals and periodicals comprising the extensive literature on organic chemistry, are those used in the World List of Scientific Periodicals.

LIST OF COMMON ABBREVIATIONS AND SYMBOLS USED

A	acid
Å	Ångström units
Ac	acetyl
a	axial
as, *asymm.*	asymmetrical
at.	atmosphere
B	base
Bu	butyl
b.p.	boiling point
C, mC and μC	curie, millicurie and microcurie
c, *C*	concentration
c.d.	circular dichroism
conc.	concentrated
crit.	critical
D	Debye unit, 1×10^{-18} e.s.u.
D	dissociation energy
D	dextro-rotatory; dextro configuration
DL	optically inactive (externally compensated)
d	density
dec. or decomp.	with decomposition
deriv.	derivative
E	energy; extinction; electromeric effect
E1, E2	uni- and bi-molecular elimination mechanisms
E1cB	unimolecular elimination in conjugate base
e.s.r.	electron spin resonance
Et	ethyl
e	nuclear charge; equatorial
f	oscillator strength
f.p.	freezing point
G	free energy
g.l.c.	gas liquid chromatography
g	spectroscopic splitting factor, 2.0023
H	applied magnetic field; heat content
h	Planck's constant
Hz	hertz
I	spin quantum number; intensity; inductive effect
i.r.	infrared
J	coupling constant in n.m.r. spectra
K	dissociation constant
k	Boltzmann constant; velocity constant
kcal.	kilocalories
L	laevorotatory; laevo configuration
M	molecular weight; molar; mesomeric effect
Me	methyl

m	mass; mole; molecule; *meta-*
ml	millilitre
m.p.	melting point
Ms	mesyl (methanesulphonyl)
[M]	molecular rotation
N	Avogadro number; normal
n.m.r.	nuclear magnetic resonance
n	normal; refractive index; principal quantum number
o	*ortho-*
o.r.d.	optical rotatory dispersion
P	polarisation; probability; orbital state
Pr	propyl
Ph	phenyl
p	*para-*; orbital
p.m.r.	proton magnetic resonance
R	clockwise configuration
S	counterclockwise config.; entropy; net spin of incompleted electronic shells; orbital state
S_N1, S_N2	uni- and bi-molecular nucleophilic substitution mechanisms
S_Ni	internal nucleophilic substitution mechanisms
s	symmetrical; orbital
sec	secondary
soln.	solution
symm.	symmetrical
T	absolute temperature
Tosyl	*p*-toluenesulphonyl
Trityl	triphenylmethyl
t	time
temp.	temperature (in degrees centigrade)
tert	tertiary
U	potential energy
u.v.	ultraviolet
v	velocity
α	optical rotation (in water unless otherwise stated)
$[\alpha]$	specific optical rotation
α_A	atomic susceptibility
α_E	electronic susceptibility
ε	dielectric constant; extinction coefficient
μ	microns (10^{-4} cm); dipole moment; magnetic moment
μ_B	Bohr magneton
μg	microgram (10^{-6} g)
λ	wavelength
υ	frequency; wave number
χ, χ_d, χ_μ	magnetic, diamagnetic and paramagnetic susceptibilities
~	about

(+)	dextrorotatory
(−)	laevorotatory
⊖	negative charge
⊕	positive charge

LIST OF ABBREVIATED NAMES OF CHEMICAL FIRMS MENTIONED IN PATENT REFERENCES

A.G.F.A., Agfa A.G.	Aktiengesellschaft für Anilinfabrikation (Berlin)
B.A.S.F.	Badische Anilin- und Soda-Fabrik (Ludwigshafen)
Bayer	Farbenfabriken vorm. Friedrich Bayer und Co. (Leverkusen)
Cassella	Leopold Cassella und Co. (Frankfurt am Main)
C.F.M.	Compagnie française des Matières Colorantes (Paris)
CIBA	Gesellschaft für chemische Industrie (Basel)
Du Pont	E.I. Du Pont de Nemours and Co. (U.S.A.)
G.A.F.	General Anilin and Film Corporation (U.S.A.)
Geigy A.G.	J. R. Geigy S.A. (Basel)
Hoechst	Hoechst A.G. (see M.L.B.)
I.C.I.	Imperial Chemical Industries, Ltd. (London)
I.G.	(= Interessen Gemeinschaft Farbenindustrie) of the principal dyestuffs manufacturers in Germany
Kalle	Kalle und Co., A.G. (Biebrich am Rhein)
M.L.B.	Farbwerke vormals Meister, Lucius und Brüning (Hoechst)
Sandoz	Sandoz A.G. Chemische Fabrik (Basel)

Chapter 20

*Six-membered Ring Compounds with One Hetero Atom: Oxygen**

R. LIVINGSTONE

Of the heterocyclic compounds containing a six-membered ring with one hetero atom, pyridine is aromatic in character having a ring containing five methine groups and a nitrogen atom, but an "aromatic formulation for pyran and thiopyran is not possible. Both 2*H*- and 4*H*-pyran contain a methylene group and lack the conjugated double bond system characteristic of benzene and pyridine. In consequence, they are *unsaturated* closed-chain compounds lacking aromatic properties. The rings are numbered as shown

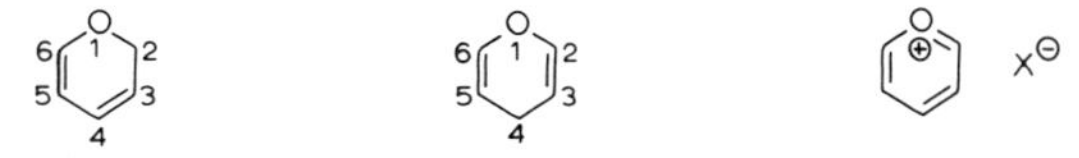

2*H*-Pyran   4*H*-Pyran   Pyrylium salt

The pyrylium salt possesses the pyrylium cation which is an aromatic system, but it is so reactive that it appears to be unstable. The pyrylium compounds are of special importance as in them oxygen exhibits high basicity and because pyrylium derivatives are found in natural pigments. Pyran and pyrylium derivatives are widely distributed in nature and are

* "Monocyclic Pyrans, Pyrones, Thiopyrans and Thiopyrones", by *J. Fried* in Heterocyclic Compounds, ed. *R. C. Elderfield*, Vol. 1, p. 343, "Some Aspects of Furan and Pyran Chemistry", by *D. G. Jones* and *A. W. C. Taylor*, Quart. Reviews, 1950, **4**, 195, "Konstitution und Vorkommen der organischen Pflanzenstoffe', by *W. Karrer*, Birkhauser Verlag, 1958, "Pyrylium Salts, Part I, Syntheses", *A. T. Balaban, W. Schroth* and *G. Fischer*, Adv. Heterocyclic Chemistry, Academic Press, New York and London, 1969, **10**, 241. "Naturally Occurring Oxygen Ring Compounds", *F. M. Dean*, Butterworths, London, 1963.

found in the carbohydrates, chromones (4*H*-benzopyran-4-ones), coumarins (2*H*-benzopyran-2-ones), flavones (2-aryl-4*H*-benzopyran-4-ones), anthocyanins, alkaloids and tocopherols.

1. Pyran and its derivatives

(a) Pyrans, pyranols and pyrylium salts

*(i) 2- and 4-Pyrans (*2H- *and* 4H-*pyrans, α- and γ-pyrans)**

Preparation. Although there is evidence for the existence of 4-pyran there is none for 2-pyran. An early attempt to prepare 4-pyran failed (*H. Normant*, Bull. Soc., chim. Fr., 1951, C113), but **4-pyran,** b.p. 80°, n_D^{20} 1.4559, is obtained by heating a mixture of glutaraldehyde and hydrochloric acid in methylene dichloride at 90°/40 mm (*J. Strating et al.*, Angew. Chem., 1962, **74,** 465). 4-Pyran has also been isolated following the pyrolysis of 2-acetoxy-3,4-dihydro-2-pyran (I) at 350° and on catalytic hydrogenation it gives tetrahydropyran (p. 48) and with 2,4-dinitrophenylhydrazine forms the 2,4-dinitrophenylhydrazone of glutaraldehyde. N.m.r. and i.r. spectral data are recorded (*S. Masamune* and *N. T. Gastellucci*, J. Amer. chem. Soc., 1962, **84,** 2452):

OAc 350°

(I)

2,6-Disubstituted pyrylium salts (p. 8) react with Grignard reagents and compounds containing an active hydrogen (nitromethane, ethyl cyanoacetate, acetylacetone) to give 2,4,6-trisubstituted 4-pyrans, which on treatment with 25% hydrochloric acid yield 1,5-diketones. The product from a 2,6-disubstituted pyrylium salt and a Grignard reagent with 70% perchloric acid gives a 2,4,6-trisubstituted pyrylium salt (*K. Dimroth* and *K. H. Wolf*, Angew. Chem., 1960, **72,** 777).

2- and 4-Pyrans are prepared in good yields from the corresponding pyrones and aliphatic and aromatic Grignard reagents (p. 17) (*R. Gompper* and *O. Christmann*, Ber., 1961, **94**, 1784).

Reactions. 4-Pyrans are dehydrogenated easily with an alkaline solution of cyanoferrate in the presence of 2,4,6-triphenylphenol. The dehydrogenation products form pyrylium salts with strong acids *(Dimroth* and *Wolf*, *loc. cit.)*. Some 4-pyran derivatives have been converted with 70% per-

* The *H* for indicated hydrogen and its locant may be omitted when no ambiguity results (I.U.P.A.C. Rule C-314).

chloric acid into naphthalene derivatives with the elimination of a ketone, *e.g.* 4-benzyl-2,4,6-triphenyl-4-pyran gave 1,3-diphenylnaphthalene (*Dimroth*, *H. Kroke* and *Wolf*, Ann., 1964, **678,** 202).

The cyclisation of 1,5-dioxopimelic acid with sulphuric acid affords 4-**pyran**-2,6-**dicarboxylic acid,** high m.p., *dimethyl ester*, m.p. 121°, *diethyl ester*, m.p. 37°, which could not be decarboxylated to 4-pyran (*E.-E. Blaise* and *H. Gault*, Bull. Soc. chim. Fr., 1907, [iv], **1,** 133). 2-**Methyl**-2-**pyran,** b.p. 106–111°, d_4^{20} 0.917, n_D 1.45421, is obtained by the action of formic acid on mannitol (*A. Windaus* and *A. Tomich*, Nachr. kgl. Ges. Göttingen, 1917, 462; *K. von Auwers*, Ann., 1920, **422,** 148). 2,4,6-**Triphenyl**-2-**pyran**, m.p. 225°, results when the pseudo-base of 2,4,6-triphenylpyrylium chloride is hydrogenated with a palladium catalyst (*W. Dilthey et al.*, J. pr. Chem., 1920, [ii], **101,** 177). δ-Diketones in the presence of acids or bases undergo cyclisation to cyclohexenones rather than 4-pyrans (*R. G. Fargher* and *W. H. Perkin Jr.*, J. chem. Soc., 1914, **105,** 1353), but diphenacyldiphenylmethane, $Ph_2C(CH_2 \cdot COPh)_2$, when boiled in xylene with phosphorus pentoxide yields 2,4,4,6-**tetraphenyl**-4-pyran, m.p. 157–158°, which with bromine gives the 3,5-*dibromo* compound, m.p. 215–216° (*A. P. de Carvalho*, Ann. Chim., 1935, [xi), **4,** 449).

Versicolin, an antibiotic from *Aspergillus versicolor*, is 4-hydroxy-2-(hydroxymethylene)-3-methyl-2*H*-pyran (*A. K. Dhar* and *S. K. Bose*, Tetrahedron Letters, 1969, 4871):

Versicolin

(ii) Pyranols

The pyranols are colourless pseudo-bases obtained by treating pyrylium salts (below) with alkali or sodium acetate, the pyrylium cation undergoing nucleophilic attack at the 2- or 4-positions. Thus 2,4,6-triphenylpyrylium chloride with alkali yields 2,4,6-triphenyl-2*H*-pyranol. The pyranols behave tautomerically either as ring compounds (II) or unsaturated open-chain δ-diketones (*e.g.*, III), which are detected by the formation of phenylhydrazones (*e.g.*, IV) (*O. Diels* and *K. Alder*, Ber., 1927, **60,** 716).

(II) (III) (IV)

Certain pyrylium salts, however, with alkali yield bluish-violet dimolecular compounds, *pyranhydrones*, which have a structure similar to that of quinhydrone (*W. Schneider* and *A. Ross*, *ibid.*, 1922, **55,** 2773). The gradual addition of alkali to a pyrylium salt containing a 4-hydroxyphenyl group in the 2-position causes a colour change, the red pyrylium salt V giving rise to a blue anhydrobase VI and eventually to the colourless pseudobase VII (*Dilthey et al., ibid.*, 1920, **53**, 252, 261). 4-(4-Hydroxyphenyl) pyrylium salts undergo attack at the 4-position. From these results *Dilthey* concluded that the positive charge is distributed round the ring-atoms and thus anticipated the modern resonance picture of the pyrylium ring:

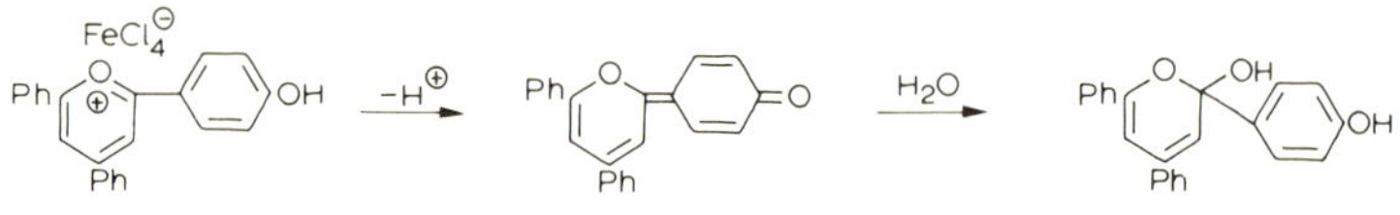

(V) (VI) (VII)

2-**Hydroxy**-2,4,6-**triphenyl**-2*H*-**pyran,** colourless needles, m.p. 119–120°, *picrate*, m.p. 226–227°, yields 2,4,6-triphenylpyridine with ammonia (*Dilthey*, J. pr. Chem., 1916, [ii], **94,** 65; *D.* and *T. Iwanov*, Ber., 1944, **77,** 180).

(iii) Pyrylium salts

Preparation. Pyrylium salts may be prepared from 4*H*-pyran-4-ones (γ-pyrones) and Grignard reagents (p. 17), but most frequently are obtained by the oxidative ring-closure of aromatic 1,5-diketones (*Dilthey*, J. pr. Chem., 1916, [ii], **94,** 53; 1917, [ii], **95,** 107). Thus the diketone VIII yields 2,4,6-triphenylpyrylium ferrichloride (IX) with acyl anhydrides as condensing agent and ferric chloride as oxidant. Ketones condense with themselves or with aromatic aldehydes to give the requisite 1,5-diketone, acetophenone (2 mols) and benzaldehyde (1 mol) or acetophenone and benzylideneacetophenone, for example, with acetic anhydride and ferric chloride affording the pyrylium salt IX:

(VIII) (IX)

Some pyrylium salts are prepared from 1,5-diketones by using chalcone as a hydride-abstracting agent and boron trifluoride etherate as a cyclisation medium (*J. A. VanAllan* and *G. A. Reynolds*, J. org. Chem., 1968, **33,** 1102). Enolisable β-diketones react with α-methylene-active ketones, in the presence of strong acids, to give poly-substituted mono- and poly-cyclic pyrylium salts (*W. Schroth* and *G. W. Fischer*, Ber., 1969, **102,** 1214). Aryl-substituted pyrylium salts may be formed in one step by the condensation of aromatic aldehydes with alkyl aryl ketones in the presence of 70% perchloric acid (*G. N. Dorofeenko* and *S. V. Krivun*, Zhur. obshcheĭ Khim., 1962, **32,** 2386; 1964, **34,** 105). The condensation of β-diketones with ketones in the presence of 70% perchloric acid also gives pyrylium salts (*Dorofeenko, Krivun* and *Zh. V. Shiyan*, *ibid.*, 1964, **34,** 167):

$$PhAc + Ac_2CH_2 \xrightarrow{HClO_4} \text{2-Ph-4,6-Me}_2\text{-pyrylium}\ ClO_4^{\ominus}$$

δ-Diketones on heating with triphenylmethyl perchlorate, $Me_3C^{\oplus}ClO_4^{\ominus}$ or perchloric acid in acetic acid, acetic anhydride, or acetic acid–acetic anhydride mixtures give pyrylium salts (*Y. Maroni-Barnaud*, Bull. Soc. chim. Fr., 1970, 1398).

Pyrylium salts may also be prepared from α,β-unsaturated ketones such as dypnone and a mixed anhydride such as $CH_3 \cdot CO \cdot O \cdot ClO_3$, obtained from perchloric acid and acetic anhydride (*Schneider et al.*, Ber., 1921, **54,** 2285; 1922, **55,** 2775; *Diels* and *Alder*, *loc. cit.*). Dypnone (X) and the mixed anhydride give 2-methyl-4,6-diphenylpyrylium perchlorate (XI):

$$PhCO{-}CH{=}C(Me)Ph\ (X) + MeC(O){-}OClO_3 \longrightarrow \text{2-Me-4,6-Ph}_2\text{-pyrylium}\ ClO_4^{\ominus}\ (XI)$$

The diacylation of olefins, for example 2-methylpropenylbenzene, yields pyrylium salts (*A. T. Balaban*, Tetrahedron Letters, 1963, 91), and depending on the catalyst employed, Friedel Crafts diacetylation of 2-methylbut-2-ene yields 2,3,4,6-tetramethyl- and/or 4-ethyl-2,6-dimethyl-pyrylium salts. The mechanism and implication of this phenomenon has been discussed (*Balaban* and *C. D. Nenitzescu*, J. chem. Soc., 1961, 3553, 3561, 3564). β-Chlorovinyl ketones react with α-methylene-active ketones, in the presence of acid catalyst, to afford pyrylium salts with an unsubstituted $C_{(4)}$ position (*Schroth, Fischer* and *J. Rottmann*, Ber., 1969, **102,** 1202). Pyrylium salts may be obtained by the cyclodeamination of oxovinylenamines of the type $R_2NCR^1{:}CR^2 \cdot CH{:}CO \cdot C_6H_4R^3$[4] (*Schroth* and *Fischer*, *ibid.*, 1969, **102,** 575). Phenyl(3-phenyl-1-propenyl)acetylene (XII) prepared by condensing 1-formyl-2-phenylacetylene with acetophenone, on treatment with perchloric acid gives 2,6-diphenylpyrylium perchlorate (*H. Stetter* and *A. Reischl*, *ibid.*, 1960, **93,** 1253):

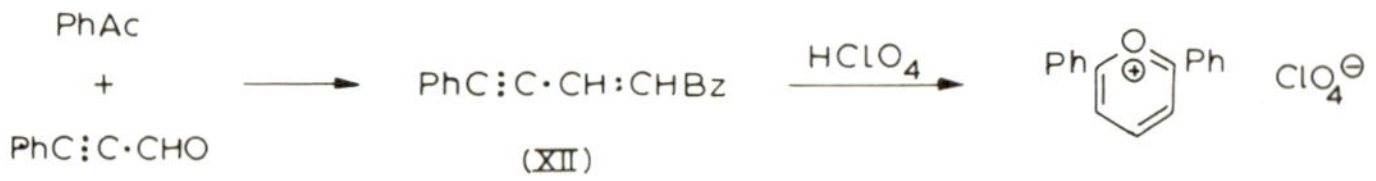

PhAc + PhC⋮C·CHO ⟶ PhC⋮C·CH:CHBz (XII) $\xrightarrow{HClO_4}$ 2,6-diphenylpyrylium $ClO_4^{\ominus}$

Pyrylium salts result when acetophenone reacts with sulphuric acid and potassium pyrosulphate and with boron trifluoride (*T. L. Davis* and *G. B. Armstrong*, J. Amer. chem. Soc., 1935, **57**, 1583; *W. C. Dovey* and *R. Robinson*, J. chem. Soc., 1935, 1389).

A pyrone and an acid may give several salts (*J. Kendall*, J. Amer. chem. Soc., 1914, **36**, 1222). 2,6-Dimethyl-4*H*-pyran-4-one (2,6-dimethyl-4-pyrone*), for example, with organic acids (Hx) forms salts $C_7H_8O_2 \cdot Hx$, $2C_7H_8O_2 \cdot 3Hx$ and $C_7H_8O_2 \cdot 2Hx$. We shall consider here those pyrylium salts with one molecule of acid to one of pyrone. *J. N. Collie* and *T. Tickle* (J. chem. Soc., 1899, **75**, 710) in their classic investigation on the formation of pyrylium salts by the action of acids on 2,6-dimethylpyran-4-one formulated these salts as oxonium compounds with a tetravalent oxygen atom; they are, in fact, pyrylium or oxonium salts (XIII). Methyl iodide and 2,6-dimethylpyran-4-one give a pyrylium salt XIV, the perchlorate of which under very mild conditions with aqueous ammonium carbonate gives 2,6-dimethyl-4-methoxypyridine (XV) (*A. von Baeyer*, Ber., 1910, **43**, 2337; Ann., 1911, **384**, 208; *Dilthey*, J. pr. Chem., 1921, [ii], **101**, 177):

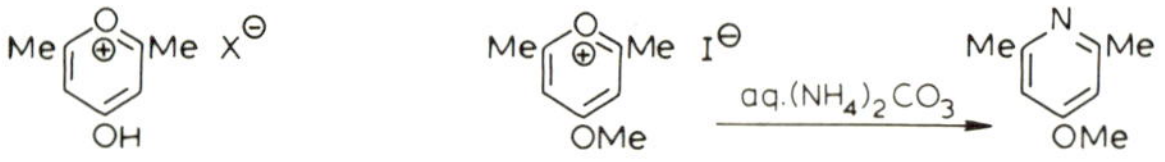

(XIII) 2,6-Me₂-4-OH-pyrylium $X^{\ominus}$; (XIV) 2,6-Me₂-4-OMe-pyrylium $I^{\ominus}$ $\xrightarrow{aq.(NH_4)_2CO_3}$ (XV) 2,6-Me₂-4-OMe-pyridine

In the pyrylium salt, therefore, the methyl group is linked to the oxygen of the carbonyl group and the pyrylium salts as suggested by Baeyer must have benzenoid and not quinonoid formulae (*e.g.* XIII and XIV). The oxonium form will be in resonance with carbonium ionic forms XVI and XVII and this accounts for the nucleophilic attack of pyrylium salts at $C_{(2)}$ and $C_{(4)}$:

(XVI) (XVII)

* See footnote p. 2.

Studies have been made on the charge density distribution of the pyrylium ion (*O. Martensson* and *C. H. Warren*, Acta Chem. Scand., 1970, **24,** 2745).

The preparation of unsubstituted pyrylium salts (XXI) has been achieved by the alkaline fission of (1-pyridinio)sulphate (XVIII) to give the sodium salt of glutacondialdehyde (XIX) (*F. Klages* and *H. Träger*, Ber., 1953, **86,** 1327). At $-20°$ this is converted by perchloric acid to the red oxonium salt of XX, which when kept for some time at 0° undergoes ring-closure to the colourless **pyrylium perchlorate** (XXI), decomposing explosively at about 275°. It is converted to pyridine by warm moist ammonium acetate.

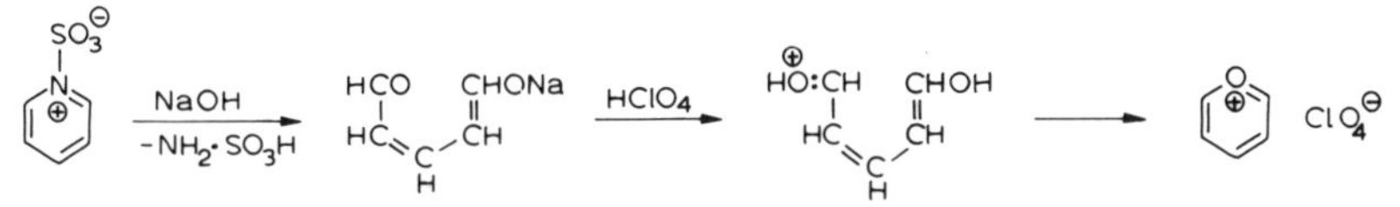

(XVIII) —NaOH, $-NH_2 \cdot SO_3H$→ (XIX) —$HClO_4$→ (XX) → (XXI)

The diketone XXIV, reacting in the mono-enol form, similarly yields 2,4,6-trimethylpyrylium perchlorate (XXIII) (*Baeyer* and *J. Piccard*, Ann., 1915, **407,** 332):

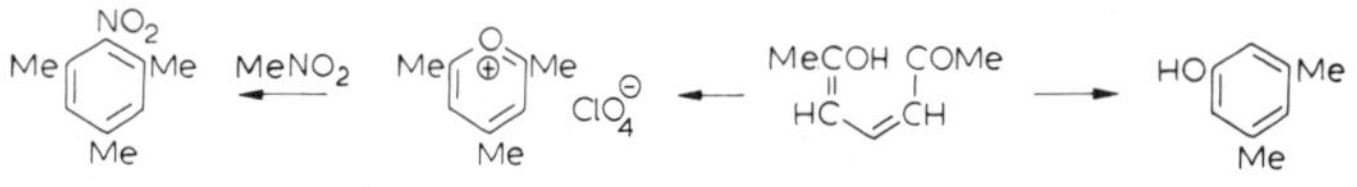

(XXII) ←$MeNO_2$— (XXIII) ← (XXIV) → (XXV)

Reactions. The sodium salt with warm alkali rapidly forms *m*-5-xylenol (XXV), and since the unsaturated ketone can be obtained from 2,6-dimethylpyran-4-one, this reaction furnishes an example of the conversion of a pyrone into a benzene derivative. *Nitromesitylene* (XXII), m.p. 41–42°, is obtained in 80–90% yield when the pyrylium perchlorate XXIII reacts with alkali and nitromethane (*K. Dimroth* and *G. Bräuniger*, Ber., 1957, **90,** 1634, 1668). The reaction between 2,4,6-triarylpyrylium salts and phenylnitromethane in the presence of bases yields mixtures of tetra-arylnitrobenzenes and tetra-arylphenols. This reaction is the first instance of an allyl migration of a nitro group (*Dimroth* and *H. Wache*, *ibid.*, 1966, **99,** 399).

2,4,6-Trialkylpyrylium salts are converted with ring contraction to alkyl 3,5-dialkyl-2-furyl ketones by oxidation with aqueous hydrogen peroxide under mild conditions (*Balaban* and *Nenitzescu*, *ibid.*, 1960, **93,** 599):

2,4,6-Triphenylpyrylium oxide (XXVII) may be obtained by the reaction between oxygen and the thiabenzene XXVI, prepared by treating 2,4,6-triphenylthiopyrylium perchlorate with phenyllithium (*G. Suld* and *C. C. Price*, J. Amer. chem. Soc., 1961, **83**, 1770; 1962, **84**, 2094):

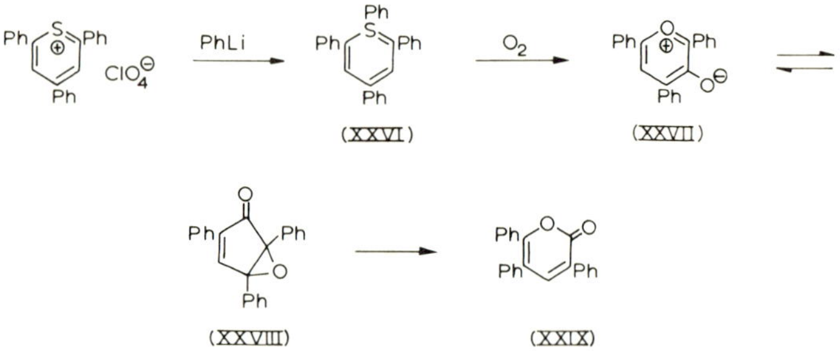

(XXVI) (XXVII)

(XXVIII) (XXIX)

Irradiation of the deep red acetonitrile solution of the oxide XXVII with light resulted in bleaching and the suggestion of the existence of a photo-equilibrium between XXVII and a tautomer tentatively assumed to be the epoxyketone XXVIII. The addition of methanol to a concentrated freshly bleached benzene solution of XXVII afforded 3,5,6-triphenyl-2*H*-pyran-2-one (XXIX) (*E. F. Ullman*, *ibid.*, 1963, **85,** 3529).

2,6-Disubstituted pyrylium salts may be converted to 2,4,6-trisubstituted 4-pyrans (p. 2). Reference has already been made to nucleophilic attack of the pyrylium cation at $C_{(2)}$ and $C_{(4)}$. Another instance is the marked reactivity of the methoxyl group in the 2,6-dimethyl-4-methoxypyrylium cation towards bases (*R. M. Anker* and *A. H. Cook*, J. chem. Soc., 1946, 117; *L. King* and *F. J. Ozog*, J. org. Chem., 1955, **20,** 448).

Unsubstituted and alkylpyrylium salts are colourless, whereas the arylpyrylium salts are coloured and fluorescent. 2,4,6-Triphenylpyrylium perchlorate and fuming nitric acid gives a tris(nitrophenyl)pyrylium perchlorate (XXX) in which the 2- and 6-phenyl groups are nitrated in the *meta*-position and the 4-phenyl group in the *para*-position (*C. G.* and *R. J. W. Le Fèvre*, J. chem. Soc., 1932, 2894).

Data on some pyrylium salts are given in Table 1.

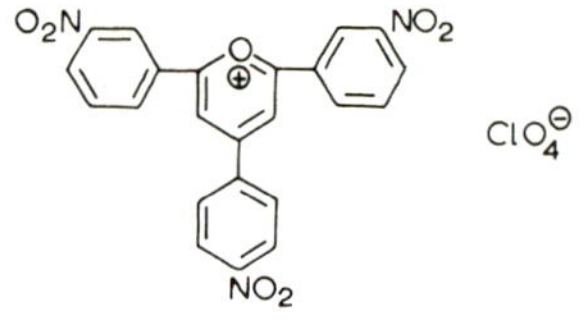

(XXX)

TABLE 1

SOME PYRYLIUM SALTS

Substituent	*Salt anion*	*M.p. (°C)*	*Yield (%)*	*Ref.*
2,4,6-Trimethyl	ClO_4	245	12	1
2,4-Dimethyl-6-phenyl	ClO_4	193–194	21	1
2,4-Dimethyl-6-(1-naphthyl)	ClO_4	262	25	1
2,4-Dimethyl-6-acenaphthenyl	ClO_4	217	21	1
2,4,6-Tri(4-methoxyphenyl)	BF_4	346–347	88	2
4-(4-Methoxyphenyl)-2,6-diphenyl	BF_4	238–240	93	2
2,6-Diphenyl	BF_4	206–207	72	2
	ClO_4	226	—	2
3-Benzyl-2,4-6-triphenyl	ClO_4	241	—	2
3-Ethoxycarbonyl-2,4,6-triphenyl	ClO_4	235	—	2
3-Ethoxycarbonyl-2-methyl-4,6-diphenyl	ClO_4	178	—	2
2-Methyl-4,6-diphenyl	ClO_4	205	—	2
3-Methyl-2,4,6-triphenyl	ClO_4	236	62	2
3-Ethyl-2,4,6-triphenyl	ClO_4	180	60	2
2,3,4,6-Tetraphenyl	ClO_4	261	—	2
2,3,6-Triphenyl-4-methyl	ClO_4	256	5	3
2,4,6-Triphenyl	ClO_4	138	89	4
4-(3-Nitrophenyl)-2,6-diphenyl	ClO_4	271	—	4
4-(4-Nitrophenyl)-2,6-diphenyl	ClO_4	302	27	4
4-(4-Dimethylaminophenyl)-2,6-diphenyl	ClO_4	360 (decomp.)	33	4
2,6-Di(4-methoxyphenyl)-4-(3-nitrophenyl)	ClO_4	195–196	26	4
2,6-Di(4-methoxyphenyl)-4-(4-dimethylaminophenyl)	ClO_4	330	30	4
2,6-Di(1-naphthyl)-4-(4-nitrophenyl)	ClO_4	330 (decomp.)	—	4

References

1 *G. N. Dorofeenko, S. V. Krivun* and *Zh. V. Shiyan*, Zhur. obshcheĭ Khim., 1964, **34**, 167.
2 *J. A. VanAllan* and *G. A. Reynolds*, J. org. Chem., 1968, **33**, 1102.
3 *A. T. Balaban*, Tetrahedron Letters, 1963, 91.
4 *Dorofeenko* and *Krivun*, Zhur. obshcheĭ Khim., 1964, **34**, 105.

(b) 2H-*Pyran-2-ones (α- or 2-pyrones) and* 4H-*pyran-4-ones (γ- or 4-pyrones)**

In the 2- and 4-pyrones the methylene group of the pyran is replaced by a carbonyl group.

* For a review see *L. F. Cavalieri*, Chem. Reviews, 1947, **41**, 525.

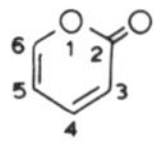

2H-Pyran-2-one 4H-Pyran-4-one

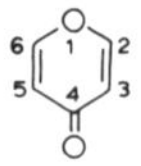

The 4-pyrones are particularly important since their study has thrown much light on the basic properties of oxygen (p. 6). The two series, though chemically similar in many respects, differ fundamentally in others, particularly in their behaviour towards acids. The two classes of compounds will therefore be considered separately.

(i) 2-Pyrones

The earliest researches into the preparation and chemistry of 2-pyrones were conducted by *H. von Pechmann* (Ann., 1891, **264**, 261), *A. Hantzsch* (*ibid.*, 1884, **222**, 1) and *R. Anschütz* (*ibid.*, 1890, **259**, 148). The 2-pyrones attracted more attention after certain derivatives were found in nature (*L. J. Haynes*, Quart. Reviews, 1948, **2**, 46). Many steroids showing cardiac activity contain a 2-pyrone side-chain, as, for example, scillaridin A (XXXI, R = a steroid nucleus).

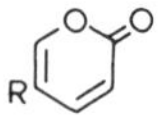

(XXXI)

Other naturally occurring 2-pyrones contain alkyl or aryl substituents or a 4-methoxy substituent.

The constitution of 2-pyrone and its derivatives follows from their synthesis and chemical properties. The coumarins are 2*H*-benzo[*b*]pyran-2-ones (benzo-α-pyrones or benzo-2-pyrones) whose derivatives include the haemorrhagic agent 3,3′-methylene-bis-4-hydroxycoumarin (p. 116).

Methods of preparation (1) Condensation-cyclisation. von Pechmann (loc. cit.) in his studies on the α-hydroxycarboxylic acids found that malic acid reacts with fuming sulphuric acid to form coumalic acid (2- or α-pyrone-5-carboxylic acid) in about 60% yield. Formylacetic acid is probably an intermediate in the reaction, the course of which can be represented as follows, the formylacetic acid reacting in the enolic form:

HOCH·CO$_2$H / CH$_2$·CO$_2$H $\xrightarrow{H_2SO_4}$ CHO / CH$_2$CO$_2$H ⇌ CHOH ‖ CHCO$_2$H

HO$_2$C·CH=CH–OH + HO–CO–CH=CHOH $\xrightarrow{-2\,H_2O}$ Coumalic acid

The formation of 2-pyrone derivatives by condensation reactions with the help of alkaline reagents is illustrated by the formation of diethyl 6-methyl-2-pyrone-3,5-dicarboxylate (diethyl 6-methylcoumalin-3,5-dicarboxylate) from acetoacetic ester and an ester-nitrile in the presence of sodium ethoxide:

MeCO·CH$_2$·CO$_2$Et + EtOCH=C(CN)·CO$_2$Et $\xrightarrow{NaOEt}$ MeCO(EtO$_2$C)CH–CH=C(CN)·CO$_2$Et ⟶ 2-imino pyran $\xrightarrow{H_2O}$ diethyl 6-methyl-2-pyrone-3,5-dicarboxylate

A somewhat similar and very useful synthesis is found in the condensation of substituted ethyl crotonates with diethyl oxalate in the presence of sodium ethoxide to give γ-oxalylcrotonic esters, ring-closure of which with hydrochloric acid yields 6-carboxy-2-pyrones (*J. Fried* and *R. C. Elderfield*, J. org. Chem., 1941, **6,** 566). Pyrolytic decarboxylation of 5-methyl-2-pyrone-6-carboxylic acid gives 3-methyl-2-pyrone, probably by a 1,5-hydrogen shift following decarboxylation (*W. H. Pirkle, H. Seto* and *W. V. Turner*, J. Amer. chem. Soc., 1970, **92,** 6984):

5-methyl-2-pyrone-6-carboxylic acid ⟶ 3-methyl-2-pyrone

Malonyl chloride condenses with benzoylacetone to yield 5-acetyl-4-hydroxy-6-phenyl-2-pyrone, which tautomerises above its melting point into the 2-hydroxy-4-pyrone (*M. A. Butt* and *J. A. Elvidge*, J. chem. Soc., 1963, 4483):

MeCO·CH:CPh(OH) + CH$_2$(COCl)$_2$ ⟶ 5-acetyl-4-hydroxy-6-phenyl-2-pyrone $\xrightarrow{\text{above } 168°}$ 5-acetyl-2-hydroxy-6-phenyl-4-pyrone

2-Pyrones may be obtained by the thermal condensation of various malonic esters with ketones (*C. Goetschel* and *C. Mentzer*, Bull. Soc. chim. Fr., 1962, 365). 6-Substituted 2-pyrones may be synthesised from 2-chlorovinyl ketones and ethoxymagnesiomalonic ester (*N. K. Kochetkov* and *L. I. Kudryashov*, Zhur. obshcheĭ Khim., 1958, **28,** 1511).

Aryl-2-pyrones can be obtained by condensing acetylenic esters with β-diketones or β-ketonic esters (*S. Ruhemann*, J. chem. Soc., 1910, **97,** 457; *G. Soliman* and *I. E.-S. El-Kholy*, *ibid.*, 1955, 2911) or acetylenic ketones and malonic ester (*E. P. Kohler*, J. Amer. chem. Soc., 1922, **44**, 379) in the presence of sodium ethoxide. Thus ethyl propionate and deoxybenzoin yield 4,5,6-triphenyl-2-pyrone:

(2) Cyclisation of glutaconic acid derivatives. Substituted glutaconic acids undergo cyclisation by loss of water to give substituted 2-pyrones. For example, when α-methylglutaconic acid is heated with acetyl chloride in a sealed tube at 100°, a mixture of products is obtained, namely, 6-hydroxy-3-methyl-2-pyrone and the corresponding 6-chloro compound (*F. B. Thole* and *J. F. Thorpe*, J. chem. Soc., 1911, **99,** 2208; see also *Thorpe* and *A. S. Wood*, *ibid.*, 1913, **103,** 1569). Dimethyl 2,4-diacetylglutaconate, with sodium, lithium or calcium methoxide in methanol–benzene is cyclised to the methyl 3-acetyl-6-methylpyrone-5-carboxylate (XXXIII) *via* the diacetylglutaconate anion XXXII. When the base used is magnesium methoxide other products are formed and the yield of the pyrone depends on the amount of magnesium methoxide employed (*L. Crombie* and *A. W. G. James*, Chem. Comm., 1966, 357):

(XXXII) (XXXIII)

(3) Other methods. (i) 4,6-Disubstituted and 4,5,6-trisubstituted 2-pyrones may be prepared by a Wittig reaction by heating the phosphorane $Ph_3P{:}CH\cdot CO_2Et$ with a β-ketone or 2-benzoylcyclohexanone (*A. K. Soerensen* and *N. A. Klitgaard*, Acta Chem. Scand., 1970, **24**, 343):

$$CH_2(COR)_2 + Ph_3P{=}CH\cdot CO_2Et \longrightarrow$$

(ii) Certain pyrazolines when heated yield 2-pyrones (*E. Buchner* and *H. Schröder*, Ber., 1902, **35,** 782, 790; *Kohler* and *L. Steele*, J. Amer. chem. Soc., 1919, **41,** 1093); thus pyrazoline (XXXIV) yields 4,6-diphenyl-2-pyrone:

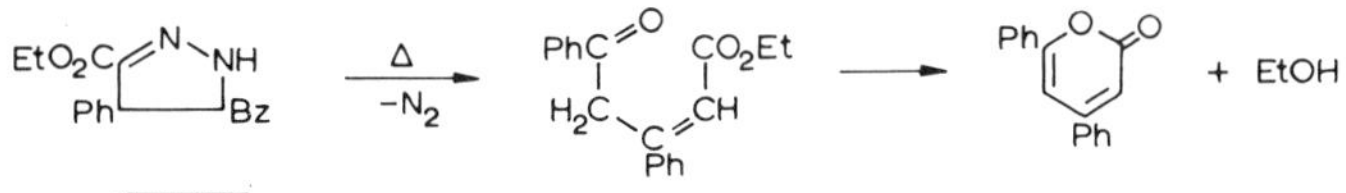

(XXXIV)

(iii) Some cyclopentadienones undergo ring-enlargement when heated with water, ethanolic sodium hydroxide, or acid (see *e.g.*, *W. Wislicenus* and *K. Schöllkopf*, J. pr. Chem., 1917, [ii], **95,** 295); in this way XXXV yields 4-ethoxy-2-pyrone-6-carboxylic acid:

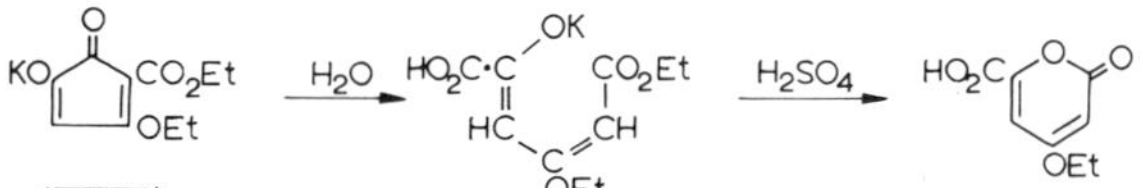

(XXXV)

(iv) β-Diketones are converted to disodio salts by treatment with excess sodamide in liquid ammonia. Suspensions of the disodio salts in ether react with carbon dioxide to yield the corresponding carboxylic acids, which are cyclised by liquid hydrogen fluoride to give the 4-hydroxy-2-pyrones in excellent yields (*T. M.* and *C. M. Harris*, J. org. Chem., 1966, **31,** 1032).

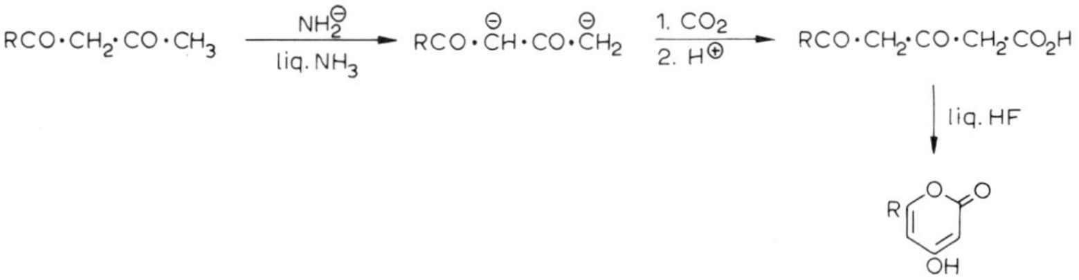

$$RCO \cdot CH_2 \cdot CO \cdot CH_3 \xrightarrow[\text{liq. } NH_3]{NH_2^{\ominus}} RCO \cdot \overset{\ominus}{C}H \cdot CO \cdot \overset{\ominus}{C}H_2 \xrightarrow[2.\ H^{\oplus}]{1.\ CO_2} RCO \cdot CH_2 \cdot CO \cdot CH_2 \cdot CO_2H \xrightarrow{\text{liq. HF}}$$

(v) Diethyl 3-phenylthiomalonate on heating with enolizable aromatic ketones at 200–240° with removal of the alcohol formed gives 3-phenylthio-4-hydroxy-2-pyrones, which on desulphuration by Raney nickel yields 4-hydroxy-2-pyrones unsubstituted in the 3-position (*A. Lefeuvre* and *Mentzer*, Compt. rend., 1963, **256,** 3316).

Reactions. (1) Diels–Alder addition. The presence of the conjugated double bonds in the 2-pyrones is clearly demonstrated by the formation of Diels–Alder adducts with maleic anhydride (*O. Diels* and *K. Alder*, Ann., 1931, 490, 257). 2-Pyrone forms the adduct XXXVI, which loses carbon dioxide to yield 1,2-dihydrophthalic anhydride. Addition of a second molecule of maleic anhydride gives the bis-adduct XXXVII as final product:

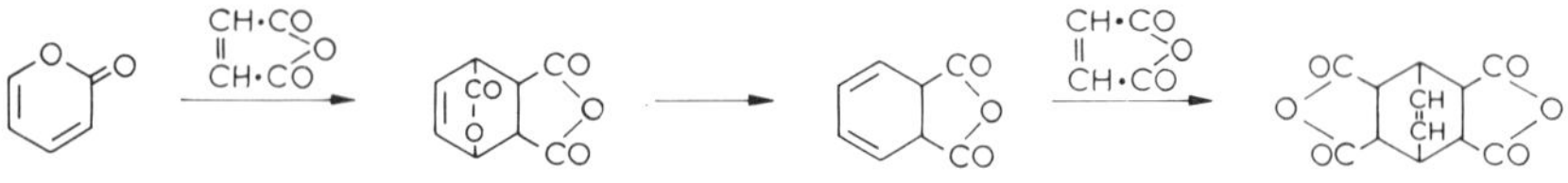

(XXXVI) (XXXVII)

(2) Hydrolysis. 2-Pyrones, being cyclic lactones, readily undergo ring-cleavage on hydrolysis, particularly in the presence of alkali. Coumalic acid in this way forms formylacetic acid:

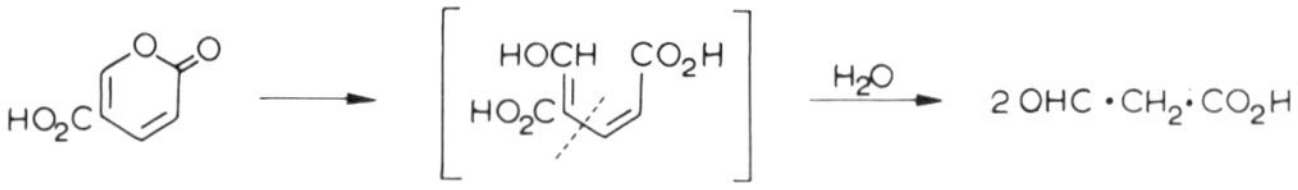

$$\xrightarrow{H_2O} 2\,OHC\cdot CH_2\cdot CO_2H$$

Ring-fission, however, may be unaccompanied by degradation of the carbon chain. 4,5,6-Triphenyl-2-pyrone with methanolic potassium hydroxide yields γ-benzoyl-β,γ-diphenyl vinylacetic acid and β-desylcinnamic acid (*Soliman* and *El-Kholy*, *loc. cit.*):

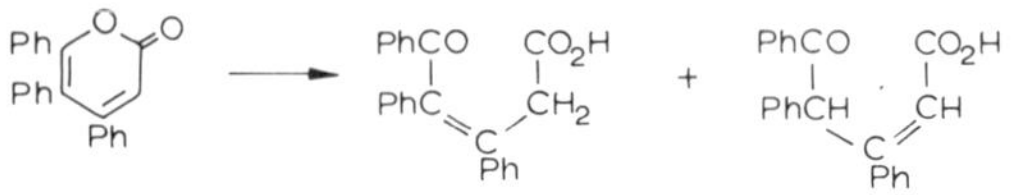

3-Hydroxy- or 3-halogeno-2-pyrone on boiling with aqueous potassium hydroxide solution yields, on acidification, furan-2-carboxylic acid; 2-pyrones having substituents in the 4-, 5-, or 6-positions give 3-, 4-, or 5-substituted furan-2-carboxylic acids (*W. Bray* and *R. Mayer*, Z. Chem., 1963, **3**, 150):

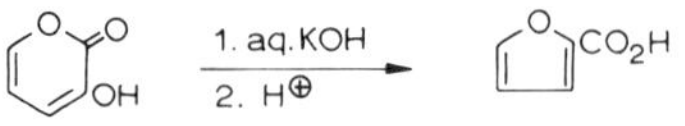

1. aq. KOH
2. H⊕

(3) The action of acids and bases. Treatment of 4-hydroxy-6-phenacyl-2-pyrone and related compounds with various acidic and basic reagents brings about cleavage of the lactone ring to form unstable intermediates which recyclise in several ways. 4-Hydroxy-6-phenacyl-2-pyrone gives 4-pyrones with strong acids, resorcinols with nucleophilic bases, and benzoylphloroglucinol with non-nucleophilic bases (*T. M. Harris* and *M. P. Wachter*, Tetrahedron, 1970, **26**, 5255).

(4) Attack by hydride ion. It has been found that the enol–lactone system of 2-pyrones undergoes attack by the hydride ion at the 6-position, with ring opening to a carboxylate anion. Thus treatment of 4,6-dimethyl-

2-pyrone with lithium tetrahydridoaluminate in ethereal solution gives *β*-methylsorbic acid (*G. Vogel*, Chem. and Ind., 1962, 268):

(5) Chloromethylation. 2-Pyrones are chloromethylated using formalin and hydrogen chloride; 3-chloromethyl-5,6-dimethyl-2-pyrone, b.p. 139–145°/6 mm, m.p. 65.5°; 3-chloromethyl-6-methyl-5-propyl-2-pyrone, b.p. 155–156°/5 mm, n_D^{20} 1.5308 (*N. P. Shusherina, N. D. Dmitrieva* and *R. Ya. Levina*, Dokl. Akad. Nauk S.S.S.R., 1962, **146,** 1113; *Dmitrieva et al.*, Zhur. obshcheĭ Khim., 1964, **34,** 2835).

(6) Reaction with ammonia and amines. A reaction of great importance, characteristic of both 2- and 4-pyrones, was discovered by *von Pechmann* and *W. Welsh* (Ber., 1884, **17,** 2391). By warming with aqueous ammonia the ring is opened, but is immediately followed by ring-closure to yield 2-pyridones. In the simplest instance, 2-pyrone forms 2-pyridone. It is doubtful, however, whether the reaction is as simple as represented:

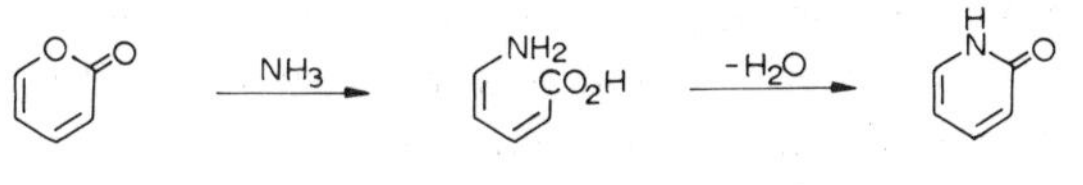

Primary amines react with 2-pyrones like ammonia, but secondary amines yield open-chain amides:

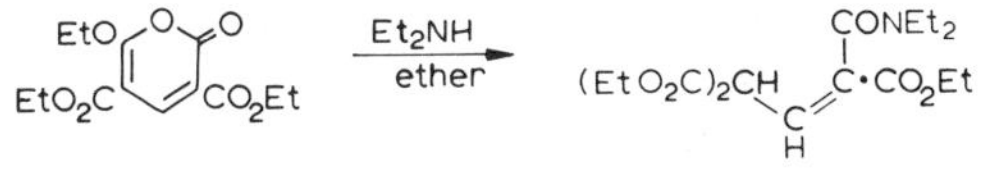

Phenylhydrazine and semicarbazide also break the 2-pyrone ring. Pyridine catalyses the condensation of hydroxylamine with 5-methoxy- and 5-aryloxy-4,6-diaryl-2-pyrones giving the corresponding 1-hydroxy-2-pyridones (*El-Kholy, F. K. Rafla* and *Soliman*, J. chem. Soc., 1961, 4490):

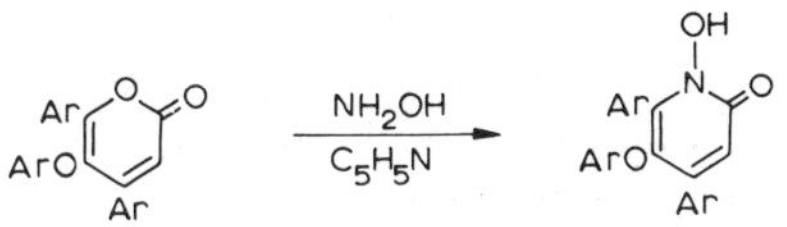

(7) Conversion to aromatic compounds. Some 2-pyrones may be converted into aromatic compounds, thus methyl 4-methoxy-2-pyrone-6-acetate on treatment with *N* NaOMe–MeOH gives methyl 2,6-dihydroxy-4-methoxybenzoate and compound XXXVIII, with *N* KOH–MeOH–H_2O (1:1)

yields 2-hydroxy-4-methoxy-6-methylbenzoic acid (*C. T. Bedford et al.*, Chem. Comm., 1968, 1091):

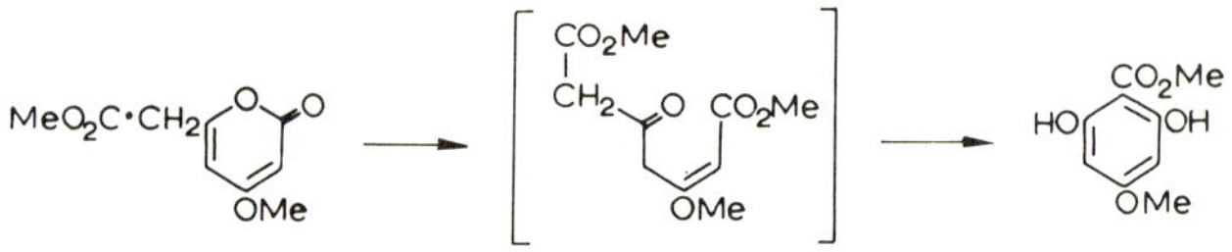

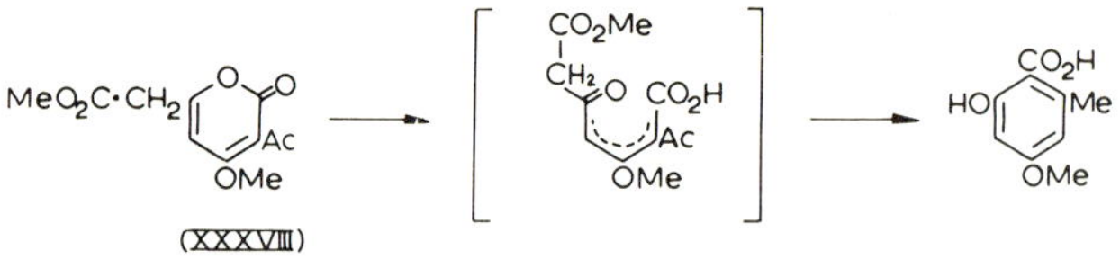

(XXXVIII)

(8) Reaction with diazomethane. 2-Pyrones with negative substituents in the 5-position are methylated by diazomethane in the 6-position; methyl coumalate reacts in this manner (*Fried* and *Elderfield*, J. org. Chem., 1941, **6**, 577).

The relative proportions of two isomeric methyl ethers XXXIX and XL produced from tautomeric pyronones by reaction with diazomethane depend on the rate of addition of diazomethane (*H. Nakata*, Bull. chem. Soc. Japan, 1960, **33**, 1688):

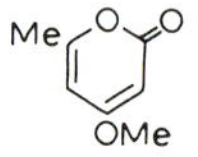

(XXXIX) m.p. 87-88°

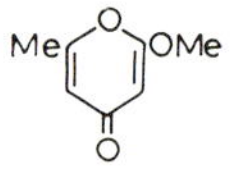

(XL) m.p. 94-95°

(9) Hydrogenation. Unsaturated lactones with hydrogen in the presence of a catalyst undergo both hydrogenation and hydrogenolysis (*W. Borsche* and *W. Peitzsch*, Ber., 1929, **62**, 360); similarly 2-pyrones yield both saturated δ-lactones and open-chain derivatives (*J. Meinwald*, J. Amer. chem. Soc., 1954, **76**, 4571; *R. H. Wiley* and *A. J. Hart*, *ibid.*, 1955, **77**, 2340). Methyl 6-methylcoumalate with hydrogen and a platinum oxide catalyst gives α-ethylglutaric acid (XLI) (after hydrolysis) and the saturated lactone methyl 5-methylpentanolide-4-carboxylate (XLII) (*Fried* and *Elderfield*, *loc. cit.*):

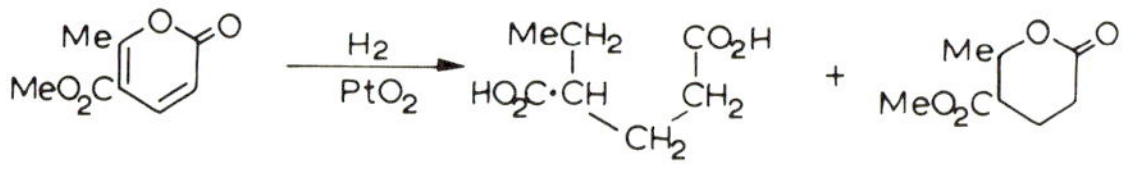

H_2 / PtO_2

(XLI) (XLII)

4,6-Dimethyl-2-pyrone with hydrogen and palladised charcoal gives an 83% yield of 4,6-dimethyltetrahydro-2-pyrone. These reactions establish without ambiguity the structure of the 2-pyrones.

(10) Halogenation. Substituted 2-pyrones react readily with halogen to give substituted products; coumalic acid yields 3-bromocoumalic acid, but unsubstituted 2-pyrone reacts with bromine and chlorine by an addition–elimination sequence rather than by direct electrophilic substitution. Several intermediate bromine addition products have been isolated (*W. H. Pirkle* and *M. Dines*, *ibid*., 1969, **34,** 2239).

(11) Grignard reagents react with 2-pyrones to give 2*H*-pyrans, but with excess reagent aromatic hydrocarbons are obtained (p. 2) (*R. Gompper* and *O. Christmann*, Ber., 1961, **94,** 1784, 1795).

(12) Friedel–Crafts reagents. 4-Hydroxy-6-methyl-2-pyrone reacted with crotonyl chloride in the presence of titanium(IV) chloride in tetrachloroethane at 100° to give 2,3-dihydro-2,7-dimethylpyrano[4,3-*b*]pyran-4,5-dione (XLIII). Under similar conditions, the crotonate of 4-hydroxy-6-methyl-2-pyrone in the presence of titanium(IV) gives similar results, but when the crotonate and aluminium chloride are boiled in carbon disulphide a mixture of XLIII and 3-crotonoyl-4-hydroxy-6-methyl-2-pyrone (XLIV) is obtained (*Y. Shizuri*, *K. Kato* and *Y. Hirata*, J. chem. Soc., C, 1969, 2774):

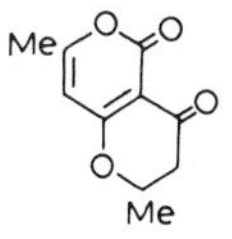

(XLIII) 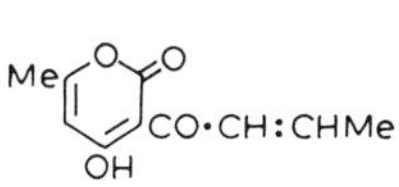 (XLIV)

2-**Pyrone,** 2H-*pyran-2-one*, *α-pyrone* or coumalin, is a liquid m.p. 5°, b.p. 206–209°, n_D^{25} 1.5272, which smells of new-mown hay and is prepared by the pyrolysis of the mercurous salt of coumalic acid (*H. von Pechmann*, Ann., 1891, **264,** 303). It is difficult to purify. It is preferable to decarboxylate 2-pyrone-6-carboxylic acid by heating with freshly reduced copper powder (*J. Fried* and *R. C. Elderfield*, J. org. Chem., 1941, **6,** 566). 2-Pyrone has been obtained in a 70–75% yield by subliming coumalic acid through copper turnings at 650–670° (*H. E. Zimmerman*, *G. L. Grunewald* and *R. M. Paufler*, Org. Synth., 1966, **46,** 101). Coumalic acid is not decarboxylated under these conditions. 2-Pyrone polymerises on standing and with alkali by ring-cleavage forms formylcrotonic acid $OHC \cdot CH_2 \cdot CH{:}CH \cdot CO_2H$.

5-**Methyl-2-pyrone,** m.p. 17–19°, has an odour of fresh hay, *maleic anhydride adduct*, m.p. 194.5–195.5°. 5-**Ethyl-2-pyrone** is a colourless oil, n_D^{25} 1.5137, *maleic anhydride adduct*, m.p. 161–162°. 3,5,6-Trimethyl-2-pyrone, b.p. 105–106°/6 mm, m.p. 37–38° (*Dmietrieva et al., loc. cit.*). 3,6-Diphenyl-4,5-disubstituted -2-pyrones may be obtained from 2,6-disubstituted-3,5-diphenyl-4-pyrans (p. 29).

Coumalic acid, α- or 2-*pyrone-5-carboxylic acid*, 2-*oxo*-2H-*pyran-5-carboxylic acid*, m.p. 205–210° (decomp.) is prepared (p. 10) by the action of fuming sulphuric acid on malic acid (*Wiley* and *N. R. Smith*, Org. Synth., 1951, **31**, 23; Coll. Vol. **4**, 201), *methyl ester*, m.p. 74°, b.p. 179°/60 mm, *ethyl ester*, m.p. 36°, b.p. 262–265°, *amide*, m.p. 244–245° (decomp.). The acid undergoes ring fission on treatment with hot alkali and yields formic acid and glutaconic acid. This reaction is characteristic of the 2-pyrone-5-carboxylic acids and is of importance not only for the structural determination of the acids, but also for the preparation of substituted glutaconic acids.

Fulvoplumierine, m.p. 151–152°, a derivative of coumalic acid, is an unstable orange-coloured pigment in the bark of *Plumiera acutifolia* and *Plumiera rubra varalba* (*H. Schmid* and *W. Bencze*, Helv., 1953, **36**, 205, 1468). It is of interest since it was the first naturally occurring fulvene to be synthesised (*G. Büchi* and *J. A. Carlson*, J. Amer. chem. Soc., 1969, **91**, 6470):

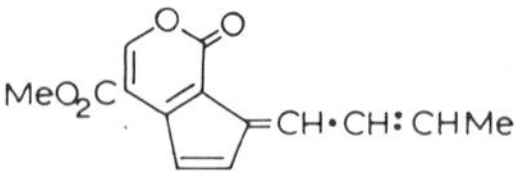

Fulvoplumierine

2-**Pyrone-6-carboxylic acid,** 2-*oxo*-2H-*pyran-6-carboxylic acid*, m.p. 228–230°, is prepared by condensing diethyl oxalate and ethyl crotonate by means of potassium in toluene and cyclising the product, diethyl α-hydroxybutadiene-α,ω-dicarboxylate, with hydrochloric acid (*A. Lapworth*, J. chem. Soc., 1901, **79**, 1279; *Wiley* and *Hart*, J. Amer. chem. Soc., 1954, **76**, 1942):

Isodehydracetic acid, 4,6-*dimethyl-2-pyrone-5-carboxylic acid*, m.p. 155°, is obtained in the form of its esters by treating acetoacetic esters with concentrated sulphuric acid or, better, dry hydrogen chloride, *methyl ester*, m.p. 67°, *ethyl ester*, m.p. 18–20°, b.p. 191°/35 mm (*Wiley* and *Smith*, Org. Synth., 1952, **32**, 76), other derivatives (*Wiley*, J. Amer. chem. Soc., 1953, **75**, 3715). The acid is decarboxylated by heating at 160° with concentrated sulphuric acid, forming 4,6-dimethyl-2-pyrone. The esters with ammonia form esters of 4,6-dimethyl-2-pyridone-5-carboxylic acid.

The pyran ring of isodehydracetic acid is broken by alkali and acetic acid and β-methylglutaconic acids are obtained. A somewhat unusual reaction is encountered when the ethyl ester of isodehydracetic acid is treated with bromine and the mono-bromo product heated with excess alkali to give methylcyclopropenedicarboxylic acid:

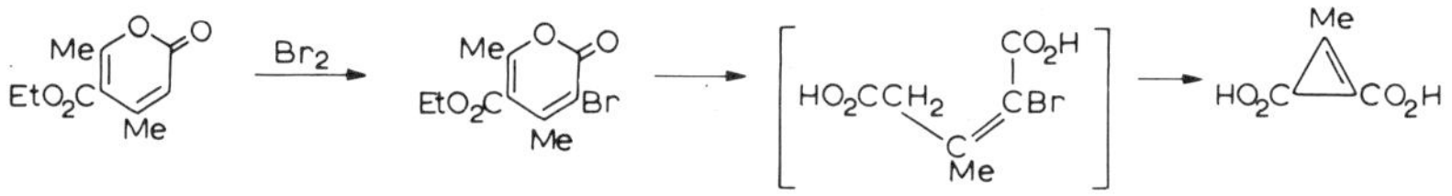

Kawain and structurally related substances are isolated from kawa resin (*W. Borsche*, Ber., 1933, **66,** 803). *Kawain, 5,6-dihydro-4-methoxy-6-styryl-2-pyrone*, m.p. 106°, $[\alpha]_D^{20}$ +105°, gives a red colour with concentrated sulphuric acid and has been used in the treatment of gonorrhoea. Its structure follows from degradation and synthesis (*E. M. F. Fowler* and *H. B. Henbest*, J. chem. Soc., 1950, 3642). It is isomerised by alkali to *kawaic acid*, m.p. 185°, $Ph(CH{:}CH)_2 \cdot C(OMe){:}CH \cdot CO_2H$ (*Borsche* and *W. Peitzsch*, Ber., 1930, **63,** 2414), the constitution of which follows from its synthesis (*D. Kostermans*, Nature, 1950, **166,** 789; *Fowles* and *Henbest*, *loc. cit.).* The absolute configurations of kawain and methysticine and their chain dihydro derivatives have been established from circular dichroism and by comparison with parascorbic acid (*G. Snatzke* and *R. Haensel*, Tetrahedron Letters, 1968, 1797. The structure of **yangonin,** m.p. 154–155°, has been established by direct synthesis (*J. D. Bu'Lock* and *H. G. Smith*, J. chem. Soc., 1960, 502):

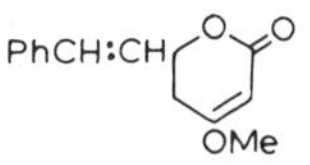

Kawain 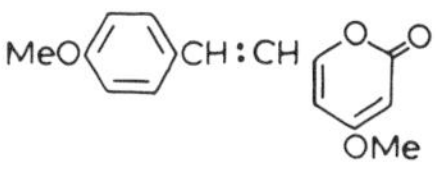 Yangonin 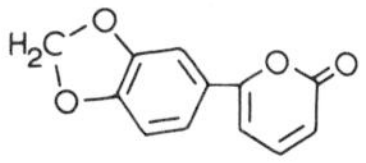 Paracotoin

Paracotoin, 6-*piperonylcoumalin*, m.p. 162° and 6-*phenylcoumalin*, m.p. 68°, are found in coto bark (*G. Ciamician* and *P. Silber*, Ber., 1894, **27,** 841); 4-*methoxyparacotoin*, m.p. 222–224°. occurs in S. American rosewood (*W. B. Mors et al.*, J. Amer. chem. Soc., 1957, **79,** 4507). The structure of 6-phenylcoumalin is established by its reduction with hydriodic acid to δ-phenylvaleric acid, its conversion by ammonia to 6-phenylpyrid-2-one, which on distillation with zinc yields 2-phenylpyridine (*J. A. Leben*, Ber., 1896, **29,** 1673). and synthesis (*J. Kalff*, Rec. Trav. chim., 1927, **46,** 594).

Alternaric acid, m.p. 138°, is a metabolite of *Alternaria solani*, having the following suggested structure on degradative and infrared evidence (*J. R. Bartels-Keith*, J. chem. Soc., 1960, 1662, 3413):

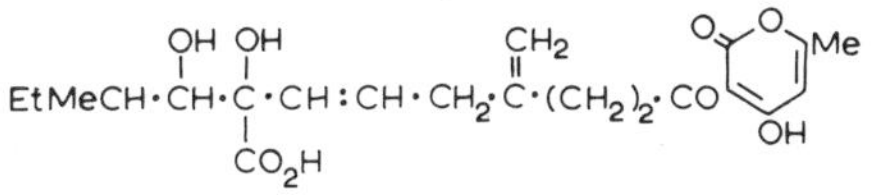

Alternaric acid

Hispidin, 4-*hydroxy-6-styryl-2-pyrone*, m.p. 259° (decomp.), occurs in the fungus produced by the basidiomycete *Polyporus hispidus* (Bull.) Fr.; its structure has been confirmed by synthesis (*R. L. Edwards*, *D. G. Lewis* and *D. V. Wilson*, *ibid.*, 1961, 4995, 5003):

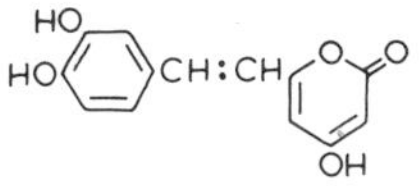

Hispidin

Opuntiol, m.p. 180–181°, a constituent of *Opuntia elatior*, is 6-*hydroxymethyl*-4-*methoxy*-2-*pyrone*; *monoacetate*, m.p. 110–111° (*A. K. Ganguly et al.*, Tetrahedron, 1965, **21**, 93).

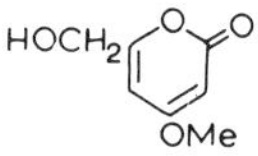

Opuntiol

Obtusifoline (see p. 283) is a derivative of 2-pyrone and flavanone.

Heating together ethyl ethoxymethylenacetoacetate, AcC(:CHOEt)CO_2Et, and ethyl sodioacetoacetate, AcCHNa·CO_2Et, gives a mixture of **diethyl xanthophanate** (XLV) and **diethyl glaucophanate** (XLVI) (*L. Crombie, D. E. Games* and *M. H. Knight*, Tetrahedron Letters, 1964, 2313):

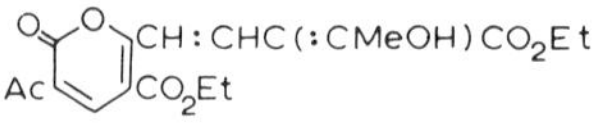

CH:CHC(:CMeOH)CO_2Et, Ac, CO_2Et

(XLV)

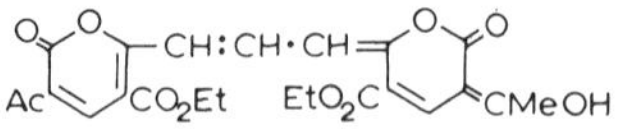

CH:CH·CH=, Ac, CO_2Et, EtO_2C, CMeOH

(XLVI)

(ii) 4-Pyrones*

Many 4-pyrone derivatives occur in nature; chelidonic acid occurs in the roots of *Chelidonium majus* (celandine) and other *Papaveraceae*, and meconic acid is found in opium. Other naturally occurring 4-pyrones are kojic acid, maltol and patulin. The 4-pyrone nucleus is also found in polycyclic substances such as chromone, flavone and xanthone (pp. 138, 162, 316). The 4-pyrones owe much of their interest to their capacity for forming stable salts with acids. These salts are of considerable theoretical importance (p. 6).

The constitution of 4-pyrone and its simple derivatives has been conclusively established by synthesis (p. 21), hydrolytic fission of the ring (p. 28), and conversion by aqueous ammonia to 4-pyridones (p. 28). There are, however, objections to the conventional formula XLVII. 4-Pyrones lack many of the characteristic properties of a carbonyl compound, failing to form phenylhydrazones, etc., and in spite of the presence of two olefinic linkages not undergoing reduction by zinc and acetic acid (*F. Feist*, Ann., 1890, **257**, 253; *J. N. Collie*, J. chem. Soc., 1904, **85**, 971; 1909, **95**, 144).

The lack of reactivity of the carbonyl and ethylenic linkages is explained by resonance theory (*F. Arndt* and *L. Lorenz*, Ber., 1930, **63**, 3121). 4-Pyrone is a resonance hybrid of this and the four contributing forms XLVIIa–d:

**J. Fried*, "Heterocyclic Compounds", John Wiley and Sons Inc., New York, 1950, Vol. I, p. 343.

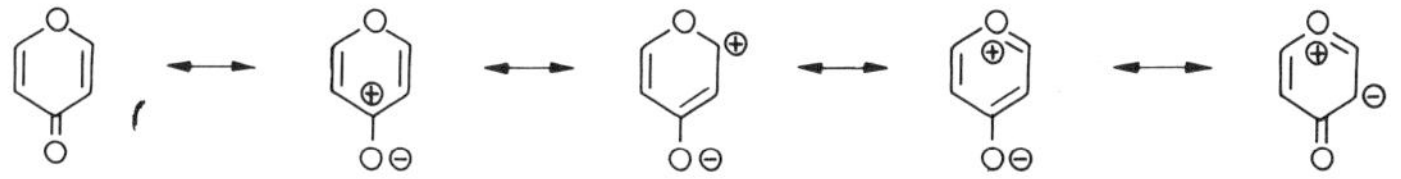

(XLVII) (XLVIIa) (XLVIIb) (XLVIIc) (XLVIId)

The polar forms arise partly from the tendency of the carbonyl group to assume the polar state by the migration of a pair of electrons from the double bond of the oxygen atom, and partly from participation of unshared electrons of the ring-oxygen to give a pseudo-aromatic system shown in formula XLVIIc (*Arndt et al.*, *ibid.*, 1924, **57,** 1903; see also *J. W. Armit* and *R. Robinson*, J. chem. Soc., 1925, **127,** 1604).

The ultraviolet absorption spectra of 2,6-dimethyl-4-pyrone (XLVIII) and 1,2,6-trimethyl-4-pyridone (XLIX) are very similar to one another and quite different from those of 2,6-dimethyl-4-methoxypyridine (L) (*R. C. Gibbs et al.*, J. Amer. chem. Soc., 1930, **52,** 4895) thereby showing the essential correctness of formulae XLVII and XLVIII. It may be noted that these spectral measurements do not distinguish between structures such as XLVII and XLVIIa. Further similarities between the 4-pyrone and the 4-pyridones may be seen in their infrared spectra where both compounds have carbonyl stretching modes at very low frequencies.

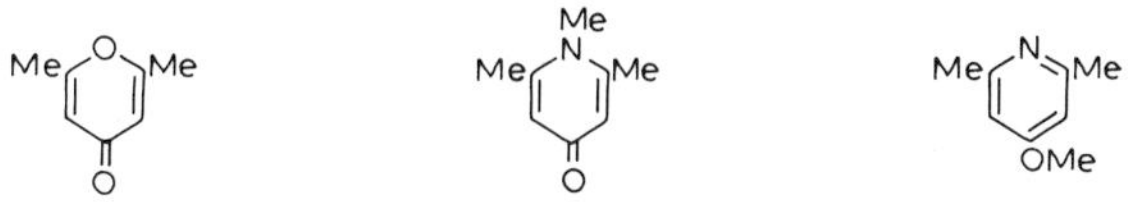

(XLVIII) (XLIX) (L)

From formulae XLVIIa–d, electrophilic substitution should occur at the 3-position and nucleophilic attack at the 2-position: actually bromination occurs at the 3- and 5-positions (p. 27) and ring-opening is initiated by alkaline reagents (nucleophilic attack) at position 2.

Molecular orbital calculations indicate that formula XLVIIa is the best single representation of the electronic structure of 4-pyrone (*R. D. Brown*, J. chem. Soc., 1951, 2670). Confirmation and refinement of the resonance picture comes from dipole measurements (*E. C. E. Hunter* and *J. R. Partington*, *ibid.*, 1933, 87; *C. G. LeFèvre* and *R. J. W. LeFèvre*, *ibid.*, 1937, 1088). 2,6-Dimethyl-4-pyrone has a moment of 4.05D (4.48D is also quoted), which is greater than the 1.75D required by structure XLVII, but considerably less than the 22.0D required by structure XLVIIc. The aromatic character of 4-pyrone is probably similar to that of 4-pyridone, since a ring current is evident from its p.m.r. spectrum (*J. Jonas*, *W. Derbyshire* and *H. S. Gutowski*, J. phys. Chem., 1965, **69,** 1).

Methods of preparation. (1) Cyclisation of open-chain compounds, typified by the ring-closure of suitable diketones. Diacetylacetone, for instance, is cyclised by phosphorus pentachloride or phosphoryl chloride to yield 2,6-dimethyl-4-pyrone:

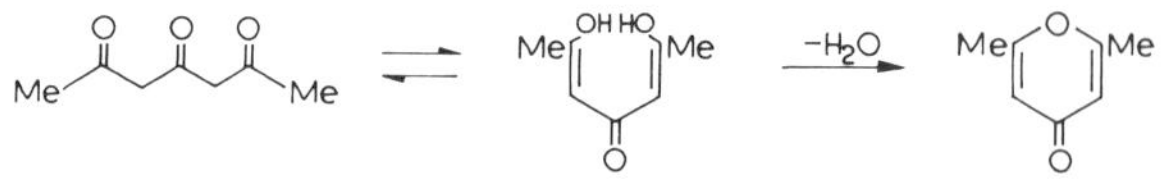

2-Methyl-4-pyrone has been obtained from acetylacetone and diethyl oxalate (*L. C. Dorman*, J. org. Chem., 1967, **32**, 4105):

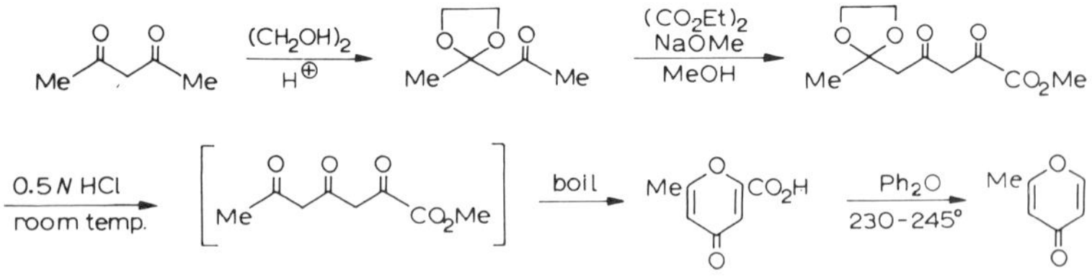

$(CH_2OH)_2$, $H^{\oplus}$; $(CO_2Et)_2$, NaOMe, MeOH; 0.5 *N* HCl, room temp.; boil; Ph_2O, 230–245°

Several β-diketones may be aroylated at the terminal methyl group with aromatic esters by using two molecular equivalents of potassium amide in liquid ammonia to form 1,3,5-triketones, which cyclise with acid to form 4-pyrones (*R. J. Light* and *C. R. Hauser*, *ibid*., 1960, **25**, 538, 1110):

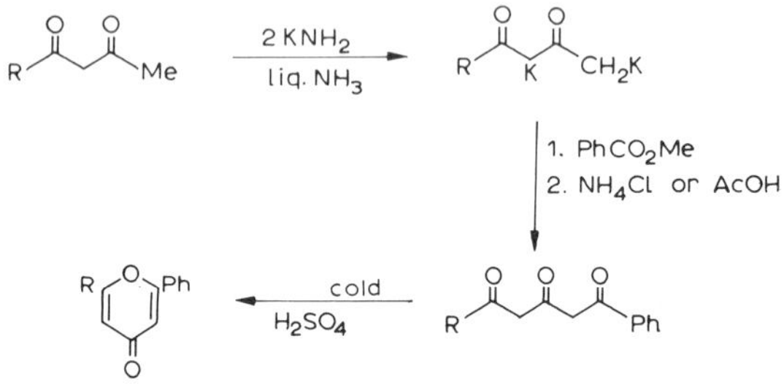

2 KNH_2, liq. NH_3; 1. $PhCO_2Me$ 2. NH_4Cl or AcOH; cold H_2SO_4

4-Hydroxy-6-phenacyl-2-pyrone on heating with hydrochloric acid and acetic acid gives a 6-phenyl-4-pyronyl-2-acetic acid (p. 14) (*T. M. Harris* and *M. P. Wachter*, Tetrahedron, 1970, **26**, 5255):

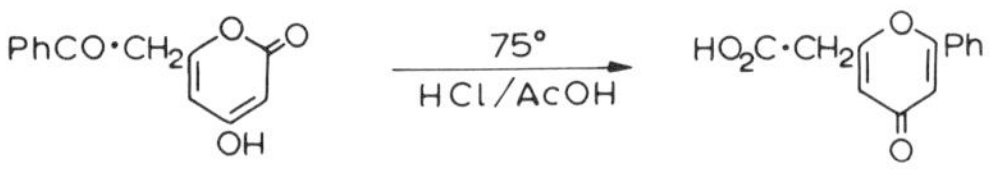

$PhCO{\cdot}CH_2$ … 75°, HCl/AcOH … $HO_2C{\cdot}CH_2$

(2) Decarboxylation of pyronecarboxylic acids. As in the pyridine series, decarboxylation of appropriate acids is extremely useful for the preparation of 4-pyrone derivatives. Thus the fairly readily accessible chelidonic acid (4-pyrone-2,6-dicarboxylic acid) when heated with copper powder decarboxylates and yields 4-pyrone together with some comanic acid (4-pyrone-2-carboxylic acid) (*L. Claisen*, Ber., 1891, **24**, 111; *R. Willstätter* and *R. Pummerer*, *ibid*., 1905, **38**, 1465):

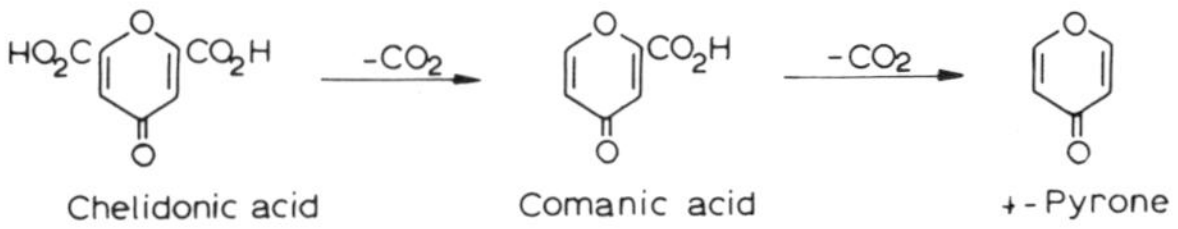

Chelidonic acid $\xrightarrow{-CO_2}$ Comanic acid $\xrightarrow{-CO_2}$ 4-Pyrone

If the mono-ethyl ester is decarboxylated only one molecule of carbon is lost and ethyl comanate is obtained. See also the conversion of dehydracetic acid into 2,6-dimethyl-4-pyrone (p. 35).

(3) By isomerisation of tetrahydropyrones. Substituted 3,5-dibenzylidene-tetrahydro-4-pyrones may be isomerised to the corresponding substituted 3,5-dibenzyl-4-pyrones in boiling diethylene glycol in the presence of palladium-on-charcoal (*N. J. Leonard* and *D. Chaudhury*, J. Amer. chem. Soc., 1957, **79**, 156):

Properties. The 4-pyrones are colourless liquids or solids, which absorb light in the ultraviolet region; 4-pyrone absorbs at 250 mμ (logε ~4.1) (*S. Aronoff*, J. org. Chem., 1940, **5**, 561). The 4-pyrones are very weak bases (*J. Walker*, Ber., 1901, **34**, 4115; *Partingon et al.*, J. chem. Soc., 1929, 1562; 1931, 86), and form solid salts with hydrogen chloride or bromide, and also with Lewis acids such as boron trifluoride (*W. Dilthey*, J. pr. Chem., 1921, [ii], **101**, 177; *E. F. Pratt* and *K. Matsuda*, J. Amer. chem. Soc., 1953, **75**, 3739). 2,6-Dimethyl-4-pyrone, for example, forms salts which are largely hydrolysed in aqueous solution (*H. N. K. Rördam*, *ibid.*, 1915, **37**, 557); has pK_b 18.7. The 4-pyrones, like the 4-pyridones, readily form hydrates.

Reactions. (1) Reactivity of the carbonyl group. Towards the usual carbonyl reagents, the 4-pyrones in general are either inert or undergo changes other than those of simple condensation. Many 4-pyrones fail to form oximes with hydroxylamine, but interaction sometimes occurs. Meconic acid forms the hydroxylamine salt and the methyl ether of comenic acid yields a 1-hydroxy-4-pyridone derivative in which the carbonyl group is still intact:

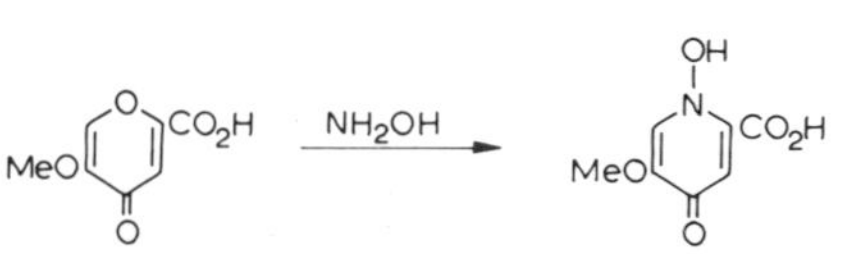

Unlike 2,6-diaryl-4-pyrones, the polycyclic 4-pyrones (LI, LII and LIII) fail to give the corresponding 1-hydroxy-4-pyridones with hydroxylamine in neutral media; in the presence of pyridine their oximes are the sole products:

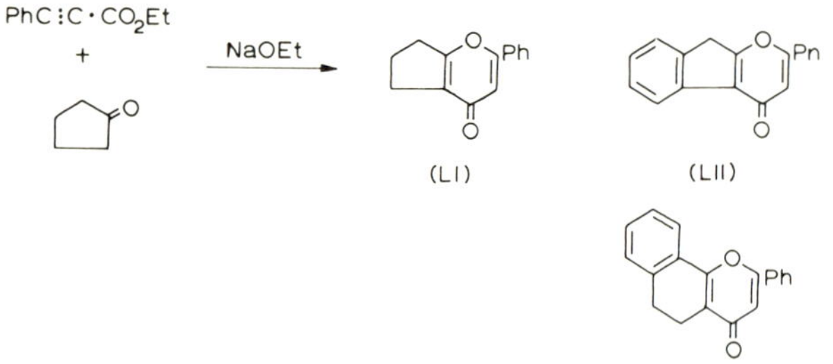

PhC⋮C·CO_2Et + cyclopentanone —NaOEt→ (LI)

(LI) (LII) (LIII)

4-Pyrones react with cyanamide in 1:1 aqueous ethanol to give 1-cyano-4-pyridones (*L. L. Woods*, Chem. and Ind., 1960, 1567):

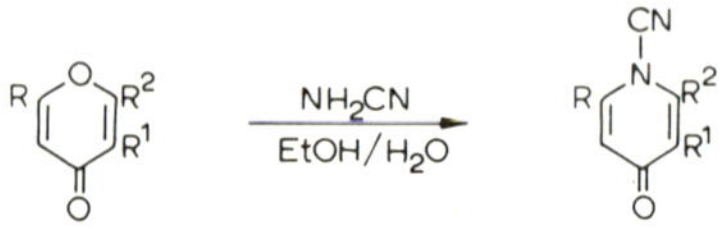

$\xrightarrow[\text{EtOH/H}_2\text{O}]{\text{NH}_2\text{CN}}$

(a) R = CH_2OH, R^1 = OH, R^2 = H
(b) R = Me, R^1 = H, R^2 = Me
(c) R = CH_2OH, R^1 = OMe, R^2 = H
(d) R = CH_2Cl, R^1 = OH, R^2 = H

Malononitrile and 2,6-dimethyl-4-pyrone when heated in acetic anhydride yield 4-(dicyanomethylene)-4*H*-pyran (*Woods*, J. Amer. chem. Soc., 1958, **80,** 1440):

NC·C·CN

It is reported that the reaction between 2,6-diphenyl-4-pyrone and hydroxylamine in pyridine gives 2,6-diphenyl-4-pyrone oxime as the main product and a small amount of 4-hydroxyamino-2,6-diphenylpyridine 1-oxide (*I. E.-S. El-Kholy, F. K. Rafla* and *G. Soliman*, J. chem. Soc., 1962, 1857):

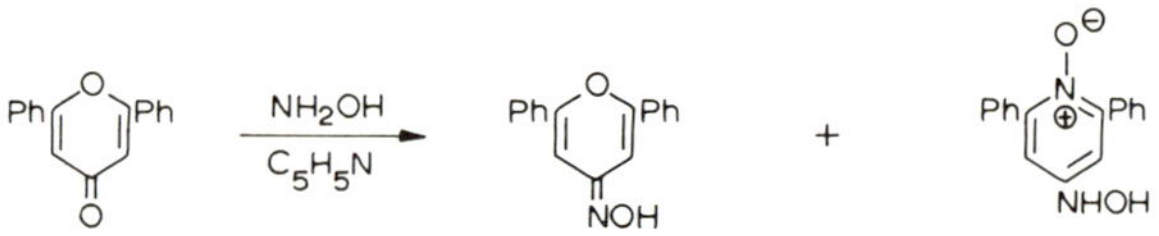

$\xrightarrow[\text{C}_5\text{H}_5\text{N}]{\text{NH}_2\text{OH}}$ NOH + NHOH

4-Pyrones in general do not form hydrazones with hydrazine and substituted hydrazines; 2,6-dimethyl-4-pyrone, for instance, fails to react with phenylhydrazine in ethanol or glacial acetic acid (*F. Feist*, Ann., 1890, **257,** 253). Diethyl chelidonate is stated to form a 4-nitrophenylhydrazone (*D. Bedekar et al.*, J. Indian chem. Soc., 1935, **12,** 465), but the claim that the free acid and 4-pyrone under the same conditions form 4-nitrophenylhydrazones of the *N*-(4-nitrophenylamino)-4-pyridone is certainly incorrect. Later work shows that 4-pyrones with hydrazine undergo ring-fission and subsequent reaction with hydrazines to form pyrazole derivatives (*R. G. Jones* and *M. J. Mann*, J. Amer. chem. Soc., 1953, **75,** 4048; *C. Ainsworth* and *Jones*, *ibid.*, 1954, **76,** 3172). Thus 4-pyrone and hydrazine give the hydrazone of pyrazole-3-acetaldehyde:

H_2O → CHO CHO / CH_2 CH_2 / CO → NH_2NH_2 → pyrazole-CH_2·CHO → NH_2NH_2 → pyrazole-CH_2·CH:N·NH_2

4-Pyrone and phenylhydrazine when heated together give the phenylhydrazone of 1-phenyl-5-pyrazoleacetaldehyde, which is reduced to 5-(β-aminoethyl)-1-phenylpyrazole and aniline. 2,6-Dimethyl- and 2,6-diphenyl-4-pyrone react similarly with phenylhydrazine.

Oximes and semicarbazones are obtained from the 4-pyran-4-thiones (4-thio-γ-pyrones) (*F. Arndt et al.*, Ber., 1924, **57,** 1910).

4-Pyrones react with Grignard reagents to give tertiary alcohols, which are generally not isolated but are converted by acids into the more stable pyrylium salts (p. 2); 2,6-dimethyl-4-pyrone for example, reacts with methylmagnesium iodide to give a product which with water yields 2,4,6-trimethyl-4*H*-pyranol (LIV), which is unstable and with perchloric acid forms 2,4,6-trimethylpyrylium perchlorate (LV) (*A. Bayer* and *J. Piccard*, Ann., 1911, **384,** 210; 1915, **407,** 332):

MeMgI → (LIV) → $HClO_4$ → (LV) $ClO_4^{\ominus}$

LIV is the pseudo-base of pyrylium salt LV and the reaction furnishes an example of the less stable alicyclic pyranol ring changing into the more stable aromatic pyrylium system.

It is reported that 2,6-dimethyl-4-pyrone reacts with three molar equivalents of methylmagnesium iodide to yield 2,4,4,6-tetramethyl-4-pyran;

similarly phenylmagnesium bromide gives 2,6-dimethyl-4,4-diphenyl-4-pyran (p. 4) (*R. Gompper* and *O. Christmann*, Ber., 1961, **94**, 1784).

A less usual method of demonstrating the presence of the carbonyl group is found in the reaction of 4-pyrone with diphenylketene at 150° to give a lactone LVI, which loses carbon dioxide and forms 4-benzhydrylidenepyran (*H. Staudinger* and *N. Kon*, Ann., 1911, **384**, 38, 129):

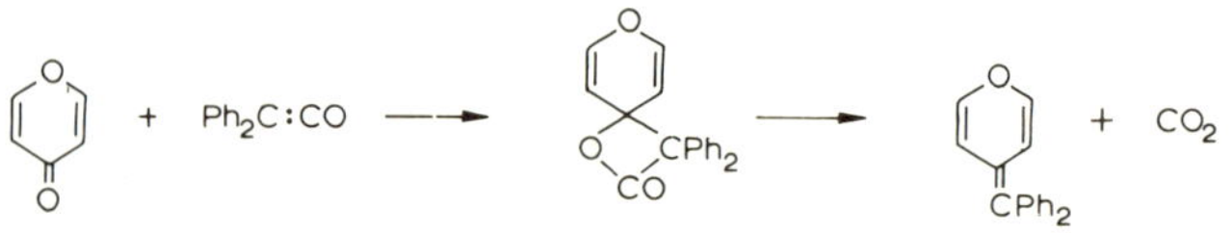

$$\text{4-pyrone} + Ph_2C{:}CO \longrightarrow \text{(LVI)} \longrightarrow \text{4-benzhydrylidenepyran } (=CPh_2) + CO_2$$

(LVI)

(2) Unsaturation of the 4-pyrone molecule. The behaviour of the 4-pyrone derivatives towards various reducing reagents is most irregular and unpredictable. Complex mixtures of unidentified products may be obtained; reduction of the ring with or without concomitant reduction of the carbonyl group may occur; or the pyrone ring may be broken.

Comenic acid with sodium amalgam undergoes reduction of the ethylenic double bonds to furnish 5-hydroxytetrahydropyran-4-one 2-carboxylic acid:

$$\text{comenic acid (HO, }CO_2H\text{)} \xrightarrow{Na/Hg} \text{5-hydroxytetrahydropyran-4-one 2-carboxylic acid}$$

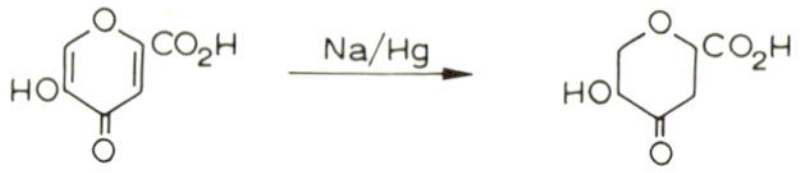

Meconic acid with the same reducing agent affords the saturated ring-compound, but, at the same time, the carbonyl group is converted to an alcohol, thus giving the cyclic glycol 3,4-dihydroxyoxane-2,6-dicarboxylic acid:

$$\text{meconic acid (}HO_2C, CO_2H, OH\text{)} \xrightarrow{Na/Hg} \text{3,4-dihydroxyoxane-2,6-dicarboxylic acid}$$

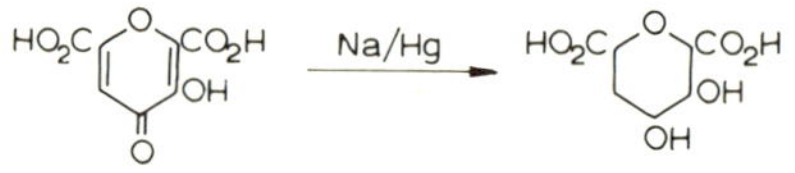

The reduction of a 4-pyrone derivative accompanied by ring-fission is exemplified by the formation of the open-chain 2,4,6-trihydroxyheptane-1,7-dioic acid, $HO_2C \cdot CHOH \cdot CH_2 \cdot CHOH \cdot CH_2 \cdot CHOH \cdot CO_2H$, from sodium chelidonate and sodium amalgam. The opening of the ring is probably caused by the alkaline nature of the reducing agent. Ring-opening may, however, occur with non-alkaline reducing agents. Diethyl chelidonate with zinc and acetic acid forms γ-oxopimelic acid; and meconic acid on hydrogenation with palladium chloride as catalyst yields the open-chain 2,3,4,6-tetrahydroxyoctane-1,8-dioic acid $HO_2C \cdot (CHOH)_3 \cdot CH_2 \cdot CHOH \cdot CO_2H$.

The most satisfactory results have been obtained by catalytic reduction. Here the products are frequently tetrahydropyrans with the carbonyl group either remaining intact or being reduced to a methanol or methylene group. 4-Pyrone with hydrogen and palladium chloride at room temperature and atmospheric pressure smoothly yields 2,3,5,6-tetrahydro-4-pyrone (*W. Borsche*, Ber., 1915, **48,** 682) whilst with hydrogen and copper chromite under high pressure at 125° it gives 23% of this derivative and 50% of the corresponding tetrahydropyranol (*R. Mozingo* and *H. Adkins*, J. Amer. chem. Soc., 1938, **60,** 669). Hydrogenation of 2,6-dimethyl-4-pyrone with a Raney nickel catalyst gives 2,6-dimethyltetrahydro-4-pyranol, whereas hydrogenation with a palladium catalyst produces at first a mixture of 2,6-dimethyldihydro-4-pyrone and 2,6-dimethyltetrahydro-4-pyrone which is eventually reduced completely to the latter product (*J. J. Vrieze*, Rec. Trav. chim., 1959, **78,** 91; *O. Riobe, V. Herault* and *L. Gouin*, Compt. rend., 1963, **256,** 1542).

Solvents sometimes play an important part in determining the course of reduction. 2,6-Dimethyl-4-pyrone on reduction with hydrogen and a platinum oxide catalyst in ethanol gives the saturated carbonyl compound, but in glacial acetic acid solution some of the tetrahydropyranol is formed and with decalin as solvent the product is almost entirely the tetrahydropyranol.

(3) Aromaticity of the 4-*pyrone nucleus.* On reaction with bromine 4-pyrone yields 3-bromo- and 3,5-dibromo-4-pyrone and the ease of substitution lies between that of benzene and pyridine. The halogen atoms attached to the 4-pyrone nucleus are inert towards most reagents. Hydroxy-4-pyrones, *e.g.*, comenic acid, brominate at the position contiguous to the carbon atom bearing the hydroxyl group. Nitration of the 4-pyrones has been effected with fuming nitric acid.

The 4-pyrones undergo mercuration (*J. R. Files* and *F. Challenger*, J. chem. Soc., 1940, 663). Comenic acid with mercuric acetate gives hydroxymercuricomenic anhydride, which with bromine affords 6-bromocomenic acid. In some instances mercuration is accompanied by decarboxylation, meconic acid giving the above hydroxymercuricomenic anhydride. It is therefore the carboxyl group in position 6 which is replaced, the same carboxyl group which is lost when meconic acid is decarboxylated to comenic acid (p. 36):

Comenic acid —2 stages→ (6-bromocomenic acid) ←2 stages— Meconic acid

2,6-Dimethyl-4-pyrone has been mercurated, but it could not be acylated using fuming stannic chloride, polyphosphoric acid, boron trifluoride etherate or hydriodic acid as catalyst. The Friedel–Crafts reaction with acetic anhydride and aluminium chloride, however, does effect acetylation in low yield (*L. L. Woods*, J. org. Chem., 1957, **22**, 341):

Acyl halides react with 4-pyrones in trifluoroacetic acid (*Woods* and *P. A. Dix*, *ibid*., 1959, **24**, 1126, 1804).

(4) Fission of the 4-pyrone ring. Cleavage of the 4-pyrone ring is not effected by acids. Alkalis, on the other hand, readily open the ring to give triketones or derived products; barium hydroxide converts 2,6-dimethyl-4-pyrone to the barium salt of diacetylacetone (*J. N. Collie* and *N. T. M. Wilsmore*, J. chem. Soc., 1896, **69**, 293), reversing method of preparation (*1*).

4-Pyrone similarly with cold sodium hydroxide yields the disodium salt of diformylacetone, which with benzoyl chloride gives the dibenzoate; at a higher temperature the diformylacetone breaks down into acetone and formic acid.

4-Pyrones, like the 2-pyrones, react under mild conditions with ammonia and primary amines to give pyridones, the intermediate amino-alcohol formed by ring-cleavage at once losing water to form the pyridone. The formation of 4-pyridones has proved most useful in determining the structures of 4-pyrones and the pyrylium salts (pp. 6, 20, 32, 35, 36):

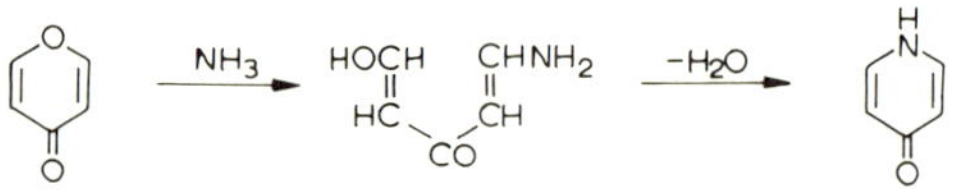

Secondary amines likewise break the ring, but the amino-alcohol cannot undergo cyclisation and is hydrolysed to acetone, etc., by the alkaline action of the amine.

4-Pyrones react with phosphorus pentasulphide in benzene to give 4*H*-pyran-4-thiones (*I. E.-S. El-Kholy*, *F. K. Rafla* and *G. Soliman*, J. chem. Soc., 1959, 2688; 1962, 1857). The conversion of 4-pyrones to 4*H*-pyran-4-thiones (and hence to 4*H*-thiopyran-4-thiones) is greatly improved by converting the pyrones to 4-methoxypyrylium salts which react readily with sodium hydrogen sulphide to give 4*H*-pyran-4-thione (*G. Traverso*, Ann. chim., Rome, 1956, **46**, 821).

An interesting degradation is the conversion of the 4-pyronone (4*H*-2,3-dihydropyran-2,4-dione) into the dimer of methylketene by the action of baryta (*R. B. Woodward* and *G. Small Jr.*, J. Amer. chem. Soc., 1950, **72,** 1297):

$$\longrightarrow$$

O=C–CHMe
MeC=COH

The irradiation of 2,6-dimethyl-4-pyrone in either the solid state or in solution gives the dimer LVII (*P. Yates* and *M. J. Jorgenson*, *ibid.*, 1963, **85**, 2956):

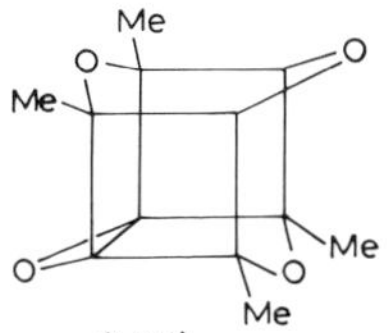

(LVII)

2,6-Disubstituted-3,5-diphenyl-4-pyrones on irradiation with a medium-pressure mercury lamp yield 3,6-diphenyl-4,5-disubstituted-2-pyrones (p. 17) (*N. Ishibe, M. Odani* and *M. Sunami*, Chem. Comm., 1971, 1034):

4-**Pyrone** (4H-*pyran*-4-one; *γ-pyrone*), colourless, m.p. 31–32°, b.p. 215°/742 mm, 101–102°/17 mm, 88.5°/7 mm. It forms a *hydrochloride*, m.p. ~139°, *acid oxalate*, m.p. 136.5° and *picrate*, m.p. 129° (*R. Willstätter* and *R. Pummerer*, Ber., 1904, **37,** 3740; 1905, **38,** 1461). It yields iodoform with iodine and potassium hydroxide (*cf. R. C. Fuson* and *B. A. Bull*, Chem. Reviews, 1934, **15,** 275; *H. Booth* and *B. C. Saunders*, Chem. and Ind., 1950, 824).

Pure 4-pyrone, first isolated independently from comanic acid by *H. Ost* (J. pr. Chem., 1884, [ii], **29,** 63) and from chelidonic acid by *L. Haitinger* and *Ad. Lieben* (Monatsh., 1884, **5,** 363), is still prepared in this way, the decarboxylation occurring smoothly in the presence of copper bronze. It is also obtained in 92% yield by heating 1-methoxy-1-penten-3-yn-5-al diethyl acetal (LVIII) with methanol and water in the presence of mercuric sulphate or sulphuric acid (*A. Dornov* and *F. Ische*, Z. angew. Chem., 1955, **67,** 653). The diacetal LIX is a possible intermediate:

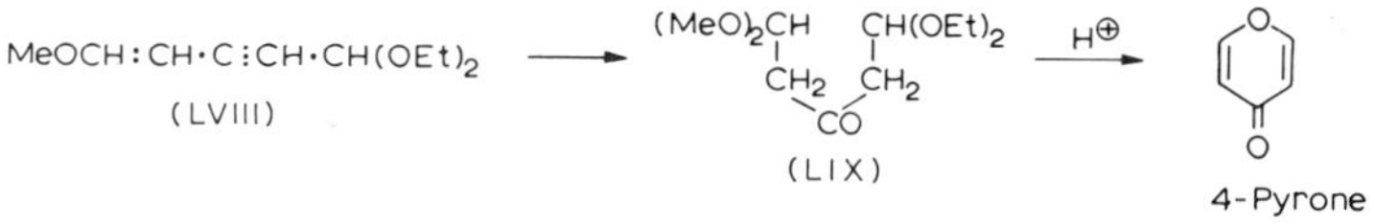

$MeOCH{:}CH{\cdot}C{:}CH{\cdot}CH(OEt)_2$ (LVIII) $\longrightarrow$ $(MeO)_2CH{\cdot}CH_2{\cdot}CO{\cdot}CH_2{\cdot}CH(OEt)_2$ (LIX) $\xrightarrow{H^{\oplus}}$ 4-Pyrone

The structure is established by the action of ammonia on 4-pyrone at 120° to give 4-pyridone.

The 2-methyl-4-pyrone nucleus is of interest since it is found in naturally occurring substances (p. 153).

2-Methylpyrones not containing a phenolic group give a violet colour with potassium hydroxide (*A. Schönberg* and *A. Sina*, J. chem. Soc., 1950, 3344).

2,6-**Dimethyl-4-pyrone,** colourless, m.p. 132°, b.p. 248–249°/713 mm, is best purified by sublimation followed by crystallisation from ether. It is soluble in water, and forms a *hydrochloride*, m.p. 83–85° *(hydrate)*, 154° (anhydrous); *oxalate*, m.p. 121° and *picrate*, m.p. 101–102°. It was first prepared by *Feist* in 1890 by heating dehydracetic acid with hydriodic acid, but it is preferable to use hydrochloric acid (*F. Arndt et al.*, Ber., 1936, **69**, 2373). It can be synthesised from diacetylacetone (p. 22) or from acetoacetic ester as indicated below:

2 COMe·CH_2·CO_2Et $\xrightarrow{COCl_2}$ EtO_2C·CH(COMe)·CO·CH(COMe)·CO_2Et $\xrightarrow{-H_2O}$ 2,6-dimethyl-3,5-di(ethoxycarbonyl)-4-pyrone $\xrightarrow{1.\ H_2O;\ 2.\ -CO_2}$ 2,6-dimethyl-4-pyrone

2,6-Dimethyl-4-pyrone is also obtained by heating acetic anhydride with sulphuric acid (*E. Philippi* and *R. Seka*, *ibid.*, 1921, **54,** 1089); as a side-product when diketene is heated with sodium phenoxide (*A. B. Steele et al.*, J. org. Chem., 1949, **14,** 460); and from hepta-2,5-diyne-4-one (*J. Chauvelier*, Ann. Chim., 1948, [xii], **3,** 393). With barium hydroxide it yields diacetylacetone (p. 28) and with ammonia 2,6-dimethyl-4-pyridone (4-lutidone). For some other reactions and derivatives of 2,6-dimethyl-4-pyrone see *Woods* (J. org. Chem., 1957, **22,** 341; 1959, **24,** 1804; Chem. and Ind., 1960, 1567; J. Amer. chem. Soc., 1958, **80,** 1440), *Woods* and *Dix (loc. cit.)*, *Woods* and *H. L. Williams* (J. org. Chem., 1960, **25,** 1052).

For spectral data: i.r., see *J. H. Looker* and *W. H. Hanneman* (*ibid.*, 1962, **27,** 381); u.v. see *L. Dorfman* (Chem. Reviews, 1953, **53**, 47), *Aronoff* (J. org. Chem., 1940, **5**, 561), *P. Franzosini et al.*, [Ann. Chim. (Phys.), 1955, **45**, 128; 1957, **47**, 346], *F. Eiden* (Arch. Pharm., Berlin, 1959, **292,** 355); and n.m.r. see *A. Terahara et al.* (Bull. chem. Soc. Japan, 1960, **33,** 1310); electronic spectra and chemical reactivity of 4-pyrone see *R. Zahradnik*, *C. Parkanyi* and *J. Koutecky* (Coll. Czech. chem. Comm., 1962, **27,** 1242).

(iii) Hydroxy-4-pyrones (hydroxy-γ-pyrones)

3-Hydroxy-4-pyrones are fully enolised and many of them give a deep red colour with ferric chloride; kojic acid and 3-hydroxy-2,6-dimethyl-4-pyrone, however, give blue and bluish-violet colours, respectively. The hydroxyl group activates the nucleus enough to allow coupling reactions with diazonium salts and, in vigorous conditions the Friedel–Crafts reaction, when acylation normally takes place at the 2-position (*L. L. Woods*, J. Amer. chem. Soc., 1953, **75,** 3608). Alkali causes a bathochromic shift in the u.v. absorption and acetylation a hypsochromic shift, in this way the 3-hydroxy-4-pyrones behave as do phenols (*J. A. Berson, W. M. Jones* and *R. L. O'Callaghan, ibid.*, 1956, **78**, 623; *E. A. Chandross* and *P. Yates*, Chem. and Ind., 1960, 149). In the case of 4-hydroxy-2-pyrones bases cause a hypsochromic shift similar to that suffered by α,β-unsaturated acids and acetylation produces a bathochromic shift.

The structures of the hydroxy-4-pyrones are established by the usual methods and especially by the formation of *O*-alkyl ethers followed by hydrolytic opening of the ring or by conversion into 4-pyridones.

Malonyl chloride condenses with acetylacetone, to yield 3-acetyl-6-hydroxy-2-methyl-4-pyrone (LX), and with benzoylacetone to yield 5-acetyl-4-hydroxy-6-phenyl-2-pyrone, which tautomerises above its melting point into 3-acetyl-6-hydroxy-2-phenyl-4-pyrone. The structures of these compounds have been investigated by deacetylation and conversion into pyridones, and by u.v., i.r. and n.m.r. spectroscopy (*M. A. Butt* and *J. A. Elvidge*, J. chem. Soc., 1963, 4483):

MeCO·CH:C(OH)Me + $CH_2(COCl)_2$ ⟶ (HO, Me, Ac, O pyrone)

(LX)

MeCO·CH:CPh·OH + $CH_2(COCl)_2$ ⟶ (Ph, Ac, OH, O 2-pyrone) —Δ, above 168°⟶ (Ph, Ac, OH, O 4-pyrone)

Aldehydes can be obtained from kojic acid, α-chloro-α-deoxykojic acid (2-chloromethyl-5-hydroxy-4-pyrone) and 2-hydroxymethyl-5-methoxy-4-pyrone by formylating the 4-pyrones using a mixture of carbon monoxide and hydrogen chloride in the presence of trifluoroacetic acid as a catalyst (*Woods* and *P. A. Dix*, J. org. Chem., 1961, **26,** 1028):

(HO, CH_2Cl, O pyrone) ⟶ (HO, CH_2Cl, CHO, O pyrone)

In the presence of trifluoroacetic acid, kojic acid and 2-hydroxymethyl-5-methoxy-4-pyrone react with lactones (*Woods* and *H. C. Smitherman*, *ibid.*, p. 2987):

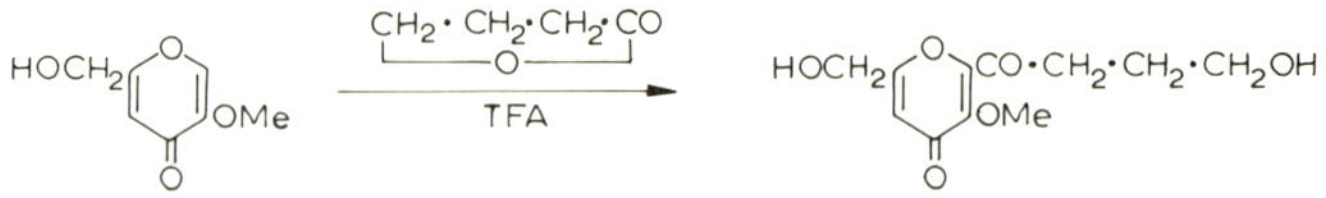

Reactions of some hydroxy-4-pyrones with aryl aldehydes under the influence of fused potassium acetate and trifluoroacetic acid have been discussed (*Woods* and *Dix*, *ibid.*, 1959, **24,** 1126).

Pyromeconic acid, 3-*hydroxy*-4-*pyrone*, m.p. 121.5°, b.p. 227–228°, is prepared by the complete decarboxylation of meconic acid, and is readily soluble in water giving a solution acid to litmus. With ferric chloride it gives a deep-red colour. It yields with acetyl chloride and diazomethane the 3-acetate and 3-methyl ether, respectively. On the other hand, the acid with ethyl nitrite forms 2-hydroxyiminopyromeconic acid.

The position of the hydroxyl group is established by alkaline hydrolysis of the methyl ether to give methoxyacetone and formic acid and by the action of ammonia on the methoxy compound to give 3-methoxy-4-pyridone:

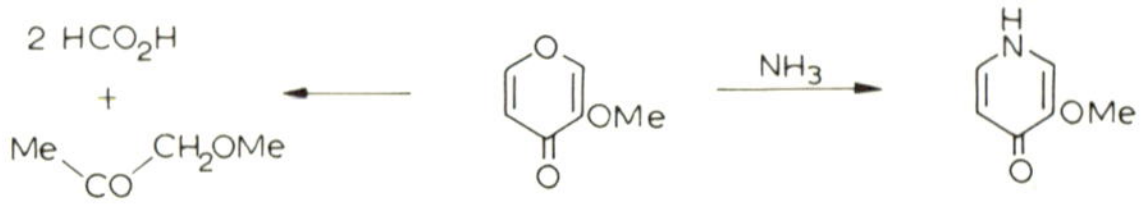

The acid is also converted into a pyridine derivative by treating the 2-hydroxyimino derivative in aqueous solution with sulphur dioxide. 1,2,3-Trihydroxy-4-pyridone is formed and by prolonged reduction 2,3,4-trihydroxypyridine (pyromecazonic acid) (*H. Ost*, J. pr. Chem., 1879, [ii], **19,** 197; 1883, **27,** 257). The formation of pyromecazonic acid must involve ring fission, reduction of the C–N linkage, and ring-closure:

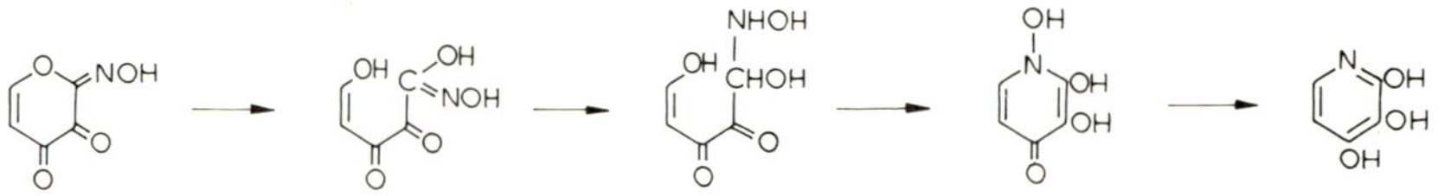

3-Hydroxy-2-iodo-4-pyrone, m.p. 175–180° (*G. A. Garkusha*, J. gen. Chem., U.S.S.R., 1946, **16,** 2025).

Maltol, 3-*hydroxy*-2-*methyl*-4-*pyrone*, *laricin*, *laricinic acid*, m.p. 162–164°, was first obtained from the bark of the larch tree (*J. Stenhouse*, Ann., 1862, **123,** 191). It also occurs in the fresh needles of the silver fir and in the plant *Corydalis ochotensis* (*R. H. F. Manske*, Canad. J. Research, 1940, **18B,** 75) and has been obtained by the dry distillation of cellulose and starch (*E. Erdmann* and *C. Schaefer*, Ber., 1910, **43,** 2398), the acid hydrolysis of soya beans, and the alkaline hydrolysis of streptomycin (*J. R. Schenk* and *M. A. Spielman*, J. Amer. chem. Soc., 1945, **67,** 2276).

Maltol in alkali gives a yellow solution, with ferric chloride gives a violet colour, and yields an O-*benzoyl* derivative, m.p. 115–116°. It reduces ammoniacal silver nitrate and Fehling's solution, and gives a positive iodoform reaction but a negative Millon's test. The constitution of maltol is indicated by the alkaline hydrolysis of the methyl ether to methoxyacetone, formic acid, and acetic acid:

Maltol → (OMe derivative) $\xrightarrow{Ba(OH)_2}$ HCO_2H + $MeCO_2H$ + $Me\text{-}CO\text{-}CH_2OMe$

These products could also be obtained from the *O*-methyl derivative of allomaltol.

The structure of maltol has been confirmed by synthesis (*Spielman* and *M. Freifelder*, *ibid.*, 1947, **69,** 2908). A Mannich reaction between comenic acid, formaldehyde and dimethylamine gives a Mannich derivative, which on reduction affords 6-methyl-comenic acid, decarboxylated to maltol by copper powder or tetrachlorobiphenyl (*I. Ichimoto, K. Fujii* and *C. Tatsumi*, Agr. biol. Chem., Tokyo, 1965, **29,** 325).

Allomaltol, 5-*hydroxy*-2-*methyl*-4-*pyrone*, m.p. 152–153°, can be prepared from kojic acid (*T. Yabuta*, J. chem. Soc., 1924, **125,** 575; *M. G. Brown, ibid.*, 1956, 2558).

Kojic acid, 5-*hydroxy*-2-*hydroxymethyl*-4-*pyrone*, colourless needles, m.p. 154–156° (*B. E. Bryant* and *W. C. Fernelius*, J. Amer. chem. Soc., 1954, **76,** 5351), was first isolated in 1907 by *K. Saito* from cultures of *Aspergillus oryzae*, a ferment used by the Japanese to produce the wine known as sake. It is formed anaerobically by various micro-organisms from carbohydrates and other substances, for example, by the action of *A. flavus* on xylose or glucose (*A. J. Moyer, ibid.*, 1931, **53,** 774), or of *A. oryzae* on disaccharides, hexoses, pentoses, tetroses, etc. (*F. Challenger et al.*, J. chem. Soc., 1929, 1498).

Its structure was established by *Yabuta (loc. cit.)*. It is the primary alcohol corresponding to comenic acid, but attempts to oxidise it to comenic acid have been unsuccessful. The methyl ether of kojic acid, however, is oxidised by potassium permanganate in acetone or by nitric acid, to the methyl ether of comenic acid (*J. W. Armit* and *T. J. Nolan*, J. chem. Soc., 1931, 3023):

Kojic acid (HO, CH_2OH) ; methyl ether (MeO, CH_2OH) → Comenic acid methyl ether (MeO, CO_2H)

The structure is confirmed by the conversion of both kojic acid and comenic acid into 4,5-dihydroxy-2-methylpyridine (p. 36); by the preparation of 5-hydroxypicoline acid from kojic acid (*K. Heyns* and *G. Vogelsang*, Ber., 1954, **87,** 13); and by the action of ammonia on kojic acid to yield 4,5-dihydroxy-2-hydroxymethylpyridine (*idem, ibid.*, p. 1377). Kojic acid is readily soluble in water and ethanol but only slightly soluble in benzene and light petroleum. It can be purified by sublimation. It is weakly acidic, reduces Fehling's solution and couples in alkaline solution with diazonium salts in the 6-position (*A. Quilico* and *C. Musante*, Gazz., 1944, **74,** 26). In ethanol or acetic acid, however, diazonium salts yield phenyl derivatives, 4-chlorobenzenediazonium sulphate for example, yielding 6-(4-*chlorophenyl*)-5-*hydroxy*-2-*hydroxymethyl*-4-*pyrone*, m.p. 194°. Kojic acid with diazomethane yields the 5-*methyl ether*, m.p. 165°, but with dimethyl sulphate and alkali the product may be either the 5- or the 7-*methyl ether*, m.p. 75–76°, or the *dimethyl ether*, m.p. 90°, according to the conditions (*e.g.*, *K. N. Campbell et al.*, J. org. Chem., 1950, **15,** 221). Kojic acid forms a *diacetate*, m.p. 101–102° and a *dibenzoate*, m.p. 136°, and with triphenylmethyl chloride gives the 7-*triphenylmethyl ether*, m.p. 185–185.5° (*A. Béelik* and *C. B. Purves*, Canad. J. Chem., 1955, **33,** 1361). U.v. absorption is recorded at 315 mμ (ε_{max} ~5000) (*M. Stacey* and *L. M. Turton*, J. chem. Soc., 1946, 661), and 269 mμ (log ε 3.90) (MeOH) (*Berson et al.*, J. Amer. chem. Soc., 1956, **78,** 622).

With vinyl cyanide and a base kojic acid reacts as a ketone to give β-(5-hydroxy-2-hydroxymethyl-4-pyrone)propionic acid (*Woods, ibid.*, 1952, **74**, 3959). The reaction of kojic acid with citric, malic and maleic acids in the presence of concentrated sulphuric acid reveals that the Pechmann reaction takes place to form pyrano-coumarins, but the products are highly contaminated compounds which are difficult to isolate and purify (*Woods* and *Dix*, J. org. Chem., 1959, **24,** 1148). Kojic acid reacts with hydrazine to give 3,6-bis(hydroxymethyl)-4-oxo-1,4-dihydropyridazine (LXI) (*A. Marxer* and *A. F. Thomas*, Chimia, Switz., 1960, **14,** 126; *Ichimoto*, *Fujii* and *Tatsumi*, Agr. Biol. Chem., Tokyo, 1967, **31,** 979):

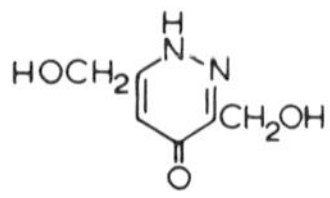

H N N HOCH$_2$ CH$_2$OH O

(LXI)

With methanesulphonyl chloride in pyridine in a short term reaction, kojic acid gave the dimethanesulphonate, but a long term reaction led to an oily product, which with hydrochloric acid gave 2-chloromethyl-5-mesyloxy-4-pyrone (*J. H. Looker et al.*, J. org. Chem., 1962, **27,** 4349):

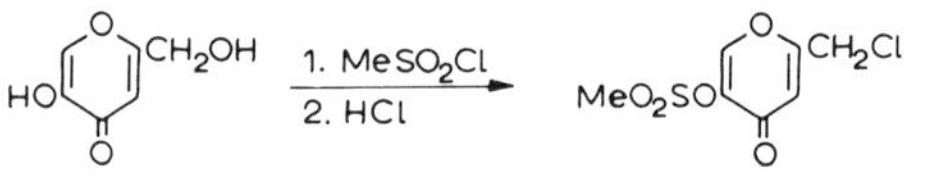

HO O CH$_2$OH O —1. MeSO$_2$Cl, 2. HCl→ MeO$_2$SO O CH$_2$Cl O

Isokojic acid, colourless plates, m.p. 183°, is formed from fructose by *Gluconoaceto-bacter roseum* along with kojic acid. It is an isomer of kojic acid and its structure has been determined from its chemical and u.v. spectral properties (*K. Aida*, *M. Fujii* and *T. Asai*, Proc. Japan. Acad., 1956, **32,** 595):

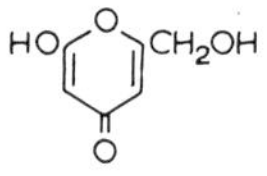

HO O CH$_2$OH O

Isokojic acid

Dehydracetic acid, m.p. 108°, has been referred to in Vol. IE, p. 213, as the cyclic lactone α,γ-diacetylacetoacetic acid. It is prepared by the self-condensation of ethyl acetoacetate under the influence of heat and sodium carbonate (*F. Arndt*, Org. Synth., 1940, **20,** 26):

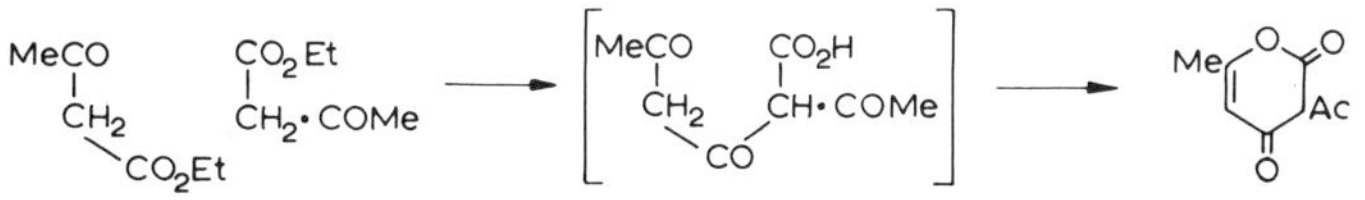

Dehydracetic acid

It is also obtained in excellent yield by the dimerisation of diketene or by the action of methylmagnesium iodide on carbon suboxide (*J. H. Billman* and *C. M. Smith*, J. Amer. chem. Soc., 1952, **74,** 3174; *I. Isoshima*, J. chem. Soc., Japan, Ind. Chem. Sect., 1954, **57,** 773):

Its structure was first suggested by *Feist*. A somewhat similar formula, LXII, was advanced by *Collie* and for some years the structure of dehydracetic acid was a subject of controversy. Feist's formula was proved to be correct by *C. F. Rassweiler* and *R. Adams* (J. Amer. chem. Soc., 1924, **46,** 2758) who showed the acid LXIII obtained by the isomerisation of dehydracetic acid with 85% sulphuric acid has indeed the structure assigned since it is converted by ammonia into 2,6-dimethyl-4-pyridone-3-carboxylic acid (LXIV), the constitution of which was established by synthesis. The formation of the isomer is readily explained by the ring-opening of dehydracetic acid to give the triketone LXV, which recyclises to LXIII. This substance would not result from LXII:

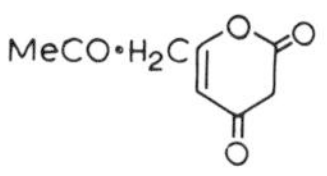

(LXII) 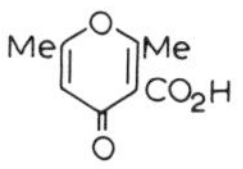 (LXIII) 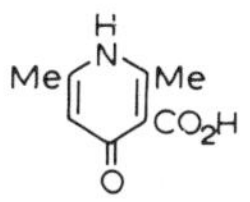 (LXIV)

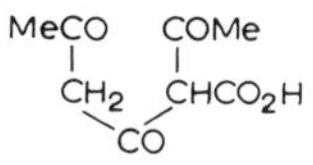

(LXV) 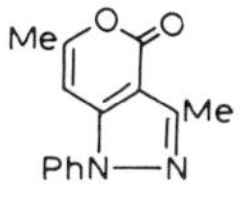 (LXVI)

Confirmation of the structure of dehydracetic acid has been provided by the demonstration that the dehydration product of the phenylhydrazone has the structure LXVI (*E. Benary*, Ber., 1910, **43,** 1070) and from a consideration of the acidic dissociation constant and u.v. spectrum of the acid (*Berson*, J. Amer. chem. Soc., 1952, **74,** 5172).

It has been reported that dehydracetic acid reacts with aromatic aldehydes via an aldol condensation to yield 3-cinnamoyl-4-hydroxy-2-pyrones (LXVII), and that it reacts with isobutyraldehyde in a similar way. If the condensation with isobutyraldehyde is carried out in the presence of a catalytic amount of piperidine in chloroform and the water removed by azeotropic distillation then 3,4-dihydro-2-isopropyl-7-methyl-2*H*,5*H*-pyrano[4,3-*b*]pyran-4,5-dione (LXVIII) is formed (*J. H. Robson* and *E. Marcus*, Chem. and Ind., 1969, 1306):

Me, O, O, COMe, OH + RCHO → Me, O, O, CO·CH:CHR, OH

(LXVII)
R=Ar or $CHMe_2$

$Me_2CH·CHO$
$(C_5H_{11}N / CHCl_3)$

Me, O, O, O, O, $CHMe_2$

(LXVIII)

(c) 4-Pyronecarboxylic acids

The 4-pyronecarboxylic acids constitute one of the most important groups of the monocyclic pyrone derivatives since many of them are accessible and can be readily decarboxylated to 4-pyrone or its substituted derivatives, as already described. Some, such as chelidonic and meconic acids, occur in nature.

Comanic acid, 4-*pyrone-2-carboxylic acid*, 4-*oxo*-4H-*pyran-2-carboxylic acid*, m.p. 250° (decomp.), absorbing at 260 mμ (ε_{max} 10,280) (*J. Attenburrow*, J. chem. Soc., 1945, 571) is conveniently prepared as the ethyl ester by the decarboxylation of monoethyl chelidonate (p. 23) (*L. Cavalieri*, Chem. Reviews, 1947, **41**, 529); *ethyl ester*, m.p. 103°.

Comenic acid, 5-*hydroxy*-4-*pyrone-2-carboxylic acid*, 5-*hydroxy*-4-*oxo*-4H-*pyran*-2-*carboxylic acid*, decomp. above 260° [m.p. 285° (decomp.)], is formed by the action of bacteria on galactose (*T. Takahashi* and *T. Asai*, C.A., 1934, **28,** 6519; 1936, **30,** 3460) and is prepared by boiling meconic acid with hydrochloric acid until the calculated quantity of carbon dioxide has been evolved (*G. A. Garkusha*, Zhur. obshcheĭ Khim., 1953, **23,** 1578). The *ethyl ester*, m.p. 135°, is prepared by the decarboxylation of the acid ester of meconic acid (see p. 38) (see *T. Yabuta*, J. chem. Soc., 1924, **125**, 575); *methyl ether*, m.p. 280–282°; *ethyl ether*, m.p. 239–240°; *benzoyl* deriv., m.p. 227–228°.

The yellow crystalline acid gives colourless aqueous solutions (compare meconic acid). The enolic form in solution is detected by the red colour given with ferric chloride, a reaction characteristic of many 3-hydroxy-4-pyrones.

The structure for comenic acid is supported by the fact that it couples with diazobenzene acetate, and is confirmed by its conversion into 4,5-dihydroxy-2-methylpyridine (5-hydroxy-2-methyl-4-pyridone), the same product being obtained from kojic acid. This somewhat difficult conversion is effected by a curious reaction involving heating the intermediate comenamic acid (*T. Reibstein*, J. pr. Chem., 1881, [ii], **24,** 284; *H. Ost*, *ibid.*, 1884, **29,** 63) with phosphorus pentachloride at 150° and treating the product first with water and then with tin and hydrochloric acid (*Yabuta, loc. cit.; P. E. Verkade*, Rec. Trav. chim., 1924, **43,** 879):

Comenic acid —2 stages→ [5-hydroxy-2-methyl-4-pyridone] ← Kojic acid; Coumenamic acid

The structure of comenic acid has been confirmed by its preparation from kojic acid (*H. D. Becker*, Acta Chem. Scand., 1962, **16,** 78):

MeO-pyrone-CH_2OH —MnO_2, C_6H_6→ MeO-pyrone-CHO —CrO_3–H_2SO_4, Me_2CO→ MeO-pyrone-CO_2H —$AlCl_3$, sealed tube 185°→ HO-pyrone-CO_2H

Chelidonic acid, 4-*pyrone*-2,6-*dicarboxylic* acid, 4-*oxo*-4H-*pyran*-2,6-*dicarboxylic acid*, m.p. 262°, first isolated in 1839 by *Probst* from *Chelidonium majus*, *dimethyl ester*, m.p. 122.5°, was thoroughly investigated by *Lieben* and *Haitinger* (1883–1885) whose researches are of great importance in the history of pyran chemistry. They deduced its structure which was later confirmed by synthesis and degradation with alkali.

The acid is prepared by the ring-closure of diethyl acetonedioxalate with hydrochloric acid, a method first worked out by *L. Claisen* and subsequently improved by *R. Willstätter* and *R. Pummerer* (Ber., 1905, **38,** 1461; see also *E. R. Riegel* and *F. Zwilgmeyer*, Org. Synth., 1937, **17,** 40).

Diethyl acetonedioxalate (enolic form) → Chelidonic acid → Comanic acid → 4-Pyrone

The ester must be completely free from sodium if the synthesis is to be successful. A modified procedure for the synthesis of chelidonic acid has been developed and a number of mono- and di-alkyl esters of chelidonic acid have been prepared (*I. Ichimoto* and *C. Tatsumi*, Bull. Univ. Osaka Prefect., 1965, **B16,** 105; C.A., 1966, **65,** 3822g).

Chelidonic acid undergoes the usual 4-pyrone reactions, with calcium or barium hydroxide giving one mol. of acetone and two mols. of oxalic acid, and with ammonia yielding 4-pyridone-2,6-dicarboxylic acid (chelidamic acid). Claisen's statement that the acid is only slightly soluble in ethanol is incorrect and the acid is purified by diluting a boiling ethanolic solution with water.

It has been reported that chelidonic acid reacts with ethylenediamine in hot ethanol to give a compound with a pyridopiperazine ring (*A. W. Schwab*, J. Amer. chem. Soc., 1954, **76**, 1189):

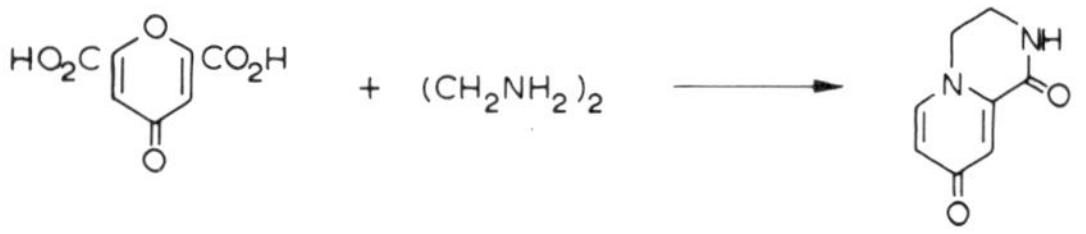

HO_2C–(4-pyrone)–CO_2H + $(CH_2NH_2)_2$ ⟶ (pyridopiperazinone)

Meconic acid, 3-*hydroxy*-4-*pyrone*-2,6-*dicarboxylic acid*, 3-*hydroxy*-4-*oxo*-4H-*pyran*-2,6-*dicarboxylic acid*, crystallises with water of crystallisation which is lost at 100°. At higher temperatures it loses carbon dioxide. It forms a *monomethyl ester*, m.p. 161°; a *monoethyl ester*, m.p. 179°; a *dimethyl ester*, m.p. 117°; a *diethyl ester*, m.p. 111.5°; an *acetyl*, m.p. 218° and a *benzoyl* derivative, m.p. 248°. It can be crystallised from warm water, but in boiling water is decarboxylated to comenic acid. Meconic acid occurs in and is obtained from opium from which it was first isolated by *Séguin* in 1814. It is colourless when pure, but is often slightly brown, due to traces of iron (*Verkade*, Rec. Trav. chim., 1924, **43**, 879). Meconic acid is a surprisingly strong acid, nearly as strong as sulphuric acid. The dissociation constants for meconic and chelidonic acids have been calculated from the neutralisation curves at 25° (*S. Miyamoto* and *E. Brochmann-Hanssen*, J. pharm. Sci., 1962, **51**, 552).

An open-chain formula for meconic acid was advanced by *W. Borsche* (Ber., 1916, **49**, 2538) since the acid on catalytic hydrogenation yields a tetrahydroxypimelic acid, but the formula given below is now accepted. The acid with absolute ethanol and hydrogen chloride gives the acid ester in good yield, decarboxylation of which gives ethyl comenate:

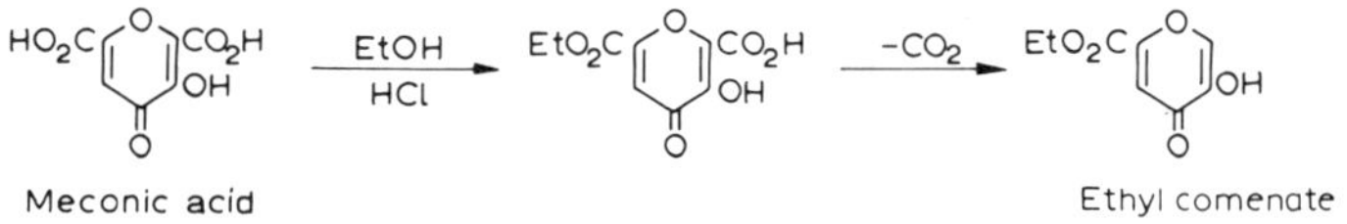

Meconic acid $\xrightarrow[HCl]{EtOH}$ (ethyl hydrogen meconate) $\xrightarrow{-CO_2}$ Ethyl comenate

Meconic acid forms a triethyl derivative which with barium hydroxide is broken down to ethoxyacetone, oxalic acid and ethanol:

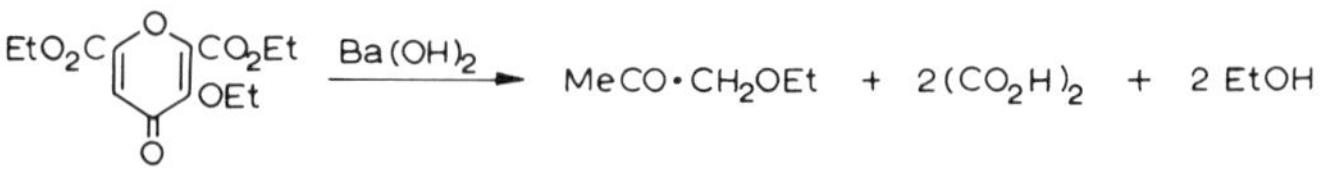

(triethyl derivative) $\xrightarrow{Ba(OH)_2}$ $MeCO{\cdot}CH_2OEt$ + $2(CO_2H)_2$ + 2 EtOH

Meconic acid is synthesised by the bromination of acetonedioxalic ester to give diethyl 3-bromo-4-pyrone-2,6-dicarboxylic acid, which by careful treatment with potassium hydroxide in aqueous acetone yields potassium meconate (*H. Thoms* and *R. Pietrulla*, Ber. deutsch. pharmaz. Gesell., 1921, **31,** 4; *J. P. Wibaut* and *R. J. C. Kleipool*, Rec. Trav. chim., 1947, **66,** 24):

$CO(CH_2{\cdot}CO{\cdot}CO_2Et)_2 \xrightarrow{Br_2}$ (EtO₂C, CO₂Et, Br-substituted 4-pyrone) $\xrightarrow{KOH}$ (KO₂C, CO₂K, OH-substituted 4-pyrone)

Evidence for the enolic form is found in the blood-red colour which the acid gives the ferric chloride, the phenolic properties of the diethyl ester and the formation of a triethyl derivative when the silver salt of the acid is treated with ethyl iodide.

Acylation of the pyrone ring of meconic acid can be carried out by the thermal rearrangement of the phenolic ester of meconic acid in which the acyl radical probably takes the position vacated by the carboxyl group, for example, benzoyl meconic acid yields 2-benzoylcomenic acid (*L. L. Woods*, J. org. Chem., 1957, **22,** 339):

(HO₂C, CO₂H, OH-4-pyrone) $\xrightarrow[145°]{PhCOCl}$ [HO₂C, CO₂H, OBz-4-pyrone] $\longrightarrow$ (HO₂C, Bz, OH-4-pyrone)

2. Hydropyrans

(a) Dihydropyrans and derivatives

(i) 2,3-Dihydro-4-pyrans

2,3-**Dihydro-4-pyran** (2,3-*dihydro*-4H-*pyran*, 2,3-*dihydro-γ-pyran*) (II), b.p. 85–86°, d_{15}^{19} 0.922, n_D^{19} 1.4402, is prepared in high yield by heating tetrahydrofurfuryl alcohol (I) over alumina at 300–350° (*R. Paul*, Bull. Soc. chim. Fr., 1933, [v], **53,** 1489; *C. H. Kline Jr.* and *J. Turkevich*, J. Amer. chem. Soc., 1945, **67,** 498; *R. L. Sawyer* and *D. W. Andrus*, Org. Synth., 1943, **23,** 25; *J. G. M. Bremner* and *D. McNeill*, B.P. 547,334/1942; *L. E. Schniepp* and *H. H. Geller*, J. Amer. chem. Soc., 1946, **68,** 1646):

(I) (tetrahydrofurfuryl alcohol, CH_2OH) $\longrightarrow$ (II)

Heat of combustion (*R. C. Cass et al.*, J. chem. Soc., 1958, 1406); critical properties and vapour pressures (*K. A. Kobe*, J. chem. Eng. Data, 1970, **15,** 182); bond lengths and angles (*E. M. Philbin* and *T. S. Wheeler*, Proc. chem. Soc., 1958, 167). It is a liquid with a penetrating ether-like odour. Containing a vinyl ether group, $CH_2{:}CHO–$, it is very reactive (*O. W. Cass*, Ind. Eng. Chem., 1948, **40**, 216; *C. H. Hurd et al.*, J. Amer. chem. Soc., 1955, **77**, 2793). Hydrogenation of 2,3-dihydro-4-pyran with a platinum black or nickel catalyst yields tetrahydropyran which with hydrogen bromide undergoes ring-fission to give 1,5-dibromopentane (*C. L. Wilson, ibid.*, 1948, **70**, 1311; *Bremner et al.*, B.P. 565,175/1944). The position of the double bond is proved by hydration with sulphuric or hydrochloric acid to 2-hydroxytetrahydropyran (*Paul*, Bull. Soc. chim. Fr., 1934, 971; *G. F. Woods Jr.*, Org. Synth., 1947, **27**, 43):

H_2O → OH

With other hydroxy compounds, such as alcohols and phenols, it gives ethers or esters, which are hydrolysed very readily by aqueous acids to give alcohols or phenols and 2-hydroxytetrahydropyran (*Schniepp* and *Geller*, *loc. cit.*; *Paul*, *loc. cit.*). This property has been utilised to "protect" hydroxyl groups attached to the benzene ring (*W. E. Parham* and *E. L. Anderson*, J. Amer. chem. Soc., 1948, **70**, 4187).

Pyrolysis of dihydropyran at 500° gives ethylene (88%) and acrylic aldehyde (85%) thus incidentally offering a convenient method for the preparation of the aldehyde (*Bremner et al.*, J. chem. Soc., 1946, 1018; cf. *Wilson*, J. Amer. chem. Soc., 1947, **69**, 3004):

$CH_2{=}CH_2$ + $CH_2{=}CH{-}CHO$

Preparation of substituted dihydropyrans. These compounds are obtained in a variety of ways, three of which are mentioned: (*1*) dehydration of the appropriate keto-alcohols, *e.g.*, 5-oxohexan-1-ol (1-hydroxyhexan-5-one) on heating gives 2-methyl-5,6-dihydro-4-pyran:

$HOCH_2CH_2CH_2CH_2COMe$ $\xrightarrow{-H_2O}$ Me

Several 5-alkyltetrahydrofurylmethanols have been dehydrated by heating with alumina at 360° to give 2-alkyl-2,3-dihydro-4-pyrans (*J. Colonge* and *A. Girantet*, Compt. rend., 1962, **254**, 498):

Me CH_2OH → Me

The stereochemistry of the catalytic dehydration of secondary tetrahydrofurfuryl alcohols has been studied and the dihydropyrans obtained are indicated below (*G. Descotes* and *A. Laily*, Bull. Soc. chim. Fr., 1967, 2989):

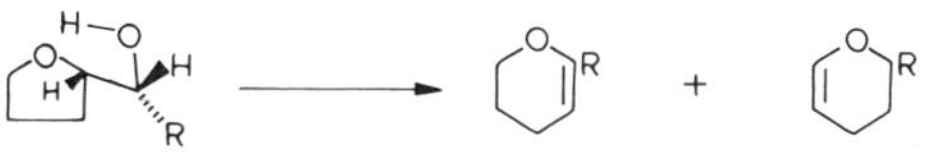

(III) (IV)

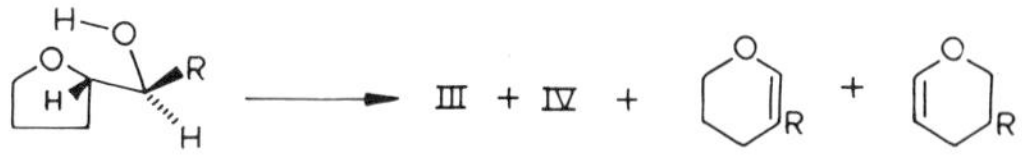

III + IV +

(*2*) Unsaturated carbonyl compounds by Diels–Alder reaction give good yields of pyran derivatives (*K. Alder et al.*, Ber., 1941, **74,** 905, 920, 926); acraldehyde in the presence of 1% of quinol affords 2,3-dihydro-4-pyran-2-carbaldehyde in good yield:

OHC·CH=CH$_2$ + O=CH–CH=CH$_2$ ⟶ OHC (2,3-dihydro-4-pyran-2-carbaldehyde)

(*3*) Derivatives of 2-alkoxy-2,3-dihydro-4-pyran may be obtained by heating ethers of α,β-unsaturated alcohols with α,β-unsaturated aldehydes under pressure at 150–300° (*R. I. Hoaglin* and *R. G. Kelso*, B.P. 739,128/1955):

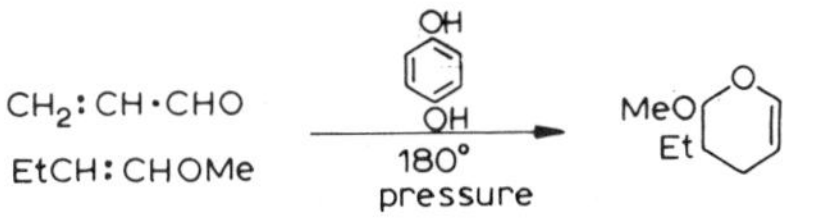

CH_2:CH·CHO + EtCH:CHOMe —(quinol, 180°, pressure)→ MeO, Et (dihydropyran)

Reactions of dihydropyrans. Halogens or hydrogen halides add to dihydro-4-pyran to give 2,3-dihalogeno- or 2-halogeno-tetrahydropyrans. The halogen atom in the 2-position is removed on distillation as hydrogen halide, *e.g.* 2,3-dibromotetrahydropyran yields 5-bromo-2,3-dihydro-4-pyran:

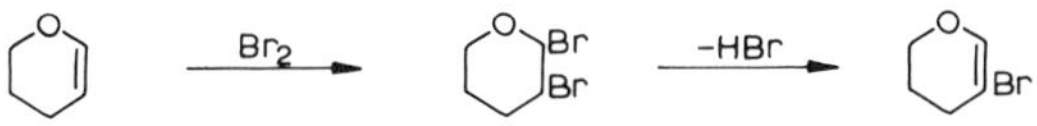

—Br_2→ (Br, Br) —−HBr→ (Br)

The 3-halogeno atom in the halogenotetrahydropyrans is relatively inert, but the 2-halogeno compounds react with Grignard reagents and with cuprous cyanide to give 2-alkyltetrahydropyrans and 2-cyanotetrahydropyrans, respectively.

Products identified from the reaction of *N*-bromosuccinimide with 2,3-dihydro-4-pyran were 3-bromo-5,6-dihydro-4-pyran, 2,3-dibromotetrahydropyran (2.5%) and both geometrical isomers of 2-succinimidyl-3-bromotetrahydropyran. The following reaction scheme accounts for the formation of the major products by a polar mechanism (*J. R. Shelton* and *C. Cialdella*, J. org. Chem., 1958, **23,** 1128):

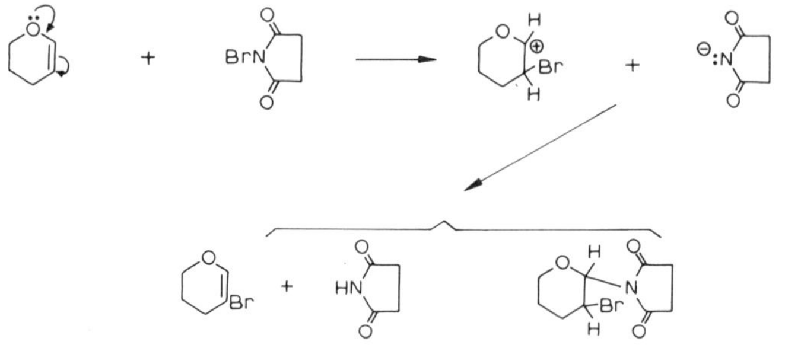

Substituted 2,3-dihydro-4-pyrans react with *N*-bromosuccinimide to give various products, depending on the substituents present in the pyran ring (*Paul* and *S. Tchelitcheff*, Bull. Soc. chim. Fr., 1956, 869, 1370; *F. Korte*, *R. Heinz* and *D. Scharf*, Ber., 1961, **94,** 825). Mechanisms of the halogenation and halogenomethoxylation of 2,3-dihydro-4-pyran have been studied by *R. U. Lemieux* and *B. Fraser-Reid* (Canad. J. Chem., 1965, **43,** 1458).

Oxidation of 2,3-dihydro-4-pyran with hydrogen peroxide in *tert*-butanol in the presence of osmium tetroxide gives tetrahydropyran-2,3-diol, which is probably in equilibrium with the open-chain aldehyde $HOCH_2 \cdot (CH_2)_2 \cdot CHOH \cdot CHO$ (*Hurd* and *C. D. Kelso*, J. Amer. chem. Soc., 1948, **70,** 1484). In consequence of this equilibrium the diol is probably a mixture of the *cis*- and *trans*-diols and not the pure *cis* isomer which would normally be expected by this method.

Oxidation with nitric acid somewhat unexpectedly gives good yields of pure glutaric acid, probably due to the initial formation of 2-hydroxy-tetrahydropyran:

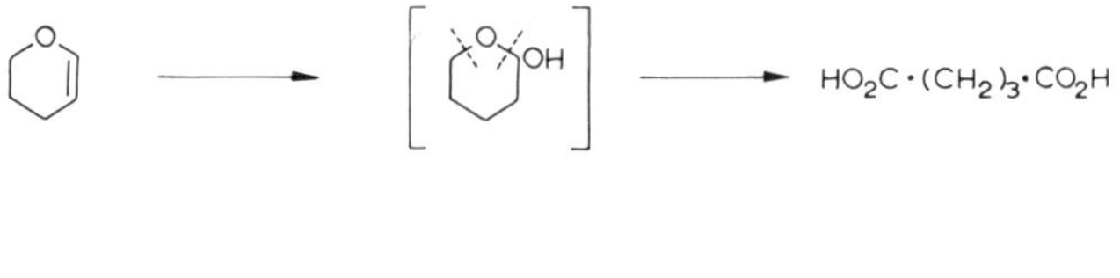

$$\longrightarrow \longrightarrow HO_2C \cdot (CH_2)_3 \cdot CO_2H$$

The ozonolysis of 2,3-dihydro-4-pyran in methanol readily affords 4-hydro-peroxy-4-methoxybutyl formate (V), which can be reduced to 4-formoxy-butyraldehyde (VI) or dehydrated to methyl 4-formoxybutyrate (VII) (*Q. E. Thompson*, J. org. Chem., 1962, **27,** 4498):

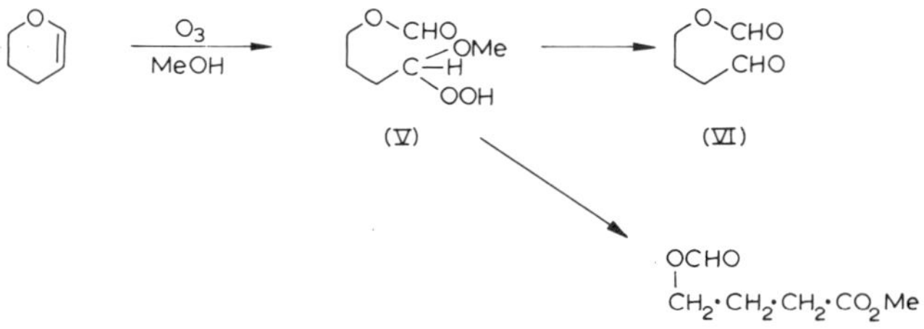

O_3 / MeOH → (V) → (VI); (V) → (VII)

(V) (VI)

OCHO
$CH_2 \cdot CH_2 \cdot CH_2 \cdot CO_2Me$

(VII)

The hydroformylation of 2,3-dihydro-4*H*-pyran takes place preferentially in the 6-position and only to a small extent in the 5-position. When there is a substituent in the 6-position and a free 5-position, hydroformylation occurs exclusively in the 5-position. 5-Alkoxycarbonyl derivatives of 2,3-dihydro-4-pyran can be hydroformylated only with difficulty (*J. Falbe* and *Korte*, Ber., 1964, **97,** 1104):

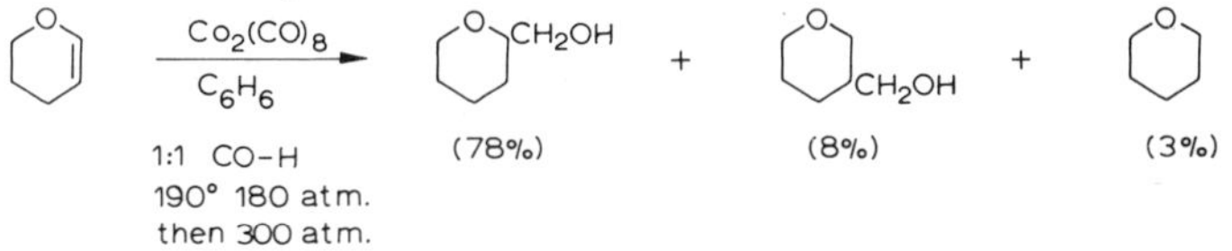

$Co_2(CO)_8$, C_6H_6; 1:1 CO–H, 190° 180 atm. then 300 atm. → (CH_2OH) (78%) + (CH_2OH) (8%) + (3%)

2,3-Dihydro-4-pyran reacts with some Schiff bases, for example, PhCH:NPh in ether in the presence of boron trifluoride etherate to give a tetrahydropyranotetrahydroquinoline (VIII) (*L. S. Pavarov et al.*, Izvest. Akad. Nauk S.S.S.R., Ser. Khim., 1964, 179):

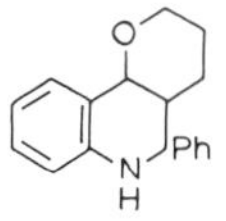

(VIII)

2,3-Dihydro-4-pyran-[2-^{14}C] isomerises over alumina at 350° to give a mixture of 2,3-dihydro-4-pyran[2-^{14}C] and 2,3-dihydro-4-pyran[6-^{14}C] (see p. 44) (*W. J. Gensler, J. E. Stouffer* and *R. G. McInnis*, J. org. Chem., 1967, **32**, 200).

Tertiary ethynylmethanols add smoothly and in high yield to 2,3-dihydro-4-pyran to form tetrahydropyranyl compounds, *e.g.* IX. The tetrahydropyranyl grouping is stable to alkali but labile to aqueous acid and exchangeable with lower alcohols by acid catalysis (*D. N. Robertson, ibid.*, 1960, **25,** 931):

$$\text{2,3-dihydro-4-pyran} + RR^1C(OH)\cdot C{:}CH \longrightarrow \text{2-tetrahydropyranyl-O-}C(R)(R^1)C{:}CH \quad (IX)$$

Diphenylacetylene reacts smoothly with an excess of 2,3-dihydro-4-pyran to give a 1:1 addition product upon irradiation at 2537Å. The product has been characterised as 7,8-diphenyl-2-oxabicyclo[4.2.0]oct-7-ene (X), on the basis of spectral evidence (*H. M. Rosenberg* and *P. Serve*, *ibid.*, 1968, **33,** 1653):

$$\text{2,3-dihydro-4-pyran} + CPh{\vdots}CPh \longrightarrow (X)$$

2-Alkoxy-2,3-dihydro-4-pyrans undergo an oxidative rearrangement to substituted tetrahydrofuran derivatives in the presence or peracids (*S. S. Hall* and *H. C. Chernoff*, Chem. and Ind., 1970, 896):

$$\text{RO-(dihydropyran)-Me} \xrightarrow[CHCl_3 \text{ or } CH_2Cl_2,\ 0°]{[3]\ Cl{-}C_6H_4{\cdot}CO_3H} \text{RO-(tetrahydrofuran)-Ac}$$

(ii) 5,6-Dihydro-2-pyran

5,6-**Dihydro-2-pyran** (*5,6-dihydro-2H-pyran*, *5,6-dihydro-α-pyran*) (XI) is obtained in 75% yield by the dehydrohalogenation of 4-bromotetrahydropyran with potassium hydroxide in glycol (*R. Paul* and *S. Tchelitcheff*, Compt. rend., 1947, **224,** 1722):

$$\text{4-bromotetrahydropyran} \xrightarrow{-HBr} (XI) \xrightarrow{600°} CH_2{:}CH{\cdot}CH{:}CH_2 + HCHO$$

$$(XI) \xrightarrow[360°]{Al_2O_3} CH_2{:}CH{\cdot}CH{:}CH_2 + CO_2$$

It forms a stable dibromide and the structure is proved by the reaction with perbenzoic acid to form a peroxide which is hydrolysed to 3,4-dihydroxytetrahydropyran, and by its pyrolysis at 600° into butadiene and formaldehyde (*cf.* 2,3-dihydro-4-pyran); under the influence of hot alumina butadiene and carbon dioxide are obtained (see p. 43) (*Gensler*, *Stouffer* and *McInnes*, *loc. cit.*).

The acid- or base-catalysed condensation of crotonaldehyde gives, besides the dimeric product, 2,6-dimethyl-5,6-dihydro-2-pyran-3-carbaldehyde, an isomer which has been identified as 2,4-dimethyl-2,3-dihydro-4-pyran-3-carbaldehyde by i.r. and n.m.r. spectroscopy (*A. Losse*, Ber., 1967, **100,** 1266):

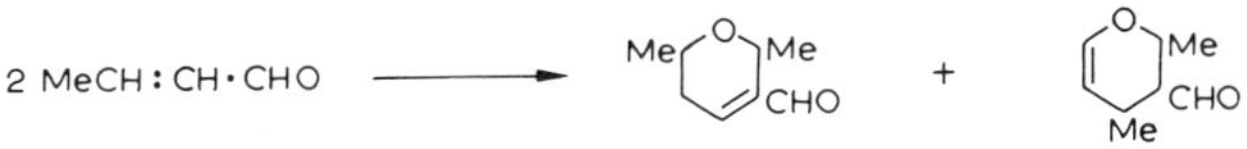

2 MeCH : CH·CHO ⟶ (Me, Me, CHO dihydropyran) + (Me, CHO, Me dihydropyran)

2-Alkoxy-5,6-dihydro-2-pyran has been subjected to addition reactions such as hydrohalogenation, hydroxyhalogenation, epoxidation and *cis*-hydroxylation, and the related 3,4-epoxide isomers, to acid hydrolysis, and addition of dimethylamine (*M. Cahu* and *G. Descotes*, Bull. Soc. chim. Fr., 1968, 2975).

(iii) Naturally occurring furo- and pyrano-dihydro derivatives

Patulin (*clavacin*), 4-*hydroxy-2-oxodihydrofuro*[3,2-c]*pyran*, m.p. 111°, *acetate*, m.p. 118–120°, *phenylhydrazone*, m.p. 149–150°, is a reduced 4-pyrone derivative first isolated from culture media of *Penicillium patulum* Bainier (*H. Raistrick et al.*, Lancet, 1943, **245,** 625) and *P. claviforme* (*F. Bergel et al.*, Nature, 1943, **152,** 750). **Clavitin,** a metabolite of *Aspergillus clavatus* is identical with patulin (*Bergel et al.*, J. chem. Soc., 1944, 415). Patulin is of great interest, partly because it is a powerful though toxic antibiotic and partly because of its exceptional chemical properties. The small molecular formula, $C_7H_6O_4$, gives no indication of the difficulty which investigators encountered in establishing the structure. The successful outcome of many researches, particularly by *Bergel* and *Raistrick*, was announced by *R. B. Woodward* and *G. Singh* and was made possible by supplementing chemical methods by careful physical measurements. Structure XII (*Woodward* and *Singh*, J. Amer. chem. Soc., 1949, **71,** 758; Experientia, 1950, **6,** 238) is now fully established (see also *I. Chmielewska*, Przemysl Chem., 1950, **29,** 46):

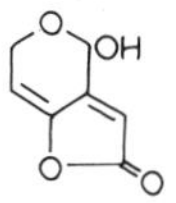

Patulin (XII) (XIII) 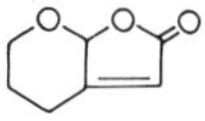 Allopatulin (XIV)

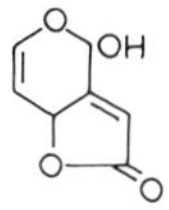

The *synthesis* of patulin involves first condensing tetrahydro-4-pyrone and ethyl mesoxalate to give the hydroxyketone XV, which after hydrolysis and heating with acetic acid and acetic anhydride yields a *cis*-acid which cyclises to XVI and the *trans*-acid XVII. Dehydration of XVI gives deoxypatulin the u.v. and i.r. spectra of which closely resemble those of patulin. Hydrogenation of the *trans*-acid gives tetrahydro-4-pyrone-3-acetic acid:

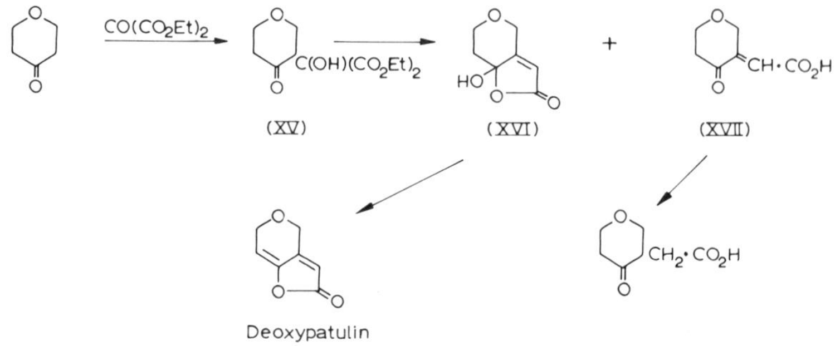

The synthesis of patulin was accomplished by brominating the acetate of the cyclised *cis*-acid XVI with *N*-bromosuccinimide (*Woodward* and *Singh*, J. Amer. chem. Soc., 1950, **72,** 1428), at position 2 ("allylic position"). Treatment with silver acetate gives the diacetate XVIII, which with a mixture of acetic acid, acetic anhydride and sulphuric acid gives patulin acetate (XIX). Hydrolysis to patulin (XII) is readily effected:

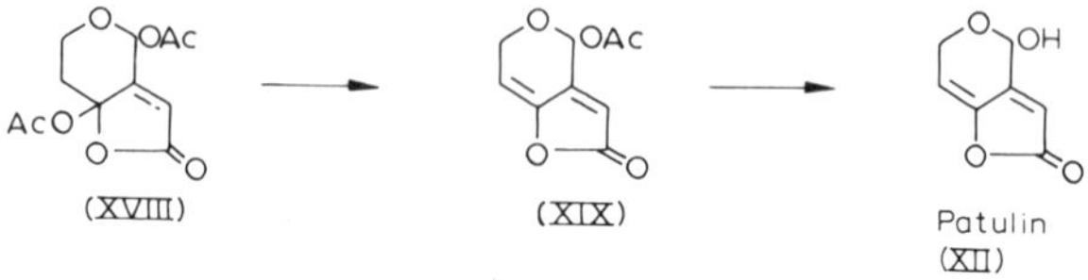

Confirmation of its structure comes from physical and chemical evidence (*F. L. Weisenborn* and *H. J. Dauben Jr.*, *ibid.*, 1949, **71,** 3853; *J. F. Grove*, J. chem. Soc., 1951, 883).

There is some spectrographic evidence that another form such as XIII may be present in small quantities. The formula XIV, which was earlier advanced for patulin (*B. G. Engel et al.*, Helv., 1949, **32,** 1166, 1752) represents another substance, **allopatulin,** m.p. 116–117°, *phenylhydrazone*, m.p. 195° (*Woodward* and *Singh*, J. Amer. chem. Soc., 1950, **72,** 5351).

Reviews of the chemistry of patulin have been given by *Woodward* and *Singh* (Experientia, 1950, **6,** 238) and *A. W. Johnson* and *H. N. Rydon* (Ann. Reports, 1950, **47,** 234); the pharmacological effects have been discussed by *A. Schweitzer* (Experimental Medicine and Surgery, 1946, **4,** 289).

Isoclavacin, m.p. 87°, is isomeric with patulin and has been synthesised from 3-ethoxalyltetrahydro-4-pyrone by opening of the ring followed successively by enolisation, lactonisation and ring-closure (*A. Cohen*, Chem. and Ind., 1949, 640):

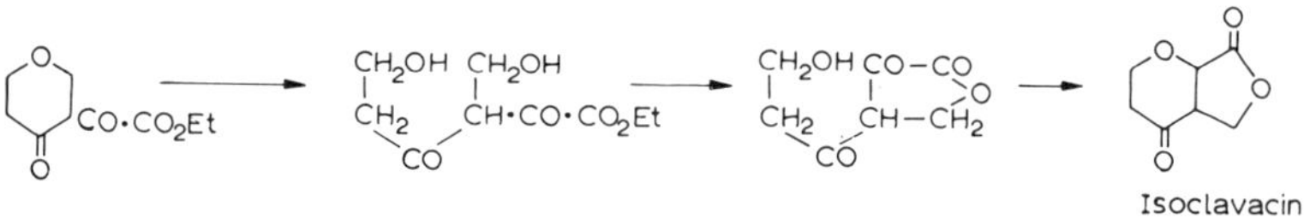

The formula is in agreement with the u.v. spectrum (λ_{max} 276 mμ, log ε 3.97) suggesting three conjugated chromophores and with i.r. data *(Grove, loc. cit.)*.

Asperuloside, m.p. 125–127°, $[\alpha]_D^{18}$ −204° (H_2O), occurs in *Galium aparine, Rubia tinctorum* and *Daphniphyllum macropodum*. Structure and absolute configuration have been assigned on the basis of chemical and spectral data (*L. H. Briggs et al.*, J. chem. Soc., 1965, 2595):

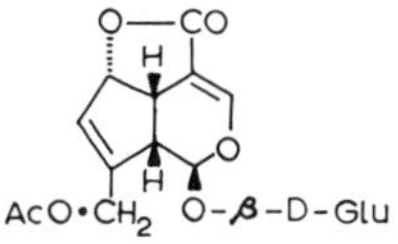

Asperuloside

Gentiopicrin *(gentiopicroside)*, $C_{16}H_{20}O_9 \cdot H_2O$, m.p. 122° (*hydrate*, m.p. 191°), $[\alpha]_D$ −196.3° (H_2O), is the bitter principle of *Gentiana lutea* and the first proposal for its structure was made by *F. Korte* (Ber., 1954, **87,** 512), and a later one by *L. Canonica et al.* (Tetrahedron Letters, 1960, 7; Tetrahedron, 1961, **16,** 192). Structure and absolute configuration have been assigned on chemical evidence and n.m.r. data (*H. Inouye et al.*, Tetrahedron Letters, 1968, 4429):

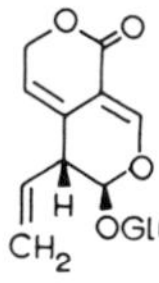

Gentiopicrin 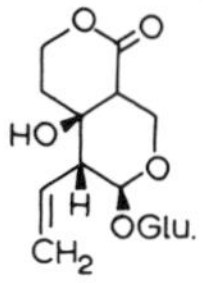 Swertiamarin 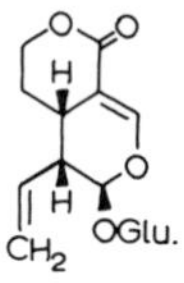 Sweroside

Swertiamarin, m.p. 103–104°, the bitter principle of *Swertia Japonica Makino*, and **sweroside,** have been assigned the structures shown above. See also opuntiol (p. 20), kawain and methysticine (p. 19).

Verbenalin, $C_{17}H_{24}O_{10}$, m.p. 182° (decomp.), $[\alpha]_D$ −185° (H_2O), is a glucoside obtained from the leaves of *Verbena officianalis* L., and its structure has been established by chemical methods and from spectral data (*G. Büchi* and *R. E. Manning, ibid.*, 1960, 5):

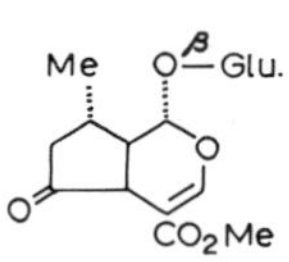

Verbenalin

(b) Tetrahydropyrans and their derivatives

(i) Tetrahydropyrans

Tetrahydropyran, *pentamethylene oxide, oxane* (XX), a liquid with a characteristic odour, b.p. 87.5–88.5°, d_4^{25} 0.8772, n_D^{25} 1.4195, $\mu = 1.87$D, is quickly and quantitatively prepared by the catalytic hydrogenation of dihydropyran over Raney nickel (*D. W. Andrus* and *J. R. Johnson*, Org. Synth., 1943, **23,** 90). It is also obtained by heating pentamethylene bromide with water and zinc oxide in a sealed tube (*H. T. Clarke*, J. chem. Soc., 1912, **101,** 1802; *J. S. Allen* and *H. Hibbert*, J. Amer. chem. Soc., 1934, **56,** 1398). Tetrahydropyran forms peroxides and is stabilised by the addition of an antioxidant. It is a good solvent for resins. The Raman spectra have been measured (*L. Kahovec* and *K. W. F. Kohlrausch*, Z. physik. Chem., 1937, **B, 35,** 29), its heat of combustion determined (*R. C. Cass et al.*, J. chem. Soc., 1958, 1406; *A. Snelson* and *H. A. Skinner*, Trans. Faraday Soc., 1961, **57,** 2125; *A. S. Pell* and *G. Pilcher*, *ibid.*, 1965, **61,** 71), and critical properties and vapour pressures are recorded (*K. A. Kobe* and *J. F. Mathews*, J. chem. Eng. Data, 1970, **15,** 182).

Tetrahydropyran is oxidised by nitric acid to succinic acid in good yield and when heated with acetic anhydride and zinc chloride at 200° gives pentane-1,5-diol diacetate (*R. Paul*, Bull. Soc. chim. Fr., 1941, [v], **8,** 369):

(XX) ⟶ $AcOH_2C$–CH_2–CH_2–CH_2–CH_2OAc

With acetic anhydride and hydrogen bromide it slowly yields 5-bromo-*n*-amyl acetate (*C. L. Wilson*, J. chem. Soc., 1945, 48), and heating it with acetyl chloride and zinc chloride affords 5-chloropentan-1-ol (*M. E. Synerholm*, J. Amer. chem. Soc., 1947, **69,** 2581). 1,5-Dibromopentane can be readily prepared by boiling tetrahydropyran with hydrobromic acid and sulphuric acid (*Andurs*, Org. Synth., 1943, **23,** 67).

Tetrahydropyran differs fundamentally from tetrahydrofuran in that it possesses a strainless ring, *cf.* the strainless non-planar structure of the pyranose ring (*T. R. R. McDonald* and *C. A. Beevers*, Acta Cryst., 1952, **5,** 654). Tetrahydropyrans may be prepared by the prototropic dehydration of the corresponding 1,7- or 1,6-allyldiols in the allyl position (*G. Ohloff, K. H. Schulte-Elte* and *B. Willhalm*, Helv., 1964, **47**, 602).

2-**Hydroxytetrahydropyrans** are known (p. 40). Molecular refractivities and boiling points indicate that they exist in the cyclic form XXI, but they exhibit ring–chain isomerism and are, at least in aqueous solution, in equilibrium with an open-chain hydroxyaldehyde form XXII. This is shown by the fact that they give positive, albeit somewhat sluggish, Schiff's and Fehling's tests, form oximes, etc., are oxidised to δ-hydroxyvaleric acids and are reduced to pentane-1,5-diols (*Paul*, Bull. Soc. chim. Fr., 1934, **1**, 971; *B. Helferich* and *T. Malkomes*, Ber., 1922, **55**, 702). With methanol and hydrogen chloride they yield hemiacetals. 5-Hydroxypentanal reacts in alkaline condensations in its tautomeric form, 2-hydroxytetrahydropyran, condensing with itself to give 2-hydroxy-3-(2-tetrahydropyranyl)tetrahydropyran (XXIII) (*J. Colonge* and *P. Corbet*, Compt. rend., 1957, **245**, 974):

(XXI) ⇌ $HO(CH_2)_4 \cdot CHO$ (XXII) (XXIII)

2-Hydroxytetrahydropyran also condenses with compounds containing an active methylene group to yield compounds of type XXIV (*Colonge*, *J. Dreux* and *M. Coblentz*, *ibid*., 1960, **250**, 3202), and with olefins to afford acetals of type XXV (*Coblentz*, *J. Royer* and *Dreux*, Bull. Soc. chim. Fr., 1963, 1279):

(XXIV): 2-($CHRR^1$)tetrahydropyran

R = Ac, Bz, CN, CO_2Et
R^1 = Ph, Ac, CO_2Et

(XXV): 2-($OCRR^1 \cdot CHR^2 \cdot COR^3$)tetrahydropyran

R =	H	H	H	Me	H	H
R^1 =	H	H	Me	Me	H	Me
R^2 =	H	Me	H	H	H	H
R^3 =	Me	Me	Me	Me	H	H

2-**Hydroxytetrahydropyran** (*δ-hydroxyvaleraldehyde, 5-hydroxypentanal*), viscous oil, b.p. 62–66°/9–10 mm, n_D^{25} 1.4513 (*G. F. Woods Jr.*, Org. Synth., 1947, **27**, 43). 2-*Hydroxy-6-methyltetrahydropyran*, b.p. 71–78°/11 mm, 80°/16 mm, d_{15}^{18} 1.055 (*Helferich* and *Malkomes, loc. cit.*).

At 100° and under pressure in the presence of Raney nickel, 2,6-dimethyl-4-pyrone (p. 30) in alcohol is hydrogenated to give 2,6-dimethyl-4-hydroxytetrahydropyran and heptan-2,4-diol (*O. Riobe*, *V. Herault* and *L. Gouin*, Compt. rend., 1963, **256**, 1542). When di-*tert*-butyl peroxide is thermally decomposed in the presence of 2-methoxytetrahydropyran rearrangement takes place to give methyl valerate (*E. S. Huyser*, J. org. Chem., 1960, **25**, 1820):

2-methoxytetrahydropyran ⟶ $Me(CH_2)_3 \cdot CO_2Me$

A number of 2-alkoxy- and 2-aryloxy-tetrahydropyrans have been hydrogenolysed in ether by lithium tetrahydridoaluminate. As the alkyl group attached to the *exo*-O atom is changed from primary to tertiary, the proportion of ring cleavage to give XXVI, to side-chain cleavage to give tetrahydropyran and the alcohol ROH, increases. Only side-chain cleavage occurs with 2-aryloxytetrahydropyrans (*U. E. Diner* and *R. K. Brown*, Canad. J. Chem., 1967, **45,** 2547):

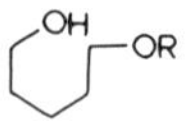

(XXVI)

A conformational analysis has been made of some 2-alkoxytetrahydropyrans (*G. O. Pierson* and *O. A. Runquist*, J. org. Chem., 1968, **33,** 2572).

6-Ethyl-2,2-dimethyltetrahydropyran-6-carboxylic acid has been synthesised, and the acid undergoes a reaction analogous to the cinenic acid rearrangement (cf. Vol. IIB, p. 205) to give 2,2-dimethyl-6-oxo-octanoic acid, the structure of which has been proved by synthesis. The transformation can be rationalised in terms of a carboxyl transfer process (*J. Meinwald* and *J. T. Ouderkirk*, J. Amer. chem. Soc., 1960, **82,** 480):

3-Alkyl-4-halogenotetrahydropyrans are obtained when α-olefins are condensed with paraformaldehyde and hydrogen halides. The best yields using hydrogen chloride are achieved at −60 to −70° (*P. R. Stapp*, J. org. Chem., 1969, **34,** 479):

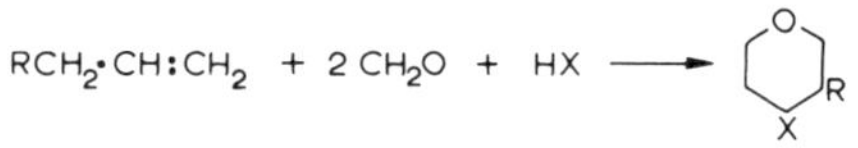

$$RCH_2 \cdot CH{:}CH_2 + 2\,CH_2O + HX \longrightarrow$$

3-Chloro-2-hydroxytetrahydropyran is prepared from 2,3-dihydropyran and hypochlorous acid (*G. Descotes* and *J. C. Soula*, Bull. Soc. chim. Fr., 1964, 2636). Alcoholates and phenolates react stereospecifically with 3-chloro-2-hydroxytetrahydropyran to give *trans*-3-hydroxy-2-alkoxy-, or *trans*-3-hydroxy-2-phenoxy-tetrahydropyrans probably through the formation of an intermediate epoxide (*G. Bakassian* and *Descotes*, Compt. rend., Ser. C, 1966, **262,** 1691):

A variety of substituted tetrahydropyrans have been found to have floral or vegetable odours (*N. P. Soloveva et al.*, Riechst. Aromen. Koerperpflegem, 1969, **19,** 252; C.A., 1969, **71,** 91208).

Tetrahydropyran-3-carboxylic acid, *ethyl ester*, b.p. 101–103°/19–20 mm, n_D^{21} 1.443 (*S. M. Sherlin et al.*, J. gen. Chem., U.S.S.R., 1938, **8,** 22); **tetrahydropyran-4-carboxylic acid,** m.p. 87°, b.p. 145–150°/18 mm, *ethyl ester*, b.p. 82.5°/12 mm (*C. S. Gibson* and *J. D. A. Johnson*, J. chem. Soc., 1930, 2527); **tetrahydropyran-2,6-dicarboxylic acid,** *cis-meso-*, m.p. 193°, *dimethyl ester*, m.p. 54°, b.p. 156°/8 mm (*W. Czornodola*, Rocz. Chem., 1936, **16,** 459); **tetrahydropyran-4,4-dicarboxylic acid**, m.p. 172–173°, *diethyl ester*, b.p. 152–155°/21 mm (*Gibson* and *Johnson, loc. cit.*).

Cladinose, $C_8H_{16}O_4$, b.p. 120–130°/0.25 mm (bath temp.), $[\alpha]_D^{25}$ −23.1° (H_2O) a hydrolysis product of the antibiotic erythromycin (*E. H. Flynn et al.*, J. Amer. chem. Soc., 1954, **76,** 3121) is 2,5-dihydroxy-4-methoxy-4,6-dimethyltetrahydropyran (*P. F. Wiley* and *O. Weaver*, *ibid.*, 1955, **77,** 3422):

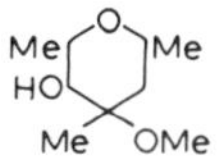

Me Me HO Me OMe

Cladinose

(ii) Hydrogenated 4-pyrones

The 4-pyrones are best reduced by catalytic methods, chemical methods generally resulting in ring-fission (p. 26).

Tetrahydro-4-pyrone, colourless liquid, b.p. 163–166°/742 mm, 67–68°/18 mm, $n_D^{24.5}$ 1.4529, is miscible with water, has a pleasant odour (*W. Borsche*, Ber., 1915, **48,** 682; *R. Cornubert* and *P. Robinet*, Bull. Soc. chim. Fr., 1933, [iv], **53,** 565; *S. R. Cawley* and *S. G. P. Plant*, J. chem. Soc., 1938, 1214), and forms a 2,4-*dinitrophenylhydrazone*, m.p. 186–187°, a *semicarbazone*, m.p. 171.5°, and *dibenzylidene* derivative, m.p. 186°. The reactivity of the 3- and 5-methylene group is utilised in synthetic work (p. 45). 2,6-**Dimethyltetrahydro-4-pyrone,** b.p. 170°/737 mm, *semicarbazone*, m.p. 192° (*Borsche, loc. cit.*); some other tetrahydro-4-pyrans (*Cornubert* and *Robinet, loc. cit.*, 619; *J. W. Baker*, J. chem. Soc., 1944, 297).

The hydrogenation of 4-pyrones sometimes yields dihydro- as well as tetrahydro-4-pyrones. Comanic acid, for instance, with hydrogen and a palladium–barium sulphate catalyst gives a 14% yield of 2,3-dihydro-4-pyrone-6-carboxylic acid (*J. Attenburrow et al.*, *ibid.*, 1945, 571). 3-**Methyl**-2,3-**dihydro**-6-**phenyl-4-pyrone,** m.p. 69°, 2,4-dinitrophenylhydrazone, m.p. 239°, is a constituent of *Myrtus bullata* (*C. W. Brandt et al.*, *ibid.*, 1954, 3245).

3. Benzo[*b*]pyrans (5,6-benzopyrans, chromenes) and their derivatives

In the older literature compounds with this ring system, and those with a saturated heterocyclic ring are called *chromenes* and *chromans*, respectively.

The now official systematic names, benzo[*b*]pyran and dihydrobenzo[*b*]-pyran, have come into general use only very recently. The former names are still acceptable trivial names according to I.U.P.A.C. rules and so will be adopted for convenience in this section.

Derivatives of benzo[*b*]pyran, 2*H*-chromene and dihydrobenzo[*b*]pyran, chroman, occur widely in nature. The oxo-compounds, *coumarin* and *chromone*, together with 2-phenylchromone, *flavone*, are the parent compounds of many naturally occurring substances. In none of these compounds are both rings aromatic but in the *benzopyrylium* or *chromylium salts*, the heterocyclic ring assumes aromatic character. Of this type are the anthocyanins and anthocyanidins. The dihydro compound *chroman* gives rise to oxygenated substances such as the chroman-4-ones and 2-phenylchroman-4-ones or *flavanones*. Many hydroxyflavanones are found in nature. Chroman and its derivatives are described in Section 4.

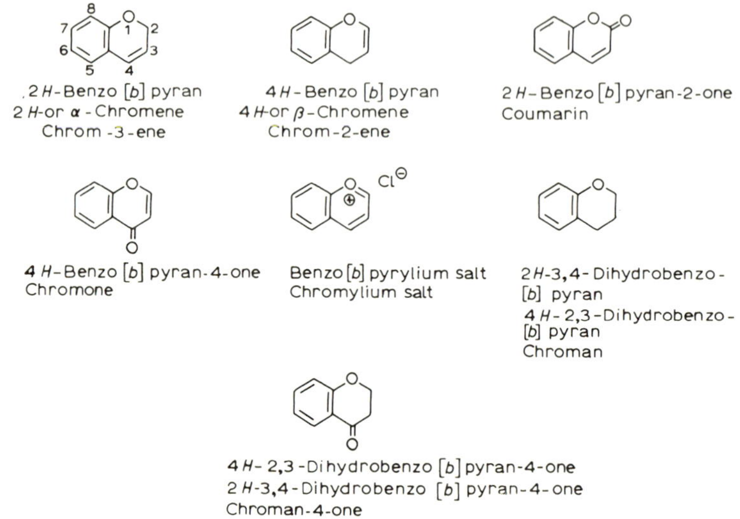

(*a*) *Chromenes and their oxidation products*

(*i*) *Chromenes, alkyl- and aryl-chromenes and flavenes*

(*1*) 2-**Chromene,** 4*H*- or *β*-**chromene,** 4*H*-**benzo**[*b*]**pyran** (I), is obtained as a liquid, b.p. 77°/9 mm, 44°/1.3 mm, n_D^{25} 1.5551, along with 2-methylcoumarone (2-methylbenzo[*b*]furan) (II), when 2-hydroxymethylcoumaran (2,3-dihydro-2-hydroxymethylbenzo[*b*]furan) is heated at 350–400° (*P. Maitte*, Ann. Chim., 1954, [xii], **9,** 460):

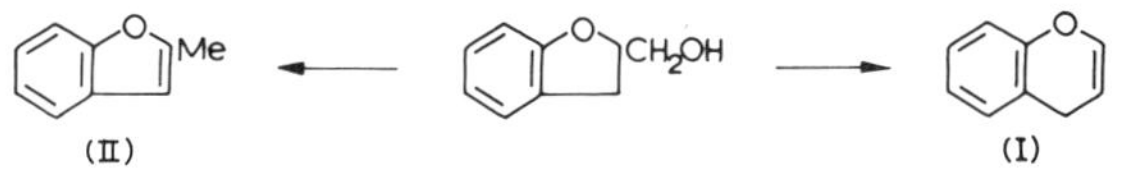

It is also obtained by the pyrolysis of 2-chromanyl acetate at 410°; i.r. and n.m.r. data have been reported (*W. E. Parham* and *L. D. Huestis*, J. Amer. chem. Soc., 1962, **84,** 813).

Reactions. 4*H*-Chromene is reduced quantitatively by hydrogen and Raney nickel to chroman, and with mercuric chloride and potassium acetate gives a mercurichloride which is easily hydrolysed by hydrochloric acid to chroman-2-ol. Bromine yields the unstable 2,3-dibromochroman, which with sodium ethoxide gives 3-*bromo*-4H-*chromene*, b.p. 119°/13 mm. The 4*H*-chromenes react at the double bond with acids and alcohols to give chroman-2-ols or their ethers, respectively; 4*H*-chromene with ethanol and hydrogen chloride yields 2-*ethoxychroman*, b.p. 111.5–112°/12 mm. The reaction between 4*H*-chromene and dichlorocarbene affords 1,1-dichloro-1a,7a-dihydro-4*H*-benzo[*b*]cyclopropa[*e*]pyran in good yield (*Parham* and *Huestis*, *loc. cit.*; *B. Graffe et al.*, Compt. rend., Ser. C, 1969, **269,** 992):

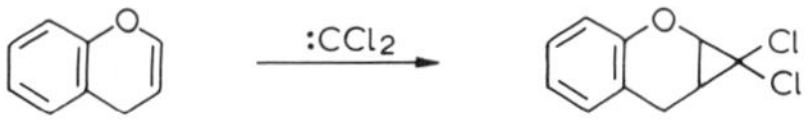

4,4-Dichloro-4*H*-chromene with nucleophilic reagents (water and aromatic amines) gives the normal products, but 3,4,4-trichloro-4*H*-chromene is attacked by water and aromatic amines at position 2 to give the corresponding ethers or amines (*V. A. Zagorevskiĭ*, *I. D. Tsvetkova* and *E. K. Orlova*, Khim. Geterotsikl Soedin., 1967, 786):

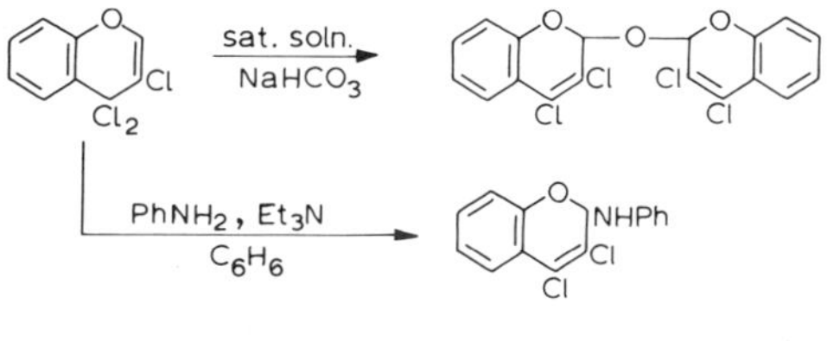

4*H*-Chromene following treatment with bromine is reacted with phenylmagnesium bromide to give 2-bromo-1-(2-hydroxyphenyl)-3-phenylpropan-3-ol, which on heating with *p*-toluenesulphonic acid yields *trans*-3-bromoflavan, converted into 2*H*-flavene by treatment with methanolic potassium hydroxide (*G. Descotes* and *D. Missos*, Synthesis, 1971, 149):

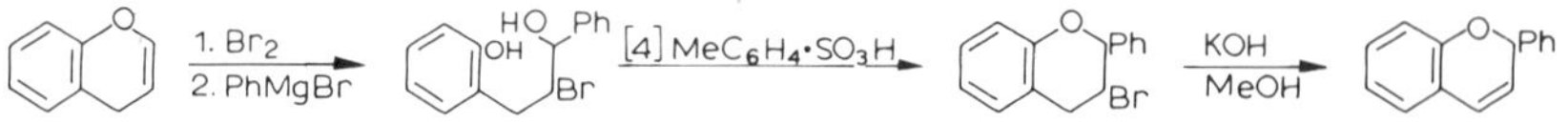

Arylcoumaranones have been cyclised using phosphorus pentoxide to the corresponding chromene derivatives (*R. B. Desai* and *J. N. Ray*, J. Indian chem. Soc., 1958, **35,** 83):

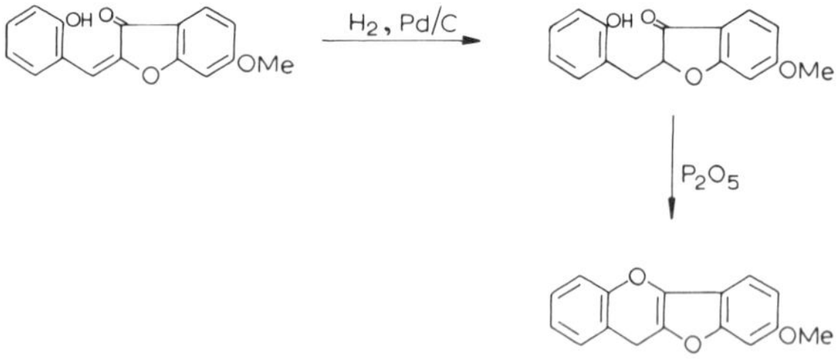

Chromone-2-carboxylic acid on boiling with excess of thionyl chloride yields 4,4-dichloro-4*H*-chromene (*Zagorevskiĭ* and *D. A. Zykov*, Zhur. obshcheĭ Khim., 1960, **30,** 3100):

(2) **Substitued** 4*H*-**chromenes**. 2,4,4-*Trimethyl*-4H-*chromene*, m.p. 104–106°/16 mm, is obtained by treating 1,1,3-trimethylindanyl hydroperoxide with sulphuric acid (*W. Webster* and *D. P. Young*, J. chem. Soc., 1956, 4785), and by dehydrating 2,4,4-trimethylchroman-2-ol in boiling benzene with oxalic acid. 4-*Methyl*-2,4-*diphenyl*-4H-*chromene*, m.p. 88–89° (*R. C. Elderfield* and *T. Piao King*, J. Amer. chem. Soc., 1954, **76,** 5439). 2,4-*Diphenyl*-4H-*chromene*, m.p. 110° (*A. Löwenbein*, Ber., 1924, **57,** 1517), is prepared by boiling with acetic acid 3-(2-hydroxyphenyl)-1,3-diphenylpropan-1-one (III), one of the products obtained on treating coumarin with phenylmagnesium bromide, the other product being 3-(2-hydroxyphenyl)-1,1-diphenylprop-2-en-1-ol (IV), which on boiling with acetic acid gives 2,2-diphenyl-2*H*-chromene (*C. S. Barnes*, *J. L. Occolowitz* and *M. I. Strong*, Tetrahedron, 1963, **19,** 839):

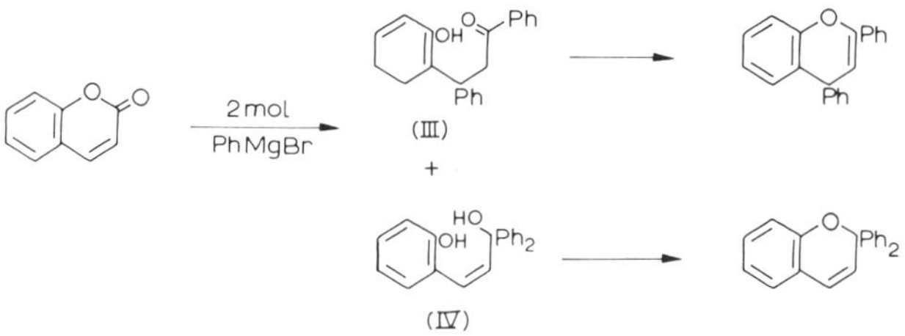

2,4-*Bis* (*2-methoxyphenyl*)-4H-*chromene*, m.p. 141–142° (*W. D. Cotterill, R. Livingstone* and *M. V. Walshaw*, J. chem. Soc., 1970, 1758).

(*3*) **Benzo-4*H*-chromenes.** The following types have been described:

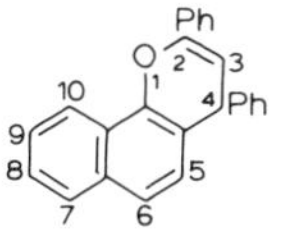

(V) 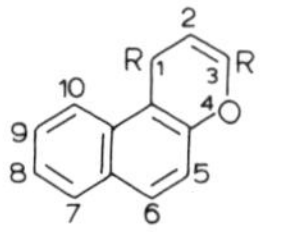 (VI) 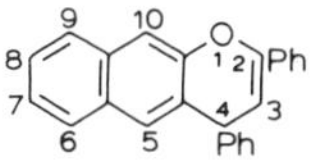 (VII)

2,4-*Diphenyl-7,8-benzo*-4H-*chrom-2-ene* (2,4-*diphenyl*-4H-*naphtho*[1,2-b]*pyran*) (V), m.p. 150–151° (*J. Cottam* and *Livingstone*, *ibid.*, 1964, 5228); 2,4-*diphenyl-5,6-benzo*-4H-*chromene* (1,3-*diphenyl*-1H-*naphtho*[2,1-b]*pyran*) (VI, R = Ph), m.p. 205° (*Livingstone, D. Miller* and *S. Morris*, *ibid.*, 1960, 5148); 2,4-*bis-(2-methoxyphenyl)-5,6-benzo*-4H-*chromene* (1,3-*bis-(2-methoxyphenyl)*-1H-*naphtho*[2,1-b]*pyran*) (VI, R = [2]-$MeOC_6H_4$), m.p. 183°; 2,4-*bis-(3-methoxyphenyl)-5,6-benzo*-4H-*chromene* (1,3-*bis-(3-methoxyphenyl)*-1H-*naphtho*[2,1-b]*pyran*) (VI, R = [3]$MeOC_6H_4$), m.p. 97°; 2,4-*bis-(4-methoxyphenyl)-5,6-benzo*-4H-*chrom-2-ene* (1,3-*bis-(4-methoxyphenyl)*-1H-*napththo*-[2,1-b]*pyran* (VI, R = [4]$MeOC_6H_4$), m.p. 196° (*Cotterill, Livingstone* and *Walshaw*, *loc. cit.*); 2,4-*diphenyl-6,7-benzo*-4H-*chrom-2-ene* (2,4-*diphenyl*-4H-*naphtho*[2,3-b]-*pyran*) (VII), m.p. 186–187° (*Cottam* and *Livingstone*, J. chem. Soc., 1965, 6646).

The colouring matters from *Croton sparsiflorus* have been studied and one of them shown probably to have structure VIII (*C. S. Pande* and *J. D. Tewari*, J. Indian chem. Soc., 1962, **39**, 545):

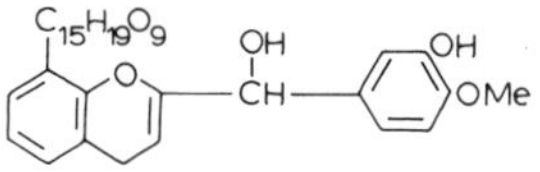

(VIII)

(*4*) 3-**Chromene,** α- or 2*H*-**chromene,** 2*H*-**benzo**[*b*]**pyran,** b.p. 92–92.5°/15 mm, d_4^{16} 1.0993, n_D^{25} 1.5883, is prepared by distilling 4-ethoxychroman, obtained from 4-bromochroman and sodium ethoxide (*Maitte, loc. cit.*), or by dehydrating chroman-4-ol (*Parham* and *Huestis, loc. cit.*). Dehydration over anhydrous copper sulphate gives a 70–80% yield of 2*H*-chromene.

Dehydration of 4-substituted chroman-4-ols in the presence of *p*-toluenesulphonic acid or anhydrous copper sulphate yields a mixture of 4-alkyl-2*H*-chrom-3-enes and 4-alkylidenechromans (*F. Baranton, G. Fontaine* and *Maitte*, Bull. Soc. chim. Fr., 1968, 4203).

Other attempts to prepare 2*H*-chromene have been unsuccessful (*R. Adams* and *R. E. Rindfuss*, J. Amer. chem. Soc., 1919, **41**, 648) but a method for the preparation of 2*H*-chromene and its derivatives involving the pyrolysis of *cis*-2-hydroxycinnamyl alcohols has been developed (*J. N. Chaterjea*, J. Indian chem. Soc., 1959, **36**, 76).

Reactions. 2*H*-Chromene is hydrogenated in the presence of Raney nickel to chroman and with chlorine and bromine it forms 3,4-*dichlorochroman*, m.p. 89–90°, and 3,4-*dibromochroman*, m.p. 127–128°, respectively (p. 224) (*Cotterill*, *Cottam* and *Livingstone*, J. chem. Soc., C, 1970, 1006).

Treatment of 2*H*-chromene with osmium tetroxide or potassium permanganate affords *cis*-chroman-3,4-diol, and with performic acid the *trans*-diol (*Baranton*, *Fontaine* and *Maitte*, Compt. rend. Ser. C, 1967, **264**, 410). The reaction between dichlorocarbene and 2*H*-chromene gives 1,1-dichloro-1a,7b-dihydro-2*H*-benzo[*b*]cyclopropa-[*d*]pyran (*Parham* and *Huestis*, *loc. cit.*):

:CCl2 → Cl Cl

Using a platinum oxide catalyst the 2*H*-chromenes are hydrogenated to the corresponding chromans, with ozone they yield salicylaldehyde, they decolorise aqueous solutions of potassium permanganate and with concentrated sulphuric acid give a red colour. The action of bromine or chlorine on 2,2-dimethylchromene and on the related dimethylbenzochromans (dimethylnaphthopyrans) results in addition or substitution products, depending on the conditions (*Livingstone et al.*, J. chem. Soc., 1962, 76; 1965, 5266; 1966, 2013; C, 1967, 1472).

The coloured photoproducts of the chromenes have been shown by spectroscopy and synthesis to be *o*-quinoneallides (*J. Kole* and *R. S. Becker*, J. phys. Chem., 1967, **71**, 4045):

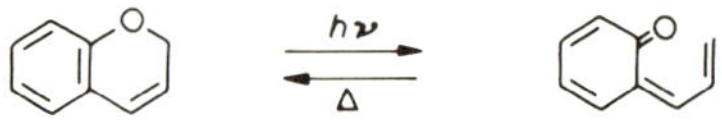

$h\nu$ ⇌ Δ

(5) **Substituted 2*H*-chromenes**. Acrolein and crotonaldehyde react with tetrachlorobenzyne to give 5,6,7,8-tetrachloro-2*H*-chromene (IX, R = H), and its 2-methyl derivative (IX, R = Me), respectively (*H. Heaney* and *J. M. Jablonski*, Chem. Comm., 1968, 1139; *Heaney et al.*, Amer. chem. Soc. Div. Petrol. chem. Prepr., 1969, **14**, C28):

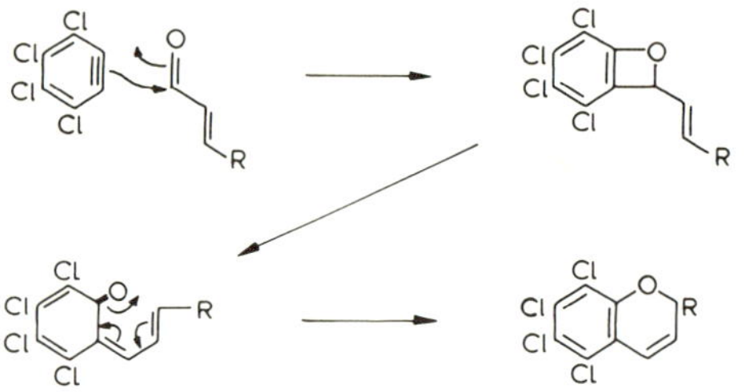

(IX)

2-Methyl-2*H*-chromene is obtained together with dihydrobenzoxepin (XI) from the aldehyde X and sodium ethoxide; dihydrobenzoxepin on boiling with the ethoxide rearranges to the chromene (*E. E. Schweizer* and *R. Schepers*, Tetrahedron Letters, 1963, 979):

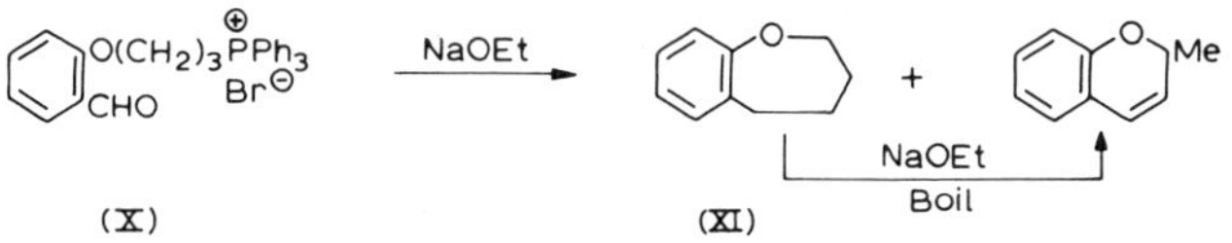

(X) → NaOEt → (XI) + 2-methylchromene; (XI) → NaOEt, Boil → 2-methylchromene

2,2-Dimethylchromenes are of particular interest since many of their derivatives are found in nature. They may be prepared by the action of methylmagnesium halide on coumarins (*L. I. Smith* and *P. M. Ruoff*, J. Amer. chem. Soc., 1940, **62**, 145) (p. 104):

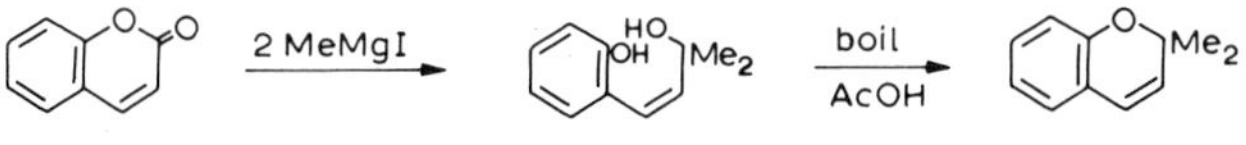

coumarin → 2 MeMgI → (XII) → boil, AcOH → 2,2-dimethylchromene

The allylic diol XII, 4-(hydroxyphenyl)-2-methylbut-3-en-2-ol, under acid conditions affords a dimer XIII of 2,2-dimethylchromene:

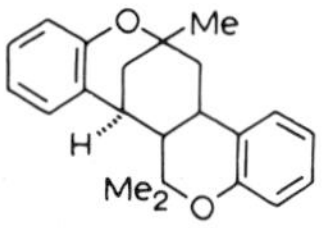

(XIII)

2,2-Dimethylchromene is also obtained by reacting coumarin with methyl-lithium (*Barnes, Occolowitz* and *Strong, loc. cit.*).

Aryl 1,1-dimethylpropargyl ethers may be rearranged to 2,2-dimethylchromenes in boiling diethylaniline:

The reaction has been used to prepare the natural products, ageratochromene, evodionol methyl ether, lapachenole, seselin, luvangetin and braylin (see next section, p. 63). A base-catalysed Wessely–Moser rearrangement is reported (*J. Hlubucek, E. Ritchie* and *W. C. Taylor*, Tetrahedron Letters, 1969, 1369; Austral. J. Chem., 1971, **24**, 2347).

2,2-Dimethylchromenes are also synthesised by the pyridine-catalysed condensation of 3-hydroxyisovaleraldehyde dimethyl acetal (or 3-methylcrotonaldehyde) with appropriate phenols; 4-acetylresorcinol heated with 3-hydroxyisovaleraldehyde dimethyl acetal (2 mol.) and pyridine gives 6-acetyl-5-hydroxy-2,2-dimethylchromene (XIV):

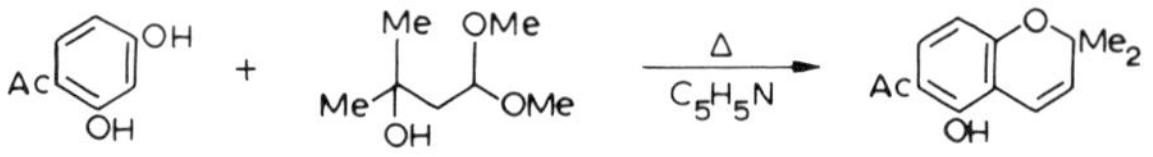

(XIV)

The method has been applied in the synthesis of lonchocarpin, evodionol 7-methyl ether, jacacreubin and acronycine (*W. M. Bandaranayake*, *L. Crombie* and *D. A. Whiting*, Chem. Comm., 1969, 970). The catalysed condensation of 3-hydroxyisovaleraldehyde dimethyl acetal with phloroacetophenone gives six characterisable products all containing the 2,2-dimethylchromene ring system, including the tricyclic pyranochromene XV (*W. J. G. Donnelly* and *P. V. R. Shannon*, *ibid*., 1971, 76):

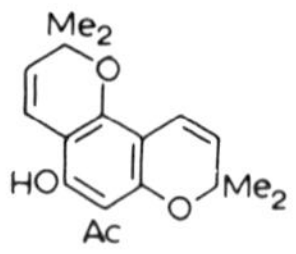

(XV)

The Wittig reaction on *o*-quinones affords a further method for the synthesis of 2,2-dimethylchromenes; 9,10-phenanthraquinone and 1,2-naphthoquinone on treatment with 3-methyl-2-butenylidenetriphenylphosphorane, $Me_2C{:}CH{\cdot}CH{:}PPh_3$ give XVI, and a mixture of XVII and XVIII, respectively (*G. Cardillo*, *L. Merlini* and *S. Servi*, Ann. Chim., Rome, 1970, **60**, 564):

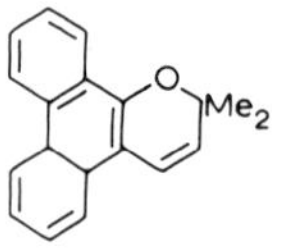

(XVI) 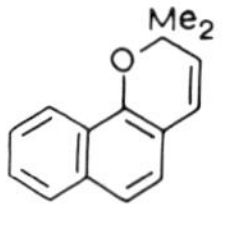 (XVII) 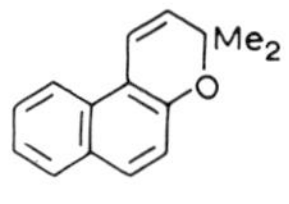 (XVIII)

2,2-*Dimethyl*-2H-*chromene*, b.p. 96–97°/15 mm, d_4^{20} 1.0163, n_D^{20} 1.5490; 2,2-*diethyl*-2H-*chromene*, b.p. 99–100°/2.8 mm; d_4^{20} 1.0049, n_D^{20} 1.5428; also *dipropyl*-, *dibutyl*- and *diamyl*-2H-*chromene* (*R. L. Shriner* and *A. G. Sharp*, J. org. Chem., 1939, **4**, 574); 5-*methoxy*-, 6-*methoxy*-, 7-*methoxy*- and 8-*methoxy*-2,2-*dimethyl*-2H-*chromene*, b.pp. 87–89°/3 mm, 132–136°/15 mm, 140–141°/16 mm and 114–115°/2 mm, respectively (*Livingstone* and *R. B. Watson*, J. chem. Soc., 1957, 1509; *J. D. Hepworth* and *Livingstone*, *ibid*., C, 1966, 2013); 4-*ethyl*-2H-*chromene*, b.p. 121°/13 mm, $d_4^{20.5}$ 1.064; 2,2,3-*trimethyl*-2H-*chromene*, b.p. 83°/1 mm, d^{25} 1.0278 (*W. R. Nummy* and *D. S.

Tarbell, J. Amer. chem. Soc., 1951, **73,** 1500); 2,2-*diphenyl*-2H-*chromene*, m.p. 93° (*A. Löwenbein*, Ber., 1924, **57,** 1517); 4-*phenyl*-2H-*chromene*, b.p. 137–138°/0.6 mm, $d_4^{21.5}$ 1.408; 3-*methyl*-2,4-*diphenyl*-2H-*chromene*, m.p. 91°; 4-*methyl*-2,2-*diphenyl*-2H-*chromene*, m.p. 89°; 4,6-*dimethyl*-2,2-*diphenyl*-2H-*chromene*, m.p. 126° (*I. M. Heilbron* and *D. W. Hill*, J. chem. Soc., 1927, 2005); 2,2,4-*triphenyl*-2H-*chromene*, m.p. 130° (*Heilbron et al.*, *ibid.*, 1931, 1701; *Smith* and *Ruoff*, *loc. cit.*).

Benzo-2*H*-chromenes. 2,2-*Diphenyl*-7,8-*benzo*-2H-*chromene* (2,2-*diphenyl*-2H-*naphtho*[1,2-b]*pyran*), m.p. 126–128° (*Cottam* and *Livingstone*, J. chem. Soc., 1964, 5228); 2,2-*dimethyl*-5,6-*benzo*-2H-*chromene* (3,3-*dimethyl*-3H-*naphtho*[2,1-b]*pyran*), m.p. 45° (*Livingstone*, *Miller* and *Watson*, *ibid.*, 1958, 2422); 2,2-*diphenyl*-5,6-*benzo*-2H-*chromene* (3,3-*diphenyl*-3H-*naphtho*[2,1-b]*pyran*), m.p. 160–161° *Livingstone*, *Miller* and *Morris*, *ibid.*, 1960, 5148); 2,2-*bis*(2-*methoxyphenyl*)-5,6-*benzo*-2H-*chromene* {3,3-*bis*(2-*methoxyphenyl*)-3H-*naphtho*[2,1-b]*pyran*}, m.p. 153°; 2,2-*bis*(3-*methoxyphenyl*)-5,6-*benzo*-2H-*chromene* {3,3-*bis*(3-*methoxyphenyl*)-3H-*naphtho*[2,1-b]*pyran*}, m.p. 101°; 2,2-*bis*(4-*methoxyphenyl*)-5,6-*benzo*-2H-*chromene* {3,3-*bis*(4-*methoxyphenyl*)-3H-*naphtho*[2,1-b]-*pyran*}, m.p. 175–176° (*Cotterill*, *Livingstone* and *Walshaw*, *ibid.*, 1970, 1758); 2,2-*diphenyl*-6,7-*benzo*-2H-*chrom*-3-*ene* (2,2-*diphenyl*-2H-*naphtho*[2,3-b]*pyran*), m.p. 165° (*Cottam* and *Livingstone*, *ibid.*, 1965, 6646). For dihalogeno addition products of the above compounds see chromans (p. 225).

Reactions. 2,2-Diarylchromenes undergo ring-fission with 50% potassium hydroxide, 4-methyl-2,2-diphenyl-2*H*-chromene, for example, yielding benzhydryl phenyl ether, m.p. 56° (*Heilbron* and *Hill*, *loc. cit.*). *Löwenbein* (*loc. cit.*) reported that 2,2-diarylchromenes isomerised under acid conditions to give coumarones or coumarans; this is incorrect and the compound XIX, which *Löwenbein* obtained is, in fact, obtained by the reaction between 2,2-diphenyl-2*H*-chromene and 2,4-diphenyl-4*H*-chromene or their intermediates, formed during the Grignard reaction between coumarin and phenylmagnesium bromide (*Cotterill et al.*, J. heterocyclic Chem., 1974, 283):

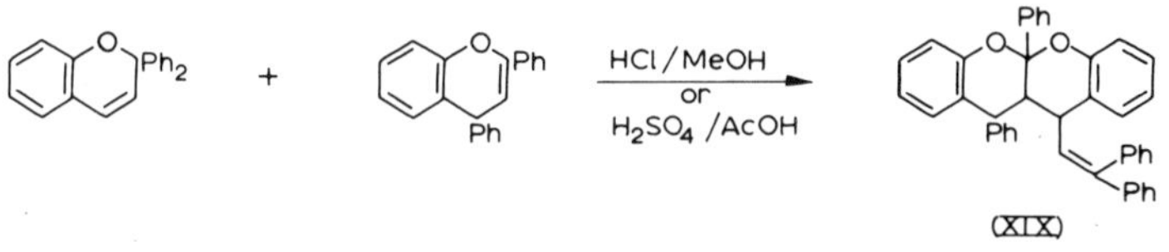

Ph₂ + Ph, Ph —HCl/MeOH or H_2SO_4/AcOH→ (XIX)

The products from the reaction between 2,2-diphenylchromene and 1,1-diarylethylenes under acid conditions have been shown by spectroscopic methods and by unambiguous synthesis to be 2,2-diaryl-4-(2,2-diphenylvinyl)chromans (XX). Similar reactions with related compounds have been carried out and a mechanism for the reaction proposed. Compound XX, with all the aryl groups phenyls, may be obtained by reacting salicylaldehyde with two mols. of diphenylethylene under similar conditions (*E. Bradley et al.*, J. chem. Soc., C, 1971, 3028):

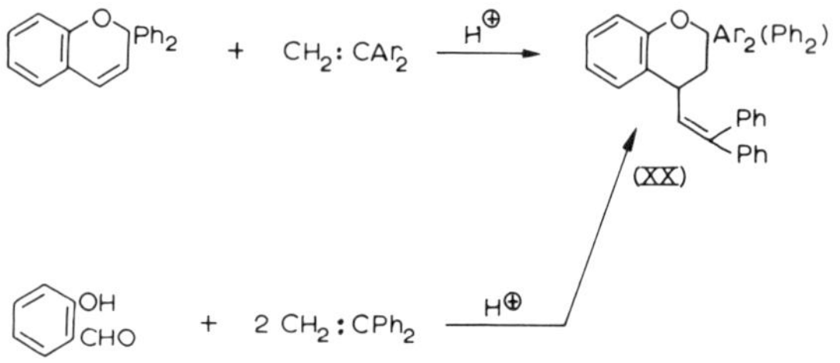

Benzofuran reacts with dichlorocarbene in hexane to form an adduct which is converted into *bis-3-chloro-2(2H-chromenyl)ether* (XXIII), m.p. 181–182°, by hydrolysis with water. The ether is thought to be formed as shown; the product XXI and/or XXII, formed prior to addition of water, decomposed upon attempted distillation or chromatography upon alumina (*Parham et al.*, J. org. Chem., 1963, **28,** 577):

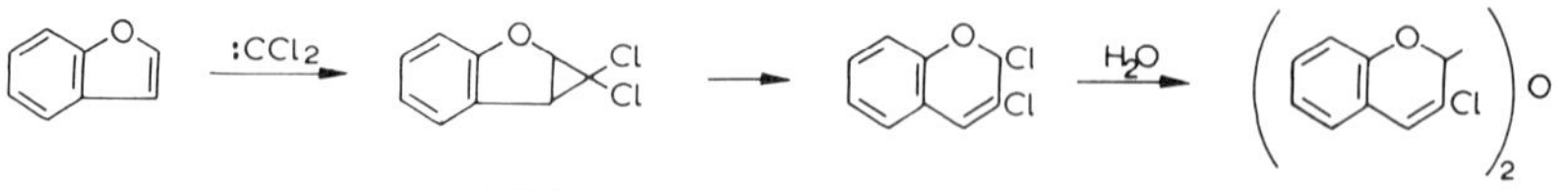

[1a,7b]Dihydro-2*H*-benzo[*b*]cyclopropa[*d*]pyran, its *cis*- and *trans*-2-methyl-, *cis*- and *trans*-2-ethyl- and 2,2-dimethyl derivatives when heated at >200° in diethylaniline rearrange to 2-alkyl-2*H*-chromenes, 2-alkyl-2*H*-benzo[*b*]pyrans (alkyl = Me, Et, Pr, $CHMe_2$) (*R. Hug et al.*, Helv., 1971, **54,** 306).

4-Ethoxy-2*H*-chromene, 4-ethoxy-2*H*-benzopyran forms an adduct with dihalogenocarbenes, which on reduction and then heating, rearranges to give 4-ethoxy-2-methyl-2*H*-chromene (*Graffe et al., loc. cit.*):

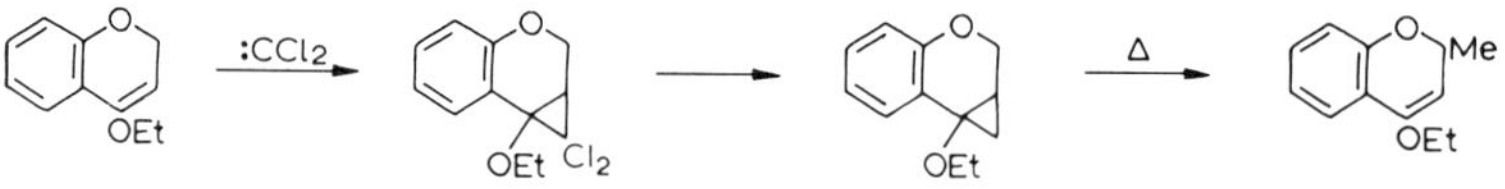

(*6*) *Flavenes.* 2′-Hydroxychalcones and α-alkoxy-2′-hydroxychalcones are converted by sodium tetrahydridoborate in isopropyl alcohol into 2*H*-flavenes (2-phenyl-2*H*-chromenes) and 3-alkoxy-2*H*-flavenes (3-alkoxy-2-phenyl-2*H*-chromenes) (XXIV). The flavenes immediately give flavylium cations in the cold when treated with acids in air, and may be reduced catalytically to flavans and 3-alkoxyflavans, the latter being obtained in the 2,3-*cis*-form (*J. W. Clark-Lewis et al.*, Chem. and Ind., 1967, 1455; Austral. J. Chem., 1968, **21,** 2247; *L. Jurd*, Chem. and Ind., 1967, 2175):

2*H*-Flavene
(2-Phenyl-2*H*-chromene)

(XXIV)
(R = alkyl)

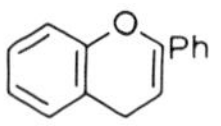

4*H*-Flavene
2-Phenyl-4*H*-chromene)

The reduction of 2′-hydroxychalcone using lithium tetrahydridoaluminate and aluminium chloride gives 1-(1-hydroxyphenyl)-3-phenyl-1- and -2-propenes; the same two propenes are obtained by reduction of 2-benzyl-2,3-dihydrobenzofuran with sodium in pyridine. If 2′-hydroxychalcone is reduced by lithium tetrahydridoaluminate alone, the products are 2*H*-flavene and 1-(1-hydroxyphenyl)-3-phenylpropan-1-ol (*T. Hase*, Acta Chem. Scand., 1968, **22,** 2845).

Some 4*H*-flavenes and 2*H*-flavenes have been synthesised by the reduction of fully methylated flavanols with lithium tetrahydridoaluminate. The two flavenes are obtained in approximately equal amounts from myricetin hexamethyl ether (XXV) and quercetin pentamethyl ether (XXVI). The 4*H*-flavenes obtained in the above preparations on warming in acetic acid isomerise quantitatively to the 2*H*-flavenes:

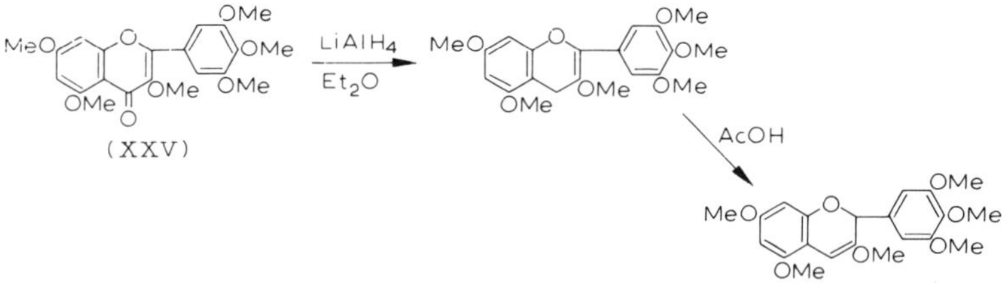

(XXV) $\xrightarrow[Et_2O]{LiAlH_4}$ → AcOH

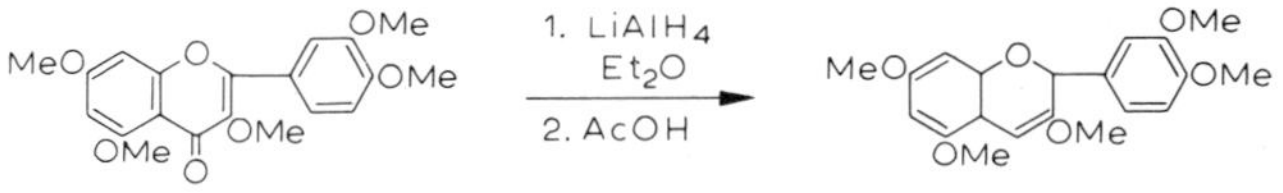

(XXVI) 1. $LiAlH_4$, Et_2O; 2. AcOH

4*H*-**Flavene** *(flav-2-ene)* has been prepared by the reduction of flavone with excess of lithium tetrahydridoaluminate, and the structure confirmed by n.m.r. spectroscopy. 2*H*-**Flavene** *(flav-3-ene)* has been obtained by the dehydrobromination of 4-bromoflavan using pyridine (*K. G. Marathe et al.*, Chem. and Ind., 1962, 1793).

4′,5,7-Trimethoxyflavone gives only 5,7-dimethoxy-4*H*-flavene, which in aqueous acetic acid rapidly hydrolyses to a dihydrochalcone (*A. C. Waiss Jr.* and *Jurd*, Chem. and Ind., 1968, 743):

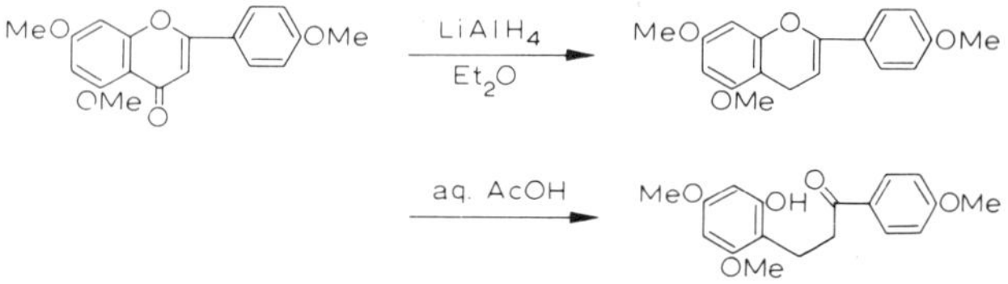

LiAlH4 / Et2O; aq. AcOH

The reaction between 2,5-dialkylphenoxymagnesium bromides and cinnamaldehyde gives the corresponding flavenes, the 2*H*-flavenes again isomerising to 4*H*-flavenes when boiled in benzene in the presence of the corresponding phenoxymagnesium bromide (*G. Casiraghi* and *G, Casnati*, Chem. Comm., 1970, 321).

o-Cinnamylphenols may be converted into 2*H*-flavenes by dehydrogenation with 2,3-dichloro-5,6-dicyanobenzoquinone (*Cardillo, R. Cricchio* and *Merlini*, Tetrahedron Letters, 1969, 907):

The photolysis of 2-hydroxychalcones in alcoholic solvents using 3500 Å light gives the corresponding 2-alkoxy-2*H*-flavenes together with small quantities of flavone (*D. Dewar* and *R. G. Sutherland*, Chem. Comm., 1970, 272):

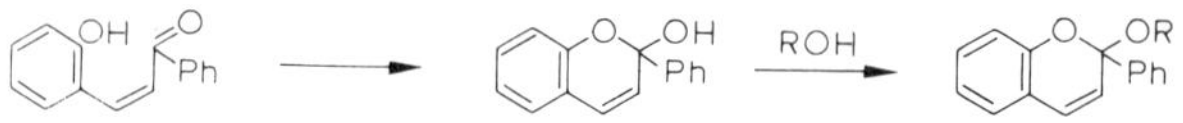

ROH

The use of 2*H*-flavenes in the synthesis of flavans (*Clarke-Lewis* and *R. W. Jemison*, Austral. J. Chem., 1970, **23,** 315) and the stereochemistry of some 2*H*-flavene derivatives have been discussed (*B. J. Bolger et al.*, Tetrahedron, 1967, **23,** 341).

4-Alkyl-3-arylcoumarins have been converted into 2,4-dialkyl-2*H*-isoflavenes (2,4-dialkyl-3-aryl-2*H*-chromenes) without the isolation of intermediate products. The scheme involves reduction of the coumarin XXVII with di-isobutylaluminium hydride to XXVIII (or a tautomeric form), which without isolation is treated with a Grignard reagent. Acid-catalysed ring closure of the resulting product, XXIX, gives the isoflavene XXX (*C. E. Cook* and *C. E. Twine Jr.*, Chem. Comm., 1968, 791):

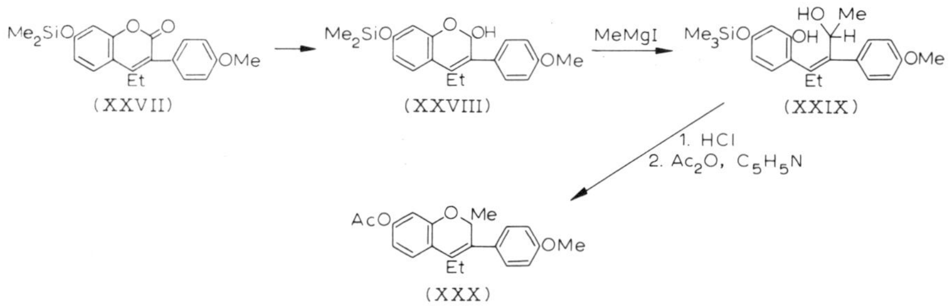
Me2SiO
OMe
Et
Me2SiO
OH
MeMgI
Me3SiO
HO Me
1. HCl
2. Ac2O, C5H5N
AcO
Me

(XXVII) (XXVIII) (XXIX) (XXX)

1,4-Diaryloxybut-2-ynes may be converted to benzofurobenzopyrans, which on boiling with diethylaniline containing *p*-toluenesulphonic acid yield benzofurobenzofurans (*B. S. Thyagarajan, K. K. Balasubramanian* and *R. B. Rao,* Chem. and Ind., 1966, 2128; Tetrahedron, 1967, **23,** 1893).

(ii) Naturally occurring derivatives of 2,2-*dimethyl*-2H-*chromene*

Ageratochromene, 6,7-*dimethoxy*-2,2-*dimethyl*-2H-*chromene,* m.p. 47.5°, λ_{max} 278, 322 mμ (log ε, 3.57 and 3.84), *dihydro* derivative, m.p. 60°, λ_{max} 293 mμ (log ε, 3.69), is isolated from *Ageratum* species (*A. R. Alertsen,* Acta Chem. Scand., 1955, **9,** 1725). Its structure is confirmed by its synthesis from 1,2,4-trimethoxybenzene, which is condensed with 3-methyl-2-butenoyl chloride in the presence of aluminium chloride to yield as one of the products, 6,7-dimethoxy-2,2-dimethylchroman-4-one. Reduction to the 4-alcohol followed by dehydration gives ageratochromene (*R. Huls,* Bull. Soc. chim. Belg., 1958, **67,** 22):

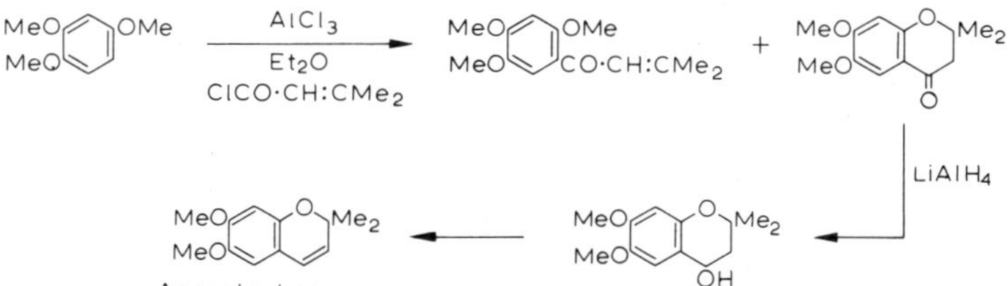
MeO
OMe
AlCl3
Et2O
ClCO·CH:CMe2
CO·CH:CMe2
Me2
LiAlH4
OH

Ageratochromene

Another synthesis of ageratochromene has been reported (p. 57).

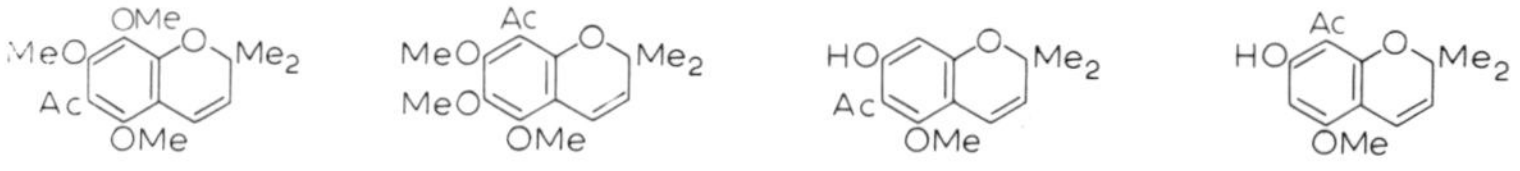
OMe
MeO
Ac
Me2
HO

Evodione Alloevodione Evodionol Alloevodionol

The volatile oil from *Evodia elleryana* contains **evodione**, 6-*acetyl*-5,7,8-*trimethoxy*-2,2-*dimethyl*-2H-*chromene*, m.p. 57°, λ_{max} 257, ~310 mμ (log ε, 4.19, 3.32) (*S. E. Wright*, J. chem. Soc., 1948, 2005) and the isomeric **alloevodione**, 8-*acetyl*-5,6,7-*trimethoxy*-2,2-*dimethyl*-2H-*chromene*, m.p. 81°, 2,4-*dinitrophenylhydrazone*, m.p. 173° (*K. D. Kirby* and *M. D. Sutherland*, Austral. J. Chem., 1956, **9**, 411). Dihydroevodione has been synthesised (*Huls* and *S. Brunell*, Bull. Soc. chim. Belg., 1959, **68**, 325; *R. Warin*, *M. Renson* and *Huls*, *ibid.*, 1960, **69**, 593). **Evodionol,** 6-*acetyl*-7-*hydroxy*-5-*methoxy*-2,2-*dimethylchromene*, m.p. 86°, *methyl ether*, m.p. 79°, *oxime*, m.p. 89°, *acetate*, m.p. 66°, is obtained from the volatile oil of *Evodia littoralis* (*F. N. Lahey*, University Queensland Papers, Chem. Dept., 1942, No. 2). **Alloevodionol,** 8-*acetyl*-7-*hydroxy*-5-*methoxy*-2,2-*dimethyl*-2H-*chromene*, b.p. 160°/0.3 mm, m.p. 71–72°, *acetate*, m.p. 74–75°, *methyl ether*, m.p. 106.5–107°, is isolated from the leaves of *Medicosma cunninghamii* (*Sutherland*, *ibid.*, 1949, 1, No. 35). Alloevodionol, evodionol methyl ether, D,L-cannabichromene, franklinone and the trimethyl ether of fleminigin C have been synthesised by cyclodehydrogenation of the corresponding isoprenylphenols with 2,3-dichloro-5,6-di-cyanobenzoquinone or *via* the chromans and dehydrogenation with the same reagent (*G. Cardillo*, *R. Cricchio* and *L. Merlini*, Tetrahedron, 1968, **24**, 4825). A variety of natural products derived from 2,2-dimethyl-2*H*-chromene have been synthesised using thermal condensation methods (*C. Mercier*, Mem. Mus. Nat. Hist. Natur., Paris, Ser. D, Sci. Phys.-Chim., 1969, **4**, 70; C.A., 1970, **72**, 121390). Six chromenes, eupatoriochromene, methyleupatoriochromene, ripariochromene A, methylripariochromene A, evodionol, methylevodionol, ripariochromene B and ripariochromene C, have been isolated from *E. riparium* and *E. glandulosum*, and their n.m.r. spectra have been recorded.

Me2 O OMe H⊕ Me2 H H H Me2 O H OMe H OMe

Lapachenole Isolapachenole

Lapachenole, 6-*methoxy*-2,2-*dimethyl*-7,8-*benzo*-2H-*chromene*, 6-*methoxy*-2,2-*dimethyl*-2H-*naphth*[1,2-b]*pyran*, $C_{16}H_{16}O_2$, m.p. 62°, *picrate*, violet needles, m.p. 141°, is isolated from the hardwood *Paratecoma alba*. Its structure follows from the synthesis of lapachenole and its dihydro derivative (*R. Livingstone et al.*, J. chem. Soc., 1955, 3631; 1956, 3701; *J. Hlubucek*, *E. Ritchie* and *W. C. Taylor*, Tetrahedron Letters, 1969, 1369; Austral. J. Chem., 1971, **24**, 2347). Lapachenole with 2,4-dinitrophenylhydrazine in ethanolic or butanolic sulphuric acid yields the dinitrophenylhydrazone of the corresponding chroman-4-one (*Livingstone* and *R. B. Watson*, J. Chem. Soc., 1957, 1509). Dinitrophenylhydrazones are similarly obtained from related 2,2-dimethylchromenes, presumably by addition of the phenylhydrazine at the double bond followed by dehydrogenation. Lapachenole under acid conditions forms a dimer; the dimerisation has been carried out in deuteriosulphuric–deuterioacetic acid with the incorporation of one atom of deuterium. A probable mechanism has been described, and this together with the n.m.r. data is compatible with the proposed structure of *isolapachenole*, m.p. 258° (*W. D. Cotterill et al.*, Tetrahedron, 1968, **24,** 1981).

Rottlerin, $C_{30}H_{28}O_8$, salmon-pink needles or plates, m.p. 201–202° (Brockmann), 212° (Robertson), λ_{max} 232, 294 and 350 mμ, is a phenol isolated from the Indian colouring matter *Kamala (Mallotus phillipinensis)*, an orange-red powder which is used as a dye for silk and as an anthelmintic. Rottlerin is responsible for both properties.

Rottlerin is a chromene derivative, the structure of which was determined by the degradative and synthetic investigations of *Brockmann* in Germany, *Robertson* in England, and their collaborators (*A. McGookin et al.*, J. chem. Soc., 1939, 1579). It forms a *penta-acetate*, m.p. 213°, and a *pentamethyl ether*, m.p. 142°; it has two double bonds, detected by hydrogenation with a palladium catalyst to *tetrahydrorottlerin* (XL), pale yellow prisms, m.p. 214°, *penta-acetate*, m.p. 188°:

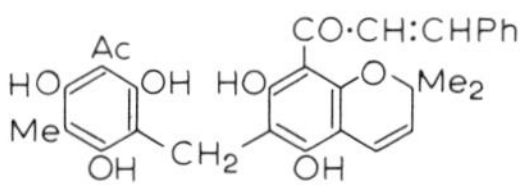

Rottlerin 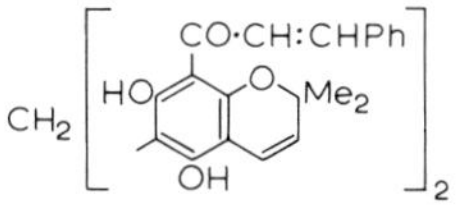 Rottlerone 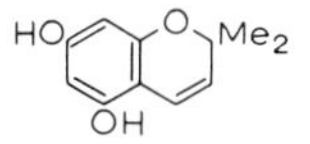 (XXXI)

When tetrahydrorottlerin is oxidised by ozone or potassium permanganate, acetone is formed, showing the presence of an isopropyl group; and with alkali it gives β-phenylpropionic acid, indicating the cinnamoyl group in the side-chain. Rottlerin with 2% sodium carbonate gives *inter alia* *C*-methylphloroglucinol, rottlerone and benzaldehyde. This shows not only that rottlerin is a derivative of phloroglucinol, but also that it is a hydroxychalcone, since it is only with this type of compound that benzylidene residues yield benzaldehyde with alkali (*R. L. Shriner*, J. Amer. chem. Soc., 1930, **52,** 2538). Tetrahydrorottlerin with hot concentrated alkali affords hydrocinnamic acid and 5,7-dihydroxy-2,2-dimethyl-2*H*-chromene (XXXI), thereby giving the essential structure of one half of the rottlerin molecule. The constitution of the "phloroglucinol half" was shown by analogy with a reaction first noted by *R. Boehm* (Ann., 1901, **318,** 253) in which methylene derivatives of phloroglucinol such as XXXII react with diazoaminobenzene to give azo compounds, with replacement of the methylene group. Rottlerin in a similar way gave the azo compound XXXIII, thus showing not only the structure of the "phloroglucinol half" in rottlerin is that given in the formula above, but also that the phloroglucinol and chromene residues are linked by a methylene group (*H. Brockmann* and *K. Maier*, Ann., 1938, **535,** 149):

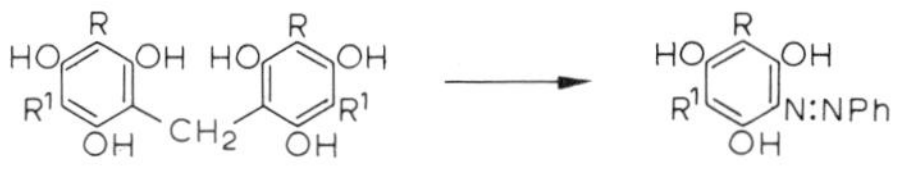

(XXXII) 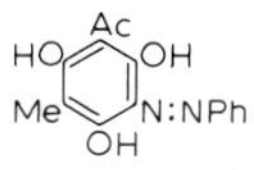 (XXXIII)

Formaldehyde condenses with a mixture of 2-acetyl-4-methylphloroglucinol and the 2,2-dimethylchroman shown to give tetrahydrorottlerin (*A. Robertson et al.*, J. chem. Soc., 1948, 113):

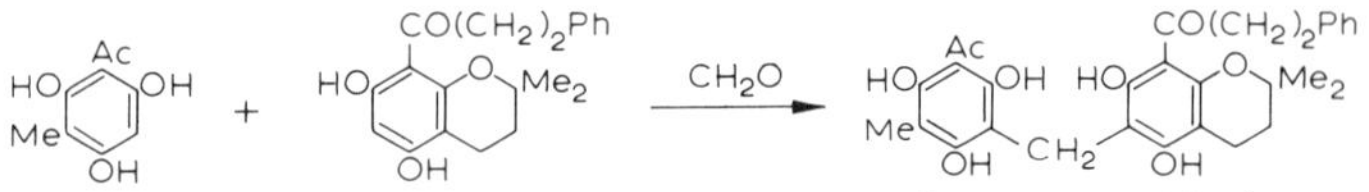

Tetrahydrorottlerin

The constitution of *rottlerone*, $C_{41}H_{36}O_8$, m.p. 236°, has been established by the synthesis of the reduction product octahydrorottlerone (*McGookin et al.*, J. chem. Soc., 1939, 1587). The tendency of rottlerin to be changed into a flavone derivative is shown by boiling it in solvents such as ethanol or toluene to give **isorottlerin,** m.p. 182°, a change which *Robertson* has formulated as a fission of the chromene ring followed by recyclisation involving the alternative hydroxyl group to form allorottlerin and thence, by a chalcone–flavanone ring closure to isorottlerin (*McGookin et al., loc. cit.*):

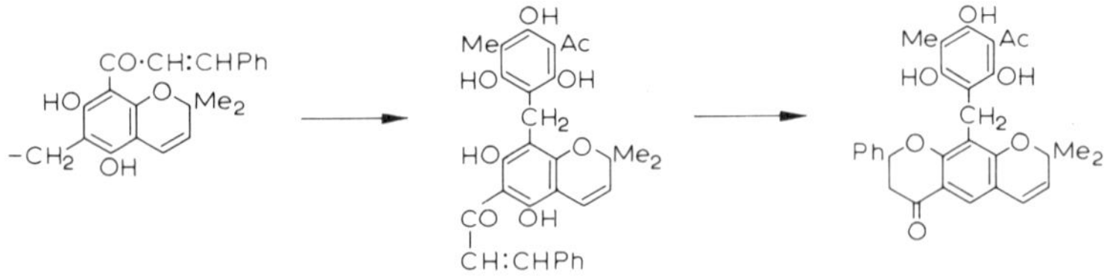

Allorottlerin

Isorottlerin

Two red pigments, 4-hydroxyrottlerin (XXXIV, $R^1 = OH$, $R^2 = H$), and 3,4-dihydroxyrottlerin (XXXIV, $R^1 = R^2 = OH$), which are minor constituents of *kamala*, have been isolated and identified (*G. Cardillo et al.*, Gazz., 1965, **95,** 725). Two other minor components have also been isolated from *kamala*; one is red *C*-methylated 5,7-dihydroxy-2,2,6-trimethylcinnamoylchromene (XXXV) related to one half of the rottlerin structure; the other (yellow) is a rearranged flavanone–chromene XXXVI, which can also be obtained from the red component by acid treatment.

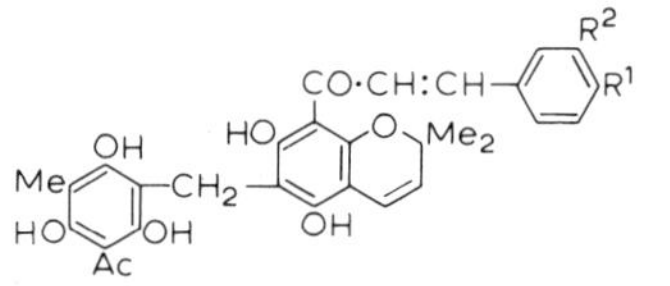

(XXXIV) 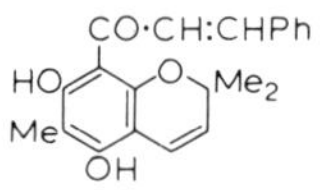 (XXXV) 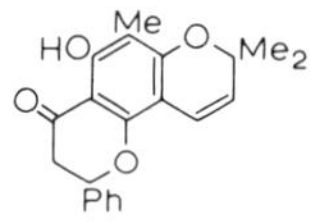 (XXXVI)

The status of these two new compounds as artefacts or natural products is debatable (*L. Crombie et al.*, J. chem. Soc., C, 1968, 2625).

Mundulone, $C_{26}H_{25}O_6$, m.p. 180°, $[\alpha]_D$ −11.5 ($CHCl_3$), occurs in the bark of *Mundulea sericea.* Although its formula has been established as an isoflavone it contains also 2,2-dimethylchromene units (*S. F. Dyke et al.*, Proc. chem. Soc., 1963, 179; *C. S. Barnes et al.*, Tetrahedron Letters, 1963, 281):

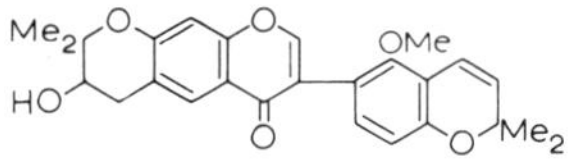

Mundulone

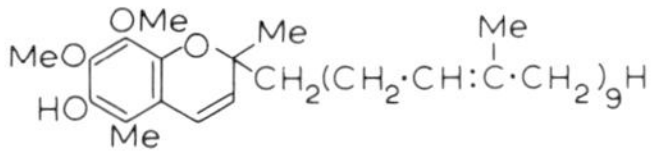

Ubichromenol

Ubichromenol, m.p. ~18°, a chromene related to the tocols (p. 226) is a yellowish compound obtained from the liver and kidney of many species of animal (*D. L. Laidman et al.*, Chem. and Ind., 1959, 1019; Biochem. J., 1960, **74,** 541). Ubiquinone left in contact with alumina overnight is partially converted into ubichromenol (*F. W. Hemming, R. A. Morton* and *J. F. Pennock, ibid.*, 1961, **80,** 445).

Quinones of the coenzyme Q group have been converted to the corresponding chromenols in good yield, by cyclisation using sodium hydride (*B. O. Linn et al.*, J. Amer. chem. Soc., 1963, **85,** 239):

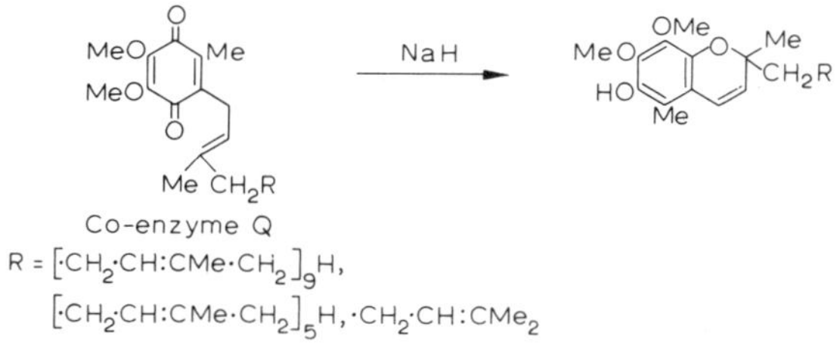

MeO, MeO, Me, Me CH$_2$R — NaH → MeO, HO, OMe, Me, Me, CH$_2$R

Co-enzyme Q

R = [·CH$_2$·CH:CMe·CH$_2$]$_9$H,

[·CH$_2$·CH:CMe·CH$_2$]$_5$H, ·CH$_2$·CH:CMe$_2$

(iii) Chromenols (5,6-benzopyranols, benzo[b]*pyranols)*

There are two types of benzo[*b*]pyranols or chromenols, 2*H*-chromenols (chromen-2-ols, α-chromenols) and 4*H*-chromenols (chromen-4-ols, β-chromenols):

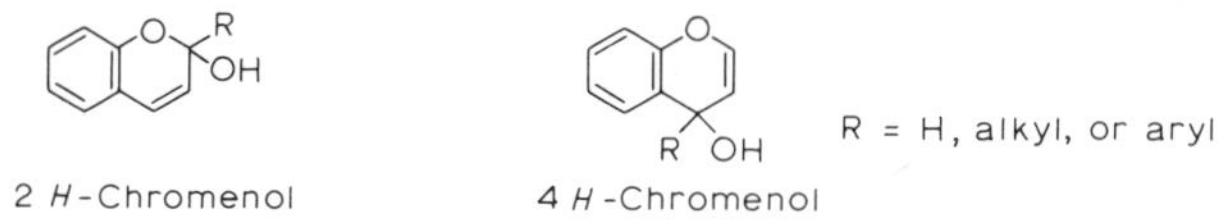

2 *H*-Chromenol   4 *H*-Chromenol   R = H, alkyl, or aryl

The chromenols are the pseudo-bases or methanol bases of the benzopyrylium (chromylium) salts and the chromen-2-ols are formed from these salts by the action of water, alkali, or acetic acid. It is stated that they are also obtained along with other products by the interaction of coumarins with Grignard reagents followed by treatment with acid. This is very doubtful and if they are obtained at all it is in very low yield (p. 105). On the other hand, chromen-4-ols are obtained from the corresponding chromones and Grignard reagents.

Many chromenols are unstable and non-crystalline, but are crystalline when the 2,3-position (and sometimes the 2,4-positions) are substituted by aryl groups (*T. A. Geissman* and *E. Baumgarten*, J. Amer. chem. Soc., 1943, **65,** 2135). 2,4-*Diphenylchromen-2-ol*, m.p. 68–71°, prepared from 2,4-diphenylbenzopyrylium ferrichloride, is unstable, colourless and insoluble in cold alkali, and is believed to revert to the yellow *trans*-chalcone XXXVII:

OH
C:CH·COPh
Ph·
(XXXVII)

whereas 2,3-diphenylchromen-2-ol, m.p. 122–122.5°, is stable, colourless, insoluble in cold alkali, and apparently exists in the cyclic form (*Geissman* and *Baumgarten*, *loc. cit.*; *H. Decker* and *P. Becker*, Ber., 1922, **55,** 375; *P. P. Hopf* and *R. J. W. LeFèvre*, J. chem. Soc., 1938, 1582). The chromenols are converted into the benzopyrylium salts by treatment with acids and when warmed with alcohols yield alkyl ethers, particularly in the presence of a trace of acid (*K. Ziegler et al.*, Ann., 1926, **448,** 249; *P. Karrer* and *C. Trugenberger*, Helv., 1945, **28,** 444). 2,3,4-*Triphenylchromen-2-ol*, m.p. 158°, for instance, gives a *methyl ether*, m.p. 139°, and *ethyl ether*, m.p. 126° (*A. Löwenbein* and *B. Rosenbaum*, Ann., 1926, **448,** 223).

Examples are known in which chromen-4-ols are converted to the -2-ols by heat, 6-*methyl*-2,3,4-*triphenylchromen*-4-*ol*, m.p. 139–140°, for instance, giving the -2-*ol*, m.p. 163° (*Ziegler et al.*, *loc. cit.*; *R. L. Shriner* and *R. B. Moffett*, J. Amer. Chem. Soc., 1941, **63,** 1694).

(iv) Benzopyrylium or chromylium salts

The chromenols are converted by acids into benzopyrylium or chromylium salts in which the heterocyclic ring becomes aromatic (compare pyrylium salts, p. 4). No single atom in the conjugated system carries all the charge; this charge is shared out over the atoms in the ring, but for convenience and also according to general practice the charge is shown on the oxygen atom. The benzopyrylium cation is best represented in terms of the theory of resonance as a resonance hybrid:

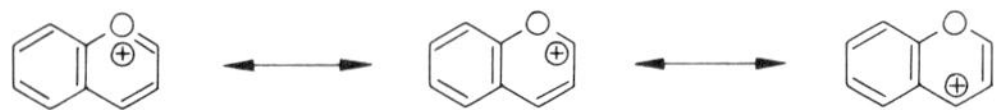

At one time there was a dispute as to whether flavylium salts were oxonium or carbenium salts, but this has now lapsed (*R. J. W. LeFèvre et al.*, J. chem. Soc., 1929, 2771; 1932, 1988; 1934, 27; *LeFèvre, ibid*., 1930, 2236; *W. Borsche* and *K. Wunder*, Ann., 1915, **411,** 38; *W. Dilthey* and *W. Hoschen*, J. pr. Chem., 1933, [ii], **138,** 42; *Shriner* and *Moffett*, J. Amer. chem. Soc., 1940, **62,** 2713; *D. W. Hill*, Chem. Reviews, 1936, **19,** 27). The presence of the hetero-atom and of a positive charge makes the heterocyclic ring extremely susceptible to attack by nucleophilic reagents, but relatively inert to attack by electrophilic reagents. The 2-phenylbenzopyrylium salts (flavylium salts) are of importance since their polyhydroxy derivatives include the anthocyanins. Benzopyrylium salts were first prepared by *Decker* and *T. von Fellenberg* (Ann., 1907, **356,** 281; 1909, **364,** 17) and *W. H. Perkin Jr.* and

R. Robinson (Proc. chem. Soc., 1907, **23,** 149; *Perkin et al.*, J. chem. Soc., 1908, **93,** 1085).

Methods of synthesis (*1*) The most important method is the cyclisation of 2-hydroxychalcones, analogous to the preparation of pyrylium salts from glutacondialdehyde (p. 7). Two methods are generally applicable. In the first, a phenol, *e.g.*, resorcinol, is condensed with a β-diketone or hydroxymethylene ketone by means of hydrogen chloride:

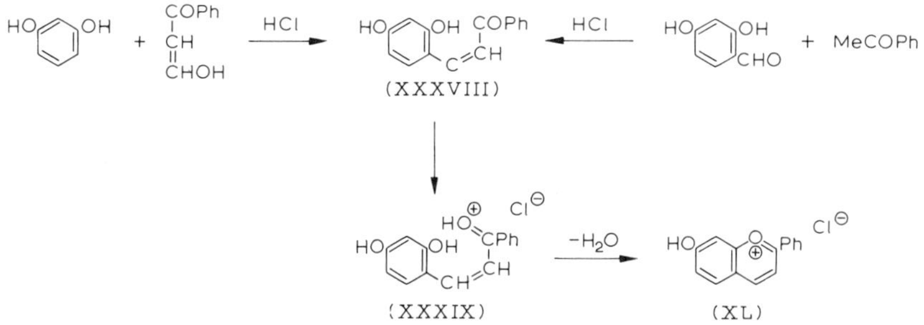

HO–C6H4–OH + COPh–CH=CHOH —HCl→ (XXXVIII) ←HCl— HO–C6H3(OH)–CHO + MeCOPh

(XXXVIII) → (XXXIX) —$-H_2O$→ (XL)

The intermediate hydroxhychalcone XXXVIII is obviously a derivative of glutacondialdehyde (enol form) and with hydrogen chloride yields an oxonium salt XXXIX which loses water to give 7-hydroxyflavylium chloride (XL) (*Robinson*, Ber., 1934, **67A,** 85; *S. Kawai* and *N. Sugiyama*, *ibid.*, 1939, **72,** 367). The isolation of pure chalcones and their conversion into benzopyrylium salts have been repeatedly effected (*D. D. Pratt* and *Robinson*, J. chem. Soc., 1922, **121,** 1577; 1923, **123,** 745).

Chalcone may be oxidatively cyclised to 3-hydroxyflavylium salts by trifluoroperoxyacetic acid in methylene chloride (*B. R. Brown, A. J. Davidson* and *R. O. C. Norman*, Chem. and Ind., 1962, 1237):

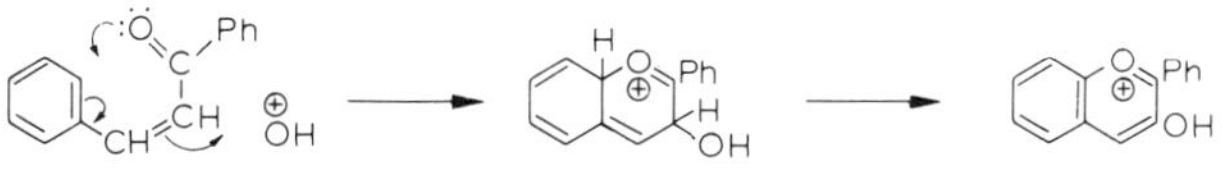

In the second, salicylaldehyde and substituted salicylaldehydes condense with aldehydes or ketones containing the $-CH_2 \cdot CO-$ grouping in the presence of hydrochloric acid, hydrogen chloride, or 70% sulphuric acid (*Decker* and *Fellenberg, loc. cit.; Perkin et al., loc. cit.*). The first synthetic benzopyrylium salt was thus obtained from helicin, the glucoside of salicylaldehyde, acetaldehyde and concentrated hydrochloric acid. The product, *benzopyrylium ferrichloride*, m.p. 126–128°, is obtained by the interaction of salicylaldehyde, acetaldehyde and hydrochloric acid followed by the addition of ferric chloride (*LeFèvre*, J. chem. Soc., 1929, 2711):

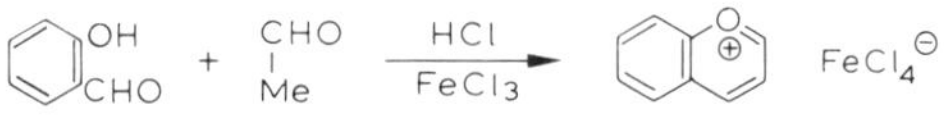

This condensation is one of the most useful methods for the preparation of benzopyrylium salts. With ketonic esters, however, the reaction sometimes takes another course. For example, ethyl benzoylacetate and salicylaldehyde with ethereal hydrogen chloride and perchloric acid give mainly 3-benzoylcoumarin and not 2-phenylbenzopyrylium (flavylium) perchlorate *LeFèvre, ibid.*, 1934, 450). Substituted β-ketonic esters under the same conditions undergo ketonic fission and the ultimate products are those obtained from salicylaldehyde and the corresponding ketone (*S. C. De*, J. Indian Chem. Soc., 1927, **23,** 137; *Löwenbein* and *W. Katz*, Ber., 1926, **59,** 1377). Methyl acetoacetic ester, and ethyl methyl ketone thus furnish the same benzopyrylium salt. 2-Benzoylcoumarin on long boiling in aqueous chloric acid (33%) affords *flavylium perchlorate* (88%), m.p. 184°, λ_{max} (in alc.) 260, 300 mμ. It is reported that this method applies generally to the synthesis of benzopyrylium salts (*M. Mercier et al.*, Compt. rend., 1957, **244,** 907; Bull. Soc. chim. Fr., 1958, 702).

2-Hydroxyacetophenone, acetophenone and 70% perchloric acid on long boiling afford 4-methylflavylium perchlorate. Several 4-methylflavylium perchlorates are obtained by this method from substituted 2-hydroxyacetophenones and the appropriate acetophenone (*G. A. Reynolds, J. A. VanAllan* and *D. Daniel*, J. heterocyclic Chem., 1970, **7,** 1395):

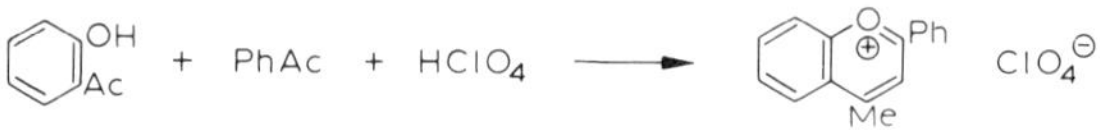

Flavylium salts have been formed by the condensation of 2-hydroxybenzaldehyde with 2-acetothienone and 3-acetothionaphthenone (*G. Bendz, O. Martensson* and *E. Nilsson*, Acta Chem. Scand., 1970, **24,** 1862):

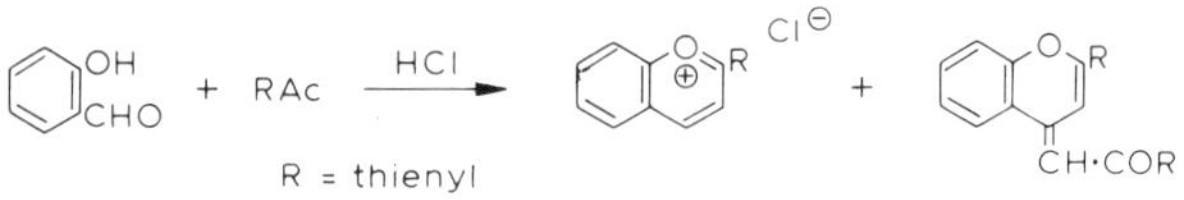

(*2*) In the Bülow synthesis, reaction occurs between resorcinol derivatives and 1,3-diketones in formic or acetic acid and hydrogen chloride (*C. Bülow* and *H. Wagner*, Ber., 1901, **34,** 1189; *J. S. Buck* and *I. M. Heilbron*, J. chem. Soc., 1923, **123,** 2521; *T. Malkin* and *Robinson, ibid.*, 1925, **127,** 1190; *H. Brockmann* and *H. Junge*, Ber., 1943, **76,** 1028; *F. Kehrmann* and *M. Rieder*, Helv., 1926, **9,** 491). Resorcinol and acetylacetone thus yield 7-hydroxy-2,4-dimethylbenzopyrylium chloride:

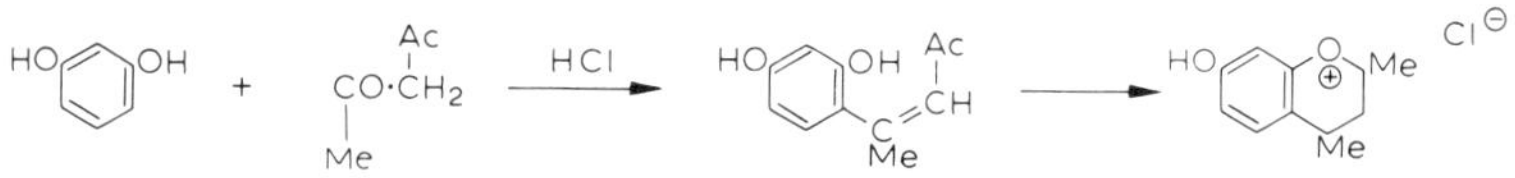

(*3*) It is stated in the literature that benzopyrones react with Grignard reagents to give chromenols (p. 67) which can be converted by acids into benzopyrylium salts. Although *Decker* and *von Fellenberg (loc. cit.)* were the first to show that coumarins can thus be converted into benzopyrylium salts, it is doubtful whether the Grignard reaction between coumarin and phenylmagnesium bromide does in fact give the 2-phenylchromen-2-ol (p. 105) along with the other products which have been isolated and identified.

The chromones react with Grignard reagents (p. 147) to give chromen-4-ols (*Heilbron* and *A. Zaki*, J. chem. Soc., 1926, 1902), which can be converted to benzopyrylium salts.

7-Methoxy-2-phenylchromone with phenylmagnesium bromide yields 7-methoxy-2,4-diphenylchromen-4-ol which with hydrochloric acid and ferric chloride forms the benzopyrylium salt (*Robinson* and *M. R. Turner*, *ibid.*, 1918, **113,** 874):

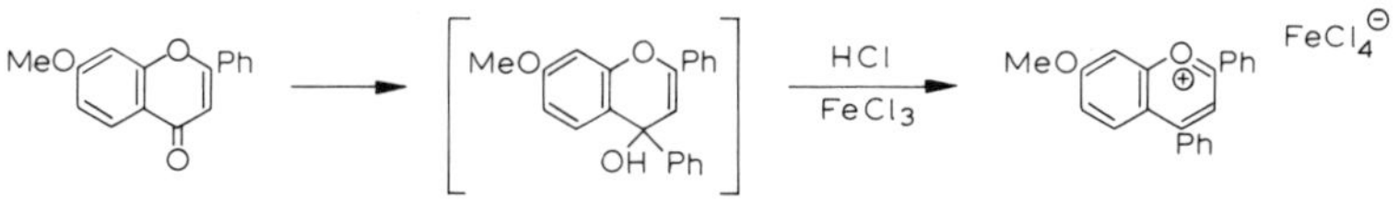

The same benzopyrylium [2,3-diphenyl-6-methyl-4-(4-tolyl)benzopyrylium] salt is obtained from 2,3-diphenyl-6-methylchromone and 4-tolylmagnesium bromide and from 6-methyl-3-phenyl-4-(4-tolyl)coumarin and phenylmagnesium bromide (*Shriner* and *Moffett*, J. Amer. chem. Soc., 1941, **63,** 1694). The formation of this benzopyrylium salt from the corresponding coumarin is probably achieved because some of the products normally expected from the Grignard reaction appear to be unlikely with the number of substituents already present in the heterocyclic ring.

4-Chromanones also react with Grignard reagents and the resulting 4-hydroxychromans readily lose water and yield 2*H*-chromenes (α-chromenes). These, on oxidation in acid solution give benzopyrylium salts (*Löwenbein*, Ber., 1924, **57,** 1517; *Löwenbein* and *Rosenbaum*, *loc. cit.)*. 2,4-*Diphenylbenzopyrylium ferrichloride*, m.p. 167–168°, from 2,4-diphenyl-4*H*-chromene and -2*H*-chromene and ferric chloride *(Löwenbein, loc. cit.)* and also from 2,4-diphenyl-2-methoxychroman (*R. Livingstone*, *D. Miller* and *S. Morris*, J. chem. Soc., 1960, 5148):

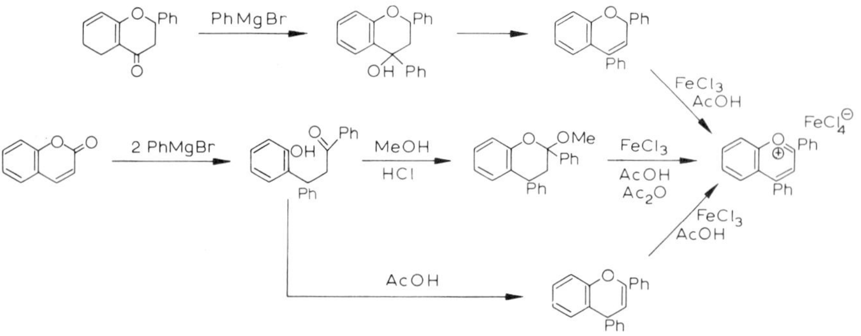

(4) Chromanones may be dehydrogenated using triphenylmethyl perchlorate to give the corresponding 4-hydroxybenzopyrylium perchlorates (*A. Schönberg* and *G. Schütz*, Ber., 1960, **93**, 1466):

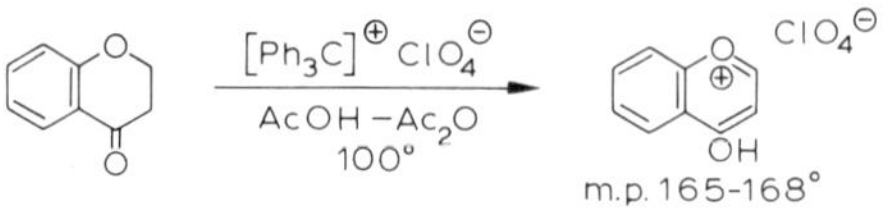

m.p. 165-168°

(5) The synthesis of benzopyrylium salts and chromans by acid-catalysed disporportionation of 2*H*-chromenes and dehydration of 4-chromanols using aqueous perchloric acid has been reported (*B. D. Tilak* and *Z. Muljiani*, Tetrahedron, 1968, **24**, 949):

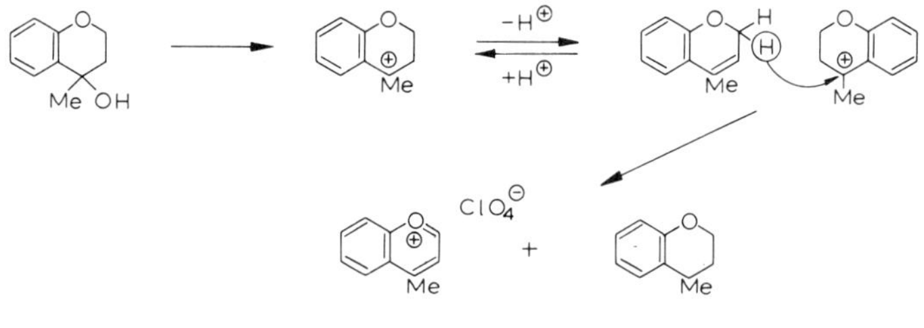

(6) Acetic acid protonated with perchloric acid results in a rapid aldol condensation of methylene compounds with aromatic ketones or aldehydes, leading to high yields of substituted benzopyrylium salts when the aromatic compounds are substituted with hydroxyl and carbonyl groups *ortho* to one another (*J. Andrieux et al.*, Compt. rend. Ser. C, 1970, **271**, 90):

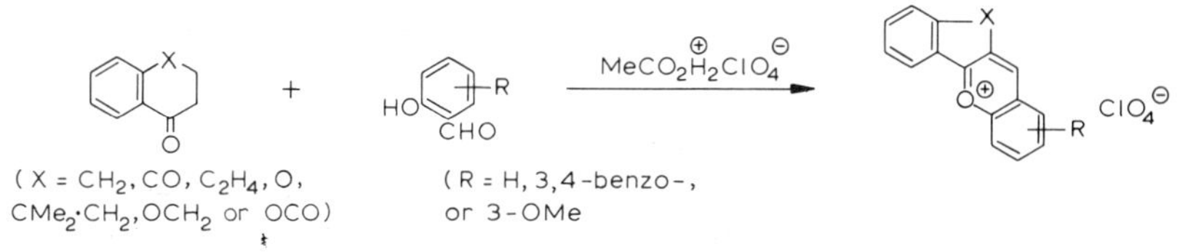

(X = CH_2, CO, C_2H_4, O, $CMe_2{\cdot}CH_2$, OCH_2 or OCO)

(R = H, 3,4-benzo-, or 3-OMe

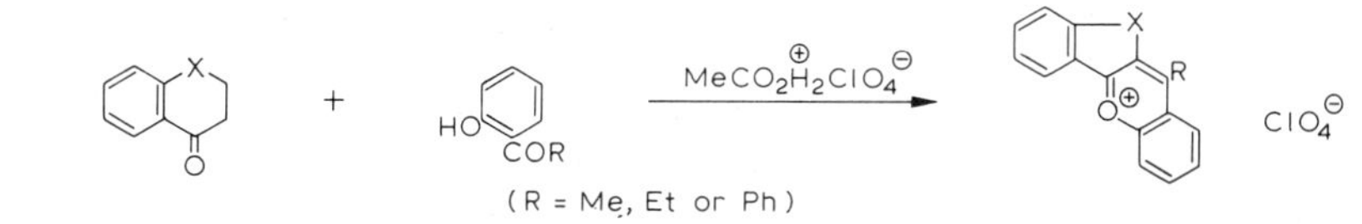

$MeCO_2H_2ClO_4$

(R = Me, Et or Ph)

(7) Other methods include the reduction of flavanols (p. 88) and the reaction of phenols with α,β-unsaturated ketones or α,β-ethynyl ketones (p. 86).

Tertiary oxonium salts of chromones react with dialkylaniline to yield 4-alkoxy-4′-dialkylaminoflavylium salts (*A. I. Tolmachev*, Zhur. obshcheĭ Khim., 1960, **30,** 2884).

Properties. The benzopyrylium salts may be isolated as crystalline double salts of gold or iron chlorides, C_9H_7O $AuCl_4$ and C_9H_7O $FeCl_4$. They are strong electrolytes as shown by their conductivity in acetonitrile (*R. R. Otter* and *Shriner*, J. Amer. chem. Soc., 1951, **73,** 887) and dipole moment measurements (*LeFèvre*, J. chem. Soc., 1936, 398). Flavylium perchlorate, for example, has a moment of approximately 8D which is comparable with that of 7.84D for lithium perchlorate.

Reactions: (1) Oxidation. The pyrylium salts are related to the pyrones as the alkylpyridinium salts are to the pyridones. Since the alkylpyridinium salts are oxidised to the pyridones it would be expected that pyrylium salts will be oxidised to pyrones. This has indeed been effected by the oxidation with chromic acid of 7-hydroxyflavylio-4-carboxylate to 7-hydroxyflavone but the yield is poor and the reaction is not general (*Bülow* and *Wagner*, Ber., 1903, **36,** 1941; *Robinson* and *G. Schwarzenbach*, J. chem. Soc., 1930, 822). By means of perphthalic acid, flavylium salts have been converted into flavanols but this reaction also is of limited application (*P. Karrer* and *W. Fatzer*, Helv., 1942, **25,** 1138). Oxidation at $C_{(2)}$ and $C_{(4)}$ has been effected. 3-Substituted flavylium salts in neutral or acid solution are oxidised by hydrogen peroxide (*Dilthey* and *E. Quint*, J. pr. Chem., 1931, [ii], **131,** 1). Unstable hydroperoxides are formed which decompose to give benzoate esters of 2-hydroxybenzyl ketones:

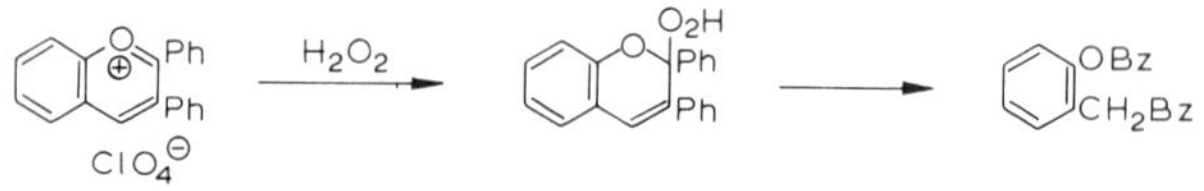

Similar results are obtained with 3-hydroxy- or 3-methoxy-flavylium salts (*Dilthey et al.*, J. pr. Chem., 1926, [ii], **114,** 179; Ber., 1931, **64,** 2082) the products being the benzoyl derivatives of 2-hydroxyphenylacetic acid or its esters. This method of ring-cleavage shows conclusively the presence of the pyrylium ring in the benzopyrylium salts.

Flavylium salts with an alkyl, alkoxy, or phenoxy group in the 3-position are oxidised by hydrogen peroxide in aqueous methanol at pH 5–7 to yield 3-acyl-, 3-alkoxycarbonyl- and 3-phenoxycarbonyl-2-phenylbenzofuran derivatives, respectively (*L. Jurd*, Chem. and Ind., 1963, 1165; J. org. Chem., 1964, **29,** 2602). In aqueous acetic acid solutions 3-alkylflavylium salts undergo a Baeyer–Villiger type oxidation with hydrogen peroxide to yield 1-(2-hydroxyphenyl)-2-benzoyloxyprop-1-ene derivatives XLI. Under slightly alkaline conditions these oxidation products readily rearrange to 2-benzoyloxybenzyl ketones XLII. Alkaline hydrolysis of both XLI and XLII give the desoxybenzoin XLIII (*idem*, Tetrahedron, 1966, **22,** 2913):

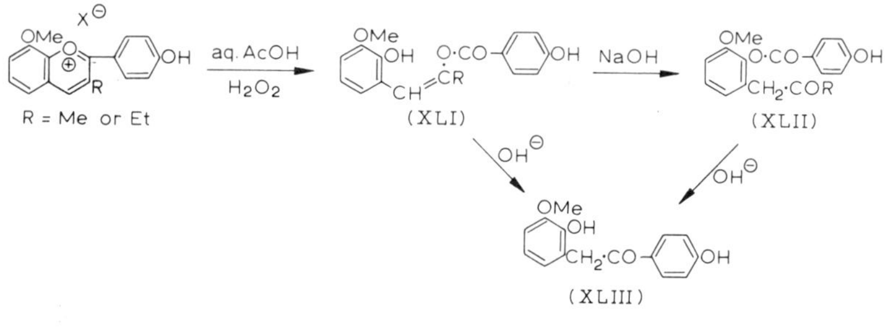

Flavylium salts, unsubstituted at the 3-position, are oxidised by hydrogen peroxide in aqueous acetic acid to (±)*threo*-dihydroxyketones (*idem, ibid.*, 1968, **24,** 4449):

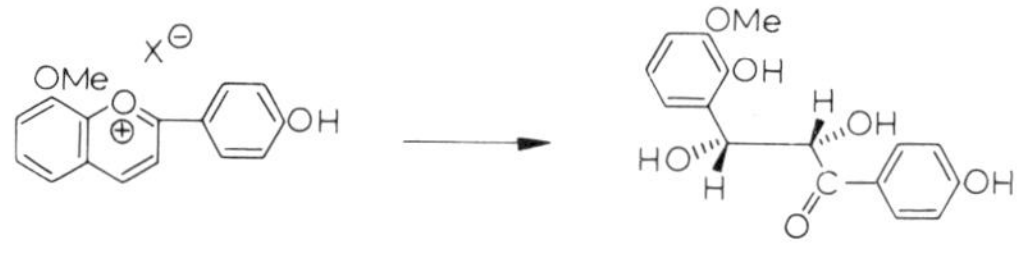

The fission of the 2,3 and 3,4 bonds in benzopyrylium salts has been accomplished by ozonisation (*Shriner* and *Moffett*, J. Amer. chem. Soc., 1940, **62,** 2711).

(2) Reduction. The products obtained by the reduction of the benzopyrylium salts depend *inter alia* on the reducing agent (*A. L. Bhalla* and *J. N. Ray*, J. chem. Soc., 1933, 288; *Karrer* and *Fatzer*, Helv., 1941, **24,** 1317; *K. Freudenberg et al.*, Ann., 1935, **518,** 37). Catalytic hydrogenation (palladium–barium sulphate, platinum black or platinum oxide) gives chromans, fisetinidin chloride thus yielding 3,7,3′,4′-tetrahydroxyflavan. Controlled reduction, however, gives chromenes or flavenes (*E. L. Fonseka*, J. chem. Soc., 1947, 1683). Lithium tetrahydridoaluminate is an excellent reducing agent for converting flavylium salts into flavans (*W. E. Elstow* and *B. C. Platt*, Chem. and Ind., 1950, 824; *Karrer* and *M. Seyhan*, Helv., 1950, **33,** 2209; *Freudenberg* and *K. Weinges*, Ann., 1954, **590,** 140). There is an interesting difference between lithium tetrahydridoaluminate and sodium tetrahydridoborate in the reduction of flavylium salts; with the former, reduction occurs at $C_{(2)}$ and with the latter at $C_{(4)}$:

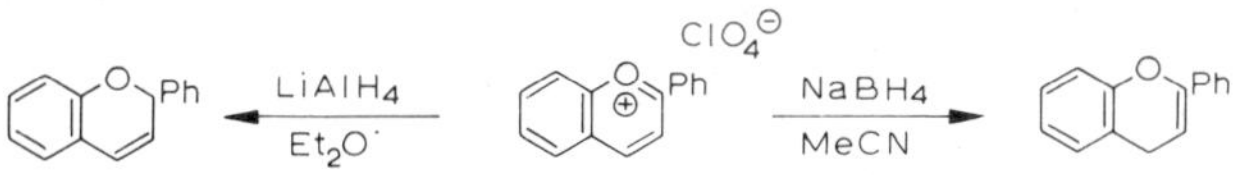

The reduction of flavylium salts and their 3-alkyl or aryl derivatives with zinc dust in acetonitrile solution gives a pair of diastereomeric dimers which are 4,4′-bis(flav-2-ene) derivatives (*Reynolds*, *VanAllan* and *T. H. Regan*, J. org. Chem., 1967, **32,** 3772):

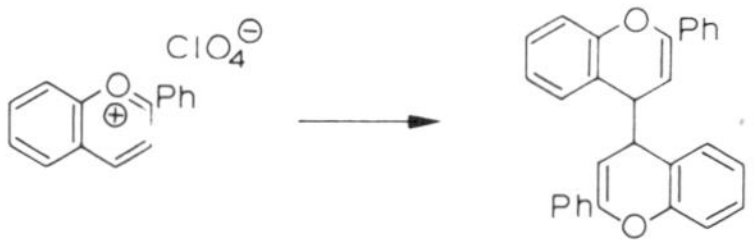

(3) Hydrolysis. The products obtained by the hydrolysis of benzopyrylium salts depend on the substituents in the molecule, the hydrolysing agent, and the conditions of hydrolysis. In most cases a chromenol or pseudo-base is probably formed initially and may then be isolated or be allowed to undergo further change. Benzopyrylium ferrichloride is converted by moist air to coumarin (*Decker* and *von Fellenberg*, Ann., 1909, **364,** 22):

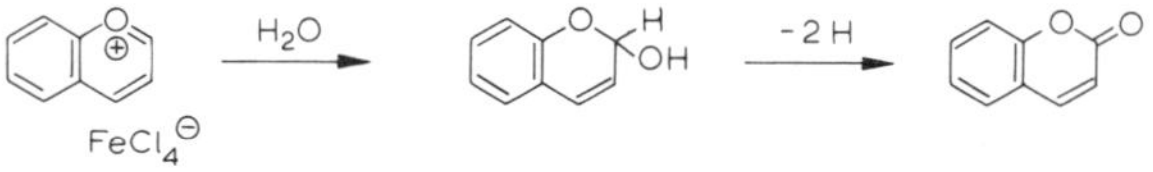

Some chromen-2-ols can be isolated, especially when the benzopyrylium salt contains a substituent in the 3-position. 2,3,4-Triphenylbenzopyrylium ferrichloride when shaken with acetone and water yields 2,3,4-triphenylchromen-2-ol (*A. Löwenbein* and *B. Rosenbaum*, Ann., 1926, **448,** 223) and 3-phenylflavylium perchlorate with water and sodium acetate gives 2,3-diphenylchromen-2-ol (*Hill* and *R. R. Melhuish*, J. chem. Soc., 1935, 1161):

Ph Ph Ph $FeCl_4^{\ominus}$ → OH Ph Ph Ph

Flavylium chloride, on the other hand, with 10% sodium hydroxide at room temperature yields flavone and the corresponding chalcone (*Hill* and *Melhuish, loc. cit.*). This is attributed to the fact that in aqueous solution the flavylium salt exists in equilibrium with the 2-phenyl-2*H*-chromen-2-ol (XLIV) and 2-phenyl-4*H*-chromen-4-ol (XLV) as well as the open chain compound, 2-hydroxychalcone (XLVI). Oxidation of XLV yields flavone. With ethanol and sodium hydrozide at low temperature it forms 4-ethoxy-4*H*-chromene.

(XLIV) ⇌ ($OH^{\ominus}$) flavylium $Cl^{\ominus}$ ⇌ ($OH^{\ominus}$) (XLV); (XLIV) ⇌ (XLVI) OH, CH:CHBz; flavylium → (NaOH, EtOH) 4-ethoxy (OEt); (XLV) → (Ox) Flavone

At higher temperatures flavylium salts with sodium hydroxide yield acetophenone and salicylaldehyde (*H. Decker* and *H. Felser*, Ber., 1908, **41,** 2997). Alkaline hydrolysis of 7-hydroxy-2,4-dimethylbenzopyrylium chloride affords 2,4-dihydroxyacetophenone and acetone.

Ring-fission by the action of alkali may take another course, the flavylium salt XLVII, for instance, with boiling aqueous potassium hydroxide yielding resorcinol, acetophenone and 4-methoxyacetophenone (*Robinson* and *J. Walker*, J. chem. Soc., 1934, 1435). Fission has obviously occurred at the dotted line giving benzoylanisoylmethane, which breaks down to the above ketones:

HO, Ph, $Cl^{\ominus}$, C_6H_4OMe (XLVII) → HO OH + $BzCH_2{\cdot}CO{\cdot}C_6H_4OMe$

When benzopyrylium salts dissolved in an alcohol are treated in the cold with sodium hydroxide, 2- and 4-alkyl chromenol ethers result *(Hill* and *Melhuish, loc. cit.)*. 3-Methoxyflavylium chloride with methanol thus gives the 2-methyl ether 2,3-dimethoxy-2-phenyl-2*H*-chromene, oxidation of which by ferric chloride or perphthalic acid gives a flavanone derivative XLVIII (*Karrer et al.*, Helv., 1942, **25,** 1129, 1138; 1943, **26,** 2116). Subsequent hydrolysis yields flavon-3-ol (XLIX). This series of reactions provides a method for converting flavylium salts into flavonols:

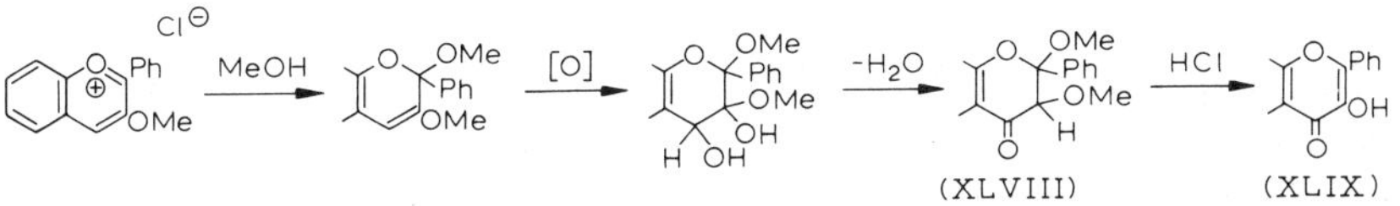

(XLVIII) (XLIX)

Changes in the visible spectra of a number of flavylium salts, unsubstituted in the 3-position, indicate that in the pH range 1–4 these salts undergo reversible, hydrolytic opening of the pyrylium ring with the formation of the corresponding 2-hydroxychalcone (*Jurd*, J. org. Chem., 1963, **28,** 987):

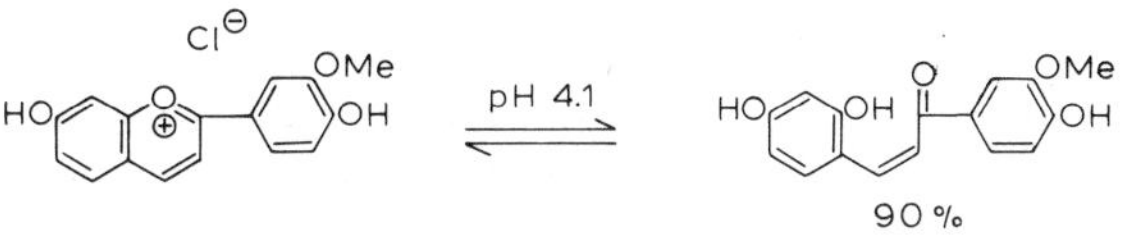

90 %

The structural transformations of flavylium salts have been examined in acid and alkaline solutions (*Jurd* and *T. A. Geissman, ibid.*, 1963, **28,** 2394), and a study made of these changes and those of 4′-hydroxy- and 4′,7-dihydroxy-flavylium perchlorates by polarography and spectroscopy at changing pH values (*K. A. Harper* and *B. V. Chandler*, Austral. J. chem., 1967, **20,** 731, 745).

(4) Condensation with reactive CH compounds. Flavylium salts react with acidic CH compounds, *e.g.* flavylium perchlorate reacts with acetylacetone to give 4-diacetylmethyl-flav-2-ene (*F. Kröhnke* and *K. Dickoré*, Ber., 1959, **92,** 46):

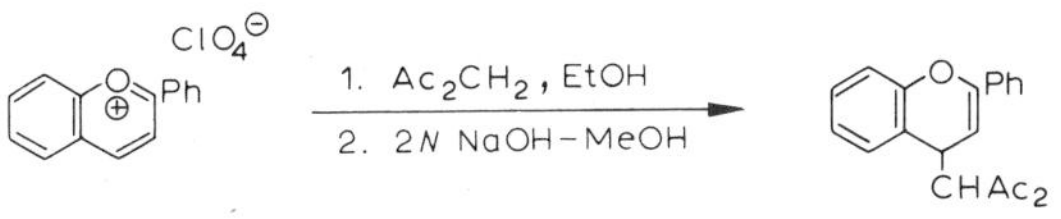

4′-Hydroxy-8-methoxy-3-methylflavylium chloride condenses with phloroglucinol, at p*H* 4–7 in aqueous methanol at room temperature to give L.

Reduction of the flavylium salt to the flav-2-ene also occurs (*M. Nakazaki* and *K. Naemura*, Chem. and Ind., 1964, 1708):

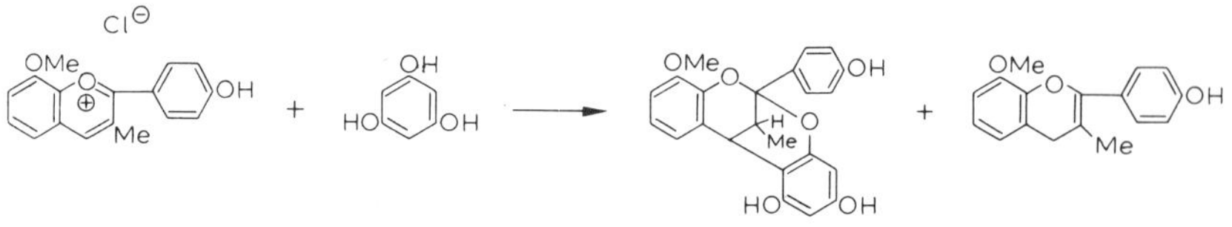

(L)

(5) Grignard reagents attack benzopyrylium salts at $C_{(2)}$ and $C_{(4)}$ (*Löwenbein*, Ber., 1924, **57**, 1517; *Löwenbein* and *Rosenbaum*, *loc. cit.*) to give chromenes, flavylium chloride and phenylmagnesium chloride yielding 2,4-diphenyl-4*H*-chromene:

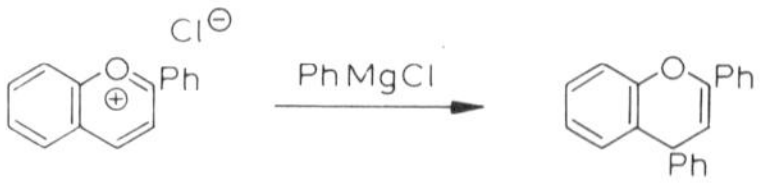

PhMgCl

(6) Anhydro-bases (colour bases). The action of aqueous alkali on a flavylium salt does not yield a flavylium hydroxide, because the hydroxide ion attaches itself to the 2- or 4-position, thus giving a hydroxychromene (*Hill* and *Melhuish*, *loc. cit.*). When there is a hydroxyl group in the 5,7,2′- or 4′-positions flavylium salts with alkali afford anhydro-bases (colour bases) (*Bülow* and *W. von Sicherer*, Ber., 1901, **34**, 2380; *E. L. Hirst*, J. chem. Soc., 1927, 2490; *F. M. Irvine* and *R. Robinson*, *ibid.*, 1927, 2086; *H. Brockmann* and *H. Junge*, Ber., 1944, **77**, 44). Thus 2′-hydroxyflavylium chloride with sodium acetate or sodium hydroxide affords the unstable violet anhydro-base, which changes very readily into the pseudo-base by hydration:

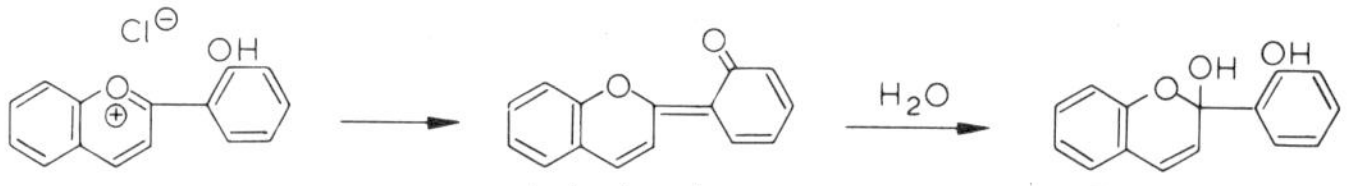

H_2O

Anhydro-base (violet)

Pseudo-base

The anhydro-base obtained from 4-phenyl-7-hydroxyflavylium chloride and sodium acetate, with methyl iodide, dimethyl sulphate and potassium carbonate, or diazomethane yields the pseudo-base and by further methylation the dimethyl ether LI (*Brockmann* and *Junge*, *loc. cit.*). This ether with acid very readily yields the methyl ether of the original flavylium salt LII, a reaction characteristic of pyranol ethers:

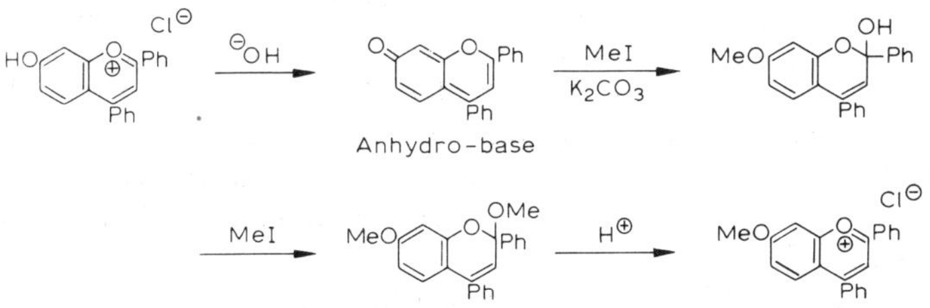

Anhydro-base

(LI) (LII)

Contrary to earlier belief it has been shown that in the formation of anhydro-bases from flavylium salts, the 7-hydroxyl group undergoes loss of a proton in preference to the 4′-hydroxyl group. It has been suggested that this gives an insight into the biogenensis of carajurin (*S. K. Arora*, *A. C. Jain* and *T. R. Seshadri*, J. Indian chem. Soc., 1962, **39**, 285). The effect of substituents on anhydro-base formation has been discussed (*C. F. Timberlake* and *P. Bridle*, J. Sci. Food Agr., 1967, **18**, 473).

Carajurin, $C_{17}H_{14}O_5$, garnet-red prisms, m.p. 205–207°, the ruby coloured pigment of the Brazilian tree *Bignonia chika* (*Robinson et al.*, J. chem. Soc., 1927, 3015) is a typical anhydro-base. It has been synthesised by the following route (*L. Ponniah* and *Seshadri*, Proc. Indian Acad. Sci., 1953, **38A,** 77):

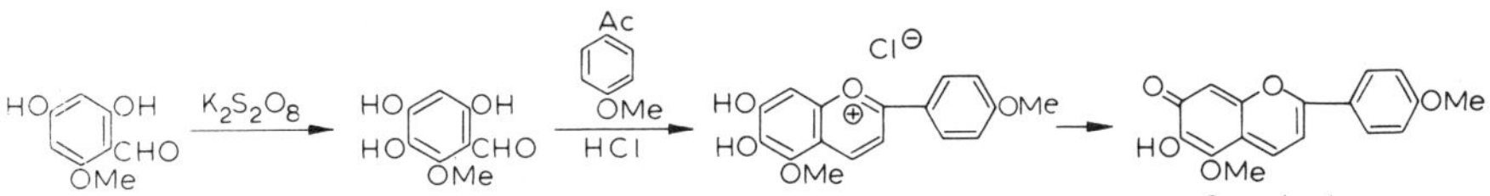

Carajurin

Dracorhodin, $C_{17}H_{14}O_3$, deep red prisma, m.p. 169°, *picrate*, m.p. 217–220°, a constituent of the resin of the dragon's blood tree, *Dracaena Draco*, is another naturally occurring anhydro-base. Its structure has been established by degradation and synthesis (*Brockmann* and *Junge*, Ber., 1943, **76,** 751; *A. Robertson* and *W. B. Whalley*, J. chem. Soc., 1950, 1882).

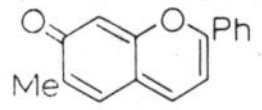

Dracorhodin

The benzopyrylium and flavylium salts do not react with ammonia to give quinoline derivatives (cf. p. 7).

Pyrylium and flavylium salts with a 2- or 4-methyl group condense in acid solution with aldehydes to give styryl derivatives, the methyl group thus resembling the methyl group in the analogous pyridine and quinoline compounds (*Dilthey* and *J. Fischer*, Ber., 1923, **56,** 1012; 1924, **57,** 1653; *J. S. Buck* and *I. M. Heilbron*, J. chem. Soc., 1923, **123,** 2521). This reactivity forms the basis of a test in which benzopyrylium salts are treated in pyridine with orthoformic ester to give an intense blue colour (*Brockmann et al.*, Ber., 1944, **77,** 529).

Brief heating of a solution of 4-methylflavylium perchlorate in pyridine gives a *dimer* (LIV), m.p. 279–280°. The dimer is readily protonated and its n.m.r. spectrum in trifluoroacetic acid shows that the protonated species LIII ($X = CF_3CO_2^\ominus$) is formed (*VanAllan* and *Reynolds*, Tetrahedron Letters, 1969, 2047):

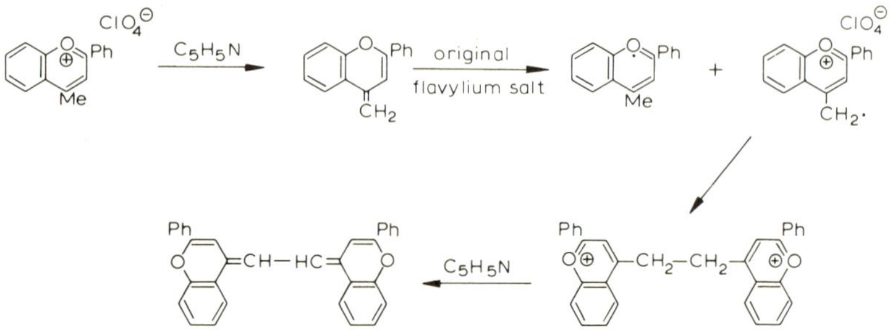

Flavylium chloride cannot be prepared by the action of salicylaldehyde on acetophenone in the presence of hydrochloric acid, but is obtained from 2-hydroxychalcone by treatment in ether with hydrogen chloride, or in acetic acid with hydrochloric acid. It forms a *perchlorate*, m.p. 178–179° and a *ferrichloride*, m.p. 138–139°. Substituted flavylium salts include 7-*hydroxyflavylium chloride*, red-brown needles, m.p. >260°, *perchlorate*, m.p. 220–221°; 5,7-*dihydroxyflavylium chloride* (**chrysinidin chloride**), red needles, m.p. >250°, *perchlorate*, red prismatic needles, m.p. 243–244° (*A. W. Johnson* and *Melhuish*, J. chem. Soc., 1947, 346; *Robinson et al., ibid.*, 1927, 1975); 5,7-*dihydroxy-4′-methoxyflavylium chloride* (**acacetinidin chloride**), red plates or needles, *perchlorate*, red needles, m.p. 278–280° (decomp.) (*D. D. Pratt*, *Robinson* and *P. N. Williams*, *ibid.*, 1924, **125,** 199); 5,7,4′-*trihydroxyflavylium chloride* (**apigenidin chloride gesneridin**), salmon-red needles (*Robinson et al., loc. cit.*). 4-*Methylflavylium perchlorate*, m.p. 212–213°; 4′-*methoxy*-4-*methylflavylium perchlorate*, m.p. 230–232°; 4′-*chloro*-4-*methylflavylium perchlorate*, m.p. 236–237°; 4,6-*dimethylflavylium perchlorate*, m.p. 278–279° (*Reynolds*, *VanAllan* and *Daniel*, *loc. cit.*).

*(v) The anthocyanins and anthocyanidins**

Anthocyanins are hydroxy derivatives of 2-phenylbenzopyrylium or flavylium salts (p. 68), and are responsible for the red, violet, and blue colours of flowers. They are also present in roots, stems, leaves, seeds and other parts of plants, *e.g.*, bracts and pollen. They exist in nature mostly as glycosides. The presence in flowers of pigments which change colour with acid or alkali has been known for centuries. The term *anthocyan* was introduced some 120 years ago (1835) by *Marquart*, and in 1836 *Hope* read a paper before the Royal Society of Edinburgh (March 21st, cf. J. pr. Chem., 1837, **10,** 269) in which he concluded from many experiments that the plant pigments were formed from faintly coloured chromogens by a variety of changes. Subsequently many unsuccessful attempts were made to isolate the pure pigments (see "The Natural Organic Colouring Matters", *A. G. Perkin* and *A. E. Everest*, Longmans, London, 1918, pp. 236 *et seq.*), but it was not until 1913 that success was eventually achieved. In that year *R. Willstätter* published his classic paper with *Everest* (Ann., 1913, **410,** 189) on the isolation and characterisation of cyanin from the corn-flower and the constitution of the anthocyanins. These authors introduced the terms which are now used; *anthocyanins* for the glycosides and *anthocyanidins* for the aglycones. Subsequent important contributions in the field have come from *P. Karrer* and especially from Sir *R. Robinson* who synthesised the most important anthocyanins and anthocyanidins.

The anthocyanins are obtained from the dried petals of flowers in which they often occur in abundance, a notable example being the blue-black viola *(viola tricolor)* whose dried petals contain as much as 24% of the pigment violanin.

Structure. An anthocyanin on hydrolysis yields a pentose or a hexose and an anthocyanidin. There is sometimes a third constituent such as 4-hydroxybenzoic acid or 4-hydroxycinnamic acid, which esterifies the hydroxyl groups of the anthocyanidin or the sugar. Anthocyanins containing such acids are termed *complex anthocyanins.*

The anthocyanidins are salts which are relatively stable in aqueous solution thus showing that they are derived from strong bases. *Willstätter* found that the salts formed from the anthocyanidins (or anthocyanins) and strong acids can be purified by dissolution in a hydroxylic solvent such as methanol followed by precipitation by means of a non-hydroxylic solvent, for

* "Uber die Synthese von Anthocyaninen", *R. Robinson*, Ber., 1934, **67,** 85; "Chemistry of the Anthocyanins", *Robinson*, Nature, 1935, **135,** 732; "The Red and Blue Colouring Matters of Plants", *Robinson*, Endeavour, 1942, **1,** 92; *S. G. P. Plant*, Ann. Reports, 1928, **25,** 163; "Naturally Occurring Oxygen Ring Compounds", Ch. 13, *F. M. Dean*, Butterworths, London, 1963.

example, ether. Purification could then be effected by crystallisation of the chloride or picrate. Other methods of purification such as chromatographic adsorption have been successfully employed (*P. Karrer* and *F. M. Strong*, Helv., 1936, **19,** 25).

Alkali fusion of the anthocyanidins yields phloroglucinol and a hydroxylated benzoic acid. These degradations taken in conjunction with what was known about flavones and flavonols, and flavylium salts led *Willstätter* to formulate the anthocyanidins as hydroxylated flavylium salts. This has been confirmed in other ways particularly by synthesis (p. 84) and from the close relationship of the anthocyanidins to the flavonols, etc. (p. 88).

The anthocyanins and anthocyanidins are flavylium salts derived for the most part from three types represented by pelargonidin chloride, cyanidin chloride and delphinidin chloride:

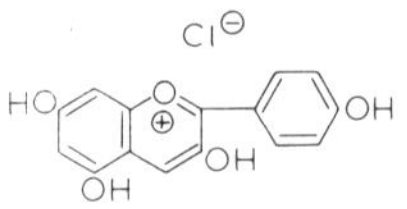

Pelargonidin chloride 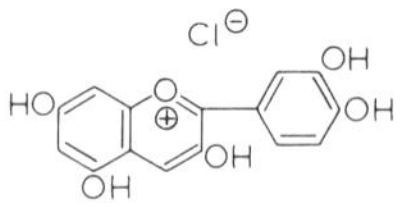 | Cyanidin chloride 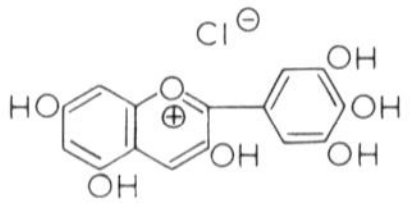 | Delphinidin chloride

The three types have in common a 3,5,7-trihydroxyflavylium system and differ by the number of hydroxyl groups on the 2-phenyl ring. This is decisively demonstrated by alkali fusion, all three compounds giving phloroglucinol, pelargonidin also yielding 4-hydroxybenzoic acid, cyanidin giving protocatechuic acid and delphinidin giving gallic acid. More detailed information is obtained by degradation with dilute barium hydroxide or 10% sodium hydroxide solution (p. 83).

Important naturally occurring derivatives of cyanidin and delphinidin are the anthocyanins peonin, petunin, malvin and hirsutin, whose aglycones are peonidin chloride, petunidin chloride, malvidin chloride and hirsutidin chloride:

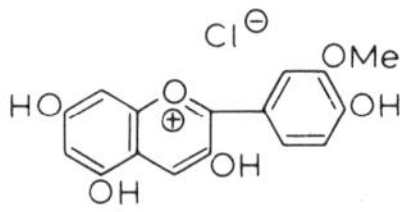

Peonidin chloride 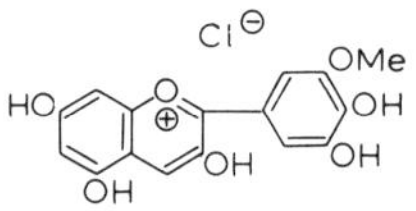 | Petunidin chloride

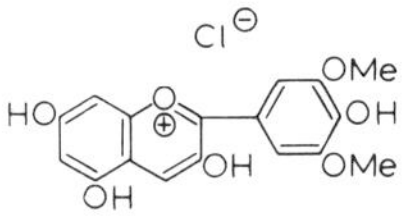

Malvidin chloride 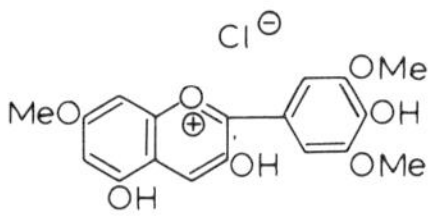 | Hirsutidin chloride 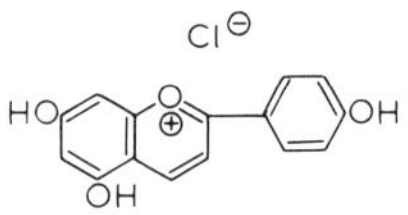 | Gesneridin chloride

A fourth type of anthocyanin has been discovered in gesnerin and the aglycone gesneridin in which the 3-hydroxyl group of the other anthocyanins is missing (*G. M.* and *R. Robinson*, Biochem. J., 1932, **26,** 1647; *G. M. Robinson* and *A. R. Todd*, J. chem. Soc., 1934, 809). Other anthocyanins which are not derived from the three main types are the nitrogenous anthocyanins such as the *betanin* group (*G. M.* and *R. Robinson*, *ibid.*, 1932, 1439; 1933, 25), and the colouring matter of the Iceland poppy (*R. Robinson et al.*, Nature, 1938, **142,** 356). These substances are to be regarded as exceptional.

The individual anthocyanins differ within the type by the number and position of the glycosidic residues and methoxyl groups. Information on these points has been obtained by methods developed by *Karrer* and his co-workers (Helv., 1927, **10,** 67, 729). The use of dilute barium or sodium hydroxide solution in an atmosphere of hydrogen breaks down the molecule but leaves the methoxyl groups intact. Thus malvidin chloride with barium hydroxide solution yields phloroglucinol and syringic acid (3,5-dimethoxy-4-hydroxybenzoic acid):

Cl⊖ OMe HO ⊕ OH OH OMe OH → HO OH OH + HO_2C OMe OH OMe

Syringic acid

The position of the sugar residue can be ascertained by methylating the anthocyanins and subsequently removing the sugar group or groups. The position of the free hydroxyl group gives the position of the sugar in the parent molecule. Anthocyanins with a free hydroxyl group in the 3-position are rapidly decolorised by ferric chloride whereas those with sugar residues in this position are stable (*R. Robinson et al.*, J. chem. Soc., 1931, 2672). Anthocyanins with sugar residues in the 3-position undergo ring-fission at the $C_{(2)}$–$C_{(3)}$ bond with hydrogen peroxide to give a sugar ester, for example, malvin yields malvone and this sugar group is easily removed by dilute ammonium hydroxide solution:

Cl⊖ OMe HO ⊕ OH OR OMe OR $\xrightarrow{H_2O_2}$ HO O·CO OMe OH OMe OR $CH_2{\cdot}CO_2R$

Malvone

$R = C_6H_{11}O_5$

If the sugar is attached to the aromatic nucleus it requires dilute hydrochloric acid for hydrolysis (*Karrer* and *G. de Meuron*, Helv., 1932, **15,** 507, 1212). As a result of these methods it has been established that the sugar residue in all the monoglycosides occupies the 3-position. It is now known that most anthocyanins fall into one of the following categories: (*a*) 3-monoglucosides and 3-monogalactosides; (*b*) 3-rhamnoglucosides and 3-pentoseglycosides; (*c*) 3-biosides; (*d*) 3,5-diglucosides and (*e*) acylated anthocyanins. The best-known belong to category *d*. Recent work has shown that other hydroxyl groups besides the 3 and 5 can be glycosidated (*W. G. C. Forsyth* and *N. W. Simmonds*, Proc. Roy, Soc. B, 1954, **142,** 549; *J. B. Harbone* and *H. S. A. Sherratt*, Experientia, 1957, **13,** 486). Anthocyanins may also be partially esterified with malonic, 4-hydroxybenzoic, or 2- or 4-hydroxycinnamic acids (*Harborne*, Phytochemistry, 1964, **3,** 151).

Methods of synthesis. (*1*) In the first general method of synthesis, *Willstätter* extended that of *Decker* and *von Fellenberg* (p. 71) to the preparation of the cyanidins. 3,5,7-Trimethoxycoumarin, for instance, yields with anisylmagnesium bromide and subsequent treatment with water the pseudo-base LV, which with hydrochloric acid gives the flavylium chloride LVI. This substance is then demethylated with hydriodic acid and subsequent treatment with silver chloride gives pelargonidin chloride. This synthesis is experimentally difficult and tends to give impure products and since the last stage involves demethylation it cannot, therefore, be adapted to the preparation of partially methylated compounds:

MeO OMe OMe + BrMg OMe → MeO OH OMe OMe OMe (LV)

→ $Cl^{\ominus}$ MeO OMe OMe OMe (LVI) → $Cl^{\ominus}$ HO OH OH OH Pelargonidin chloride

(*2*) *Robinson* modified the second benzopyrylium synthesis (p. 70). Phloroglucinaldehyde monobenzoate, from phloroglucinaldehyde and benzoyl chloride, is a particularly useful starting material, and when condensed with an ω-acetoxyacetophenone (*e.g.* LVII) in ethyl acetate by hydrogen chloride it yields a benzoylated anthocyanidin (LVIII, R = COPh), condensation being accompanied by deacetylation. Treatment with barium hydroxide removes the benzoyl residue but opens the pyrylium ring to give a chalcone as the barium salt. Finally ring-closure to malvidin chloride (LVIII, R = H) is effected by adding sulphuric acid equivalent to the amount of barium salt and then hydrochloric acid:

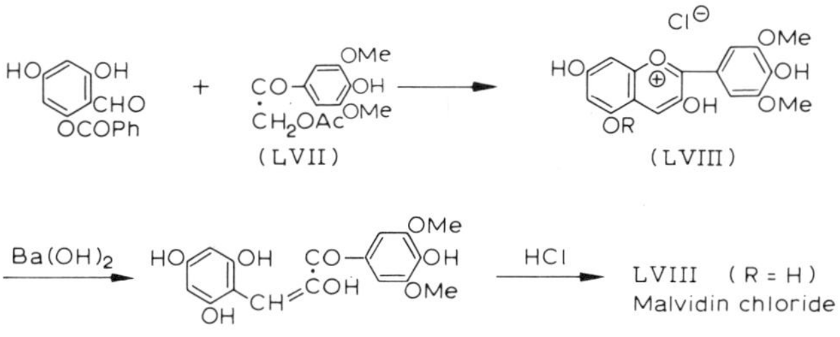

HO, OH, CHO, OCOPh + CO, CH2OAc, OMe, OH, OMe (LVII) → Cl⊖ HO, OR, OH, OMe, OH, OMe (LVIII)

$\xrightarrow{Ba(OH)_2}$ HO, OH, OH, CH=COH, CO, OMe, OH, OMe $\xrightarrow{HCl}$ LVIII (R = H) Malvidin chloride

This method is most convenient and widely applicable and by it *Robinson* has synthesised all the important anthocyanidins. By the use of suitable glucosated intermediates it is possible to prepare the isomeric β-glucosides of any particular anthocyanidin. Phloroglucinaldehyde with acetobromoglucose gives 2-*O*-tetra-acetyl-β-glucosidophloroglucinaldehyde, which in the presence of ethyl acetate and hydrogen chloride condenses with ω-*O*-tetra-acetyl-β-glucosidoxy-4-acetoxyacetophenone to give the acetylated pelargonin (LIX) from which by suitable treatment pelargonin chloride is obtained:

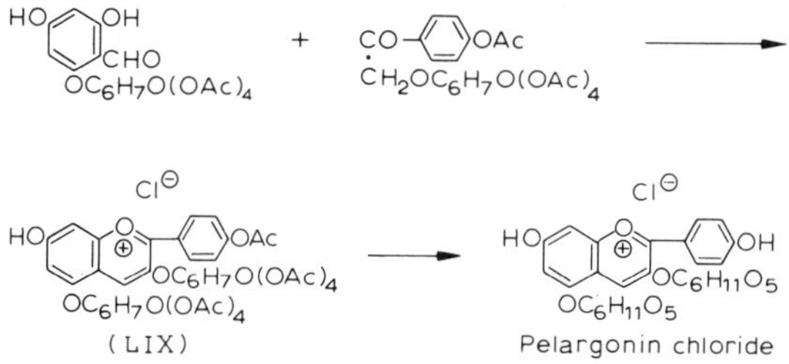

HO, OH, CHO, $OC_6H_7O(OAc)_4$ + CO, OAc, $CH_2OC_6H_7O(OAc)_4$ →

Cl⊖ HO, OAc, $OC_6H_7O(OAc)_4$, $OC_6H_7O(OAc)_4$ (LIX) → Cl⊖ HO, OH, $OC_6H_{11}O_5$, $OC_6H_{11}O_5$ Pelargonin chloride

Robinson and his co-workers by such methods synthesised the four β-glucosides of pelargonidin and showed that the 3-glucoside is identical with *Willstätter*'s *callistephin chloride.* This constituted the first synthesis of a flower pigment.

(*3*) In a related synthesis reactive phenols are condensed with hydroxymethylene derivatives of acetophenone in the presence of hydrogen chloride (*C. von Bülow* and *W. von Sicherer*, Ber., 1901, **34**, 3889):

Galanginidin 3-methyl ether

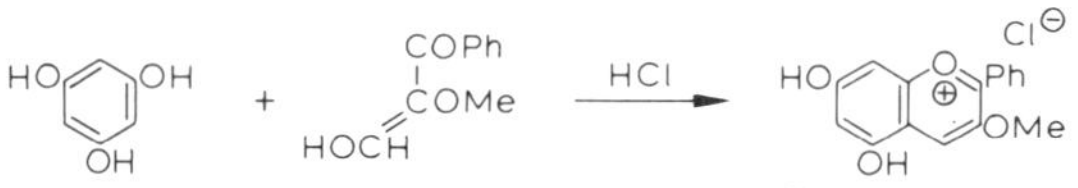

The 3-methyl ether of galanginidin chloride has been synthesised in this way (*T. Malkin* and *Robinson*, J. chem. Soc., 1925, **127,** 1190).

(*4*) The condensation of α,β-unsaturated ketones with phenols can be effected in the presence of hydrogen chloride and the mild oxidising agent chloranil (*Robinson* and *J. Walker*, *ibid.*, 1934, 1435; 1935, 941). In this way chrysinidin chloride has been synthesised:

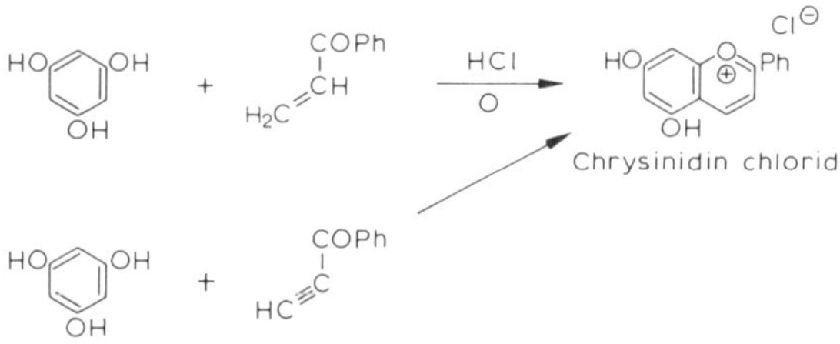

Chrysinidin chloride

α,β-Ethynyl ketones likewise condense with phenols in the presence of acids to give flavylium salts, no oxidising agent being required (*A. W. Johnson* and *R. R. Melhuish*, *ibid.*, 1947, 346). Chrysinidin and related chlorides are readily prepared by this method.

Isomeric rearrangements during the synthesis and purification of 5,6,7- and 5,7,8-trihydroxyanthocyanidins have been discussed and spectral evidence has been presented in support of the 3,4′,5,6,7-pentahydroxyflavylium structure for aurantinidin (*L. Jurd* and *Harborne*, Phytochemistry, 1968, **7,** 1209). The syntheses of anthocyanins have been reviewed (*Robinson*, Beitr. Biochem. Physiol. Naturstoffen, Festschr., 1965, 11).

Properties of the anthocyanins and anthocyanidins. The anthocyanins are crystalline substances soluble in water, but insoluble in non-hydroxylic solvents such as ether and benzene. They are amphoteric and their salts with strong acids such as hydrochloric acid are more stable than the salts obtained from the 4-pyrones and chromones. The importance of this amphoteric character is demonstrated by the variation of colour of an anthocyanin or anthocyanidin with pH. For instance, cyanin is red at pH 3, violet at pH 8.5 and blue at pH 11.0. These colour changes are accounted for by the structural changes shown below in which cyanidin is shown as the red flavylium salt (a), the violet colour base (b) and the blue salt of the colour base (d). The methanol base (pseudo-base) (c) is colourless:

NaOAc / HCl

(a) Red ⇌ (b) Violet

H_2O

HCl / H_2O

Na_2CO_3

(c) Colourless

(d) Blue

Support for these formulae comes from the colours of certain other anthocyanins and anthocyanidins. Galanginidin chloride (p. 86), on account of the lack of a hydroxyl group in the 4′-position, cannot change into the quinonoid form and in consequence does not form a blue or a violet salt with alkali. Hirsutidin chloride (p. 82) has no free hydroxyl group in the 3′- or 5′-positions and can form a violet colour base but no blue alkali salt. For the formation of a pure blue colour a quinonoid structure and a salt-forming auxochrome are evidently required. Confirmation is provided by the blue colour of the sodium salt of malvin which has a hydroxyl group in the 7-position and the violet colour of the sodium salt of hirsutin with a 7-methoxyl group (*Robinson* and *A. R. Todd*, *ibid*., 1932, 2293).

Willstätter noted that blue flowers contained rather large proportions of certain metals, chiefly sodium and potassium, and many workers have regarded blue flowers as containing metal salts of anhydro-bases and red flowers as containing pyrylium salts, for example, cyanin is present in the cornflower as the blue potassium salt. In spite of this, the hypothesis that there is a simple relationship between the presence of metals and the colour of flowers must be regarded as inadequate (*E. M. Chenery*, J. roy. hort. Soc., 1937, **62,** 304; *K. Shibata*, *K. Hayashi* and *T. Isaka*, Acta phytochim., Tokyo, 1949, **15,** 17, 41; *H. Kikkawa*, *Z. Oggta* and *S. Fujito*, Kagaku, 1955, **25,** 139; *A. Storck*, Angew. Bot., 1942, **24,** 397). Also, the idea of any simple connection between acidity and the colour of flowers must be discounted, because the pH of flower petals does not vary very much: it usually lies between 3.8 and 5.8 (*G. M. Robinson*, J. Amer. chem. Soc., 1939, **61**, 1606). The colour of the anthocyanins in acid solution is deepened by the presence of co-pigments which include tannins and flavones (*E. Bayer et al*., Ber., 1958, **91,** 1115; 1959, **92,** 1062; 1960, **93,** 2871). The co-pigments play an important role in the colouring of flowers.

Genuine anthocyanins, corresponding to pelargonin, cyanin and violanin, have been isolated from flowers of cornflower, rose and pansy, respectively, without the use of mineral acid, and identical anthocyanins have also been prepared *in vitro* from the chlorides of the corresponding anthocyanins by chemical means. The basic structure of all genuine anthocyanins seems to correspond to a free state of oxonium compound instead of to a hydroxylated form of the flavylium skeleton. E.s.r. spectra of genuine anthocyanins, both natural and dechlorinated, indicated the presence of some free radicals within the molecular structures (*K. Takeda, N. Saito* and *Hayashi*, Proc. Jap. Acad., 1968, **44,** 352).

The flavanols, catechins or epicatechins and anthocyanidins are structurally closely related and should be interconvertible by oxidation and reduction (p. 240). Such conversions have been effected though frequently in very poor yields.

Quercetin $\xrightarrow{H_2}$ → $\xrightarrow[-H_2O]{H^\oplus}$ Cyanidin; Cyanidin → Epicatechin; Cyanidin $\xrightarrow{Zn,\ C_5H_5N,\ AcOH}$ Leucoanthocyanidin; Leucoanthocyanidin $\xrightarrow{O_2}$ Cyanidin

Quercetin is reduced to cyanidin chloride by magnesium and methanolic hydrogen chloride (*R. Willstätter* and *H. Mallinson*, Ber., 1914, **47,** 2874) and quercetin pentamethyl ether with titanium chloride yields cyanidin pentamethyl ether (*P. Karrer et al.*, Helv., 1930, **13,** 1308). *R. Robinson* and *R. Mirza* (Nature, 1950, **166,** 997) have shown that the flavonol–anthocyanidin change may be readily effected by means of lithium tetrahydridoaluminate and in this way kaempferol gives a 32% yield of pelargonidin chloride. Reductive acetylation of flavanols with zinc and anhydrous sodium acetate in acetic anhydride gives acetates of undetermined structure having one acetyl less than would be expected. Hot mineral acid converts the acetates into the corresponding anthocyanidins smoothly and in good yield; quercetin gives cyanidin chloride, fisetin gives fisetinidin chloride, and kaempferol gives pelargonidin chloride (*H. G. C. King* and *T. White*, J. chem. Soc., 1957, 3901):

Quercetin $\xrightarrow[Ac_2O \quad 2.\ HCl]{1.\ Zn/NaOAc}$ Cyanidin chloride

Cyanidin chloride is reduced by zinc and pyridine with a few drops of acetic acid to a leucoanthocyanidin (*R. Kuhn* and *A. Winterstein*, Ber., 1932, **65,** 1742). It is readily oxidised by air to cyanidin. Cyanidin chloride has been reduced to epicatechin (p. 240; see also *L. Reichel* and *W. Burkart*, Ann., 1938, **536,** 164).

Some dihydrohydroxyflavonols, taxifolin, aromadendrin and dihydrorobinetin on heating with acetic anhydride and sodium acetate or pyridine at 150–160° give, after acid hydrolysis, the corresponding cyanidin chloride. No change occurs on similar treatment of the related methyl ethers (*H. G. Krishnamurty, T. R. Seshadri* and

B. Venkataramani, J. Sci. Ind. Res., India, 1960, **19B,** 115). The conversion of 5,7,3′,4′-tetramethoxyflavan-3,4-diol (LX) into 5,7,3′,4′-tetramethylcyanidin chloride by 10*M* hydrochloric acid in 2-propanol has been examined. It was concluded that the mechanism must be oxidative, since the alternative of disproportionation was ruled out (*R. V. Robertson*, Canad. J. Chem., 1959, **37,** 1946):

(LX) —10 *M* HCl→ (flavylium chloride)

Only one example of the oxidation of a catechin to an anthocyanidin is known (p. 240). The oxidation of flavylium salts to flavones and flavonols has already been mentioned (p. 76).

A great variety of rapid tests are used to detect and identify anthocyanins (*G. M. Robinson* and *R. Robinson*, Biochem. J., 1931, **25,** 1687, 1704; 1932, **26,** 1647; *R. Robinson*, J. Soc. chem. Ind., 1933, **52,** 742). Colour tests are also valuable. Extraction with amyl alcohol, addition of sodium acetate, and finally a trace of ferric chloride results in a change of the violet alcoholic solution to a pure blue if cyanidin or delphinidin is present. In the oxidation test 10% sodium hydroxide solution is added to the solution of the anthocyanidin; perunidin and delphinidin are destroyed immediately whereas other anthocyanidins are relatively stable.

Spectral methods for characterising anthocyanins have been described (*Harborne*, Biochem. J., 1958, **70,** 22), and the i.r. spectra of various anthocyanins have been examined (*P. Ribéreau-Gayon* and *M. L. Josien*, Bull. Soc. chim. Fr., 1960, 934).

Structural transformation of anthocyanidins and anthocyanins in solution have been studied by spectral and pH measurements, and the effect of substituents on anhydro-base formation has been discussed (*C. F. Timberlake* and *P. Bridle*, J. Sci. Food Agr., 1967, **18,** 473, 479).

Gesneridin chloride, $C_{15}H_{11}O_4Cl$, 5,7,4′-trihydroxyflavylium chloride, pale orange-yellow prisms, occurs as the 5-glucoside in the flowers of *Gesneria fulgens* (*G. M. Robinson, R. Robinson* and *Todd*, J. chem. Soc., 1934, 809). Its structure has been confirmed by synthesis (*D. D. Pratt et al., ibid.*, 1927, 1975; *A. Robertson* and *R. Robertson, ibid.*, 1927, 1710; 1928, 1526, 1533, 1537; *Y. Asahina* and *M. Inubuse*, Ber., 1928, **61**, 1646). **Apigenidin chloride,** obtained by the reduction of apigenin (p. 180), is identical with gesneridin chloride. Acacetinidin chloride is the 4′-methoxy compound.

Luteolinidin (5,7,3′,4′-tetrahydroxyflavylium) occurs as a glycoside with gesnerin in the petals of *Gesnera cardinalis*, and its structure has been established on spectral and chromatographic evidence (*J. B. Harborne*, Chem. and Ind., 1960, 229). *Robinson et al.* (J. chem. Soc., 1934, 809) have suggested previously that what is here called luteolinidin was the malvidin analogue (3′,5′-dimethoxy-5,7,4′-trihydroxyflavylium).

Pelargonidin chloride and related substances. **Pelargonidin chloride,** $C_{15}H_{11}OCl$, 3,5,7,4′-*tetrahydroxyflavylium chloride*, m.p. $>350°$, separates from water as the mono-hydrate. Its constitution follows from its breakdown with alkali to 4-hydroxybenzoic acid and phloroglucinol (*Willstätter* and *E. K. Bolton*, Ann., 1915, **408,** 59) and synthesis (*Willstätter et al.*, Ber., 1924, **57,** 1938; *Robertson et al.*, J. chem. Soc., 1928, 1533; *L. Bauer*, *A. J. Birch* and *W. E. Hillis*, Chem. and Ind., 1954, 433).

Callistephin chloride, $C_{21}H_{21}O_{10}Cl$, the 3-β-glucoside of pelargonidin chloride, was first isolated along with chrysanthemin from the red aster (*Callistephus chinensis*, N. ab E.) (*Willstätter* and *Ch. L. Burdick*, Ann., 1917, **412,** 149). It also occurs in carnations. Callistephin forms fine orange-red needles, and a picrate, red needles. With hydrochloric acid it is hydrolysed to perlargonidin chloride. Its synthesis constitutes the first artificial production of a flower pigment (*Robertson* and *R. Robinson*, J. chem. Soc., 1928, 1460). **Fragarin chloride,** $C_{21}H_{21}O_{10}Cl$, is the 3-β-galactoside, and can be isolated from the wild strawberry *(Fragaria vesca)* as dark red needles. It has been synthesised (*P. V. Nair* and *Robinson*, *ibid.*, 1934, 1611). **Pelargonin chloride,** $C_{27}H_{31}O_{15}Cl$, $4H_2O$, occurs in the scarlet pelargonium and the red cornflower *(Willstätter* and *Bolton*, *loc. cit.)*. It forms scarlet-red needles, sintering at 178° and melting at 187°. It is the 3,5-di-D-glucoside and with 20% hydrochloric acid is hydrolysed to pelargonidin and glucose. It has been synthesised (*Robinson* and *Todd*, J. chem. Soc., 1932, 2488). **Monardin** and **salvinin** are identical with pelargonin (*P. Karrer* and *R. Widmer*, Helv., 1927, **10**, 729; *Robinson* and *Todd*, *loc. cit.*).

Monardaein chloride (salvianin chloride), occurs in "golden balm" *(Monarda didyma)* *(Karrer* and *Widmer*, Helv., 1928, **11,** 837; 1929, **12,** 292). It is hydrolysed to 4-hydroxycinnamic acid, malonic acid and pelargonidin chloride and partial hydrolysis yields pelargonin chloride and 4-hydroxycinnamic acid. Hydrolysis with 20% hydrochloric acid for 3 minutes at 100° liberated malonic acid. The constitution of some *O*-acylanthocyanins related to salvianin have been established (*L. Birkofer et al.*, Z. Naturforsch., 1965, **20b,** 424).

Cyanidin chloride and related substances. **Cyanidin chloride,** $C_{15}H_{11}O_6Cl$, 3,5,7,3′,4′-pentahydroxyflavylium chloride, is obtained as violet-red needles, and is broken down by alkaline fission into phloroglucinol and protocatechuic acid. Its constitution follows from this and its synthesis (*Willstätter et al.*, Ber., 1924, **57,** 1938; *Pratt* and *Robinson*, J. chem. Soc., 1925, **127,** 166; *Robertson* and *Robinson*, *ibid.*, 1928, 1526). Quercetin and dihydroquercetin on reductive acetylation give the same leucocyanidin (flav-3-en-3-ol) acetate, which has been converted into cyanidin chloride (*K. R. Laumas* and *Seshadri*, Proc. Indian Acad. Sci., 1959, **49A,** 47). (−)-Epicatechin penta-acetate when oxidised with *N*-bromosuccinimide gives cyanidin bromide, which can be converted to the chloride (*Seshadri* and *R. K. Trikha*, Indian J. Chem., 1971, **9,** 626).

Cyanin chloride, $C_{27}H_{31}O_{16}Cl$, cyanidin 3,5-di-β-glucoside, crystallises as the trihydrate, which after drying still contains 0.75 mol. water. It is obtained as brownish crystals with a greenish-bronze lustre, m.p. 205° (decomp.), from the red rose *(Rose gallica)*, blue cornflower *(Centaurea cyanus)*, red poppy and red dahlia. It has been synthesised *(Robinson* and *Todd*, *loc. cit.)*.

Mecocyanin chloride, $C_{27}H_{31}O_{16}Cl$, is isolated in dark red needles from the flowers of the red poppy *(Papaver rhoeas)* (*Willstätter* and *F. Weil*, Ann., 1917, **412,** 231) and is purified through the sparingly soluble ferrocyanide. It is cyanidin 3-cellobioside (*J. B. Harborne* and *H. S. A. Sherratt*, Biochem. J., 1961, **78,** 298), although for a long time it was believed to be the 3-gentiobioside. It is partially hydrolysed to the monoglucoside and both cyanidin 3-cellobioside and 3-gentiobioside have been synthesised (*K. E. Grove et al.*, J. chem. Soc., 1934, 1608).

Antirrhinin chloride, keracyanin chloride, prunicyanin chloride, $C_{27}H_{31}O_{15}Cl$, yellow needles, occurs in the sweet cherry *(Prunus avium)*. It is cyanidin 3-rhamnoglucoside, being hydrolysed to cyanidin, glucose and rhamnose (*Willstätter* and *E. H. Zollinger*, Ann., 1917, **412,** 164; *K. Hayashi, T. Noguchi* and *Y. Abe*, Bot. Mag., Tokyo, 1955, **68,** 129; Pharm. Bull. Japan, 1954, **2,** 41).

Chrysanthemin chloride, $C_{21}H_{21}O_{11}Cl$, bronze red prisms, is obtained from the summer aster *(Aster chinensis)* or winter aster (*Chrysanthemum indicum* Linn.) (*Willstätter* and *Bolton, loc. cit.; Robinson* and *H. Smith*, Nature, 1955, **175,** 634; *K. C. Li* and *A. C. Wagenknecht, ibid.*, 1958, **182,** 657; *B. V. Chandler, ibid.*, p. 933). It is the 3-β-glucoside and is hydrolysed to cyanidin and glucose. It has been synthesised (*S. Murakami et al.*, J. chem. Soc., 1931, 2665).

Idaein chloride, $C_{21}H_{21}O_{11}Cl$, H_2O, brownish-green prisms, m.p. 210° (decomp.), has been isolated from cranberries (*Vaccinium vitis idaea*, L.) (*Willstätter* and *Mallinson*, Ann., 1915, **408,** 15; *C. E. Sando*, J. biol. Chem., 1935, **117,** 45). It is cyanidin 3-galactoside and has been synthesised (*Grove* and *Robinson*, J. chem. Soc., 1931, 2722).

Delphinidin chloride and related substances. **Delphinidin chloride,** $C_{15}H_{11}O_7Cl$, 3,5,7,3′,4′,5′-*hexahydroxyflavylium chloride*, is obtained in brownish-red plates, m.p. >350°. Hydrates containing 1, 1.5, 2, 3.5 and 4 mols. of water have been prepared (*Willstätter* and *W. Mieg*, Ann., 1915, **408,** 61; *Willstätter* and *F. J. Weil, ibid.*, 1916, **412,** 178). Alkaline fission yields phloroglucinol, gallic acid and pyrogallol. This together with its synthesis established the constitution (*Pratt* and *Robinson, loc. cit.; W. Bradley et al.*, J. chem. Soc., 1930, 793).

Delphin chloride, $C_{27}H_{31}O_{17}Cl$, $3H_2O$, m.p. 202–203° (decomp.) is obtained from the blue flowers of *Salvia patens*. Hydrolysis with boiling 10% hydrochloric acid gives delphinidin chloride and glucose. It is the 3,5-diglucoside of delphinidin chloride and is thus the analogue of cyanin. It has been synthesised (*T. M. Reynold et al., ibid.*, 1934, 1235).

Delphinin chloride, $C_{41}H_{39}O_{21}Cl$, $2H_2O$, m.p. 200–203° (decomp.), is a diglycoside obtained from the wild purple larkspur *(Willstätter* and *Mieg, loc. cit.)*. With 20% hydrochloric acid it yields delphinidin chloride, glucose (2 mol) and 4-hydroxybenzoic acid (2 mol).

Empetrin chloride, myrtillin-b **chloride,** obtained from Japanese crowberry *(Empetrum nigrum)*, is believed to be the 3-galactoside of delphinidin (*Reynold* and *Robinson*, J. chem. Soc., 1934, 1039; *Hayashi, G. Suzushino* and *K. Ouchi*, Proc. Japan Acad., 1951, **27,** 430).

Vicin chloride I and **II,** $C_{21}H_{21}O_{12}Cl$, $2H_2O$, occurs in scarlet flowered vetch (*P. Karrer* and *R. Widmer*, Helv., 1927, **10,** 67). Vicin is a mixture of the 3-glucoside and the 3-rhamnoside and is degraded by alkali to phloroglucinol and by concentrated hydrochloric acid to delphinidin, glucose and rhamnose.

Gentianin chloride, $C_{30}H_{27}O_{14}Cl$, is obtained from the dwarf gentian *(Gentiana acaulis)* *(Karrer* and *Widmer, loc. cit.)*. Fission with 2*M* sodium hydroxide gives 4-hydroxycinnamic acid and with 20% hydrochloric acid affords this acid, delphinidin and glucose. It gives a violet-blue colour with ferric chloride in ethanol. 1,7-Dihydroxy-

3-methoxyxanthone obtained from the roots of *Gentiana lutea* is also called gentianin (*R. Binaghi* and *P. Falgui*, Ann. chim. applicata, 1925, **15,** 386; *G. V. Rao* and *Seshadri*, Proc. Indian Acad. Sci., 1953, **37A,** 710).

Violanin chloride, $C_{36}H_{37}O_{18}Cl$, is obtained from the black pansy (*Viola tricolor*) (*Willstätter* and *Weil*, Ann., 1917, **412,** 178). With dilute hydrochloric acid it gives delphinidin, rhamnose, glucose and 4-hydroxycinnamic acid. It is a 3-rhamnoglucoside with a molecule of 4-hydroxycinnamic acid attached to the sugar residue (*Karrer* and *G. de Meuron*, Helv., 1933, **16,** 292; *Harborne*, Arch. Biochem. Biophys., 1962, **96,** 171; *Takeda, Abe* and *Hayashi*, Proc. Japan Acad., 1963, **39,** 225). It gives a blue colour with ferric chloride.

Peonidin chloride and related substances. **Peonidin chloride, paeonidin chloride,** $C_{16}H_{13}O_6Cl$, 3,5,7,4′-tetrahydroxy-3′-methoxyflavylium chloride, is obtained as the monohydrate in dark brown prismatic needles or in two modifications with 1.5 mol. of water of crystallisation. It dissolves in aqueous sodium carbonate to give a blue solution. Its constitution follows from its degradation with boiling 10% sodium hydroxide to phloroglucinol and vanillic acid (*Karrer* and *Widmer*, Helv., 1927, **10,** 5) and from its synthesis (*Robinson et al.*, J. chem. Soc., 1926, 1968; 1928, 1537).

Peonin chloride, $C_{28}H_{33}O_{16}Cl$, is obtained as the mono- or penta-hydrate or in the anhydrous state. It occurs in the deep violet-red peony (*Paeonia arborea*) (*Willstätter* and *T. J. Nolan*, Ann., 1915, **408,** 136). Its constitution as the 3,5-di-β-glucoside was established by hydrolytic breakdown to peonidin chloride and glucose (2 mols.) and by synthesis (*Robinson* and *Todd, loc. cit.*). **Rosinidin,** the principal pigment in flowers of *Primula rosea*, is 7-methylpeonidin (7,3′-dimethoxy-3,5,4′-trihydroxyflavylium). Its identity has been established on spectral and chromatographic evidence (*Harborne*, Chem. and Ind., 1960, 229).

Oxycoccicyanin chloride, $C_{22}H_{23}O_{11}Cl.2H_2O$, is obtained in dark brown needles from the bilberry and cranberry. It is the 3-β-glucoside of peonidin chloride and was synthesised before being isolated from natural sources (*L. F. Levy* and *R. Robinson*, *ibid.*, 1931, 2715; *Grove* and *Robinson, loc. cit.*).

Malvidin chloride and related substances. **Malvidin chloride (syringidin chloride),** $C_{17}H_{15}O_7Cl$, is obtained in dark brown prisms. Its formulation as the 3′,5′-dimethyl ether of delphinidin chloride, *i.e.*, as 3,5,7,4′-tetrahydroxy-3′,5′-dimethoxyflavylium chloride (*E. S. Gatewood* and *Robinson*, *ibid.*, 1926, 1959; *T. Endo*, Bor. Mag., Tokyo, 1959, **72,** 10) was confirmed by fission with sodium or barium hydroxide to yield phloroglucinol and 4-hydroxy-3,5-dimethoxybenzoic acid (*Willstätter* and *Mieg*, Ann., 1915, **408,** 122; *Karrer* and *Widmer, loc. cit.*) and by synthesis (*Bradley* and *Robinson*, J. chem. Soc., 1928, 1541).

Malvin chloride, $C_{29}H_{35}O_{17}Cl$, is isolated in dark brown needles from the flowers of wild mallow (*Malva sylvestris*), primulas (*Primula viscosa*) (*Willstätter* and *Mieg*, *loc. cit.*) and many other sources (*G. M. Robinson* and *R. Robinson*, Biochem. J., 1931, **25,** 1687). It is the 3,5-di-β-glucoside of malvidin chloride and has been synthesised (*R. Robinson* and *A. R. Todd*, J. chem. Soc., 1932, 2299).

Oenin chloride (cyclamin chloride), $C_{23}H_{25}O_{12}Cl$, is obtained from black grapes (*Willstätter* and *Zollinger*, Ann., 1915, **408,** 83; 1916, **412,** 195; *Endo, loc. cit.*; *K. Kondo*, Chem. Ztbl., 1930, **I,** 3191; 1931, **II,** 450), and from *Cyclamen persicum* (*Karrer* and

Widmer, loc. cit.) and was the first monoglucoside anthocyanin to be isolated from natural sources. It is malvidin 3-β-glucoside and on hydrolysis with 25% hydrochloric acid yields glucose and malvidin (oenidin) chloride. Oenin and cyclamin are identical (*J. C. Bell* and *Robinson*, J. chem. Soc., 1934, 813). It has been synthesised (*Levy et al., ibid*., 1931, 2701).

Hirsutidin chloride and related substances. **Hirsutidin chloride,** $C_{18}H_{17}O_7Cl$, 3,5,4′-trihydroxy-7,3′,5′-trimethoxyflavylium chloride (*Karrer* and *Widmer*, Helv., 1927, **10,** 758; *W. G. C. Forsyth* and *N. W. Simmonds*, Nature, 1957, **180,** 247), was shown to be 7,3′,5′-trimethyl ether of delphinidin chloride (7-methylmalvidin chloride) by synthesis (*Bradley et al*., J. chem. Soc., 1930, 793, 808).

Hirsutin chloride, $C_{30}H_{37}O_{17}Cl$, is isolated from primulas *(Primula hirsuta)* *(Karrer* and *Widmer, loc. cit.)* and has been synthesised (*Robinson* and *Todd*, J. chem. Soc., 1932, 2293). It is the 3,5-diglucoside of hirsutidin and is unique among the anthocyanins in containing a single free hydroxyl group.

The constitution of acyl anthocyanins from *Petunia hybrida, Salvia splendens* and *Raphanus sativus* have been established (*Birkofer et al*., Z. Naturforsch., 1965, **20b,** 424). It has been shown that in acylated anthocyanins the acyl groups are attached to the sugar groups in the 3-position of the anthocyanins (*Harborne*, Phytochem., 1964, **3,** 151).

Leucoanthocyanins and leucoanthocyanidins.* The origin of anthocyanins and catechins in plants and their structural relationship to the flavonoids are of great interest and importance. Colourless anthocyanin precursors, which with hydrochloric acid yield anthocyanins, occur in plant tissues. To these *Rosenheim* gave the name *leucoanthocyanins* implying that they are glucosides and have a structural relationship to the leuco forms of other dyestuffs, whereas in fact they are not necessarily glycosides and are not related to other leuco-dyestuffs. These leucoanthocyanins are widely distributed in the plant kingdom, examples being a substance in the gum of *Butea frondosa* which can be converted into cyanidin chloride (*G. M. Robinson*, J. chem. Soc., 1937, 1157) and the colourless crystalline peltogynol, containing an extra ring to the normal leucoanthocyanins, but related to flavan-3,4-diol the parent compound of the leucoanthocyanins and leucoanthocyanidins. Peltogynol may be regarded as the first known leucoanthocyanidin.

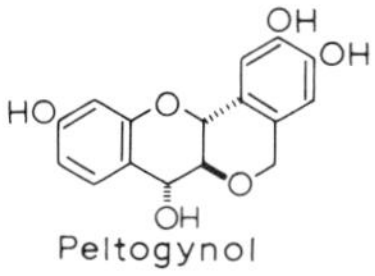

Peltogynol

Peltogynol, $C_{16}H_{14}O_6$, decomp. ~240°, $[\alpha]_D^{21}$ +273° (ethyl acetate), *tetra-acetate*, m.p. 173°, *tetrabenzoate*, m.p. 244°, which occurs in *Peltogyne porphyrocardia* and *P. pubescens* (*R. Robinson* and *G. M. Robinson*, *ibid*., 1935, 744). It forms a pyrylium salt

* *T. Swain* and *E. C. Bate-Smith*, The Chemistry of Vegetable Tannins, p. 109 (Society of Leather Trades' Chemists, London, 1956).

under mild conditions with hydrochloric acid and its structure follows from a number of reactions including the formation of 2,4,6-trinitroresorcinol and 4,5-dinitroveratrol by the action of nitric acid on peltogynol and its *trimethyl ether*, m.p. 198°, respectively, and its oxidation with manganese dioxide to a ketone which with sodium tetrahydridoborate regenerates peltogynol (*W. R. Chan et al., ibid.*, 1958, 3174). It has been shown to have an equatorial 4-hydroxy group, and **peltogynol B,** which accompanies it in the extract of *Peltogyne porphyrocardia* has an axial 4-hydroxyl group. (±)-Peltogynol trimethyl ether and its *cis,cis*-stereoisomer have been synthesised (*J. W. Clark-Lewis* and *M. M. Mahandru*, Chem. Comm., 1970, 1287; Austral. J. Chem., 1971, **24**, 549). A number of flavan-3,4-diols (p. 249) have been prepared and their chemical behaviour furnishes support for the theory that leucoanthocyanins belong to this class of compound.

Reduction of quercetin penta-acetate by lithium tetrahydridoaluminate (chosen because of its alliance to reducing coenzymes) followed by treatment with hydrochloric acid gives a good yield of cyanidin chloride and an almost colourless gum (*L. Bauer et al.*, Chem. and Ind., 1954, 433). The gum also yields cyanidin chloride but only with hydrochloric acid in hot propan-2-ol. It is reasonable to suppose that dihydroguercetin, a form of the colourless base of cyanidin, is formed and readily yields cyanidin chloride with hydrochloric acid, while the gum is the leucoanthocyanidin and is a 3,4-dihydroxyflavan, particularly since ethanolic hydrochloric acid is a recognised reagent for converting leucoanthocyanidins into anthocyanidins (*G. M Robinson, loc. cit.*):

Dihydroquercetin —HCl→ Cyanidin chloride ←HCl, hot Me_2CHOH— (gum)

Hot mineral acid will dehydrate the flavan-3,4-diol to a 3-oxoflavan (*A. J. Birch*, Ann. Reports, 1950, **47**, 184) which will disproportionate to a flavylium salt and a reduced product such as catechin (LXI):

(LXI)

This has been substantiated by a leucoanthocyanin from cocoa beans with hydrochloric acid yielding cyanidin chloride, (−)-epicatechin, and "phlobaphene" (p. 245) (*W. G. C. Forsyth*, Nature, 1953, **172,** 726; *M. E. Kieser et al.*, Chem. and Ind., 1953, 1260). Naturally occurring flavan-3,4-diols which behave as leucoanthocyanidins have been isolated. **Melacacidin,** 3,4,7,8,3′,4′-hexahydroxyflavan, is obtained from *Acacia melanoxylon* wood and when heated with hydrochloric acid yields the corresponding 3,7,8,3′,4′-pentahydroxyflavylium chloride and a polymer (*F. E. King* and *W. Bottomley*, J. chem. Soc., 1954, 1399; *Bottomley*, Chem. and Ind., 1954, 516). The *tetramethyl ether* of melacacidin, m.p. 145–146°, $[\alpha]_D$ −84.4° (EtOH), was synthesised by hydrogenating (Raney nickel) the corresponding tetramethylflavonol (*King* and *Clark-Lewis*, *ibid.*, 1954, 757). The absolute configuration of (−)-melacacidin has been established as 2*R*, 3*R*, 4*R* by conversion of its tetramethyl ether into (−)-7,8,3′,4′-tetramethoxy-2,3-*cis*-flavan-3-ol, and analogue of (−)-epicatechin tetramethyl ether; (−)-**tetracacidin,** 3,4,7,8,4′-pentahydroxyflavan obtained from *Acacia intertexta*, 4′,7,8-*trimethyl ether*, m.p. 159°, $[\alpha]_D^{18}$ −65° (EtOH), is assigned the 2*R*, 3*R*, 4*R*-configuration by analogy with (−)-melacacidin (*Clark-Lewis* and *G. F. Katekar*, J. chem. Soc., 1962, 4502). **Isomelacacidin,** the 4-epimer of melacacidin has been found in the heartwoods of three Australian *Acacia* species (*A. excelsa*, *A. harpophylla* and *A. melanoxylon*) (*Clark-Lewis* and *P. I. Mortimer*, *ibid.*, 1960, 4106). **Isoteracacidin,** obtained from *Acacia intertexta* appears to be related to teracacidin in a similar way (*Clark-Lewis*, *Katekar* and *Mortimer*, *ibid.*, 1961, 499). The crystalline leucocyanidin **mollisacacidin,** 3,4,7,3′,4′-*pentahydroxyflavan*, $C_{15}H_{14}O_6$, m.p. 125–130° (decomp.), $[\alpha]_D^{18}$ +12.6° (MeOH), *trimethyl ether*, m.p. 129°, is obtained from the heartwood of *Acacia mollissima* (black wattle) and is prepared by the catalytic hydrogenation of fustin (*H. H. Keppler*, *ibid.*, 1957, 2721). The 2*R*-, 3*S*-configuration of (+)-mollisacacidin has been confirmed by conversion of its trimethyl ether into (−)-7,3′,4′-trimethoxy-2,3-*trans*-flavan-3-ol, an analogue of (+)-catechin tetramethyl ether (*Clark-Lewis* and *Katekar*, *loc. cit.*; *B. R. Brown* and *J. A. H. MacBride*, Chem. and Ind., 1963, 1037). For the absolute configuration of free phenolic forms of diastereoisomeric flavan-3,4-diols related to (+)-mollisacacidin, see *S. E. Drewes* and *Roux*, *ibid.*, 1964, 1555. The inversion of $C_{(2)}$ and $C_{(4)}$ of (+)-mollisacacidin and its three diastereoisomers has been studied by two-dimensional paper chromatography (*idem.*, Chem. Comm., 1965, 282). **Leucofisetinidin,** 3,4,7,3′,4′-pentahydroxyflavan, found in the heartwoods of *Schinopsis balansae* and *S. quebracho-colorado* (*S. lorentzii*) is enantiomorphous with (+)-mollisacacidin (*Clark-Lewis* and *Roux*, Chem. and Ind., 1958, 1475; J. chem. Soc., 1959, 1402). These compounds have the 2,3-*trans*-3,4-*trans* arrangements of substituents, and the absolute configurations 2*R*, 3*S*, 4*R* and 2*S*, 3*R*, 4*S*, respectively (*Clark-Lewis* and *L. R. Williams*, Austral. J. Chem., 1963, **16**, 869; *S. E. Drewes* and *Roux*, Biochem. J., 1964, **90**, 343; Chem. and Ind., 1964, 1799):

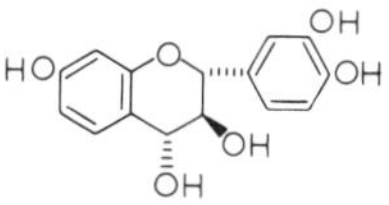

Mollisacacidin

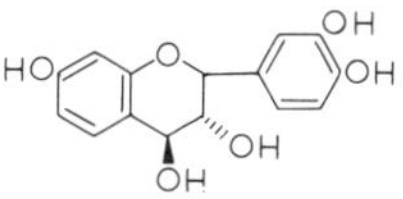

Leucofisetinidin

Optically active diastereoisomers of (+)-mollisacacidin have been obtained by epimerisation (*Drewes* and *Roux*, Biochem. J., 1965, **94,** 482). The stereochemistry of a 4,6-linked (+)-bileucofisetinidin from *Acacia mearnsii*, has been established by mass and n.m.r. spectroscopy (*J. Feeney* and *S. H. Eggers*, Chem. Comm., 1966, 368). Other leucoanthocyanidins: **guibourtacacidin,** obtained from *Guibourtia coleosperma*, *G. demeusei* and *G. tessmannii* (*Roux*, Nature, 1959, **183**, 890; *D. M. Phatak* and *A. B. Kulkarni*, Current Sci., 1959, **28**, 328; *Roux* and *G. C. de Bruyn*, Biochem. J., 1963, **87**, 439); **leucoperargonidin,** obtained from the gum of *Eucalyptus calophylla* (*T. R. Seshadri et al.*, J. sci. Ind. Res., 1958, **17B,** 44, 167, 168; 1961, **20B,** 615; J. chem.

Soc., 1961, 2787); **leucocyanidins** (*Seshadri et al., loc. cit.;* Tetrahedron, 1958, **3,** 225; 1959, **6,** 21; *T. Swain*, Chem. and Ind., 1954, 1144); **leucodelphinidins** (*Seshadri et al., loc. cit.; D. E. Hathway*, Biochem. J., 1958, **70,** 34); **margicassidin,** m.p. 320°, $[\alpha]_D^{25}$ $-93°$ (acetone), obtained from the flowers of *Cassia marginata* (*D. Adinarayana* and *Seshadri*, Indian J. Chem., 1960, **4,** 73); **teracacidin,** (−)-7,8,4′-*trihydroxy*-2,3-cis-*flavan*-3,4-cis-*diol*, an analogue of melacacidin, has been isolated from the heartwood of *Acacia intertexta* (*Clark-Lewis, Katekar* and *P. I. Mortimer*, J. chem. Soc., 1961, 499). The structures of *Acacia* leucoanthocyanidins have been reviewed, together with the structures of optically active flavonoids (*Clark-Lewis*, Rev. Pure Appl. Chem., Australia, 1962, **12,** 32, 97). A survey of leucoanthocyans from various natural sources suggestes that they are mostly of the leucoanthocyanidin and leucoanthodelphinidin type (*E. C. Bate-Smith*, Biochem. J., 1954, **58,** 122). It is noteworthy that the most commonly encountered hydroxyflavans in nature are the catechins and gallocatechins (p. 238).

Taxifolin (p. 284), is reduced by sodium tetrahydridoborate to a non-crystalline 3,4-dihydroxyflavan, which is an unstable leucoanthocyanidin and with hot dilute hydrochloric acid in an open vessel gives a deep red solution of cyanidin chloride and a "phlobaphene" (*Swain, loc. cit.*). The fact that leucoanthocyanidins can yield catechins and "phlobaphene" indicates that they may be precursors of the condensed tannins (p. 244).

Although it may be concluded that flavan-3,4-diols are true leuco compounds, not all leuco compounds are necessarily flavan-3,4-diols. Flavan-2,3-diols obtained by hydroxylating flav-2-ens with osmium tetroxide, on treatment with acid afford anthocyanidins (*J. W. Gramshaw, A. W. Johnson* and *T. J. King*, J. chem. Soc., 1958, 4040):

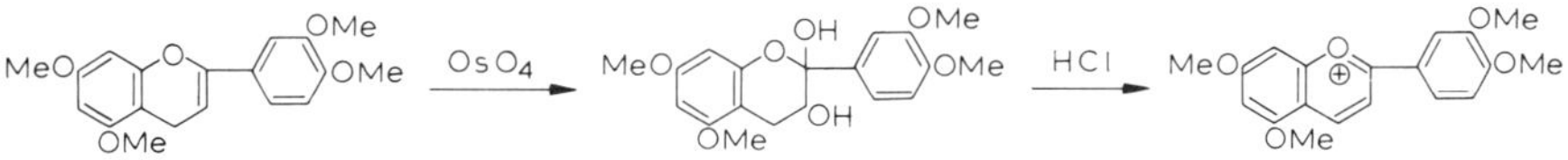

MeO OMe OMe OMe $\xrightarrow{OsO_4}$ MeO OMe OH OH OMe OMe $\xrightarrow{HCl}$ MeO OMe ⊕ OMe OMe

*(vi) Coumarin**, *(*2H-*benzo*[b]*pyran-2-one*, 5,6-*benzo-2-pyrone*, 5,6-*benzo-α-pyrone) and its derivatives*

The benzo[*b*]pyrones are of two types, 2*H*-benzo[*b*]pyran-2-ones or coumarins, and 4*H*-benzo[*b*]pyran-4-ones or chromones (p. 138):

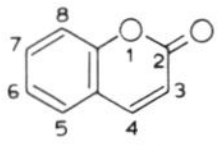

Coumarin

Since in the major part of the literature 2*H*-benzo[*b*]pyran-2-ones are referred to as coumarins, the latter name will be used for naming compounds in this section.

S. M. Sethna* and *N. M. Shah*, Chem. Reviews, 1945, **36, 1; *S. Wawzonek*, Heterocyclic Compounds (John Wiley and Sons Inc., New York, 1951), Vol. 2, p. 173.

Coumarins are found in the vegetable kingdom either in the free or combined state, particularly in plants of the *Orchidaceae*, *Leguminoceae*, *Rutaceae* and *Umbelliferae* species (*E. Späth*, Ber., 1937, **70A,** 83). Coumarin itself was first isolated in 1820 from tonquin beans.

Coumarin is the lactone of 4-hydroxycinnamic acid and this formulation expresses satisfactorily the chemical behaviour of the coumarins.

The natural coumarins* may be broadly divided into the following groups: (1) coumarin and simple derivatives such as hydroxy- and methoxy-coumarins and their glycosides; (2) furocoumarins; (3) coumarins fused to a 2,2-dimethylpyran ring.

Synthesis of coumarin and its derivatives. (1) Perkin synthesis. This classic synthesis was discovered by *W. H. Perkin Sr.* (J. chem. Soc., 1868, 53; 1877, 388) who prepared coumarin by heating salicylaldehyde with acetic anhydride and anhydrous sodium acetate. It was supposed that a 2-hydroxy-cinnamic acid was first formed and lactonised to the coumarin.

The mechanism of the reaction has given rise to considerable differences of opinion (see *J. R. Johnson*, Org. Reactions, Vol. 1, p. 210). It seems certain, firstly, that the aldehyde reacts with acetic anhydride while the sodium acetate functions as a basic catalyst; secondly, that 2-hydroxy-cinnamic acid is improbable as an intermediate since the *cis*-acids are encountered only rarely in the reaction (cf. *e.g. D. S. Breslow* and *C. R. Hauser*, J. Amer. chem. Soc., 1939, **61,** 786). It has been suggested that coumarin and its substituted derivatives are formed by one or other of two reactions (*M. Crawford* and *J. A. M. Shaw*, J. chem. Soc., 1953, 3435). In both cases the hydroxyaldehyde is acylated to yield the product *O*-acyl derivative, which in the presence of the base sodium acetate undergoes an internal aldol condensation to the cyclic compound, from which subsequent removal of acetic acid gives the coumarin:

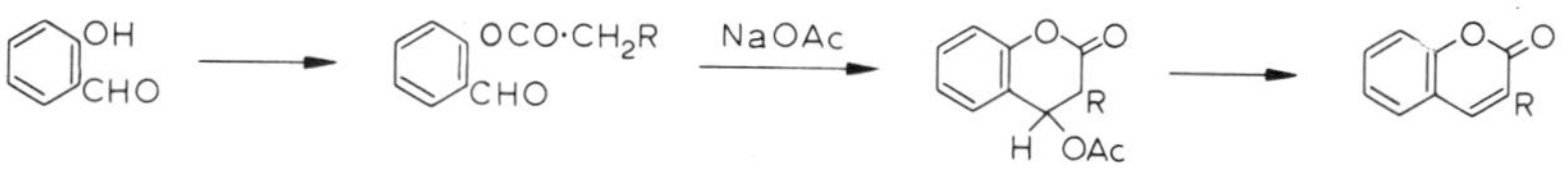

Salicyclaldehyde, phenylacetic anhydride and sodium phenylacetate thus yield 3-phenylcoumarin (R = Ph), but this mechanism does not apply to the formation of coumarin itself, since 2-acetylsalicylaldehyde gives only a trace of coumarin when heated with sodium acetate. The acetyl compound, however, with acetic anhydride readily gives coumarin probably through con-

* *E. Späth*, Monatsh., 1936, 69, 75; *F. M. Dean*, "Progress in the Chemistry of organic Natural Products", Springer Verlag, Vienna, 1953, Vol. 9, p. 225; "Naturally occurring Oxygen Ring Compounds", Butterworths, London, 1963, p. 176.

densation of the acetic anhydride with the aldehydic group and ring-formation to the cyclic compound (R = H).

The Perkin reaction has been widely applied (*Späth et al.*, Ber., 1937, **70**, 83) but has its limitations. The required 2-hydroxybenzaldehydes are not always easily obtained and the yields of coumarins are generally low although it is claimed that they are improved by the addition of iodine (*H. Yanagisawa* and *H. Kondo*, J. pharm. Soc. Japan, 1921, **472,** 498). Sometimes coumaric acids are formed but not coumarins; for example, *o*-vanillin undergoes the Perkin reaction to yield *trans*-2-hydroxy-3-methoxycinnamic acid which fails to give 8-methoxycoumarin (*B. B. Dey* and *V. A. Kutti*, Proc. Nat. Inst. Sci. India, 1940, **6,** 641). The *trans*-acids may be isomerised by u.v. irradiation to the *cis*-coumaric acid (coumarinic acid) and hence to the coumarin. 2-Alkoxycinnamic acids have been converted to coumarins by reacting with thionyl chloride and then heating the resulting product (*F. D. Popp* and *W. Blount*, J. org. Chem., 1961, **26,** 2108):

OMe; CH:CMe·CO_2H —(1. $SOCl_2$; 2. Distil 170–172°/60 mm)→ Me

A reaction somewhat similar to the Perkin reaction is that of *E. Knoevenagel* (Ber., 1898, **31,** 2585; 1904, **37,** 4461; see also *E. C. Horning et al.*, Org. Synth., 1948, **28,** 24), 2-hydroxybenzaldehydes condense with substance containing reactive methylene groups such as ethyl acetoacetate in the presence of organic bases to yield coumarins. Thus salicyladehyde and malonic acid yield coumarin-3-carboxylic acid:

OH; CHO + CO_2H; $CH_2{\cdot}CO_2H$ —($-H_2O$)→ CO_2H

The syntheses of five 3,3′-biscoumarins by a modification of the Perkin reaction have been reported (*L. L. Woods* and *S. M. Shamma*, J. chem. Eng. Data, 1970, **15,** 355).

(*2*) *Pechmann–Duisberg reaction**. Coumarins are prepared by the condensation of phenols with *β*-ketonic esters such as acetoacetic ester in the presence of sulphuric acid. The method has been widely used to prepare coumarins substituted in both the aromatic and pyrone nuclei.

Very occasionally with sulphuric acid as condensing agent chromones are formed, 4-chloro-3,5-xylenol giving only the chromone and no coumarin (*R. Adams* and *J. W. Mecorney*, J. Amer. chem. Soc., 1944, **66,** 802). With

---

* The method has been reviewed by *S. Sethna* and *R. Phadke* (Org. Reactions, Vol. 7, p. 1) and *S. M. Sethna* and *N. M. Shah* (Chem. Reviews, 1945, **36,** 10).

other condensing agents such as phosphorus pentoxide the reaction takes another course and forms chromones (Simonis reaction, p. 142).

Replacement of the sulphuric acid by liquid hydrogen fluoride often leads to increased yield of coumarin (*O. Dann* and *G. Mylius*, Ann., 1954, **587,** 1), *e.g.* from phenol and *p*-cresol with acetoacetic ester. *p*-Cresol and ethyl benzoylacetate with sulphuric acid give a small yield of 4-phenyl-6-methylcoumarin (*A. Robertson* and *W. F. Sandrock*, J. chem. Soc., 1932, 1180), but with hydrogen fluoride a mixture of 6-methylflavone (24%) and 4-phenyl-6-methylcoumarin (16%) is obtained. Polyphosphoric acid has been recommended as condensing agent (*J. Koo*, Chem. and Ind., 1955, 445). Resorcinol dimethyl ether and ethyl acetoacetate fail to react in the presence of polyphosphoric acid, but with pyrophosphoric acid 7-methoxy-4-methylcoumarin (35%) is obtained along with some ethyl *trans*-β-methyl-2,4-dimethoxycinnamate (*M. Narayana, J. F. Dash* and *P. D. Gardner*, J. org. Chem., 1962, **27,** 4704):

Several mechanisms appeared reasonable for the conversion of resorcinol dimethyl ether to the coumarin; some were excluded because it was found that methyl *cis*-2-methoxycinnamate on treatment with pyrophosphoric acid at 30° gave a 79% yield of coumarin:

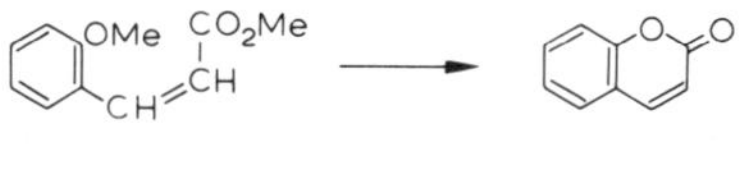

A plausible mechanism of the reaction postulates 2-hydroxycinnamic acids as intermediates since such acids readily yield coumarins by treatment with sulphuric acid (*Robertson et al.*, J. chem. Soc., 1932, 1681). The controlling step is probably a condensation of the phenol with acetoacetic ester in the form of a carbonium ion to give the 2-hydroxycinnamic acid (or ester) (R = H or alkyl). Ring-closure then yields the coumarin (*R. N. Lacey*, *ibid.*, 1954, 854):

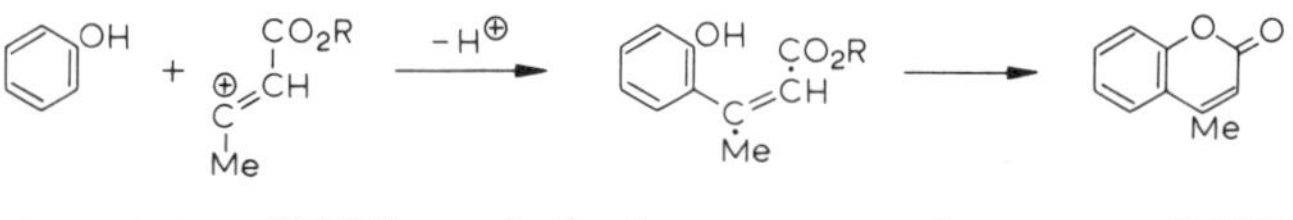

Aryl acetoacetates (LXII) and β-phenoxycrotonic esters (LXIII) (with rearrangement) with sulphuric acid undergo ring-closure to coumarins (*Lacey, loc. cit.; Dann* and *G. Illing*, Ann., 1957, **605,** 158):

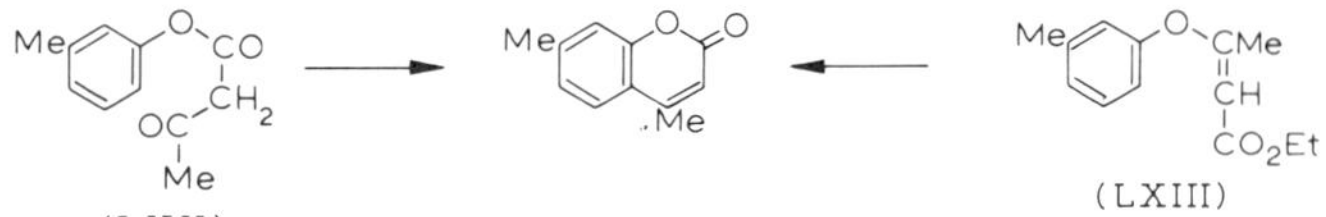

(LXII) (LXIII)

The reaction of ethyl acetoacetate with hot sodium ethoxide affords 3-acetyl-6-ethoxycarbonyl-5-hydroxy-4,7-dimethylcoumarin (*K. Anderton* and *R. W. Rickards*, J. chem. Soc., 1965, 2543).

Reactive phenols in the presence of cation-exchange resins have been condensed with β-ketonic esters to give coumarins (*K. N. Trivedi*, J. Indian chem. Soc., 1965, **42,** 273). Appropriate phenols when boiled with diethyl α-acetylglutarate or diethyl α-acetylsuccinate in trifluoroacetic acid afford highly substituted coumarins (*Woods* and *D. Johnson*, Texas J. Sci., 1965, **17,** 340; C.A., 1966, **64,** 15826).

Benzoylacetonitrile and acetoacetonitrile undergo carbon–carbon condensation with phenols providing a route to coumarins. The aluminium chloride catalysed reaction of *meta-* and *para-*substituted phenols with benzoylacetonitrile in the presence of dry hydrogen chloride affords the corresponding iminocoumarins and/or coumarins:

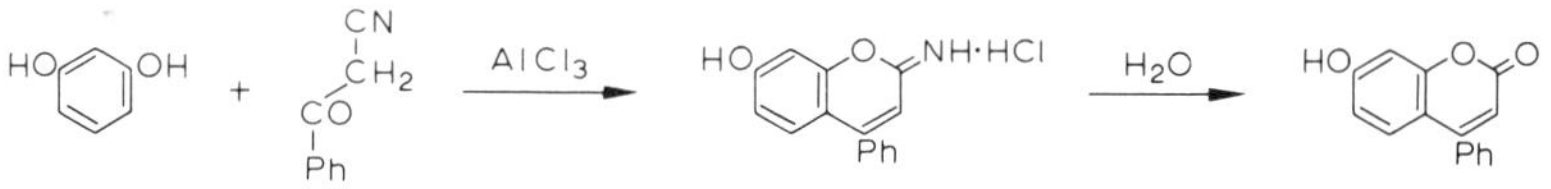

$AlCl_3$ H_2O

Phenol and *o*-cresol yield predominantly β-(4-hydroxyphenyl)cinnamonitriles. In contrast, acetoacetonitrile reacts with phenols in polyphosphoric acid or its ethyl ester to give 4-methylcoumarins in appreciable yields (*K. Sato* and *T. Amakasu*, J. org. Chem., 1968, **33,** 2446). Benzoylacetic acid with phenols in the presence of polyphosphoric acid affords 4-phenylcoumarins (*L. Reichel* and *G. Proksch*, Ann., 1971, **745,** 59).

(*3*) Methylated quinones react with sodium enolates in benzene solution to give hydroxycoumarins. Duroquinone and ethyl sodiomalonate, for example, yield ethyl 6-hydroxy-5,7,8-trimethylcoumarin-3-carboxylate (LXV) (*L. I. Smith* and *F. J. Dobrovolny*, J. Amer. chem. Soc., 1926, **48,** 1693; *Smith*, *ibid.*, 1938, **60,** 676; 1946, **68,** 887). The reaction is formulated as a Michael condensation of the quinone in the enolic form LXIV with the sodium enolate to give a quinol derivative; this is cyclised to the 3,4-dihydrocoumarin which is dehydrogenated to the hydroxycoumarin LXV:

(LXIV) (LXV)

(*4*) Coumarins may be prepared by condensing malononitrile with aldehydes or ketones in the presence of piperidine and hydrolysing the product with hydrochloric acid (*Woods* and *J. Sapp*, J. org. Chem., 1965, **30,** 312):

$$\text{2-HOC}_6\text{H}_4\text{CHO} + CH_2(CN)_2 \xrightarrow[\text{2. HCl}]{\text{1. } C_5H_{11}N} \text{coumarin-3-}CO_2H$$

Coumarins have been prepared by the action of cyanoacetic acid on certain polyhydroxyaryl acids (*Woods* and *J. Sterling*, Texas J. Sci., 1963, **15,** 200), and by reacting 2-hydroxyaryl acids with malonic acid in the presence of trifluoroacetic acid. The latter method affords 4-hydroxycoumarin-3-carboxylic acids (*Woods* and *Johnson*, J. org. Chem., 1965, **30,** 4343):

$$\text{2-HOC}_6\text{H}_4CO_2H + CH_2(CO_2H)_2 \longrightarrow \text{4-OH-coumarin-3-}CO_2H$$

(*5*) The compounds obtained by reacting phenoxides with polyhalogenoacroleins, after treatment with formic acid give esters, which are cyclised with sulphuric acid at 0° to yield 3-halogenocoumarins (*M. LeCorre* and *E. Levas*, Compt. rend., 1963, **257,** 1622):

$$Cl_2C{:}CX{\cdot}CHO + ArONa \longrightarrow ArOCCl{:}CX{\cdot}CHO \xrightarrow{HCO_2H} ArOCO{\cdot}CHX{\cdot}CHO \rightleftharpoons ArOCO{\cdot}CX{:}CHOH \xrightarrow{H_2SO_4} \text{3-X-coumarin}$$

(X = Cl or Br; Ar = phenyl or naphthyl)

Some coumarins may be prepared by heating phenolic compounds with dimethyl acetylenedicarboxylate and zinc chloride (*Woods* and *V. Hollands*, Texas J. Sci., 1969, **21,** 91; C.A., 1970, **72,** 12486):

$$\text{HO-C}_6\text{H}_4\text{-OH} + C{\cdot}CO_2Me \equiv C{\cdot}CO_2Me \xrightarrow{ZnCl_2} \text{7-HO-coumarin-4-}CO_2Me$$

m.p. 183–185°

(*6*) Coumarin-4-carboxylic acids may be obtained by boiling coumarin-2,3-diones in acetic anhydride containing a catalytic amount of pyridine (*M. Kawai*, *T. Matsuura* and *R. Nakashima*, Bull. chem. Soc., Japan, 1970, **43,** 1590):

$$\text{coumaran-2,3-dione} \xrightarrow[C_5H_5N]{Ac_2O} \text{coumarin-4-}CO_2H$$

(7) Coumarin may be obtained in a high yield by oxidising benzopyrylium perchlorate with manganese dioxide in chloroform or acetonitrile (*I. Degani* and *R. Fochi*, Ann. chim., Rome, 1968, **53**, 251):

$ClO_4^{\ominus}$; MnO_2 ; $CHCl_3$ or MeCN ; (89–93%)

(8) On photolysis 4-phenylchroman-3-one rearranges to 4-phenyldihydrocoumarin (*P. K. Grover* and *N. Anand*, Chem. Comm., 1969, 982):

Ph ; Ph

Properties of coumarins. The coumarins, as lactones, are readily hydrolysed by alkali to the salts of the corresponding coumarinic acids, which usually cannot be isolated as the acids revert immediately to the coumarins (for exceptions see below). The coumarinic acids are *cis*-acids and are converted by prolonged treatment with alkali into the salts of the stable *trans*-acids (coumaric acids):

NaOH ; HCl ; Coumarinic acid *cis*- ; Coumaric acid *trans*-

The coumaric and coumarinic acids and their relationships to coumarin have been discussed in Vol. IIIE, pp. 208 *et seq.* See also *A. J. Harle* and *L. E. Lyons* (J. chem. Soc., 1950, 1575); *M. Crawford et al.* (*ibid.*, 1953, 3435; 1956, 2155); *T. R. Seshadri* (Chem. and Ind., 1954, 308). The hydrolysis of coumarin has been studied spectrophotometrically (*B. N. Mattoo*, Trans. Faraday Soc., 1957, **53**, 760; *E. Cingolani*, Gazz., 1959, **89**, 985; *E. R. Garrett*, *B. Chippold* and *J. B. Mielck*, J. pharm. Sci., 1971, **60**, 396), and the rate coefficients have been measured for the alkaline ring fission of substituted coumarins in 70% v/v dioxane–water at several temperatures by a conductimetric method (*K. Bowden*, *M. J. Hanson* and *G. R. Taylor*, J. chem. Soc., B, 1968, 174).

A sensitive test for coumarin is based on its ring-opening by alkali and the photo-catalysed conversion of coumarinic acid to coumaric acid (*F. Feigl et al.*, J. Amer. chem. Soc., 1955, **77**, 4162). Hydrolysis to a coumarinic acid is the decisive test for a coumarin and the best procedure is to hydrolyse the coumarin and simultaneously prevent ring-closure by methylation with alkali and dimethyl sulphate to a 2-methoxycinnamic acid (*F. W. Canter* and *A. Robertson*, J. chem. Soc., 1931, 1875; *N. M. Shah* and *R. C. Shah*, J. Univ. Bombay, 1938, **7,** 213; *R. Adams et al.*, J. Amer. chem. Soc., 1943, **65,** 356), or with sodium methoxide and methyl iodide (*S. Rangaswami* and *Seshadri*, Proc. Indian Acad. Sci., 1936, **5A,** 249); 3-phenylcoumarin yields *cis*-2-methoxy-α-phenylcinnamic acid, m.p. 131.5° (*Crawford* and *G. W. Moore*, J. chem. Soc., 1955, 3445), while with sodium hydroxide and mercuric oxide it yields *trans-2-hydroxy-α-phenylcinnamic* acid, m.p. 195°, converted by dimethyl sulphate and sodium hydroxide into the methoxy acid, m.p. 184°:

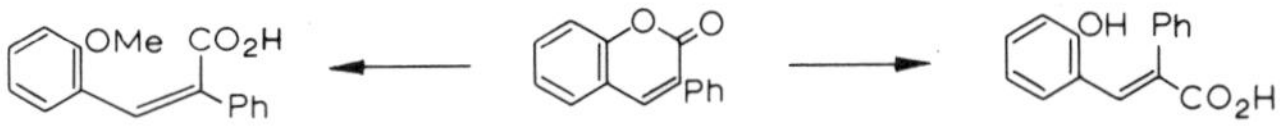

OMe CO₂H Ph ← Ph → OH Ph CO₂H

Hydrolysis of substituted coumarins may take another course and yield a mixture of a phenol and a ketone (*W. Baker*, *ibid.*, 1925, **127,** 2349; *S. M. Sethna* and *R. C. Shah*, J. Indian chem. Soc., 1940, **17,** 211); 7-hydroxy-4-methyl-3-phenylcoumarin, with 33–50% potassium hydroxide gives methyl benzyl ketone and resorcinol as the main products:

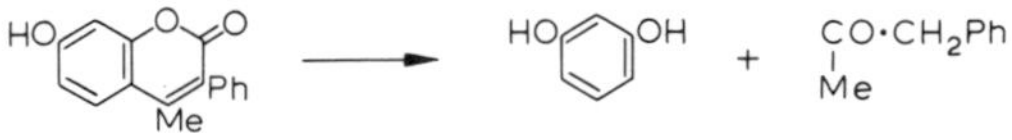

HO Ph Me → HO OH + CO·CH_2Ph Me

Fission of the heterocyclic ring of the coumarin nucleus can be effected by oxidation (p. 109); by fusion with potassium hydroxide, 4,7-dimethylcoumarin yielding 4-methylsalicylic acid; by catalytic hydrogenolysis in presence of copper chromite at 250° and by reduction with zinc and alkali (Table 2). Coumarins can be characterised by reduction to dihydrocoumarins which are hydrolysed to acids, *e.g.* coumarin yields melilotic acid, which does not cyclise at room temperature. These acids on oxidation with potassium permanganate afford succinic acid. The test can be carried out with 0.02 g of coumarin:

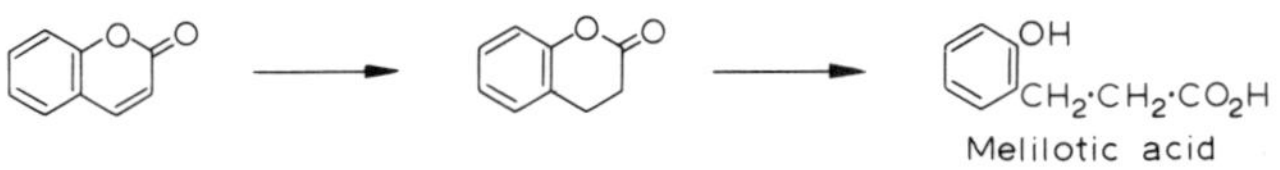

→ → OH CH_2·CH_2·CO_2H Melilotic acid

Reduction products of coumarin by different reducing agents are shown in Table 2.

TABLE 2

REDUCTION PRODUCTS OF COUMARIN

Method	*Product*	*Ref.*
Hydrogen/$CuCrO_2$ at 150°	Dihydrocoumarin	1
Hydrogen/Raney nickel at 85° or 100°	Dihydrocoumarin	1,2
Hydrogen/Pd–C	Dihydrocoumarin	3
Hydrogen/PtO_2	Dihydrocoumarin	4
Hydrogen/$CuCrO_2$ at 250°	3-(2-hydroxyphenyl)propanol	1
Sodium and ethanol	3-(2-hydroxyphenyl)propanol	5
NaOH and zinc dust	Melilotic acid and tetrahydro-bicoumarinyls*	6
Zinc and acetic acid	Dihydrocoumarin and tetrahydro-4,4′-bicoumarinyls*	4
Diborane	*o*-Allylphenol	7
Polarographic	Tetrahydro-4,4′-bicoumarinyls*	8,9

* The constitution of these substances is somewhat uncertain. See ref. 4.

References
1 *P. L. deBenneville* and *R. Connor*, J. Amer. chem. Soc., 1940, **62,** 283, 3067.
2 *L. Palfray* and *S. Sabetay*, Bull. Soc. chim. Fr., 1938, [v], **5,** 1423.
3 *E. Späth* and *O. Pesta*, Ber., 1933, **66,** 754.
4 *L. I. Smith* and *R. O. Denyes*, J. Amer. chem. Soc., 1936, **58,** 304.
5 *F. W. Semmler*, Ber., 1906, **39,** 2851.
6 *K. Fries* and *G. Fickewirth*, Ann., 1908, **362,** 30.
7 *K. M. Biswas* and *A. H. Jackson*, J. chem. Soc., C, 1970, 1667.
8 *A. J. Harles* and *L. E. Lyons*, *ibid.*, 1950, 1575.
9 *R. Patzak* and *L. Neugebauer*, Monatsh., 1951, **82,** 662.

The double bond in the pyrone ring is olefinic. It adds bromine and resembles the double bond in α,β-unsaturated carbonyl compounds. Coumarin, for instance with cyanoacetamide and alkali yields 3,4-dihydrocoumarin-4-cyanoacetamide (*Seshadri*, J. chem. Soc., 1928, 166). With 2,3-dimethylbutadiene by a Diels–Alder addition coumarin forms the compound LXVI *(Adams et al., loc. cit.)*:

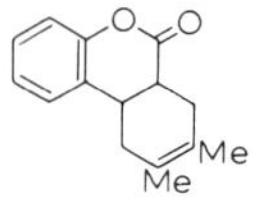

(LXVI)

Coumarins with Grignard reagents are reported to give products which may be benzopyrylium salts, dialkyl- or diaryl-chromenes, 3-(2-hydroxyphenyl)-1,3-diarylpropan-1-ones, chromanols or open-chain tertiary alcohols [3-(2-hydroxyphenyl)-1,1-diarylprop-2-en-1-ols] (*H. Decker* and *T. von*

Fellenberg, Ann., 1907, **356,** 281; *R. Willstätter et al.*, Ber., 1924, **57,** 1938, 1945; *I. M. Heilbron et al.*, J. chem. Soc., 1931, 1701). In general coumarins with alkylmagnesium bromide or iodide afford tertiary alcohols, which easily dehydrate on boiling with acetic acid to give 2,2-dialkylchromenes (*R. L. Shriner* and *A. G. Sharp*, J. org. Chem., 1939, **4,** 575; *L. I. Smith* and *P. M. Ruoff*, J. Amer. chem. Soc., 1940, **62,** 145; *R. Livingstone et al.*, J. chem. Soc., 1958, 2422; 1966, 2013):

1. RMgI; 2. aq. NH_4Cl → *tert*-alcohol; Δ, AcOH → 2,2-Dialkylchromene

The tertiary alcohol obtained from coumarin and methylmagnesium iodide on treatment with acetic acid containing a small amount of sulphuric acid or methanol saturated with hydrogen chloride gives a dimer of 2,2-dimethylchromene (LXVII) (*C. A. Barnes, J. L. Occolowitz* and *M. I. Strong*, Tetrahedron, 1963, **19,** 839):

(LXVII)

A. Löwenbein (Ber., 1924, **57,** 1517) reported that the reaction between coumarin and phenylmagnesium bromide gives 2,2-diphenylchromene and 2,4-diphenylchroman-2-ol; in fact the products of the Grignard reaction are 3-(2-hydroxyphenyl)-1,1-diphenylprop-2-en-1-ol and 3-(2-hydroxyphenyl)-1,3-diphenylpropan-1-one *(Barnes, Occolowitz* and *Strong, loc. cit.)*, the former compound on boiling with acetic acid yields 2,2-diphenyl-2*H*-chromene (LXVIII) and the latter, 2,4-diphenyl-4*H*-chromene (LXIX) (p. 54):

1. PhMgBr; 2. NH_4Cl; Δ, AcOH

(LXVIII) (LXIX)

Although the literature indicates that warm solutions with excess of Grignard reagent afford chromenes or chroman-2-ols (*J. Houben*, Ber., 1904, **37,** 489; *Heilbron* and *D. W. Hill*, J. chem. Soc., 1927, 2005; 1931,

1701; *Löwenbein* and *B. Rosenbaum*, Ann., 1926, **448,** 223; *Shriner* and *Sharp, loc. cit.)*, in general the reaction between a coumarin and an aryl-magnesium bromide gives a mixture of all or some of the following products: the tertiary alcohol, the 2,2-diphenylchromene, the 2,4-diphenyl-chromene and the open-chain ketone; 2*H*-naphtho[1,2-*b*]pyran-2-one (7,8-benzocoumarin) (*J. Cottam* and *Livingstone*, J. chem. Soc., 1964, 5228) and 2*H*-naphtho[2,3-*b*]pyran-2-one (6,7-benzocoumarin) (*idem, ibid.*, 1965, 6646) react with phenylmagnesium bromide to give, besides other products, the open-chain ketone and no diphenyldihydronaphthopyran-2-ol (diphenyl-benzochroman-2-ol); 3*H*-naphtho[2,1-*b*]pyran-3-one (5,6-benzocoumarin) with arylmagnesium bromide yields, besides other products, the corresponding diphenyldihydronaphthopyran-3-ols (5,6-benzochroman-2-ol) and no open-chain ketones (*idem, ibid.*, 1964, 5228; *W. D. Cotterill, Livingstone* and *M. V. Walshaw, ibid.*, 1970, 1758).

It has been found that ethyl-, isopropyl- and cyclohexyl-magnesium bromides undergo 1,4-additon to 4-substituted coumarins, best in tetrahydrofuran, to yield 4,4-disubstituted 3,4-dihydrocoumarins. This novel mode of addition is encouraged by $-I$ substituents in the 4-position, discouraged by a 4-methyl group $(+I)$, unaffected by a 5-hydroxy group, and not apparently subject to steric effects from the 4-substituent or the Grignard reagent. Methyl- and phenyl-magnesium bromide undergo normal 1,2-addition to the 4-substituted coumarins, except that exclusive 1,4-addition occurs with a 4-(2-pyridyl)coumarin (*J. A. Elvidge, R. W. Tickle* and *T. Melton*, J. chem. Soc., Perkin I, 1974, 569).

Coumarin, alkyl- and aryl-coumarins. **Coumarin,** rhombic prisms, m.p. 68°, b.p. 303°, occurs in woodruff, tonka beans, lavender oil and is found in more than 60 species. It sublimes without decomposition, is easily volatile in steam, is very soluble in boiling water, and can be purified by crystallisation from light petroleum. Coumarin has the pleasant odour of asperula and is used in perfumery for the preparation of asperula essence. Coumarin is non-fluorescent, but in alkaline solution develops a green fluorescence in ultra-violet light (*B. N. Mattoo*, Trans. Faraday Soc., 1956, **52,** 1184). It absorbs in the u.v. region λ_{max} 274.5 mμ (logε 4.06) and 311 mμ (logε 3.77) (87% ethanol). The spectrum is radically altered by the introduction of hydroxyl groups, in the 5-, 6- and 7-positions (*Mattoo, loc. cit.; J. A. Berson*, J. Amer. chem. Soc., 1953, **75,** 3521; *V. C. Farmer* and *R. H. Thomson*, Chem. and Ind., 1957, 113). Methyl groups in the coumarin molecule do not alter the absorption curves significantly (*B. K. Ganguly* and *P. Bagchi*, J. org. Chem., 1956, **21,** 1415). The i.r. spectrum of coumarin (*R. S. Rasmussen* and *R. R. Brattain*, J. Amer. chem. Soc., 1949, **71,** 1073), and the Raman spectrum in the solid state and in solution (*T. R. Seshadri*, Proc. Indian Acad. Sci., 1939, **8A**, 519) have been reported; the carbonyl frequency is considerably lower than expected. The dipole moment of coumarin is 4.51D (*M. A. G. Rau*, Current Sci., 1936, **5**, 132).

On prolonged illumination in the solid state or in solution coumarin is converted into a *dimer* LXX, m.p. 262° (272–274°) (*A. W. K. de Jong*, Rec. Trav. chim., 1924, **43,** 316; *F. von Wessely* and *I. Plaichinger*, Ber., 1942, **75,** 971), with a head-to-head *cis*-cyclobutane structure (*R. Anet*, Chem. and Ind., 1960, 897; Canad. J. Chem., 1962, **40,** 1249). The *trans*-head-to-head *dimer* LXXI, m.p. 179°, is obtained by irradiating coumarin in ethanol in the presence of benzophenone; when the solvent is benzene, besides dimer LXXI, traces of the *trans*-head-to-tail dimer LXXII is also formed. Some details of the mechanism of the direct and sensitised photodimerisation of coumarin have been worked out (*G. S. Hammond, C. A. Stout* and *A. O. Lamola*, J. Amer. chem. Soc., 1964, **86,** 3103):

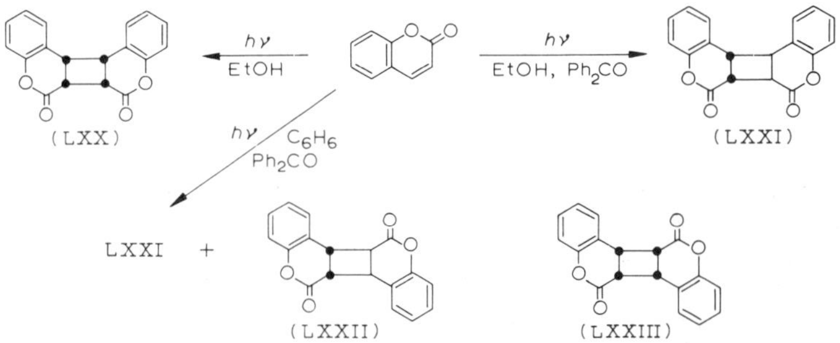

(LXX) ← hν, EtOH — coumarin — hν, EtOH, Ph_2CO → (LXXI)

coumarin — hν, C_6H_6, Ph_2CO → LXXI + (LXXII)

(LXXIII)

The *cis*-head-to-tail *dimer* LXXIII, m.p. 204–206°, has also been isolated and it has been reported that the irradiation of coumarin in various solvents yields in all cases simultaneously the dimers LXX, LXXI and LXXIII. The structures of LXX, LXXI, LXXII and LXXIII have been confirmed by their n.m.r. spectra (*C. H. Krauch, S. Farid* and *G. O. Schenck*, Ber., 1966, **99**, 625). The photodimerisation of various substituted coumarins has been studied (*A. Mustafa, M. Kamel* and *M. Ali Allam*, J. org. Chem., 1957, **22,** 888).

With diazotised anilines coumarins yield 3-arylcoumarins (*H. Meerwein et al.*, J. pr. Chem., 1938, [ii], **152,** 237) or 6-benzeneazo derivatives (*W. Borsche*, Ber., 1904, **37,** 346). Other cases of arylation of coumarin in the 3-position have been investigated (*O. Vogel* and *C. S. Rondestvedt Jr.*, J. Amer. chem. Soc., 1955, **77**, 3067). Aqueous-ethanolic potassium cyanide and coumarin give a product which on hydrolysis yields (2-hydroxyphenyl)succinic acid. The Michael addition of cyanoacetamide, $NC \cdot CH_2 \cdot CONH_2$ and malononitrile, $CH_2(CN)_2$, to coumarin yields 3-cyano-3,4-dihydro-4-acetamido-coumarin (LXXIV) and 2-amino-3-cyano-4-acetamido-4*H*-benzo[*b*]pyran (LXXV), respectively (*H. Junek* and *H. Sterk*, Monatsh., 1967, **98,** 144):

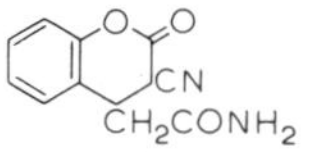

(LXXIV) CN, CH_2CONH_2

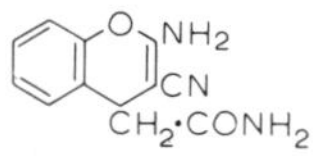

(LXXV) NH_2, CN, $CH_2 \cdot CONH_2$

Coumarin condenses with dimethyl homophthalate to give compound LXXVI (*E. Eisenhuth, H. B. Renfroe* and *H. Schmid*, Helv., 1965, **48,** 375):

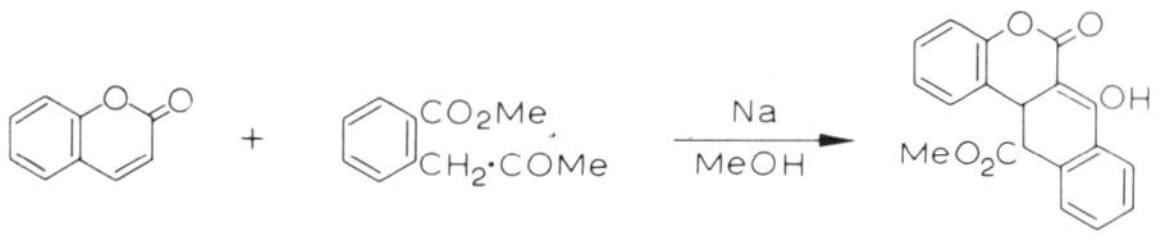

(LXXVI)

and on heating with *o*-phenylenediamine it affords 2-(2-hydroxyphenyl)benzimidazole (LXXVII) (*F. Sparatore* and *P. Schenone*, Ann. Chim., Rome, 1964, **54**, 602):

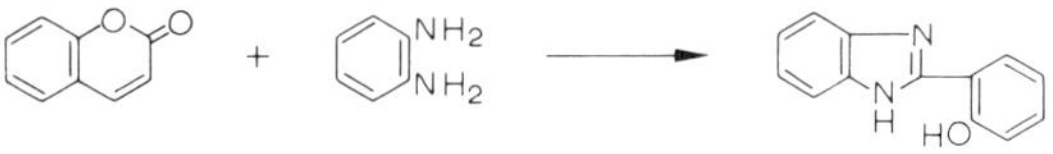

(LXXVII)

Coumarin with hydroxylamine does not form an oxime, but undergoes ring fission (*T. Posner* and *R. Hess*, Ber., 1913, **46**, 3816). The *oxime*, m.p. 131° (decomp.), and the *semicarbazone*, m.p. 218–220°, are obtained from 2-iminochrom-3-ene, an unstable substance obtained as the hydrochloride by the action of hydrogen chloride on 2-hydroxycinnamonitrile (*J. Houben* and *E. Pfankuch*, *ibid.*, 1926, **59**, 1594). The imine with phenylhydrazine gives a pyrazoline derivative LXXVIII, m.p. 146–147°, also obtained by the action of phenylhydrazine on 2-thionocoumarin.

Coumarin with hydrazine yields a product, m.p. 128–129°, probably *β-hydrazino-β-(2-hydroxyphenyl)propionhydrazide* (LXXIX) (*A. Darapsky*, J. pr. Chem., 1936, [ii], **147**, 145):

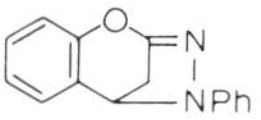

(LXXVIII) (LXXIX)

Structural parameters and thermodynamic quantities for the protonation of coumarin in fluorosulphonic acid at 25° have been obtained (*E. M. Arnett*, *R. P. Quirk* and *J. W. Larsen*, J. Amer. chem. Soc., 1970, **92**, 3977). Diazomethane and coumarin react to give *3-methoxycarbonyl-4-(2-methoxyphenyl)pyrazoline*, yellow crystals, m.p. 94.5° (*E. Y. Spencer* and *G. F. Wright*, *ibid.*, 1941, **63**, 2017). The hydroboration of coumarin gives chroman-3-ol, whereas 3-methylcoumarin affords 3-methylchroman-4-ol (*B. Kirkiacharian* and *D. Raulais*, Bull. Soc. chim. Fr., 1970, 1139):

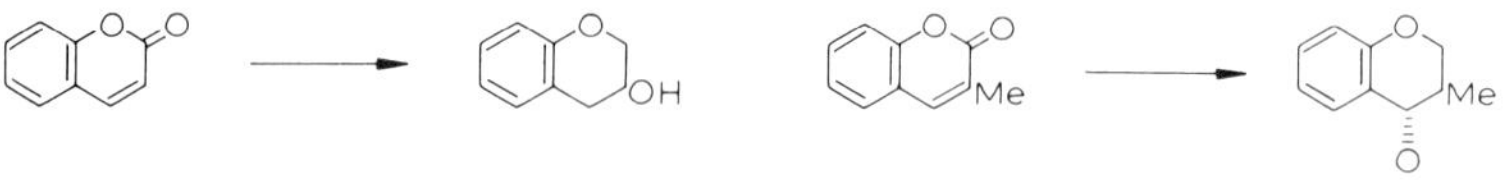

Nitration of coumarin w.th mixed acids gives mainly 6-**nitrocoumarin,** m.p. 183° (*A. Clayton*, J. chem. Soc., 1910, **97,** 2106; *Clayton* and *W. Godden*, *ibid.*, 1912, **101,** 210) and to a much smaller extent 8-**nitrocoumarin,** m.p. 191° (*B. B. Dey* and *P. Krishnamurti*, J. Indian chem. Soc., 1927, **4,** 197). 7-**Nitrocoumarin**, m.p. 198–200°, may be obtained by an indirect method (*D. Libermann*, *A. Desnoës* and *L. Hengl*,

Compt. rend., 1951, **232**, 2027). Benzoyl nitrate is stated to give 5-nitrocoumarin in quantitative yield (*F. Francis*, Ber., 1906, **39**, 3803) but the product may well be 6-nitrocoumarin. Further nitration of the 6- and 8-nitrocoumarins proceeds sluggishly, 6-nitrocoumarin giving 3,6-**dinitrocoumarin,** m.p. 180–181°, and 8-nitrocoumarin giving 6,8-**dinitrocoumarin,** m.p. 158° (*Clayton*, J. chem. Soc., 1910, **97**, 1388). Oxidation of the nitrocoumarins gives nitrosalicylic acids, while treatment of the 3-nitrocoumarins with warm sodium or ammonium hydroxide solution gives salicylaldehydes, ω-nitrostyrene derivatives being intermediates (*Clayton, loc. cit.*):

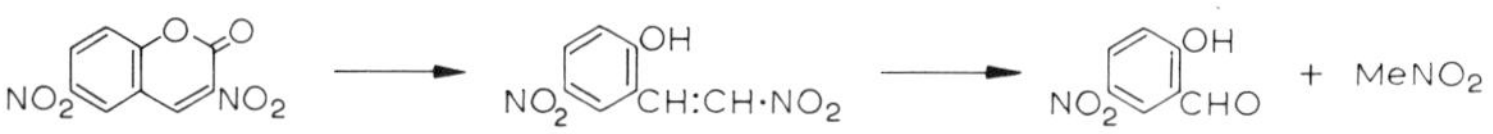

For nitro derivatives of 7-methyl-, 4,7- and 6,7-dimethyl-, 4,6,7- and 4,6,8-trimethylcoumarin see *Clayton, loc. cit.* Sulphonation affords **coumarin-6-sulphonic** and -3,6-**disulphonic acids.**

With halogen, addition to the pyrone ring occurs first, coumarin with bromine yielding *coumarin dibromide*, m.p. 105°, which with alkali gives 3-**bromocoumarin,** m.p. 183°, obtained also from coumarin and *N*-bromosuccinimide (*D. Molho* and *C. Mentzer*, Compt. rend., 1946, **223,** 1141). 3,4-Dichlorocoumarins on treatment with phenylmagnesium bromide are converted to 3-chloroflavones (*M. S. Newman* and *J. L. Ferrari*, Tetrahedron Letters, 1962, 199):

$R = R^1 = H$; $R = Me, R^1 = H$;
$R = Cl, R^1 = H$; $R = Me, R^1 = Bz$

4-**Chlorocoumarin**, m.p. 95–95.5°, obtained by heating 4-hydroxycoumarin with phosphoryl chloride (*V. A. Zagorevskiĭ* and *D. A. Zykov*, Zhur. obshcheĭ Khim., 1960, **30,** 1378).

4-Chlorocoumarin in liquid ammonia in the presence of copper gives almost exclusively 4-**aminocoumarin,** m.p. 161.5–162°, but with concentrated ammonium hydroxide in dioxane at room temperature it yields 4-aminocoumarin (25%) and (2-hydroxyphenyl)propiolamide (52%) (*Zagorevskiĭ* and *N. V. Dudykina*, *ibid.*, 1962, **32,** 2384):

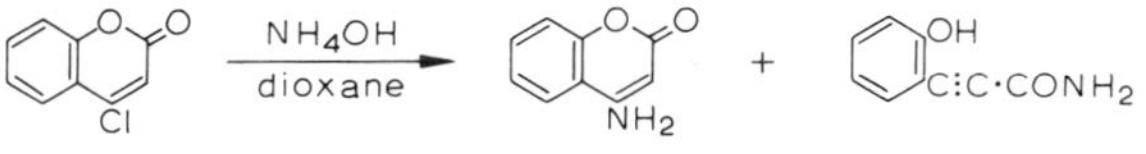

Several 3,3′- and 8,8′-bicoumarinyls have been synthesised by the Ullmann reaction on iodocoumarins (*S. S. Lele, M. G. Patel* and *S. Sethna*, J. chem. Soc., 1961, 969). A number of 6- and 7-halogenocoumarins have been synthesised including 7-**bromocoumarin,** m.p. 105°, 7-*bromo*-4-*methylcoumarin*, m.p. 138°, 7-*chloro*-4-*methyl*-3-*phenylcoumarin*, m.p. 140°, 7-*bromo*-4-*methyl*-3-*phenylcoumarin*, m.p. 129°, 6-*bromo*-4-*methylcoumarin*, m.p. 142° and 6-*bromo*-4-*methyl*-3-*phenylcoumarin*, m.p. 189° (*N. V. S. Subba*

Rao and *V. Sundaramurthy*, Proc. Indian Acad. Sci., 1961, **54,** 105). A number of coumarin derivatives have been chloromethylated and the structures of the chloromethyl derivatives established; 3-**chloromethylcoumarin**, m.p. 111°; 3-*chloromethyl-4-methylcoumarin*, m.p. 139–140° and 6-*chloromethyl-7-methoxy-4-methylcoumarin*, m.p. 246° (*Lele, N. G. Savant* and *Sethna*, J. org. Chem., 1960, **25,** 1713). The crystal dimensions of 7-chloro-4-methylcoumarin have been determined (*K. V. K. Rao* and *P. V. Rao*, Z. Krist., 1965, **122,** 318). Several new heterocycles have been prepared by the action of difunctional reagents on 3,4-dichloro- and 3,4,6-trichloro-coumarins. The products have a heterocyclic ring fused at the 3,4-positions of the coumarin ring (*Newman* and *C. Y. Perry*, J. org. Chem., 1963, **28,** 116).

2-**Thiocoumarin** (2*H*-**chromene**-2-**thione**), m.p. 101°, yellow colour and unpleasant smell, *picrate*, yellow needles, m.p. 118° (*B. N. Ghosh*, J. chem. Soc., 1915, **107,** 1588):

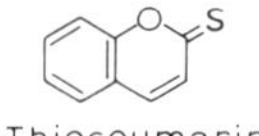

2-Thiocoumarin

The *trinitrobenzene* derivative, m.p. 87° (*A. Kent et al., ibid.*, 1939, 1858), is prepared by heating coumarin with phosphorus pentasulphide or phosphorus and sulphur (*H. Kitagawa*, J. pharm. Soc., Japan, 1950, **70,** 730; C.A., 1951, **45,** 7113). In contrast to coumarin it forms an *oxime*, m.p. 131°.

3-**Methylcoumarin,** m.p. 91°, has the odour of coumarin and is destroyed by potassium permanganate (*H. Simonis*, Ber., 1915, **48,** 1583; *F. D. Popp* and *W. Blount*, J. org. Chem., 1961, **26,** 2108), whereas 4-*methylcoumarin*, m.p. 83–84° is odourless and is not affected by permanganate (*H. von Pechmann* and *E. von Kraft*, Ber., 1901, **34,** 421; *E. H. Woodruff*, Org. Synth., 1944, **24,** 69); 5-*methylcoumarin*, m.p. 65.8°, has a feeble odour of coumarin (*P. Chuit* and *F. Bolsing*, Bull. Soc. chim. Fr., 1906, **35,** 76); 6-*methylcoumarin*, m.p. 74° (*T. J. Thompson* and *R. H. Edee*, J. Amer. chem. Soc., 1925, **47,** 2556); 7-*methylcoumarin*, m.p. 126°, and 8-*methylcoumarin*, m.p. 109–110° (*Posner* and *Hess, loc. cit.*); 3-**ethylcoumarin,** m.p. 72° (*Popp* and *Blount, loc. cit.*); 4-*ethylcoumarin*, m.p. 70° (*W. Cocker et al.*, J. chem. Soc., 1953, 2355).

3,4-**Dimethylcoumarin,** m.p. 115° (*Simonis* and *F. Peters*, Ber., 1908, **41,** 830; *I. M. Heilbron et al.*, J. chem. Soc., 1933, 1263); 4,6-, m.p. 150° and 4,8-*dimethylcoumarin*, m.p. 114° (*D. Chakravarti* and *N. Dutta*, J. Indian chem. Soc., 1940, **17,** 65); 4,7-*dimethylcoumarin*, m.p. 132° (*K. Fries* and *W. Volk*, Ann., 1911, **379,** 101): 5,8-, m.p. 122–123°, 6,7-, m.p. 148–149°, and 6,8-*dimethylcoumarin*, m.p. 95° (*Clayton*, J. chem. Soc., 1908, **93,** 2018). The catalysed hydrolysis of coumarin and some halogeno- and methyl-coumarins by hydroxyl ions has been examined and the mechanisms discussed (*E. R. Garrett, B. C. Lippold* and *J. B. Mielck*, J. pharm. Sci., 1971, **60,** 396). The alkaline hydrolysis of 6- and 7-substituted coumarins has also been studied (*T. Nakabayashi*, Yakugaku Zasshi, 1957, **77,** 523).

3-**Phenylcoumarin**, m.p. 140–141°, μ 4.3D (*C. G.* and *R. J. W. Le Fèvre*, J. chem. Soc., 1937, 1088) is prepared by heating sodium phenylacetate and acetylsalicylaldehyde (*M. Crawford* and *J. A. M. Shaw, ibid.*, 1953, 3435). It is reduced by sodium amalgam to 2′-hydroxydibenzyl-α-carboxylic acid [2-(hydroxyphenyl)-1-phenylethane-1-carboxylic acid] and in sunlight gives a *dimer*, m.p. ~242° (*A. Schönberg et al., ibid.*, 1950,

374). 3-Arylcoumarins have been prepared by the arylation of a number of substituted coumarins by 4-substituted arenediazonium chloride in acetone at pH 2–4 in the presence of cupric chloride (Meerwein arylation reaction) (*P. C. Taunk, J. K. Jain* and *R. L. Mital*, Ann. Soc. Sci. Bruxelles, Ser. 1, 1971, **84,** 383). Contrary to simple coumarins, 3-phenylcoumarins do not undergo inversion into the corresponding coumaric acids to any large extent (*N. R. Krishnaswamy, Seshadri* and *B. R. Sharma*, Indian J. Chem., 1964, **2,** 182). 4-*Phenylcoumarin*, m.p. 105° (*R. Stoermer* and *E. Friderici*, Ber., 1908, **41,** 340; *F. Bergmann et al.*, J. Amer. chem. Soc., 1948, **70,** 1612); a number of 4-phenylcoumarins with substituents in the benzene ring have been prepared (*S. K. Mukerjee, T. Saroja* and *Seshadri*, Indian J. Chem., 1969, **7,** 671). 6-*Phenylcoumarin*, m.p. 118° (*F. D. Cramer* and *H. Windel*, Ber., 1956, **89,** 354); a number of 8-*phenylcoumarins* have been prepared (*Lele, Patel* and *Sethna, loc. cit.*). Synthesis of cyclohexenocoumarins, *e.g.* 3-hydroxy-6-oxo-7,8,9,10-tetrahydro-6*H*-dibenzo[*b*,*d*]-pyran (*M. G. Parekh* and *K. N. Trivedi*, Austral. J. Chem., 1970, **23,** 407):

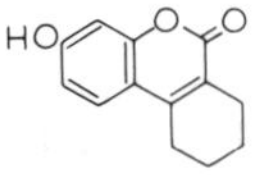

HO

Several coumarins and their methyl derivatives have been synthesised, and their u.v. (*Ganguly* and *Bagchi, loc. cit.; A. Mangini* and *R. Passerini*, Gazz., 1957, **87,** 243; *C. F. Wheelock*, J. Amer. chem. Soc., 1959, **81,** 1348; *R. S. Shah* and *S. L. Bafna*, Indian J. Chem., 1963, **1,** 400; *J. Mendez* and *M. I. Lojo*, Microchem. J., 1968, **13,** 506) and i.r. spectra recorded (*B. K. Sabata* and *M. K. Rout*, J. Indian chem. Soc., 1964, **41,** 74); i.r. spectra of 3-acylcoumarins (*P. Bassignana* and *C. Cogrossi*, Tetrahedron, 1964, **20,** 2859). The reduction of some 3-acylcoumarins with sodium tetrahydridoborate gives the corresponding 3-acyl-3,4-dihydrocoumarin (*H. Wamhoff, G. Schorn* and *F. Korte*, Ber., 1967, **100,** 1296).

Coumarin-8-carboxylic acid, m.p. 245–247° (decomp.) (*E. Cingolani, G. Bellomonte* and *A. Sordi*, Gazz., 1965, **95,** 413).

(vii) Hydroxycoumarins

Besides 7-hydroxycoumarin, umbelliferone, most naturally occurring monohydroxycoumarins are its derivatives; only a small group including ferulenol are not derived from umbelliferone.

3-**Hydroxycoumarin,** m.p. 152°, is obtained by cyclisation of (2-hydroxyphenyl)pyruvic acid (*E. Erlenmeyer Jr.* and *W. Stadlin*, Ann., 1904, **337,** 283; *H. A. Offe* and *H. Jatzkewitz*, Ber., 1947, **80,** 469). It is tautomeric with chroman-2,3-dione. A number of derivatives of 3-hydroxycoumarin have been prepared, including 6-bromo-, 6,8-dibromo-, 4-bromo-, 4,6-dibromo-, 4,6,8-tribromo-, 6-nitro-, 8-nitro-, 6,8-dinitro-, 5-hydroxy-6-methoxycarbonyl-, 5-hydroxy-6-acetyl- (*K. N. Trivedi* and *S. Sethna*, J. org. Chem., 1960, **25,** 1817), 5-methyl-, and other derivatives (*D. Chakravarti* and *R. Das*, J. Indian chem. Soc., 1971, **48,** 371). 3-Hydroxycoumarins undergo the

Wanzlick oxidative coupling reaction with catechol in the presence of potassium ferricyanide to give 6*H*-benzofuro[2,3-*c*]benzopyran-6-ones (*J. N. Chatterjea, K. Achari* and *N. C. Jain*, Tetrahedron Letters, 1969, 3337):

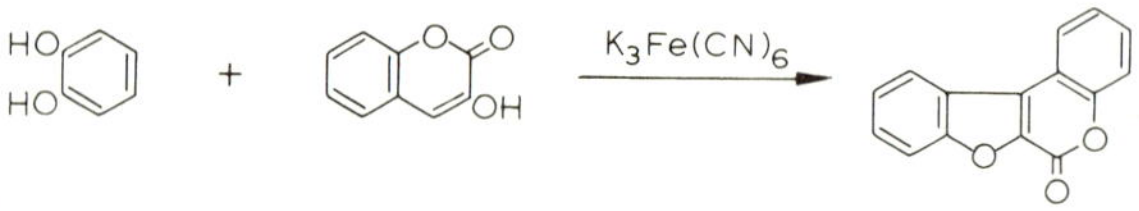

4-Hydroxycoumarin, m.p. 210–212°, can be prepared by the internal condensation of methyl acetylsalicylate using sodium in liquid paraffin at 240° (*M. A. Stahmann et al.*, J. Amer. chem. Soc., 1943, **65,** 2285); by the condensation of 2-hydroxyacetophenone with ethyl carbonate in the presence of sodium (*J. Boyd* and *A. Robertson*, J. chem. Soc., 1948, 174); by treating 2-acetoxybenzoic acid with phosphorus pentachloride in nitrobenzene, and then with aluminium chloride followed by diethyl malonate (*H. Kaneyuki*, Bull. chem. Soc. Japan, 1962, **35,** 519); by cyclodehydration of (2-hydroxybenzoyl)acetic acid with tetraphosphoric acid (*A. Resplandy*, Compt. rend., 1965, **260,** 6479); from 2-hydroxyacetophenone by a modified Kolbe–Schmidt reaction (*P. Da Re* and *E. Sandri*, Ber., 1960, **93,** 1085); in 85% yield by cyclising diphenyl malonate by using aluminium chloride (*E. Ziegler* and *H. Junek*, Monatsh., 1955, **86,** 29):

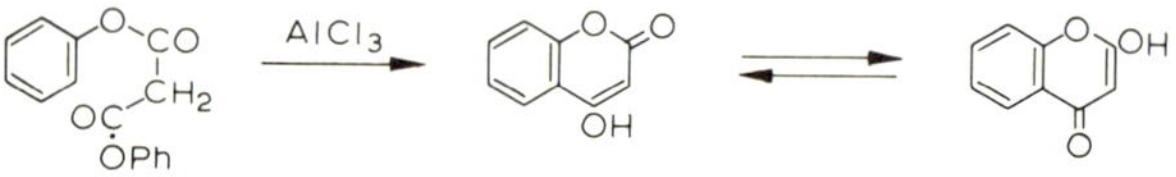

and by treating phenol with an equimolar proportion of malonic acid in the presence of a mixture of 2–3 mols. each of anhydrous zinc chloride and phosphoryl chloride at 60–75° (*R. C.* and *V. R. Shah* and *J. L. Bose*, J. org. Chem., 1960, **25,** 677; Chem. and Ind., 1960, 623). The last two methods can also be applied to the preparation of substituted 4-hydroxycoumarins.

4-Hydroxycoumarin exists in solution as 2-hydroxychromone (4*H*-benzo[*b*]pyran-4-one) (*F. Arndt et al.*, Ber., 1951, **84,** 319; *C. F. Huebner* and *K. P. Link*, J. Amer. chem. Soc., 1945, **67**, 99; *J. F. Garden et al.*, J. chem. Soc., 1957, 3315); 4-hydroxycoumarin with diazomethane yields both 4-methoxycoumarin and 2-methoxychromone. 4-Hydroxycoumarin with aniline gives 4-anilinocoumarin and 2-anilinomalonylphenol (*A. Mustafa et al.*, Ann., 1968, **712**, 107), and with phenylhydrazine it forms a pyrazolone derivative in contrast to 3-phenyl-4-hydroxycoumarin which gives a phenylhydrazone. With aliphatic acid chlorides in pyridine and a trace of piperidine it yields 3-acyl-4-hydroxycoumarins; acetyl chloride giving 3-*acetyl-4-hydroxycoumarin*, m.p. 134° (*T. Ukita et al.*, J. Amer. chem. Soc., 1950, **72**, 5143; *J. Klosa*, Arch. Pharm., 1955, **288**, 356); but with aromatic acid chlorides it forms the corresponding 4-esters (*H. R. Eisenhauser* and *Link, ibid.*, 1953, **75**, 2044, 2046). Treatment of 4-hydroxycoumarin with acetic anhydride gives 4-acetoxycoumarin below 155° and 3-acetyl-4-hydroxycoumarin above 180°; the former product on heating above 180° rearranges to yield the latter product. The rearrangement is thought to be intramolecular (*N. Matzet*,

H. Wamhoff and *F. Korte*, Ber., 1969, **102**, 3122). The reaction between 6-substituted 4-hydroxycoumarins and α,β-unsaturated acyl chlorides has been carried out in pyridine or in tetrachloroethane containing titanium(IV) chloride. In the case of reactions carried out in pyridine, the ratio of the two isomers formed is related to the substituents at the β-position of the α,β-unsaturated acyl chlorides and at the 6-position of the 4-hydroxycoumarin (*Y. Shizuri et al.*, J. chem. Soc., C, 1969, 2774). 4-Acyloxy- or 4-aroyloxy-coumarins on heating in trifluoroacetic acid rearrange to give 5-acyl- or 5-aroyl-4-hydroxycoumarin, respectively, in good yield (*L. L. Woods* and *M. Fooladi*, J. org. Chem., 1968, **33**, 2966):

OAc TFA Ac OH

The arylation of 4-hydroxycoumarin with arenediazonium salts has been carried out under Meerwein arylation conditions; under alkaline conditions 2,3,4-trioxochroman 3-arylhydrazones (LXXX) are formed, while under acid conditions a mixture of 3-aryl-4-hydroxycoumarins, 2,3,4-trioxochroman 3-arylhydrazones, and salicylic acids is obtained (*V. V. Bhat* and *Bose*, C.A., 1968, **68,** 95624x):

N·NHAr

(LXXX)

A number of 3-aryl-4-hydroxycoumarins have been prepared (*B. van Zanten*, *U. A. Th. Brinkman* and *W. Th. Nauta*, Rec. Trav. chim., 1960, **79,** 1223). Arylmethylation occurs when 4-hydroxycoumarin is boiled with anisole and trioxymethylene in propionic acid. The product 3-(4-methoxybenzyl)-4-hydroxycoumarin is obtained in a 20% yield following the removal of bis(4-hydroxycoumarinyl)methane (*C. Deschamps-Vallet* and *C. Mentzer*, Compt. rend., 1963, **256**, 1807):

OH PhOMe CH_2 OMe OH

The condensation of 4-hydroxycoumarin with primary phenol alcohols in chloroform gives in the presence of hydrogen chloride or phosphoryl chloride, 3-benzyl substituted 4-hydroxycoumarins (*Ziegler* and *U. Rossmann*, Monatsh., 1957, **88,** 25); with glyceraldehyde in an aqueous medium it yields compound LXXXI, whereas in dioxane compound LXXXII is formed (*M. Eckstein* and *H. Pazdro*, Rocz. Chem., 1964, **38,** 1115):

OH CH_2OH $CH(OH)CH_2OH$

(LXXXI) (LXXXII)

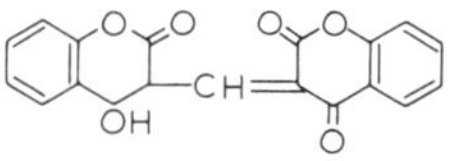

(LXXXIII)

and with orthoformic esters it gives product LXXXIII (*S. Checchi, L. P. Vettori* and *G. Renzi*, Farmaco, Ed. Sci., 1970, **25,** 54). 4-Hydroxycoumarin reacts with substituted salicylaldehydes, 2,6-dichloro- or chloronitrobenzaldehydes to yield unstable 3,3′-benzylidenebis(4-hydroxycoumarins), intermediates which undergo ring closure with the hydroxyl group of one of the 4-hydroxycoumarin moieties to give derivatives of 7-(4-hydroxy-3-coumarinyl)benzopyrano[3,2-*c*]coumarins (*F. Litvan* and *W. G. Stoll*, Helv., 1959, **42,** 878):

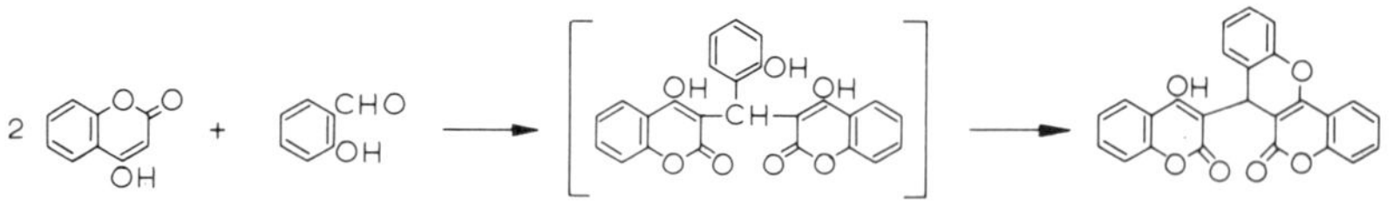

With tetracyanoethylene 4-hydroxycoumarin affords an adduct (*Junek*, Monatsh., 1965, **96,** 1421):

$+ (NC)_2C:C(CN)_2 \longrightarrow$ $C(CN)_2{\cdot}CH(CN)_2$

and with mesityl oxide it gives two products, LXXXIV and LXXXV, the former being converted into a methyl ether of a cyclic tertiary alcohol by the action of methanolic hydrogen chloride (*D. W. Hutchinson* and *J. A. Tomlinson*, Tetrahedron Letters, 1968, 5027):

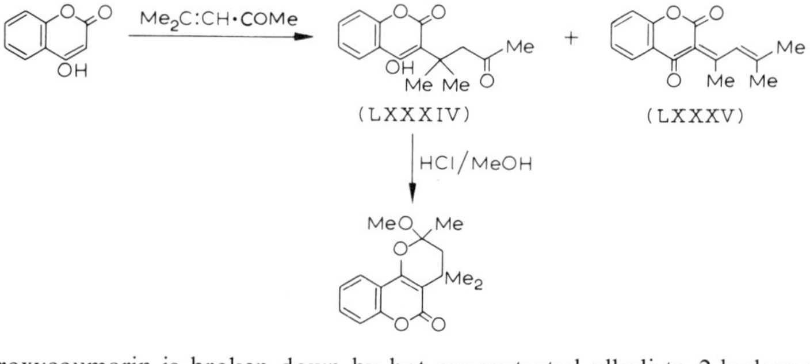

$Me_2C:CH{\cdot}COMe$

(LXXXIV) (LXXXV)

HCl/MeOH

4-Hydroxycoumarin is broken down by hot concentrated alkali to 2-hydroxyacetophenone, and with phosphorus pentachloride yields 4-*chlorocoumarin*, m.p. 92°, also obtained along with a yellow compound derived from three mols. of hydroxycoumarin by reaction with phosphoryl chloride (*V. A. Zagorevskiĭ* and *D. A. Zykov*, Zhur. obshcheĭ Khim., 1960, **30,** 1378; *A. R. Knight* and *J. S. McIntyre*, Canad. J. Chem., 1968, **46,** 2495; *Checchi* and *Vettori*, Chim Ind., 1969, **51,** 292):

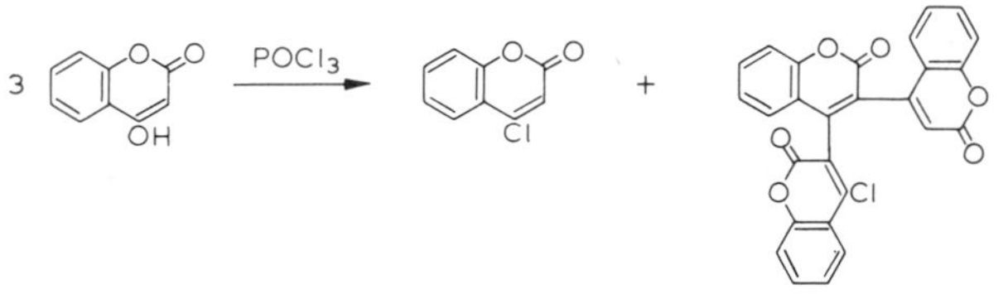

With bromine it affords 3-*bromo*-4-*hydroxycoumarin*, m.p. 190–192° (183°) (*P. Meunier et al.*, Helv., 1946, **29,** 1291); for the synthesis of 6-bromo-4-hydroxycoumarin see *M. Dezelic* and *M. Trkovnik*, J. med. Chem., 1964, **7,** 284. 4-Hydroxycoumarin has λ_{max} 290 mμ (log ε 4.10) and λ_{mfl} 301 mμ (log ε 4.08) (*R. H. Thomson et al.*, J. chem. Soc., 1956, 3315; Chem. and Ind., 1957, 112); the u.v. spectra of a number of 4-hydroxycoumarins have been reported (*V. C. Farmer*, Spectrochim. Acta, 1959, 870); and the structure of 4-hydroxycoumarin monohydrate has been determined by 3-dimensional X-ray analysis (*J. Gaultier* and *C. Hauw*, Compt. rend., 1965, **260,** 5787; Acta Cryst., 1966, **20,** 646).

4-*Hydroxy*-3-*phenoxycoumarin*, m.p. 216°, is prepared by condensing acetylsalicyloyl chloride and diethyl sodiophenoxymalonate, NaC(OPh)($CO_2Et)_2$, in toluene (*B. Kirkiacharian, D. Billet* and *Mentzer*, Compt. rend., 1963, **257,** 2676). Aurone epoxides on treatment with boron trifluoride etherate give the corresponding 4-hydroxy-3-phenylcoumarins (*M. Geoghegan, W. I. O'Sullivan* and *E. M. Philbin*, Tetrahedron, 1966, **22,** 3209; *A. C. Jain, V. K. Rohatgi* and *T. R. Seshadri*, Tetrahedron Letters, 1966, 2701). 4,5-Dihydroxy-, 4,5-dihydroxy-6-methoxy-, 4,7,8-trihydroxy- (*N. J. Desai* and *Sethna*, J. org. Chem., 1957, **22,** 388) and 5,7-dimethoxy-4-hydroxy-3-(2,4,5-trimethoxyphenyl)coumarin (*W. J. Bowyer, Robertson* and *W. B. Whalley*, J. chem. Soc., 1957, 542) have been prepared. The condensation of 3-formyl-4-hydroxycoumarin with ethyl cyanoacetate or benzoyl acetonitrile affords ethyl or phenyl 2,5-dioxopyrano-[3,2-*c*]benzopyran-3-carboxylate, respectively (*Checchi* and *Vettori*, Gazz., 1965, **95,** 1502). Treatment of the silver salt of 4-hydroxy-3-methylcoumarin with methyl iodide gives a mixture of 4-methoxy-3-methylcoumarin, 2-methoxy-3-methyl-4-oxo-4*H*-chromene and 3,3-dimethyl-2,4-dioxochroman (*D. Anker* and *J. Massicot*, Bull. Soc. chim. Fr., 1969, 2181).

3-Alkyl- or 3-aryl-4-hydroxycoumarins on heating with 5% aqueous sodium hydroxide in a sealed tube yield a hydroxyketone, which after treatment with sodium in benzene reacts with dimethyl carbonate to reform the coumarin. This synthetic route offers a way of inserting an isotopically labelled $C_{(2)}$ atom into the coumarin (*L. A. Goding* and *B. D. West*, J. org. Chem., 1968, **33,** 437):

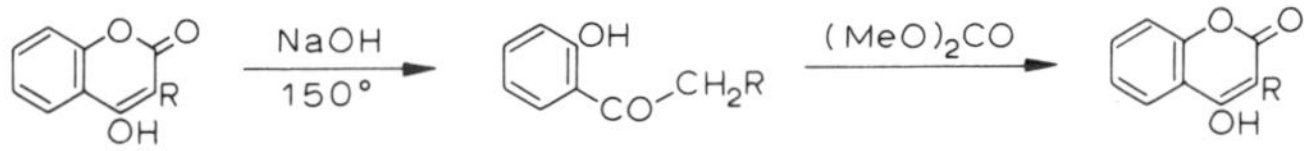

3-Acyl-4-hydroxycoumarins with aniline give the corresponding imines (*A. Mustafa et al.*, Ann., 1968, **712,** 107).

Warfarin, coumoxin, 1-(4-hydroxycoumarin-3-yl)-3-oxo-1-phenylbutane, m.p. 161°, widely used as a rodenticide, is prepared by the Michael condensation of benzylideneacetone with 4-hydroxycoumarin. Anticoagulants of the warfarin type have been prepared by the condensation of alkynols with 4-hydroxycoumarin (*J. Reisch*, Arch. Pharm., 1966, **299**, 806). The Beckmann rearrangement of warfarin oxime and the action of hydrazoic acid on warfarin (Schmidt reaction) both give 3-[α-(acetamidomethyl)benzyl]-4-hydroxycoumarin (*C. Wiener et al.*, J. org. Chem., 1962, **27**, 3086):

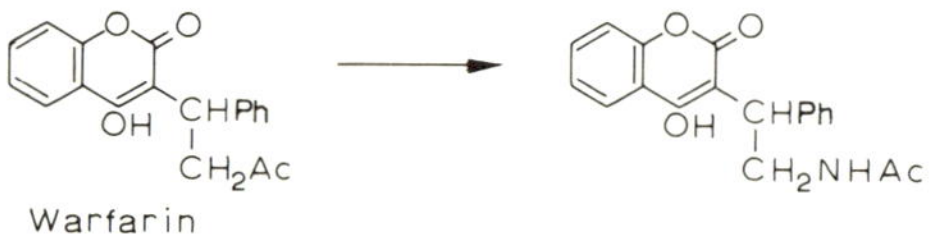

OH CHPh CH$_2$Ac → OH CHPh CH$_2$NHAc

Warfarin

Ferulenol, m.p. 64–65°, isolated from *Ferula communis*, is a member of the small group of natural coumarins not derived from umbelliferone (p. 117) (*S. Carboni*, *V. Malaguzzi* and *A. Marsili*, Tetrahedron Letters, 1964, 2783):

OH

Ferulenol

The crystal and molecular structure of **marcoumar,** 3-(1-phenylpropyl)-4-hydroxycoumarin, a synthetic antagonist of vitamin K has been determined by X-ray diffraction studies (*G. Bravic*, *J. Gaultier* and *Hauw*, Compt. rend. Ser. C, 1971, **272,** 1112).

3,3′-**Methylenebis-4-hydroxycoumarin,** commonly termed *dicoumarin* or *dicoumarol*, m.p. 288–289°, *diacetate*, m.p. 250–252°, is the agent responsible for the haemorrhagic "sweet clover disease" of cattle (*H. A. Campbell* and *Link*, J. biol. Chem., 1941, **138,** 21). It is readily prepared by condensing formaldehyde with two molecules of 4-hydroxycoumarin (*R. Anschütz*, Ber., 1903, **36,** 463), and its structure is confirmed by its breakdown by 10% aqueous sodium hydroxide solution to 1,3-bis(2-hydroxybenzoyl)propane (*Stahmann et al.*, J. biol. Chem., 1941, **138,** 513):

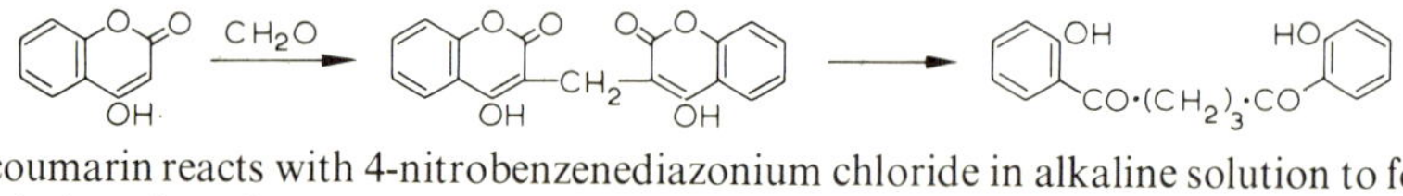

OH →(CH_2O) OH $-CH_2-$ OH → OH HO $CO\cdot(CH_2)_3\cdot CO$

Dicoumarin reacts with 4-nitrobenzenediazonium chloride in alkaline solution to form the 4-nitrophenylhydrazone of 2,3,4-trioxochroman and formaldehyde (*Huebner* and *Link*, J. Amer. chem. Soc., 1945, **67,** 99; *Ziegler* and *Junek*, *loc. cit.*). This nitrophenylhydrazone can likewise be obtained from 4-hydroxycoumarin.

Other coumarin derivatives which show haemorrhagic activity are ayapanin (p. 117) and ayapin (p. 123). *Mentzer* concludes that the ·C(OH):CR·CO· grouping in hydroxycoumarin derivatives is essential for anticoagulant activity (Bull. Soc. chim. biol., 1948, **30,** 872).

5-**Hydroxycoumarin,** m.p. 229°, *acetate*, m.p. 88–89°, is obtained by the Perkin reaction on 2,6-dihydroxybenzaldehyde (*H. Böhme*, Ber., 1939, **72,** 2130); for other syntheses see *H. A. Shah* and *R. C. Shah* (J. chem. Soc., 1938, 1832) and *Th. Nogradi*, *J. Swoboda* and *F. Wessely* (C.A., 1963, **59,** 2796). In aqueous ethanol it gives with ferric chloride a bluish-green colour and with diazomethane forms 5-*methoxycoumarin*, m.p. 85–87°. In addition to dehydroacetic acid, 3,6-diacetyl-4,7-dimethyl-5-hydroxycoumarin and 3,8-diacetyl-4,7-dimethyl-5-hydroxycoumarin have been isolated as condensation products of diketone (*E. Marcus* and *J. K. Chan*, J. org. Chem., 1967, **32,** 2881).

6-**Hydroxycoumarin,** m.p. 250°, is prepared from quinol by the Pechmann synthesis *(Böhme, loc. cit.)*, by decarboxylation of the acid obtained by hydrolysis of *ethyl 6-hydroxycoumarin-3-carboxylate*, m.p. 192°, formed by reacting 2,5-dihydroxybenzaldehyde with diethyl malonate in the presence of a few drops of piperidine (*F. D. Cramer* and *H. Windel*, Ber., 1956, **89,** 354), and in a 77% yield by treating 2,5-dimethoxycinnamic acid with boron tribromide in chloroform (*J. I. DeGraw* and *P. Tsakotellis*, J. chem. Eng. Data, 1969, **14,** 509). The latter method is potentially useful for the preparation of other hydroxycoumarins. Some of the reactions of 6-hydroxycoumarin have been studied (*M. G. Patel* and *Sethna*, J. Indian chem. Soc., 1962, **39,** 511); it is non-fluorescent in u.v. light except in concentrated sulphuric acid. Nitration of 6-hydroxy-4-methylcoumarin with a mixture of nitric and sulphuric acids at 0–5° affords 6-hydroxy-4-methyl-5,7-dinitrocoumarin, m.p. 220° (decomp.) (*G. S. Mewada* and *N. M. Shah*, Ber., 1956, **89,** 2209).

Umbelliferone, 7-**hydroxycoumarin, shimmetin, hydrangin,** m.p. 225–228° (235°, 240°) is found in the bark of *Daphne mezereum*; the distillates of *umbellifera* resins; as the 7-*glucoside, skimmin*, m.p. 220° (*E. Späth* and *O. Neufeld*, Rec. Trav. chim., 1938, **57,** 535); and in the roots of *Hydrangea arborescens* (*K. H. Palmer*, Canad. J. chem., 1963, **41**, 2387). It is prepared by the Perkin synthesis from 2,4-dihydroxybenzaldehyde (*F. Tiemann* and *L. Lewy*, Ber., 1877, **10**, 2216); by the decarboxylation of umbelliferone-6-carboxylic acid (*R. C. Shah et al.*, J. Indian chem. Soc., 1937, **14**, 717); and by the Pechmann synthesis from resorcinol, malic acid and sulphuric acid (Ber., 1884, **17,** 932). It crystallises from water; has a smell like that of coumarin; gives a brilliant blue fluorescence in u.v. light and its formation and consequent fluorescence from resorcinol and a hydroxydicarboxylic acid such as malic acid constitutes a test for this type of acid. 7-Methoxycoumarin-3-carbonyl chloride may be used as a fluorescent reagent for the chromatographic separation of hydroxy- and amino-compounds (*W. Baker* and *C. B. Collie*, J. chem. Soc., 1949, S12). 7-Hydroxycoumarin forms an *acetate*, m.p. 140°, and a *methyl ether*, **herniarin (ayapanin),** m.p. 117–118°, which occurs in the leaves of *Herniaria hirsuta* L., and has been synthesised from resorcinol monomethyl ether and malic acid (*B. B. Dey et al.*, J. Indian chem. Soc., 1934, **11,** 743; 1935, **12,** 140). Herniarin does not give an oxime, but the 2-thiono-compound readily furnishes an *oxime*, m.p. 138°, and a *phenylhydrazone*, m.p. 115°. With iminodiethanol 7-hydroxycoumarin gives a Mannich base (*R. C. Elderfield* and *A. C. Mehta*, J. med. Chem., 1967, **10,** 921).

Umbelliprenin, $C_{24}H_{30}O_3$, *umbelliferone farnesyl ether*, m.p. 61–63°, occurs in the roots of *Angelica archangelica* L. (*Späth* and *F. Vierhapper*, Ber., 1938, **71,** 1667; Monatsh., 1938, **72,** 179; *R. B. Bates, J. H. Schauble* and *M. Souček*, Tetrahedron Letters, 1963, 1683).

Marmin, m.p. 123–124°, $[\alpha]_D$ +25° (EtOH), has been isolated from the trunk bark of *Aegle marmelos* Correā *(Rutaceae)* and its structure proposed on the evidence of m.s. and n.m.r. data (*A. Chatterjee et al.*, Tetrahedron Letters, 1967, 471):

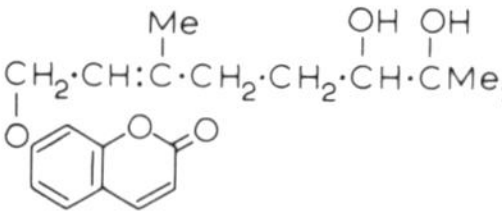

Me OH OH
$CH_2\cdot CH:C\cdot CH_2\cdot CH_2\cdot CH\cdot CMe_2$

Marmin

The structure of a number of complex naturally occurring ethers containing the umbelliferone nucleus have been determined; **samarcandin** and **samarcandone** from the roots of *Ferula samarcandica* (*N. P. Kir'yalov* and *S. D. Movchan*, Khim. Prir. Soedin., 1968, **4,** 73; *A. Ban'kovskii et al., ibid.*, 1970, **6,** 173; *V. Yu. Bagirov, Kir'yalov* and *V. I. Sheichenko, ibid.*, p. 465); **badrakemin** (*Ban'kovskii et al., loc. cit.; Bagirov et al., ibid.*, 1970, **6,** 466); **geiparvarin,** from the leaves of *Geijera parviflora* (*F. N. Lahey* and *J. K. MacLeod*, Austral. J. Chem., 1967, **20,** 1943); **coladin** and **coladonin** (*Ban'kovskii et al., loc. cit.*). **Resocyanin,** 4-*methylumbelliferone*, 7-*hydroxy*-4-*methylcoumarin*, m.p. 185°, is obtained from resorcinol and acetoacetic ester (*A. Russell* and *J. R. Frye*, Org. Synth., 1941, **21,** 22; *N. Andric*, Ann. Chim., 1960, **5,** 1373), *acetate*, m.p. 150–151°; 7-hydroxy-4-methylcoumarins have been prepared in 68–84% yield, from di- or triphenols with OH groups in *meta* position, and acetoacetic ester in the presence of trichloroacetic acid (*L. Reichel* and *W. Hampel*, Ann., 1970, **733,** 1); 4-methyl-, 3,4-dimethyl- and 6-ethyl-4-methyl-7-hydroxycoumarin react with thionyl chloride and aluminium chloride to give the corresponding 8-*sulphinic acids* (*R. N. Usgaonkar* and *G. V. Jadhav*, J. Indian chem. Soc., 1958, **35,** 251); 6-*acetylumbelliferone*, 6-*acetyl*-7-*hydroxycoumarin*, m.p. 175°, may be obtained by reacting 3,4-dihydroumbelliferone with acetic anhydride in the presence of aluminium chloride and dry hydrogen chloride, followed by dehydrogenation of the product (*D. K. Chatterjee* and *K. Sen. ibid.*, 1969, **46,** 275). Nuclear methylation of umbelliferone occurred on treatment with methyl iodide and potassium hydroxide in boiling methanol to give 7-methoxy-8-methylcoumarin. Under similar conditions 4-methylumbelliferone afforded a mixture of 7-hydroxy- and 7-methoxy-4,8-dimethylcoumarin (*V. N. Gupta* and *T. R. Seshadri*, J. sci. ind. Research India, 1957, **16B,** 257). The sulphonation of 7-hydroxy-3,4-dimethylcoumarin, its methyl ether, 7-hydroxy-4-methyl-6-ethylcoumarin, its methyl ether, 7-hydroxy-3,4-dimethyl-6-ethylcoumarin and its methyl ether, with chlorosulphonic acid has been described and the structure of the sulphonated products established (*J. R. Merchant* and *R. C. Shah*, J. org. Chem., 1957, **22,** 884). **Geijerin,** 7-*methoxy*-8-*isovalerylcoumarin*, $C_{15}H_{16}O_4$, m.p. 121–123°, 2,4-*dinitrophenylhydrazone*, m.p. 181°, is isolated from the bark of *Geijera salicifolia* Schott and was at first believed to be 7-methoxy-6-isovalerylcoumarin (*Lahey* and *D. J. Wluka*, Austral. J. Chem., 1955, **8**, 125). It has been synthesised by *L. G. Shah et al.* (J. sci. ind. Research India, 1956, **15B**, 580) who indicated that it was 7-methoxy-8-isovalerylcoumarin. **Osthol,** 7-methoxy-8-(2-isopentenyl)coumarin, $C_{15}H_{16}O_{31}$, m.p. 85° (*Späth et al.*, Ber., 1942, **75**, 1623) and **ostruthin,** 6-(3,7-*dimethyl*-2,6-*octadienyl*)-7-*hydroxycoumarin*, $C_{19}H_{22}O_3$, m.p. 119° (*Späth* and *K. Klager, ibid.*, 1934, **67**, 859) both occur in *Imperatoria ostruthium* and angelica. **Osthenol,** 7-*hydroxy*-8-(2-*isopentenyl*)*coumarin*, $C_{14}H_{14}O_3$, m.p. 124–125°, occurs in angelica and is methylated by diazomethane to osthol (*Späth* and *J. Bruck, ibid.*, 1937, **70**, 1023). It has been synthesised by first converting 7-hydroxycoumarin to its 1,1-dimethylallyl ether, by reaction with 3-chloro-3-methylbutyne, followed by partial hydrogenation. The ether on pyrolysis gave mainly the $C_{(8)}$ Claisen rearrangement product, osthenol (*R. D. H. Murray, M. M. Ballantyne* and *K. P. Mathai*, Tetrahedron, 1971, **27**, 1247):

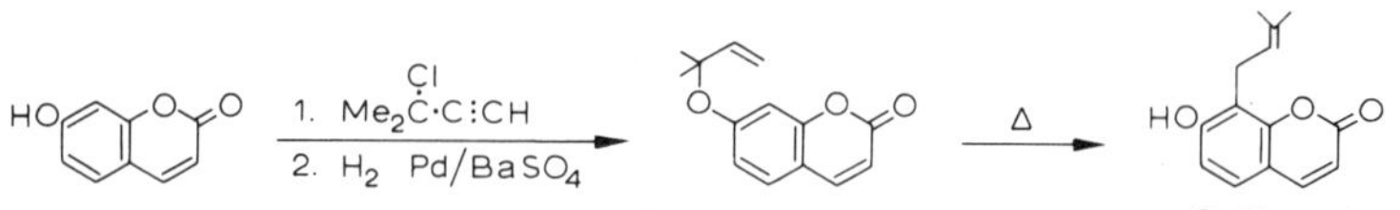

1. $Me_2\overset{Cl}{C}\cdot C:CH$
2. H_2 Pd/$BaSO_4$
Δ
Osthenol

Pinnaterin, 3-(1,1-dimethylallyl)-6-hydroxymethyl-7-methoxycoumarin, m.p. 130–132, obtained from the roots of *Ruta pinnata* L. (*A. G. Gonzàlez, R. Estévez* and *I. Jaraiz*, Phytochemistry, 1971, **10**, 1971).

Auraptenol, 7-methoxy-8-(2-hydroxy-3-methyl-3-butenyl)coumarin, has been isolated from bitter (Seville) orange oil and evidence for its structure obtained from n.m.r., i.r. and analytical data (*W. L. Stanley et al.*, Tetrahedron, 1965, **21,** 89). **Calcicolin,** *7-hydroxy-8-(β-hydroxy-β-isopropoxyethyl)coumarin*, m.p. 172°, obtained from *Angelica saxicola Yoshinagae* (*T. Nakabayashi*, Nippon Kagaku Zasshi, 1962, **83,** 182); **micropubescin,** 7-methoxy-8-(1-oxo-3-methyl-3-butenyl)coumarin from *Micromelum pubescens* (*A. Chatterjee, C. P. Dutta* and *S. Bhattacharyya*, Sci. Cult., 1968, **34,** 366); **pranferin,** 7-methoxy-8-(2-acetoxy-3,3-dimethylbutyl)coumarin from *Prangos ferulaceae* roots (*A. Z. Abyshev et al.*, Khim. Prir. Soedin., 1970, **6,** 675). The structure of a number of natural coumarins containing the umbelliferone nucleus with a complex $C_{(6)}$ side-chain have been determined; **micromelin,** from *Micromelum minutum* (*J. A. Lamberton, J. R. Price* and *A. H. Redcliffe*, Austral. J. Chem., 1967, **20,** 973); **dehydrogeijerin,** from the leaves of *Geijera parviflora* (*Lahey*, and *MacLeod, loc. cit.*); **peucenol, peumorisin,** m.p. 146°, from the roots of *Peucedanum morisoni* (*Yu. N. Sheinker et al.*, Khim. Prir. Soedin., 1969, **5,** 361); **ulopterol,** from *Prangos uloptera* (*Abyshev et al., ibid.*, 1970, **6,** 300).

Thamnosin, m.p. 244–247°, a minor component from *Thamnosma montana* Torr. and Frem., represents a novel dimeric coumarin. The coumarin nuclei possess a 7-methoxy group (*J. P. Kutney, T. Inaba* and *D. L. Dreyer*, Tetrahedron, 1970, **26,** 3171):

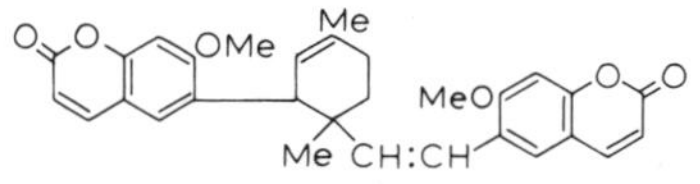

OMe Me MeO Me CH:CH

Thamnosin

8-**Hydroxycoumarin,** m.p. 160°, obtained by demethylating 8-methoxycoumarin (*Böhme, loc. cit.*), gives in aqueous ethanol a deep blue colour with ferric chloride, and with diazomethane yields *8-methoxycoumarin*, m.p. 90–91°, which can be synthesised from *o*-vanillin by the Perkin reaction (*F. Mauthner*, J. pr. Chem., 1939, [ii], **152,** 23), *acetate*, m.p. 131°.

The cyanoethylation of some hydroxycoumarins has been investigated and the cyanoethyl derivatives obtained have been hydrolysed to the corresponding acids (*Merchant* and *J. R. Patell*, J. chem. Soc., C, 1969, 1544). It has been found that the

γ,γ-dimethylallyl ethers of 5-, 6-, 7- and 8-hydroxycoumarin are degraded to phenolic coumarins and isoprene under Claisen rearrangement conditions (*B. Chaudhury, S. K. Saha* and *A. Chatterjee*, J. Indian chem. Soc., 1962, **39,** 783). 2-(1-Naphthyl)coumarino-4-pyrones have been synthesised from 8-acetyl-7-hydroxycoumarin and 1-naphthoyl chloride (*K. A. Thaker* and *N. R. Manjaramkar, ibid.*, 1971, **48,** 221). The u.v. (*Böhme* and *T. Severin*, Arch. Pharm., 1957, **290,** 448; *K. Sen* and *P. Bagchi*, J. org. Chem., 1959, **24,** 316), fluorescence (*B. N. Mattoo*, Trans. Faraday Soc., 1956, **52,** 1184), u.v. and i.r. (*M. E. Perel'son* and *Sheinker*, Zhur. Prikl. Spektroskopii, Akad. Nauk Belorussk., 1966, **5,** 104) and n.m.r. and i.r. spectra (*D. W. Hutchinson* and *J. A. Tomlinson*, Tetrahedron, 1969, **25,** 2531) of some hydroxycoumarins and related compounds have been discussed. The magnetic susceptibilities of several hydroxycoumarins and their derivatives have been determined (*R. M. Mathur*, Trans. Faraday Soc., 1958, **54,** 1609; 1960, **56,** 325).

(viii) Dihydroxycoumarins

Dihydroxycoumarins with hydroxyl groups in the 5,7-, 6,7- and 7,8-positions are found in plants, both in the free state or as glucosides or methyl ethers.

Ammoresinol, $C_{24}H_{30}O_4$, m.p. 109°, is obtained from the resin of *Dorema ammoniacum* Don. and has the structure given below (*K. Kunz*, J. pr. Chem., 1934, [ii], **141,** 350; *E. Späth et al.*, Ber., 1936, **69,** 1656; *H. Raudnitz et al.*, Ber., 1936, **69,** 1957); *acetate*, m.p. 129°; *diacetate*, m.p. 102–103°. It is a 4-hydroxycoumarin, examples of which are rare:

HO, O, O, OH, $CH_2(CH{:}\underset{Me}{C}{\cdot}CH_2{\cdot}CH_2)_2CH{:}CMe_2$

Ammoresinol

The structure of the coumarin moiety of the antibiotic novobiocin is 3-amino-4,7-dihydroxy-8-methylcoumarin (*C. H. Stammer et al.*, J. Amer. chem. Soc., 1958, **80,** 137).

Limettin, citropten, 5,7-*dimethoxycoumarin*, $C_{11}H_{10}O_4$, m.p. 145–146°, is obtained from some essential oils of the citrus species (*W. A. Tilden* and *H. Burrows*, J. chem. Soc., 1902, **81,** 508; *A. G. Caldwell* and *E. R. H. Jones, ibid.*, 1945, 540; *T. Matsuno*, J. pharm. Soc. Japan, 1959, **79,** 540). It is synthesised by methylating 5,7-**dihydroxycoumarin,** m.p. 140°, obtained by the Perkin reaction on phloroglucinaldehyde (*R. G. Heyes* and *A. Robertson*, J. chem. Soc., 1936, 1831). Partial alkylation of 5,7-diacetoxycoumarin gives the 5-alkyl ether, but in the case of 5,7-dihydroxy-4-methylcoumarin the 7-alkyl ether is obtained (*V. K. Ahluwalia et al.*, Indian J. Chem., 1969, **7,** 115). **Toddaculine,** 5,7-*dimethoxy*-6-(2-*isopentenyl*)*coumarin*, m.p. 95° (*G. Combes, R. Pernet* and *R. Pierre*, Bull. Soc. chim. Fr., 1961, 1609) and **toddalolactone,** 5,7-*dimethoxy*-6-(2,3-*dihydroxyisopentyl*)*coumarin*, m.p. 132–132.5°, *diacetate*, m.p. 119°, have been isolated from the root-bark of *Toddalia aculeata* Pers. (*B. B. Dey* and *P. P. Pillay*, Arch. Pharm., 1935, **273,** 223; *Späth, Dey* and *E. Tyray*, Ber., 1939, **72,** 53; *Combes*,

Pernet and *Pierre, loc. cit.*). **Glabralactone,** 5,7-*dimethoxy*-8-(3,3-*dimethylacryloyl*)-*coumarin*, m.p. 130°, from *Angelica glabra* (*K. Hata* and *Y. Tanaka*, Yakugaku Zasshi, 1957, **77,** 937, 941; *T. Kariyone* and *Hata*, J. pharm. Soc. Japan, 1956, 649, 666):

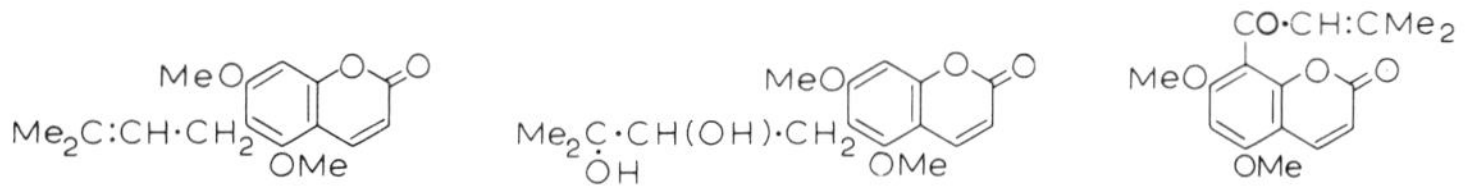

Toddaculine Toddalolactone Glabralactone

Coumurrayin, 5,7-*dimethoxy*-8-(2-*isopentenyl*)*coumarin*, m.p. 157°, isolated from the ripe fruits of *Murraya paniculata* (L.) Jack (Rutaceae) (*E. Ramstad et al.*, Tetrahedron Letters, 1968, 811). It has been synthesised by a simple three-step method for introducing a 3,3-dimethylallyl unit *ortho* to a phenolic hydroxyl group (*R. D. H. Murray, M. M. Ballantyne* and *K. P. Mathai*, Tetrahedron, 1971, **27,** 1247). **Pinnarin,** 5,7-*dimethoxy*-8-(1,1-*dimethylallyl*)*coumarin*, m.p. 166–167°, from the roots of *Ruta pinnata* (*R. E. Reyes* and *A. G. González*, Phytochemistry, 1970, **9,** 833):

Coumurrayin Pinnarin

It has been synthesised from 7-hydroxy-5-methoxycoumarin *via* a Claisen rearrangement of 5-methoxy-7-*O*-(3,3-dimethylallyl)coumarin (*Murray* and *Ballantyne*, Tetrahedron, 1970, **26,** 4667). **Sesibiricin,** 8-*isopentenyl*-5-*isopentenyloxy*-7-*methoxycoumarin*, m.p. 120–122°, has been isolated from the roots of *Seseli sibiricum* (*T. R. Seshadri* and *T. Vishwapaul*, Indian J. Chem., 1970, **8,** 202) and synthesised (*Murray et al.*, Tetrahedron Letters, 1971, 3317):

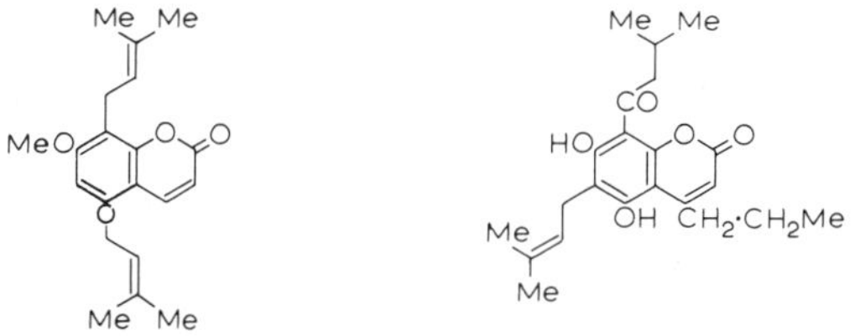

Sesibiricin Mammein

Mammein, 5,7-*dihydroxy*-6-*isopentenyl*-8-*isovaleroyl*-4-*propylcoumarin*, m.p. 129°, obtained from the seeds of *Mammea americana* L. shows insecticidal properties; some coumarins related to mammein have been synthesised (*C. Djerassi et al.*, J. org. Chem., 1960, **25,** 2164, 2169); **isomammein** is 5,7-dihydroxy-8-isopentenyl-6-isovaleroyl-4-propylcoumarin (*Djerassi et al.*, Tetrahedron Letters, 1959 (March), 10). **Mammeisin,** 5,7-*dihydroxy*-6-*isovaleryl*-8-*isopentenyl-phenylcoumarin*, m.p. 98–109°, *diacetate*, m.p. 122–124°, *dimethyl ether*, 86–89°, obtained from the peelings of the fruit of *Mammea americana* L. (*R. A. Finnegan et al.*, J. org. Chem., 1961, **26,** 1180; 1965, **30,** 2342).

Ferruol A, m.p. 126.5°, is a 4-alkylcoumarin isolated from the trunk bark of *Mesua ferrea* L. (*T. R. Govindachari et al.*, Tetrahedron, 1967, **23**, 4161):

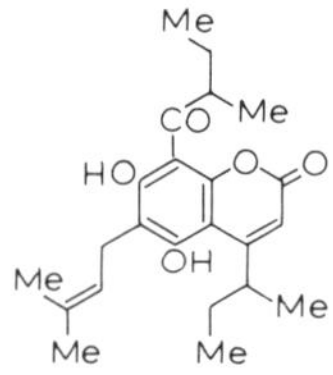

Ferruol A

The reaction between phloroglucinol and ethyl acetoacetate in the presence of a cation exchange resin, gives 5,7-dihydroxy-4-methylcoumarin (*N. Andric*, Ann. Chim., Paris, 1960, **5**, 1373); 5,7-diacetoxy-4-methylcoumarin reacts with *N*-bromosuccinimide in the presence of benzoyl peroxide to give 5,7-*diacetoxy*-4-*(bromomethyl)coumarin*, m.p. 164–165°, which may be converted to 5,7-*diacetoxy*-4-*(acetoxymethyl)coumarin*, m.p. 176–177° and hence to 5,7-*dihydroxy*-4-*(hydroxymethyl)coumarin*, m.p. 253–255° (*J. M. Sehgal* and *Seshadri*, J. sci. ind. Research India, 1957, **16B,** 12); 5,7-*dimethoxy*-4,6-*dimethylcoumarin*, m.p. 158–159°; 4,6,6-*trimethyl*-5,7-*dioxo*-5,6,7,8-*tetrahydrocoumarin*, m.p. 255–256° (*V. N. Gupta* and *Seshadri*, *ibid*., p. 257). Some complex 4-alkyl-5,7-dihydroxycoumarins have been isolated from the bark of *Mammea africana* G. Don. (*I. Carpenter, E. J. McGarry* and *F. Scheinmann*, Tetrahedron Letters, 1970, 3983).

Aesculetin, esculetin, 6,7-**dihydroxycoumarin,** $C_9H_6O_4$, m.p. 276° (decomp.), *diacetate*, m.p. 133–134°, occurs in the bark of the horse chestnut *Aesculus hippocastanum*. It has a pale blue fluorescence in dilute aqueous alkaline solution and its structure follows from its Perkin synthesis from 2,4,5-trihydroxybenzaldehyde (*L. Gattermann* and *M. Köbner*, Ber., 1899, **32,** 287; *H. von Pechmann* and *E. von Krafft*, *ibid*., 1901, **34,** 423). The irradiation of 3,4-dihydroxycinnamic acid in the presence of oxygen gives aesculetin (*J. Kagan*, J. Amer. chem. Soc., 1966, **88,** 2617). It gives a green colour with ferric chloride and reduces Fehling's solution and ammoniacal silver nitrate. With diazomethane it forms 6-*hydroxy*-7-*methoxy*-, m.p. 185–186° and 6,7-*dimethoxycoumarin*, m.p. 144° (*R. Seka* and *P. Kallir*, Ber., 1931, **64,** 909). A study has been made of the partial alkylation (methylation, benzylation and allylation) of aesculetin (*V. K. Ahluwalia et al.*, India J. Chem., 1969, **7,** 59). **Aesculin, esculin,** $C_{15}H_{16}O_9 \cdot 2H_2O$, m.p. 160°, 205° (anhydrous), $[\alpha]_D^{15}$ −146° (MeOH), the 6-β-glucoside of aesculetin, occurs in the bark of the horse chestnut and in the roots of the wild jasmin, *Gelsemium sempervirens*. Its aqueous solutions show a blue fluorescence. The position of the sugar residue was established by methylating aesculin and hydrolysing the product to 6-*hydroxy*-7-*methoxycoumarin*, m.p. 185° (*F. S. H. Head* and *Robertson*, J. chem. Soc., 1930, 2434; *A. K. Macbeth*, *ibid*., 1931, 1288; *Seka* and *Kallir*, *loc. cit.)*. It has been synthesised by attaching the sugar residue to the 7-benzyl ether of aesculetin and then removing the protecting group by selective hydrogenation (*K. W. Merz* and *W. Hagemann*, Naturwiss., 1941, **29,** 650). A number of coumarins, which are derivatives of aesculetin have been isolated from the heartwood of *Ptaeroxylon obliquum* (*F. M.*

Dean et al., Tetrahedron Letters, 1967, 2147), and some of them have been synthesised (*Ballantyne, P. H. McCabe* and *Murray*, Tetrahedron, 1971, **27,** 871).

Cichoriin, 6-*hydroxy-7-β-glucosidocoumarin*, $C_{15}H_{16}O_2 \cdot 2H_2O$, m.p. 213–214°, $[\alpha]_D$ —104.5° (dioxane), is isomeric with aesculin, *penta-acetate*, m.p. 218°, and is obtained from *Cichotium intybus* L. It is non-fluorescent in solution and has been synthesised (*Head* and *Robertson*, J. chem. Soc., 1939, 1266). **Scopoletin,** 7-*hydroxy-6-methoxycoumarin*, $C_{10}H_8O_4$, m.p. 204°, *acetate*, m.p. 177°, occurs in the free state and as the glucoside **scopolin,** $C_{22}H_{28}O_{14}$, m.p. 218°, in *Solanaceae* and *Scopolia* species. It reduces Fehling's solution and in aqueous or ethanolic solution especially when alkaline exhibits a striking blue fluorescence. It has been synthesised from 2,4-dihydroxy-5-methoxybenzaldehyde (*Seka* and *Kallir*, *loc. cit.; Head* and *Robertson*, J. chem. Soc., 1931, 1241); isovanillin (*D. G. Crosby*, J. org. Chem., 1961, **26,** 1215); aesculin (*H. D. Braymer, M. R. Shetlar* and *S. H. Wender*, Biochim. et Biophys. Acta, 1960, **44,** 163); 2,4-dihydroxyanisole and ethyl 3,3-diethoxypropionate (*Crosby* and *R. V. Berthold*, J. org. Chem., 1962, **27,** 3083); and 6,7-diacetoxycoumarin (*Ahluwalia, loc. cit.).* **Fabiatrin,** $C_{21}H_{26}O_{13} \cdot 2H_2O$, the 7-*β*-primveroside of scopoletin, m.p. 236–238°, $[\alpha]_D^{20}$ −140° (H_2O), *hexaacetate*, m.p. 172°, occurs in *Fabiana imbricata* and has been synthesised (*D. N. Chaudhury, R. A. Holland* and *Robertson*, J. chem. Soc., 1948, 1671; *F. Kenji* and *N. Mitsuru*, Bull. chem. Soc. Japan, 1962, **35**, 1321; *Crosby* and *Berthold, loc. cit.*). **Ayapin,** 6,7-methylenedioxycoumarin, $C_{10}H_6O_4$, m.p. 231–232° (evacuated tube) occurs in the leaves of *Eupatorium ayapana* Vent. and has been obtained by the action of methylene iodide on aesculetin (*Späth et al.*, Ber., 1937, **70,** 702). **Scoparone,** *aesculetin dimethyl ether*, m.p. 146–147°, has been isolated from the leaves of *Zanthoxylum setosum* (*I. Araki* and *Y. Miyashita*, J. pharm. Soc. Japan, 1928, **48,** 437); the fruit of *Artemisia capillaris* (*S. Sera* and *C. Shibuye*, J. agr. chem. Soc. Japan, 1930, **6,** 600); the heartwood of *Fagara macrophylla* (olon) (*F. E. King, J. R. Housley* and *T. J. King*, J. chem. Soc., 1954, 1392) and *Artemisia scoparia* (*G. Singh et al.*, Chem. and Ind., 1954, 1294; J. sci. ind. Research India, 1956, **15B,** 190). It has been synthesised (*Crosby* and *Berthold, loc. cit.).*

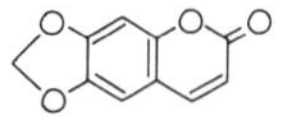

Ayapin

MeO MeO

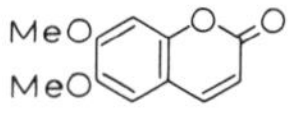

Scoparone

Dalbergin, 6-*hydroxy-7-methoxy-4-phenylcoumarin*, m.p. 209–210°, is the chief constituent of the heartwood of *Dalbergia sissoo* with its O-*methyl ether*, m.p. 145–146°, as a minor constituent (*Ahluwalia* and *Seshadri*, J. chem. Soc., 1957, 970), and was the first natural 4-phenylcoumarin to be isolated. It has been synthesised (*Ahluwalia, A. C. Mehta* and *Seshadri*, Tetrahedron, 1958, **4,** 271), and the methyl ether which had been synthesised earlier, is oxidised to 2-hydroxy-4,5-dimethoxybenzophenone, and with dimethyl sulphate, methanol and potassium hydroxide yields 2,4,5-*trimethoxy-β-phenylcinnamic acid*, m.p. 170–171°.

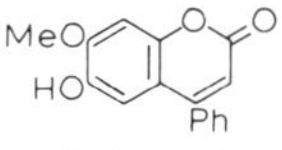

Dalbergin

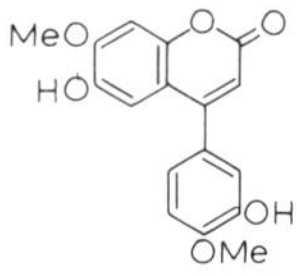

Melannein

The properties of the methyl ether and related 4-phenylcoumarins have been compared with those of ordinary coumarins (*Ahluwahlia et al.*, Proc. Indian Acad. Sci., 1959, **49,** 104). The synthesis of dalbergin has been reviewed (*G. Bargellini*, Gazz., 1958, **88,** 3). **Melannein,** 6-*hydroxy*-(3-*hydroxy*-4-*methoxyphenyl*)-7-*methoxycoumarin*, m.p. 221–223°, isolated from the heartwood of *Dalbergia baroni* Baker, *dimethyl ether*, m.p. 220–221° (*B. J. Donnelly, D. M. X. Donnelly* and *A. M. O'Sullivan*, Tetrahedron, 1968, **24,** 2617). It has been synthesised (*S. K. Mukerjee, T. Saroja* and *Seshadri*, Indian J. Chem., 1969, **7,** 844).

Lasiocephalin, 6-(7-*coumaroxymethyl*)-7-*methoxycoumarin*, m.p. 215°, has been isolated from the root bark of *Lasiosiphon eriocephalus* Decne. (*A. K. Bhattacharya* and *S. C. Das*, Chem. and Ind., 1971, 885):

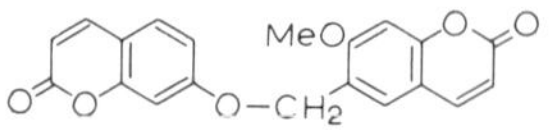

Lasiocephalin

Daphnetin, 7,8-*dihydroxycoumarin*, $C_9H_6O_4$, m.p. 256°, *dimethyl ether*, m.p. 119–121°, *diacetate*, m.p. 137° (128–130°), is found in plants of the *Thymelaecae* order, and gives a green colour with ferric chloride. The structure follows from the hydrolysis of the diethyl ether by means of sodium hydroxide to the diethyl ether of daphnetinic acid, which after conversion to the triethyl ether is oxidised to 2,3,4-triethoxybenzoic acid. The benzoic acid derivative on decarboxylation gives pyrogallol triethyl ether:

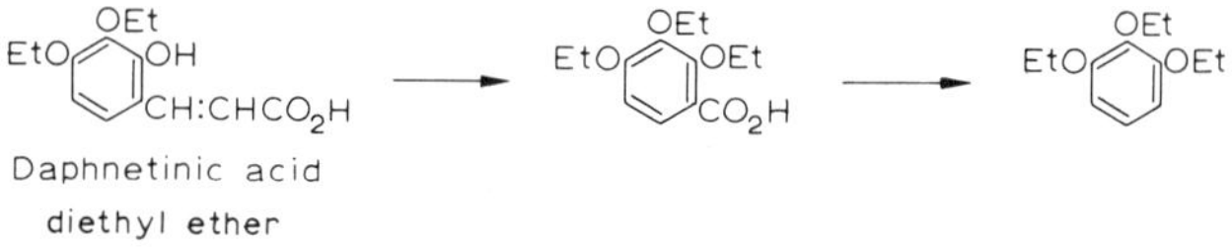

Daphnetinic acid diethyl ether

The structure is confirmed by synthesis from pyrogallolcarbaldehyde (*L. Gattermann* and *M. Köbner*, *loc. cit.*) or from pyrogallol and malic acid (*H. von Pechmann*, Ber., 1884, **17,** 929). 7,8-Dihydroxy-4-methylcoumarin, 4-methyldaphnetin, has been synthesised (*Andric*, Ann. Chim., Paris, 1960, **5,** 1373) and a study made of its partial alkylation (*Ahluwahlia et al., loc. cit.*). Daphnetin occurs in nature as the 7-*β-glucoside*, **daphnin,** m.p. 228–229°, $[\alpha]_D$ −115°, which is hydrolysed by boiling dilute acids to daphnetin and glucose. It is the 7-glucoside since methylation and subsequent removal of the sugar group gives 7-hydroxy-8-methoxycoumarin (*F. Wessely* and *K. Sturm*, Ber., 1930, **63,** 1299). Its hydrolysis by emulsion shows it to be a *β*-glucoside. The selective glycosidation and acetylation of daphnetin has been studied by *A. Grandini* (Gazz., 1940, **70,** 611) who found that daphnin could be prepared by attaching the sugar residue to 8-acetoxydaphnetin and then removing the protective acetyl group. **Sabandinone,** *7,8-methylenedioxycoumarin*, m.p. 154–155°, isolated from the ripe fruits of *Ruta pinnata*. Its structure has been confirmed by i.r., u.v., n.m.r. and mass spectroscopic data (*González, R. Estévez* and *I. Jaraiz*, An. Quim., 1970, **66,** 1017). 3-(1,1-*Dimethylallyl)daphnetin dimethyl ether* (LXXXVI), m.p. 85–87° and *8-methoxygravelliferone*, 3-(1,1-*dimethylallyl*)-7-*hydroxy*-6-*isopentenyl*-8-*methoxycoumarin*, m.p. 131–133°, have been isolated from *Ruta graveolens* L. (*J. Reisch et al.*, Tetrahedron Letters, 1970, 4305):

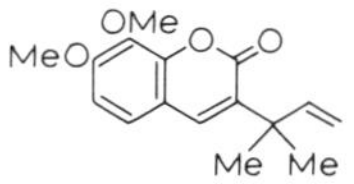

(LXXXVI)

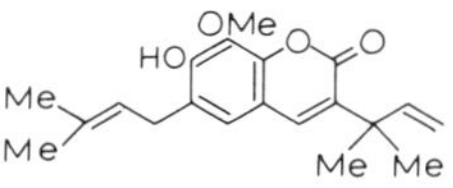

8-Methoxygravelliferone

Ellagic acid, m.p. >360°, is one of the most common yellow pigments in the plant world (*A. G. Perkin* and *M. Nierenstein*, J. chem. Soc., 1905, **87,** 1412; *F. Stitt et al.*, J. Amer. chem. Soc., 1959, **81,** 4615; *L. Judd, ibid.*, pp. 4606 and 4610). It undergoes *C*-benzylation giving **ellaborugin,** m.p. 220°, as one of several products *(Judd, loc. cit.)*:

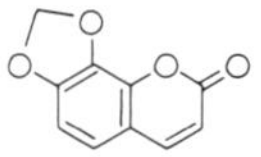

Sabandinone

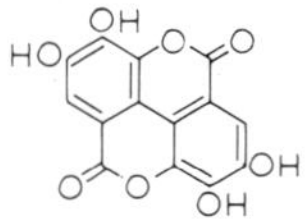

Ellagic acid

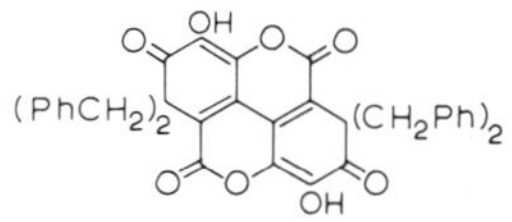

Ellagorubin

(ix) Trihydroxycoumarins and derivatives

Daphnoretin, *3-coumaroxy-7-hydroxy-6-methoxycoumarin*, m.p. 244–247° (crystals change at 220° approx.), isolated from *Daphne* spp. and identified by degradation and synthesis (*R. Tschesche, U. Schacht* and *G. Legler*, Ann., 1963, **662,** 113; *B. Kirkiacharian* and *C. Mentzer*, Bull. Soc. chim. Fr., 1966, 770):

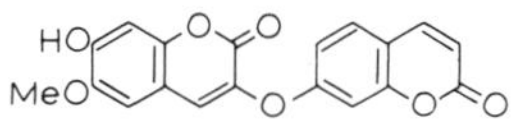

Daphnoretin

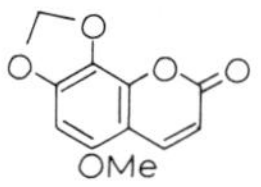

Sabandinin

Sabandinin, *5-methoxy-7,8-methylenedioxycoumarin*, m.p. 192–194°, constituent of the roots of *Ruta pinnata* L. (*González* and *Estévez*, An. Quim., 1971, **67,** 207; *González, Estévez* and *Jaraiz*, Phytochemistry, 1971, **10,** 1621). **Exostemin,** 5,7-dimethoxy-8-hydroxy-4-(4-methoxyphenyl)coumarin, isolated from *Exostemma caribaeum* (*F. Sanchez-Viesca, ibid.*, 1969, **8,** 1821).

Fraxetin, *7,8-dihydroxy-6-methoxycoumarin*, $C_{10}H_8O_5$, m.p. 227–228°, *acetate*, m.p. 192–193°, occurs in *Fraxinus excelsior* as the 9-glucoside, **fraxin,** m.p. 205° (*F. Wessely* and *E. Demmer*, Ber., 1929, **62,** 120). Fraxetin gives a bluish-green colour with ethanolic ferric chloride. Its structure has been confirmed by synthesis (*E. Späth* and

E. Dobrovolny, *ibid.*, 1938, **71**, 1831; *K. Aghoramurthy* and *T. R. Seshadri*, J. chem. Soc., 1954, 3065; *V. K. Ahluwalia et al.*, J. sci. ind. Research India, 1960, **19B**, 345). Also obtained from the bark of *Fraxinus excelsior* L. are **fraxidin,** 6,7-*dimethoxy-8-hydroxycoumarin*, m.p. 196–197°; **isofraxidin,** 6,8-*dimethoxy-7-hydroxycoumarin*, m.p. 148–149°, and **fraxinol,** 5,7-*dimethoxy-6-hydroxycoumarin*, m.p. 171–172°, *acetate*, m.p. 140–141° (*Späth et al.*, Ber., 1937, **70,** 698, 1019) which has been synthesised from bergopten (*A. Schönberg et al.*, J. Amer. chem. Soc., 1955, **77,** 5390). Fraxidin (*Ahluwalia*, *V. N. Gupta* and *Seshadri*, Tetrahedron, 1959, **5,** 90) and isofraxidin have been synthesised and some ring isomeric changes noted in related coumarins during the Dakin reaction (*Ahluwalia et al., loc. cit.).*

4-Methyl-5,6,7,8-tetrahydroxycoumarin and its partially and completely methylated derivatives have been synthesised (*R. M. Naik* and *V. M. Thakor*, J. org. Chem., 1957, **22,** 1626).

*(x) Furocoumarins**

Examples of the furocoumarins are found in psoralene and angelicin (in some literature called isopsoralene) which differ from each other in the points of attachment of the furan ring to the benzene nucleus of coumarin, the former being fused in a linear structure more common than the angular structures of which the latter is one example. Certain fish poisons contain the furocoumarin fragment:

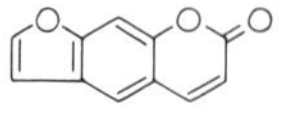

Psoralene 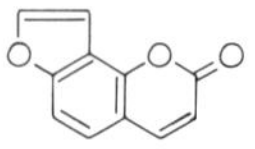 Angelicin

Some furocoumarins are of medicinal value, for example, the powdered fruit of *Ammi majus* L. has been used since olden times in the treatment of leukodermia and has been found to contain as principles xanthotoxin and imperatorin in the ratio 2:1 (*A. Schönberg* and *A. Sina*, J. Amer. chem. Soc., 1950, **72,** 4826). These substances are now administered in the pure form.

Much of our knowledge of the chemistry of these substances is due to the work of *E. Späth* and his collaborators (Monatsh., 1936, **69,** 75). Furocoumarins derived both from psoralene and angelicin had previously been synthesised by *D. B. Limaye* (Ber., 1932, **65,** 375; 1934, **67,** 12). In recent years a number of coumarins containing a reduced furan ring have been isolated from natural sources.

Psoralene, ficusin, 2H-*furo*[3,2-*g*]*chromen-2-one*, $C_{11}H_6O_3$, m.p. 171° (161–162, 165.5), was first isolated from an Indian plant, *Psoralea coryfolia* L. and later from the

---

A. Mustafa*, "Furopyrans and Furopyrones", The Chemistry of Heterocyclic Compounds, Interscience, New York, 1967, **23, 14; *F. M. Dean*, "Naturally Occurring Oxygen Ring Compounds", Butterworths, London, 1963, p. 198.

leaves of *Ficus carica*. It also occurs in most species of *Coronilla* in the form of the furocoumarinic acid, m.p. 125° (*A. Stoll et al.*, Helv., 1950, **33,** 1637). Its structure was determined by *Späth et al.*, (Ber., 1936, **69,** 1087).

Oxidation of psoralene affords furan-2,3-dicarboxylic acid (LXXXVII) and 2,4-dihydroxybenzene-1,5-dicarboxylic acid (LXXXVIII) isolated as the dimethoxy compound.

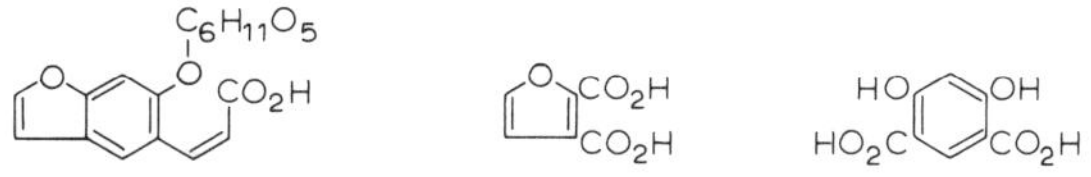

$C_6H_{11}O_5$ CO_2H Furocoumarinic acid; CO_2H CO_2H (LXXXVII); HO OH HO_2C CO_2H (LXXXVIII)

Confirmation of structure is afforded by the synthesis from 6-hydroxy-2,3-dihydrobenzo[*b*]furan (6-hydroxydihydrocoumarone) by treatment with malic acid and sulphuric acid to give the coumarin derivative, which by dehydrogenation yields psoralene:

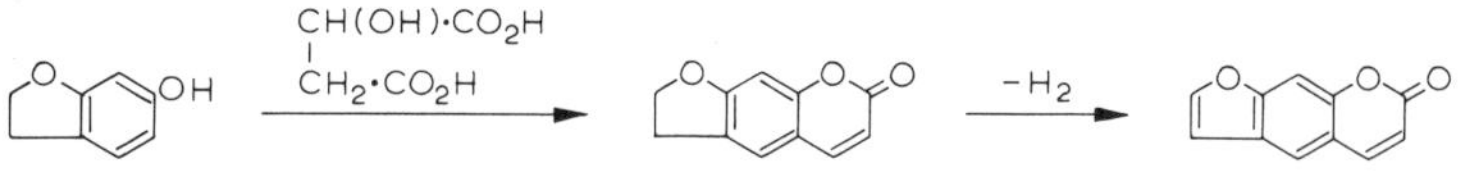

OH $\xrightarrow{CH(OH)\cdot CO_2H,\ CH_2\cdot CO_2H}$ → $\xrightarrow{-H_2}$

Psoralene has also been synthesised by another method related to the above synthesis (*E. C. Horning* and *D. B. Reisner*, J. Amer. chem. Soc., 1950, **72,** 1514); by a route involving the conversion of 6-dimethylallyl-7-hydroxycoumarin into psoralene (*R. Aneja, S. K. Mukerjee* and *T. R. Seshadri*, Tetrahedron, 1958, **4,** 256); and from *β*-resorcylaldehyde (*L. R. Worden et al.*, J. org. Chem., 1969, **34,** 2311).

3-*Methylpsoralene*, m.p. 235–235.5° (*Worden et al., loc. cit.*). A number of methyl- (*K. D. Kaufman*, J. org. Chem., 1961, **26,** 117), alkyl- and alkyl-4-phenyl-substituted psoralenes (*N. H. Pardanani* and *K. N. Trivedi*, J. Inst. Chem., Calcultta, 1970, **42,** 153; J. Indian chem. Soc., 1971, **48,** 339) have been synthesised.

Angelicin, 2H-*furo*[2,3-h]*chromen-2-one*, $C_{11}H_6O_3$, m.p. 139–140°, isomeric with psoralene and in some cases referred to as isopsoralene, was first isolated from *Angelica archangelica* L. and the wax of *Psoralea corylifolia* and has a characteristic odour. The presence of the furan and coumarin rings was shown by the oxidation of angelicin with alkaline hydrogen peroxide to furan-2,3-dicarboxylic acid, and with potassium permanganate to umbelliferone-8-carboxylic acid and thence to umbelliferone (7-hydroxycoumarin) (*Späth* and *O. Pesta*, Ber., 1934, **67**, 853). It can also be degraded to 2,4-dimethoxybenzene-1,3-dicarboxylic acid. The structure was confirmed by synthesis (*Späth* and *M. Pailer*, *ibid.*, 1934, **67,** 1212; 1935, **68,** 940; *Aneja, Mukerjee* and *Seshadri, loc. cit.; Y. Kawase, M. Nakayama* and *H. Tamatsukuri*, Bull. chem. Soc. Japan, 1962, **35,** 149). 4-*Methylangelicin*, m.p. 190–192° (*Kawase, Nakayama* and *Tamatsukuri, loc. cit.*); methyl- (*Kaufman, loc. cit.*), alkyl- and 3-phenyl-alkyl-substituted angelicins (*Pardanani* and *Trivedi, loc. cit.*).

A variety of psoralene and angelicin derivatives found in nature have for long been employed as fish poisons by the natives of Indonesia. The fish so poisoned are fit for human consumption.

Xanthotoxin, *9-methoxypsoralene, 9-methoxy-2H-furo[3,2-g]chromen-2-one*, $C_{12}H_8O_4$, m.p. 146°, occurs in the pods of *Fagara xanthoxyloides* (*H. Thoms*, Ber., 1911, **44,** 3325; *Späth* and *Pailer, ibid.*, 1936, **69,** 767), and has been found in domestic celery, which has developed pink-rot disease, caused by the fungus *Sclerotinia sclerotiorum* (*L. D. Scheel et al.*, Biochemistry, 1963, **2,** 1127). It is demethylated by aniline hydrochloride to **xanthotoxol,** m.p. 251–252° (sealed tube) (*Schönberg* and *G. Aziz*, J. Amer. chem. Soc., 1953, **75,** 3265; *C. Lagercrantz*, Acta Chem. Scand., 1956, **10,** 647), which on methylation with diazomethane or methyl iodide and potassium carbonate regenerates xanthotoxin (*Späth* and *F. Vierhapper*, Ber., 1937, **70,** 248; *Schönberg* and *A. Sina, loc. cit.*). It has been synthesised from 2,3-dihydro-6,7-dihydroxybenzo[*b*]furan-3-one (6,7-dihydroxycoumaran-3-one) (*Späth* and *Pailer, loc. cit.; Lagercrantz, loc. cit.*) and from 7-hydroxy-8-methoxycoumarin (*G. Rodighiero* and *C. Antonello*, Ann. chim., Rome, 1956, **46**, 960). Reactions of xanthotoxin including oxidation, reduction, nitration, halogenation, sulphonation and demethylation have been studied (*M. E. Brokke* and *B. E. Christensen*, J. org. Chem., 1958, **23,** 589; 1959, **24,** 523). Xanthotoxin on irradiation in dioxane or benzene gives an *anti*-head-head-$C_{(4)}$-cyclodimer LXXXIX (*C. H. Krauch* and *S. Farid*, Ber., 1967, **100,** 1685):

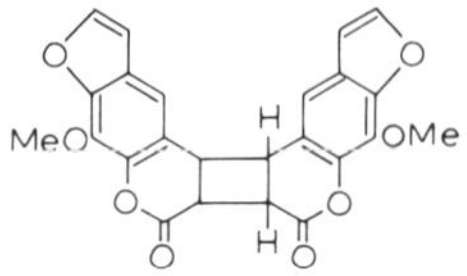

MeO H OMe H

(LXXXIX)

Some reactions and derivatives of xanthotoxol have been discussed (*E. A. Abu-Mustafa, B. A. H. El-Tawil* and *M. B. E. Fayez*, Indian J. Chem., 1967, **5,** 283).

Imperatorin, ammidin, *9-(3-methyl-2-butenyloxy)psoralene, 9-(3-methyl-2-butenyloxy)-2H-furo[3,2-g]chromen-2-one*, $C_{16}H_{14}O_4$, m.p. 102°, is found in the roots of *Imperatoria ostruthium*, and has been shown to have the assigned structure by degradative experiments such as the breakdown by sulphuric acid in acetic acid to xanthotoxol and 3-methylbut-2-en-1-ol (*Späth* and *H. Holzen*, Ber., 1933, **66,** 1137). It has been synthesised by the interaction of the sodium salt of xanthotoxol and γ,γ-dimethylallyl bromide (*Späth* and *Holzen, ibid.*, 1935, **68,** 1123; *Späth* and *Vierhapper, loc. cit.*). When distilled it undergoes the Claisen rearrangement to give the phenol, *alloimperatorin*, m.p. 233°. Some reactions and derivatives of imperatorin have been described (*Abu-Mustafa, El-Tawil* and *Fayez, loc. cit.*).

Bergapten, *5-methoxypsoralene, 5-methoxy-2H-furo[3,2-g]chromen-2-one*, m.p. 190–191°, is isomeric with xanthotoxin (*Thoms* and *E. Baetcke*, Ber., 1912, **45,** 3705; *Späth* and *S. Rashka, ibid.*, 1934, **67,** 62) and has been synthesised (*Späth et al., ibid.*, 1937, **70,** 478; *W. N. Howell* and *A. Robertson*, J. chem. Soc., 1937, 293; *G. Caporale*, Ann. chim., Rome, 1958, **46,** 650; *V. K. Ahluwalia et al.*, Indian J. Chem., 1969, **7,** 831) along with some of its derivatives (*H. M. N. El-Din* and *H. M. Safwat*, U.A.R.J. pharm. Sci., 1970, **11,** 45). The demethylated compound *bergaptol*, $C_{11}H_6O_4$, m.p. 281°, occurs in bergamot oil (*Späth* and *L. Socias*, Ber., 1934, **67,** 59). It yields bergapten when methylated with diazomethane and has been synthesised (*Späth* and *G.*

Kubiczek, *ibid.*, 1937, **70**, 1253). 2,9-Dimethyl-5-methoxypsoralen (*V. N. Dholakia* and *Trivedi*, J. Indian chem. Soc., 1970, **47**, 1058):

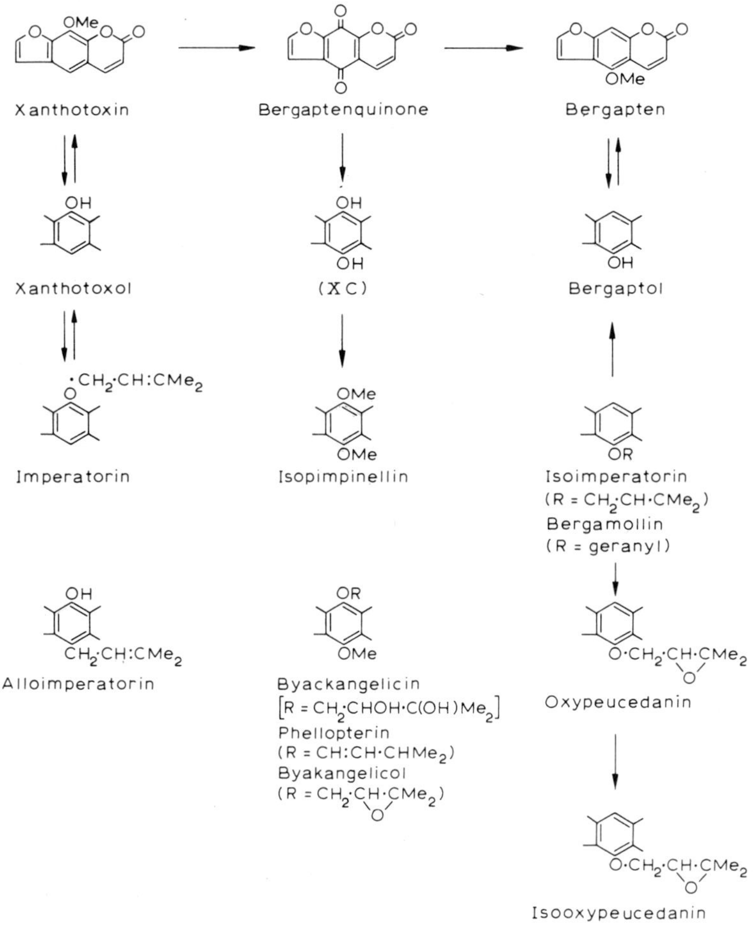

Xanthotoxin → Bergaptenquinone → Bergapten

Xanthotoxol (XC) Bergaptol

Imperatorin ($O\cdot CH_2\cdot CH{:}CMe_2$) Isopimpinellin Isoimperatorin ($R = CH_2{:}CH\cdot CMe_2$) Bergamollin (R = geranyl)

Alloimperatorin ($CH_2\cdot CH{:}CMe_2$) Byackangelicin [$R = CH_2\cdot CHOH\cdot C(OH)Me_2$] Phellopterin ($R = CH{:}CH\cdot CHMe_2$) Byakangelicol ($R = CH_2\cdot CH\cdot CMe_2$, epoxide) Oxypeucedanin

Isooxypeucedanin

Isoimperatorin, 5-(3-*methyl*-2-*butenyloxy*)-2H-*furo*[3,2-g]*chromen*-2-*one*, $C_{16}H_{14}O_4$, m.p. 109°, isolated from *Imperatoria ostruthium*, and whose constitution was shown by hydrolytic breakdown with sulphuric acid in acetic acid to bergaptol and an unsaturated alcohol which on hydrogenation yielded isoamyl alcohol (*Späth* and *L. Kahovec*, Ber., 1933, **66**, 1146) and by synthesis (*Späth* and *E. Dobrovolny*, *ibid.*, 1939, **72**, 52). **Auraptin,** obtained from orange oil has been shown to be isoimperatorin (*T. Matsuno*, J. pharm. Soc. Japan, 1956, **76**, 1136). Closely related to isoimperatorin is **oxypeucedanin,** $C_{16}H_{14}O_5$, m.p. 142–143°, which occurs in *Peucedanum officinale* (*Späth* and *K. Klager*, Ber., 1933, **66,** 914). Its structure was confirmed by synthesis by the oxidation of isoimperatorin by perbenzoic acid (*Späth* and *Kahovec*, Arch. Pharm., 1935, **273**, 223; *Späth* and *Dobrovolny*, *loc. cit.*). Oxypeucedanin is isomerised by phosphorus pentoxide in boiling toluene to **isooxypeucedanin,** m.p. 148°, *oxime*, m.p. 186°, the epoxy side-chain being converted into a ketonic grouping (*Späth* and *Klager*, Ber., 1933, **66,** 914, 921). **Bergamollin,** *bergaptyl geranyl ether*, 5-*geranyloxy*-2H-*furo*[3,2-g]*chromen*-2-*one*, $C_{21}H_{22}O_4$, m.p. 59–61°, obtained from bergamot oil occurring in *Citrus bergamia* Risso (*Späth* and *P. Kainrath*, *ibid.*, 1937, **70,** 2272). It has been synthesised from bergaptol and geranyl chloride in ethanol in the presence of sodium ethoxide (*A. Chatterjee* and *B. Chaudhury*, J. chem. Soc., 1961, 2246).

Isopimpinellin, 5,9-*dimethoxypsoralene*, 5,9-*dimethoxy*-2H-*furo*[3,2-g]*chromen*-2-*one*, $C_{13}H_{10}O_5$, yellow crystals, m.p. 148° (151°), which occurs in *Pimpinella sacifraga* L. and in *Heracleum sphondylium* (*F. Wessely et al.*, Monatsh., 1932, **59,** 161; 1932, **60,** 141) and has been obtained from West Indian lime oil (from *Citrus aurantifolia*, Swingle) (*A. G. Caldwell* and *E. R. H. Jones*, J. chem. Soc., 1945, 540). It can be obtained by oxidising the amino derivative of xanthotoxin or bergapten to bergaptenquinone, reducing this to the quinol XC, and finally methylating. **Halkendin,** 3,4-*dimethoxypsoralene*, m.p. 173–173.5°, has been found in the bark of *Halfordia kendack* (*Nakayama et al.*, Austral. J. Chem., 1971, **24,** 209). **Halfordin,** 3,5,9-*trimethyoxypsoralene*, m.p. 136–137°, has been isolated from the bark of *Halfordia scleroxyla* (*M. P. Hegarty* and *P. N. Lahey*, *ibid.*, 1956, **9,** 120: *S. Bhanu et al.*, Indian J. Chem., 1971, **9,** 380).

The seeds of the Mexican tree *Casimiroa edulis* Llave et Lex. contain 9-*hydroxy*-5-*methoxypsoralene*, 9-*hydroxy*-5-*methoxy*-2H-*furo*[3,2-g]*chromen*-2-*one*, yellow needles, m.p. 223–224°, *acetate*, m.p. 181–182°, *benzoate*, m.p. 203–205° (*F. A. Kincl et al.*, J. chem. Soc., 1956, 4163). It is oxidised by chromic acid to bergaptenquinone and is methylated to isopimpinellin. It has also been prepared from 9-aminobergapten by the action of nitrous acid and by the hydrolysis of **byakangelicin,** $C_{17}H_{18}O_7 \cdot H_2O$, pale yellow needles, m.p. 117–118°, 125–126° (anhydrous), $[\alpha]_D$ +24.62° (pyridine), which occurs in *Angelica glabra* Makino (*T. Noguchi* and *M. Kawanami*, Ber., 1938, **71,** 344; *Noguchi*, Reports Jap. Assocn. Advancement of Science, 1943, **17,** 234; C.A., 1950, **44,** 3989) along with **phellopterin,** $C_{17}H_{16}O_5$, m.p. 102° (*Noguchi* and *Kawanami*, J. pharm. Soc. Japan, 1940, **60,** 57; 1941, **61,** 77) and **byakangelicol,** m.p. 106°, $[\alpha]_D$ +35.8° (pyridine) (*Noguchi* and *Kawanami*, Ber., 1938, **71,** 344; 1939, **72,** 483).

Derived from angelicin are isobergapten and pimpinellin both of which occur in *Pimpinella saxifraga* (*Wessely*, *loc. cit.*). **Isobergapten,** 5-*methoxyangelicin*, m.p. 222°, has been synthesised from bergaptol (*Späth* and *Kubiczek*, Ber., 1937, **70**, 1253; *Ahluwahlia et al.*, Indian J. Chem., 1969, **7**, 115); **pimpinellin**, 5,6-*dimethoxyangelicin*, m.p. 119° (*Wessely et al.*, *loc. cit.*). **Allobergapten,** 8-*methoxy*-2H-*furo*[2,3-f]*chromen*-2-*one*, m.p. 207° (*Späth*, *Wessely* and *Kubiczek*, Ber., 1937, **70,** 243) has been synthesised (*R. T. Foster*, *Howell* and *Robertson*, J. chem. Soc., 1939, 930; *Ahluwahlia et al.*, *loc. cit.*). **Archangelin,** $C_{21}H_{22}O_4$, m.p. 132°, has been isolated from the root of *Angelica archangelica* L. (*Chatterjee* and *S. S. Gupta*, Tetrahedron Letters, 1964, 1961):

Archangelin

Sphondin, 6-*methoxyangelicin*, m.p. 189–191°, and the so-called sphondylin have been isolated from *Heracleum sphondylium* L. (*Späth* and *A. F. J. Simon*, Monatsh., 1936, **67,** 344; *Späth* and *H. Schmid*, Ber., 1941, **74,** 595). "Spondylin" is a mixture of sphondin and bergapten (*Wessely* and *J. Kotlan*, Monatsh., 1955, 86, 430):

Sphondin

Heraclenin
(R = $CH_2{\cdot}CH{\cdot}CMe_2$ with O bridging CH and CMe₂)

Heraclenol
[R = $CH_2{\cdot}CH(OH){\cdot}C(OH)Me_2$]

Heraclenin, prangenin, 8-(2,3-*epoxy*-3-*methylbutoxyl*)*psoralene*, m.p. 111°, $[\alpha]_D^{32}$ +22 (pyridine), u.v. spectrum similar to that of imperatorin, has been isolated from the roots of *Prangos pabularia* and may be obtained by reacting imperatorin with perbenzoic acid. On treatment with sulphuric acid in acetic acid it gives xanthotoxol (*Späth* and *H. Holzen*, Ber., 1935, **68,** 1123; *G. A. Kuznetsova* and *G. V. Pigulevskii*, Zhur. obshcheĭ Khim., 1961, **31,** 323; *Y. N. Sharma*, *A. Zaman* and *A. R. Kidawi*, Tetrahedron, 1964, **20,** 87). **Heraclenol,** 8-(2,3-dihydroxy-3-methylbutoxy)psoralen, obtained from *Heracleum candicans* (*Sharma et al.*, Naturwiss., 1964, **51,** 537). 5-(2,3-*Epoxy*-3-*methylbutyl*)-9-*methoxypsoralene*, m.p. 103–104°, has been isolated from *Thamnosma montana* (*J. P. Kutney*, *R. N. Young* and *A. K. Verma*, Tetrahedron Letters, 1969, 1845).

Some naturally occurring furocoumarins contain substituents in the furan ring. **Peucedanin,** 7-*isopropyl*-6-*methoxypsoralene*, 7-*isopropyl*-6-*methoxy*-2H-*furo*[3,2-g]-*chromen*-2-*one*, $C_{15}H_{14}O_4$, m.p. 109°, is obtained from the roots of *Peucedanum officinale* (*Späth* and *Klager*, Ber., 1933, **66,** 749). It is an enol-ether and in consequence is readily demethylated by acids giving **oreoselone,** 6,7-*dihydro*-7-*isopropyl*-2H-*furo*-[3,2-g]*chromen*-2,6-*dione*, $C_{14}H_{12}O_4$, m.p. 177–178°, which also occurs in *P. officinale* and whose structure has been established by degradative oxidation. With potassium permanganate one of the products obtained is 4,6-dihydroxyisophthalic acid, suggesting that oreoselone is a 7-hydroxycoumarin derivative. Oxidation of dihydro-oreoselone gives succinic acid thus confirming the presence of a coumarin nucleus:

Peucedanin ⇌ Oreoselone

Oreoselone has been synthesised (*F. v. Bruchhausen* and *H. Hoffmann*, Ber., 1941, **74,** 1584; *K. N. Gaind et al.*, J. Indian chem. Soc., Ind. News Edit., 1946, **23,** 370). It is only slightly enolised in solution and is methylated to peucedanin only by the unusual reagent methanolic aluminium chloride (*H. Schmid* and *A. Ebnöther*, Helv., 1951, **34,** 1982).

Oroselone, kvanin, 8-*isopropenylangelicin*, 8-*isopropenyl*-2H-*furo*[2,3-h]*chromen*-2-*one*, $C_{14}H_{10}O_3$, m.p. 188–189° (evacuated tube), $[\alpha]_D^{22}$ +96°, reduced by hydrogen in the presence of palladium to the *dihydro* compound, m.p. 142°, which on ozonolysis yields umbelliferone-8-carbaldehyde and isobutyric acid (*Späth et al.*, Ber., 1940, **73,** 709; 1941, **74,** 595); and *oroselol*, $C_{14}H_{12}O_4$, are found with athamantin in *Athamanta oreoselinum* L.

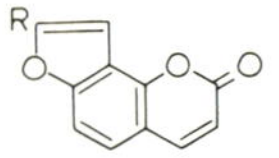

Oroselone (R = CH_2:ĊMe)
Oroselol [R = ·C(OH)Me_2]

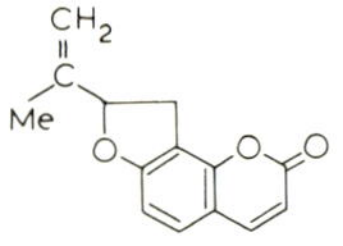

Majurin

There are a number of naturally occurring furocoumarins which contain a dihydrofuran ring. **Majurin,** 8,9-*dihydro*-8-*isopropenyl*-2H-*furo*[2,3-h]*chromen*-2-*one*, $C_{14}H_{12}O_3$, m.p. 95–96°, has been isolated from the fruits of *Ammi majus* L. (*Aby-Mustafa, F. K. A. El-Bay* and *Fayez*, Tetrahedron Letters, 1971, 1657). **Athamantin,** 8,9-*dihydro*-9-*hydroxy*-8-(1-*hydroxy*-1-*methylethyl*)-2H-*furo*[2,3-h]*chromen*-2-*one diisovalerate*, $C_{24}H_{30}O_7$, $[\alpha]_D^{22}$ +96° (MeOH) (*Späth* and *Schmid*, Ber., 1940, **73,** 1309; *Schmid et al.*, Helv., 1957, **40,** 758). Its structure follows from degradations, hydrochloric acid and methanol giving oroselone and 2 mol, of isovaleric acid, and sodium methoxide, depending on the conditions, affording either the methyl ester of the coumaric acid XCI or the furocoumarin XCII:

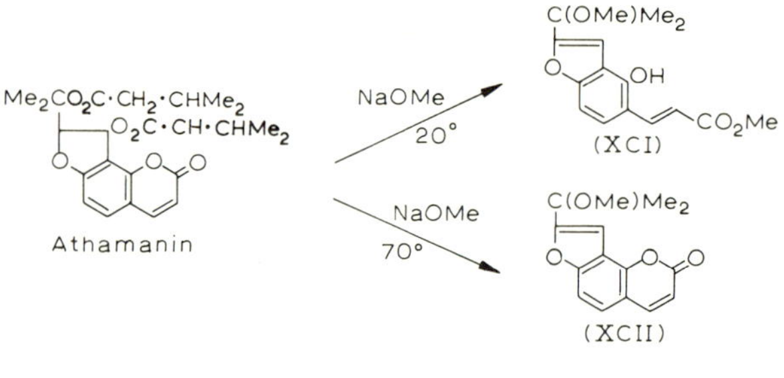

Athamanin — NaOMe, 20° → (XCI); NaOMe, 70° → (XCII)

Edultin, 8-(1-*acetoxyisopropyl*)-9-*angeloyloxy*-2H-*furo*[2,3-h]*chromen*-2-*one*, $C_{21}H_{22}O_7$, m.p. 136–142°, $[\alpha]_D^{10.7}$ 41.5° (pyridine), isolated from the root of *Angelica edulis.* On treatment with 7.5% methanolic sodium hydroxide it gives angelicin, oroselone, oroselol methyl ether, acetic acid and angelic acid (*H. Mitsuhashi* and *T. Itoh*, Chem. and pharm. Bull., Tokyo, 1961, **9,** 170; 1962, **10,** 511). The absolute configurations of athamantin and edultin have been discussed (*M. Nakazaki, Y. Hirose* and *K. Ikematsu*, Tetrahedron Letters, 1966, 4735):

Me_2CO_2CMe Me
$O_2C{\cdot}\dot{C}{:}CHMe$

Edultin

Columbianadin, 8-(1-*angeloyloxy*-1-*methylethyl*)-8,9-*dihydro*-2H-*furo*[2,3-h]*chromen*-2-*one*, $[\alpha]_D^{27}$ +26.5° (dioxane) and **columbianin,** 8,9-*dihydro*-8-(1-*glucosyloxy*-1-*methylethyl*)-2H-*furo*[2,3-h]*chromen*-2-*one*, $[\alpha]_D^{23}$ +118° (H_2O), have been isolated from *Lomatium columbianum* and on alkaline hydrolysis they both afford the same alcohol, **columbianetin** and tiglic acid (angelic acid readily isomerises to tiglic acid) and glucose, respectively (*R. E. Willette* and *T. O. Soine*, J. pharm. Sci., 1964, **53,** 275). *Columbianadin oxide* (R = $-\overset{Me}{\dot{C}}-CHMe$, with O bridge) m.p. 97°, $[\alpha]_D^{28}$ 305° (MeOH), along with other coumarins has been obtained from the fruits of *Peucedanum palustre* (*B. E. Nielsen* and *J. Lemmich*, Acta Chem. Scand., 1965, **19,** 1810).

Me_2COCO_2R → Me_2COH

Columbianadin
(R = $-\underset{Me}{\dot{C}}{:}CHMe$)
Columbianin
(R = D-glucosyl)

Columbianetin

(±)-Columbianetin has been synthesised (*M. Shipchandler, Soine* and *P. T. Gupta*, J. pharm. Sci., 1970, **59,** 67). Archangelicin, 9-*angeloxyloxy*-8-(1-*angeloyloxy*-1-*methylethyl*)-8,9-*dihydro*-2H-*furo*[2,3-h]chromen-2-one, $[\alpha]_D^{26}$ +112.7° (MeOH), has been obtained from the root of *Angelica archangelica* subsp. *litoralis* (*Nielsen* and *Lemmich*, Acta Chem. Scand., 1964, **18,** 932), and on treatment with sodium methoxide it gave oroselone, oroselol methyl ether and the coumaric ester XCI, also obtained by similar treatment of athamantin. A study has been made of the configuration of archangelicin and related coumarins (*idem, ibid.*, 1964, **18,** 2111):

$Me_2CO_2C{\cdot}\overset{Me}{\dot{C}}{:}CHMe$
$O_2C{\cdot}\underset{Me}{\dot{C}}{:}CHMe$

Archangelicin

A number of coumarins with a complex structure and containing a 6,7-dihydro-2*H*-furo[2,3-*f*]chromen-2-one ring system have been isolated from the seeds of *Mammea americana* L. (*L. Crombie et al.*, Tetrahedron Letters, 1970, 3975).

Nieshoutaol, $C_{15}H_{16}O_5$, m.p. 143–144°, *acetate*, m.p. 126–128°, obtained from

Ptaeroxylan utile (sneezewood) (*R. D. H. Murray* and *M. M. Ballantyne*, *ibid.*, 1969, 4031):

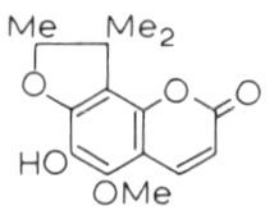

Nieshoutol

3-*Methyl*-2H-*furo*[3,2-c]*chromen*-2-*one* (XCIII), m.p. 154–156°, is prepared by boiling 4-acetonyloxycoumarin, polyphosphoric acid, and a drop of phosphoryl chloride in sodium-dried xylene (*F. M. Dean et al.*, J. chem. Soc., C, 1967, 2232). Its mass spectrum has been compared with that of the corresponding pyronocoumarin:

(XCIII)

Furocoumarins have been synthesised by thermal or acid-catalysed condensation of β-dicarboxyl or β-carboxycarbonyl compounds with 4-, 5-, 6- or 7-hydroxy-2,3-dimethylbenzofurans (*J. P. Lechartier et al.*, Bull. Soc. chim. Fr., 1966, 1716); angular furocoumarins have been obtained from 6-hydroxy-4-methylcoumarin carrying a substituent in the 5-position, and linear isomers by blocking the reactive 5-position and from 5-hydroxy-4-methylcoumarin, which eliminates the need for a block group (*K. D. Kaufman et al.*, J. org. Chem., 1962, **27**, 2567); 4-hydroxy-2*H*-furo[2,3-*h*]-chromen-2-one and its 3-phenyl- and 3-(2-methoxyphenyl) derivatives have been prepared (*Y. Kawase, M. Nanbu* and *H. Yanagihara*, Bull. chem. Soc. Japan, 1968, **41**, 1201); 9-methyl-, 9-ethyl- and 9-propyl-2*H*-furo[3,2-*g*]chromen-2-ones have been obtained by condensing the appropriate 6-acetoxy-7-alkyl-2,3-dihydrobenzofuran with acrylonitrile in the presence of zinc chloride and hydrochloric acid, and then dehydrogenating the product over a 10% palladium/carbon catalyst (*D. K. Chatterjee* and *K. Sen*, Indian J. Chem., 1971, **9**, 31):

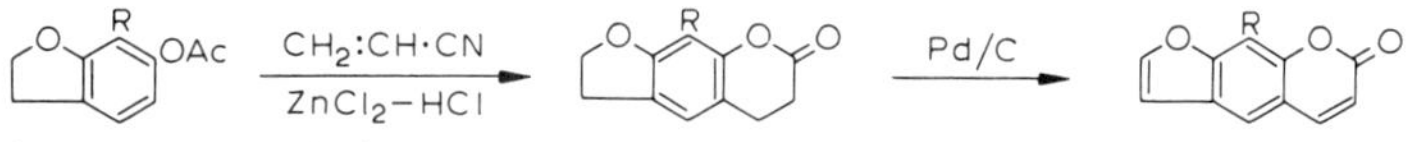

CH_2:CH·CN, $ZnCl_2$–HCl; Pd/C

(R = Me, Et or Pr)

Various methods used to synthesise furocoumarins have been discussed (*Trivedi*, Symp. Syn. Heterocyclic Compounds Physiol. Interest, Hyderabad, India, 1964, 17). The i.r. (*T. V. Bukreeva* and *G. V. Pigulevskii*, Zhur. prikl. Khim., 1966, **39**, 1541), u.v. (*Bukreeva*, *ibid.*, p. 1653), and i.r., u.v. and n.m.r. (*K.-H. Lee* and *Soine*, J. pharm. Sci., 1969, **58**, 681) spectra of a number of furocoumarins have been reported and discussed.

(xi) Pyranocoumarins

The 2,2-dimethylpyran ring occurs in a variety of naturally occurring compounds (p. 63) and three coumarin derivatives containing this ring are **xanthyletin,** 8,8-*di*-

methyl-2H,8H-*benzo*[1,2-b:5,4-b]*dipyran-2-one*, $C_{14}H_{12}O_3$, m.p. 128–128.5°, **xanthoxyletin,** $C_{15}H_{14}O_4$, m.p. 133° (*J. C. Bell* and *A. Robertson*, J. chem. Soc., 1936, 1828; *Robertson* and *T. S. Subramaniam*, *ibid*., 1937, 286; *Bell et al*., 1937, 1542; *E. Späth* and *R. Hillel*, Ber., 1939, **72,** 2093; *L. H. Briggs* and *R. H. Locker*, J. chem. Soc., 1951, 3131), and **alloxanthoxyletin,** $C_{15}H_{14}O_4$, pale yellow prisms, m.p. 115.5° (*Robertson* and *Subramaniam*, *ibid*., 1937, 1545). They occur in *Xanthoxylum americanum* and are characterised by the comparative ease with which they yield acetone by alkaline fission. Xanthyletin has been synthesised along with its 3,4-dihydro- and 3,4,6,7-tetrahydro derivatives (*A. K. Das Gupta* and *K. R. Das*, J. chem. Soc., C, 1969, 33; *P. W. Austin* and *T. R. Seshadri*, Indian J. Chem., 1968, **6**, 412):

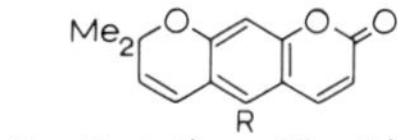

Xanthyletin (R = H)
Xanthoxyletin (R = OMe) Alloxanthoxyletin 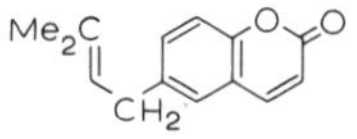 Suberosin

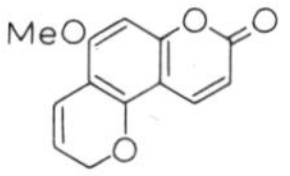

Structurally related is **suberosin,** $C_{15}H_{16}O_3$, m.p. 87.5°, which is obtained from the bark of *Z. suberosum* (*J. Ewing et al*., Austral. J. Sci. Res., 1950, **3A**, 342) and *Z. flavum* (*F. E. King et al*., J. chem. Soc., 1954, 1392). With hydrobromic acid and phosphorus it yields dihydroxanthyletin. Suberosin and 7-*dimethylsuberosin*, m.p. 131–133°, have been synthesised (*Austin* and *Seshadri*, *loc. cit.*). 4,6-*Dimethylxanthyletin* (XCIV), m.p. 172–173°, has been obtained by the condensation of 7-hydroxy-2,2,4-trimethyl-2*H*-chromene with ethyl acetoacetate in nitrobenzene in the presence of aluminium chloride; 6,7-*dihydro*-4,6-*dimethylxanthyletin* (XCV), m.p. 155–156° (*F. N. Faquih* and *R. N. Usgaonkar*, Curr. Sci., 1970, **39,** 61):

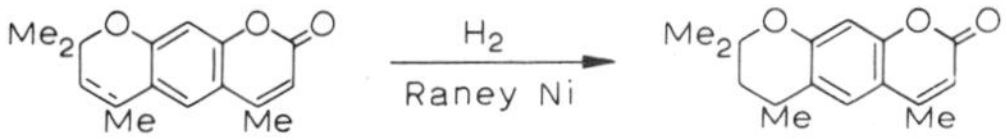

(XCIV) —H_2, Raney Ni→ (XCV)

The fruit of *Luvanga scandens*, an important drug in Hindu medicine, contains xanthotoxin, zanthyletin and **luvangetin,** $C_{15}H_{14}O_4$, m.p. 108–109°, which has been shown by synthesis and degradation to be isomeric with xanthoxyletin (*Späth et al*., Ber., 1940, **73,** 1361; 1941, **74,** 193):

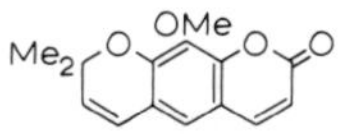

Luvangetin 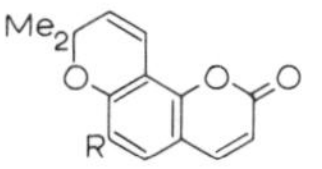 Seselin (R = H)
Braylin (R = OMe)

Seselin, amyrolin, $C_{14}H_{12}O_3$, m.p. 120°, obtained from the fruits of *Seseli indicum*, possesses an angular ring system (*Späth et al*., Ber., 1939, **72,** 821, 2093); its crystal structure has been determined (*K. Kato*, Acta Crystallogr., Sect. B, 1970, **26,** 2022); *dihydroseselin*, m.p. 103–104° (*Austin* and *Seshadri*, *loc. cit.*; *K. H. Boltze* and *H. D. Dell*, Angew. Chem. internat. Edn., 1966, **5,** 415). **Braylin,** m.p. 150°, *dibromide*, m.p.

196° (decomp.), occurs in the bark of *Flindersia brayleyana* F. Muell. and is a methoxy-seselin (*F. A. L. Anet, G. K. Hughes* and *E. Ritchie*, Austral. J. Sci. Res., 1949, **2A,** 608).

Mammeigin, $C_{25}H_{24}O_5$, m.p. 144–146°, has been isolated from the seed oil of *Mammea americana* L. and shown to be 5-*hydroxy*-6-*isovaleryl*-8,8-*dimethyl*-4-*phenyl*-2H,8H-*pyrano*[2,3-h]*chromen*-2-*one*. The structure is assigned not only on spectroscopic evidence, but also on a chemical interrelation of mammeigin with mammeisin (p. 121) (*R. A. Finnegan* and *W. H. Mueller*, J. org. Chem., 1965, **30**, 2342):

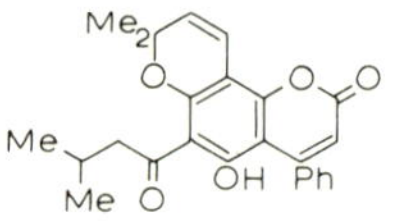

Mammeigin

A number of naturally occurring coumarins containing a 3,4-dihydro-2,2-dimethyl-pyran ring have been isolated; **decursin,** 6,7-*dihydro*-8,8-*dimethyl*-7-*senecioyloxy*-2H,8H-*pyrano*[3,2-g]*chromen*-2-*one*, $C_{19}H_{20}O_5$, m.p. 110–111°, $[\alpha]_D^{15}$ +172.9° ($CHCl_3$), from the root of *Angelica decursiva* (*K. Hata* and *K. Sano*, Tetrahedron Letters, 1966, 1461):

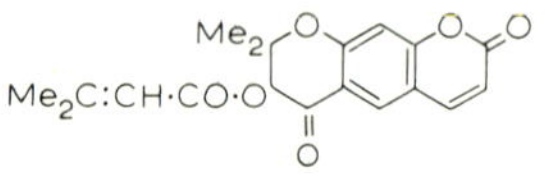

Decursin 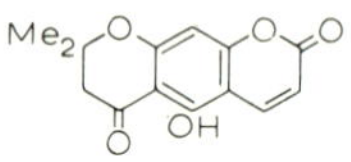 Clausenin 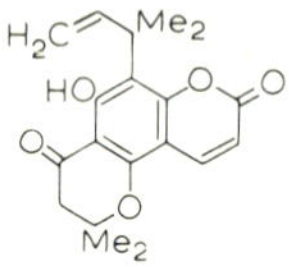 (XCVI)

Clausenin, $C_{14}H_{12}O_5$, m.p. 156–157°, and **clausenidin,** $C_{19}H_{20}O_5$, m.p. 136–137°, from the root of *Clausena heptaphylla*, the structure of the former has been confirmed by synthesis and one, XCVI, suggested for the latter compound (*B. S. Joshi* and *V. N. Kamat*, *ibid.*, p. 5767); **peuformosin,** from the root of *Peucedanum formosanum* (*Hata, M. Kozawa* and *K. Y. Yen*, Chem. pharm. Bull., Tokyo, 1966, **14,** 442. The following structures have been suggested for **lomatin,** (+)-9,10-dihydro-9-hydroxy-seselin, $C_{14}H_{14}O_4$, m.p. 187–188° (183°), $[\alpha]_D^{25}$ +74.8° (EtOH), from *Lomatium nuttallii* (*T. O. Soine* and *F. H. Jawad*, J. pharm. Sci., 1964, **53,** 990; *J. Lemmich* and *B. E. Nielsen*, Tetrahedron Letters, 1969, 3):

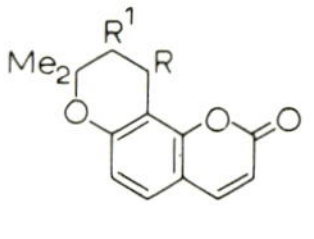

Lomatin (R = H, R[1] = OH)
Peuformosin (R = O·CO·CH:CMe2, R[1] = O·CO·CMe:CHMe)
Anomalin (R = R[1] = O·CO·CMe:CHMe)
Floroselin (R = O·CO·CH:CHSMe, R[1] = O·CO·CMe:CHMe)
Jatamansin (R = H, R[1] = O·CO·CMe:CHMe)
Samidin (R = OAc, R[1] = O·CO·CH:CMe)

Anomalin, m.p. 173–174°, $[\alpha]_D$ −78.4° (EtOH), from *Angelica anomala* (*Hata, Kozawa* and *Y. Ikeshiro*, Chem. Pharm. Bull. Tokyo, 1966, **14,** 94); **selinidin,** m.p. 97–98°, $[\alpha]_D^{29}$ +20.3° (dioxane), from the root of *Selinum vaginatum*, on hydrolysis with alkali affords tiglic acid and selinetin, which has been shown to be identical with lomatin

(*T. R. Seshadri et al.*, Tetrahedron Letters, 1964, 3367); **floroselin,** from the root of *Seseli sessiliflorum* (*A. A. Savina et al.*, Khim. Prir. Soedin., 1970, **6,** 517); **jatamansin,** $C_{19}H_{20}O_5$, m.p. 97–98°, from the oil obtained from the root of *Nardostachys jatamansi* (*S. N. Shanbhag et al.*, Tetrahedron, 1964, **20,** 2605); and *calophyllolide*, m.p. 158–160°, from the nut of *Calophyllum inophyllum*, alkaline hydrolysis gives 5-*hydroxy*-7-*methoxy*-4-*phenylcoumarin*, m.p. 205–207°, and other products (*J. Polonsky et al.*, Bull. Soc. chim. Fr., 1953, 546; 1954, 925; 1955, 541; 1956, 914; 1957, 929). **Samidin,** $C_{21}H_{22}O_7$, m.p. 135°, **visnadin,** $C_{21}H_{22}O_7$, m.p. 87°, and **dihydrosamidin,** $C_{31}H_{24}O_7$, m.p. 113°, are similar compounds obtained from *Ammi visnega* L. and like some of the above compounds are diesters of khellactone (*E. Smith et al.*, J. Amer. chem. Soc., 1957, **79,** 3534; *H. D. Schroeder et al.*, Ber., 1959, **92,** 2338):

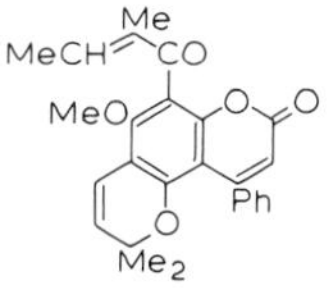

Calophyllolide

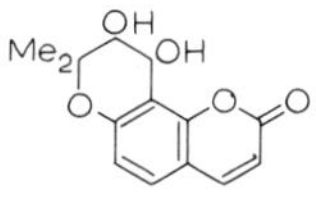

Khellactone

The stereochemistry of natural coumarins containing the 3-hydroxy-2,2-dimethylchroman system has been discussed (*Lemmich* and *Nielsen, loc. cit.*).

The Pechmann condensation of malic acid with 7-hydroxy-2,2-dimethylchroman-4-one gives 7,8-dihydro-6,6-dimethyl-2*H*,6*H*-pyrano[2,3-*f*] chromene-2,8-dione (*A. S. Mujumdar* and *R. N. Usgaonkar*, Indian J. Chem., 1971, **9,** 294):

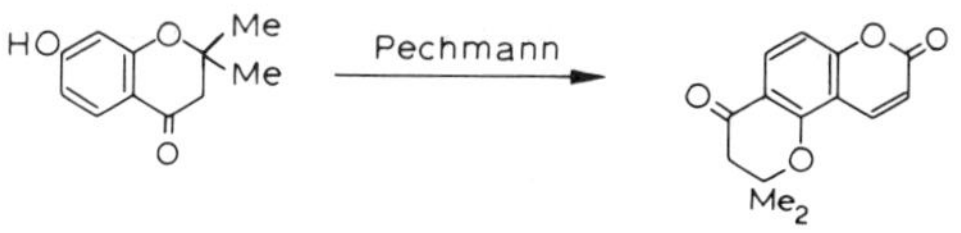

Related compounds have been prepared, and some of them have been converted to pyranocoumarins by treatment with sodium tetrahydridoborate followed by sulphuric acid. Some 2*H*,5*H*-pyrano[3,2-*c*]chromene-2,5-diones are obtained by the Pechmann condensation of 4-hydroxycoumarins with ethyl acetoacetate or other *β*-ketonic esters (*J. Payell* and *Usgaonkar*, J. Indian chem. Soc., 1966, **43,** 536). The mass spectrum of 4-methyl-2*H*,5*H*-pyrano[3,2-*c*]chromene-2,5-dione (XCVII) has been determined and compared with that of the related furocoumarin XCVIII (*F. M. Dean et al.*, J. chem. Soc., C, 1967, 2232):

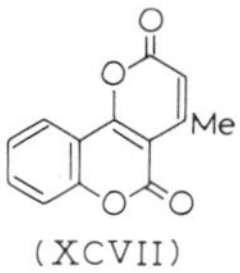

(XCVII) 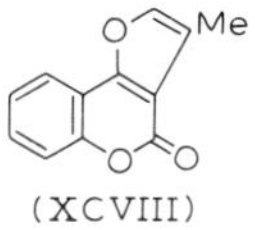 (XCVIII)

The reaction between 8-hydroxycoumarin-7-carbaldehyde, acetic anhydride and sodium acetate gives 2H,9H-*pyrano*[3,2-h]*chromene*-2,9-*dione*, m.p. 378–380°, and between the coumarin and diethyl malonate in the presence of a few drops of pyridine affords *ethyl* 2H,9H-*pyrano*[3,2-h]*chromene-2,9-dione-3-carboxylate*, m.p. 255–256°.

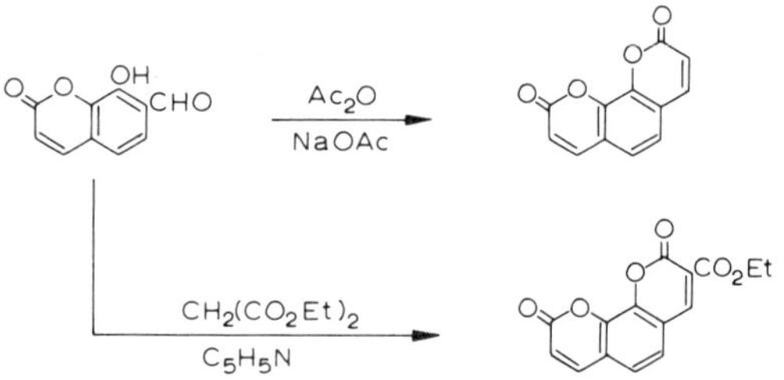

The reaction between 7-acetyl-8-hydroxycoumarin, benzoic anhydride and sodium benzoate yields 3-*benzoyl-2-phenyl*-4H,9H-*pyran*[3,2-h]*chromene-2,9-dione*, m.p. 253–254° (*R. H. Mehta*, Indian J. Chem., 1965, **3,** 186):

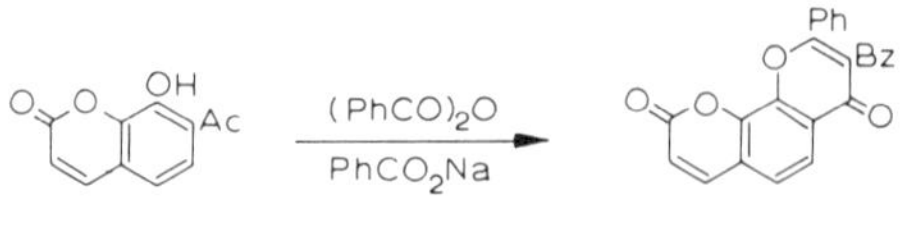

The Kostanecki–Robinson acylation of 7-hydroxy-6-acyl-8-methyl-4-alkyl- or aryl-coumarins gives the corresponding 2*H*,6*H*-pyrano[3,2-*g*]chromene-2,6-dione (XCIX), while 7-hydroxy-6-benzoyl-4,8-dimethylcoumarin yields 4,10-dimethyl-6-phenyl-2*H*,9*H*-pyrano[3,2-*g*]chromene-2,9-dione (C) (*K. Pardanani et al.*, J. Indian chem. Soc., 1970, **47,** 36):

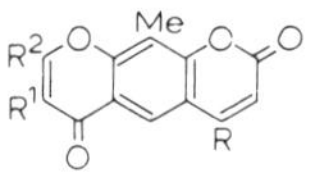

(XCIX) (C)

R = Me or Ph
R^1 = Ac, Me or H
R^2 = Me or Ph

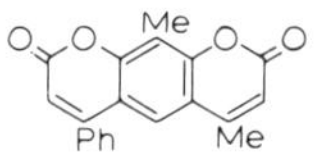

(xii) Chromones (4H-*benzo*[b]*pyran-4-one*, 5,6-*benzo-4-pyrone*, 5,6-*benzo-γ-pyrone*)*

4*H*-Benzo[*b*]pyran-4-one or chromone, the name given by *von Kostanecki*, is the parent coumpound of important vegetable colouring matters, which are derived from *flavone*, 2-phenylchromone, 2-phenyl-4*H*-benzo[*b*]-

* *S. Wawzonek*, "Heterocyclic Compounds", Vol. 2, p. 228 (J. Wiley and Sons, Inc., New York, 1951). "Recent Advances in the Chemistry of Naturally Occurring Pyrones" (Sci. Proc. roy. Dublin Soc., 1956, **27,** 75). *F. M. Dean*, "Naturally Occurring Oxygen Ring Compounds" Ch. 8, p. 251 (Butterworths, 1963).

pyran-4-one; *flavonol*, 3-hydroxy-2-phenyl-4*H*-benzo[*b*]pyran-4-one; *flavanone*, 2,3-dihydro-2-phenyl-4*H*-benzo[*b*]pyran-4-one; and *isoflavone*, 3-phenyl-4*H*-benzo[*b*]pyran-4-one. In consequence these phenylchromones have been more systematically studied that the simple chromones, although a number of 2-methylchromones have been obtained from natural sources (p. 153), and they resemble the coumarins in being found as hydroxy derivatives or as furochromones (p. 155).

Much of our fundamental knowledge of the chromones and flavones is due to the classical researches of *St. von Kostanecki* (1860–1910) (see *J. Tambor*, Ber., 1912, **45,** 1683) and *A. G. Perkin* (see *R. Robinson*, J. chem. Soc., 1938, 1738).

Methods of synthesis

(*1*) By cyclisation of β-diketones of the type $RCO \cdot CH_2 \cdot COR^1$, in which R is 2-hydroxyphenyl and R^1 is alkyl, aryl, OH etc. Thus 2-hydroxybenzoylpyruvic acid gives chromone, decarboxylation accompanying ring-closure:

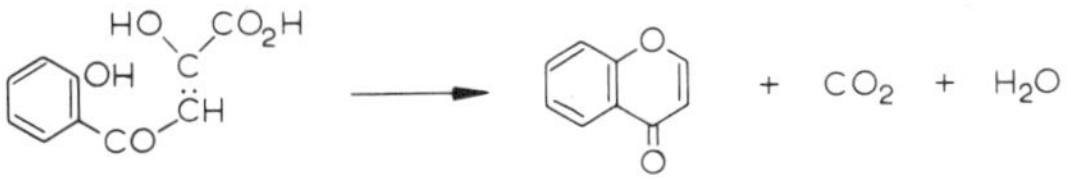

HO CO₂H / OH C / ĊH / CO → Chromone + CO_2 + H_2O

Chromone

2-Methoxydibenzoylmethanes with hydrogen iodide undergo simultaneous demethylation and ring-closure to give flavones (*von Kostanecki et al.*, Ber., 1900, **33,** 471, 1998; 1901, **34,** 2942).

The β-diketones for these syntheses are obtained (*a*) by Claisen condensation of 2-hydroxyacetophenone with esters in presence of sodium or (*b*) by the Baker–Venkataraman transformation of *O*-acyl- or *O*-aroyl-2-hydroxy acetophenones.

(*a*) 2-Hydroxyacetophenone and ethyl propionate give the diketone 1-(2-hydroxyphenyl)pentane-1,3-dione cyclised by acetic and hydrochloric acids to 2-ethylchromone (*M. Blumberg* and *Kostanecki*, Ber., 1903, **36,** 2191; *P. F. Wiley*, J. Amer. chem. Soc., 1952, **74,** 4329):

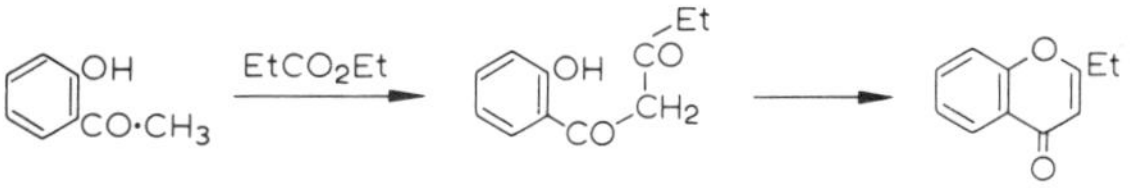

OH / $CO \cdot CH_3$ —$EtCO_2Et$→ OH / CO–CH_2–CO–Et → 2-ethylchromone (Et)

This condensation can be effected in pyridine or xylene or by means of sodium hydride (*R. Mozingo*, Org. Synth., 1941, **21,** 42; *J. Schmutz et al.*, Helv., 1952, **35,** 1168).

(*b*) An *O*-acyl- or *O*-aryl-2-hydroxyacetophenone is heated, *e.g.*, in benzene or toluene with sodium, sodamide or potassium carbonate (*W. Baker*, J. chem. Soc., 1933, 1381; *H. S. Mahal* and *K. Venkataraman*, *ibid.*, 1934, 1767; 1935, 868; *T. S. Wheeler et al.*, Proc. roy. Dublin Soc., 1948, **24,** 291). Thus 2-Benzoyloxyacetophenone yields 1-(2-hydroxyphenyl)-3-phenylpropane-1,3-dione, which cyclises with sulphuric acid to flavone (*Wheeler*, Org. Synth., 1962, **32,** 72):

OBz, CO·CH₃ ⟶ OH, COPh, CH₂, CO ⟶ O, Ph, O

Flavone

The migration provides a general method for the preparation of flavones and other 2-substituted chromones (*Baker et al.*, J. chem. Soc., 1939, 956, 1922; *V. V. Virkar* and *Wheeler*, *ibid.*, 1939, 1679, 1681). There is evidence that the Baker–Venkataraman transformation is a base-catalysed intramolecular Claisen condensation, but it may be noted that at sufficiently high temperatures no catalyst is required (*Wheeler et al.*, *ibid.*, 1950, 1252, 1925; 1956, 4411; *H. Schmid* and *K. Banholzer*, Helv., 1954, **37,** 1706). Various phenolic compounds have been condensed with ethyl acetoacetate and ethyl benzoylacetate in boiling diphenyl ether to give the corresponding 2-methylchromones and flavones in good yield (*K. B. Desai*, *K. N. Trivedi* and *S. Sethna*, C.A. 1958, **52,** 11030g).

In the Kostanecki–Robinson synthesis, 2-hydroxyacetophenones are heated with the sodium salt and the anhydride of an aliphatic acid to form 2-alkylchromones (*Kostanecki et al.*, Ber., 1901, **34,** 102, 2942), or with sodium salts and anhydrides of aromatic acids to give flavones (*J. Allan* and *Robinson*, J. chem. Soc., 1924, **125,** 2192). Resacetophenone when heated with acetic anhydride and sodium acetate yields 7-acetoxy-3-acetyl-2-methylchromone:

HO, OH, Ac $\xrightarrow[\text{AcONa}]{Ac_2O}$ AcO, O, Me, Ac, O

Alkaline hydrolysis removes the acetyl groups to give 7-hydroxy-2-methylchromone.

The mechanism of the reaction probably involves a Baker–Venkataraman transformation (*Baker*, *ibid.*, 1933, 1381).

The essential steps are given below:

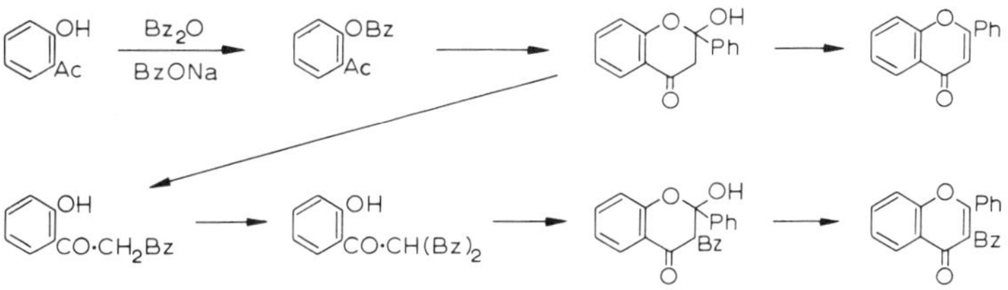

The product obtained, depending on the conditions including temperature, may be either a chromone or a 3-acylchromone. Since the 3-acylchromones are β-diketones the 3-acyl group is readily removed by alkaline hydrolysis. Formation of 3-aroylflavones can be avoided by heating 2-aroyloxyacetophenones in anhydrous glycerol at 250° when flavones are obtained (*Wheeler et al., ibid.*, 1950, 1252).

When aliphatic anhydrides and salts derived from acids of the general type $RCH_2 \cdot CO_2H$ are used the products may be coumarins (*T. C. Chada et al., ibid.*, 1933, 1459; *I. M. Heilbron et al., ibid.*, 1934, 1311; 1936, 295; *S. M. Sethna* and *N. M. Shah*, Chem. Reviews, 1945, **36**, 8). This is explained by the *O*-acyloxyketone CI first formed losing water in different ways to give either a chromone CII, or a coumarin CIII, or a mixture of the two (*G. Wittig et al.*, Ann., 1925, **446,** 155):

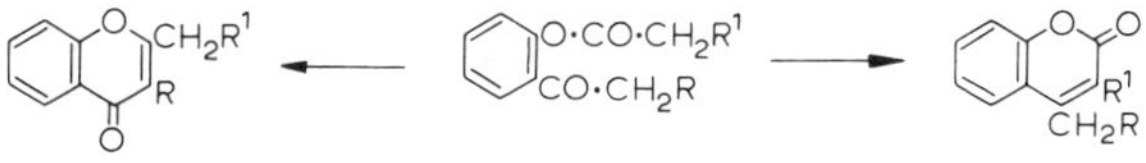

(CII) (CI) (CIII)

The Baker–Venkataraman reaction has been extended to ω-methyl, ω-methoxy and ω-phenyl derivatives of 2-acyloxyacetophenone, 3-substituted chromones being obtained (*H. M. Lynch et al.*, J. chem. Soc., 1952, 2063; *W. D. Ollis* and *D. Weight, ibid.*, 1952, 3826).

(*2*) 2-Hydroxychalcones are converted into flavones by acetylation and addition of bromine, the resultant dibromides with ethanolic potassium or sodium hydroxide losing hydrogen bromide to give flavones:

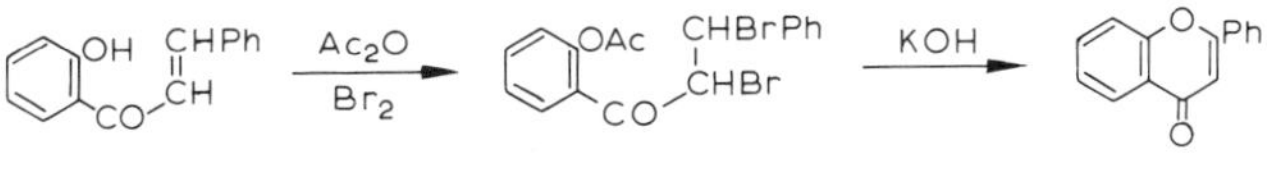

The method, by which the first synthesis of a flavone was achieved (*T. Emilewicz* and *Kostanecki*, Ber., 1898, **31,** 696), is of wide application, but in certain cases yields 2-benzylidenecoumaran-3-ones (aurones) (*Wheeler et al.*, J. chem. Soc., 1937, 1798; 1939, 91; Chem. and Ind., 1952, 130; Chem. Comm., 1966, 351). Certain of the dibromo intermediates, which with alkali gives benzylidenecoumaranones, yield the corresponding flavones with ethanolic potassium cyanide (*D. M. Fitzgerald et al.*, J. chem. Soc., 1955, 860; *J. E. Gowan, E. M. Philbin* and *Wheeler*, Sci. Proc. roy. Dublin Soc., 1956, **27,** 185). It has been established that benzylidenecoumaranones are intermediates in the reaction and under the action of potassium cyanide or other weak bases undergo ring expansion:

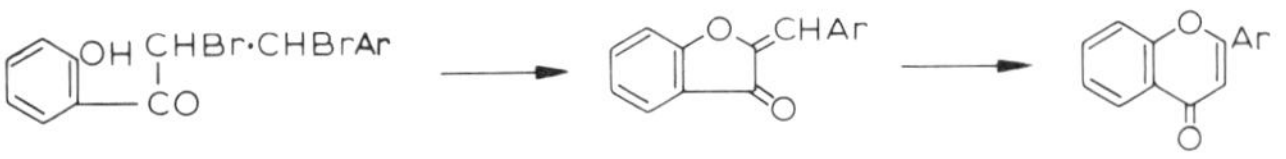

Oxidative ring-expansion occurs when aurones are treated with alkaline hydrogen peroxide and thus yield flavanols along with aurone epoxides (*W. E. Fitzmaurice et al.*, Chem. and Ind., 1955, 652).

o-Hydroxychalcones and flavanones (2,3-dihydro-2-phenyl-4*H*-benzo[*b*]-pyran-4-ones) are oxidised by selenium dioxide to give good yields of flavones (*Mahal et al.*, J. chem. Soc., 1935, 866; 1936, 569; *G. Bargellini* and *G. B. Marini-Bettolo*, Gazz., 1940, **70,** 170):

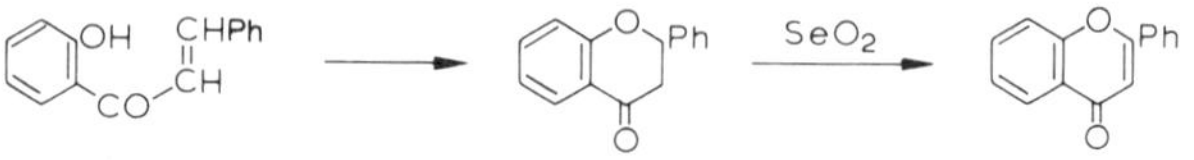

Flavanones are also dehydrogenated to flavones by phosphorus pentachloride in bezene (*A. Löwenbein*, Ber., 1924, **57,** 1515).

(*3*) Flavanones can be converted into flavones by bromination in the 3-position and subsequent dehydrobromination (*Kostanecki et al.*, *ibid.*, 1904, **37,** 2634). Bromination is effected by *N*-bromosuccinimide or pyridinium perbromide (*N. B. Lorette et al.*, J. org. Chem., 1951, **16,** 930; *N. R. Bannerjee* and *T. R. Seshadri*, Proc. Indian Acad. Sci., 1952, **36A,** 134; *S. Hishida et al.*, J. chem. Soc. Japan, 1953, **74,** 697); with *N*-bromosuccinimide flavanone undergoes partial dehydrogenation to yield a mixture of flavone and 3-bromoflavanone (*R. Bognar* and *M. Rakosi*, Chem. and Ind., 1955, 773).

(*4*) *The Simonis method.* Alkylacetoacetic esters condense with phenols in the presence of sulphuric acid to yield coumarins (Pechmann reaction, p. 98), but when phosphorus pentoxide is used poor yields of chromones are frequently obtained (*H. Simonis et al.*, Ber., 1913, **46,** 2014; 1914, **47,** 692, 2229). Thus phenol and methylacetoacetic ester give 2,3-dimethylchromone:

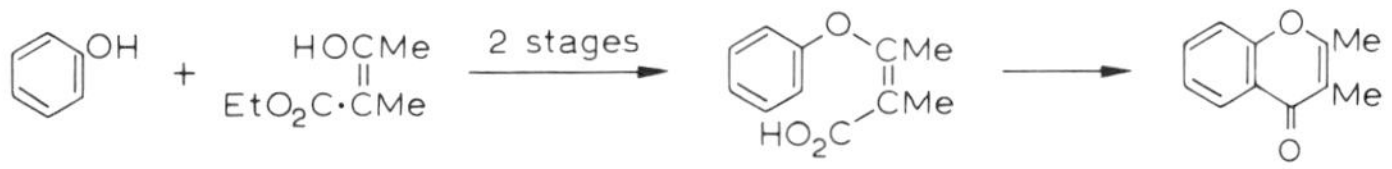

With phosphorus pentoxide, however, coumarins sometimes result and may lead to confusion (*Baker* and *Robinson*, J. chem. Soc., 1925, **127**, 1981, 2349; *D. Chakraverti*, J. Indian chem. Soc., 1931, **8**, 129). *p*-Cresol, for example, with ethyl acetoacetate and phosphorus pentoxide yields 4,6-dimethylcoumarin (*A. Robertson et al.*, J. chem. Soc., 1931, 1877, 2426; 1932, 1180). Heating the reactants in the absence of a condensing agent gives chiefly chromones (*C. Mentzer et al.*, Compt. rend., 1951, **232**, 1488).

The Simonis reaction probably involves the formation of a β-phenoxyacrylic acid, which undergoes ring-closure as shown above (*Robertson et al.*, J. chem. Soc., 1932, 1681). This is comparable to the preparation of chromones and flavones by the ring-closure of phenoxyfumaric and β-phenoxycinnamic acids (*S. Ruhemann et al.*, *ibid.*, 1900, **77**, 1119, 1179; Ber., 1913, **46**, 2188). Other theories have been advanced (*R. N. Lacey*, J. chem. Soc., 1954, 854).

(*5*) 3-Methylchromones may be prepared from 2-hydroxypropiophenones by condensation with diethyl oxalate in the presence of sodium (*Kostanecki et al.*, Ber., 1901, **34**, 2475; *K. von Auwers*, Ann., 1920, **421**, 1). The product undergoes ring-closure in the presence of hydrochloric acid to form 3-methylchromone-2-carboxylic acid which on heating gives 3-methylchromone. The method gives good yields of 3-methylchromones with substituents in the benzene ring, but only poor yields of 3-methylchromone itself (*Mentzer* and *P. Meunier*, Bull. Soc. chim. Fr., 1944, [v], II, 302):

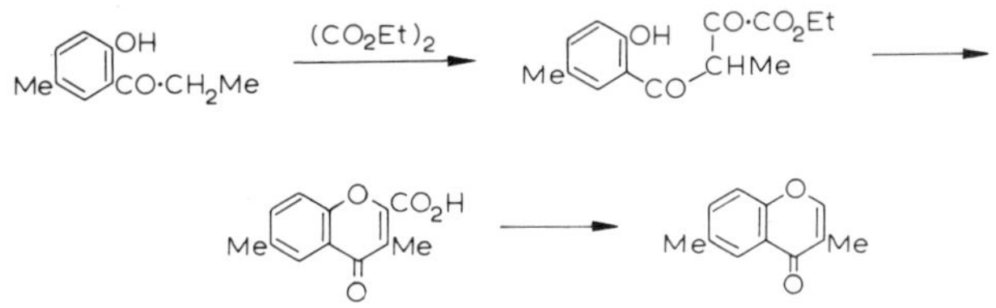

(*6*) ω-Halogeno-2-hydroxyacetophenones (*e.g.*, CIV) and aromatic aldehydes condense in the presence of alkali to give 2-arylidenecoumaran-3-ones (CV) (*G. Woker et al.*, Ber., 1903, **36**, 4235), but if the reaction is carried out in the cold with excess alkali present the corresponding flavonols (CVI) are obtained (*Wheeler et al.*, J. chem. Soc., 1955, 862; *D. B. Limaye*, Rasayanam, 1955, **2**, No. 3, 53):

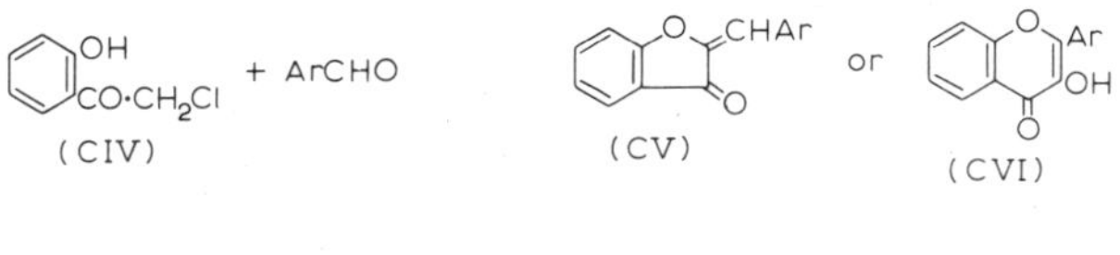

(7) 2-Hydroxyacetophenone reacts with the dimethyl acetal of dimethylformamide at 120° in xylene to give 2-(2-hydroxybenzoylvinyl)dimethylamine (CVII), which on heating with dilute sulphuric acid at 100° gives chromone (*B. Foehlisch*, Ber., 1971, **104,** 348):

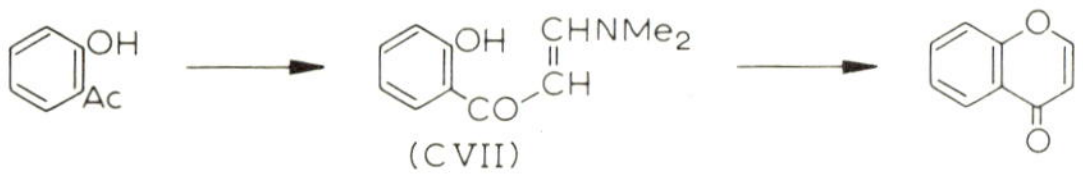

(CVII)
(80 %)
(71 %)

(8) Salicylaldehyde reacts with 1-morpholino-1-phenylethene in benzene solution to form a non-cyrstalline adduct, which on oxidation with chromium trioxide-pyridine yields flavone (*L. A. Paquette* and *H. Stucki*, J. org. Chem., 1966, **31,** 1232):

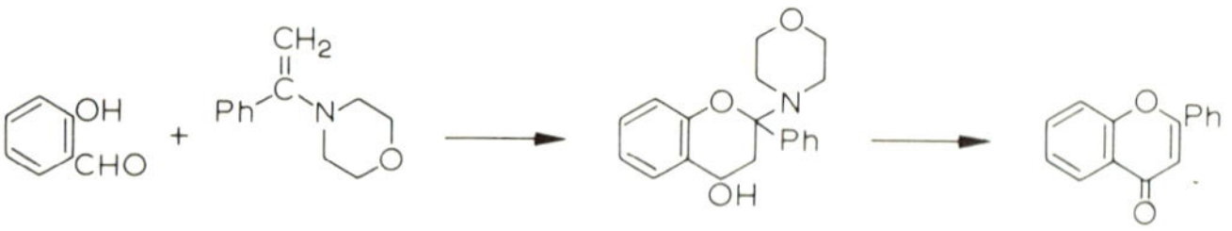

(9) 7-Methoxy-2-methylchroman-4-one on treatment with vinylmagnesium bromide under the Normant reaction conditions, besides giving the expected 4-hydroxy-7-methoxy-2-methyl-4-vinylchroman, also affords 7-methoxy-2-methylchromone (*S. R. Ramadas* and *J. Radhakrishnan*, Chem. Comm., 1970, 529):

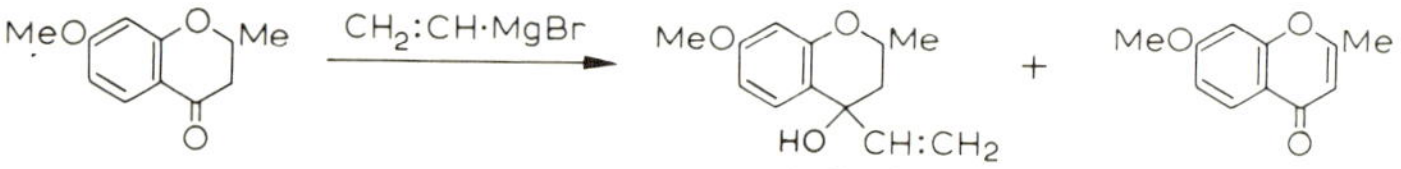

$CH_2{:}CH{\cdot}MgBr$
(50 %)
(12 %)

(*10*) Benzopyrylium perchlorates may be converted to chromones by treatment with aqueous sodium hydrogen carbonate solution; the benzopyrylium perchlorates being obtained by reacting chromanones in acetic acid with triphenylmethyl perchlorate. Chromanone itself gives almost 100% of chromone (*A. Schönberg* and *G. Schütz*, Ber., 1960, **93,** 1466).

The methods of synthesising chromones have been reviewed (*M. Vandewalle*, Ind. Chim. Belge, 1959, **24,** 1037, 1207).

Structure and properties. The chromones are crystalline substances with properties very similar to those of the 4-pyrones. They react as α,β-unsaturated ketones.

The structures of the chromones follow from their synthesis and from *ring-fission* of the pyrone nucleus by treatment with alkali.

The 2-hydroxyphenyl-β-diketones thus formed are seldom isolated but undergo typical β-diketone cleavage. The β-diketone CVIII from 2,3-dimethylchromone may undergo hydrolysis at the points marked by arrows yielding either salicylic acid and ethyl methyl ketone or 2-hydroxypropiophenone and acetic acid:

Me, Me, O → OH, CO·CHMe·COMe (CVIII) → OH, CO_2H + $MeCH_2$·COMe; OH, CO·CH_2Me + $MeCO_2H$

The products obtained depend both on the hydrolytic agent used and on the substituents in the chromone molecule (*Simonis*, Ber., 1915, **50,** 779). Concentrated alkali gives a mixture of the four products, while sodium ethoxide gives mainly the 2-hydroxyphenyl ketones and fatty acids (*W. Feuerstein* and *Kostanecki*, *ibid*., 1898, **31,** 1762; *E. Petschek* and *Simonis*, *ibid*., 1913, **46,** 2015). Molar sodium hydroxide yields mainly the salicylic acids and aliphatic ketones (*Simonis* and *C. A. Lehmann*, *ibid*., 1914, **47,** 692). The introduction of substituents into the 5-position results in slow fission by normal alkali and the formation of an alkyl 2-hydroxyphenyl ketone. Chromone yields salicylic acid with normal alkali, and 2-hydroxyacetophenone and formic acid with sodium ethoxide. The chromones thereby differ from the coumarins, which with alkali merely undergo delactonisation (cf. however, p. 102) and yield salicylic acid only when fused with alkali.

Differentiation from coumarins. Chromenes can be distinguished from coumarins also by their formation of orange-yellow *oxonium salts* with hydrochloric acid, perchloric acid, etc. (*Wittig et al*., Ann., 1925, **446,** 157).

Hydrogen chloride, for instance, passed into an ethereal solution of a chromone and a coumarin forms an insoluble oxonium salt of the chromone, while the coumarin remains in solution.

Chromones in acetic acid with bromine yield insoluble perbromides (which regenerate the chromones when treated with sulphurous acid) whereas coumarins form insoluble 3-bromocoumarins (*R. D. Desai*, Rasayanam, 1938, **1,** 155; C.A., 1939, **33,** 1697).

The u.v. absorption spectra of the chromones differ markedly from those of the coumarins (*B. K. Ganguly* and *P. Bagchi*, J. org. Chem., 1956, **21,** 1415); i.r. spectra (*J. H. Looker* and *W. H. Hanneman*, *ibid*., 1962, **27,** 381; *R. D. Murray* and *P. H. McCabe*, Tetrahedron, 1969, **25,** 5819). The dielectric constants have been measured over a range of temperatures at various concentrations in benzene solution and the dipole moments calculated for a number of chromones (*S. K. K. Jatkar* and *C. M. Deshpende*, J. Indian chem. Soc., 1960, **37,** 69).

Certain 5,8-dialkoxychromones when heated with hydriodic acid undergo both dealkylation and the Wessely–Moser rearrangement (p. 176) (*D. K. Chakravorty et al.*, Proc. Ind. Acad. Sci., 1952, **35A**, 37); thus 5,7,8-trihydroxy-3-methoxy-2-methylchromone when heated in phenol with hydriodic acid at 170° for 1 hour gives 3,5,6,7-tetrahydroxy-2-methylchromone (*D. M. Donnelly et al.*, J. chem. Soc., 1956, 4409):

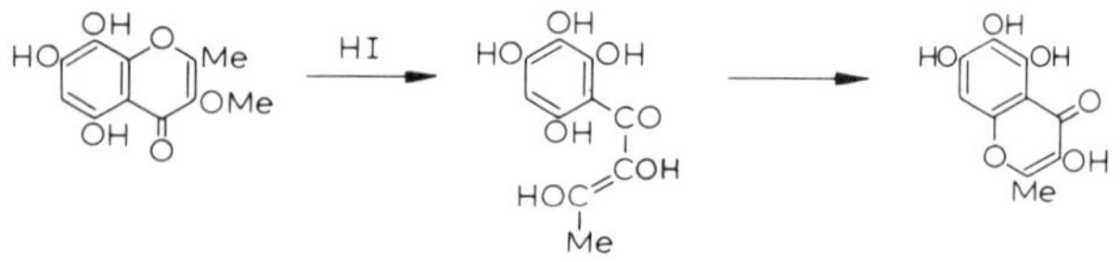

Boiling 3-acetoacetyl-2-methylchromone in 48% hydrobromic acid gives a Wessely–Moser rearrangement to a product tentatively identified as 3-methyl-1*H*,4*H*-xanthene-1,9-dione (*H. Abu-Shady* and *H. Darwish* U.A.R.J. Pharm. Sci., 1970, **11,** 289):

Oxidation is frequently used in structural determination, 3,7-dimethoxychromone, for instance, with potassium permanganate and acetic acid yielding 2-hydroxy-4-methoxybenzoic acid.

The valuable method whereby pyrones are converted into pyridones finds no analogy in the chromone series.

Carbonyl reactions. Chromone is a resonance hybrid as shown:

and many characteristic ketonic reactions are masked.

2,8-Dimethylchromone (or the corresponding 4-thionochromone) in neutral solution gives the *oxime*, m.p. 146°, and in alkaline solution the dioxime of 2-acetylaceto-6-methylphenol (*Wittig* and *F. Bangert*, Ber., 1925, **58,** 2636). Chromone or 4-thionochromone with hydrazine gives 3(5)-(2-*hydroxyphenyl*)*pyrazole*, m.p. 96° (*Baker et al.*, J. chem. Soc., 1952, 1303) and 2-methylchromone reacts similarly (*E. Königs* and *J. Freund*, Ber., 1947, **80,** 143; *Schmutz* and *R. Hirt*, Helv., 1953, **36,** 132; *C. Alberti*, Gazz., 1957, **87,** 781). Phenylhydrazine is stated not to react with chromones (*Simonis* and *S. Rosenberg*, Ber., 1914, **47,** 1232), but 2-methylchromone is reported to react with phenylhydrazine and hydroxylamine to give a pyrazole and an isoxazole, respectively (*Alberti, loc. cit.*). Phenylhydrazones can be obtained from the more reactive thionochromones.

The carbonyl group in the chromones reacts normally with Grignard reagents to give the corresponding tertiary alkanols. (*Heilbron et al.*, J. chem. Soc., 1936, 1380). Treatment of chromone with ammonia, *n*-butylamine, benzylamine or piperidine gives a benzene derivative CIX, (R = NH_2, BuNH, $PhCH_2NH$, or piperidino) (*V. A. Zagorevskiĭ et al.*, Khim. Geterotsikl. Soedin., 1970, 1024):

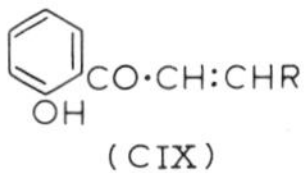

(CIX)

Nuclear methylation with methyl iodide and alkali is encountered in the chromone series (*A. C. Jain* and *Seshadri*, Quart. Reviews, 1956, **10,** 169), 2-methyl-5,7-dihydroxychromone for instance, yielding 2,6-dimethyl-5-hydroxy-7-methoxychromone (*R. Iengar et al.*, J. sci. ind. Research, India, 1953, **12B,** 119; Proc. Indian Acad. Sci., 1953, **38A,** 208) and noreugenin yielding eugenitin (p. 153) (see also *T. P. C. Mulholland* and *G. Ward.*, J. chem. Soc., 1953, 1642).

Colour reactions. 2-Methylchromones like the 2-methyl-4-pyrones, give a reddish-violet colour with concentrated potassium hydroxide (*Schönberg* and *A. Sina*, J. Amer. chem. Soc., 1950, **72,** 1611; J. chem. Soc., 1950, 3344), but colour formation is prevented by substituent hydroxyl groups (*A. Bolleter et al.*, Helv., 1951, **34,** 186). 5-Hydroxychromones give a violet-blue colour with ferric chloride (*Bolleter al., loc. cit.*) and 5,7-dihydroxychromones give a blue colour with alkaline hydrogen peroxide (*E. Späth* and *H. Eiter*, Ber., 1941, **74,** 1851). 2-Methylchromone with sodium nitroprusside in alkaline solution gives a yellow-red coloration which changes to red-violet on acidification with acetic acid (*M. M. Sidby* and *M. R. Mahren*, Arch. Pharm., 1963, **296,** 569).

Hydrogenation of the chromones with platinum or palladium catalysts reduces the olefinic linkage to give chromanones, which readily undergo further hydrogenation to chromans (*F. Prillinger* and *Schmid*, Monatsh., 1938, **72,** 427; *P. Pfeiffer* and *J. Grimmer*, Ber., 1917, **50,** 911). With hydrogen and a copper chromite catalyst reduction of the olefinic linkage, the carbonyl group, or both may occur (*Mozingo* and *H. Adkins*, J. Amer. chem. Soc., 1938, **60,** 669); at about 110° 2-ethylchromone gives 2-ethylchroman-4-one, 2-ethylchrom-2-en-4-ol, 2-ethylchroman-4-ol, and 2-ethylchroman; at higher temperatures only 2-ethylchroman is obtained.

The *sulphonation* of 2,3-dimethylchromone gives the 6-sulphonyl chloride and the 6-sulphonic acid; 5-hydroxy- and 7-hydroxy-2-methylchromone gives the 8-sulphonic acid and the 6,8-disulphonic acids, whereas 5-methoxy- and 7-methoxy-2-methylchromone affords only the 8-sulphonic acid (*D. V. Joshi, J. R. Merchant* and *R. C. Shah*, J. org. Chem., 1956, **21,** 1104).

Methyl group activity. The methyl group in 2-methylchromones is reactive and in the presence of sodium ethoxide benzylidene compounds are readily formed (*I. M. Heilbron et al.*, J. chem. Soc., 1923, **123**, 2559; *K. C. Gulati et al., ibid.*, 1934, 1765; *A. Zaki* and *R. C. Azzam, ibid.*, 1943, 434; *Schönberg* and *Sina, loc. cit.*). The presence of hydroxyl or methoxyl groups may inhibit reactivity, 7-methoxy-2,3-dimethylchromone, for instance, failing to condense with aromatic aldehydes. The related pyrylium salt, 7-methoxy-2,3-dimethyl-4-phenylchromylium chloride, however, condenses without a catalyst to give the 2-styrylchromylium salt (*Heilbron* and *A. Zaki*, J. chem. Soc., 1926, 1902):

$Cl^{\ominus}$ MeO, Me, Me, Ph —PhCHO→ $Cl^{\ominus}$ MeO, CH:CHPh, Me, Ph

(Compare the enhanced reactivity of the methyl group in 2-methyl-1-ethylquinolinium salts, Vol. IV F, p. 262 *et seq.*) 2-Ethylchromones also yield styryl derivatives though less readily than the methyl compounds (*Heilbron et al., ibid.*, 1934, 1311). 2-Methylchromone is oxidised by selenium dioxide to chromone-2-carboxylic acid *Schmutz et al.*, Helv., 1952, **35,** 1168).

Oxonium salts of chromones react with dialkylaniline to yield 4-alkoxy-2-(4-dialkylaminophenyl)benzo[b]pyrylium salts (4-alkoxy-4′-dialkylaminoflavilium salts) (*A. I. Tolmachev*, Zhur. obshcheĭ Khim., 1960, **30,** 2884).

Dimerisation. Chromone is dimerised with sodium ethoxide to 1-(3-*chromonyl*)-2-salicyloylethylene (CX), which can be isomerised reversibly to 2-(3-*chromonyl*)*chromanone* (CXI), m.p. 154–155°, by boiling in acetic acid containing a small amount of sulphuric acid (*Schönberg* and *E. Singer*, Ber., 1963, **96,** 1529):

—NaOEt→ OH (CX) —Δ, H_2SO_4/AcOH→ (CXI)

Depending on the conditions and solvent used during the crystallisation of ω-formyl-2-hydroxyacetophenone, the compound itself, chromone, or dimer

CXI may be obtained (*K. Kostka*, Rocz. Chem., 1967, **41,** 71). 6-Methylchromone is also dimerised by sodium ethoxide.

The **thionochromones** containing the C:S group are prepared by heating the chromones with phosphorus pentasulphide and are obtained as deep red needles (*W. Baker et al.*, J. chem. Soc., 1952, 1303). These react with hydrazine, phenylhydrazine, etc., to give either hydrazones or hydroxyphenylpyrazoles depending on the substituents in the molecule. 4-Thionoflavone, for instance, with phenylhydrazine yields the phenylhydrazone (p. 169), while 2-methylthionochromone affords 5-(2-hydroxyphenyl)-3-methyl-1-phenylpyrazole probably by nucleophilic attack at $C_{(2)}$ to give the intermediates shown in brackets followed by successive ring-fision and ring-closure (*Baker et al.*, *ibid.*, 1954, 998):

2-Methylthionochromone → [NH·NHPh, Me, O, S intermediate] → [OH, N·NHPh, CMe, C=CH, SH intermediate] → [OH, Ph, N, N, Me pyrazole]

The methyl group in the 2-methyl-4-thionochromones is reactive (*Schönberg et al.*, J. Amer. chem. Soc., 1954, **76,** 5115).

2-**Methyl-**, m.p. 96–97°, *benzylidene-*, m.p. 84°, and 4-*dimethylaminophenylimino* deriv., green crystals, m.p. 236° (*idem, loc. cit.*); 2,3-**dimethyl-**, m.p. 120°; 2,8-**dimethyl-**, m.p. 134°; 2,3,5-**trimethyl-**, m.p. 116°; 2,3,6-**trimethyl-**, m.p. 125°; and 2,3,8-**trimethyl-4-thionochromone,** m.p. 139° (*Simonis* and *Rosenberg*, Ber., 1914, **47,** 1232). Like 2-methyl-4-thionochromone all the above substances with phenylhydrazine form pyrazole derivatives.

(*xiii*) *Chromone and alkyl-chromones*

Chromone, needles, m.p. 59°, λ_{max} 297–298 mμ (log ε 3.82) (96% EtOH) is prepared by the thermal decomposition of *chromone-2-carboxylic acid*, m.p. 252° (*R. Heywang* and *St. von Kostanecki*, *ibid.*, 1902, **35,** 2887; *K. Kostka*, Rocz. Chem., 1967, **41,** 71); by the action of sulphuric acid on phenoxyfumaric acid (*M. Gomberg* and *L. H. Cone*, Ann., 1910, **376,** 228); or preferably from 2-hydroxyacetophenone and ethyl formate in the presence of sodium (*Schönberg* and *Sina*, J. Amer. chem. Soc., 1950, **72,** 3396). It gives a colourless solution in cold concentrated sulphuric acid with a violet-blue fluorescence and forms a crystalline *hydrobromide*, m.p. 175°, and *hydrochloride*, m.p. 101–102°.

The m.p.'s and absorption in the u.v. region of a number of alkylchromones are given in Table 3.

TABLE 3

ALKYLCHROMONES

Compound	*M.p. °C*	*λ_{max} (mμ); log ε in brackets*[a]		*Ref.*
2-Methylchromone	71	225(4.35)	295(3.91)	1,2
3-Methylchromone	69	225(4.33)	304(3.87)	3,4
6-Methylchromone	88–89			5
7-Methylchromone	72–73			6
8-Methylchromone	84–85			6
2-Ethylchromone	(b.p. 124–126°/ 2 mm)			7
2,3-Dimethylchromone	97	225(4.39)	299–304(3.94)	8
2,6-Dimethylchromone	103	225(4.42)	303(3.88)	5
2,7-Dimethylchromone	98	225(4.34)	294–295(3.91)	8,9
2,8-Dimethylchromone	115	225(4.41)	299(3.87)	8
3,6-Dimethylchromone	66			3,5
2,3,6-Trimethylchromone	105–106			10
2,3,7-Trimethylchromone	89			9
2,5,8-Trimethylchromone	81			11
2,6,8-Trimethylchromone	125			12
2-Styrylchromone	131			13

[a] In 96% ethanol (ref. 8).

References
1 *G. G. Babcock et al.*, J. Amer. chem. Soc., 1950, **72,** 3396.
2 *S. Biniecki* and *E. Kesler*, Acta Polon. Pharm., 1956, **13,** 503.
3 *Ch. Mentzer* and *P. Meunier*, Bull. Soc. chim. Fr., 1944 [v], **11,** 302.
4 *A. Schönberg* and *A. Sina*, J. chem. Soc., 1950, 3344.
5 *K. von Auwers*, Ann., 1920, **421,** 1.
6 *S. Ruhemann* and *H. W. Bausor*, J. chem. Soc., 1901, **79,** 470.
7 *R. Mozingo*, Org. Synth., 1941, **21,** 42.
8 *B. K. Ganguly* and *P. Bagchi*, J. org. Chem., 1956, **21,** 1415.
9 *A. Zaki* and *R. C. Azzam*, J. chem. Soc., 1943, 434.
10 *A. Robertson et al., ibid.*, 1931, 2426.
11 *I. Goodall* and *Robertson, ibid.*, 1936, 426.
12 *D. G. Flynn* and *Robertson, ibid.*, 1936, 215.
13 *G. Aziz*, J. org. Chem., 1962, **27,** 2954.

2-**Ethylchromone** on treatment with sodium ethoxide is converted to a stable dimeric red enol CXII (*Schönberg* and *Singer*, Ber., 1963, **96,** 3062):

2 [2-ethylchromone] —NaOEt / Et_2O→ (CXII)

2-Ethyl-6-methylchromone behaves similarly.

2,3-Dihydro-*N*,*N*-dimethyl-*p*-toluidine obtained by the reduction of *N*,*N*-dimethyl-

p-toluidine by sodium and alcohol in liquid ammonia, on reacting with diketene gives 7,8-dihydro-2,6-dimethylchromone (*B. B. Millward*, J. chem. Soc., 1960, 26):

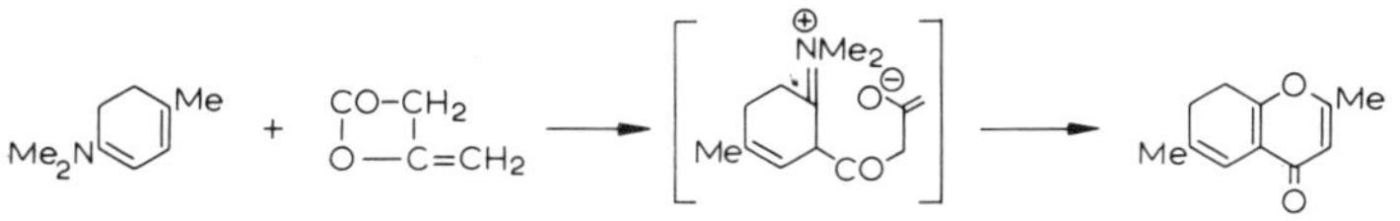

Chromone-2-carboxylic acid may be converted to 4,4-dichlorochromone-2-carbonyl chloride, 4-chlorocoumarin and ethyl and butyl chromone-2-carboxylates (*V. A. Zagorevskiĭ* and *D. A. Zykov*, Zhur. obshcheĭ. Khim., 1960, **30,** 3100); esters of chromone-2-carboxylic react with some secondary amines at normal temperatures in water-free medium to give *β*-(2-hydroxybenzoyl)acrylic acids (*Z. Jezzmanowska* and *Kostka*, Rocz. Chem., 1957, **31,** 1071); a number of chromone-2-carboxylic acids have been converted to 4-chlorocoumarins (*Zagorevskiĭ, Zykov* and *E. K. Orlova*, Zhur. obshcheĭ Khim., 1961, **31,** 568).

(xiv) Hydroxychromones

ω-Formyl-1-hydroxyacetophenone and benzaldehyde in aqueous alcoholic solution of potassium hydroxide gives 3-**chromonylphenylmethanol** (CXIII), which on oxidation yields 3-*benzoylchromone* (CXIV), m.p. 132°. The latter on heating with hydroxylamine hydrochloride in a mixture of pyridine, ethanol, and acetic anhydride yields 3-cyanoflavone (*F. Eiden* and *H. Haverland*, Naturwiss., 1965, **52,** 513; Ber., 1967, **100,** 2554):

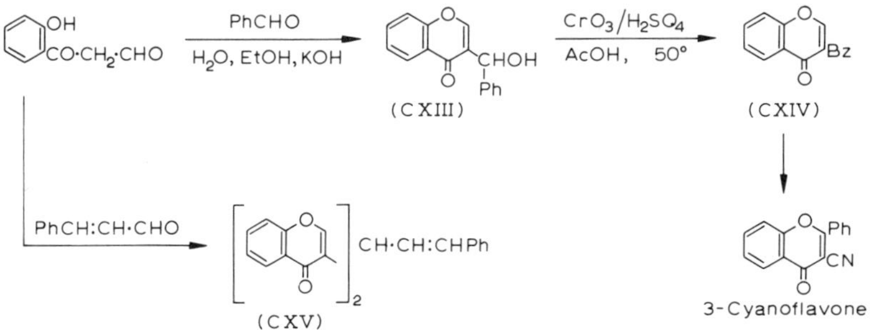

(CXIII) (CXIV) (CXV) 3-Cyanoflavone

The reaction between *ω*-formyl-1-hydroxyacetophenone and cinnamaldehyde yields styrylbis(chromon-3-yl)methane (CXV).

3,7-**Dihydroxychromone,** m.p. 271°, with a blue fluorescence in concentrated sulphuric acid, is obtained by the aerial oxidation of brazilin; the 7-*methyl ether*, m.p. 171–172°, has been synthesised (*P. Pfeiffer* and *J. Oberlin*, Ber., 1924, **57,** 208), and the *dimethyl ether*, m.p. 169–170°, with ethanolic potassium hydroxide gives fisetol dimethyl ether (*W. Feuerstein* and *St. von Kostanecki*, *ibid.*, 1899, **32,** 1025):

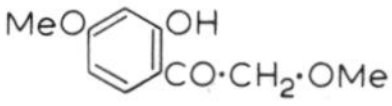

5,7-**Dihydroxychromone,** m.p. 272°, 7-*methyl ether*, m.p. 117–118°, 5,7-*dimethyl ether*, m.p. 131–132° (*Kostanecki* and *J. C. de R. de Wildt*, *ibid.*, 1902, **35,** 861; *N. Narasimhachari et al.*, J. Sci. Ind. Research, India, 1953, **12,** 287); 7,8-**dihydroxychromone,** m.p. 262°, *dimethyl ether*, m.p. 124° (*E. David* and *Kostanecki*, Ber., 1903, **36,** 128); Friedel–Crafts acylation of 7-hydroxychromones gives the 8-acyl derivatives, which are smoothly oxidised with hydrogen peroxide to the corresponding 7,8-dihydroxychromones (*S. M. Parikh* and *V. M. Thaker*, J. Indian chem. Soc., 1959, **36,** 841); the application of the Mannich reaction to 7-hydroxychromones gives rise to the corresponding 8-aminomethyl derivatives (*P. Da Re*, *L. Verlicchi* and *I. Setnikar*, J. org. Chem., 1960, **25,** 1097):

HO, O, Me, Me, O — 1. CH_2O, $NHMe_2$, EtOH; 2. HCl/EtOH → $CH_2NMe_2{\cdot}HCl$, HO, O, Me, Me, O

Chromones are not affected normally by treatment with alkali under mild conditions although they undergo fission of the pyrone ring when boiled in alkali; treatment of 5-methoxy-2-methyl-8-nitrochromone with boiling 10% aqueous sodium carbonate yields 5-hydroxy-2-methyl-6-nitrochromone (27%) and 2,6-dihydroxy-3-nitroacetophenone (50%). The rearrangement is of the Wessely–Moser type which occurs normally in acid conditions, although it has been seen to occur in alkali with certain flavones and flavanones (*W. Marlow*, Chem. and Ind., 1969, 1838):

NO_2, O, Me, OMe, O — 10% Na_2CO_3 → O, Me, O_2N, OH, O + OH, Ac, OH, NO_2

The bromination of 6-**hydroxy**-2-**methylchromone** with bromine in sulphuric acid at 25° gives 5-*bromo*-6-*hydroxy*-2-*methylchromone*, m.p. 235–236°, whereas with bromine in acetic acid at 45–50° 3,5-*dibromo*-6-*hydroxy*-2-*methylchromone*, m.p. 275–276° is obtained; nitration with nitric acid in acetic acid at 25° affords 6-*hydroxy*-2-*methyl*-5-*nitrochromone*, m.p. 291–292°, and at 45° 6-*hydroxy*-2-*methyl*-5,7-*dinitrochromone*, m.p. 190–191° (*G. Barker* and *G. P. Ellis*, J. chem. Soc. C, 1970, 2230).

The behaviour of 2,3-**dimethyl**-8-**hydroxychromone** on substitution has been studied; most substituents, for example bromine, enter the 7-position, the nitro group enters the 5- and 7-positions, and the sulphonic acid and diazo groups the 5-position (*Da Re* and *L. Cimatoribus*, Ber., 1962, **95,** 2912).

Alginetin, 3,8-*dihydroxy*-2-*methylchromone*, m.p. 235°, *diacetate*, m.p. 125°, is obtained by heating alginic acid at 160° and 2-oxoglutaraldehyde is believed to be an intermediate (*K. Aso et al.*, J. agr. chem. Soc., Japan, 1934, **10,** 1189; C.A., 1935, **29,** 1067; see also 1957, **51,** 8658).

The introduction of a 6-formyl group into 2-**methyl**-5,7,8-**trihydroxychromone** may be effected *via* the Gattermann reaction, but not *via* the hexamine method; the 8-methyl ether is formylated satisfactorily *via* the hexamine method (*V. V. S. Murti* *et al.*, Proc. Indian Acad. Sci., 1959, **50A,** 192).

The simplest naturally occurring hydroxychromone* is **eugenin,** $C_{11}H_{10}O_4$, m.p. 119–120°, which is obtained along with eugenitin, eugenitol, isoeugenitol and isoeugenitin, from wild cloves, *Eugenia caryophyllata* L. Thung.; *acetate*, m.p. 152.5–153.5°. Its constitution, *5-hydroxy-7-methoxy-2-methylchromone*, follows from its demethylation to 5,7-dihydroxy-2-methylchromone, its preparation by the methylation of that compound, and its degradation with alkali to 2,6-dihydroxy-4-methoxyacetophenone (*Th. M. Meijer* and *H. Schmid*, Helv., 1948, **31,** 1603).

Eugenitin, $C_{12}H_{12}O_4$, m.p. 162–163°, has been shown by synthesis to be *5-hydroxy-7-methoxy-2,6-dimethylchromone* (*Schmid*, *ibid.*, 1949, **32,** 813; *Schmid* and *A. Bolleter*, *ibid.*, 1950, **33,** 917). It is obtained in good yield by the action of methyl iodide in methanol on 5,7-dihydroxy-2-methylchromone, methylation of the hydroxyl group being accompanied by nuclear methylation (*W. B. Whalley*, J. Amer. chem. Soc., 1952, **74**, 5795; cf. p. 116); *5,7-dihydroxy-2-methylchromone*, m.p. 289°, may be obtained in a 50% yield along with *7-hydroxy-5-methoxy-2-methylchromone*, m.p. 277°, by the de-acylation of 3-acetyl-7-hydroxy-5-methoxy-2-methylchromone on boiling with aqueous sodium carbonate solution (10%) (*M. Davies* and *H. J. Smith*, Chem. and Ind., 1965, 1597). **Eugenitol,** *5,7-dihydroxy-2,6-dimethylchromone*, $C_{11}H_{10}O_4$, m.p. 290–292°, *diacetate*, m.p. 200–202° (*S. K. Mukerjee, T. R. Seshadri* and *S. Varadarajan*, Proc. Indian Acad. Sci., 1953, **37A**, 127; Chem. and Ind., 1955, 1009). **Isoeugenitol,** *5,7-dihydroxy-2,8-dimethylchromone*, $C_{11}H_{10}O_4$, m.p. 237–237.5° (*Schmid* and *Belleter*, Helv., 1949, **32,** 1358) is obtained by heating eugenitin with hydriodic acid, demethylation being accompanied by rearrangement (p. 116) (*Schmid*, *loc. cit.*; *Mukerjee*, *Seshadri* and *Varadarajan loc. cit.*). Isoeugenitol and diazomethane affords **isoeugenitin,** *5-hydroxy-7-methoxy-2,8-dimethylchromone*, $C_{12}H_{12}O_4$, m.p. 147–148°, *acetate*, m.p. 136–137°,

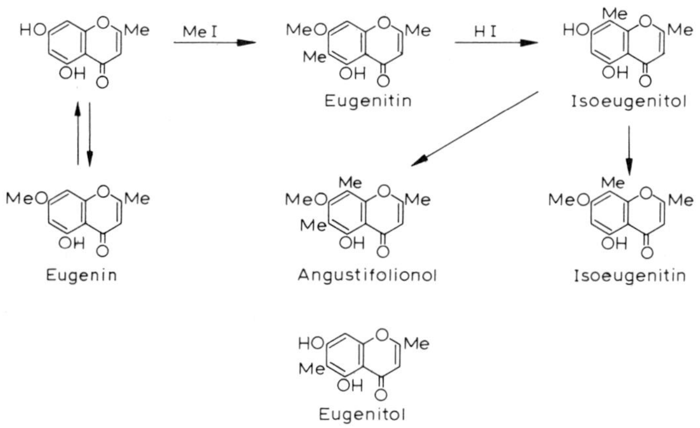

* For a review see *H. Schmid*, Fortschritte der Chemie organischer Naturstoffe, Vol. XI, p. 124 (1954).

which has been synthesised (*Schmid* and *Bolleter*, Helv., 1950, **33,** 1770; *Mukerjee*, *Seshadri* and *Varadarajan*, *loc. cit.*).

Angustifolionol, 5-*hydroxy*-7-*methoxy*-2,6,8-*trimethylchromone*, $C_{13}H_{14}O_4$, m.p. 119°, is isolated from *Backhousia angustifolia*. Alkaline degradation affords 2,6-*dihydroxy*-4-*methoxy*-3,5-*dimethylbenzoic acid*, m.p. 163°; and it has been synthesised by the nuclear methylation of isoeugenitol (*A. J. Birch et al.*, Austral. J. Chem., 1954, **7,** 169; 1955, **8,** 409).

Peucenin, $C_{15}H_{16}O_4$, m.p. 210–211° (vac. tube), 7-*methyl ether*, m.p. 108–109°, was first isolated from *Peucedanum Ostruthium* Koch (*E. Späth* and *K. Eiter*, Ber., 1941, **74,** 1851). The structure assigned from degradative experiments has been confirmed by synthesis (*Bolleter et al.*, Helv., 1951, **34,** 186; *A. C. Jain*, *P. Lal* and *Seshadri*, Indian J. Chem., 1969, **7,** 1072):

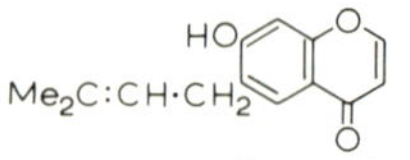

HO
Me$_2$C:CH·CH$_2$
O
O
Peucenin

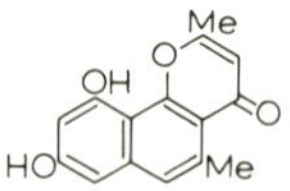

Me
O
OH
O
HO
Me
Eleutherinol

A number of compounds connected with peucenin have been synthesised (*C. Mercier*, Ann. Chim., Paris, 1970, **5,** 373).

Eleutherinol, $C_{15}H_{12}O_4$, decomp. >290–300°, *diacetate*, m.p. 188–189°, from *Eleutherine bulbosa* Mill., was the first naphthopyrone to be isolated from nature. The *dimethyl ether*, m.p. 186–187° is degraded by alkali to 1-*hydroxy*-6,8-*dimethoxy*-3-*methylnaphthalene*, m.p. 82° (*Birch* and *F. W. Donovan*, Austral. J. Chem., 1953, **6,** 373) and the structure assigned is supported by spectral data and synthesis (*H. Frei* and *Schmid*, Ann., 1957, **603,** 169).

Flavasperone, $C_{16}H_{14}O_4$, m.p. 203–204°, is obtained from the dried mycelium of *Aspergillus niger*, van Tiegham (*B. W. Bycroft*, *T. A. Dobson* and *J. C. Roberts*, J. chem. Soc., 1962, 40):

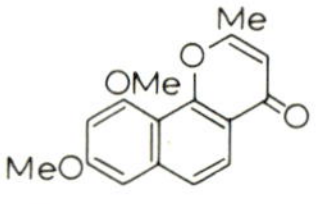

Me
O
OMe
O
MeO
Flavasperone

8-*Chloro*-5,7-*dihydroxy*-2,6-*dimethylchromone*, m.p. 265–266°, obtained from *Lecanora rupicola* has been synthesised (*S. Huneck* and *J. Santesson*, Z. Naturforsch. B, 1969, **24,** 750).

Two isomeric dimethyl ethers CXVII and CXVIII of **dicoumarol** have been obtained, which suggest for it a coumarin–chromone structure CXVI (*I. Chmielewska* and *J. Cieślak*, Tetrahedron, 1958, **4,** 135):

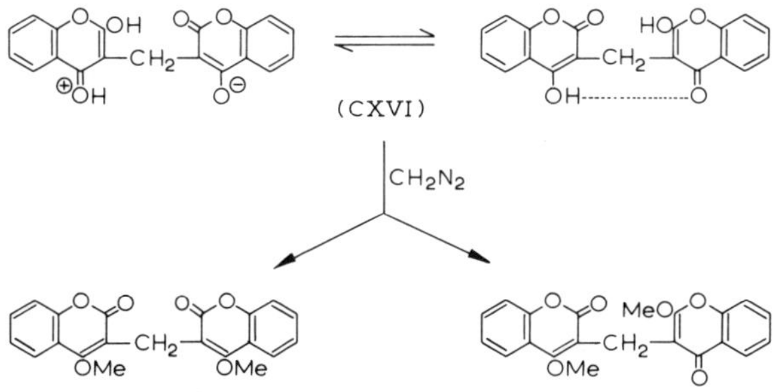

(CXVI)

CH_2N_2

(CXVII) (CXVIII)

Hydroxy derivatives of 8,8′- and 6,6′-**bichromonyl** have been prepared by a variation of the Wessely–Moser rearrangement (*B. Franck* and *G. Baumann*, Ber., 1963, **96**, 3209):

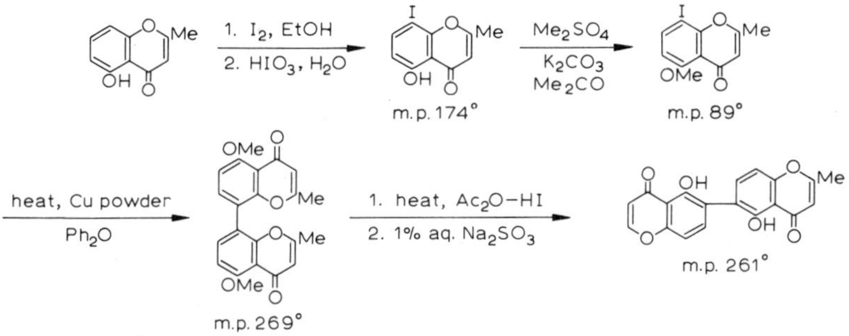

1. I_2, EtOH 2. HIO_3, H_2O — m.p. 174° — Me_2SO_4, K_2CO_3, Me_2CO — m.p. 89° — heat, Cu powder, Ph_2O — m.p. 269° — 1. heat, Ac_2O–HI 2. 1% aq. Na_2SO_3 — m.p. 261°

Leptorumol, 5,7-*dihydroxy*-6,9-*dimethylchromone*, the first natural chromone without a 2-methyl group and with a fully substituted benzene ring has been synthesised by two methods, one from 5,7-dimethoxy-8-methylchromone and the other from 5,7-dihydroxy-8-methylchromone by the Duff reaction with hexamine followed by catalytic reduction of the resulting formylchromone (*Mukerjee, S. Raychaudhuri* and *Seshadri*, Indian J. Chem., 1969, **7**, 1070):

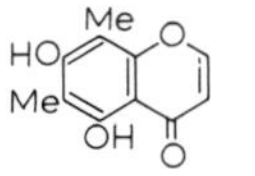

Leptorumol

*(xv) Furochromones**

Linear furochromones, whose structures recall those of the furocoumarins

---

* *C. P. Huttrer* and *E. Dale*, Chem. Reviews, 1951, **48**, 543; *H. Schmid*, Proc. roy. Dublin Soc., 1956, **27**, 145; *A. Mustafa* "Furopyrans and Furopyrones", Interscience Publishers, New York, 1967, Chap. 3, p. 102.

(p. 126) are known. Of these the following, which occur in the fruit and seeds of *Ammi visnaga* L., call for particular mention: khellin (kellin, visammin), visnagin (visnagidin), khellinin (kellinin, kellol glycoside), ammiol, khellinol (kellinol), and visamminol. These are other crystalline constituents of the plant have been isolated by *K. Samaan* and other workers, and although they are limited in number, many other furochromones both of the linear and angular type have been obtained synthetically. The *Ring Index* naming system 5*H*-furo[3,2-*g*]benzopyran-5-one (CXIX) is not commonly used, and although several nomenclature systems are in use along with trivial names, in the following discussion furo [3,2-*g*] chromone (CXX) will be regarded as the parent compound.

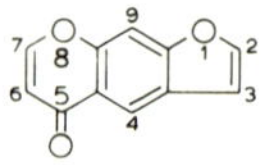

(CXIX) 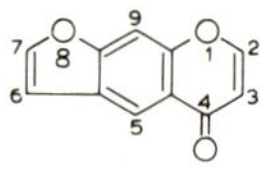 (CXX)

Khellin, 5,9-*dimethoxy-2-methylfuro*[32-g]*chromone*(4,9-*dimethoxy-7-methyl-*5H-*furo*[3,2-g]*benzopyran-5-one*), $C_{13}H_{10}O_5$, m.p. 154°, is sparingly soluble in cold water and has a bitter taste. A drop of an aqueous solution of the substance added to solid potassium or sodium hydroxide produces a rose-red colour. Compounds believed to be oxonium salts, have been obtained by dissolving khellin in concentrated sulphuric acid, when a deep orange coloured solution is formed. Khellin has pronounced physiological activity and has clinical potentialities in the treatment of anginal syndrome and other kinds of muscle spasm. It is used in cases of bronchial asthma and whooping cough.

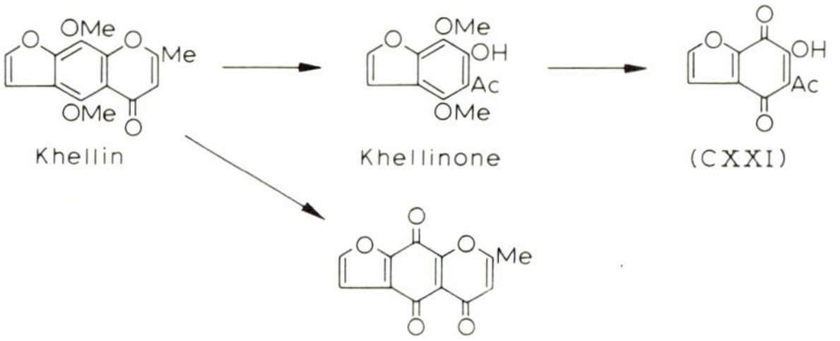

Khellin Khellinone (CXXI)

(CXXII)

The structure of khellin, first advanced by *E. Späth* and *W. Gruber* (Ber., 1938, **71,** 106) was established by degradation and synthesis. It contains two methoxyl groups and the presence of the furan ring is shown by oxidation with alkaline hydrogen peroxide to furan-2,3-dicarboxylic acid. Less violent degradation is effected by alkali which opens the ring to give the ketone *khellinone*, m.p. 95°. A detailed study has been made of this reaction (*C. Musante*, Gazz., 1958, **88,** 910). Khellinone is oxidised by nitric acid to 3-acetyl-2-hydroxyfuro[3,2-l]benzo-1,4-quinone, $C_{10}H_6O_5$, (CXXI) thus showing that the two methoxyl groups in khellin occupy the two *para*-positions. A similar conclusion is drawn from the oxidation of khellin with nitric acid to the quinone CXXII. Reduction of this quinone with sulphurous acid gives demethylated khellin, which with methyl iodide yields khellin. The position of the methyl group is confirmed by the condensation of khellin with aromatic aldehydes, anisaldehyde, for example, yielding 2-(4-methoxybenzylidene)khellin (cf. p. 148).

The structure of khellin was finally established by synthesis (*J. R. Clarke* and *A. Robertson*, J. chem. Soc., 1949, 302; *R. A. Baxter et al.*, *ibid.*, 1949, 30; *V. V. S. Murti* and *T. R. Seshadri*, Proc. Indian Acad. Sci., 1949, **30,** 107; *T. A. Geissman* and *T. G. Halsall*, J. Amer. chem. Soc., 1951, **73,** 1280; *O. Dann* and *G. Illing*, Ann., 1957, **605,** 146; *R. Aneja, S. K. Mukerjee* and *Seshadri*, J. Sci. Ind. Res. India, 1958, **17B,** 382; Ber., 1960, **93,** 297; *Dann* and *H. G. Zeller*, *ibid.*, 1960, **93,** 2829). Acetylation of 2,3-dihydro-4,6,7-trimethoxybenzo[*b*]furan to *dihydrokhellinone*, m.p. 104°, constitutes an improved method for the prepation of this compound, an intermediate in the synthesis of khellin (*W. J. Horton* and *E. G. Paul*, J. org. Chem., 1959, **24,** 2000):

OMe, OMe, OMe — BF_3, AcOH, Ac_2O → OMe, OH, Ac, OMe
Dihydrokhellinone

It is manufactured in high yield from 3,5,7-triacetoxybenzofuran (*Dann*, U.S. 3,099,660/1963) and attempts to obtain it from 2-methyl-5, 7-dihydroxychromone have been reported (*M. M. Badawi* and *M. B. E. Fayez*, Acta Chim., Budapest, 1968, **55,** 397).

Primary aliphatic amines react with khellin and its derivatives with opening of the 4-pyrone ring to give a series of derivatives of 6-hydroxy-4,7-methoxybenzofuran with the amino group in a side chain (*Musante* and *A. Stener*, Gazz., 1956, **86,** 297). Oxidation of khellin with selenium dioxide in boiling ethyl acetate afforded a mixture of 5,9-*dimethoxyfuro*[3,2-g]*chromone-2-carbaldehyde* (4,9-*dimethoxy-5-oxo-5H-furo*[3,2-g]-*benzopyran-7-carbaldehyde*), 2,4-*dinitrophenylhydrazone*, m.p. 320–322° and the corresponding *carboxylic acid*, m.p. 270–272° (*G. Renzi* and *P. Perini*, Farmaco, Ed. Sci., 1969, **24,** 1073; C.A. 1970, **72,** 78917):

OMe, Me, OMe, O — SeO_2 → OMe, CHO, OMe, O + OMe, CO_2H, OMe, O

The hydrogenation products obtained following the reduction of khellin with platinum in methanol, palladium on barium sulphate in acetone, and lithium tetrahydridoaluminate in tetrahydrofuran have been described (*Dann* and *G. Volz*, Ann., 1965, **685,** 167). Khellin forms a crystalline addition compound with thiourea (*N. A. Strakowsky*, Egypt. J. Chem., 1959, **2,** 111). Crystallographic data have been presented for khellin (*S. Morgante* and *A. Damiani*, Periodico Mineral., Rome, 1962, **31,** 99). Colour reaction for khellin (*M. M. Sidky* and *M. R. Mahran*, Arch. Pharm., 1963, **296,** 569).

Ammiol, m.p. 211°, has been synthesised from khellin (*A. Mustafa, Starkowsky* and *T. I. Salama*, J. org. Chem., 1961, **26,** 886:

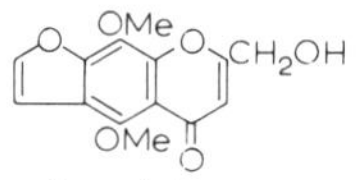

Ammiol 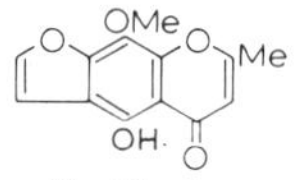 Khellinol

Khellinol, *demethylkhellin*, $C_{13}H_{10}O_5$, m.p. 203–203.5°, a yellow crystalline substance, is identical with 5-norkhellin obtained by the demethylation of khellin (*W. Bencze et al.*, Helv., 1956, **39,** 923; *Mustafa et al.*, Monatsh., 1967, **98,** 310).

Visnagin, 5-*methoxy*-2-*methylfuro*[3,2-g]*chromone* (4-*methoxy*-7-*methyl*-5H-*furo*[3,2-g]*benzopyran*-5-*one*), $C_{13}H_{10}O_4$, m.p. 144°, is hydrolysed to *visnaginone*, $C_{11}H_{10}O_4$, m.p. 109–111°, which gives a green colour with ferric chloride, forms an *acetate*, m.p. 64.5–65.5°, and has been synthesised (*Geissman* and *E. Hinreiner*, J. Amer. chem. Soc., 1951, **73,** 782). The structure of visnagin has been established by synthesis (*Clarke et al.*, J. chem. Soc., 1948, 2260; *J. S. H. Davies* and *W. L. Norris*, *ibid.*, 1950, 3195; *Gruber* and *K. Horvath*, Monatsh., 1950, **81,** 819; *Badawi* and *Fayez*, Tetrahedron, 1965, **21,** 2925); *demethylvisnagin*, m.p. 156° (*Mustafa et al., loc. cit.*). Visnagin forms crystalline addition compounds with thiourea and urea (*Strakowsky, loc. cit.*), and gives a characteristic colour reaction (*Sidky* and *Mahran, loc. cit.*):

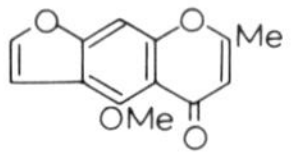

Visnagin 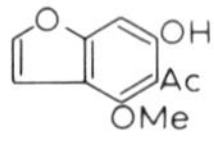 Visnaginone 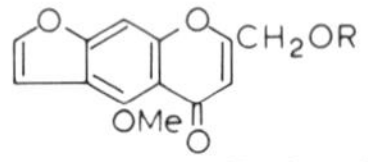 Khellinin (R = $C_6H_{11}O_5$)

Khellol glucoside, *khellinin*, (CXXIII, R = $C_6H_{11}O_5$), $C_{19}H_{20}O_{10} \cdot 2H_2O$, m.p. 142–144°, anhyd., m.p. 175°, $[\alpha]_D$ −33° (pyridine), is a glucoside of hydroxyvisnagin (*Späth* and *Gruber*, Ber., 1941, **74,** 1549), which exerts a stimulating action on the heart and increases the coronary flow. Alkali converts it into visnaginone, but acid gives **khellol** (CXXIII, R = H), $C_{13}H_{10}O_5$, m.p. 176–178°, which has been synthesised (*Geissman* and *J. W. Bolger*, J. Amer. chem. Soc., 1951, **73,** 5875) and also obtained from visnagin (*Mustafa, Starkowsky* and *Salama, loc. cit.*).

Visamminol, a dihydrovisnagin derivative, $C_{15}H_{16}O_5$, m.p. 160–160.5°, $[\alpha]_D^{18}$ +93° ($CHCl_3$), *diacetate*, m.p. 148–149°, *methyl ether*, m.p. 141.8–142.3° (*E. Smith et al.*, Science, 1952, **115,** 520; *Bencze et al., loc. cit.*); its chemistry has been reviewed (*H. Schmid*, Proc. roy. Dublin Soc., 1956, **27,** 145):

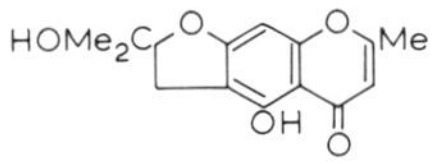

Visamminol

Hydroxyfurochromones and related substances give a colour reaction with uranyl acetate solution (*Mustafa, Starkowsky* and *E. Zaki*, J. org. Chem., 1960, **25,** 794).

Linear and angular furochromones can be synthesised by thermal or acid-catalysed condensation of β-dicarboxyl or β-carboxycarbonyl compounds with 4-, 5-, 6- or 7-hydroxy substituted 2,3-dimethylbenzofurans (*J. Pierre et al.*, Bull. Soc., chim. Fr., 1966, 1716).

Certain derivatives of **furochromone,** on treatment with aqueous alkali, are converted to benzodifurans (*Musante*, Ann. Chim., Rome, 1956, **46,** 768):

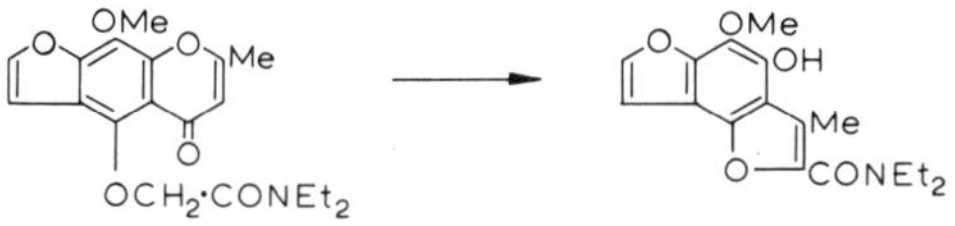

OMe, Me, $OCH_2{\cdot}CONEt_2$ → OMe, OH, Me, $CONEt_2$

Derivatives of benzofuro[3,2-*b*]benzopyran-6-one have been synthesised (*R. Bryant*, J. chem. Soc., 1965, 5140):

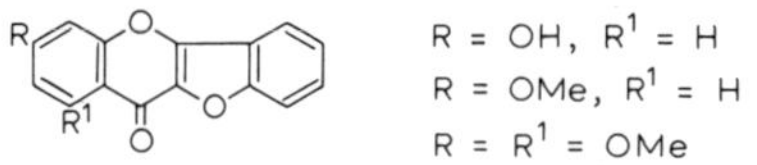

R = OH, R[1] = H
R = OMe, R[1] = H
R = R[1] = OMe

Lisetin, $C_{21}H_{18}O_7$, m.p. 283–286°, has been isolated from the root bark of Jamaican Dogwood, *Piscidia erythrina* and its structure elucidated (*C. P. Falshaw et al.*, Tetrahedron Suppl., 1966, **7,** 333):

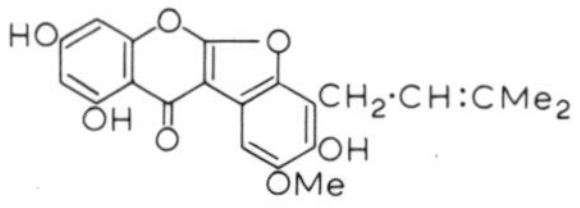

HO, OH, $CH_2{\cdot}CH{:}CMe_2$, OH, OMe

Lisetin

(xvi) Pyranochromones

Hamaudol, $C_{15}H_{16}O_5$, m.p. 197–197.5°, *diacetate*, m.p. 185–185.5°, is a constituent of the roots of *Angelica japonica*, which on long boiling in benzene with *p*-toluenesulphonic acid affords *anhydrovisamminol*, m.p. 117–118°. Its structure has been suggested on chemical and spectroscopic evidence (*A. Nitta*, Yakugaku Zasshi, 1965, **85,** 55):

Hamaudol   Allopteroxylin

Allopteroxylin, $C_{15}H_{16}O_4$, m.p. 156°, from *Ptaeroxylon obliquuom* has been synthesised (*C. Mercier, C. Mentzer* and *D. Billet*, Compt. rend. Ser. C, 1967, **265,** 945; *Mercier*, Ann. Chim., Paris, 1970, **5,** 373). Although the above two compounds possess a dihydropyran ring, a number of similar pyranochromones have been isolated from *Spathelia sorbifolia* containing both linear and angular ring systems (*D. R. Taylor* and

J. A. Wright, Rev. Latinoamer. Quim., 1971, **2,** 84). The *methyl ether* of allopteroxylin, m.p. 155–157°, has been obtained from the heartwood of *Cedrelopsis grevei* and synthesised (*F. M. Dean* and *M. L. Robinson*, Phytochem., 1971, **10,** 3221; *B. S. Bajwa, P. Lal*, and *T. R. Seshadri*, Indian J. Chem., 1971, **9,** 17).

Sorbifolin, $C_{20}H_{22}O_5$, m.p. 143–145°, is obtained from the roots and stems of *Spathelia sorbifolia* L. Its structure is supported by spectral data and simple chemical transformations, and confirmed by synthesis of the tetrahydroanhydro derivative (*W. R. Chan, Taylor* and *C. R. Willis*, J. chem. Soc. C, 1967, 2540):

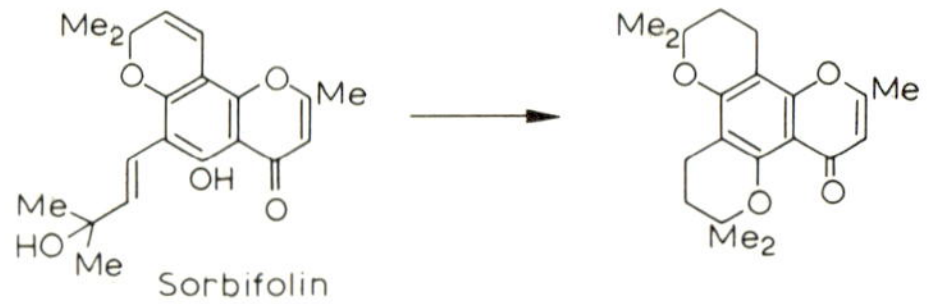

Sorbifolin

Fulvic acid, $C_{14}H_{12}O_8$, m.p. 246° is a yellow metabolite produced by fungi including *Carpenteles brefeldianum* Dodge (*A. E. Oxford et al.*, Biochem. J., 1935, **29,** 1102), *ethyl ester*, m.p. 230–234°, *trimethyl ether* m.p. 192°. On dehydration it yields anhydrofulvic acid, $C_{14}H_{10}O_7$, isomeric with citromycetin (below). Its structure was determined by *Dean et al.* (J. chem. Soc., 1957, 3497; 1961, 798):

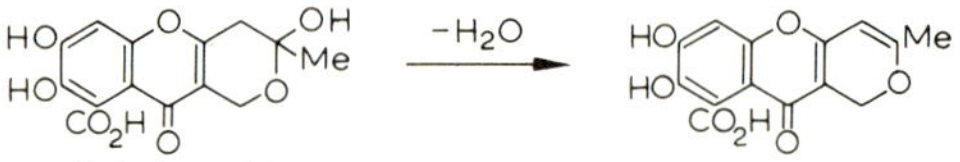

Fulvic acid

Citromycetin, $C_{14}H_{10}O_7 \cdot 2H_2O$, yellow, m.p. 285°, *diacetate*, m.p. 224°, is a metabolic product of various *Citromyces* species. Its structure was established by a study of its degradation products (*A. Robertson et al., ibid.*, 1949, 848; 1950, 903, 1031; 1951, 2013). Decarboxylation of citromycetin gives *citromycin*, $C_{13}H_{10}O_5$, m.p. 290°, *diacetate*, m.p. 224°. Alkaline hydrolysis of di-*O*-dimethylcitromycin affords 4,5-dimethyoxy-2-hydroxyacetophenone, and oxidation yields di-*O*-methylcitromycinol (CXXIV, R = H, R′ = OH) and di-*O*-methylcitromycinone (CXXIV, RR′ = O). Acid hydrolysis of this ketone breaks the middle ring to give a pyrone derivative (CXXV) (*G. W. K. Cavill et al., ibid.*, 1950, 1031). Methyl di-*O*-methylcitromycetin (CXXVI) is oxidised with lead tetra-acetate and chromic acid to give methyl di-*O*-methylcitromycetinol (CXXVII, R = H, R′ = OH) and methyl di-*O*-methylcitromycetinone (CXXVII, RR′ = O), respectively. Aqueous alkali hydrolysis of ketone (CXXVII, RR′ = O) yields 2-acetyl-7-hydroxy-4,5-dimethoxyindane-1,3-dione (CXXVIII), a degradation product also common to methyl di-*O*-methylanhydrofulvate (*Dean, D. R. Randell* and *G. Winfield, ibid.*, 1961, 792):

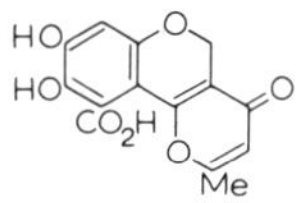

Citromycetin  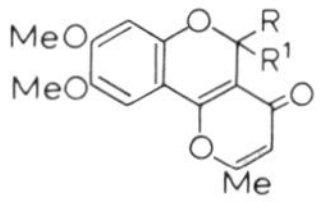 (CXXIV) 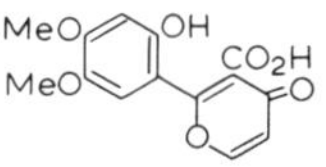 (CXXV)

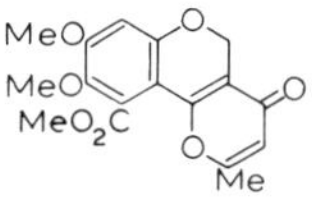

(CXXVI) 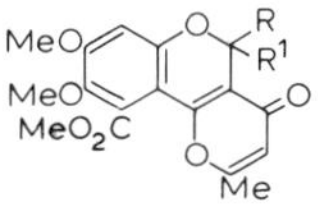 (CXXVII) 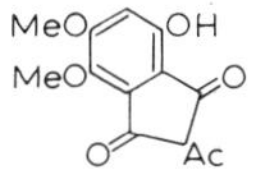 (CXXVIII)

Some of the details of citromycetin chemistry given in the early literature have been revised because of evidence obtained from spectroscopic data and other techniques (*Dean, Randell* and *Winfield, loc. cit.; Dean* and *Randell*, J. chem. Soc., 1961, 798).

Distemonanthin, $C_{17}H_{10}O_9$, yellow needles, m.p. 351° (decomp.), contains a chromonoisocoumarin nucleus, *tetra-acetate*, m.p. 225–227°, *tetrabenzoate*, m.p. 239–240°; *tetramethyl ether*, m.p. 272° (*F. E. King et al., ibid.*, 1954, 4594). The structural similarity to peltogynol (p. 93) is striking:

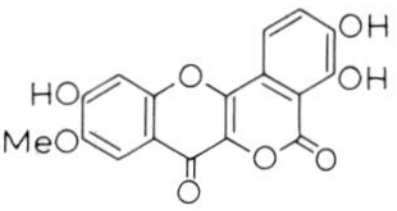

Distemonanthin 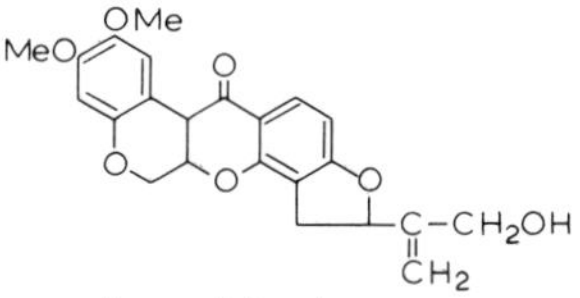 Amorphigenin

The hydrolysis of amorphin gives glucose, arabinose and **amorphigenin,** the structure of which is supported by chemical and spectral evidence (*L. Crombie* and *R. Peace*, Proc. chem. Soc., 1963, 246). For the biosynthesis of amorphigenin see *Crombie, P. M. Dewick* and *D. A. Whiting*, Chem. Comm., 1970, 1469; 1971, 1182, 1183.

6,7-*Dihydropyrano*[3,2-g]*chromone-2-carboxylic acid* (6,7-*dihydro*-4H,8H-*pyrano*-[3,2-g]*benzopyran-4-one-2-carboxylic acid*) (CXXIX, R=CO_2H), m.p. 278°, *ethyl ester*, (CXXIX, R=CO_2Et), m.p. 129–130°, *methyl ester* (R=CO_2Me), m.p. 132°; 6,7-*dihydropyrano*[3,2-g]*chromone* (CXXIX, R=H), m.p. 116.5° (*P. Naylor* and *G. R. Ramage*, J. chem. Soc., 1960, 1956).

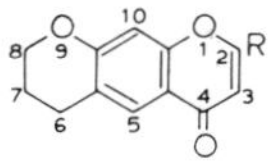

(CXXIX) 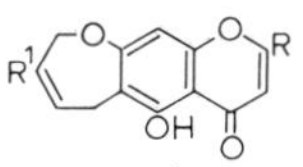 Desoxykarenin (R = R^1 = Me)

Desoxykarenin, $C_{15}H_{14}O_4$, m.p. 133–135°, and **karenin** (R=CH_2OH, R′=Me), $C_{15}H_{14}O_5$, m.p. 204–205°, have been obtained from the heartwood of *Ptaeroxylon obliquum* (Thunb.) Radlk., more commonly known as sneezewood, and their constitution has been elucidated by conversion to a derivative of peucenin (p. 154) (*P. H. McCabe*, *R. McCrindle* and *R. D. H. Murray*, *ibid.*, C, 1967, 145).

(xvii) Flavones (2-phenylchromones, 2-phenyl-4*H*-benzo[*b*]pyran-4-ones)*

Flavone (from the latin *flavus*, yellow), 2-phenylchromone, 2-phenyl-4*H*-benzo[*b*]pyran-4-one, is the parent substance of a number of important mordant dyes found in plants, leaves, fruits, and flowers. These substances all contain hydroxyl groups and occur naturally in the uncombined state or as glycosides. Some flavones have been used as dyestuffs, but most of them have been replaced by synthetic dyes. In the literature, up to recent times the name flavone is extensively used, but now the studies of flavones as a class are indexed under this heading, whereas there is a tendency to derive the name of individual flavones from the parent name 2-phenyl-4*H*-benzo-[*b*]pyran-4-one. When the former naming is used it is necessary to indicate differently the numberings of the phenyl group in position 2. This does not arise when the latter naming is applied.

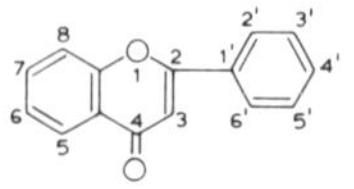

Flavone 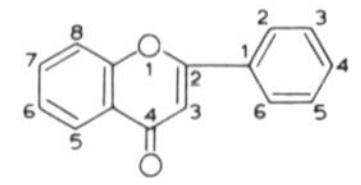   2-Phenyl-4*H*-benzo[*b*]pyran-4-one

In this section compounds will be given names which are generally used for them in the literature. The polyhydroxyflavones, which form a group of naturally occurring substances, are described in a separate section (p. 174).

An alternate nomenclature of flavonoids based on polyhydroxyflavans with well-known modified trivial names has been suggested (*K. Freudenberg* and *K. Weinges*, Tetrahedron, 1960, **8**, 336).

Preparation. A number of general methods for the preparation of flavones have been discussed (p. 139); they have been obtained by method (*1*)(*b*) (p. 140) (*L. Reichel* and *H. G. Henning*, Ann., 1959, **621**, 72; *A. V. R. Rao*, *S. A. Telang* and *P. M. Nair*, Indian J. chem., 1964, **2**, 431) and by a modified Baker–Venkataraman transformation (*G. Srimannarayana* and *N. V. S. Rao*, *ibid.*, 1968, **6**, 696; 1969, **7**, 940); by method (*2*) by the action of pyridine on chalcone dibromides (*P. N. Wadodkar*, *ibid.*, 1963, **1**, 163) and by treating the chalcone with iodine chloride in acetic acid, followed by reacting the product with pyridine, ethanolic sodium hydroxide or silver acetate in acetic acid (*idem*, *ibid.*, p. 122); 8-hydroxyflavone has been prepared

* *T. A. Geissman* and *E. Hinreiner*, The Botanical Review, 1952, **18**, 120; *F. M. Dean*, "Naturally Occurring Oxygen Ring Compounds", Butterworths, 1963, Ch. 10, p. 280.

from the chalcone obtained from 2,3-dihydroxyacetophenone (*W. I. Awad, M. F. El-Neweiky* and *S. F. Selim*, J. org. Chem., 1969, **25,** 1333); by method (*3*), 7-hydroxyflavanone is acetylated before treatment (*K. Nakagawa* and *H. Tsukahima*, Nippon Kagaku Zasshi, 1954, **75,** 485); by method (*4*), increased yields of flavones are obtained by carrying out the condensation between the di- or poly-phenol and ethyl aroylacetate in nitrobenzene under nitrogen (*R. Teoule*, Bull. Soc. chim. Fr., 1959, 423), by heating the reactants together under reduced pressure (*Teoule et al.*, *ibid.*, 1961, 546), and the thermal condensation of monosubstituted monophenols and ethyl benzoylacetate or 3,4,5-trimethoxybenzoylacetate (*A. Vialard-Goudou* and *N. Blanchecotte*, Compt. rend., 1962, **255,** 953; 1965, **260,** 6930).

β-Morpholino-2-chlorochalcone obtained by the addition of morpholine to 2-chlorophenylbenzoylacetylene, cyclises when heated to yield 4-morpholinoflavylium chloride, which is readily hydrolysed to flavone (*P. L. Southwick* and *J. R. Kirchner*, J. Amer. chem. Soc., 1957, **79,** 689):

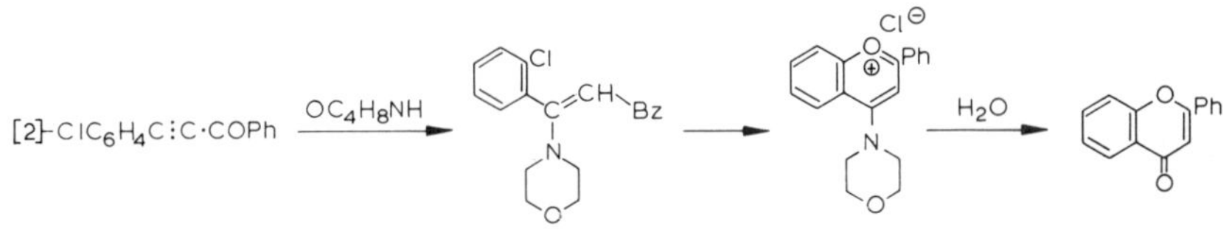

[2]-$ClC_6H_4C{:}C{\cdot}COPh$ $\xrightarrow{OC_4H_8NH}$ → H_2O →

Flavones have been synthesised by heating the appropriate acetylene compound, *e.g.* CXXX in acetic acid containing polyphosphoric acid (*Y. Okajima*, Yakugaku Zasshi, 1960, **80,** 322):

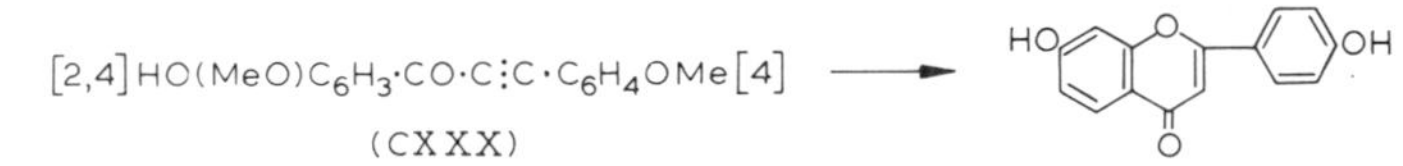

[2,4] $HO(MeO)C_6H_3{\cdot}CO{\cdot}C{:}C{\cdot}C_6H_4OMe$ [4] →

(CXXX)

The condensation of phenylacetylenecarboxylic acid (phenylpropiolic acid) with phenols gives substituted flavones (*N. Hasebe*, Nippon Kagaku Zasshi, 1957, **78**, 1102; *Reichel* and *G. Proksch*, Ann., 1971, **745**, 59).

Phenyl epoxycinnamate, besides undergoing photochemical cleavage to phenylcarbene, also undergoes a Fries rearrangement giving 2′-hydroxyepoxychalcone, which is partially photolysed further into the diketone CXXXI, easily converted to flavone (*V. T. Ramakrishnan* and *J. Kagan*, J. org. Chem., 1970, **35,** 2898):

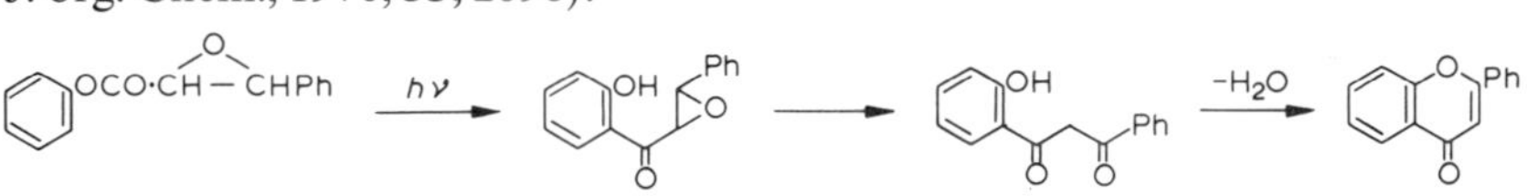

$h\nu$ → → $-H_2O$ →

(CXXXI)

Salicylaldehyde reacts with 1-morpholino-1-phenylethylene in benzene solution to form a non-crystalline adduct (CXXXII), which is oxidised to flavone with chromium trioxide-pyridine (*L. A. Paquette* and *H. Stucki*, *ibid.*, 1966, **31,** 1232):

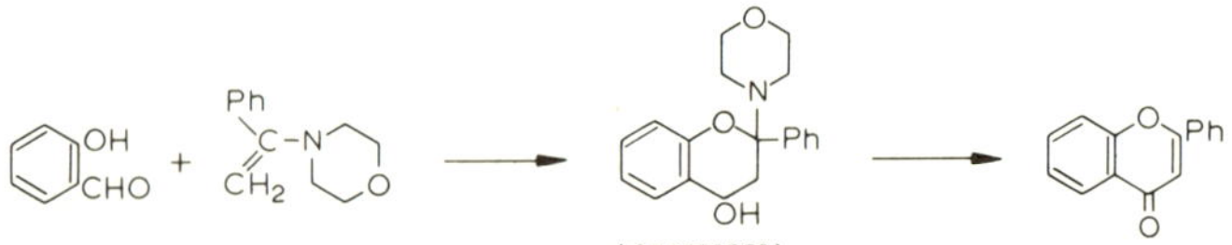

(CXXXII)

A number of flavonecarboxylic acids and aminoflavones, with the carboxylic acid group and the amino group substituents in the 2-phenyl group have been synthesised (*A. I. Tolmachev*, *Zh. N. Belaya* and *L. M. Shulezhko*, Zhur. obshcheĭ Khim., 1968, **38,** 1139), *flavone-2'-carboxylic acid*, m.p. 220°, *methyl ester*, m.p. 128–130°, dinitrophenylhydrazone, m.p. 288° (decomp.) (*G. Wurm* and *H. Loth*, Arch. Pharm., 1970, **303,** 413); preparation of 4',5,6,7-oxygenated flavones (*M. G. Stout*, *H. Reich* and *M. H. Huffman*, J. pharm. Sci., 1964, **53,** 192); review of *R. Robinson*'s work on the synthesis of flavones (Beitr. Biochem. Physiol. Naturstoffen, Festschr., 1965, 11); other reviews on the synthesis and chemistry of flavones (*N. V. Bringi et al.*, Sci. Proc. roy. Dublin. Soc., 1956, **27,** 93; *T. H. Simpson*, *ibid.*, p. 111; *O. Isaac*, Pharm. Ztg., 1957, **102,** 71).

2-(2-Furyl)chromones have been prepared by the oxidation of furfurylidenechalcones with selenium dioxide in boiling isoamyl alcohol (*S. S. Kumari et al.*, Curr. Sci., 1967, **36,** 430):

Properties and structure. The structure of flavones is determined chiefly by means of alkaline degradation. The final products (compare chromones, p. 144) obtained by *ring-fission* with alkali depend on the nature of the substituent (R) in position 3 and the nature of the hydrolysing agent employed. Anhydrous ethanolic potassium hydroxide causes fission at *A*, whereas aqueous potassium hydroxide does so mainly at *B* (cf. p. 145). There

is evidence that both in acid and alkaline solution an equilibrium between the flavone and the diketone exists, *i.e.*, ring-fission may occur in acid solution and ring-closure in alkaline solution (see, *e.g.*, *S. K. Mukerjee* and *T. R. Seshadri*, Chem. and Ind., 1955, 271).

Ring-fission accompanied by *reduction* can be effected by heating flavones with tetralin and palladium/charcoal (*D. Pillon*, Bull. Soc. chim. Fr., 1955, 39), 7-hydroxy-5-methoxyflavone thus yielding 4,6-dihydroxy-2-methoxy-dihydrochalcone. Flavanones have also been obtained by this reduction (*C. Mentzer* and *J. Massicot*, *ibid.*, 1956, 144). Reduction with lithium tetra-hydridoaluminate gives a mixture of flavenes and flaven-4-ols (*C. G. Joshi* and *A. B. Kulkarni*, Chem. and Ind., 1954, 1421), but it is reported that flav-2-ene is obtained when the reaction is carried out using an excess of lithium tetrahydridoaluminate (*K. G. Marathe*, *E. M. Philbin* and *T. S. Wheeler*, *ibid.*, 1962, 1793). Sodium amalgam reduces the carbonyl group and the double bond at $C_{(2)}-C_{(3)}$ to give 4-hydroxyflavans, the corresponding chalcones, and coloured products (*Y. Asahina et al.*, Ber., 1928, **61,** 1646; 1929, **62,** 3016; *T. A. Geissman* and *R. O. Clinton*, J. Amer. chem. Soc., 1946, **68,** 697, 700, 706). It is claimed that hydrogenation with an Adams platinum oxide catalyst gives flavanones and by this means luteolin was converted into eriodictyol, the first example of the conversion of a naturally occuring flavone into the corresponding naturally occurring flavanone (*Geissman* and *Clinto*, *loc. cit.*, p. 697). When the hydrogenation of 5-hydroxy-, 7-hydroxy-, and 5,7-dihydroxy-flavone with a platinum catalyst is stopped when absorption of hydrogen becomes slow after 3 moles have been taken up, it is found that the aromatic group at the 2-position is hydrogenated preferentially (*M. Suzuki*, *H. Mizuno* and *M. Takai*, Nippon Kagaku Zasshi, 1967, **88,** 675):

3 moles H_2 / Pt

R	R[1]	M.p.(°C)	Yield (%)
OH	H	102-102.5	74
H	OH	200-201	54
OH	OH	216-217	74
H	MeO	84-85	60
H	AcO	101.5-102	50

It is reported that high pressure hydrogenation of flavone gives flavan-4-ol, flaven-4-ol, and flavanone depending on the conditions and the catalyst (*R. Mozingo* and *H. Adkins*, J. Amer. chem. Soc., 1938, **60,** 669). 7-Methoxy-flavone has been hydrogenated at room temperature and under ordinary pressure using various catalysts (*Suzuki* and *T. Oda*, Nippon Kagaku Zasshi, 1968, **89,** 878).

The flavones are colourless or yellow crystalline substances, soluble in water and ethanol, and can be isolated by the insoluble double salts they form with calcium chloride (*D. W. Hill* and *R. R. Melhuish*, J. chem. Soc., 1935, 1161). They are moderately strong oxygen bases (*C. T. Davis* and *Geissman*, J. Amer. chem. Soc., 1954, **76,** 3507), soluble in acids. The oxonium salts thus formed are in general more highly coloured than the parent flavones, and are unstable to water (compare the stable oxonium salts of the anthocyanidins). The flavones dissolve in alkali to give yellow solutions.

i.r. Spectra (*L. Henry* and *D. Molho*, Colloq. intern. Centre natl. recherche sci., Paris, 1955, **64,** 341; C.A., 1960, **54,** 10516; *G. E. Inglett*, J. org. Chem., 1958, **23,** 93; *L. H. Briggs* and *L. D. Colebrook*, Spectrochim. Acta, 1962, **18,** 939; *J. H. Looker* and *W. W. Hanneman*, J. org. Chem., 1962, **27,** 381); u.v. spectra (*Briggs* and *R. H. Locker*, J. chem. Soc., 1951, 3136; *Geissman*, "Modern Methods of Plant Analysis", Vol. III, p. 487 (Springer-Verlag, 1955); *J. J. Chirikdjian* and *W. Bleier*, Sci. Pharm., 1971, **39,** 65) and n.m.r. spectra (*T. J. Batterham* and *R. J. Highet*, Austral. J. Chem., 1964, **17,** 428; *A. Grouiller*, Bull. Soc. chim. Fr., 1966, 2405). Flavones have been tested for use as u.v. filters (*H. Ikeda* and *K. Toba*, J. Soc. Sci. Phot. Japan, 1956, **18,** 110; C.A. 1957, **51,** 2395). Dielectric constants and dipole moments over a range of temperatures have been reported for a number of flavones (*S. K. K. Jatkar* and *C. M. Deshpande*, J. Indian chem. Soc., 1960, **37,** 69).

Carbonyl group reactivity. The inertness of the carbonyl group in the chromones is also exhibited by the flavones although flavone does form a 2,4-dinitrophenylhydrazone. Flavone condenses with 9-fluorenylsodium to give 2-hydroxy-β-fluorenylidene-β-phenylpropiophenone (*A. Schönberg* and *E. Singer*, Ber., 1961, **94,** 241), and with some compounds containing a reactive methylene group (*F. Eiden*, Arch. Pharm., 1962, **295,** 127).

The application of the *Mannich reaction* to flavones gives rise to the corresponding 8-aminomethyl derivatives (*P. Da Re*, *L. Verlicchi*, and *I. Setnikar*, J. org. Chem., 1960, **25,** 1097).

Substitution occurs in the 2-phenyl group, flavone, for instance, on nitration yielding a mixture of 2′-, 3′-, and 4′-nitroflavones (*M. T. Bogert* and *J. K. Marcus*, J. Amer. chem. Soc., 1919, **41,** 83), but may occur in the Bz ring if this contains hydroxyl groups.

Thus 5-hydroxyflavone, for instance, depending on the conditions yields 5-*hydroxy*-6-*nitro*-, m.p. 237–238°, or -8-*nitro-flavone*, m.p. 225–226° (*P. E. McCusker*, *Philbin* and *Wheeler*, J. chem. Soc., 1963, 2374); 7-hydroxyflavone gives 7-*hydroxy*-8-*nitro*-, m.p. 300°, and 7-*hydroxy*-6,8-*dinitro-flavone*, m.p. 288° (*A. M. Mehta et al.*, Proc. Indian Acad. Sci., 1949, **29A,** 314); a number of hydroxynitroflavones have been synthesised (*S. Seshadri* and *P. L. Trivedi*, J. org. Chem., 1958, **23,** 1735). Monobromination of 5-hydroxyflavone yields 8-*bromo*-5-*hydroxyflavone*, m.p. 179–180°, *acetate*, m.p. 212–214°; 8-*bromo*-5-*methoxyflavone*, m.p. 189–190°; 6,8-*dibromo*-5-*hydroxyflavone*, m.p. 250–251°, *acetate*, m.p. 187–189° (242°); 6,8-*dibromo*-5-*methoxyflavone*, m.p. 226–228° (242°) (*McCusker, Philbin* and *Wheeler, loc. cit.).* Boiling flavone in benzene with thionyl chloride yields 3-*chloroflavone*, m.p. 125° (*J. R. Merchant* and *D. V. Rege*, Tetrahedron Letters, 1969, 3589); sulphuryl chloride reacts with flavone and 3-chloroflavone to give 2,3,3-*trichloroflavanone*, m.p. 114–115° (*idem*, Chem. Comm., 1970, 380); 3-*chloro*-6-*methylflavone*, m.p. 135°; 3-*chloro*-6-*dichloromethylflavone*, m.p. 182–183°; 3-*chloro*-6-*formylflavone*, m.p. 198°; 3-*chloro*-7-*methoxyflavone*, m.p. 119–120°; 3,8-*dichloro*-7-*methoxyflavone*, m.p. 208° (*idem*, Tetrahedron, 1971, **27,** 4837). 6- and 7- Hydroxyflavone have been chloromethylated (*U. K. Jagwani*, J. Indian chem. Soc., 1970, **47,** 119).

Oxidation. Flavone has been directly oxidised by Udenfriend reagent (a buffered solution of ascorbic acid and ferrous ion chelate) in position 3′ and 4′. Under optimum conditions 3-hydroxyflavone is formed by a non-radical complex process (*B. Winicki et al.*, Compt. rend., 1960, **251,** 103; Bull. Soc. chim. Fr., 1962, 1695).

3,7-Dimethoxyflavone undergoes oxidative photocyclisation to a chromonoisochroman (CXXXIII, R = OMe) and a chromonoisocoumarin (CXXXIV, R = OMe); 3-methoxyflavone yields the chromonoisocoumarin (CXXXIV, R = H), and under similar conditions 3-methoxy-5-hydroxyflavones are unreactive (*T. Matsuura* and *H. Matsushima*, Tetrahedron, 1968, **24,** 6615):

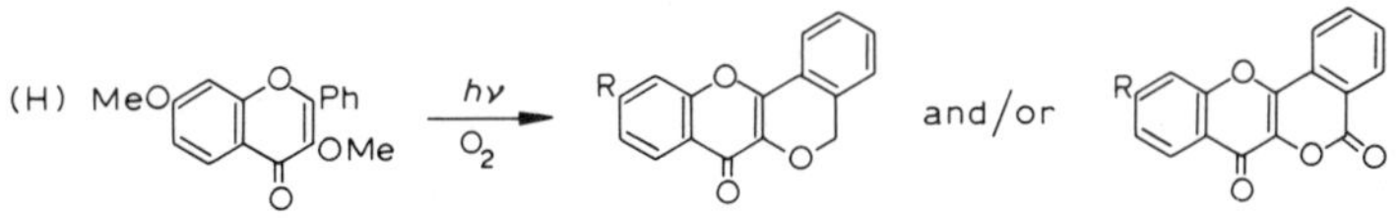

(H) MeO ... Ph, OMe —$h\nu$, O_2→ (CXXXIII) and/or (CXXXIV)

Flavones, unsubstituted in the 3- and 5-positions, form yellow, dimeric bisflavenylidenes CXXXV under the condition of *reductive acetylation*, which on treatment with mineral acids are mono-protonated to yield substituted 4-flavenylflavylium salts (CXXXVI) (*B. J. Bergot* and *L. Jurd*, Tetrahedron, 1965, **21,** 657):

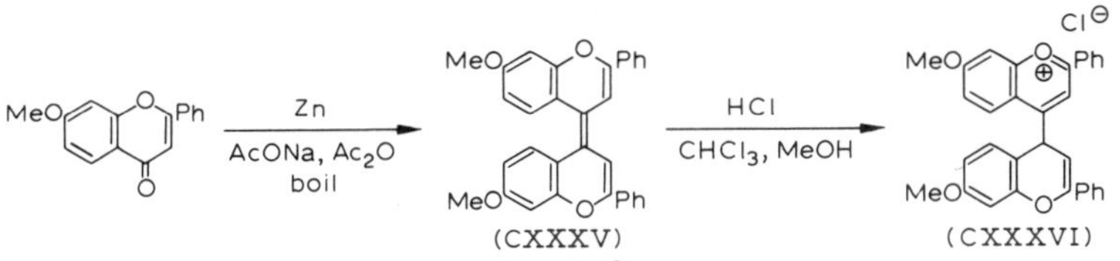

MeO ... Ph —Zn, AcONa, Ac_2O, boil→ (CXXXV) m.p. 223° —HCl, $CHCl_3$, MeOH→ (CXXXVI) $Cl^{\ominus}$

Hydroxy groups are acetylated during the above reaction. 3-Acetylflavone, its 4′-hydroxy and 4′-nitro derivatives and 3-nitroflavone react with amines to yield the corresponding 2-(β-aminocinnamyl)phenol derivative (*Z. Jerzmanowska* and *B. Podwinski*, Monatsh., 1967, **98**, 1395). 3-Cyanoflavone has been obtained from 3-benzoylchromone (p. 151).

Read of chromonoisocoumarin (CXXXVII) with lithium tetrahydridoaluminate in ether gives a spiran, 3-hydroxycoumaran-2-spiro-3′-isochroman (CXXXVIII), which with acid undergoes cleavage to the isochromene, as expected from its acetal structure, and subsequent ring closure and oxidation occurs readily to yield a flavylium salt (*J. W. Clark-Lewis* and *D. C. Skingle*, Tetrahedron Letters, 1966, 4199):

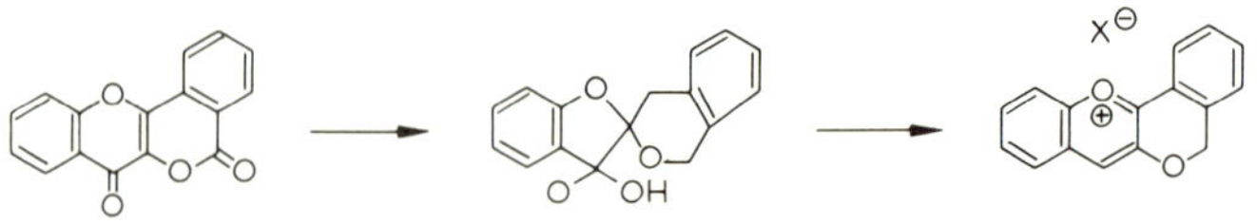

(CXXXVII) (CXXXVIII)

Flavone, 2-phenylchromone, 2-phenyl-4H-benzo[b]pyran-4-one, $C_{15}H_{10}O_2$, needles, m.p. 100°, 2,4-*dinitrophenylhydrazone*, m.p. 282°, occurs almost pure in the dust found on the stalks and leaves of primulas (*H. Müller*, J. chem. Soc., 1915, **107**, 872). It is prepared by method 1b, p. 140 (*Wheeler*, Org. Synth., 1952, **32**, 72; Coll. Vol. **4**, 478). It dissolves in sulphuric acid to give solutions, variously described as colourless or yellow, which exhibit a blue fluorescence. With fused or aqueous alkali it gives salicylic acid, 2-hydroxyacetophenone, benzoic acid, and acetophenone, and affords with methanolic barium hydroxide the intermediate 2-hydroxydibenzoylmethane. Flavone with hydroxylamine hydrochloride in pyridine yields 3-(2-*hydroxyphenyl*)-5-*phenyl-iso-oxazole*, m.p. 234° (*W. Baker et al.*, J. chem. Soc., 1952, 1303). Flavone does not give a colour reaction with sodium nitroprusside in alkali solution, cf. methylchromones (p. 147) (*M. M. Sidky* and *M. R. Mahran*, Arch. Pharm., 1963, **296**, 569).

3-**Methylflavone,** m.p. 72–74° (*A. T. M. Dunne et al.*, J. chem. Soc., 1950, 1252); 5-*methyl*-, m.p. 130° (*A. Robertson et al.*, *ibid.*, 1932, 1681); 6-*methyl*-, m.p. 122–123° (*S. Ruhemann*, Ber., 1913, **46**, 2193); 7-*methyl*-, m.p. 120°, (*Robertson et al., loc. cit.*); 8-*methyl*-, m.p. 170° (*Ruhemann, loc. cit.*); 3,6-*dimethyl*-, m.p. 93° (*G. Wittig*, Ann., 1925, **446**, 197); and 5,8-*dimethyl-flavone*, m.p. 171° (*I. Goodall* and *Robertson*, J. chem. Soc., 1936, 426).

5-**Hydroxyflavone,** m.p. 156–157° (*H. Simonis* and *S. Danshevski*, Ber., 1926, **59**, 2914; *S. Sugasawa*, C.A. 1934, **28**, 6717); 6-*hydroxy*-, m.p. 231–232° (*St. von Kostanecki et al.*, Ber., 1899, **32**, 331; *N. Hasebe*, Nippon Kagaku Zasshi, 1962, **83**, 119); 7-*hydroxy*-, m.p. 240° (*Kostanecki et al.*, Ber., 1899, **32**, 312), alkyl ethers (*J. Klosa*, J. pr. Chem., 1963, **22**, 259); 8-*hydroxy*-, m.p. 249–250° (*V. K. Ahluwalia et al.*, Proc. Indian Acad. Sci., 1953, **38A**, 480); 2′-*hydroxy*-, m.p. 249–250° (*Bogert* and *Marcus*, J. Amer. chem. Soc., 1919, **41**, 95); 3′-*hydroxy*-, m.p. 208° (*Kostanecki et al.*, Ber., 1901, **34**, 1692); 4′-*hydroxy-flavone*, m.p. 269–270° (*idem, ibid.*, 1900, **33**, 2516; *S. Hattori*, C.A., 1926, **20**, 2162).

Diflavone, 2,8-*diphenyl*-4H, 6H-*benzo* [1,2-b; 5,4-b′]*dipyran*-4,6-*dione*, $C_{24}H_{14}O_4$, needles, m.p. 280–281°, is prepared by the action of bromine on the bis-chalcone CXXXIX and treatment of the resulting tetrabromide with ethanolic potassium hydroxide (*H. Ryan* and *P. O'Neill*, Proc. roy. Irish Acad., 1915, **32B,** 48, 167); by the action of hydriodic acid on the corresponding dimethoxy compound (*J. Algar* and *K. J. Hanway*, *ibid*., 1934, **42B,** 9); and by heating the dibenzoate of 4,6-diacetylresorcinol in glycerol (*H. M. Lynch et al.*, J. chem. Soc., 1952, 2063):

(CXXXIX)

Diflavone

The diflavone, 3,9-diphenyl-6-hydroxy-1*H*,7*H*-benzo[1,2-*b*; 4,3-*b'*]dipyran-1,7-dione has been prepared by heating 8-acetyl-5,7-dihydroxyflavone with benzoic anhydride and sodium benzoate at 180° (*Y. Omoto*, *Y. Takizawa*, and *N. Sugiyama*, Bull. chem. Soc., Japan, 1971, **44,** 1160):

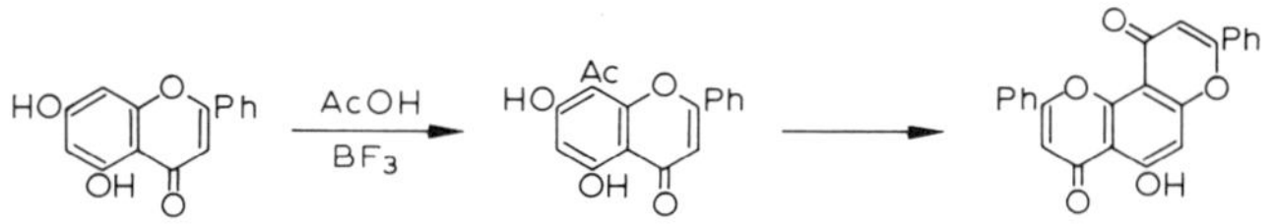

4-Thionoflavone, red needles, m.p. 87° (p. 149), *phenylhydrazone*, m.p. 155°, 2,4-*dinitrophenylhydrazone*, m.p. 282°, forms a maroon coloured *methiodide* m.p. 220–222°, which with boiling water yields flavone and methanethiol. The methiodide readily yields flavone-imines (R=alkyl or aryl) when treated with amines, e.g. *flavonebenzylimine*, m.p. 95–96°, and analogously *flavone-oxime* (R=OH), m.p. 184–186°, -*semicarbazone*,

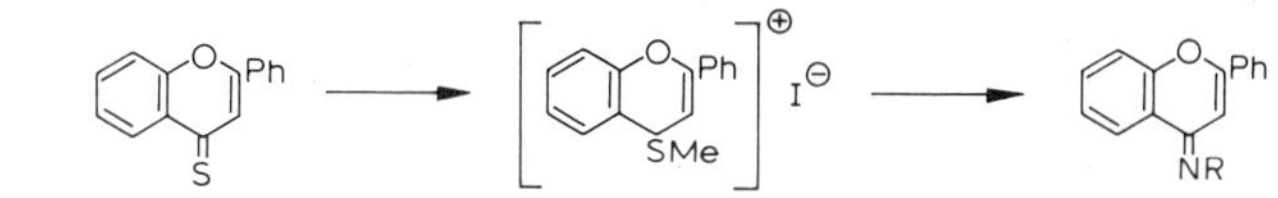

4-Thionoflavone

R=$NHCONH_2$), yellow prisms, m.p. 245° (decomp.), and -*hydrazone* (R=NH_2), yellow needles, m.p. 136° and deep yellow needles, m.p. 121.5–122.5° (*Baker et al.*, J. chem. Soc., 1952, 1303).

3-*Methyl*-, purple plates, m.p. 128–129°; 7-*methoxy*-, red or orange needles, m.p. 134–136°; 4'-*methoxy*-, red needles, m.p. 137°, and 5,7-dimethoxy-4-*thionoflavone*, dark green prisms, m.p. 184° (*Baker et al.*, *ibid*., 1954, 998).

5,6,7,8-**Tetrahydroflavone,** m.p. 121–123°, prepared from 2-acetylcyclohexanone via the triketone, 2-benzoylacetylcyclohexanone, (*R. J. Light* and *C. R. Hauser*, J. org. Chem., 1960, **25,** 538):

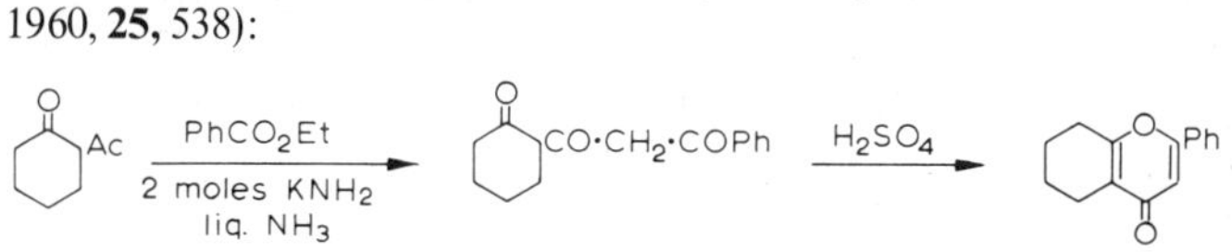

4'-Hydroxyflavones and 4'-halogenoflavones have been reacted together to give **biflavonyl ethers** (*C. L. Huang*, *T. Weng* and *F. C. Chen*, J. Heterocycl, chem., 1970, **7,** 1189):

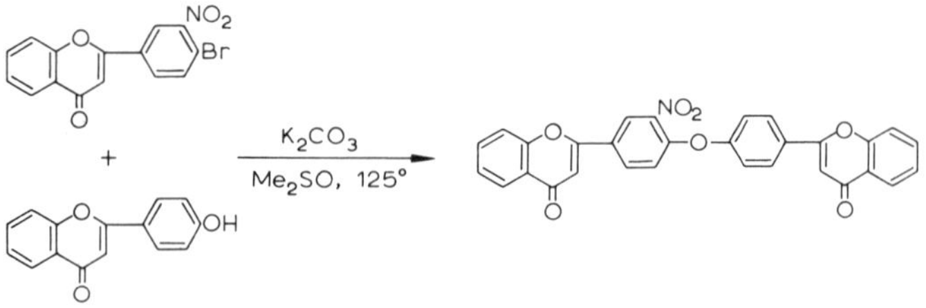

3'-Nitro-4',4'''-biflavonyl ether, m.p. 257–258°; *3'-amino-4',4'''-biflavonyl ether*, m.p. 228–230°; *4',4'''-biflavonyl ether*, m.p. 236–238°.

u.v. Absorption spectra in ethanol of flavone and some hydroxyl derivatives are given in Table 4. The data show the bathochromic effect of hydroxyl groups, particularly in the 5- and 8- positions (*D. Pillon*, Bull. Soc. Chim. Fr., 1954, 9).

TABLE 4

ABSORPTION SPECTRA OF HYDROXYFLAVONES (IN ETHANOL)

Substance	*λ_{max} (mμ);*	*Log ε in brackets*
Chromone		297–298 (3.82)
Flavone	250 (4.07)	297 (4.20)
5-Hydroxyflavone	271.5	340
7-Hydroxyflavone	250	310
5,7-Dihydroxyflavone (Chrysin)	270 (4.42)	330 (3.90)
5,7-Diacetoxyflavone	255 (4.18)	302.5 (4.43)
5-Hydroxy-7-methoxyflavone (Tectochrysin)	270 (4.40)	330 (3.88)
3,5,7-Trihydroxyflavone (Galangin)	267.5 (4.23)	360 (4.07)
3,7,8-Trihydroxyflavone	282.5	365
5,7,4'-Trihydroxyflavone (Apigenin)	265 (4.25)	340 (4.31)
5,7,4'-Trihydroxyflavone	265 (4.25)	325 (4.33)

*(xviii) Flavonols (3-hydroxyflavones, 3-hydroxy-2-phenyl-4H-benzo[*b*]-pyran-4-ones)*

Many of the ubiquitous yellow colouring materials found in nature are derivatives of flavonol, 3-hydroxyflavone, 3-hydroxy-2-phenyl-4*H*-benzo-[*b*]pyran-4-one. A few, however, are colourless, *e.g.*, morin.

Preparation. Flavonols are prepared from flavanones by reaction with nitrous acid to form 3-hydroxyiminoflavonones followed by hydrolysis with dilute mineral acid (*St. von Kostanecki et al.*, Ber., 1904, **37**, 773, 1402; *A. Sonn*, *ibid.*, 1925, **58**, 1103). Naturally occurring flavonols such as quercetin have been synthesised by this method:

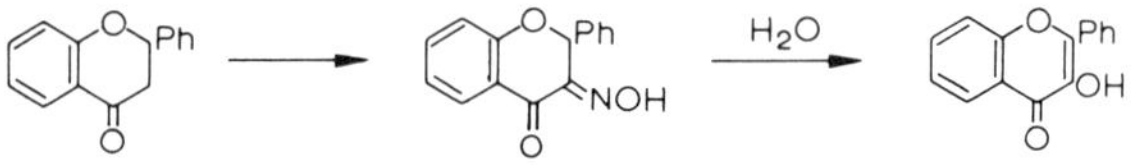

Flavonols are also prepared by a modification of the Allan–Robinson method (p. 140; J. chem. Soc., 1924, **125**, 2192), in which ω-methoxy-2-hydroxyacetophenones are heated with the anhydride and sodium salt of an aromatic acid such as benzoic acid. The resulting 3-methoxyflavones on hydrolysis yield flavonols.

Related to this method are the thermal cyclisation of aroyl esters of ω-methoxyphloracetophenones to flavonol 3-methyl ethers (*H. M. Lynch et al.*, J. chem. Soc., 1952, 2063) and the base-catalysed rearrangement of ω-methoxy derivatives of *o*-acyloxyacetophenone (*W. D. Ollis* and *D. Weight*, *ibid*., 1952, 3826).

Good yields of flavonols (CXL) are obtained by oxidising solutions of 2-hydroxychalcones in hot ethanolic potassium hydroxide solution with hydrogen peroxide (*J. Algar* and *J. P. Flynn*, Proc. roy. Irish Acad., 1934, **42B**, 1; *T. Oyamada*, Bull. chem. Soc. Japan, 1935, **10**, 182), dihydroflavonols being intermediate in this reaction. If a methoxyl or a methyl substituent is present in the 6′-position in the chalcone, aurones (CXLI) rather than flavonols are obtained, provided the 2- or 4-position does not carry a hydroxyl group. Conflicting statements in the literature, led to a re-investigation of the above reactions. Besides flavonols and aurones, 2-benzyl-2-hydroxydihydrobenzofuran-3-ones (CXLII) and 2-arylbenzofuran-3-carboxylic acids (CXLIII) are sometimes formed (*B. Cummins et al.*, Tetrahedron, 1963, **19**, 499):

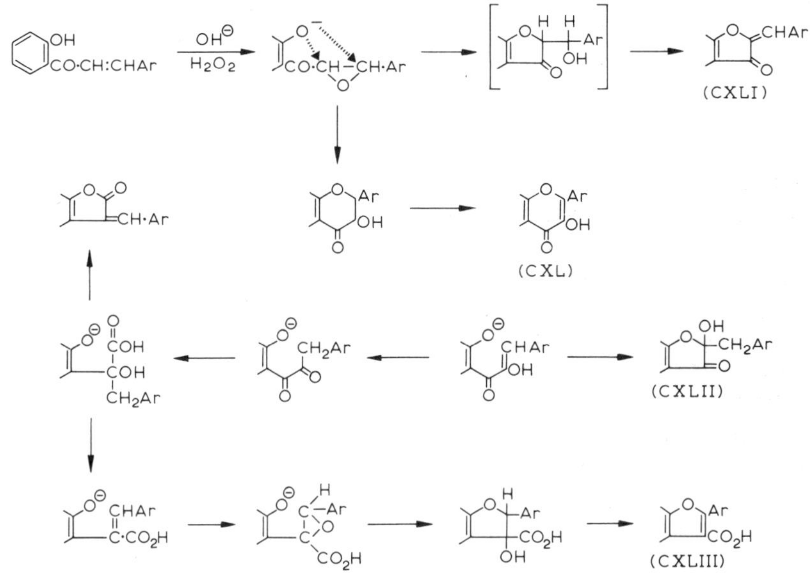

(CXLI)

(CXL)

(CXLII)

(CXLIII)

2-(α-Hydroxybenzyl)-2-methoxycoumaran-3-ones are major products in the oxidation of 2′-hydroxy-α-methoxychalcones containing a phloroglucinol type ring, *e.g.* CXLIV, with alkaline hydrogen peroxide, while flavonols predominate in the oxidation products of chalcones without a 6′-substituent. Under similar oxidation conditions, α-methyl-substituted chalcones give similar results. Stereochemistry has been assigned to the 3-methylflavan-3-ols on the basis of chemical and n.m.r. evidence (*W. P. Cullen et al.*, J. chem. Soc., C, 1971, 2848):

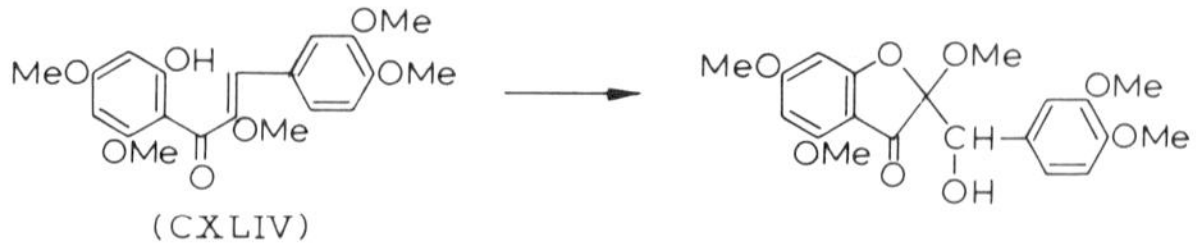

(CXLIV)

Methods for the oxidative enlargement of aurones to flavonols and other products have been discussed (*W. E. Fitzmaurice et al.*, Chem. and Ind., 1955, 652; *J. E. Gowan, E. M. Philbin* and *T. S. Wheeler*, Sci. Proc. roy. Dublin Soc., 1956, **27**, 185).

For oxidation of 3-hydroxyflavanones to flavonols see p. 276. Flavonol and 6-methyl-3′,4′-dimethoxyflavonol have been obtained by autoxidation (*K. G. Marathe*, Science and Culture, India, 1956, **22**, 175).

2,4-Dihydroxy-5-bromoacetophenone has been condensed with aromatic aldehydes and the resulting chalcones converted to flavonols (*B. J. Ghiya* and *M. G. Marathey*, J. Indian. chem. Soc., 1960, **37,** 739).

3-Hydroxyflavanone may be converted to 4-amino-3-hydroxyflavylium chloride, the diacetyl derivative of which may be hydrolysed to give flavonol (*G. Janzso, F. Kallay*, and *I. Koczor*, Tetrahedron Letters, 1965, 2269).

Chemical properties. Reduction. Flavonol with hydrogen and a copper oxide catalyst affords flavan-3,4-diol. Flavonols with sodium dithionite not only undergo reduction to flavanonols, but also suffer ring-contraction to give 2-benzylcoumaranones (*T. A. Geissman* and *H. Lischner*, J. Amer. chem. Soc., 1952, **74,** 3001). Thus quercetin yields dihydroquercetin and 4,6,3′,4′-tetrahydroxy-2-benzylcoumaran-3-one. A qualitative analysis has been made of the electrolytic reduction products of flavonols (*J. P. Haluk* and *M. Metche*, Chim. Anal., Paris, 1970, **52,** 12). The reduction of 6-methyl-3,4′-dimethoxyflavone with lithium tetrahydridoaluminate yields 6-*methyl-3,4′-dimethoxyflav-2-ene*, m.p. 108° (*A. B. Kulkarni* and *C. G. Joshi*, J. sci. ind. Research, India, 1957, **16B,** 249):

Reductive acetylation of flavonols gives acetates of undetermined structures having one acetyl group less than would be expected. Hot mineral acid converts the acetates into the corresponding anthocyanidins (p. 81) (*H. G. C. King* and *T. White*, J. chem. Soc., 1957, 3901).

Oxidation. Flavonols are oxidised by periodic acid in aqueous dioxane to form dihydrobenzofuran-3-ones (CXLV), a ring–chain tautomer of hydroxy-1,3-diphenylpropanetrione, and the corresponding flavandione (*M. A. Smith*, J. org. Chem., 1963, **28**, 933):

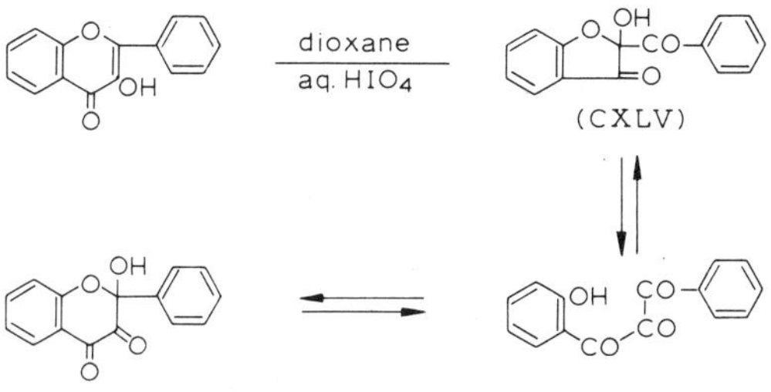

dioxane

aq. HIO_4

(CXLV)

When the oxidation of flavonol is carried out in methanol the product formed is a 3-hemiacetal of the flavan-3,4-dione (CXLVI). Sublimation of the hemiacetal gives the flavandione (*Smith*, *L. A. Webb* and *L. J. Cline*, *ibid*., 1965, **30**, 995):

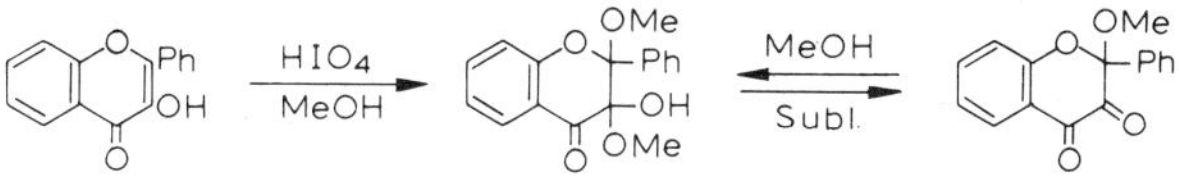

HIO_4 MeOH

MeOH Subl.

(CXLVI)

Irradiation of a solution of flavonol in isopropyl alcohol–benzene with a high pressure mercury lamp through Pyrex gives an isomer, 3-hydroxy-3-phenylindan-1,2-dione (*T. Matsuura*, *T. Takemoto* and *R. Nakashima*, Tetrahedron Letters, 1971, 1539):

Flavonol, 3-hydroxyflavone, *3-hydroxy-2-phenyl-4H-benzo[*b*]pyran-4-one*, $C_{15}H_{10}O_3$, needles, m.p. 169° (171–172°), *acetate*, m.p. 110–111°, *methyl ether*, m.p. 114°, strong violet fluorescence in sulphuric acid, is hydrolysed by boiling ethanolic potassium hydroxide to benzoic acid and 2-hydroxybenzoylmethanol. The flavonols are profoundly decomposed by oxidation and hydrolysis is often effected in an atmosphere of hydrogen (*S. Hattori* and *K. Hayashi*, Ber., 1933, **66,** 1279).

Diflavonol, 2,8-*diphenyl*-4H,6H-*benzo*[1,2-b;5,4-b′]*dipyran*-3,7-*diol*-4,6-*dione*, yellow prisms, m.p. 323°, is formed from the appropriate chalcone by treatment with hydrogen peroxide and alkali (*J. Algar* and *D. E. Hurley*, Proc. roy. Irish Acad., 1936, **43B,** 83); for 2,8-diaryl-4*H*,6*H*-benzo[1,2-b;5,4-*b*′)dipyran-3,7-diol-4,6-diones, see *L. Reichel* and *H. W. Doering*, Ann., 1971, **745,** 71:

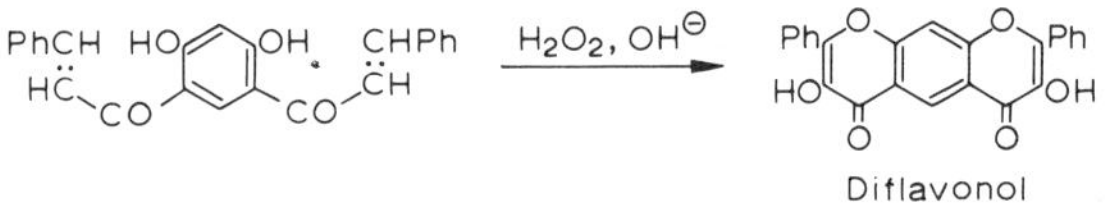

PhCH HO OH CHPh; HC CO CO CH $\xrightarrow{H_2O_2,\ OH^{\ominus}}$ Diflavonol

Diflavonol

Flavonols have been methylenated with methylene iodide to afford bi(flavonyloxy)-methanes (CXLVII):

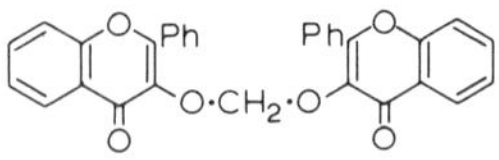

O·CH$_2$·O

(CXLVII)

Similarly methination using iodoform yields tris(flavonyloxy)methanes (*D. D. Berge* and *M. M. Bokadia*, J. India chem. Soc., 1970, **47,** 941).

(xix) Flavone and flavonol pigments

Hydroxyflavones and anthocyanins (p. 81) are the main pigments in fruits, flowers and trees. A number of naturally occurring hydroxyflavones such as quercetin, luteolin, fisetin, etc., were used as yellow dyestuffs for many centuries and their structures were first established by *St. von Kostanecki.* The pigments are polyhydroxyflavones and flavonols which occur in nature mostly as glycosides and in some cases as methyl ethers. Many are 5,7-dihydroxy derivatives.

Synthesis. Naturally occurring flavone glycosides may be synthesised from fully acetylated flavanone glycosides by bromination and treatment of the 3-bromo product with alkali (p. 143); thus hesperidin gives diosmin (*G. Zemplén* and *R. Bognár*, Ber., 1943, **76,** 452, 776). *N. Narasimhachari* and *T. R. Seshadri* (Proc. Indian Acad. Sci., 1949, **30,** 151) showed that flavanone glycosides, preferably those containing a free phenolic group in position 5, are smoothly oxidised to flavone glycosides by iodine in boiling ethanol in the presence of sodium acetate. The flavone glycosides are also obtained by the interaction of hydroxyflavones with reagents such as tetra-acetyl-α-glycosyl bromide followed by deacetylation, and from 2-glucosyloxyaroylacetophenones by intramolecular rearrangement and ring-closure (*W. Baker et al.*, J. chem. Soc., 1952, 1505; *J. Chopin, A. Durix* and *M. L. Bouillant*, Tetrahedron Letters, 1966, 3657). Both methods are of limited scope. The conditions of formation of flavones and flavone glucosides have been discussed (*L. Reichel* and *H. G. Henning*, Ann., 1959, **621,** 72) and the synthetic routes to some flavones containing a 7-hydroxy group have been compared (*M. Leonte*, *M. Beschia* and *V. Nica*, Steroids Lipids Res., 1970, **4,** 185).

The oxidation of flavones with a 5-hydroxyl group by potassium persulphate in alkaline aqueous pyridine to products with an additional hydroxyl group in the 8-position is used to synthesise naturally occurring flavones (*K. V. Rao* and *Seshadri*, Proc. Indian Acad. Sci., 1947, **26,** 417). Thus chyrsin, 5,7-dihydroxyflavone, is oxidised to norwogin, 5,7,8-trihydroxyflavone. By this method 8- as well as 5-hydroxyflavones yield 5,8-dihydroxyflavones (*Baker et al.*, J. chem. Soc., 1949, 1560). On occasions, two hydroxyl groups may be introduced, 5-hydroxy-4′-methoxyflavone, for example, yielding both 5,8-dihydroxy-4′-methoxyflavone and 5,6,8-trihydroxy-4′-methoxyflavone.

Chemical properties. The hydroxyflavones can be methylated, acetylated, etc., but chelation between the 5-hydroxy group and the contiguous carbonyl group results in greatly decreased reactivity of this hydroxyl group. 5,8-Dihydroxyflavone (primetin) for instance, is methylated only in the 8-position. Diazomethane often fails to methylate 5-hydroxyflavones (*V. C. Farmer et al.*, *ibid.*, 1956, 3600; cf. however *O. Kubota* and *A. G. Perkin*, *ibid.*, 1925, **127,** 1889). This chelation of the 5-hydroxy group with the 4-carbonyl group is reflected in the chromatographic behaviour of 5-hydroxyflavones (*T. H. Simpson* and *L. Garden*, *ibid.*, 1952, 4638), their i.r. spectra (*B. L. Shaw* and *Simpson*, *ibid.*, 1955, 655; *H. L. Hergert* and *E. F. Kurth*, J. Amer. chem. Soc., 1953, **75,** 1622; *G. E. Inglett*, J. org. Chem., 1958, **23,** 93), the small difference between their dry and wet melting-points (*K. M. Gallagher et al.*, J. chem. Soc., 1953, 3770), and the shift of the u.v. absorption maxima to shorter wave-lengths when the hydroxyl group is methylated.

Methyl iodide in ethanolic alkali can methylate the ring, quercetin, for instance, yielding *6-methylquercetin-3,7,3′,4′-tetramethyl ether*, m.p. 183–184° (*A. C. Jain* and *Seshadri*, J. sci. ind. Res. India, 1953, **12B,** 564), chrysin affording *5-hydroxy-7-methoxy-6-methylflavone*, m.p. 170–172° (*R. Iengar et al.*, *ibid.*, p. 119). Completely acetylated polyhydroxyflavones are preferentially alkylated at the 7-position (*L. Jurd*, Chem. and Ind., 1957, 1452). Nuclear methylation does not occur using dimethyl sulphate, aqueous potassium hydroxide, and acetone (*Baker* and *R. Robinson*, Jchem. Soc., 1928, 3115).

The position of the sugar residue in the flavone pigments can be determined by methylation and subsequent hydrolysis, the glucoside galuteolin, for example, thus giving 5-hydroxy-7,3′,4′-trimethoxyflavone, thereby showing that the sugar attachment is at $C_{(5)}$. u.v. Absorption measurements can also be used (*G. H. Mansfield et al.*, Nature, 1953, **172,** 23).

The demethylation of methoxyflavones is accomplished by hydriodic acid, hydrobromic acid in acetic acid, or aluminium chloride in nitrobenzene, but with the first two reagents rearrangement may also occur (*F. Wessely* and *G. H. Moser*, Monatsh., 1930, **56,** 97; 1932, **60,** 26; *R. C. Shah et al.*, J. chem. Soc., 1938, 1555; *V. D. N. Sastri et al.*, Proc. Indian Acad. Sci., 1946, **24A,** 238).

5,8-Dimethoxyflavone with hydrobromic acid and acetic acid yields 5,6-dihydroxyflavone (*Baker et al.*, J. chem. Soc., 1939, 956, 1922). The value of such rearrangements in synthesis is exemplified by the preparation of 5,6,7-trihydroxyflavones from 5,7,8-trihydroxyflavones, 5,8-dimethoxy-7-hydroxyflavone with hydriodic acid affording 5,6,7-trihydroxyflavone (baicalein), and 7-hydroxy-5,84′-trimethoxyflavone yielding 5,6,7,4′-tetrahydroxyflavone (scutellarein) (*L. H. Briggs* and *R. H. Locker*, *ibid.*, 1949, 2157; *Rao et al.*, Proc. Indian Acad. Sci., 1949, **29A,** 72).

The rearrangement of 5,8-dimethoxyflavone is probably the result of demethylation and ring-fission to an intermediate *β*-diketone, which can undergo ring-closure to yield either 5,8- or 5,6-dihydroxyflavone (*Shah et al., loc. cit.*):

OMe, Ph, OMe, O → OH, OH, $CO{\cdot}CH_2{\cdot}COPh$, OH → (OH), Ph, (HO), OH, O

support for this comes from the rearrangement of 2′-methoxyflavones when heated with hydriodic acid at 170–180° (*T. S. Wheeler et al.*, J. chem. Soc., 1953, 3770; 1955, 4249). 2′,5′-Dimethoxyflavone, for example, yields 6,2′-dihydroxyflavone:

A, B, MeO, OMe, O → A, OH, HO, B, $CO{\cdot}CH_2{\cdot}CO$, OH → HO, B, HO, A, O

Rearrangement of 5,8-dihydroxyflavones to the 5,6-isomers has been reviewed and the rearrangement of 2′-hydroxyflavones has been discussed (*E. M. Philbin* and *Wheeler*, Colloq. intern. Centre natl. Rech. Sci., Paris, 1955, **55,** C.A., 1963, **58,** 6781).

Polymethoxyflavones can be partially demethylated by aluminium chloride in nitrobenzene. The rates of demethylation of methoxy groups by hydrobromic acid lie in the order $3' > 4' > 7$ (*Simpson* and *J. L. Beton*, *ibid.*, 1954, 4065), 7,3′-dimethoxyflavone thus yielding 3′-hydroxy-7-methoxyflavone. The 5-methoxyl group is readily demethylated by hydrobromic acid at room temperature in acetic acid, a reagent which also demethylates 3-methoxyflavone (*Shah et al.*, J. Indian chem. Soc., 1942, **19,** 135). Demethylation of 6-acetyl-5,7,4′-trimethoxyflavone with hydrochloric acid – acetic acid gives 6-*acetyl*-7,4′*dimethoxy*-5-*hydroxyflavone*, m.p. 220°, and with aluminium chloride in nitrobenzene *6-acetyl-5,4′-dimethoxy-7-hydroxyflavone*, m.p. 225°, is obtained (*K. Nakazawa*, *S. Tsubouchi* and *Y. Enokida*, J. pharm. Soc. Japan, 1956, **76,** 1204). When 2′,6′-dimethoxyflavonol is treated with either hydriodic acid under pressure at 185–200° or boiled with sulphuric acid (70%) the expected product 2′,6′- or the 5,2′-5,2′-dihydroxyflavonol is not obtained. The resulting product which failed to give a positive test with ethanolic ferric chloride possessed a u.v. spectrum similar to that of tri-*O*-methylwedelolactone (CXLVIII) *Philbin*, *Wheeler* and *F. Ó. Cinnéide*, Chem. and Ind., 1961, 715):

(CXLVIII)

Boron trichloride readily demethylates methoxy groups *ortho* to a carbonyl group; quercetin pentamethyl ether gives 5-hydroxy-3,7,3′,4′-tetramethoxyflavone (*F. M. Dean et al.*, Tetrahedron Letters, 1966, 4153).

Methoxy groups have been located in flavone and flavonol aglycones and some glycosides by converting them to their trimethylsilyl ethers and measuring the n.m.r. spectra in carbon tetrachloride and benzene. Both methoxy and trimethylsilyl groups show diagnostic benzene-induced resonance shifts (*E. Rodriguez*, *N. J. Carman* and *T. J. Mabry*, Phytochem. 1972, **11,** 409).

7-Acetoxy- and 5,7-diacetoxyflavones are very resistant to alkaline hydrolysis (*Baker* and *V. S. Butt*, J. chem. Soc., 1949, 2142).

The flavonols, unlike the flavones, are oxidised by air in alkaline solution.

3- and 5-Hydroxyflavones give intense colours with ferric chloride, the former giving invariably a brown and the latter almost invariably a green colour. Colours are also given by 8-hydroxyflavones, but not by 6-, 7-, or 4′-hydroxyflavones (*L. H. Briggs* and *R. H. Locker*, J. chem. Soc., 1951, 3136). 5-Hydroxyflavones give a yellow colour with boric acid, the test being indicative of chelation; thus quercetin gives a positive and fisetin a negative test (*C. W. Wilson*, J. Amer. chem. Soc., 1939, **61,** 2303). The addition of cadmium acetate or cupric acetate solution to a dilute methanolic solution of the glycosides of 3- or 5-hydroxyflavones gives a colour or coloured precipitate which is decolourised by acetic acid; the coloured solution or precipitate obtained from 3- and 5-hydroxyflavones is stable towards acetic acid (*R. Neu*, Z. anal. Chem., 1956, **153,** 95). Fluorescence is characteristic of 3-hydroxy- but not 5-hydroxy-flavones.

The u.v. spectra have been determined for a number of flavonols before and after the addition of $Al_2Cl_6 \cdot 12H_2O$. The following showed shifts (56–62 mμ) towards lower wavelength by the addition of Al_2Cl_6; 3-hydroxy-, 3,7-dihydroxy-, 3-hydroxy-7-methoxy-, 3,5,7-trihydroxy-, and 3,5,7,2′,4′-pentahydroxy-flavone; no shift occurred with those flavonols possessing a 3-methoxy, a 5-hydroxy or a 7-hydroxy group, for example, 5,7-dihydroxy-3-methoxy-, 5-hydroxy-, 3,7,2′,4′-tetramethoxy-, and 7-hydroxy-3-methoxy-flavone. Hence the 3-hydroxy group in flavonol is essential for effecting changes in the u.v. spectra (*K. Hayashiya*, Nippon Nogeikagaku Kaishi, 1959, **33,** 1063). Hydroxyl substitution in the 3-position causes a marked bathochromic shift in the u.v. absorption band at 300–450 mμ when the spectra are recorded in 0.01 *M* NH_4OH–EtOH; similarly, a 7-hydroxy group causes a bathochromic shift in the band at 240–300 mμ compared with the spectra in NaOAc–EtOH (*F. Tomas, O. Carpena*, and *J. Mataix*, An. Quim., 1972, **68,** 123). A number of mono- and di-*O*-methylflavonols have been prepared and their i.r. and u.v. spectra recorded (*H. Pacheco* and *A. Grouiller*, Bull. Soc. chim. Fr., 1965, 779).

Some flavonols have been acetylated by reaction with acetic anhydride containing a small amount of magnesium perchlorate at 60° (*V. A. Bandyukova* and *V. D. Ponomarev*, Khim. Prir. Soedin., 1970, **6,** 418).

Treatment of 7-hydroxy- and 3-methoxy-7-hydroxy-flavone with methylene iodide and potassium carbonate in boiling acetone and dioxane–acetone, respectively, gives di(7-flavonyloxy)- (CXLIX, R = H) and di(3-methoxy-7-flavonyloxy)-methane (CXLIX, R = OMe) (*S. K. Grover, Jain* and *Seshadri*, Tetrahedron, 1964, **20,** 555):

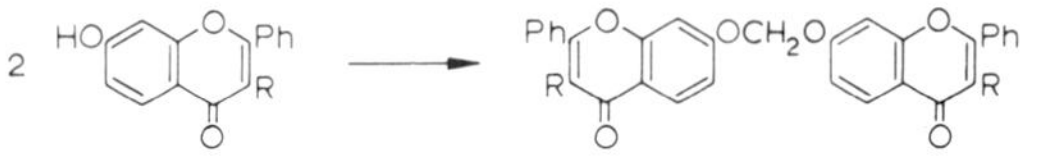

(CXLIX)

The i.r. (*L. Henry* and *D. Molho*, C.A., 1960, **54,** 10516; *I. P. Kovalev* and *V. I. Litvinenko*, Khim. Prirodn. Soedin., Akad. Nauk Uz. SSR, 1965, 233; *J. H. Looker et al.*, J. Heterocycl. Chem., 1966, **3,** 55), u.v. (*J. J. Chirikdjian* and *W. Bleier*, Sci. Pharm., 1971, **39,** 65), and n.m.r. spectra (*T. J. Batterham* and *R. J. Highet*, Austral. J. Chem., 1964, **17,** 428) of a number of flavones and flavonols have been reported. The redox potentials of some 3′,4′-dihydroxy flavone and flavonol derivatives have been determined (*H. Loth* and *H. Diedrich*, Arch. Pharm., 1968, **301,** 103). The reduction of some flavonoid glycosides by magnesium in ethanol gives products the u.v. absorption intensities of which have been used to calculate their molecular weights with a 0.3 to 2.0% error (*E. T. Oganesyan et al.*, Khim. Prir. Soedin., 1972, **8,** 57).

Flavone *O*-glycosides and *C*-glycosides may be identified by c.d. spectra (*O. Oster et al.*, Proc. Conf. appl. phys. chem., 1971, **1,** 213; C.A., 1972, **76,** 46442).

Hydroxyflavones with no 3-hydroxyl group

Chrysin, 5,7-*dihydroxyflavone*, $C_{15}H_{10}O_4$, yellow plates, m.p. 275°, *diacetate*, m.p. 192°, was discovered in 1864 by J. Piccard in poplar buds (*Pinus nigra., P. pyramidalis*) and was first synthesised by *St. von Kostanecki* (Ber., 1899, **32,** 2445). It is degraded by potassium hydroxide to phloroglucinol, benzoic acid, acetic acid and acetophenone. Methylation of chrysin by methyl iodide in alkali gives 5-*hydroxy*-7-*methoxy*-6-*methylflavone*, m.p. 170–171° and 6-*methylchrysin (strobochrysin)*, m.p. 312–314°; prolonged boiling of these products with dimethyl sulphate in acetone yields 5,7-*dimethoxy*-6-*methylflavone*, m.p. 170–171° (*J. Chopin* and *M. Justin*, Compt. rend., 1959, **248,** 3307); nuclear methylation may be achieved by means of Mannich bases, 8-*methylchrysin*, m.p. 250–253° (*J. Aknin* and *D. Molho*, Bull. Soc. chim. Fr., 1963, 604). Nuclear allylation of chrysin takes place in the 6- and 8-positions with equal ease and is accompanied by *O*-allylation in the 7-position; 6,8-diallylation also takes place, but not together with *O*-allylation in the 7-position (*V. K. Ahluwalia*, *G. P. Sachdev* and *T. R. Seshadri*, Indian J. Chem., 1967, **5,** 97). **Tectochrysin**, *chrysin* 7-*methyl ether*, $C_{16}H_{12}O_4$, yellow prisms, m.p. 165.5–166°, *acetate*, m.p. 149°, occurs in poplar buds and has been synthesised (*Kostanecki*, *loc. cit.*; *R. Robinson* and *K. Venkataraman*, J. chem. Soc., 1926, 2344). Chrysin 7-glucoside, **toringin,** m.p. 242°, is found in the bark of *Pinus toringo* and has been synthesised (*S. Hattori* and *M. Shimokoriyama*, Acta Phytochim., Japan, 1943, 109; *G. Zemplén et al.*, Ber., 1944, **77,** 99).

Stobochrysin, 5,7-*dihydroxy*-6-*methylflavone*, $C_{16}H_{12}O_4$, m.p. 285–288°, pale yellow needles, *acetate*, m.p. 190–191°, *dimethyl ether*, m.p. 170–171°, is isolated from the heartwood of *Pinus strobus* (*G. Lindstedt* and *A. Misiorny*, Acta Chem. Scand., 1951, **5,** 1). Chrysin with methyl iodide and sodium hydroxide yields *strobochrysin* 7-*methyl ether*, m.p. 170–172°, methylation of the hydroxyl group being accompanied by nuclear methylation (*D. Pillon* and *C. Mentzer*, Bull. Soc. chim. Fr., 1954, **30;** *S. K. Mukerjee* and *Seshadri*, Proc. Indian Acad. Sci., 1953, **38A,** 208).

Tachrosin, 5,7-*dimethoxy*-8-(2,3-*dihydro*-2,2-*dimethyl*-3-*oxo*-4-*furyl*)*flavone*, $C_{23}H_{20}O_6$, m.p. 226–227°, isolated from the leaves and stems of *Tephrosia polystachyoides* (*T. M. Smalberger*, *R. Vleggaar* and *H. L. DeWaal*, J.S. Afr. chem. Inst., 1971, **24,** 1):

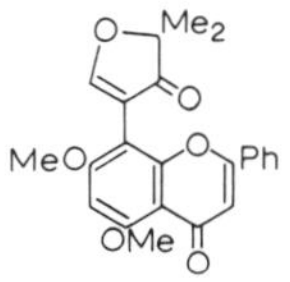

Me2 O O MeO O Ph OMe O

Tachrosin

Primetin, 5,8-*dihydroxyflavone*, $C_{15}H_{10}O_4$, yellow prisms, m.p. 230–231°, is obtained from the leaves of *Primula modesta* and forms a *dimethyl ether*, m.p. 146°, 8-*methyl ether*, m.p. 210°, and *diacetate*, m.p. 189° (*Hattori* and *W. Nagai*, J. chem. Soc. Japan, 1930, **51,** 162). The structure (*W. Baker*, J. chem. Soc., 1939, 956) was confirmed by the synthesis of primetin 8-methyl ether (*Baker et al.*, *ibid.*, 1939, 1922; *Z. Horii*, J. pharm. Soc. Japan, 1939, **59,** 209) and later by that of primetin itself (*Pillon*, Bull. Soc. chim. Fr., 1954, 9). Primetin and ferric chloride give a green colour, generally indicative of vicinal hydroxyl groups.

Baicalein, 5,6,7-*trihydroxyflavone*, $C_{15}H_{10}O_5$, yellow prisms, m.p. 264–265° (decomp.), *triacetate*, m.p. 190–192°, *trimethyl ether*, m.p. 168–169°, occurs in *Scutel-*

laria baicalensis, as the *7-glucuronate baicalin*, $C_{21}H_{18}O_{11}$, m.p. 223° (*K. Shibata* and *Hattori*, Acta Phytochim., 1930, **5**, 117) and in the root bark of *Oroxylum indicum* Vent. It gives a greenish-brown colour with ethanolic ferric chloride and is detected by its reddish-brown solution in alkali which then deposits green flocks. It has been synthesised by oxidising 5,6,7-trimethoxyflavanone with selenium dioxide followed by demethylation (*A. Oliverio et al.*, Gazz., 1948, **78**, 363), by demethylation and rearrangement of 7-hydroxy-5,8-dimethoxyflavone (*R. C. Shah et al.*, J. chem. Soc., 1938, 1555), from visnagin by three successive oxidation reactions, using selenium dioxide, chromic acid, and hydrogen peroxide (*A. Schönberg*, *N. Badran* and *N. A. Starkowsky*, J. Amer. chem. Soc., 1955, **77**, 5390), and by Dakin oxidation of 6-acetyl-5,7-dihydroxyflavone (*Y. S. Agasimundin* and *S. Siddappa*, J. chem. Soc. Perk. I, 1973, 503). For synthesis of the baicalein group see *M. Krishnamurti* and *Seshadri* (Chem. and Ind., 1954, 542). *Baicalein-7-β-L-rhamnofuranoside* has been isolated from *Scutellaria galericulata* (*E. V. Gella*, Khim. Prir. Soedin., 1972, 242).

Oroxylin-A, *5,7-dihydroxy-6-methoxyflavone*, $C_{16}H_{12}O_5$, yellow needles, m.p. 231–232°, *diacetate*, m.p. 131–132°, *7-benzoate*, m.p. 210°, occurs in the root bark of *Oroxylum indicum* Vent. (*Shah et al., loc. cit.*; 1936, 591). It has been synthesised (*V. V. S. Murti*, and *Seshadri*, Proc. Indian Acad. Sci., 1949, **29A,** 1; *T. H. Simpson*, Proc. roy. Dublin Soc., 1956, **27,** 111; *P. S. Sarin* and *Seshadri*, J. sci. ind. Research, Indian, 1960, **19B,** 117; *Molho* and *M. C. Gerphagnon*, Bull. Soc. chim. Fr., 1963, 607; *J. Vardy*, Tetrahedron Letters, 1965, 4281; *P. Rivaille* and *Mentzer*, Compt. rend., 1965, **260,** 2243) along with some of its derivatives (*K. Fukui*, *M. Nakayama* and *T. Horie*, Bull. chem. Soc. Japan., 1970, **43,** 1524).

Norwogonin, *5,7,8-trihydroxyflavone*, $C_{15}H_{10}O_5$, yellow needles, m.p. 258–260°, (227–228°) *triacetate*, m.p. 227–228° (216–217°) is obtained by careful demethylation of wogonin (*Hattori*, Ber., 1939, **72,** 1914) and 5,7,8-trimethoxyflavone (*V. D. N. Sastri*, and *Seshadri*, Proc. Indian Acad. Sci., 1946, **24A,** 243), and from chrysin (*Pillon*, Bull. Soc. chim. Fr., 1954, 9). **Wogonin,** *5,7-dihydroxy-8-methoxyflavone*, $C_{16}H_{12}O_5$, deep yellow needles, m.p. 200–201°, *7-methyl ether*, m.p. 183°, *5,7-dimethyl ether*, m.p. 167–168°, is obtained from the roots of *Scutellaria baicalensis* Georgi (*Hattori*, Acta Phytochim. Japan, 1923, **1,** 105; 1930, **5,** 114, 219; Ber., 1933, **66,** 1279). It is obtained by the partial demethylation of 7-hydroxy-5,8-dimethoxyflavone (*Shah et al.*, J. chem. Soc., 1938, 1555; *Rivaille* and *Mentzer*, Compt. rend. Ser. C., 1969, **268,** 2213) and when demethylated with hydriodic acid yields either norwogonin or baicalein (*Hattori*, Ber., 1939, **72,** 1914). 7-Methylwogonin has been converted to 7-methyloroxylin A (*L. Farkas*, *A. Major* and *J. Strelisky*, *ibid.*, 1963, **96,** 1684) and 7-benzyl- and 7-methylwogonin have been prepared from wogonin and isomerised to almost 100% of 7-benzyl- and 7-methyl-oroxylin A, respectively (*Varady, loc. cit.*).

7-Hydroxy-4'-methoxyflavone, $C_{16}H_{12}O_4$, pale yellow needles, m.p. 263–264° (rapid heating), *acetate*, dimorphic, m.p. 176–177° (stable), 167–168° (unstable), has been synthesised (*Robinson* and *Venkataraman*, J. chem. Soc., 1926, 2344; *Baker*, *ibid.*, 1933, 1381; *D. R. Nadkarni* and *T. S. Wheeler*, *ibid.*, 1938, 1321), but has not been obtained from natural sources although under the name pratol it was thought to occur in red clover (*Robinson*, Chem. and Ind. 1953, 1317).

Apigenin, *5,7,4'-trihydroxyflavone*, $C_{15}H_{10}O_5$, yellow needles, m.p. 347–348°, *tri-*

acetate, m.p. 185–187°, *tribenzoate*, m.p. 210–212°, *trimethyl ether*, m.p. 156°, occurs in yellow dahlia (*L. Schmid* and *A. Waschkau*, Monatsh., 1928, **49,** 83) and is prepared by hydrolysing apiin. It is broken down by alkali to phloroglucinol, 4-hydroxyacetophenone, and 4-hydroxybenzoic acid (*A. G. Perkin*, J. chem. Soc., 1897, **71,** 805) and its structure has been confirmed by synthesis (*Kostanecki* and *J. Tamber*, Ber., 1904, **37,** 792; *W. A. Hutchins* and *Wheeler*, J. chem. Soc., 1939, **91**; *M. O. Farocq et al.*, Arch. Pharm., 1959, **292,** 792). The 5-acetoxy group in apigenin triacetate and 5,4′-diacetoxyflavone can be selectively deacetylated by heating in a mixture of acetic acid, acetic anhydride and silver acetate (*J. H. Looker* and *M. J. Holm*, J. heterocycl. Chem., 1965, **1,** 290).

Vertiaflavone, *apigenin 5-methyl ether*, m.p. 325–327°, *diacetate*, m.p. 199–201°, has been obtained from the seeds of *Thevetia peruviana* and its structure established by mass, i.r. and n.m.r. spectroscopy (*H. W. Voigtlaender* and *G. Balsam*, Arch., Pharm., 1970, **303,** 792), and by synthesis (*J. Kraemer*, *ibid.*, p. 1013). **Cephalotaxoside,** *apigenin 5-rhamnosylglucoside*, has been isolated from the dry leaves of *Cephalotaxus drupacea* (*V. Plouvier*, Compt. rend. Ser. D., 1966, **263,** 1529).

Apiin, *apigenin 7-apiosylglucoside*, $C_{26}H_{28}O_{14} \cdot H_2O$, m.p. 228°, found in parsley and celery, is hydrolysed to apiose and cosmetin by 0.25*M*-sulphuric acid. Further hydrolysis with emulsin or acid gives apigenin and D-glucose (*C. G. Nordström et al.*, Chem. and Ind., 1953, 85; *R. Hemming* and *W. D. Ollis*, *ibid.*, p. 85). A synthesis of 4′-*O*-methylapiin has been described, which unequivocally establishes the correctness of the structure proposed for apiin and the β-configuration for the D-apiosyl unit (*A. D. Ezekiel*, *W. G. Overend* and *N. R. Williams*, J. chem. Soc., C, 1971, 2907). **Cosmetin, cosmosiin,** *apigenin 7-D-glucoside* $C_{21}H_{20}O_{10} \cdot H_2O$, m.p. 178°, anhyd., m.p. 227–230°, $[\alpha]_D^{17}$ −65.7 (pyridine–water), occurs in the flowers of *Cosmos bipinatus* (*T. Nakaoki*, J. pharm. Soc. Japan, 1935, **55,** 173, 967; 1940, **60,** 502; *M. Nogradi et al.*, Tetrahedron Letters, 1967, 1453) and along with its 4′-glucoside in the *Dahlia variabilis* (*Nordsyröm* and *T. Swain*, J. chem. Soc., 1953, 2764). Cosmosiine and **tilianin** have been synthesised *via* the application of a transacylation reaction (*Nogradi et al.*, Ber., 1967, **100,** 2783). **Rhoifolin,** *apigenin 7-rhamnoglucoside*, $C_{27}H_{30}O_{14} \cdot 3H_2O$, m.p. 263–265° (decomp.) is obtained from the latter source from *Rhus succedanea* and from the ripe fruit peel of Japanese bitter orange (*Citrus aurantium*) (*Hattori et al.*, Arch. Biophys., 1952, **37,** 85; J. Amer. chem. Soc., 1952, **74,** 3614). **Terniflorin,** *apigenin 7-(p-coumaroyl)-β-D-glucoside*, m.p. 266–267°, occurs in the flowers of *Clematis terniflora var. robusta.* Boiling with 50% potassium hydroxide solution gives *p*-coumaric acid, 4-hydroxybenzoic acid and phloroglucinol, while mild hydrolysis with 1% potassium hydroxide solution affords apigenin-7-β-D-glucoside and *p*-coumaric acid (*M. Aritomi*, Chem. pharm. Bull. Tokyo, 1963, **11,** 1225).

Acacetin, 5,7-*dihydroxy-4′-methoxyflavone*, *apigenin 4′-methyl ether*, $C_{16}H_{12}O_5$, colourless needles, m.p. 261°, *diacetate*, m.p. 203°, *dibenzoate*, m.p. 201°, *dimethyl ester*, m.p. 204°, occurs in *Robinia pseudacacia* Linn. (*Perkin*, J. chem. Soc., 1900, **77,** 430) and is prepared by the hydrolysis of linarin. It has been synthesised (*Robinson* and *Venkataraman*, *ibid.*, 1926, **128,** 2344; *Zemplén* and *R. Bognár*, Ber., 1943, **76,** 452). **Acaciin,** *acacetin 7-rhamnoglucoside*, $C_{28}H_{32}O_{13} \cdot 4H_2O$, colourless needles, m.p. 263°, $[\alpha]$ −85.3° (pyridine), −99.5° (acetic acid), from the leaves of *Robinia pseudacacia*

[*Hattori*, Acta Phytochim. Tokyo, 1925, **2,** 99; *Zemplén* and *L. Mester*, Magy. Kemi. Foly. (Ung. Z. Chem.), 1950, **56,** 2]. **Fortunellin,** obtained from the peel of the fruit of kumquat is also believed to be acacetin 7-rhamnoglucoside (*T. Matsuno*, J. pharm. Soc. Japan, 1958, **78,** 1311; *T. Nakabayashi*, Nippon Nogei Kagaku Kaishi, 1961, **35,** 45; C.A., 1963, **59,** 14232). **Linarin,** *acacetin 7-β-rutinoside*, $C_{28}H_{32}O_{14} \cdot H_2O$, monohydrate, decomp., 256° (rapid heating), anhydrous m.p. 261–263°, occurs in *Linaria vulgaris*. Its structure follows from its breakdown with alkali to acacetin, D-glucose, and L-rhamnose and synthesis (*Zemplén* and *Bognár*, Ber., 1941, **74,** 1818). *Buddleflavonoloside*, isolated from *Buddleia variabilis* is identical with linarin (*Baker et al.*, J. chem. Soc., 1951, 691). *6,8-Di-C-β-D-glucopyranosylacacetin* and its *monoacetate*, have been isolated from the seeds of *Trigonella corniculata* (*Seshadri*, *A. R. Sood* and *I. P. Varshney*, Indian J. Chem., 1972, **10,** 26).

Genkwanin, *apigenin 7-methyl ether*, $C_{16}H_{12}O_5$, yellow needles, m.p. 282°, is isolated from *Daphne genkwa* or the Chinese drug *yuen-hua* (*M. Nakano* and *K. Tseng*, J. pharm. Soc. Japan, 1932, No. 602, 343; 1933, No. 608, 905) and the bark of *Prunus puddum* (*D. Chakravarti* and *R. P. Gosh*, J. Indian chem. Soc., 1944, **21,** 171; *M. Ohta* and *S. Nishikawa*, J. pharm. Soc. Japan, 1947, **67,** 40; *M. Hasegawa* and *T. Shirato*, J. Amer. Chem., Soc., 1952, **74,** 6114). It has been synthesised (*Tseng*, J. pharm. Soc. Japan, 1935, No. 636, 30; *H. S. Mahal* and *Venkataraman*, J. chem. Soc., 1936, 569). Alkaline dimethyl sulphate yields *acacetin 7-methyl ether*, m.p. 171°. *6-Methoxygenkwanin*, m.p. 255–257°, obtained from *Teucrium polium*; methylation of the free 4′-OH group gives salvigenin, 5-hydroxy-6,7,4′-trimethoxyflavone, isolated from *Salvia triloba* (*C. H. Brieskorn* and *W. Biechele*, Tetrahedron Letters, 1969, 2603). **Swertisin,** m.p. 243°, isolated from *Swertia japonica*, and its acid-converted isomer, **isoswertisin,** m.p. 295°, have been shown to be *6-C-β-D-glucopyranosyl-* and *8-C-β-D-glucopyranosylgenkwanin*, respectively (*M. Komatsu*, *T. Tomimori* and *M. Ito*, Chem. pharm. Bull., Tokyo, 1967, 1967, **15,** 263).

Apigenin 5,7-dimethyl ether is easily obtained by selective methylation of 7,4′-diacetoxy-5-hydroxyflavone (*S. Heitz* and *Mentzer*, Compt. rend., 1961, **252,** 4214). Oxidative coupling of apigenin 7,4′-dimethyl ether with ferric chloride in boiling dioxane yields a dimer. Its methyl ether is *8,8″-biapigeninyl hexamethyl ether* (*S. Natarajan*, *Murti* and *Seshadri*, Indian J. Chem., 1971, **9,** 383).

Vitexin, $C_{21}H_{20}O_{10}$, m.p. 265°, $[\alpha]_D$ −14.5 (pyridine), obtained from *Vitex littoralis*; the original suggested structure has now been revised to 8-C-β-D-glucopyranosylapigenin on the basis of n.m.r. spectral data (*R. M. Horowitz* and *B. Gentili*, Chem. and Ind., 1964, 498). It has been shown that *orientoside* is identical with vitexin (*L. Hörhammer et al.*, Arch. Pharm., 1959, **292,** 380). **Isovitexin,** m.p. 239°, accompanies vitexin in many plants and equilibrates with it in acid solution, it is believed to be *6-C-β-D-glucopyranosylapigenin*. C.d. studies of *C*-glycosylflavones have shown that a position Cotton effect at 250–275 mμ indicates that the glycosyl residue is linked to C-6 (*e.g.*, isovitexin) while a negative Cotton effect at 250–275 mμ indicates it is linked at C-8 (*e.g.*, vitexin) (*W. Gaffield* and *Horowitz*, Chem. Comm., 1972, 648). **Bayin,** $C_{21}H_{20}O_9 \cdot 2H_2O$, m.p. 220° (decomp.), obtained from the wood of *Castanospermum australe* Cunn. et. Fras. ("Moreton Bay Chestnut") is believed to be 5-deoxyvitexin (*R. A. Eade*, *I. Salasoo* and *J. J. H. Simes*, Chem. and Ind., 1962, 1720). **Swertisin,**

6-C-*β*-D-*glucopyranosyl genkwanin*, $C_{22}H_{22}O_{10}$, pale yellow needles, m.p. 243° (decomp.), $[\alpha]_D^{20}$ −10.0 (*c*, 0.9 pyridine) isolated from the whole herb of *Swertia japonica*; methylation with diazomethane gives *swertisin dimethyl ether*, m.p. 302° (*Komatsu* and *Tomimori*, Tetrahedron letters, 1966, 1611).

Quinqueloside, m.p. 265–267°, isolated from the blooming plants of *Leonuris quinquelobatus*; on acid hydrolysis it gives apigenin, D-glucose, and *p*-coumarinic acid, and on hydrolysis with potassium hydroxide solution it yields **cosmosine**, m.p. 220–222° and *p*-coumarinic acid (*V. V. Petrenko*, Khim. Prirodn. Soedin., Akad. Nauk Uz.SSR, 1965, 414):

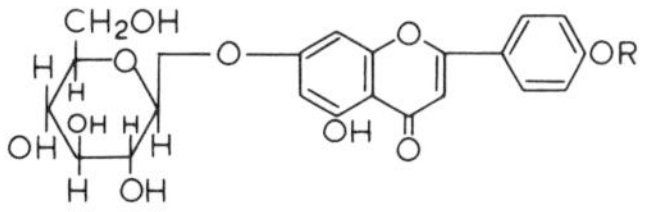

Quinqueloside (R =[4] $OHC_6H_4 \cdot CH{:}CHCO$)
Cosmosine (R = H)

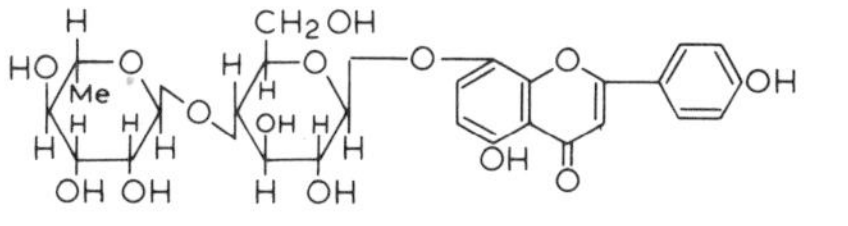

Cratenacin

Cratenacin, $C_{27}H_{30}O_{14}$, m.p. 215° u.v. λ_{max} 270 and 335 mμ (log ε 4.155 and 4.123) *monoacetate*, m.p. 254° u.v., λ_{max} 270 and 335 mμ (log ε 4.178 and 4.160) has been obtained from *Crataegus curvisepala* (*V. S. Batyuk, N. V. Chernobrovaya* and *A. P. Prokopenko*, Khim. Prirodn. Soedin., Akad. Nauk Uz. SSR, 1966, **2,** 90).

Eucalpytin, *7,4′-dimethoxy-6,8-dimethyl-5-hydroxyflavone*, $C_{19}H_{18}O_5$, pale yellow needles, m.p. 198.5–200°, is found in small amounts in the leaf waxes of *Eucalyptus globulus, E. cinerea*, and *E. risdoni*. Its structure was readily determined from its n.m.r. spectrum. Eucalyptin is obtained by boiling methyl matteucinol with iodine and potassium acetate in acetic acid (*D. H. S. Horn* and *J. A. Lamberton*, Chem. and Ind., 1963, 691). *5-Hydroxy-7,4′-dimethoxy-6-methylflavone*, m.p. 183–184°; *5,7,4′-trimethoxy-6-methylflavone*, m.p. 188–192°; *5,7,4′-trihydroxy-6-methylflavone (6-methylapigenin)*, m.p. 361°; *5,7,4′-trimethoxy-8-methylflavone*, m.p. 230–231°; *5,7,4′-trihydroxy-8-methylflavone* (*8-methylapigenin*, m.p. 348°; also reported the u.v. spectra of the above compounds (*Chopin* and *M. Chadenson*, Compt. rend., Ser. C, 1966, **262**, 662).

Tithonine, *3′-hydroxy-7,4′-dimethoxyflavone*, $C_{17}H_{15}O_5$, m.p. 193–194°, *acetate*, m.p. 178°, *methyl ether*, m.p. 182–183°, isolated from *Tithonia tubaeformis* (*J. Correa* and *M. L. Cervera*, Bull. Soc. chim. Fr., 1971, 475). It has been synthesised *via* the aldol condensation of 2-hydroxy-4-methoxyacetophenone (*Correa, Cervera* and *R. M. Maniero*, Rev. Latinoamer. Quim., 1971, **2**, 67).

Luteolin, *5,7,3′,4′-tetrahydroxyflavone*, $C_{15}H_{10}O_6 \cdot H_2O$, orange-yellow crystals, m.p. 329–330°, *tetramethyl ether*, m.p. 192°, *tetraethyl ether*, m.p. 153–155°, *tetra-acetate*, m.p. 225–226, the colouring matter of weld (*Reseda luteola*) and dyer's broom (*Genista tinctoria* L.), is said to be the oldest known European dyestuff dating back to the time of Julius Caesar. It is degraded by fusion with alkali to phloroglucinol and protocatechuic acid and by 50% potassium hydroxide to phloroglucinol and 3,4-dihydroxyacetophenone (*Perkin* and *L. H. Horsfall*, J. chem. Soc., 1900, **77,** 314), and has been synthesised (*S. Fainberg* and *Kostanecki*, Ber., 1904, **37,** 2625; *Hutchins* and *Wheeler*, J. chem., Soc., 1939, 91). **Galuteolin,** *luteolin 5-glucoside*, $C_{21}H_{20}O_{11} \cdot 3H_2O$, yellow needles, m.p. 260–263°, is obtained from the seeds of *Galega officinalis* (*G. Barger* and *F. D. White*, Biochem. J., 1923, **17,** 836) or the stems of *Equisetum arvense* L. (*H. Nakamura* and *G. Hukuti*, J. pharm. Soc. Japan, 1940, **60,** 449). **Glucoluteolin,** *luteolin 7-glucoside*, pale yellow needles, m.p. 254°, is isolated from the leaves of *Humulus japonicus* Siebold, *Sophora angustifolia* Siebold, *Digitalis purpurea* (*Hattori* and *H. Matsuda*, J. Amer. chem. Soc., 1954, **76,** 5792), and *Achillea millefolium*. It has been synthesised (*L. Hörhammer et al.*, Magy. Kem. Foly., 1964, **70,** 392; C.A., 1964, **65,** 14769). **Bignonoside,** *luteolin* 7-(6-O-p-*cumaroyl*)-*β*-D-*glucoside*, obtained from the leaves of *Catalpa bignoniodes* (*L. Birkofer*, *C. Kaiser* and *F. Becker*, Z. Naturforsch., 1965, **20b,** 923). **Orientin,** *luxtexin*, $C_{21}H_{20}O_{11}$, m.p. 265–267° (257°), obtained from *Spartium junceum* L., *Polygenum orientale*, and *Hordeum vulgare* L., and **homo-orientin,** *lutonaretin*, $C_{21}H_{22}O_{12}$, m.p. 235°, from *Aspalathus acuminatus* have been shown by their reactions and n.m.r. spectra to be 8- and 6-substituted luteolin *C*-glycosides (*B. H. Koeppen, ibid.*, 1964, **19b,** 173; Biochem. J., 1965, **97,** 444). **Adonivernite,** 6-O-*β*-D-*xylopyranosylorientin*, m.p. 205°, $[\alpha]_D$ −20.0° (ethanol), obtained from swollen duckweed (*A. I. Tikhonov*, C.A., 1972, **76,** 46453); **lucenin,** 6,8-di-C-*β*-D-*glucopyranosulluteolin*, m.p. 210–220° (decomp.), $[\alpha]_D$ +52.0° (DMF), from common duckweed (*Tikhonov* and *P. E. Krivenchuk*, C.A., 1972, **76,** 46454). 6-*Methoxyluteolin*, m.p. 264–266°, *tetra-acetate*, m.p. 202.5–203° (*K. Fukui*, *M. Nakayama* and *T. Horie*, Experientia, 1969, **25,** 355). **Swertiajaponin,** 6-C-*β*-D-*glucopyranosylluteolin-7-methyl ether*, $C_{22}H_{22}O_{11} \cdot \frac{1}{2}H_2O$, m.p. 265° (decomp.), is obtained from *Swertia japonica* (*Komatsu* and *Tomimori*, Tetrahedron Letters, 1966, 1611).

Diosmetin, *luteolin 4′-methyl ether*, $C_{16}H_{12}O_6$, yellow needles, m.p. 253–254°, *triacetate*, m.p. 195–196°, has been synthesised (*A. Lovecy et al.*, J. chem. Soc., 1930, 817; *R. Téoule et al.*, Bull. Soc. chim. Fr., 1959, 854) and occurs in *Scrophularia nodosa*, *Hyssopus officinalis*, and the bark of *Zanthoxylum nitidum* as **diosmin,** 5,7,3′-*trihydroxy-4′-methoxyflavone 7-rhamnoglucoside dihydrate*, $C_{28}H_{32}O_{15} \cdot 2H_2O$, m.p. 280° (decomp. vac. tube) (*O. A. Oesterle* and *G. Wander*, Helv., 1925, **8,** 519; *H. R. Arthur et al.*, J. chem. Soc., 1956, 632). Diosmin has been synthesised (p. 174). The 5-acetoxy group in diosmetin triacetate may be selectively deactylated (*Looker* and *Holm*, *loc. cit.*). 6-*Methoxy-diosmetin*, *demethoxycentaureidin*, m.p. 264–266°, *triacetate*, m.p. 185–187° (*Fukui*, *Makayama* and *Horie*, *loc. cit.*). **Flavoyadorinin B,** *luteolin*, 7,3′ *dimethyl ether* 4′-D-*glucoside* (7,3′-*di*-O-*methyl-luteolin*-4′-O-*mono*-D-*glucoside*) and **homoflavoyadorinin,** *luteolin* 7,3′-*dimethyl ether* 4′-D-*glucoapioside* (7,3′-*di*-O-*methyl-luteolin*-4′-O-D-*glucoapioside*) have been isolated from the leaves of *Viscum album* (mistletoe (*N. Ohta* and *K. Tagishita*, Agric. biol. Chem., 1970, **34,** 900).

5-*Hydroxy*-7,3′-*dimethoxy*-4′-(3,6-*dimethylhept-2-enyloxy*)-*flavone* has been obtained from *Melicope sarcococca* (A); the presence the unusual C-9 carbon atom alkenyloxy group is supported by n.m.r. spectral evidence and by degradation studies (*W. Brune* and *T. A. Geissman*, Austral. J. Chem., 1965, **18,** 1649).

Scutellarein, 5,6,7,4′-*tetrahydroxyflavone*, $C_{15}H_{10}O_6$, yellow leaflets, m.p. >340°, *tetramethyl ether*, two forms, m.p. 142 and 161°, occurs in the leaves and flowers of *Scutellaria* as **scutellarin,** $C_{21}H_{18}O_{12} \cdot 2.5H_2O$, yellow needles, m.p. >310°j $[\alpha]_D^{18}$ −140° (aq. pyridine), which is hydrolysed by the enzyme baicalinase to scutellarein and D-glucuronic acid (*T. Miwa*, Acta Phytochim., 1932, **6,** 155). The structure of scutellarein was proved by degradation and synthesis (*G. Goldscmiedt* and *E. Zerner*, Monatsh., 1910, **31,** 439; *Robinson* and *G. Schwarzenbach*, J. chem. Soc., 1930, 822; *F. Wessely* and *G. H. Moser*, Monatsh., 1930, **56,** 97; *A. Major*, Periodica Polytech., 1963, **7,** 117; *M. Jouanne* and *Mentzer*, Compt. rend. Ser. C, 1966, **263,** 1022). It is formed by the demethylation and rearrangement of 7-hydroxy-5,8,4'-trimethoxyflavone (*Wessely* and *F. Kallab*, Monatsh., 1932, **60,** 26). **Plantaginin,** m.p. 214°, obtained from *Plantago asiatica* is *scutellarein 7-glucoside* (*Nakaoki et al.*, Yakugaku Zasshi, 1961, **81,** 1697), and **dinatin,** m.p. 274–276°, from the leaves of *Digitalis lanata* is *scutellarein 6-methyl ether* (*S. Rangaswami* and *E. V. Rao.*, Proc. Indian Acad. Sci., 1961, **54A,** 51; *D. K. Bhardwaj, S. Neelakantan* and *Seshadri*, Indian J. Chem., 1966, **4,** 173).

Pectolinarigenin, *5,7-dihydroxy-6,4'-dimethoxyflavone*, $C_{17}H_{14}O_6$, yellow needles, m.p. 215–216°, *dimethyl ether*, m.p. 162°, *diacetate*, m.p. 151°, has been synthesised (*Zemplén* and *Farkas*, Ber., 1943, **76,** 937; *Murti* and *Seshadri*, Proc. Indian Acad. Sci., 1949, **30A,** 78; *Farkas* and *Strelisky*, Tetrahedron Letters, 1970, 187). It occurs in the flowers of *Linaria vulgaris* Linn. as **pectolinarin,** $C_{29}H_{34}O_{15}$, the *7-rutinoside*, m.p. 252–253° (decomp.) which has been synthesised (*Zemplén* and *Farkas*, Ber., 1941, **74,** 1818). Pectolinarin has also been isolated from *Cirsium oleraceum* (*H. Wagner, L. Hörhammer* and *W. Kirchner*, Arch. Pharm., 1960, **293,** 1053).

Salvigenin, *5-hydroxy-6,7,4'-trimethoxyflavone*, $C_{18}H_{16}O_6$, has been isolated from the air-dried leaves and stems of *Salvia triloba* and its structure established from spectral data, degradation products and partial methylation of pectolinarigenin (*A. Ulubelen, S. Ozturk* and *S. Isildatici*, J. pharm. Sci., 1968, **57,** 1037). **Stachyflaside,** *5,6,7,4'-tetrahydroxyflavone-7-O-β-D-glucopyranosyl-(2→1)-O-β-D-mannopyranoside*, m.p. 220–224°, $[\alpha]_D^{20}$ −60° (methanol), obtained from the grass *Stachys annua* (*I. P. Sheremet* and *N. F. Komissarenko*, Khim. Prir. Soedin., 1971, **7,** 721). *5,6,7,2'-Tetramethoxyflavone* has been prepared and converted to the 5-hydroxy derivative by partial demethylation in boiling 20% hydrochloric acid solution; complete demethylation occurs in boiling hydriodic acid (*Murti, P. V. Raman* and *Seshadri*, Current Sci., 1965, **34,** 398).

Zapotinin, *5-hydroxy-6,2'-6'-trimethoxyflavone*, $C_{18}H_{16}O_6$, m.p. 224–225°, and **zapotin,** *5,6,2',6'-tetramethoxyflavone*, $C_{19}H_{18}O_6$, m.p. 150–151° have been isolated from *Casimiroa edulis* Llave et Lex. (*F. Sondheimer* and *A. Meisels*, Tetrahedron, 1960, **9,** 139; *B. R. Pai, P. S. Subramaniam* and *V. Subramanyam*, *ibid.*, 1965, **21,** 3573; *Farkas* and *Norgradi*, Ber., 1965, **98,** 164; *Farkas, A. Gottsegen* and *Nogradi*, Tetrahedron Letters, 1968, 3993) and their structure confirmed by synthesis (*P. S. Phadke et al.*, Indian. J. chem., 1968, **6,** 177; *S. C. Dalta et al.*, *ibid.*, 1969, **7,** 746). The Sayre relations have been used to determine the molecular structure of *5,6,3',5'-tetramethoxyflavone*, isolated from the fruit of *Sargentia greggii*, a citrus plant found in Mexico (*H.-Y. Ting, W. H. Watson* and *X. A. Dominguez*, Acta Crystallogr. B., 1972, **28,** 1046).

Artocarpesin, *6-isopentenyl-5,7,2',4'-tetrahydroxyflavone*, $C_{20}H_{18}O_6$, m.p. 250°, and **norartocarpetin,** *5,7,2',4'-tetrahydroxyflavone*, $C_{15}H_{10}O_6$, m.p. 330° (decomp.) have been isolated from relatively young heartwood of *Artocarpbs heterophyllus* (*P. V. Radhakrishnan, A. V. R. Rao* and *Venkataraman*, Tetrahedron Letters, 1965, 663); **artocarpetin** is *artocarpesin 7-methyl ether*:

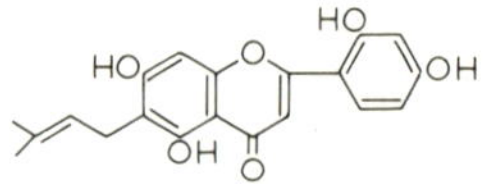

Artocarpesin

Pilloin, 5,3′-*dihydroxy*-7,5′-*dimethoxyflavone*, $C_{17}H_{14}O_6$, m.p. 236.5–237.5°, isolated from *Ovidia pillo-pillo* Meisner (*J. Núñez-Alarcón*, J. org. Chem., 1971, **36,** 3829). 5,3′-*Dihydroxy*-7,4′-*dimethoxyflavone* has been prepared (*Fa-C. Ch'en* and *T. Weng*, Tai-Wan K'o Hsueh, 1972, **26**, 46).

Velutin, 5,4′-*dihydroxy*-7,3′-*dimethoxyflavone*, $C_{17}H_{14}O_6 \cdot \frac{1}{2}H_2O$, *diacetate*, m.p. 207°, has been isolated from the leaves of *Ceanothus velutinus* and its structure established on chemical and spectral information (*K. C. Das, W. J. Farmer* and *B. Weinstein*, J. org. Chem., 1970, **35,** 3939).

Three flavones isolated from *Lindera lucida* have been shown by degradation and synthesis to be 5,6,7,8-*tetramethoxyflavone*, m.p. 112–113° (117–118°), 3′,4′-*methylenedioxy*-5,6,7,8-*tetramethoxyflavone* m.p. 171–172°, and 5,7-*dihydroxy*-3′,4′,-*methylenedioxy*-6,8-*dimethoxyflavone* **(lucidin)**, m.p. 255–257°. The effect of sodium acetate on the u.v. absorptions of lucidin, 5,7-dihydroxy-6,8-dimethoxyflavone, and their corresponding 5-methyl ethers has been discussed (*H. H. Lee* and *C. H. Tan*, J. chem., Soc., 1965, 2743). 5,6,8,4′-*Tetramethoxyflavone*, m.p. 183–184° (*A. C. Jain, S. K. Mathur* and *Seshadri*, Indian, J. Chem., 1965, **3,** 351).

Cytisoside, *β*-D-8-*glucopyranosyl*-5,7-*dihydroxy*-4′-*methoxyflavone*, $C_{34}H_{34}O_{16}$, has been obtained from the leaves and flowers of *Cytisus laburnum*; heating under acidic conditions gives rise to isomerisation, yielding a mixture of cytisoside and isocytisoside [*Chopin, M. L. Bobillant* and *A. Durix*, Compt. rend., 1965, **260,** (Group 13), 4850].

Hypolaetin, 5,7,8,3′,4′-*pentahydroxyflavone*, $C_{15}H_{10}O_7$, has been found in *Hypolaena fastigiata*; on treatment with acid, it isomerises to 5,6,7,3′,4′-*pentahydroxyflavone*, 6-*hydroxyluteolin* (*J. B. Harborne* and *H. T. Clifford*, Phytochem., 1969, **8,** 2071); 7-O-*glucosyl*, yellow needles, m.p. 243–245° is a constituent of *Juniperus macropoda* (*S. A. Siddiqui* and *A. B. Sen, ibid.*, 1971, **10,** 434).

Nodifloretin, isolated from *Lippia nodiflora* (*A. K. Barua, P. Chakrabarti* and *P. K. Sanyal*, J. Indian chem. Soc., 1969, **46,** 271) and **batatifolin** obtained from *Mikania batataefolia* have been shown to be 3′-*methoxy*-5,6,7,4′-*tetrahydroxyflavone*, $C_{16}H_{12}O_7$, m.p. 250–253° λ_{max} 265, 274 and 252 mμ, *tetra-acetate*, m.p. 201–203°; methylation furnishes 5,6,7,3′,4′-*pentamethoxyflavone (sinensetin)*, m.p. 166–168° (*W. Hertz et al.*, Tetrahedron Letters, 1969, 3419). It has been synthesised from sinensetin (*idem*, Ber., 1970, **103,** 1822).

Eupafolin, isolated *Eupatorium cuneifolium* (TOURN.) L. has been shown to be 6-*methoxy*-5,7,3′,4′-*tetrahydroxyflavone*, $C_{16}H_{12}O_7$, yellow needles, m.p. 271–273°, λ_{max}^{alc} 253, 272 and 342 mμ (ε 10,200, 10,100, and 16,300). The earlier assignment of this structure for pedalitin was incorrect and the alternative 7-*methoxy*-5,6,3′,4′-*tetrahydroxyflavone* structure has been proposed for **pedalitin** (*S. M. Kupchan et al.*, Tetrahedron, 1969, **25,** 1603).

Tricin, 3′,5′-*dimethoxy*-5,7,4′-*trihydroxyflavone*, $C_{17}H_{14}O_7$, pale yellow needles, m.p.

291°, *diacetate*, m.p. 211–213°, *triacetate*, m.p. 251–254°, *trimethyl ether*, m.p. 192–193°, is the colouring matter of "Khapli" wheat, *Triticum dicoccum* (*J. A. Anderson* and *Perkin*, J. chem. Soc., 1931, 2624) and is responsible for the ruminant disease "bloat" (*W. S. Ferguson et al.*, Nature, 1950, **166,** 116). Its structure has been established by synthesis (*K. C. Gulati* and *Venkataraman*, J. chem. Soc., 1933, **942,** 1644; *Pillon*, Bull. Soc. chim. Fr., 1953, 538; *E. Owada* and *M. Mieno*, Nippon Kagaku Zasshi, 1970, **91,** 1002) and by demethylation with hydriodic acid to **tricetin,** 5,7,3′,4′,5′-*pentahydroxyflavone*, $C_{15}H_{10}O_7 \cdot H_2O$, m.p. >330° (decomp.), *pentacetate*, m.p. 241–242° (*I. C. Badhwar et al.*, J. chem. Soc., 1932, 1107), which has been synthesised along with 3′-*hydroxy*-5,7,4′,5′-*tetramethoxyflavone* (*E. Malcher* and *E. Lamer-Zarawska*, Diss. Pharm. Pharmacol., 1970, **22,** 237; C.A., 1971, **74,** 3469). 5,7,4′-*Trihydroxy*-8,3-*dimethoxyflavone* (*K. Fukui, M. Nakayam* and *Horie*, Bull. chem. Soc. Japan, 1969, **42,** 2327).

Pedaliin, 5,3′,4′-*trihydroxy-7-methoxyflavone* 6-*glucoside*, $C_{22}H_{22}O_{12} \cdot 2.5H_2O$, m.p. 254° (decomp.), $[\alpha]_D^{19}$ +27.2° (1:9 pyridine–ethanol) is obtained from the dried leaves of *Sesamum indicum* (*N. Morita*, Chem. pharm. Bull., Tokyo, 1960, **8,** 59). **Pedalitin,** 5,6,3′,4′-*tetrahydroxy-7-methoxyflavone*, $C_{16}H_{12}O_7$, yellow needles, m.p. 300–301° (decomp.), *tetra-acetate*, m.p. 242–243°, *tetramethyl ether*, m.p. 172–173°, aglycone pedaliin (*S. M. Kupchan et al.*, Tetrahedron, 1969, **25,** 1603). Pedalitin has been prepared in one step from 5,6,7,3′,4′-penta-acetoxyflavone (*S. Mahey, Mukerjee* and *Seshadri*, Indian J. Chem., 1970, **8,** 1052).

Eupatorin, 5,3′-*dihydroxy*-6,7,4′-*trimethoxyflavone*, $C_{18}H_{16}O_7$, yellow needles, m.p. 196–198°, λ_{ma}^{alcx} 243, 254, 274 and 342 mμ (ε, 17,400, 19,300, 18,800 and 27,700) and **eupatilin,** 5,7-*dihydroxy*-6,3′,4′-*trimethoxyflavone*, $C_{18}H_{16}O_7$, yellow rhombohedral plates, m.p. 234–236°, have been isolated from *Eupatorium semiserratum* DC. (*Kupchan et al., loc. cit.*). Eupatorin has been synthesised by the alkali ring isomerisation of 5,3′-*dihydroxy*-7,8,4′-*trimethoxyflavone* and by the Baker–Venkataraman rearrangement (*Farkas, Strelisky* and *B. Vermes*, Ber., 1969, **102,** 112). Eupatilin has also been synthesised (*Horie et al.*, Bull. chem. Soc. Japan, 1971, **44,** 3198). 5-*Hydroxy*-7,8,3′,4′-*tetramethoxyflavone*, a pigment from bergamot oil has been synthesised (*Nakayama et al.*, Bull. chem. Soc. Japan., 1972, **45,** 2202).

Xanthomicrol, 5,4′-*dihydroxy*-6,7,8-*trimethoxyflavone*, $C_{18}H_{16}O_7$, m.p. 227–230°, has been isolated from yerbabuena, *Micromeria chamissonis* and its structure confirmed by synthesis (*G. H.* and *V. F. Stout*, Tetrahedron, 1961, **14,** 296).

Wightin, 5,3′-*dihydroxy*-7,8,2′-*trimethoxyflavone*, $C_{18}H_{16}O_7$, slender yellow needles, m.p. 188–189°, has been obtained from *Andrographis wightiana* (*T. R. Grovindachari et al., ibid.*, 1965, **21,** 3237) and its structure confirmed by synthesis (*Dalta, Murti* and *Seshadri*, Indian J. Chem., 1969, **7,** 110). Wightin appears to be the first example of a naturally occurring flavone with a unique 2′,3′-oxygenation pattern instead of the normal 3′,4′-pattern.

Demethoxysudachitin, 6,8-*dimethoxy*-5,7,4′-*trihydroxyflavone*, $C_{16}H_{14}O_7$, yellow needles, m.p. 271–273°, *triacetate*, m.p. 188–190°, *triethyl ether*, m.p. 128–129°, is obtained from the skin of unripened fruit of *Citrus sudachi* (*Horie et al.*, Nippon Kagaku Zasshi, 1962, **83,** 602).

Tangeritin, tangeretin, 5,6,7,8,4′-*pentamethoxyflavone*, $C_{20}H_{20}O_7$, m.p. 152°, has been isolated from tangerine oranges, *Citrus nobilis deliciosa* (*E. K. Nelson*, J. Amer. chem. Soc., 1934, **56,** 1392) and its structure established by synthesis (*J. M. Sehgal et al.*, Proc. Indian Acad. Sci., 1955, **42A,** 252; *L. J. Goldsworthy* and *Robinson*, Chem. and Ind., 1957, 47). Tangeritin has been converted to nobiletin (*L. J. Swift*, U.S.P. 3,598,841/1971). **Ponkanetin,** from the peels of *Citrus deliciosa* var. Tenore, appears to be tangeritin (*A. Bellino, P. Venturella* and *S. Cusmano*, Ann. chim., Rome, 1962, **52,** 795). From the peel of *Citrus jambhiri* lush., a lemon from Nagaland besides tangeritin, 5-O-*desmethyltangeritin*, 5-*hydroxy*-6,7,8,4′-*tetramethoxyflavone*, $C_{19}H_{18}O_7$, m.p. 176–177°, has been isolated (*B. P. Chaliha, G. P. Sastry* and *P. R. Rao*, Tetrahedron, 1965, **21,** 1441). 6,3′-*Dimethoxy*-5,7,4′-*triethoxyflavone* has been prepared by heating 5,3′-dimethoxy-2-hydroxy-4,6,4′-triethoxydibenzoylmethane, obtained from 4,6-diethoxy-2-(4-ethoxy-3-methoxybenzoyloxy)-5-methoxyacetophenone and sodamide, with a mixture of acetic acid and sulphuric acid (*Horie et al.*, Bull. chem. Soc. Japan, 1968, **41,** 1460).

Sudachitin, 5,7,4′-*trihydroxy*-6,8,3′-*trimethoxyflavone*, $C_{18}H_{16}O_8$, m.p. 239.5–240.5°, *triacetate*, m.p. 167.5–168.5°, *trimethyl ether*, m.p. 135.5–136.5°, *triethyl ether*, m.p. 143–144°, isolated from *Citrus sudachi* gives a greenish brown colour with ferric chloride solution and a reddish orange colour with magnesium and hydrochloric acid. The trimethyl ether and nobiletin are identical. Its structure has been established by synthesis (*Horie et al.*, Bull. chem. Soc. Japan, 1961, **34,** 1547; Nippon Kagaku Zasshi, 1962, **83,** 468). **Majoranin,** obtained from the leaves of *Majorana hortensis* is the same compound (*S. Subramanian et al.*, Current Sci., 1972, **41**, 202). 5,7-Dihydroxy-8,3′,4′,5′-tetramethoxyflavone, 3′-methylisoirigenin has been synthesised (*Farkas* and *Varady*, Magy. Kem. Foly., 1961, **67,** 431).

Hymenoxin, 5,7-*dihydroxy*-6,7,3′,4′-*tetramethoxyflavone*, $C_{19}H_{18}O_8$, m.p. 211–213°, has been isolated from the leaves of *Hymenoxys scaposa* (Compositae) and its structure confirmed by synthesis (*M. B. Thomas* and *T. J. Mabry*, J. org. Chem., 1967, **32,** 3254).

Mandarin, isolated from *Citrus reticulata* is not 5,4′-*dihydroxy*-6,7,8,3′-*tetramethoxyflavone* (*H. Wagner et al.*, Ber., 1971, **104,** 3357).

Serpyllin, 5-*hydroxy*-7,8,2′,3′,4′-*pentamethoxyflavone*, $C_{20}H_{20}O_8$, yellow fluffy needles, m.p. 170°, obtained from *Andrographis serpyllifolia* gives a green colour with alcoholic ferric chloride and a reddish orange colour with magnesium and hydrochloric acid. Its structure has been established on the basis of spectral and synthetic evidence. 5,6,7,2′,3′,4′-*Hexamethoxyflavone*, prisms, m.p. 141–142°, has also been synthesised (*Govindachari* and *P. C. Kalyanaraman*, Tetrahedron, 1968, **24,** 7027). 5-*Hydroxy*-6,7,2′,3′,6′- and 5-*hydjoxy*-7,8,2′,3′,6′-*pentamethoxyflavone* (*idem*, Indian J. Chem., 1970, **8,** 27).

Nobiletin, 5,6,7,8,3′,4′-*hexamethoxyflavone*, $C_{21}H_{22}O_8$, yellow crystals, m.p. 134°, is obtained from the oil which results when the Chinese drug chen-pi (*Citrus nobilis*) is extracted with cold methanol. It is degraded by ethanolic potassium hydroxide to acetoveratrone (*Robinson* and *K.-F. Tseng*, J. chem. Soc., 1938, 1004) and its structure has been established by synthesis (*Murti* and *Seshadri*, Proc. Indian Acad. Sci., 1948, **27A,** 217; 1949, **30,** 12).

Scaposin, 6,8,3′,4′-*tetramethoxy*-5,7,5′-*trihydroxyflavone*, $C_{19}H_{18}O_9$, fine yellow needles, m.p. 210–212°, *triacetate*, m.p. 179–180°, has been isolated from *Hymenoxys* *scaposa* (Compositae) and its structure confirmed by synthesis (*Thomas* and *Mabry*, Tetrahedron, 1968, **24,** 3675).

Gardenin A, 5-*hydroxy*-6,7,8,3′,4′,5′-*hexamethoxyflavone*, $C_{21}H_{22}O_9$, m.p. 163–164°, is the principal pigment from *Gardenia lucida*. Its structure has been confirmed by synthesis (*A. V. R. Rao et al.*, Indian J. Chem., 1970, **8,** 398; *M. Kamalam and Rao, ibid.*, p. 573). This compound had been given previously the structure 5-hydroxy-3,6,8,3′,4′,5′-hexamethoxyflavone (*P. K. Bose* and *R. Nath*, J. Indian chem. Soc., 1945, **22,** 233; *V. K. Ahluwalia et al.*, J. chem. Soc., 1954, 3988; *Rao* and *Venkataraman*, Indian J. Chem., 1968, **6,** 677), **Gardenin B,** 5-*hydroxy*-6,7,8,4′-*tetramethoxyflavone*, 5-O-*demethyltangeritin;* **gardenin C,** 5,3-*dihydroxy*-6,7,8,4′,5′-*pentamethoxyflavone*, m.p. 179–180°; **gardenin D,** 5,3′-*dihydroxy*-6,7,8,4′-*tetramethoxyflavone*, m.p. 190–192° have also been obtained from *G. lucida (Rao et al., loc. cit.)*. For the synthesis of some hydroxyflavone derivatives see *A. Arcoleo, Bellino* and *Venturella*, Ann. chim. Rome, 1957, **47,** 66; 8-*benzyl*-5,7-*dihydroxy*-6-*methoxyflavone* derivatives (*Horie et al.*, Nippon Kagaku Kaishi, 1972, 773).

Luteolin, tricin, and 8-acetyl-5,7-dihydroxy-3′,4′-dimethoxyflavone, are more resistant to u.v. radiation in the presence of air than the flavonols quercetin, morin and arthraxin (*M. Kaneta* and *N. Sugiyama*, Bull. chem. Soc. Japan, 1971, **44,** 3211).

Hydroxyflavones with a 3-hydroxyl group, hydroxyflavonols

Galangin, 5,7-*dihydroxyflavonol*, $C_{15}H_{10}O_5 \cdot H_2O$, yellowish needles, m.p. 215°, *triacetate*, m.p. 142°, *tribenzoate*, m.p. 177°, *trimethyl ether*, m.p. 197–198°, a constituent of galanga root, *Alpina officinarum*, Hance, is synthesised from ω-benzoyloxyphloroacetophenone, benzoic anhydride, and sodium benzoate (*T. Heap* and *R. Robinson*, J. chem. Soc., 1926, 2336). The 3-*methyl ether*, pale yellow needles, m.p. 299°, *diacetate*, m.p. 175–176°, is also obtained from galanga root and has been synthesised (*J. Kalff* and *Robinson*, *ibid.*, 1925, **127,** 182; *J. J. Chavan* and *Robinson*, *ibid.*, 1933, 368; *H. M. Lynch et al.*, *ibid.*, 1952, 2063). The results of nuclear allylation of galangin 3-methyl ether run parallel to those obtained with chrysin. Therefore substitution in the 3-position as in flavonols does not affect the pattern of nuclear allylation, but the yields of the allylated products are comparatively higher (*V. K. Ahluwalia, G. P. Sachdev* and *T. R. Seshadri*, Indian J. Chem., 1967, **5,** 243).

Izalpinin, *galangin 7-methyl ether*, $C_{16}H_{12}O_5$, m.p. 194–195°, is obtained from *Alpinia japonica* and *A. chinensis* (*Y. Kimura* and *M. Hoshi*, J. pharm. Soc. Japan, 1935, **55,** 229) and the heartwood of *Pinus clausa*, Vasey (*G. Lindstedt*, Acta Chem. Scand., 1950, **4,** 1043). Its structure follows from its preparation from pinobanksin 7-methyl ether by dehydrogenation with iodine and ethanolic sodium acetate (*V. B. Mahesh* and *Seshadri*, Proc. Indian Acad. Sci., 1955, **41,** 210) and from synthesis (*Z. V. Rao* and *Seshadri*, *ibid.*, 1945, **22A,** 383).

Datiscetin, 5,7,2′-*trihydroxyflavonol*, $C_{15}H_{10}O_6$, pale yellow needles, m.p. 276°, *tetra-acetate*, m.p. 141°, is isolated from the leaves of the Bastard Hemp, *Datisca cannabina*, Linn. Its constitution follows from its breakdown with alkali to phloroglucinol and salicylic acid (*J. Leskiewicz* and *L. Marchlewski*, Ber., 1914, **47,** 1599) and its synthesis (*Kalff* and *Robinson*, J. chem. Soc., 1925, **127,** 1968; *T. H. Simpson* and *W. B. Whalley*, *ibid.*, 1955, 166). The *rutinoside*, **datiscin,** $C_{27}H_{30}O_{15} \cdot 4H_2O$, m.p.

192–193°, occurs in the roots of *Datisca* (*C. Charaux*, Compt. rend., 1925, **180,** 1419).

Kaempferol, campherol, *kampherol, populnetin, robigenin, rhamnolutin,* 5,7,4'-*trihydroxyflavonol*, $C_{15}H_{10}O_6$, yellow needles, m.p. 276–278°, *tetra-acetate*, m.p. 116°, resolidifying at higher temperatures and then melting at 181°, *tetramethyl ether*, m.p. 165–166°, present in *Ranunculaceae*, *Leguminosae*, etc., gives a yellow solution with aqueous alkali, a green colour with ethanolic ferric chloride, and a purple colour with ferric ammonium sulphate. Its structure follows from its degradation with potassium hydroxide to phloroglucinol, 4-hydroxybenzoic acid, and acetic acid, and its synthesis from phloroacetophenone dimethyl ether and anisaldehyde *via* the 3-hydroxyiminotrimethoxyflavanone (p. 170) (*St. von Kostanecki et al.*, Ber., 1904, **37**, 2096). The 3,4'-*dimethyl ether*, m.p. 233–234°, is obtained in 40% yield by the thermal rearrangement and cyclisation of ω-methoxyphoroacetophenone tri-*p*-anisoate (*Lynch et al., loc. cit.*). **Kaempferide,** *kaempferol* 4'-*methyl ether*, $C_{16}H_{12}O_6 \cdot H_2O$, m.p. 227–229° (anhyd.), *diacetate*, m.p. 188–189°, *triacetate*, m.p. 193–195°, *tribenzoate*, m.p. 177–178°, is a component of galanga root and has been synthesised (*Heap* and *Robinson, loc. cit.; Rao* and *Seshadri*, J. chem. Soc., 1947, 122). 6-Methoxykaempferol, 5,7,4'-trihydroxy-6-methoxyflavonol, has been isolated from *Prunus avium* var *juliana* (cherry) (*P. Lebreton et al.*, Compt. rend. Ser. C, 1971, **272,** 1529). **Rhamnocitrin,** *kaempferol* 7-*methyl ether*, yellow needles, m.p. 220–221°, *triacetate*, m.p. 200–201°, occurs in the fruit of *Rhamnus catharticus*, Linn., and has been obtained by the partial demethylation of 4'-hydroxy-3,5,7-trimethoxyflavone (*Rao* and *Seshadri, loc. cit.*). Some Kaempferol methyl ethers have been synthesised (*K. Y. Sim*, Phytochem., 1969, **8,** 1597). **Isoanhydroicaritin,** 8-*isopentenylrhamnocitrin*, m.p. 275°, and **noranhydroicaritin,** 8-*isopentenylkaempferol*, m.p. 226°, have been isolated from the root of *Sophora angustifolia* (*M. Komatsu et al.*, Yakugaku Zasshi, 1970, **90,** 463). **Populnin,** *kaempferol* 7-*glucoside*, $C_{21}H_{20}O_{11}$, m.p. 228–230°, occurs in *Thespesia populnea*; it gives a pale green colour with ferric chloride (*P. R. Rao* and *Seshadri*, Proc. Indian Acad. Sci., 1946, **24A,** 456). **Tiliroside,** $C_{30}H_{26}O_7$, golden-yellow plates, m.p. 247–256°, obtained from *Tilia argentea* has been shown to be kaempferol 3-(*p*-coumaroylglucoside) and not a 7-*p*-coumaroyl derivative as first reported (*J. B. Harborne*, Phytochem., 1964, **3,** 151). **Petunoside,** kaempferol 3-[2-(*O*-feruloyl)-*O*-β-D-glucopyranosyl-(1→2)-β-D-glucopyranoside] has been obtained from *Petunia hybrida* (*L. Birkofer, C. Kaiser* and *H. Kosmol*, Z. Naturforsch., 1965, **20b,** 605).

Kaempferitrin, *kaempferol* 3,7-*dirhamnoside*, $C_{27}H_{30}O_{14} \cdot 5H_2O$, m.p. 190–192°, is obtained from the leaves of *Indigofera arrecta* (*A. G. Perkin*, J. chem. Soc., 1907, **91,** 435) and is identical with **lespedin,** which occurs in *Lespedeza cyrtobotyra* (*S. Hattori* and *M. Hasegawa*, Science, Japan, 1951, **21,** 475; C.A., 1952, **46,** 2541); the *dimethyl ether*, m.p. 173°, is hydrolysed to *kaempferol* 5,4' *dimethyl ether*, m.p. 275°. **Jaranol,** *kaempferol* 3,7-*dimethyl ether*, 5,4'-*dihydroxy*-3,7-*dimethoxyflavone*, $C_{17}H_{14}O_6$, m.p. 252–253°, is obtained from *Cistus ladanifera* var. *maculatus* and *C. ladanifera* var. *albiflorus* and its structure determined by degradation and synthesis (*C. P. Bahl, M. R. Parthasarthy* and *Seshadri*, Current Sci., 1966, **35,** 281; *J. De Pascual Teresa, C. P. Marcos* and *I. S. Bellido*, An. Quim., 1968, **64,** 623).

Gnaphaliin, 5,7-*dihydroxy*-3,8-*dimethoxyflavone*, $C_{17}H_{14}O_6$, m.p. 174–175°, *diacetate*, m.p. 135–138°, and 5-*hydroxy*-3,7,8-*trimethoxyflavone*, $C_{18}H_{16}O_6$, m.p. 176–

178°, *monoacetate*, 155–158°, have been obtained from *Gnaphalium obtusifolium* (*R. Hänsel* and *D. Ohlendorf*, Tetrahedron Letters, 1969, 431; *H. Wagner et al.*, Ber., 1971, **104,** 2381). **Isognaphaliin** from *Achyrocline satureoides* is 5,8-*hydroxy*-3,7-*dimethoxyflavone*. **Robinin,** $C_{33}H_{40}O_{19}$, m.p. 195–197° and 249–250° (two forms), is a diglycoside with L-rhamnose in the 7-position and robinobiose in the 3-position (*G. Zemplén* and *R. Bognár*, *ibid.*, 1941, **74,** 1783) which is isolated from the flowers of *Robinia pseudacacia*. With enzymes it is broken down into kaempferol L-rhamnoside and robinobiose, an L-rhamnosideo-D-galactose. **Primflasine,** *kaempferol* 3-O-*β*-D-*glucopyranosyl*-(1→4)-α-L-*arabofuranosyl*-(1→2)-α-L-*arabinopyranoside*, $C_{31}H_{36}O_{19}\cdot 2H_2O$, m.p. 207–211°, has been isolated from *Primula algida* (*A. M. Zakharov, V. N. Glyzin* and *A. I. Ban'kovskii*, Khim. Prir. Soedin., 1971, **7,** 832).

Vogelin, 6,7,4′-trihydroxyflavonol 4′-*O*-*β*-rhamnoside, vogeletin-4′-*O*-*β*-rhamnoside, has been isolated from *Polygonum recumbens* as its penta-acetate. Its structure was established by u.v., i.r. and n.m.r. data and by chemical reactions (*N. K. Sen et al.*, Ber., 1971, **104,** 3425).

Fisetin, 7,3′,4′-*trihydroxyflavanol*, $C_{15}H_{10}O_6$, yellow prisms, m.p. 330°, *tetra-acetate*, m.p. 201.5°, *tetramethyl ether*, m.p. 180°, occurs in the wood of your fustic (*Rhus cotinus* and *Quebracho colorado*). The alkaline solution in air yields protocatechuic acid and resorcinol, the latter being also produced by reductive fission with sodium amalgam (*O. Gerngross* and *H. Hübner*, *ibid.*, 1927, **60,** 2094). The structure first assigned by Herzig (1891) was confirmed by synthesis (*Kostanecki et al.*, *ibid.*, 1904, **37,** 784; *J. Allan* and *Robinson*, J. chem. Soc., 1926, 2334). Fisetin 3-methyl ether has been synthesised (*Sim, loc. cit.*).

Galetin, probably 5,6,7,4′-*tetrahydroxyflavonol*, $C_{15}H_{10}O_7$, m.p. 105–107°, *penta-acetate*, m.p. 192–194°, obtained following the hydrolysis of **galein,** *galetin* 3,6-*dirhamnoside*, $C_{27}H_{30}O_{15}$, m.p. 182–184°, has been isolated from the flowers of *Galega officinalis* (*N. P. Maksyutina* and *V. I. Litvinenko*, Doklady Akad. Nauk S.S.S.R., 1964, **154,** 1123). **Vogeltin,** 6,7,4′-*trihydroxy*-5-*methoxyflavonol*, m.p. 284–285° (*Wagner et al.*, Ber., 1966, **99,** 2430). **Mikanin,** 5-*hydroxy*-6,7,4′-*trimethoxyflavonol*, $C_{18}H_{16}O_7$, m.p. 222–224°, *diacetate*, m.p. 187–189°, occurs in *Mikania cordata*. Reaction with diazomethane or one mole of dimethyl sulphate gives a *monomethyl ether*, m.p. 174–175°; prolonged heating with excess of dimethyl sulphate gives a *dimethyl ether*, m.p. 157–158° (*A. K. Kiang, Sim* and *J. Goh*, J. chem. Soc., 1965, 6371). Mikanin has been synthesised along with *combretol*, 5-*hydroxy*-3,7,3′,4′,5′-*pentamethoxyflavone*, m.p. 145–147°, and **ptaeroxylol,** 5,7-*dihydroxy* 2′-*methoxyflavonol*, m.p. 258–260° (*Sim*, J. chem. Soc. C, 1967, 976).

Penduletin, 5,4′-*dihydroxy*-3,6,7-*trimethoxyflavone*, $C_{18}H_{16}O_7$, m.p. 216–217°, *diacetate*, m.p. 157–158°, isolated as its glucoside, **pendulin,** $C_{24}H_{26}O_{12}$, m.p. 178–179°, $[\alpha]_D^{20}$ −34° (pyridine), which occurs in *Brickelia pendula*. The glucose is attached at the 4′-position (*S. E. Flores* and *J. Herrán*, Tetrahedron, 1958, **2,** 308). Penduletin (*Flore, Herwn* and *H. Menchaca*, *ibid.*, 1958, **4,** 132) and pendulin (*L. Farkas, B. Vermes* and *M. Nogradi*, Magy. Kem. Foly., 1966, **72,** 482) have been synthesised.

Quercetin, 5,7,3′,4′-*tetrahydroxyflavanol*, $C_{15}H_{10}O_7\cdot 2H_2O$, yellow needles, m.p. 313–314° (anhyd.) (316–317° is also recorded), the most important and widely distributed pigment of the flavone series, is obtained from quercitron bark

(Quercus tinctoria) and occurs in the form of glycosides and methyl ethers (see below). It reduces Fehling's solution, gives a green colour with ferric chloride, a green fluorescence in concentrated sulphuric acid, a yellow solution in alkali, and is a mordant dye. Quercetin forms a *penta-acetate*, m.p. 195°, and with diazomethane or dimethyl sulphate and alkali gives *5-hydroxy-3,7,3′,4′-tetramethoxyflavone*, m.p. 159–160°, which with dimethyl sulphate and finely divided potassium hydroxide affords *3,5,7,3′,4′-pentamethoxyflavone*, m.p. 152° (*A. S. Gomm* and *M. Nierenstein*, J. Amer. chem. Soc., 1931, **53**, 4408). With methanesulphonyl chloride and acetic acid in pyridine it yields the *5-acetoxytetramethoxyflavone*, m.p. 169–170° (*J. H. Looker* and *F. C. Ernest*, *ibid.*, 1954, **76**, 294). Quercetin when fused with potassium hydroxide gives phloroglucinol and protocatechuic acid and its structure has been confirmed by synthesis *(Allan* and *Robinson, loc. cit.)*. A completely acetylated polyhydroxy flavone reacts with excess alkyl halide in acetone containing potassium carbonate with replacement of the 7-acetyl group by an alkyl group, for example, quercetin penta-acetate gives 80% of rhamnetin tetra-acetate (*L. Jurd*, Chem. and Ind., 1957, 1452; J. Amer. chem. Soc., 1958, **80**, 5531). Methylation of a fully acetylated flavonol in acetone containing methanol, however, preferentially methylates hydroxyls in the 7-, 4′- and 3′- positions (*idem*, J. org. Chem., 1962, **27**, 1294). Quercetin 3,7,3′,4′-tetra-acetate and rhamnetin 3,3′,4′-triacetate react with methyl iodide and potassium carbonate in acetone to give 5-*O*-methylquercetin tetra-acetate and 5,7-di-*O*-methylquercetin triacetate (*Jurd* and *L. A. Rolle*, J. Amer. chem. Soc., 1958, **80**, 5527). Reductive acetylation of quercetin pentamethyl ether with zinc and anhydrous sodium acetate in acetic anhydride gives a decamethoxybis-flavenylidene (CL); flavone and 7-methoxyflavone react similarly (*J. S. Chadha*, Chem. and Ind., 1965, 1263):

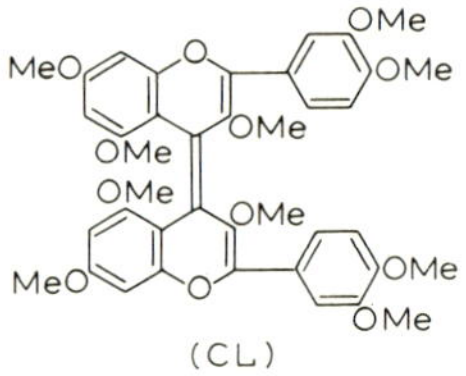

(CL)

u.v. irradiation of quercetin pentamethyl ether in methanol gives four photoproducts; lumimethylquercetin (CLI), α-photomethylquercetin (CLII), β-photomethylquercetin (CLIII) and methoxy-β-photomethylquercetin (CLIV). The structures of these compounds have been established by spectroscopic and chemical means and possible mechanisms have been proposed (*A. C. Waiss et al.*, J. Amer. chem. Soc., 1967, **89**, 6213):

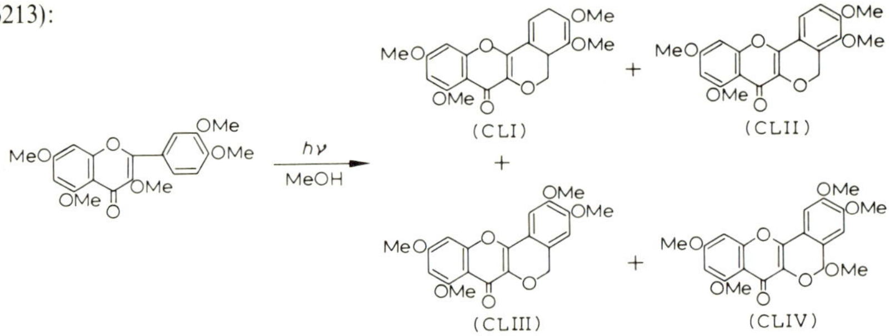

$\xrightarrow[\text{MeOH}]{h\nu}$ (CLI) + (CLII) + (CLIII) + (CLIV)

Quercetin may be obtained by treating taxifolin with sodium hydrogen sulphite (*A. Grouiller et al.*, Bull. Soc. chim. Fr., 1969, 2889):

The autoxidation of quercetin in aqueous solution has been investigated (*C. G. Nordström, R. Hakalax* and *L. J. Karumaa*, Suomen kemistilehti, 1963, **36B,** 102; C.A., 1963, **59,** 7470). 6-C-*Allylquercetin* 3,7,3′,4′-*tetra-allyl ether*, m.p. 152–153°; 7-O-*allylquercetin* 3,5,3′,4′-*tetramethyl ether*, m.p. 132°, is given by a Claisen rearrangement 8-*C*-allylquercetin 3,5,3′-4′-tetramethyl ether (*W. Heimann* and *H. Baer*, Ber., 1965, **98,** 114).

Quercitrin, *quercetin* 3-L-*rhamnoside*, $C_{21}H_{20}O_{11} \cdot 2H_2O$, pale yellow needles, m.p. 168°, 182–185° (air dried), 250–252° (anhyd.), occurs in oak bark (*Zemplén et al.*, Ber., 1928, **61,** 2486). **Isoquercitrin,** *quercetin* 3-*β*-D-*glucoside*, $C_{21}H_{20}O_{12} \cdot 4H_2O$, m.p. 233°, occurs along with quercetin in grapes, *Vitis vinifera* (*B. L. Williams* and *S. H. Wender*, J. Amer. chem. Soc., 1952, **74,** 4372) and when methylated and hydrolysed yields 5,7,3′,4′-*tetramethoxyflavonol*, m.p. 193–195°. It has been synthesised (*C. H. Ice* and *Wender*, *ibid.*, 1952, **74,** 4606). **Hyperin,** *quercetin* 3-D-*galactoside*, $C_{21}H_{20}O_{12}$, m.p. 237–238° (decomp.), $[\alpha]_D^{24}$ −59° (ethanolic pyridine), is obtained in various hydrated forms and occurs in Grimes Golden apples, *Hypericum performatum* L. (*Z. Jerzmanowska*, Wisdomosci farm., 1937, **64,** 475; C.A., 1939, **33,** 7299), leaves of *Betula verrucosa* (*P. Casparis*, Pharm. Acta Helv., 1946, **21,** 341), and the leaves and fruit of *Crataegus oxyacantha* (*U. Fiedler*, Naturwiss., 1953, **40,** 226). Hyperin yields a *tetramethyl ether*, m.p. 219–221°.

Avicularin, *quercetin* 3-α-L-*arabinoside*, $C_{20}H_{18}O_{11}$, m.p. 216–217°, occurs in *Polygonum aviculare* and is said to be the first flavonol pentoside to be found in nature (*T. Ohta*, Z. physiol. Chem., 1940, **263,** 221; *Ice* and *Wender*, J. Amer. chem. Soc., 1953, **75,** 50; *G. H. Mansfield, T. Swain* and *Nordström*, Nature, 1953, **172,** 23). It has also been isolated from *Polygonum polystachum* along with **polystachoside,** *quercetin* 3-*β*-L-*arabinoside*, $C_{20}H_{18}O_{11}$, yellow needles, m.p. 246–247° (*L. Hörhammer et al.*, Arch. Pharm., 1955, **288,** 419). **Isohyperoside,** *quercetin* 3-*β*-D-*galactofuranoside*, $C_{21}H_{20}O_{12}$, m.p. 235–238°, is obtained from *Betula middendorfii* (*V. I. Glyzin* and *Ban'kovskii*, Khim. Prir. Soedin., 1971, 662). **Quercimeritrin,** *quercetin* 7-*β*-D-*glucoside*, $C_{21}H_{20}O_{12} \cdot 3H_2O$, pale yellow needles, m.p. 247–249° (anhyd.), is found in cotton (*Gossypium*) *Perkin*, J. chem. Soc., 1909, **95,** 2181; *C. F. Attree* and *Perkin*, *ibid.*, 1927, 234; *K. Neelakantan, R. H. Rao* and *Seshadri*, Chem. Ztbl., 1936, **I,** 3518). **Spiraeoside,** *quercetin* 4′-*glucoside*, $C_{21}H_{20}O_{12}$, m.p. 210–212°, has been isolated from the flowers of Spireaea *ulmaria* L. and *Hamamelis japonica* (*E. Steinegger* and *P. Casparis*, Pharm. Acta Helv., 1945, **20,** 154; *Hörhammer* and *R. Griesinger*, Naturwiss., 1959, **44,** 427). **Quercituron,** $C_{21}H_{18}O_{13}$, pale yellow crystals, m.p. 192°, occurs in bean leaves, *Phaseolus vulgaris* and is hydrolysed to quercetin and glucuronic acid (*G. Endres et al.*, Ann., 1939, **537,** 205; *C. A. Marsh*, Nature, 1955, **176,** 176). **Rutin,** *quercetin* 3-*rutinoside*, $C_{27}H_{30}O_{16} \cdot 2H_2O$, m.p. 190°, is found in rue, *Ruta graveolens* (*Perkin*, J. chem. Soc., 1910, **97,** 1776; 1927, 234; *Zemplén* and *A. Gerecs*, Ber., 1935, **68,** 1318) and has been synthesised (*G. I. Samokhvalov, M. K. Shakhova* and *N. A. Preobrahenskiĭ*, Doklady Akad. Nauk., S.S.S.R., 1958, **123,** 305; *Shakhova et al.*, Zhur., obshcheĭ, Khim., 1962, **32,** 390; *Hörhammer et al.*, Ber., 1968, **101,** 1183). Isoquercitrin and quercimeritrin are β-glucosides since they are hydrolysed by emulsin but not by an α-glucosidase (*C. W. Nystrom et al.*, J. Amer. chem. Soc., 1954, **76,** 1950).

Quercetin methyl ethers are found in nature. **Xanthorhamin,** $C_{34}H_{42}O_{20}$, yellow needles, $[\alpha)_D$ +3.75°, is obtained from *Rhamnus tinctoria* and on hydrolysis yields rhamnose, galactose, and rhamnetin. The sugar residues are attacked at $C_{(3)}$ since xanthorhamnin when methylated with diazomethane and then hydrolysed affords 5,7,3′,4′-tetramethoxyflavonol (*Attree* and *Perkin, loc. cit.*). **Azaleatin,** *quercetin 5-methyl ether, 5-methoxy-7,3′,4′-trihydroxyflavonol*, $C_{16}H_{12}O_7 \cdot H_2O$, bright yellow needles, m.p. about 320°, obtained from **azalein,** *azaleatin 3-rhamnoside*, $C_{22}H_{22}O_{11} \cdot H_2O$, m.p. 181–185°, is isolated from *Rhododendron mucronatum* G. Don (*E. Wada*, J. Amer. chem. Soc., 1956, **78,** 4725). Its structure has been confirmed by the application of spectral procedures (*Jurd* and *R. M. Horowitz*, J. org. Chem., 1957, **22,** 1618). **Rhamnetin,** *quercetin 7-methyl ether*, $C_{16}H_{12}O_7$, yellow needles, m.p. 294–295°, *tetra-acetate*, m.p. 185–187°, occurs in the fruit of *Rhamnus cathartica*, Linn., and other species and has been synthesised (*R. Kuhn* and *I. Löw*, Ber., 1944, **77,** 211; *N. Kawano, H. Miura* and *E. Matsuishi*, Chem. pharm. Bull., Tokyo, 1967, **15,** 711). **Isorhamnetin,** *quercetin 3′-methyl ether*, $C_{16}H_{12}O_7$, greenish-yellow crystals, m.p. 305° (decomp.), occurs in senna leaves and in *Typha augustata*; in alkaline solution it is degraded by air to phloroglucinol and 4-hydroxy-3-methoxybenzoic acid (*M. Fukuda*, Bull. chem. Soc., Japan, 1928, **3,** 52). Isorhamnetin has been synthesised (*Heap* and *Robinson*, J. chem. Soc., 1926, 2336; *Kuhn* and *Löw*, Ber., 1944, **77,** 196). **Pasternoside,** *isorhamnetin 3-O-β-D-glucofuranosyl-(1→4)-α-L-rhamnopyranoside*, m.p. 235–237°, $[\alpha]_D^{20}$ −120° (cl. 3, pyridine); its structure was elucidated from its u.v. and i.r. spectra , analysis of molar rotations, and by chemical methods (*Maksyutina* and *Litvinenko*, Khim. Prirodn., Soedin., Akad. Nauk Uz.S.S.R., 1966, **2,** 20). **Narcissin,** *isorhamnetin 3-rutinoside*, $C_{28}H_{32}O_{16} \cdot 2H_2O$, m.p. 180–182°, isolated from the flowers of *Narcussus tazetta* or the pollen of *Lilium auratum* (*T. Kubota* and *T. Hase*, Nippon Kagaku Zasshi, 1956, **77,** 1059; *M. Kotake* and *H. Arakawa*, Bull. chem. Soc. Japan, 1957, **30,** 862). **Dactylin,** *isorhamnetin 3,4′-diglucoside*, $C_{27}H_{30}O_{17} \cdot H_2O$, m.p. 188–189° (decomp.), $[\alpha]_D^{20}$ −85° (M/10 NaOH) occurs in the pollens of timothy and orchard grass (*G. E. Inglett*, Nature, 1956, **178,** 1346; *J. org. Chem.*, 1957, **22,** 189) and *Crocus*, Sir John Bright (*Kuhn* and *Löw, loc. cit.*). **Tamarixin,** $C_{16}H_{11}O_7 \cdot SO_3K$, m.p. 316–317°, has been isolated from the epigeous part of *Tamarix laxa* and fresh leaves of *T. troupii*, and its structure determined (*L. M. Utkin*, Khim. Prirodn. Soedin., Akad. Nauk Uz.S.S.R., 1966, **2,** 162):

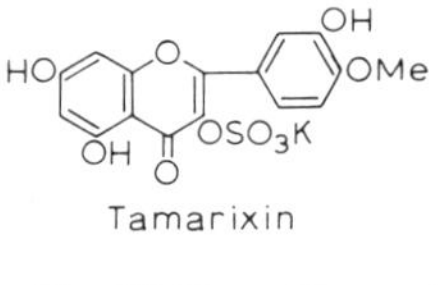

Tamarixin

Quercetin 4′-methyl ether, $C_{16}H_{12}O_7 \cdot H_2O$, golden yellow prisms, m.p. 259–260°, has been obtained from tamarixin and synthesised (*S. R. Gupta* and *Seshadri*, J. chem. Soc., 1954, 3063).

Rhamnazin, *quercetin* 7,3′-*dimethyl ether*, $C_{17}H_{14}O_7$, yellow prisms, m.p. 216–218°, *triacetate*, m.p. 154°, *tribenzoate*, m.p. 205–207°, is found in the fruits of *Rhamnus infectoria*, Linn. Its constitution (*A. G. Perkin* and *J. R. Allison*, *ibid.*, 1902, **81,** 470) was confirmed by its synthesis by the partial demethylation of quercetin 3,5,7,3′-tetramethyl ether (*K. V. Rao* and *Seshadri*, *ibid.*, 1946, 771; *Kuhn* and *Löw*, *loc. cit.)*. **Flavoyadorinin A,** rhamnazin-3-*O*-mono-D-glucoside (7,3′-di-*O*-methyl-quercetin-3-*O*-mono-D-glucoside) has been isolated from the leaves of *Viscum album* (*N. Ohta* and *K. Tagishita*, Agric. biol. Chem., 1970, **34,** 900). **Ombuoside,** rutin 7,4′-dimethyl ether, has been isolated from the leaves of *Phytolacca dioica* L. and Ombù *(Ciencia e invest)* (*V. Deulofeu*, *B. Noir* and *E. Hug*, Gazz., 1952, **82,** 726) and synthesised *(Hörhammer et al.*, *loc. cit.)*.

Ayanin, *quercetin* 3,7,4′-*trimethyl ether*, $C_{18}H_{16}O_7$, m.p. 172–173°, *diacetate*, m.p. 176–177°, is obtained from the African tree *Distemonanthus Benthamianus* (*F. E. King et al.*, J. chem. Soc., 1952, 92). Also isolated from this tree are **oxyayanin-A,** 6′-*hydroxyayanin*, yellow needles, m.p. 229–230°, *triacetate*, m.p. 183–184° and **oxyayanin-B,** 6-*hydroxyayanin*, pale yellow needles, m.p. 208–209°, *triacetate*, m.p. 183–184° (*King et al.*, *ibid.*, 1954, 4587). Oxyayanin-B has been synthesised (*R. N. Goel et al.*, *ibid.*, 1956, 1369). 5,2′,5′-Trihydroxy-3,7,4′-trimethoxyflavone, and its *acetate*, m.p. 160–161°, have been synthesised (*A. C. Jain*, *S. K. Mathur* and *Seshadri*, Indian J. Chem., 1965, **3,** 418). **Lotoflavin,** isolated from *Lotus arabicus*, is quercetin containing a small proportion of kaempferol (*M. L. Duporto et al.*, J. chem. Soc., 1955, 4249).

Pinoquercetin, 6-*methylquercetin*, $C_{16}H_{12}O_7$, yellow needles, m.p. 289–290°, *pentaacetate*, m.p. 200–201°, is found in the bark of *Ponderosa* pine and is obtained by demethylating 6-*methylquercetin* 3,3′,4′-*trimethyl ether*, m.p. 264° (*K. Venkataraman et al.*, J. sci. ind. Research, India, 1956, **15B,** 139, 490). On heating with hydriodic acid, pinoquercetin rearranged to 8-methylquercetin (*Jain* and *Seshadri*, Quart. Reviews, 1956, **10,** 169).

Morin, 5,7,2′,4′-*tetrahydroxyflavonol*, $C_{15}H_{10}O_7$, pure morin is obtained in almost colourless needles, m.p. 286–288°, and when crystallised from acids contains water of crystallisation, one molecule of which is retained even when morin is heated at 130° in a high vacuum. It occurs in the wood of *Morus tinctoria* (dyers' mulberry) partly in the free state and partly as a calcium derivative and was isolated from this source by Chevreul. It is also found in *Chlorophora tinctoria* Gaudich (old fustic) and in the Indian dye "jack-wood" *(Artocarpus integrifolia)*. In aqueous alkali it gives a yellow solution and reduces warm alkaline copper sulphate. With alumina mordants, morin gives beautiful bright yellow shades. It gives a yellow colour with boric acid in the presence of tartaric acid and this is the basis of a method for the quantitative determination of boron (*A. Jewsbury* and *G. H. Osborn*, Analyt. Chim. Acta, 1949, **3,** 481).

The structure of morin was determined by *H. Bablich* and *Perkin*, (J. chem. Soc., 1896, **69**, 792). With methyl iodide and alkali it forms a *tetramethyl ether*, m.p. 131–132° (inert 5-hydroxy group, see p. 175) and a *pentamethyl ether*, m.p. 155–157°, by the action of a great excess of sodium hydroxide and dimethyl sulphate (*J. Herzig* and *Br. Hofmann*, Ber., 1909, **42,** 155; *Perkin* and *E. R. Watson*, J. chem. Soc., 1915, **107,** 199). Alkaline fission gives phloroglucinol and 2,4-dihydroxybenzoic acid. The structure was confirmed by synthesis (*St. von Kostanecki et al.*, Ber., 1906, **39,** 625; *R. Robinson* and

Venkataraman, J. chem. Soc., 1929, 61). The nuclear methylation of morin and related compounds has been studied (*Jain*, *P. D. Sarpal* and *Seshadri*, Indian J. Chem., 1966, **4**, 223).

Robinetin, 7,3′,4′,5′-*tetrahydroxyflavonol*, $C_{15}H_{10}O_7$, forms greenish yellow needles, decomposing at 325–330°, *penta-acetate*, m.p. 233°, and is obtained from *Robinia pseudacacia* and *Gleditschia monosperma*, Walt. (*D. G. Roux* and *E. Paulus*, Biochem. J., 1962, **82,** 324). Its structure (*L. Schmid* and *F. Tadros*, Ber., 1932, **65,** 1689) was confirmed by synthesis (*E. H. Charlesworth* and *Robinson*, J. chem. Soc., 1933, 26B; *K. Brass* and *H. Kranz*, Ann., 1932, **499,** 175), and it is an effective fluorigenic spray reagent for the detection and quantitative determination of some organic compounds on a cellulose matrix, for example, the organothiophosphorus insecticide **proban** (*V. Mallet* and *R. W. Frei*, J. Chromatogr., 1971, **60,** 213). **Kanugin,** 3,7,5′-*trimethoxy*-3′,4′-*methylenedioxyflavone*, $C_{19}H_{16}O_7$, m.p. 207–208°, is found in the roots of *Pongamia glabra* along with *demethoxykanugin*, 3,7-*dimethoxy*-3′,4′-*methylenedioxyflavone*, $C_{18}H_{14}O_6$, m.p. 146–147° (*S. Rangaswami et al.*, Proc. Indian Acad. Sci., 1942, **16A,** 319; *S. Rajagopalan et al.*, *ibid.*, 1946, **23A,** 60, 147; *O. P. Mittal* and *Seshadri*, J. chem. Soc., 1956, 2176). The methylenedioxy group is detected by the green colour afforded by the flavone with gallic acid in sulphuric acid. *Norkanugin* is identical with robinetin. **Betuletol,** 5,7-*dihydroxy*-6,4′-*dimethoxyflavonol*, $C_{17}H_{14}O_7$, yellow-orange crystals, m.p. 223–225°, and its 3-*methyl ether*, yellow-orange crystals, m.p. 159–161°, have been isolated from the buds of *Betula ermanii* Cham. (*E. Wollenweber* and *Lebreton*, Biochimie, 1971, **53,** 935).

Herbacetin, 5,7,8,4′-*tetrahydroxyflavonol*, $C_{15}H_{10}O_7$, m.p. 278–280°, *penta-acetate*, m.p. 189–191°, occurs as the 7-*glucoside*, **herbacitrin,** $C_{21}H_{20}O_{12}$, m.p. 247–249°, in certain cotton plant flowers and has been synthesised (*L. J. Goldsworthy* and *Robinson*, J. chem. Soc., 1938, 56). **Prudomestin,** 5,7-*dihydroxy*-8,4′-*dimethoxyflavonol*, $C_{17}H_{14}O_7$, m.p. 209–210°, has been isolated from the heartwood of *Prunus domestica* along with kaempferol and dihydrokaempferide (*G. R. Nagarajan* and *Seshadri*, Phytochem., 1964, **3,** 477). *Herbacetin* 3,7,8-*trimethyl ether*, m.p. 267–268°, has been synthesised (*Bahl*, *Parthasarthy* and *Seshadri*, Current Sci. India, 1966, **35,** 281).

Myricetin, 5,7,3′,4′,5′-*pentahydroxyflavonol*, $C_{15}H_{10}O_8$, yellow needles, m.p. 357–360° (decomp.), *hexa-acetate*, m.p. 214–215°, occurs with its 3-*rhamnoside*, **myricitrin,** $C_{21}H_{20}O_{12}$, m.p. 197–198°, in the bark of Box myrtle, *Myrtle nagi* (*S. Hattori* and *K. Hayashi*, Acta Phytochim., 1931, **5**, 213). It forms oxonium salts with mineral acids in glacial acetic acid, and reacts characteristically with aqueous alkali giving a green colour which in air turns to bluish-violet. With ethyl iodide the flavonol gives the hexaethyl ether, m.p. 149–150°, while methyl iodide yields the *pentamethyl ether*, m.p. 167–170° (*Perkin*, J. chem. Soc., 1902, **81,** 203; 1911, **99,** 1720). Its structure was confirmed by synthesis (*Kalff* and *Robinson*, *ibid.*, 1925, **127,** 181). **Combretol,** *myricetin* 3,7,3′,4′,5′-*pentamethyl ether*, $C_{20}H_{20}O_8$, yellow needles, m.p. 144°, has been isolated from *Combretum quadrangulare* (*S. Mongkolsuk*, *F. M. Dean* and *L. E. Houghton*, *ibid.*, C, 1966, 125). *Myricetin* 7-*methyl ether*, m.p. 267–268°, *penta-acetate*, m.p. 220–222°, has been prepared by partial demethylation of myricetin hexamethyl ether (*Kawano*, *Miura* and *Matsuishi*, *loc. cit.*). **Betmidin,** *myricetin* 3-α-L-*arabinofuranoside*, $C_{20}H_{18}O_{12}$, m.p. 233–234° (*Glyzin* and *Ban'kovskii*, Khimm. Prir. Soedin., 1971, 662).

Pinomyricetin, 6-*methylmyricetin,* $C_{16}H_{12}O_8$, yellow needles, m.p. >330° (decomp.), *hexa-acetate,* m.p. 229–230°, occurs in the bark of *Ponderaso* pine and is obtained by the action of aluminium bromide on 6-*methyl-myricetin* 3,3′,4′,5′-*tetramethyl ether,* m.p. 272° (*Venkataraman et al.,* J. sci. ind. Research, India, 1956, **15B,** 139, 490). Cold dimethyl sulphate and alkali yield the *hexamethyl ether,* m.p. 178°, which with boiling ethanolic potassium hydroxide gives gallic acid trimethyl ether and 2-*hydroxy*-4,6-*dimethoxy*-5-*methylphenyl methoxymethyl ketone,* m.p. 92°. The structure has been confirmed by synthesis.

Centaureidin, 5,7,3′-trihydroxy-3,6,4′-trimethoxyflavone and 5,7,3′-trihydroxy-3,8,4′-trimethoxyflavone have been synthesised (*T. Horie,* Bull. chem. Soc. Japan, 1969, **42,** 2701). **Centaurein,** *centaureidin* 7-*glucoside,* m.p. 208–209°, has been obtained from the roots of *Centaurea jacea* and synthesised (*Farkas et al.,* Ber., 1964, **97,** 1666).

Eupatoretin, 3′-hydroxy-5,6,74′-tetramethoxyflavonol, and **eupatin,** 5,3′-dihydroxy-6,7,4′-trimethoxyflavonol have been synthesised (*K. Fukui, T. Matsumoto* and *S. Iami,* Bull. chem. Soc. Japan, 1971, **44,** 1698; *F. C. Chen* and *T. Uong,* J. Chin. chem. Soc., Taipei, 1972, **19,** 87).

Casticin, 5,3′-*dihydroxy*-3,6,7,4′-*tetramethoxyflavone,* $C_{19}H_{18}O_8$, yellow prisms, m.p. 186–187°, is isolated from the seeds of *Vitex agnus-castus.* Its structure has been confirmed by synthesis (*Hörhammer et al.,* Tetrahedron Letters, 1964, 323; Ber., 1964, **97,** 2857).

A lipophilic flavone 4,5′-dihydroxy-3,6,7-3′-tetramethoxyflavone has been isolated from camomile (*Matricaria chamomilla*) (*R. Hänsel, H. Rimpler* and *K. Walther,* Naturwissensch., 1966, **53,** 19).

Limocitrin, 5,7,4′-*trihydroxy*-8,3′-*dimethoxyflavonol,* $C_{16}H_{14}O_8$, bright yellow rosettes, m.p. 274–275° (subl.) (*R. M. Horowitz* and *B. Gentili,* J. org. Chem., 1961, **26,** 2899), **limocitril,** 5,7,4′-*trihydroxy*-6,8,3′-*trimethoxyflavonol,* $C_{18}H_{16}O_9$, m.p. 221–222°, and **isolimocitril,** 5,7,3′-*trihydroxy*-6,8,4′-*trimethoxyflavonol,* $C_{18}H_{16}O_9$, yellow needles, m.p. 236–238°, occur in *Citrus limon* as the 3-*β*-D-glucosides (*Gentili* and *Horowitz,* Tetrahedron, 1964, **20,** 2313).

3,8-Dimethoxy-5,7,3′,4′-tetrahydroxyflavone and 5,3′,4′-trihydroxy-3,7,8-trimethoxyflavone obtained from *Ricinocarpos muricatus* have been synthesised (*Farkas* and *Nogradi,* Ber., 1968, **101,** 3897).

Gossypetin, 5,7,8,3′,4′-*pentahydroxyflavonol,* $C_{15}H_{10}O_8$, yellow needles, m.p. 310–314° (decomp.), *hexa-acetate,* m.p. 229–230°, isolated from the flowers of the Indian cotton plant (*Perkin,* J. chem. Soc., 1913, **103,** 650), occurs as **gossypitrin,** the 7-*glucoside,* $C_{21}H_{20}O_{13}$, m.p. 242–244° (*V. V. Murti* and *Seshadri,* Proc. Indian. Acad. Sci., 1948, **27A,** 258) and is obtained by heating meliternin (p. 200) with hydriodic acid and phenol at 150–160° (*L. H. Briggs* and *R. H. Locker,* J. chem. Soc., 1949, 2157). *Gossypetin hexamethyl ether,* m.p. 170–172°, when fused with potassium hydroxide yields the dimethyl ether of protocatechuic acid and *gossypitol tetramethyl ether,* m.p. 115–116°:

OMe
MeO OH
CO·CH_2OMe
OMe

The structure of gossypetin follows from this and from its synthesis (*W. Baker et al., ibid.*, 1929, 74). Gossypetin 3,3′,4′-trimethyl ether when heated with hydriodic acid and phenol at 180–190° undergoes the Wessely–Moser rearrangement (p. 176) and yields quercetagetin (*D. M. Donnelly et al., ibid.*, 1956, 4409). *Gossypetin 3,7,8,4′-tetramethyl ether*, m.p. 184–185 , has been synthesised (*M. Krishnamurti, Seshadri* and *P. R. Shankaran*, Current Sci., 1965, **34,** 559). Gossypetin-7-*β*-D-glucopyranoside on heating for 6h with dimethyl sulphate and potassium carbonate in acetone gives a tetra-*O*-methyl derivative, which contains a free hydroxyl group in the phenyl ring; methylation for 30h with excess dimethyl sulphate and potassium carbonate yields gossypetin 3,5,8,3′,4′-pentamethyl ether with elimination of the sugar residue; methylation of gossypetin gives gossypetin hexamethyl ether, which on acetylation with $Ac_2O - NaOAc$ affords *acetylated gossypetin hexamethyl ether*, m.p. 169 , previously described as acetylated gossypetin pentamethyl ether (*Z. P. Pakudina et al.*, Khim. Prir. Soedin., 1972, 170). **Tambutetin,** yellow needles, m.p. 269–270°, green colour with ethanolic ferric chloride, found in the seeds of *Zanthoxylum acanthopodium* and believed previously to be herbacetin 8-methyl ether (*K. J. Balakrishna* and *Seshadri*, Proc. Indian. Acad. Sci., 1947, **26,** 234) has been shown to be a *glucoside* of *gossypetin* 7 (or 8), 4′-*dimethyl ether* (*J. B. Harborne et al.*, Phytochem., 1971, **10,** 883). A second compound **tambalin,** obtained from the same seeds, appears from mass spectral data to be a gossypetin trimethyl ether. *Gossypetin 8-methyl ether, 5,7,3′,4′-tetrahydroxy-8-methoxyflavonol*, m.p. 275–277°, has been synthesised from gossypetin and shown to be identical with the aglycone of the glycoside **ranupin,** obtained from the flowers of *Ranunculus repens* and supposed previously to be gossypetin 4′-methyl ether (*Wagner et al.*, Tetrahedron Letters, 1970, 2813). **Gossytrin,** *gossypetin 3-glucoside*, $C_{21}H_{20}O_{13}$, m.p. 182–184°, is obtained from the petals of *Hibiscus sabdariffa* (*Seshadri* and *R. S. Thakur*, J. Indian chem. Soc., 1961, **38,** 649).

Quercetagetin, 5,6,7,3′,4′-*pentahydroxyflavonol*, $C_{15}H_{10}O_8 \cdot 2H_2O$, yellow needles, m.p. 316° (decomp.), *hexa-acetate*, m.p. 210°, *hexamethyl ether*, (dimorphic), m.p. 141–142° and 157°, is obtained from **quercetagitrin,** the 7-*glucoside* $C_{21}H_{20}O_{13}$, m.p. 236–238° (decomp.), which occurs in African marigold, *Tagetes erecta* (*P. S. Rao* and *Seshadri*, Proc. Indian. Acad. Sci., 1941, **14A,** 289; *Rajagopalan* and *Seshadri, ibid.*, 1948, **28A,** 31), and has been synthesised (*Baker et al., loc. cit.; L. R. Row* and *Seshadri*, Proc. Indian Acad. Sci., 1946, **23A,** 23). The *hexamethyl ether*, m.p. 141–142°, has also been synthesised (*Jain et al.*, J. chem. Soc., 1955, 3908).

Patuletin, *quercetagetin 6-methyl ether*, $C_{16}H_{12}O_8$, yellow needles, m.p. 262–264°, *penta-acetate*, m.p. 170–172°, when methylated gives 3,6,7,3′,4′-*pentamethoxy-*, m.p. 158–159°, and 3,5,6,7,3′,4′-*hexamethoxyflavone*, m.p. 141–142° (*Seshadri et al.*, Proc. Indian Acad. Sci., 1945, **22A,** 215; 1946, **23A,** 23, 140), and is obtained from the flowers of *Tagetes patula* (French marigold) along with its 7-glucoside, **patuletrin,** $C_{22}H_{22}O_{13}$, which when methylated and hydrolysed gives 7-*hydroxy*-3,5,6,3′,4′-*pentamethoxyflavone*, m.p. 234–235°, also obtained from quercetagitrin by the same process (*N. R. Bannerjee et al. ibid.*, 1956, **44A,** 284). Patuletin and **spinacetin,** *quercetagetin 6,3′-dimethyl ether*, fine yellow needles, m.p. 235–236°, have been isolated from spinach leaves (*Spinacia oleracea*) (*A. Zane* and *S. H. Wender*, J. org. Chem., 1961, **26,** 4718). Patuletin has been partially synthesised from quercetagitrin (*D. K. Bhardwaj, S.*

Neelakantan and *Seshadri*, Indian J. Chem., 1966, **4,** 417). **Axillarin,** *quercetagetin 3,6-dimethyl ether*, $C_{17}H_{18}O_8$, m.p. 207–209° (217–218°), has been obtained from the leaves of *Xanthium pensylvanicum* (*A. O. Taylor* and *E. Wong*, Tetrahedron Letters, 1965, 3675) and synthesised (*Fukui, M. Nakayama* and *T. Horie*, Experientia, 1968, **24,** 769). *Quercetagetin 3,3'-dimethyl ether*, $C_{17}H_{14}O_8$, m.p. 214–215° is isolated from *Parthenium tomentosum* (*E. Rodriguez et al.*, Phytochem., 1972, **11,** 1507).

Artemetin, Artemisetin, *5-hydroxy-3,6,7,3',4'-pentamethoxyflavone*, $C_{20}H_{20}O_8$, m.p. 161.5° (168°), *monoacetate*, 161° (173°), has been obtained from *Artemisia absinthium* L. (*P. Tunmann* and *O. Isaac*, Z. angew. Chem., 1955, **67,** 708), and *Kuhnia eupatorioides* L. var. *pyramidalis* (*W. Herz*, J. org. Chem., 1961, **26,** 3014). **Polycladin,** $C_{19}H_{18}O_8$, m.p. 203°, isolated from *Lepidophyllum quadrangulare* (*G. B. Marini-Bettòlo, S. Chiavarelli* and *C. Giulio*, Gazz., 1957, **87,** 1185), has been shown by synthesis not to be *5,4'-dihydroxy-3,6,7,3'-tetramethoxyflavone*, m.p. 181–182° (*Hörhammer et al.*, Magy. Kem. Foly., 1965, **71,** 203), which is identical with **chrysosplenetin** isolated from *Chrysosplenium japonicum* Makino and believed previously to be *3,5,4'-trihydroxy-6,7,3'-trimethoxyflavone*; **chrysosplenin** is chrysosplenetin 4'-glucoside (*Fukui, T. Matsumoko* and *S. Tanaka*, Bull. chem. Soc. Japan, 1969, **42,** 1398; *G. K. Nikonov, L. N. Safronich* and *M. G. Pimenov*, Khim. Prir. Soedin., 1970, **6,** 268). **Chrysosphenol** C, *5,6,4'-trihydroxy-3,7,3'-trimethoxyflavone*,, m.p. 210–211° (*Nakayama et al.*, Nippon Kagaku Zasshi, 1970, **91,** 739).

Calycopterin, thapsin, *5,4'-dihydroxy-3,6,7,8-tetramethoxyflavone*, $C_{19}H_{18}O_8$, yellow prisms, m.p. 225–226°, *4'-methyl ether*, m.p. 120°, *dimethyl ether*, m.p. 131°, is obtained from *Calycopteris floribunda* (along with 4'-methyl ether) and *Digitalis thapsi*, Linn. (Spanish fox-glove) (*R. C. Shah et al.*, J. Indian chem. Soc., 1942, **19,** 135; *Rodriguez et al.*, Phytochem., 1972, **11,** 2311; *A. V. R. Rao* and *M. Varadan*, Indian J. Chem., 1973, **11,** 403). *5,3'-Dihydroxy-3,7,8,4'-tetramethoxyflavone* and related flavones isolated from *Ricinocarpos stylosus* have been synthesised (*Farkas, Nogradi* and *Vermes*, Ber., 1967, **100,** 22296).

Hibiscetin, *5,7,8,3',4',5'-hexahydroxyflavonol*, $C_{15}H_{10}O_9$, m.p. 350° (decomp.), *heptaacetate*, m.p. 242–244°, *heptamethyl ether*, m.p. 198–199° isolated from *Hibiscus sabdariffa* (*Perkin*, J. chem. Soc., 1909, **95,** 1855), has been synthesised (*P. R. Rao et al.*, Proc. Indian Acad. Sci., 1944, **19A,** 88). **Hibiscitrin,** *hibiscetin 3-glucoside*, $C_{21}H_{20}O_{14}$, m.p. 238–240°, is isolated from the same source (*Seshadri et al.*, *ibid.*, 1942, **15A,** 148; 1948, **27A,** 209).

Apuleisin, 5,6,2',3'-tetrahydroxy-3,7,4'-trimethoxyflavone and **apuleitrin,** 5,6,3'-trihydroxy-3,7,4',5'-tetramethoxyflavone have been isolated from the heartwood of *Apuleia leiocarpa* along with oxyayanin B and ayanin (*R. Braz Filho*, An. Acad. Brasil. Cienc., 1970, **42,** (Supl.), 55).

Apuleirin, *6,3'-dihydroxy-3,5,7,4',5'-pentamethoxyflavone*, $C_{20}H_{20}O_9$, m.p. 218–220°, *diacetate*, m.p. 174–175° and **apuleidin,** *5,2',3'-trihydroxy-3,7,4'-trimethoxyflavone*, $C_{18}H_{16}O_8$, yellow plates, m.p. 154–156°, *triacetate*, m.p. 198–200°, have been obtained from the heartwood of *Apuleia leiocarpa* Macbr. (*Braz Filho* and *O. R. Gottlieb*, Phytochem., 1971, **10,** 2433). *5,3',5'-Trihydroxy-3,6,7,4'-tetramethoxyflavone*, m.p. 172–173°, has been isolated from *Eremophila fraseri* (*P. R. Jefferies, J. R. Knox* and *E. J. Middleton*, Austral. J. Chem., 1962, **15,** 532).

Digicitrin, *5,3'-dihydroxy-3,6,7,8,4',5'-hexamethoxyflavone*, $C_{21}H_{22}O_{10}$, yellow-orange crystals, m.p. 181–182°, *diacetate*, m.p. 145–146°, has been isolated from the leaves of *Digitalis purpurea* (red fox-glove) (*W. Meier* and *A. Fuerst*, Helv., 1962, **45,** 232) and synthesised (*Farkas, Nogradi* and *J. Strelisky*, Magy. Kem. Foly., 1966, **72,** 485; Ber., 1966, **99,** 3218). **Exoticin,** 3,5,6,7,8,3',4',5'-octamethoxyflavone, digicitrin dimethyl ether is obtained from the leaves of *Murraya exotica* (*B. S. Joshi* and *V. N. Kamat*, Indian J. Chem., 1969, **7,** 636).

The following flavonols have been isolated from the small New Zealand tree *Melicope ternata* (*Briggs* and *Locker*, J. chem. Soc., 1949, 2157; 1951, 3131). **Meliternatin,** *3,5-dimethoxy-6,7,3',4'-dimethylenedioxyflavone*, colourless crystals, m.p. 198–198.5°, *perchlorate*, m.p. 223°; **ternatin,** *5,4'-dihydroxy-3,7,8,3'-tetramethoxyflavone*, yellow needles, m.p. 210–210.5°; **meliternin,** 3',4'-methylenedioxy-*3,5,7,8-tetramethoxyflavone*, colourless prisms, m.p. 185.5–186°; and **wharangin,** possibly 5,3',4'-trihydroxy-3-methoxy-7,8-methylenedioxyflavone. **Melisimplin,** *5-hydroxy-3',4'-methylenedioxy-3,6,7-trimethoxyflavone*, light yellow needles, m.p. 234–235°, and **melisimplexin**, *3,5,6,7-tetramethoxy-3',4'-methylenedioxyflavone*, colourless needles, m.p. 185.5° have been isolated from *Melicope simplex* and have been synthesised (*Briggs* and *Locker*, *ibid.*, 1950, 2376, 2379; *Jain et al.*, *ibid.*, 1955, 3908). Meliternatin and melisimplexin have been isolated along with **melibentin,** *3,5,6,7,8-pentamethoxy-3',4'-methylenedioxyflavone*, m.p. 134–135°, from the wood and bark of *Melicope broadbentiana* (*E. Ritche, W. C. Taylor* and *S. T. K. Vautin*, Austral. J. Chem., 1965, **18,** 2021).

Racemic peltogynol trimethyl ether (CLVII) has been synthesised from (±)-peltogynone trimethyl ether (CLVI) obtained indirectly from 2'-hydroxymethyl-7,4',5'-trimethoxyflavonol (CLV) (*J. W. Clark-Lewis* and *M. M. Mahandru*, Chem. Comm., 1970, 1287); the flavonol has also been converted into the racemic *cis, cis*-isomer of peltogynol trimethyl ether (CLVIII):

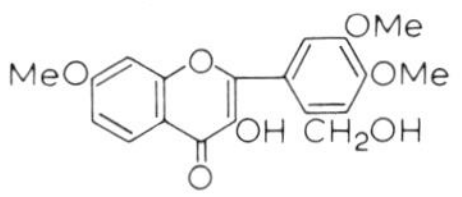

(CLV) 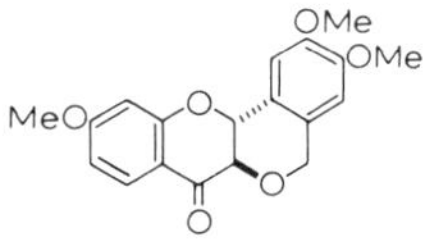 (CLVI)

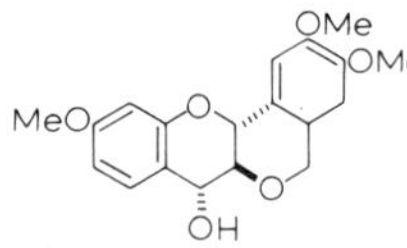

(CLVII) 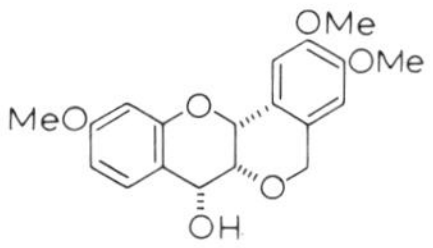 (CLVIII)

The autoxidation of flavonols in aqueous solution (*Nordström*, Suomen Kemistilehti, 1965, **B38,** 239; C.A., 1966, **64,** 9539), and the photosensitised oxygenation of some flavonols (*T. Matsuura, M. Matsushima* and *R. Nakashima*, Tetrahedron, 1970, **26,** 435) have been investigated.

The following flavonol 3-methyl ethers have been synthesised: 5,7,4'-trihydroxy-3,8-dimethoxyflavone, a pigment from *Cyanostegia angustifolia* (*Fukui et al.*, Experientia,

1969, **25**, 349; *Fukui, Matsumoto* and *Tanaka*, Bull. chem. Soc. Japan, 1969, **42**, 2380), 3,6-(**axillarin**), and 5,7,3′,4′-tetrahydroxy-3,8-dimethoxyflavone (*Fukui, Nakayama* and *Horie, ibid.*, 1969, **42,** 1649), *5,7,4′-trihydroxy-3,8,3′-trimethoxyflavone*, m.p. 213.5–215° (*idem*, Experientia, 1968, **24,** 417; *Fukui et al.*, Bull. chem. Soc. Japan, 1968, **41,** 2805); 5,7,3′-trihydroxy-3,8,4′,5′-tetramethoxyflavone and related compounds (*idem, ibid.*, 1970, **43,** 3276); the synthesis of certain partial *O*-methylated flavonols, rhamnetin, quercetin 3-methyl ether, quercetin 3,3′-dimethyl ether, ombuin (5,3′-dihydroxy-7,4′-dimethoxyflavonol), gossypetin 3,7-dimethyl ether and reso-oxyayanin-A (*N. K. Anand et al.*, J. sci. ind. Research, India, 1962, **21B,** 322), has been effected.

(xx) Biflavonyls

Ginkgetin, $C_{32}H_{22}O_{10}$, yellow crystals m.p. 344°, is isolated from the autumn leaves of the maidenhair tree *(Ginkgo biloba)*. Its structure has been determined (*K. Nakazawa* J. pharm. Soc. Japan, 1941, **61,** 174, 228; *T. Kariyone et al., ibid.*, 1956, 448, 451, 453, 457; 1958, 1010, 1013, 1016; 1959, 1182; *N. Kawano*, Chem. and Ind., 1959, 368; *W. Baker et al.*, Proc. chem. Soc., 1959, 91, 269) and confirmed by synthesis (*Nakazawa* and *M. Ito*, Tetrahedron Letters, 1962, 317). **Isoginkgetin,** m.p. 355°.

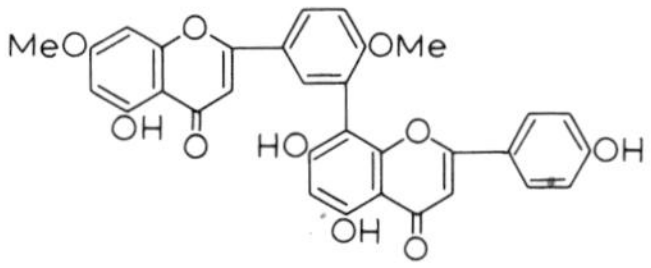

Ginkgetin 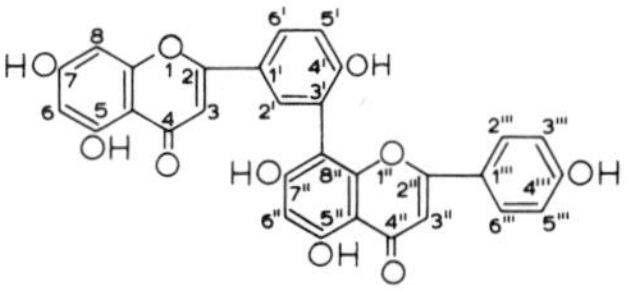 (CLIX)

Sciadopitysin, *7,4′,4‴-trimethyl ether* of CLIX, $C_{33}H_{24}O_{10}$, m.p. 287°, found in *Sciadopitys verticillata* Sieb. et Zucc. and in the leaves of *Taxus cuspidata*, is also found in the leaves of *Torreya nucifera* along with **kayaflavone**, *4′,7″,4‴-trimethyl ether* of CLIX, m.p. 314° (decomp.) *(Kawano, loc. cit.)*. **Sotetsuflavone,** *7″-methyl ether* of CLIX, $C_{31}H_{30}O_{10}$, m.p. 265° (decomp.), is found in the leaves of *Cycas revoluta* and *Cycas japonica* var. *araucarioides* (*Kawano* and *M. Yamada*, J. Amer. chem. Soc., 1960, **82,** 1505; J. pharm. Soc. Japan, 1960, **80,** 1576).

Hinokiflavone, $C_{30}H_{18}O_{10} \cdot H_2O$, m.p. >340° (decomp.), obtained from the dried leaves of *Chamaecyparis obtusa* and *Cryptomeria japonica* is a diflavonyl ether (*Kariyone* and *T. Sawada*, Yakugaku Zasshi, 1958, **78,** 1020; *Y. Fukui* and *Kawano*, J. Amer. chem. Soc., 1959, **81,** 6–31:

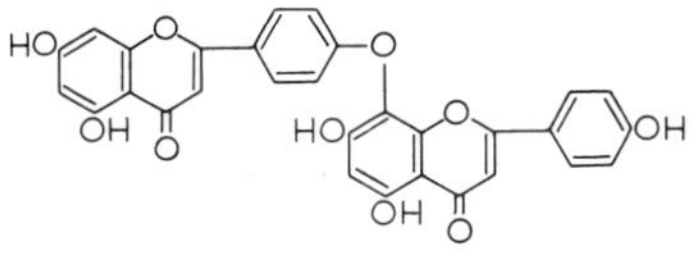

Hinokiflavone

The partial demethylation of hinokiflavone pentamethyl ether gives *isocryptomerin* (*hinokiflavone 7″-methyl ether*), *neocryptomerin* (*hinokiflavone 7-methyl ether*) and hinokiflavone 7,7″-*dimethyl ether* (*H. Miura* and *Kawano*, Chem. pharm. Bull. Tokyo, 1968, **16**, 1838). A Wessely–Moser rearrangement occurs during the Ullmann condensation of CLX and CLXI to give CLXII instead of CLXIII (*S. Natarajan* and *V. V. S. Murti*, Indian J. Chem., 1970, **8**, 116), as indicated previously (*Natarajan*, *Murti* and *T. R. Seshadri*, *ibid.*, 1968, **6**, 549).

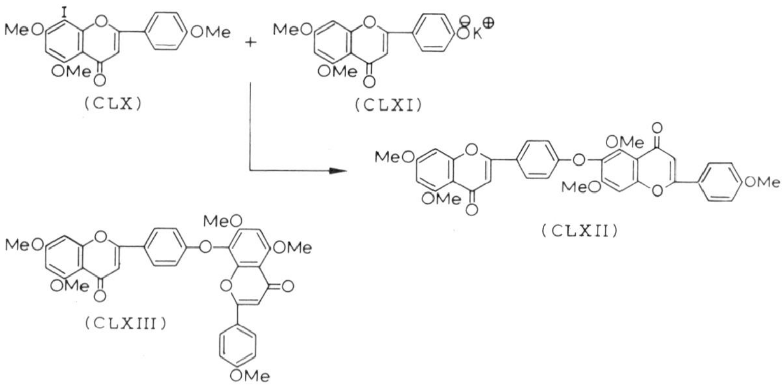

(CLX) (CLXI) (CLXII) (CLXIII)

Two 3,8″-biflavonyls **talbotaflavone, volkensiflavone,** 5,7,4′-*trihydroxy*-3-(5,7,4′-*trihydroxy*-8-*flavonyl*)*flavanone*, $C_{30}H_{20}O_{10}$, m.p. 300° (decomp.), and **morelloflavone, fukugetin,** 5,7,4′-*trihydroxy*-3-(5,7,3′,4′-*tetrahydroxy*-8-*flavonyl*)*flavanone*, $C_{30}H_{20}O_{11}$, m.p. 244–245°, 3′-*methyl ether*, m.p. 290–291° have been isolated from the roots of *Garcinia talboti* and their structures derived by spectral and chemical methods (*B. S. Joshi et al.*, Phytochem., 1970, **9**, 881):

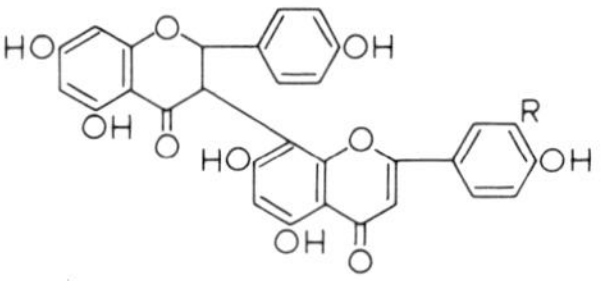

Talbotaflavone (R = H)
Morelloflavone (R = OH)

Morelloflavone 7″-*β*-*glucoside*, *fukugiside*, $C_{36}H_{30}O_{16}$, m.p. 242–243° (decomp.), is a constituent of the bark of *Garcinia spicata Hook* (*M. Konoshima* and *Y. Ikeshiro*, Tetrahedron Letters, 1970, 1717). Besides talbotaflavone and morelloflavone, a new flavanone–flavone whose constituent units are naringenin and apigenin has been isolated from *Garcinia volkenii* (*G. A. Herbin et al.*, Phytochem., 1970, **9**, 221).

Amentoflavone, 5,7,4′,5″,7″,4‴-*hexahydroxy*-8,3‴-*biflavonyl*, $C_{30}H_{18}O_{10}$, $[\alpha]_D^{40}$ +9° (pyridine), is isolated from the leaves of *Podocarpus gracilior*. **Heveaflavone,** 5,5″,4‴-

trihydroxy-7, 4′,7″-trimethoxy-8,3‴-biflavonyl, $C_{33}H_{24}O_{10}$, yellow rods, m.p. 300°, *tri-acetate*, m.p. 229–231°, $[\alpha]_D^{22}$ −15° (pyridine), *trimethyl ether*, m.p. 219–220°, is obtained from the foliage of the rubber tree (*Heavea braseliensis* Euphorbeacea) (*R. Madhav*, Tetrahedron Letters, 1969, 2017; *N. Chandramouli et al.*, Indian J. Chem., 1971, **9**, 895). **Sequoiaflavone,** *7″-methoxy 5,7,4′,5″,4‴-pentahydroxy-8,3‴-biflavonyl*, m.p. 340–349° (decomp.), *penta-acetate*, m.p. 245–247°, and amentoflavone have been isolated from the dried leaves of *Sequoia sempervirens* and *Cunninghamia lanceolata* var *Konishii*. Sequoiaflavone may be obtained by the partial demethylation of sciadopitysin, 5,7,5″-trihydroxy-4′,7″,4‴-trimethoxy-8,3‴-biflavonyl (*Miura* and *Kawano*, Yakugaku Zasshi, 1968, **88**, 1489).

Compound W13, (+)-*amentoflavone 7,4′,7″,4‴-tetramethyl ether*, $C_{34}H_{26}O_{10}$, m.p. 273°, *diacetate*, m.p. 220–225° has been isolated from *Araucaria cookii* and characterised from its u.v. and n.m.r. spectra. The diacetate and a *dimethyl ether*, (±)-*amento-flavone*, m.p. 217–218°, have also been isolated from the same source (*A. Pelter et al.*, Experientia, 1969, **25**, 350).

Agathisflavone, *5,7,4′,5″,7″,4‴-hexahydroxy-6,8″-biflavonyl*, *6,8″-bis-(5,7,4′-trihy-droxyflavone)*, $C_{30}H_{18}O_{10}$; the *7-methyl ether*, $C_{31}H_{20}O_{10}$, m.p. >320°, $[\alpha]_D^{34}$ −50° (pyridine–ethanol), *penta-acetate*, m.p. 165–168°, *7,7″-dimethyl ether*, $C_{32}H_{22}O_{10}$, have been obtained from *Araucaria bidwillii*; *7,4‴-dimethyl ether*, m.p. 212–213°, $[\alpha]_D^{34}$ −55° (pyridine–ethanol); *hexamethyl ether*, m.p. 162–164°; the *7-methyl ether* and *7,4‴-di-methyl ether*, m.p. 212–213°, $[\alpha]_D^{34}$ −55° (pyridine–ethanol), have been isolated from *Agathis palmerstoni* (*idem, ibid.*, p. 351; *N. U. Khan et al.*, Tetrahedron Letters, 1970, 2941; *S. Moriyama, M. Okigawa* and *Kawano, ibid.*, 1972, 2105; *Khan et al.*, Tetra-hedron, 1972, **28**, 5689). *Agathisflavone hexamethyl ether*, m.p. 262–265°, has been synthesised (*Moriyama, Okigawa* and *Kawano*, Tetrahedron Letters, 1972, 2105).

Cupressuflavone, *8,8″-biapigeninyl*, $C_{30}H_{18}O_{10} \cdot H_2O$, small yellow needles, m.p. >360°, *hexa-acetate*, m.p. 252–254°, *hexamethyl ether*, m.p. 295–297°, has been obtained from the leaves of *Cupressus torulosa* and *C. sempervirens*. Its structure has been deduced on the basis of colour reactions, u.v. i.r. and n.m.r. spectral data and degradation (*Murti, P. V. Raman* and *Seshadri*, Tetrahedron, 1967, **23**, 397):

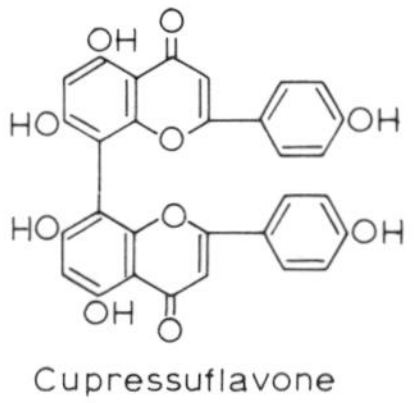

OH O
HO O OH
HO O OH
OH O

Cupressuflavone

Cupressuflavone hexamethyl ether and *7,7″,4′,4‴-tetramethyl ether*, m.p. 259–261° have been obtained *via* 8-iodo-5,7,4′-trimethoxyflavone (*Murti, Raman* and *Seshadri*, *loc. cit.*; *K. Nakazwa* and *Ito*, Tetrahedron Letters, 1962, 317; *Natarajan, Murti* and *Seshadri*, Indian J. Chem., B, 1970, 113), and starting from 2,4,6-trimethoxyiodobenzene (*S. Ahmad* and *S. Razaq*, Tetrahedron Letters, 1971, 4633):

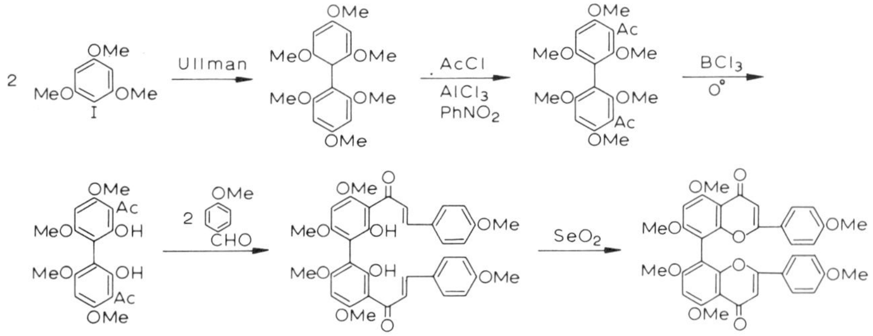

A more recent synthesis of the hexamethyl ether is described by *Moriyama, Okigawa* and *Kawano (loc. cit.)*.

The first two optically active biflavonyls WBI, (–)-4,7,7″,4‴-*tetra-O-methyl-cupressuflavone*, $C_{34}H_{26}O_{10}$, m.p. 151°, *dimethyl ether*, m.p. 161°, *diacetate*, m.p. 156°, and W11, 4,4‴-*di-O-methylcupressuflavone*, $C_{32}H_{22}O_{10}$, m.p. 301–304°, have been isolated from *Araucaria cunninghamii* and *Araucaria cookii* (*M. Ilyas et al.*, Tetrahedron Letters, 1968, 5515).

3,3-**Bipigeninyl,** 3,3-*bis*-(5,7,4′-*trihydroxyflavone*), 5,7,4′,5″,7″,4‴-*hexahydroxy*-3,3″-*bi-flavonyl*, pale yellow needles, m.p. 315–317°, *hexa-acetate*, m.p. 163–164°, *hexamethyl ether*, m.p. 186–188° (decomp.), and 3,3′-*bipigeninyl*, 5,7,4′,5″,7″,4‴- *hexahydroxy*-3,3‴-*biflavonyl*, pale yellow plates, m.p. 328° (decomp.), *hexa-acetate*, m.p. 195–197°, *hexa-methyl ether*, m.p. 264–265°, have been obtained by phenolic oxidative coupling of apigenin (p. 180), using alkaline potassium ferricyanide as the oxidising agent (*R. J. Molyneux, A. C. Waiss, Jr.*, and *W. F. Haddon*, Tetrahedron, 1970, **26**, 1409).

The u.v. spectra for the following **biflavonyls** have been reported; 6,6″-, m.p. 312–313°; 7,7″-, m.p. 346°; 8,8″-, m.p. 290–291°; 3,3‴-, m.p. 285–287°; 4,4‴- m.p. 325°; 6,4‴-, m.p. 283–284°; 7,4‴-, m.p. 304° (*C. H. Lin, K. K. Hsu* and *F. C. Chen*, J. Chinese chem. Soc. (Taiwan) Ser. II, 1961, **8**, 126; C.A., 1962, **57**, 10664; *B. Kanaklakshmi*, J. Indian chem. Soc., 1969, **46**, 279).

A 4′,4‴-diflavonyl ether has been prepared by condensing 4′-bromo-3′-nitroflavone with 4′-hydroxyflavone; following this reaction the nitro group may be converted to an amino group, which may be replaced by hydrogen by decomposition of the diazonium salt (*C. L. Huang, T. Weng* and Chen, J. Heterocycl. Chem., 1970, **7**, 1189). 6,6″-Dimethoxy-5,5″-, 7,7″-dimethoxy-6,6″-, and 7,4′,7″,4‴-tetramethoxy-6,6″-biflavonyl have been obtained (*U. K. Jagwani*, J. Indian chem Soc., 1972, **49**, 293).

(xxi) Furoflavones

Kellinone (p. 156) with aromatic aldehydes yields styryl ketones which with alcoholic phosphoric acid give the corresponding furoflavanones (*J. R. Clarke* and *A. Robertson*, J. chem. Soc., 1949, 302). Benzoylated kellinone undergoes the Baker–Venkataraman reaction (p. 140) to give a ketone which with acetic acid and hydrochloric acid is cyclised to 5,9-dimethoxy-furo[3,2-*g*]flavone (CLXIV, R = Me), yellow prisms, m.p. 180°.

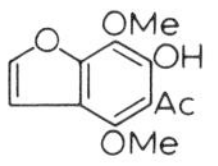

Kellinone 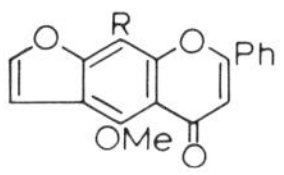 (CLXIV)

Furoflavones are found in nature; **karanjin,** 3-*methoxyfuro*[2,3-h]*flavone*, $C_{18}H_{12}O_4$, m.p. 158.5°, is obtained from the seed oil of *Pongamia glabra* and the roots of *Pongamia pinnata* L. (*L. R. Row*, Austral. J. Sci. Res., 1952, **5A**, 754). Its structure was determined by degradation (*B. L. Munjunath et al.*, Ber., 1939, **72**, 93); its breakdown to **karanjic acid,** m.p. 218° (decomp.), *acetate*, m.p. 173° (*R. T. Foster* and *Robertson*, J. chem. Soc., 1948, 115); and synthesis (*T. R. Seshadri* and *V. Venkateswarlu*, Proc. Indian Acad. Sci., 1943, **17A,** 16; *R. Aneja, K. Mukerjee* and *Seshadri*, Tetrahedron, 1958, **2,** 203). The acid chloride of acetylkaranjic acid on treatment with diazomethane and copper bronze gives *karanjin ketone*, m.p. 96–97°, which when subjected to Allan–

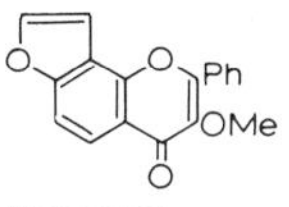

Karanjin 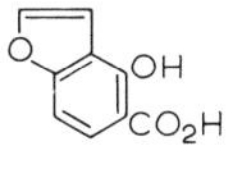 Karanjic acid

Robinson benzoylation yields karanjin (*Row* and *D. V. Rao*, J. sci. ind. Research, India, 1958, **17B,** 199). 8-*Methylkaranjin*, m.p. 121–122°, and related compounds have been synthesised (*S. S. Chibber et al.*, Proc. Indian Acad. Sci., 1957, **46A,** 19). **Pongapin,** 3-*methoxy*-3′,4′-*methylenedioxyfuro* [2,3-h]*flavone*, 3′,4′-*methylenedioxykaranjin*, $C_{19}H_{12}O_6$, pale yellow needles, m.p. 190–191°, is found in the bark of *Pongamia pinnata (Row. loc. cit.)* and has been synthesised *(Aneja, Mukerjee* and *Seshadri, loc. cit.).* In sulphuric acid it gives a bright yellow colour changing to blue. **Kanjone,** 6-*methoxyfuro*[2,3-h]*flavone*, $C_{18}H_{12}O_4$, colourless cubes, m.p. 190–191°, is obtained from the seeds of *Pongamia glabra* and its structure confirmed by synthesis (*Aneja, R. N. Khanna* and *Seshadri*, J. chem. Soc., 1963, 163). **Pongaglabrone,** 3′,4′-*methylenedioxyfuro*[2,3-h]*flavone*, $C_{18}H_{10}O_5$, m.p. 233°, is a constituent of the seeds of *Pongamia*

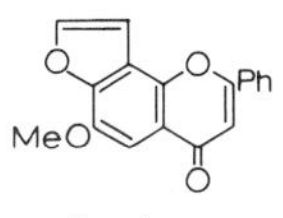

Kanjone 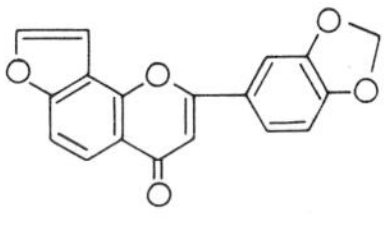 Pongaglabrone

glabra and with sulphuric acid it gives a violet colour (*Khanna* and *Seshadri*, Tetrahedron, 1963, **19,** 219). **Atanasin,** 8,9-*dihydro*-5,6-*dihydroxy*-3,9,4′-*trimethoxy-furo*-[2,3-h]*flavone*, $C_{20}H_{18}O_8$, m.p. 220–221°, isolated from *Brickelia squarrosa*, gives a diacetate and drastic demethylation with hydriodic acid affords 5,6,7,4′-tetrahydroxyflavonol (*S. E. Flores* and *J. Herrán*, Chem. and Ind., 1960, 291).

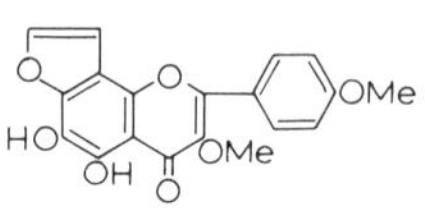

Atanasin 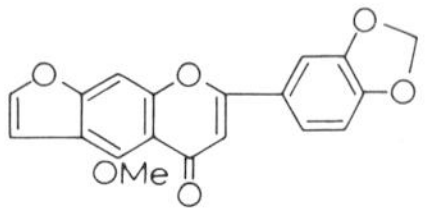 Gamatin

Comparable with the furochromones kellin and visnagin (p. 158) are **pinnatin,** *5-methoxyfuro*[3,2-g]*flavone*, $C_{18}H_{12}O_4$ (CLXIV, R = H), m.p. 180–181°, and **gamatin,** *5-methoxy-3′,4′-methylenedioxyfuro*[3,2-g]*flavone*, $C_{19}H_{12}O_6$, m.p. 233–234°, which occur in *Pongamia pinnata* and have been synthesised from visnaginone (*S. K. Pavanaram* and *Row*, Austral. J. Sci. Res., 1956, **9,** 132), and from 6-acetyl-5,7-dihydroxyflavone (*Y. S. Agasimundin* and *S. Siddappa*, J. chem. Soc., Perk. I, 1973, 503).

Linear, (*Clarke* and *Robertson*, J. chem. Soc., 1949, 302; *Row et al.*, J. sci. ind. Res. India, 1955, **14B,** 157; 1958, **17B,** 199; Indian J. Chem., 1963, **1,** 521; *A. Stener*, *Farmaco, Pavia*, Ed. Sci., 1960, **15,** 642; C.A., 1963, **58,** 497; *S. Alford* and *C. Mentzer*, Compt. rend., 1964, **258,** 2854) and angular (*Seshadri et al.*, Proc. Indian Acad. Sci., 1939, **11A,** 206; 1941, **13A,** 43; 1957, **46A,** 19; *S. D. Limaye* and *K. G. Marathe*, Rasayanam, 1950, **2,** 6; *M. G. Marathy*, Sci. Cult., Calcutta, 1951, **16,** 86) furoflavones have been synthesised.

(xxii) Pyranoflavones

Arthraxin, 8-(2-*hydroxypropyl*)-10-*oxo*-5,9,3′,4′-*tetrahydroxy*-10H-*pyrano*[2,3-h]*flavone*, $C_{21}H_{16}O_9$, pale yellow micro-needles, m.p. 336°, *penta-acetate*, m.p. 233–234°, *pentamethyl ether*, m.p. 204°, has been isolated from *Arthraxon hispidus* Makino, and its structure assigned on the basis of chemical and spectral data. The pentamethyl ether has been synthesised (*M. Kaneta* and *N. Sugiyama*, J. chem. Soc., C, 1971, 1982).

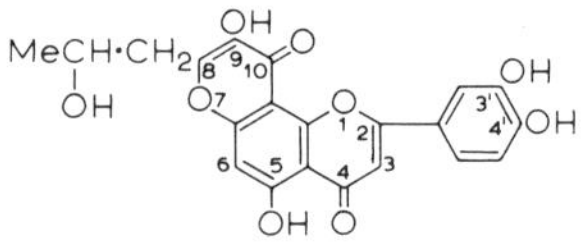

Arthraxin

Cycloartocapesin and **oxydihydroartocarpesin** have been isolated from the heartwood of *Artocarpus heterophyllus* and structures assigned to them on the bases of n.m.r. spectral data (*P. C. Parthasarathy et al.*, Indian J. Chem., 1969, **7,** 101):

Cycloartocarpesin Oxydihydroartocarpesin

(xxiii) Isoflavones, 3-phenylchromones, 4-oxo-3-phenyl-4H-benzo*[b]*pyrans*

3-Phenylchromone, 4-oxo-3-phenyl-4*H*-benzo[*b*]pyran is known as isoflavone, a number of derivatives of which have been found in nature either in the free state or as glycosides.

The structure of isoflavone derivatives is proved both by alkaline degradation and by synthesis. In the former process the pyrone ring is ruptured. Whilst, however, flavones yield polyhydric phenols and polyhydroxybenzoic acids, isoflavones afford either benzyl hydroxyphenyl ketones or polyhydric phenols and hydroxyphenylacetic acids.

Syntheses. (*1*) Benzyl 2-hydroxyphenyl ketones react with ethoxalyl chloride in pyridine to give 2-ethoxycarbonylisoflavones, hydrolysis and decarboxylation of which gives isoflavones. In the example shown, from 4-methoxybenzyl phloroglucinyl ketone 2-ethoxycarbonyl-5,7-dihydroxy-4′-methoxyisoflavone (60%) is formed, which by hydrolysis gives the acid (98%). Decarboxylation at the melting point then yields biochanin-A (81%):

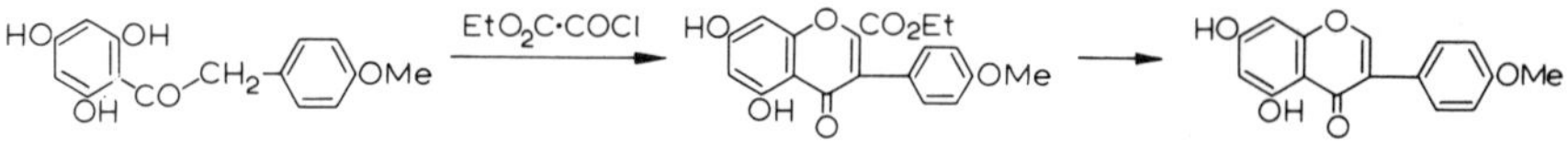

Biochanin-A

Good yields are obtained at all stages and the method is particularly suitable for the preparation of partially alkylated polyhydroxyisoflavones (*W. Baker* and *W. D. Ollis*, Nature, 1952, **169,** 706). Genistein and ψ-baptigenin and other naturally occurring isoflavones have been synthesised by this method (*Baker et al.*, J. chem. Soc., 1953, 1852, 1860). 2-Hydroxyisoflavanones are intermediates in the synthesis (see also *M. L. Wolfrom et al.*, J. Amer. chem. Soc., 1941, **63,** 1248, 1253). In a modification of the method 2-hydroxybenzoins are boiled with formamide; isoflavone is obtained in this way in 30% yield (*J. E. Gowan et al.*, Chem. and Ind., 1954, 1201).

Ethoxalylation of 4-methoxybenzyl 3-methyl-2,4,6-trihydroxyphenyl ketone gives a mixture of *ethyl* 5,7-*dihydroxy*-4′-*methoxy*-6-*methyl*-, m.p. 201–203°, and *ethyl* 5,7-*dihydroxy*-4′-*methoxy*-8-*methyl-isoflavone*-2-*carboxylate*, m.p. 199–201° (*W. Rahman* and *Kh. T. Nasim*, Tetrahedron Letters, 1961, 628):

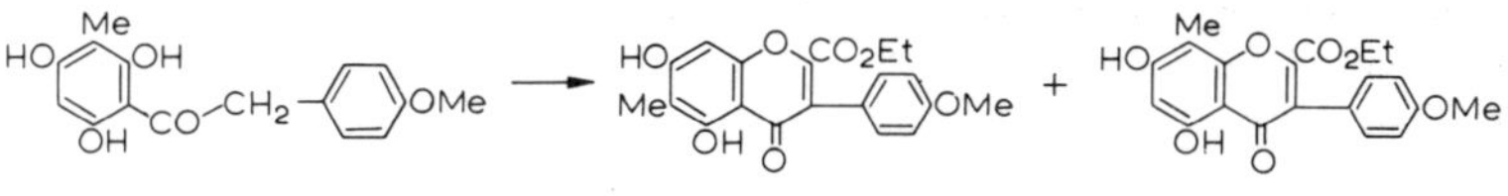

5,7-*Dihydroxy*-2,6-*dimethyl*-, m.p. 224–225° (249–251°), 5,7-*dihydroxy*-2,8-*dimethyl-isoflavone*, m.p. 256–257° (*idem*, J. org. Chem., 1962, **27,** 4215).

* *W. K. Warburton*, Quart. Reviews, 1954, **8,** 67; *W. Baker* and *W. D. Ollis*, Proc. roy. Dublin Soc., 1957, **27,** 119.

(*2*) Small yields of isoflavones are obtained by heating benzyl 2-hydroxyphenyl ketones with ethyl formate and sodium at 100° in sealed tubes (*E. Späth* and *E. Lederer*, Ber., 1930, **63,** 743); better yields are obtained at room temperature (*P. C. Joshi* and *K. Venkataraman*, J. chem. Soc., 1934, 513, 1120, 1796); thus isoflavone is obtained from benzyl 2-hydroxyphenyl ketone:

OH HCO_2Et Na → Isoflavone
$CO \cdot CH_2 \cdot Ph$

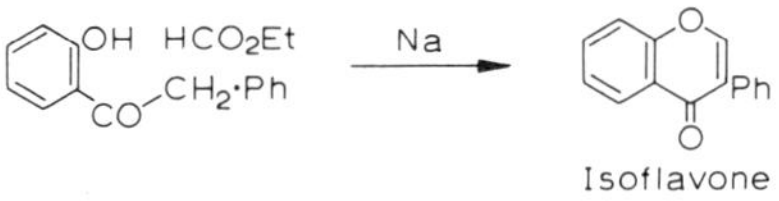

It has been reported that methyl formate does not give isoflavones, but 2-hydroxyisoflavanones (*N. Narasimhachari et al.*, J. Sci. ind. res. India, 1953, **12B,** 287).

The above two methods do not necessarily yield the same products. For instance, the substituted deoxybenzoin CLXV by the ethyl formate method gives a 5,6,7-substituted isoflavone, whereas the deoxybenzoin CLXVI, the structure of which is very closely related to that of CLXV wtih ethoxalyl chloride gives a 5,7,8-substituted isoflavone (*Baker et al.*, Chem. and Ind., 1953, 277):

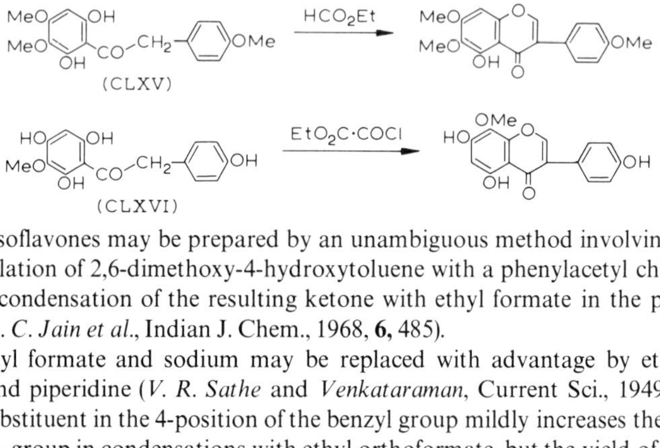

(CLXV) $\xrightarrow{HCO_2Et}$

(CLXVI) $\xrightarrow{EtO_2C \cdot COCl}$

6-Methylisoflavones may be prepared by an unambiguous method involving Friedel–Crafts acylation of 2,6-dimethoxy-4-hydroxytoluene with a phenylacetyl chloride, followed by condensation of the resulting ketone with ethyl formate in the presence of sodium (*A. C. Jain et al.*, Indian J. Chem., 1968, **6,** 485).

The ethyl formate and sodium may be replaced with advantage by ethyl orthoformate and piperidine (*V. R. Sathe* and *Venkataraman*, Current Sci., 1949, **18,** 373). A nitro-substituent in the 4-position of the benzyl group mildly increases the reactivity of the CH_2 group in condensations with ethyl orthoformate, but the yield of isoflavone depends more on the nature and position of the substituents in the 2-hydroxyphenyl half of the deoxybenzoin (*S. S. Karmarkar*, J. sci. ind. Research, India, 1961, **20B,** 334).

(*3*) 2-Styrylisoflavones prepared either from benzyl 2-hydroxyphenyl ketones, cinnamic anhydride and sodium cinnamate or by the action of benzaldehyde on 2-methylisoflavones are oxidised to the isoflavone-2-carboxylic acids provided the hydroxyl groups are first protected. Hydrolysis and decarboxylation yield the isoflavones (*Baker et al.*, J. chem. Soc., 1937, 805). The yields are generally poor.

(*4*) ω-Phenoxyacetophenone cyanohydrins undergo ring-closure with zinc chloride and hydrogen chloride in ethereal solution. Hydrolysis of the resulting ketimine hydrochlorides yields isoflavones (*idem, ibid.*, 1929, 1468; 1933, 374):

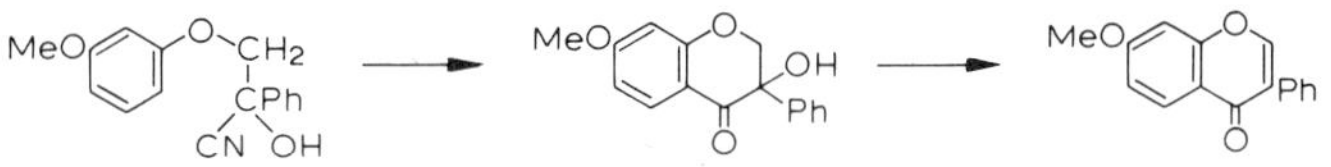

(*5*) Isoflavones have been synthesised from chalcones by the use of thallium(III) acetate. The relationship of this synthesis to the biosynthesis of isoflavones has been discussed (*Ollis et al.*, J. chem. Soc., C, 1970, 125):

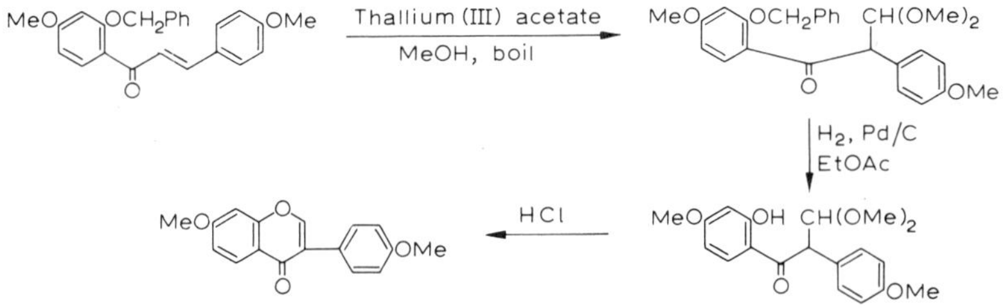

Using thallium(III) nitrate in methanol the rearrangement of simple chalcones is quantitative at room temperature (*L. Farkas et al.*, Chem. Comm., 1972, 825).

Isoflavones have been synthesised in good yields by subjecting chalcone epoxides to aryl migration using boron trifluoride-etherate (*S. K. Grover, A. C. Jain* and *Seshadri*, Indian J. Chem., 1963, **1**, 517). This method gives mainly dihydroflavonols (flavanols) if ring B in the formula shown is unsubstituted, but isoflavones are formed if methoxy groups are present in positions 4 or 3 and 4 (*S. C. Bhrara, Jain* and *Seshadri*, Tetrahedron, 1965, **21**, 963):

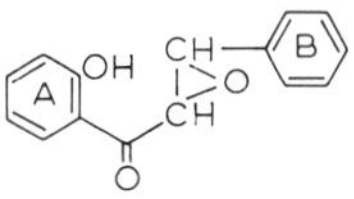

(*6*) Several 7-hydroxyisoflavones have been prepared by treating benzyl 2-hydroxyphenyl ketones with zinc cyanide in the presence of hydrogen chloride, hydrolysing the β-oxoaldimines formed to β-oxocarbaldehydes and α-hydroxymethylene ketones and cyclising (*Farkas et al.*, Periodica Polytech., 1958, **2**, 231; C.A., 1960, **54**, 1510). 7-Hydroxy- and 7-methoxyisoflavone have also been obtained *via* the reaction between the appropriate benzyl 2-hydroxyphenyl ketone and formimidoyl chloride in dry ether, in the presence of zinc chloride (*Farkas*, Ber., 1957, 90, 2940; Chem. and Ind., 1957, 1212).

(*7*) Isoflavones may be obtained by the dehydrogenation of isoflavanones, by first brominating with *N*-bromosuccinimide in the presence of perbenzoic acid and then dehydrobrominating with potassium acetate and acetic acid; for example, 7-methoxyisoflavanone affords *7-methoxyisoflavone*, m.p. 153–155° in 60% yield (*N. Inoue*, Sci. Repts. Tohoku Univ., First Serv., 1961, **45,** 63; C.A., 1963, **58,** 5619).

(*8*) Isoflavones have been prepared by oxidation of the product obtained on boiling *N*-styrylmorpholine and salicylaldehyde in benzene, with the chromium trioxide-pyridine (*L. A. Paquette* and *H. Stucki*, J. org. Chem., 1966, **31,** 1232):

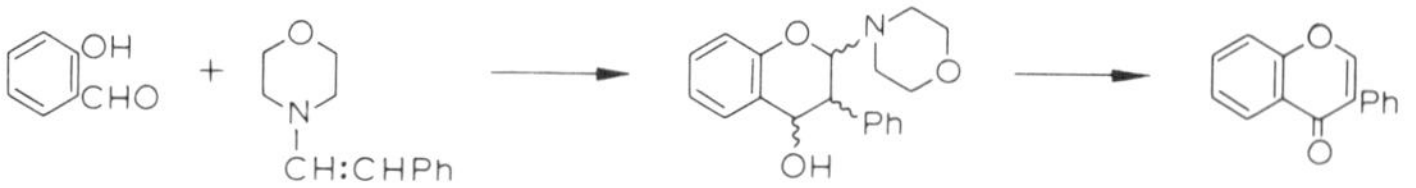

(*9*) 2′,4′,5′-Trihydroxyisoflavones have been synthesised by condensing chromanones with 4,5-dimethoxy-1,2-benzoquinone (*C. A. Weber-Schilling* and *H. W. Wanzlick*, Ber., 1971, **104,** 1518). The synthesis of isoflavones has been reviewed (*N. V. Bringi et al.*, Sci. Proc. roy. Dublin Soc., 1956, **27,** 93; *Baker* and *Ollis*, *ibid.*, p. 119), and recently developed methods for the synthesis of isoflavones have been evaluated (*Jain*, Symp. Syn. heterocycl. Compounds Physiol. Interest, Hyderabad, India, 1964, 31).

(*10*) Isoflavone glucosides are synthesised by treating hydroxyisoflavones with acetobromoglucose in the presence of aqueous potassium hydroxide (*G. Zemplén* and *Farkas*, Ber., 1943, **76,** 1110); genistin, genistein 7-glucoside, has been synthesised in this way.

Isoflavone is a colourless, crystalline substance, m.p. 150° (148°, 138°), the stability of which is shown by its inertness to cold sulphuric acid or boiling acetic anhydride. Its synthesis has already been described (methods *2* and *8*).

5-*Hydroxy*-, m.p. 102° (*S. S. Karmaker, K. H. Shah* and *Venkataraman*, Proc. Indian Acad. Sci., 1952, **36A,** 552); 7-*hydroxy*-, m.p. 215° (205–206°) (*Baker et al.*, Chem. Soc., 1953, 1852; *Farkas et al., loc. cit.*; *S. A. Kagal, P. M. Nair* and *Venkataraman*, Tetrahedron Letters, 1962, 593); 8-*methoxyisoflavone*, m.p. 138–139° (*Paquette* and *Stucki, loc. cit.*); 7-*hydroxy-6-methyl*-, m.p. 250°, *acetate*, m.p. 137–138°; 7-*hydroxy-2,6-dimethyl-isoflavone*, m.p. 245–246°, *acetate*, m.p. 161° (*Zemplén, Farkas* and *T. Sattler*, Acta Chim. Acad. Sci. Hung., 1960, **22,** 449); 7-*hydroxy-2′-methoxy*-, m.p. 187–189°; 7-*hydroxy-2′-methoxy-2-methyl-isoflavone-8-carbaldehyde*, m.p. 166° (*Y. Kawase et al.*, Bull. chem. Soc. Japan, 1960, **33**, 1240); 7,4′-*dimethoxyisoflavone*, m.p. 161° (*Ollis et al., loc. cit.*). 5,6-*Dihydroxyisoflavone*, yellow cubes, m.p. 168–169°; 5,6,7-*trihydroxyisoflavone*, yellow prisms, m.p. 189–190°; in the synthesis, the isoflavone unsubstituted in the 5-position, is treated with hexamine in acetic acid to give the 5-

carbaldehyde, which is oxidised by hydrogen peroxide to the corresponding 5-hydroxyisoflavone (*K. Aghoramurthy*, *T. R. Seshadri* and *G. B. Venkatasubramanian*, J. sci. ind. Research India, 1956, **15B,** 11). 7-*Hydroxy*-5,8-*dimethoxyisoflavone*, m.p. 251–252°, methylation gives 5,7,8-*trimethoxyisoflavone*, m.p. 167° (*Farkas* and *J. Várady*, Acta Chim. Acad. Sci. Hung., 1959, **20,** 169). 7,4′-*Dihydroxy*-5,8-*dimethoxyisoflavone* may be converted to 4′-*hydroxy*-5,7,8-*trimethoxyisoflavone*, m.p. 247–248°, then to 5,4′-*dihydroxy*-7,8-*dimethoxyisoflavone*, m.p. 181°, and finally to 5,4′-*dihydroxy*-6,7-*dimethoxyisoflavone*, m.p. 226–227°; 7,4′-*dihydroxy*-5,6-*dimethoxyisoflavone*, 5-*methyltektorigenine*, m.p. 264° (*Farkas*, *Várady* and *A. Gottsegen*, Magy. Kem. Foly., 1964, **70,** 349).

Properties. The isoflavones are colourless crystalline substances. *Ring-fission with alkali* yields a deoxybenzoin and formic acid (usually quantitative); thus formononetin gives the dihydroxymethoxydeoxybenzoin shown (*F. Wessely et al.*, Monatsh., 1931, **57,** 395; 1933, **63,** 201). Reduction of formononetin by hydrogen with a specially prepared catalyst (*Wessely* and *F. Prillinger*, *ibid.*, 1938, **72,** 197; see also *R. Mozingo* and *H. Adkins*, J. Amer. chem. Soc., 1938, **60,** 669) gives 7-hydroxy-4′-methoxyisoflavan:

HO OH CO CH$_2$ OMe ← OH$^{\ominus}$ — HO O OMe O → H$_2$ → HO O OMe

Formononetin

Hydrogenation of isoflavones under atmospheric pressure and room temperature in acetic acid using a platinum oxide catalyst gives isoflavanones (*N. Inoue*, Sci. Repts. Tohoku Univ., First Ser., 1961, **45,** 63; C.A., 1963, **58,** 5619). *Reduction* with sodium tetrahydridoborate in ethanol gives 80–85% yields of isoflavan-4-ols (*L. R. Row*, *A. S. R. Anjaneyulu* and *C. S. Krishna*, Current Sci., 1963, **32,** 67):

MeO O Ph O → MeO O Ph OH

2-Alkylisoflavones when subjected to the Clemmensen reduction give 2-alkylisoflav-3-enes; 2-unsubstituted isoflavones afford an isoflavene characterised as a mixture of the 2- and 3-enes (*K. H. Dudley et al.*, J. org. Chem., 1967, **32,** 2317):

AcO O Me OMe O → Zn/Hg, EtOH, 6*M* HCl → AcO O Me OMe

As in the flavone series (p. 176) the *demethylation* of methoxyisoflavones may be accompanied by rearrangement; thus 5,8-dimethoxyisoflavones yield 5,6-dihydroxyisoflavones.

Heating 5,7,8-trimethoxy-2-methylisoflavone with hydrobromic acid produces 5,6,7-trihydroxy-2-methylisoflavone (*W. Baker et al.*, Chem. and Ind., 1953, 277; see also *W. B. Whalley*, J. chem. Soc., 1953, 3366); on the other hand 5,7,8-trimethoxyisoflavone and its 2-methyl derivative have been demethylated by hydriodic acid without rearrangement (*N. Narasimhachari et al.*, Proc. Indian Acad. Sci., 1952, **35A**, 46; *A. Ballio* and *F. Pocchiari*, Gazz., 1949, **79**, 913; *L. Farkas* and *J. Várady*, Acta Chim. Acad. Sci. Hung., 1960, **24**, 225; see, however, *Whalley*, Chem. and Ind., 1954, 1230). Demethylation with aluminium chloride may also be accompanied by rearrangement (*Whalley*, *ibid.*, 1953, 277; see also *D. M. Donnelly et al.*, *ibid.*, 1953, 567). 5,6,7-Substituted isoflavones have been prepared by the isomerisation of the more readily available 5,7,8-substituted isoflavones (*M. L. Dhar* and *T. R. Seshadri*, Tetrahedron, 1959, **7**, 77). During demethylation fision of the isoflavone molecule at the $C_{(1)}$–$C_{(2)}$ bond may occur (*Whalley*, Chem. and Ind., 1954, 1230).

Certain isoflavones undergo isomerisation when heated with ethanolic potassium hydroxide, 7,8-dimethoxy-5-hydroxyisoflavone, for example, yielding besides benzyl 2,6-dihydroxy-3,4-dimethoxyphenyl ketone, a small quantity of *5-hydroxy-6,7-dimethoxyisoflavone*, m.p. 204–205° (*U. B. Mahesh* and *Seshadri*, J. sci. ind. Research, India, 1955, **14B**, 671).

5-Hydroxy-7-methoxy-8-methylisoflavone on treatment with potassium ethoxide rearranges to give *5-hydroxy-7-methoxy-6-methylisoflavone*, m.p. 170–172° (*Farkas*, *Várady* and *A. Gottsegen*, Tetrahedron Letters, 1962, 889):

Me MeO O Ph OH O ⟶ MeO Me O Ph OH O

Similarly 7-benzyloxy-5-hydroxy-8,2′,4′,5′-tetramethoxyisoflavone has been isomerised to 7-benzyloxy-5-hydroxy-6,2′,4′,5′-tetramethoxyisoflavone (*Farkas* and *Várady*, Ber., 1961, **94**, 2501), 5-hydroxy-7,8-2′,5′-tetramethoxyisoflavone to 5-hydroxy-6,7-2′,5′-tetramethoxyisoflavone (*idem*, Acta Chim. Acad. Sci. Hung., 1963, **38**, 283), and 5-hydroxy-7,8,4′-trimethoxyisoflavone to 5-hydroxy-6,7,4′-trimethoxyisoflavone (*idem*, *ibid.*, 1960, **24**, 225). Polyhydroxyisoflavones, *e.g.* as CLXVII, on boiling in alcohols of suitable b.p. in the presence of potassium carbonate under anhydrous conditions are directly isomerised to the isomers CLXVIII (*Várady*, Tetrahedron Letters, 1965, 4273):

OMe MeO O OH OH O (CLXVII) ⟶ MeO MeO O OH OH O (CLXVIII)

As in the flavone series a hydroxyl group in the 5-position is resistant to *methylation* with methyl iodide and alkali. 5,7-Diacetoxyisoflavone is selectively methylated with methyl iodide in the 7-position, whereas 5,7,4′-triacetoxyisoflavone affords the 5-methyl ether (*S. Heitz*, *J. Massicot* and *C. Mentzer*, Compt. rend., 1963, **257**, 3054).

The methoxyl group in the 7-position is comparatively stable, a property utilised in the conversion of *7,4′-dimethoxyisoflavone* into *isoformononetin*, *4′-hydroxy-7-methoxyisoflavone*, m.p. 218° (*K. Aghoramurthy et al.*, Proc. Ind., Acad. Sci., 1951, **33A,** 257), 7,3′,4′-trimethoxy- into 3′,4′-dihydroxy-7-methoxyisoflavone, and 5,7-dimethoxy- into 5-hydroxy-7-methoxyisoflavone (*Dhar et al.*, J. sci. ind. Research, India, 1955, **14B**, 73).

The first flavonoid to be *C*-methylated was genistein (*A. G. Perkin et al.*, J. chem. Soc., 1899, **75,** 836; 1900, **77,** 1311, 1317; *Baker* and *R. Robinson, ibid.*, 1926, 2713), the product being 5-hydroxy-7,4′-dimethoxy-6-methylisoflavone (*A. C. Mehta* and *Seshadri, ibid.*, 1954, 3823). 2-Methylgenistein likewise is methylated in the 6-position. Nuclear prenylation of 5,7-dihydroxyisoflavones with prenyl bromide in the presence of methanolic methoxide gives a mixture of 6,8-di-*C*-prenyl (25%), 6-*C*-prenyl (15%) and 7-*O*-prenyl (2%) derivatives. Allylation with allyl bromide in the presence of methanolic potash yields a mixture of 6-*C*-allyl (17%), 6-*C*-allyl-7-*O*-allyl (5%), 8-*C*-allyl (12%), 8-*C*-allyl-7-*O*-allyl (3%), 6,8-di-*C*-allyl (2%), and 7-*O*-allyl (3%) derivatives (*A. C. Jain, P. Lal* and *Seshadri*, Tetrahedron, 1970, **26,** 1977).

7-Methoxyisoflavone has been chloromethylated using formaldehyde and hydrochloric acid to give *8-chloromethyl-7-methoxyisoflavone*, m.p. 143–144°, which has been converted to *8-acetoxymethyl-7-methoxyisoflavone*, m.p. 196–198°, *7-methoxy-8-methoxymethylisoflavone*, m.p. 142–143°, *8-iodomethyl-7-methoxyisoflavone*, m.p. 166–167°, and *8-hydroxymethyl-7-methoxyisoflavone*, m.p. 186–187° (*Y. Kawasw, M. Nakayama* and *S. Matsutani*, Bull. chem. Soc. Japan, 1962, **35,** 1369).

7-Hydroxy-2-methyl- and 7-hydroxy-2-phenylisoflavone on treatment with chlorosulphonic acid sulphonates first in the 8-position, and then in the 6,8-positions giving also the corresponding disulphonyl chlorides (*D. V. Joshi, J. R. Merchant* and *R. C. Shah*, J. org. Chem., 1956, **21,** 1104).

The reaction of 7-methoxyisoflavone with *dimethylsulphoxonium methylide* gives as the major product the cyclopropane derivative CLXIX. The other product is 2-phenyl-2-vinylcoumaran-3-one (CLXX) (*G. A. Caplin, W. D. Ollis* and *I. O. Sutherland*, Chem. Comm., 1967, 575):

MeO ... $\xrightarrow{Me_2S(=O)=CH_2}$ MeO ... Ph (CLXIX) + MeO ... H Ph (CLXX)

Isoflavones generally give the same *colour tests* as flavones (p. 177). They show maximum absorption at 262–270 mμ and sometimes at 320–360 mμ (less intense) (*E. D. Walter*, J. Amer. chem. Soc., 1941, **63,** 3273; *Baker et al.*, J. chem. Soc., 1953, 1852) and their absorption spectra are sufficiently different from those of the flavones to distinguish between the two classes of compound. In the isoflavones a free 7-hydroxy group can be detected by u.v. spectral changes observed on the addition of sodium acetate, while a free 5-hydroxyl group can be detected by the addition of aluminium chloride

(*R. M. Horowitz* and *L. Jurd*, J. org. Chem., 1961, **26,** 2446). For n.m.r. spectra see *A. Grouiller*, Bull. Soc. chim. Fr., 1966, 2405.

(xxiv) Naturally occurring isoflavones

Daidzein, 7,4′-*dihydroxyisoflavone*, $C_{15}H_{10}O_4$, colourless needles, m.p. 320–328°, *diacetate*, m.p. 189°, *dimethyl ether*, m.p. 162–164°, is obtained from the 7-*glucoside*, **daidzin,** $C_{21}H_{20}O_9$, m.p. 234–236, $[\alpha]_D^{20}$ −36.4° (aq.KOH), which occurs in soya bean. Daidzein is broken down by alkali to 2,4-dihydroxyphenyl 4-hydroxybenzyl ketone and has been synthesised (*F. Wessely et al.*, Ber., 1933, **66,** 685; *H. S. Mahal et al.*, J. chem. Soc., 1934, 1769; *W. Baker et al., ibid.*, 1953, 1860; *L. Farkas et al.*, Periodica Polytech., 1958, **2,** 231; C.A., 1960, **54,** 1510; *J. L. Bose* and *N. L. Dutta*, J. sci. ind. Res. India, 1958, **17B,** 266; *S. K. Grover et al.*, Indian J. Chem., 1963, **1,** 517). Daidzin has also been synthesised (*Farkas* and *J. Várady*, Ber., 1959, **92,** 819).

Formononetin, 7-*hydroxy*-4′-*methoxyisoflavone*, $C_{16}H_{12}O_4$, colourless needles or plates, m.p. 265°, *acetate*, m.p. 166°, occurs in *Ononis spinosa* L. as the 7-*glucoside*, **ononin,** $C_{22}H_{22}O_9 \cdot \frac{3}{4}H_2O$, m.p. 214°, 245° (anhyd.), $[\alpha]_D^{23}$ −24.2° (pyridine), and in the flowers of red clover species (*Trifolium pratense* L., *T. subterraneum* L.) (*F. B. Power* and *A. H. Salway*, J. chem. Soc., 1910, **97,** 231; *R. B. Bradbury* and *D. E. White*, *ibid.*, 1951, 3447; *E. C. Bate-Smith et al.*, Chem. and Ind., 1953, 1127). It had previously been surmised that the substance from thse sources was the corresponding flavone, pratol (*R. Robinson* and *K. Venkataraman*, J. chem. Soc., 1926, 2344). Formononetin has also been isolated from *Castranosperumum australe* (*R. A. Eade, H. Hinterberger* and *J. J. H. Simes*, Austral. J. Chem., 1963, **16,** 188) and *Pterocarpus indicus* heartwood (*R. G. Cooke* and *I. D. Rae*, *ibid.*, 1964, **17,** 379). Formononetin (*Baker et al.*, J. Chem. Soc., 1953, 1856; *Farkas et al.*, *loc. cit.*; Ber., 1958, **91,** 2858; *S. A. Kagal, P. M. Nair* and *Venkataraman*, Tetrahedron Letters, 1962, 593), and ononin (*G. Zemplén*, Ber., 1944, **77B,** 452; *Farkas* and *Várady*, *loc. cit.)* have been synthesised. 3′-*Hydroxyformononetin* (*A. C. Jain, P. Lal* and *T. R. Seshadri*, Indian J. Chem., 1969, **7,** 305), and **isoformononetin,** 4′-*hydroxy*-7-*methoxyisoflavone*, m.p. 233–234° *(Bose* and *Dutta*, *loc. cit.)* have been synthesised. A compound believed to be 3′-hydroxyformononetin, isolated from *Pterocarpus dalbergioides* heartwood (*M. R. Parthasarathy et al.*, Indian J. Chem., 1969, **7,** 118) and **calycosin** have been shown to be 7,3′- dihydroxyisoflavone (*Farkas et al.*, Magy. Kem. Foly., 1972, **78,** 252).

Pseudobaptigenin, 7-*hydroxy*-3′,4′-*methylenedioxyisoflavone*, $C_{16}H_{10}O_5$, colourless crystals, m.p. 298–299° (292° and 303–304° are also reported), *acetate*, m.p. 176°, *methyl ether*, m.p. 178–179°, is found in *Baptisia tinctoria* as the 7-*rhamnoglucoside*, **pseudobaptisin,** $C_{28}H_{30}O_{14} \cdot 3H_2O$ (*E. Späth et al.*, Monatsh., 1929, **53/54** 454; Ber., 1930, **63,** 743). It has been synthesised (*Baker et al.*, J. chem. Soc., 1953, 1852; *Farkas et al., loc. cit.;* Ber., 1958, **91,** 2858).

Genistein, prunetol, 5,7,4′-*trihydroxyisoflavone*, $C_{15}H_{10}O_5$, colourless needles, m.p. 301–302° (decomp.) [296–298° (decomp.)], *triacetate*, m.p. 205–206°, *tribenzoate*, m.p. 239°, occurs along with luteolin in dyers broom, *Genista tinctoria*, Linn (*A. G. Perkin* and *F. G. Newbury*, J. chem. Soc., 1899, **75,** 830); as the 7-*glucoside*, **genistin,** $C_{21}H_{20}O_{10}$, m.p. 254–256°, $[\alpha]_D^{21}$ −27.7° (aq. methanol), in *Soja hispida* (*E. Walz*, Ann., 1931, **489,** 118; *E. D. Walter*, J. Amer. chem. Soc., 1941, **63,** 3273; *Zemplén* and *Farkas*, Ber.,

1943, **76,** 1110) and in the fruits of *Sophora Japonica* L. as the 4′-*glucoside*, **sophoricoside** (*Zemplén et al., ibid.*, Ber., 1943, **76,** 267; *R. Bognár* and *V. Szabo*, Chem. and Ind., 1954, 518). It gives on alkaline fusion phloroglucinol, formic acid, and 4-hydroxyphenylacetic acid and has been synthesised (*Baker* and *Robinson*, J. chem. Soc., 1928, 3115; *R. L. Shriner* and *C. J. Hull*, J. org. Chem., 1945, **10,** 288). It was by an investigation of a genistein derivative that *Baker* and *Robinson* proved for the first time that isoflavones occur in nature (J. chem. Soc., 1926, 2713). Genistein is a pro-oestrogen, responsible for most of the oestrogenic activity of subterranean clover (*R. B. Bradbury* and *D. E. White, ibid.*, 1953, 871). Genistein triacetate on treatment with methyl iodide in acetone in the presence of potassium carbonate is selectively methylated in the 5-position to give 7,4′-*diacetylgenistein* 5-*methyl ether*, m.p. 173°, which on boiling with 2% alcoholic potassium hydroxide affords *genistein 5-methyl ether*, m.p. 300–305° (*S. Heitz* and *C. Mentzer*, Compt. rend., 1959, **248,** 3575; *Heitz, J. Massicot* and *Mentzer, ibid.*, 1963, **257,** 3054; also see *Farkas et al.*, Acta Chim., Budapest, 1969, **60,** 293). Genistein 5-methyl ether has been obtained from fresh *Laburnum anagyroides* and *Cytisus laburnum* sapwood (*J. Chopin, M. L. Bouillant* and *P. Lebreton*, Compt. rend., 1963, **256,** 5653; Bull. Soc. chim. Fr., 1964, 1038).

Biochanin-A, *genistein 4′-methyl ether, 5,7-dihydroxy-4′-methoxyisoflavone*, $C_{16}H_{12}O_5$, m.p. 215–216°, dimorphous, *diacetate*, m.p. 192.5°, is obtained from germinated Chana grain, (*S. Siddiqui*, J. sci. ind. Res. India, 1945, **4,** 68), *Cicer arietinum*, the heartwood of *Verreira spectabilis*, and red clover. It has oestrogenic activity approximately equal to that of genistein (*G. S. Pope et al.*, Chem. and Ind., 1953, 1092). It has been synthesised (*Farkas et al.*, Acta Chim., Budapest, 1969, **60,** 293).

Prunetin, *genistein 7-methyl ether, 5,4′-dihydroxy-7-methoxyisoflavone*, $C_{16}H_{12}O_5$, m.p. 237–238° (242°), 4′-*acetate*, m.p. 190°, *diacetate*, m.p. 222.5°, 4′-*methyl ether*, m.p. 144°, is isolated from the bark of a wild cherry related to *Prunus emarginata* (*H. Finnemore*, Pharm. J., 1910, **31,** 604) and the commercial timber muninga, *Pterocarpus angolensis* (*F. E. King* and *L. Jurd.*, J. chem. Soc., 1952, 3211). It gives an intense brown-purple colour with ferric chloride indicating a chelated 5-hydroxyl group. Prolonged methylation gives *genistein trimethyl ether*, m.p. 161.5°, and demethylation yields genistein (*Baker* and *Robinson, ibid.*, 1925, **127,** 1981; 1926, 2713). Its structure follows from its synthesis (*R. N. Iyer et al.*, Current Sci., 1949, **18,** 404; Proc. Indian Acad. Sci., 1951, **33A,** 116, 228; *Baker et al.*, J. chem. Soc., 1953, 1852; *Farkas et al.*, Acta Chim., Budapest, 1969, **60,** 293) and its difference from the 5- and 4′-methyl ethers of genistein (*Shriner* and *Hull, loc. cit.; King* and *Jurd, loc. cit.*). Prunetin dimethyl ether on hydrogenation over palladium–carbon gives the isoflavanone, believed to be padmakastein (p. 286), which on oxidation with selenium dioxide yields the starting material (*S. Ramanujam* and *Seshadri*, Proc. Indian Acad. Sci., 1958, **48A,** 175).

Prunitrin, *prunetin 4′-glucoside*, $C_{22}H_{22}O_{10}\cdot 4H_2O$, m.p. 235–236°, $[\alpha]_D^{20}$ −15.4° (pyridine) occurs in *Prunus Serotina* L. and is synthesised by methylating sophoricoside (*Zemplén* and *Farkas*, Ber., 1957, **90,** 836).

Tlatlancuayin, *5,2′-dimethoxy-6,7-methylenedioxyisoflavone*, $C_{18}H_{14}O_6$, m.p. 147–148°, obtained from *Iresin celosioides* L., was the first naturally occuring isoflavone to be isolated containing a sole 2′-methoxy substituent in the phenyl ring. It constituted also the first instance where a methylenedioxy grouping was observed attached to the chromone ring (*P. Crabbé, P. R. Leeming* and *C. Djerassi*, J. Amer. chem. Soc., 1958, **80,** 5258).

Tectorigenin, 6-*methoxy*-5,7,4′-*trihydroxyisoflavone*, $C_{16}H_{12}O_6$, m.p. 227°, occurs as the 7-*glucoside*, **tectoridin** in *Iris tectorum* (*B. Shibata*, J. pharm. Soc., Japan, 1927, **47,** 380). Its *dimethyl ether* m.p. 188° (*Shriner* and *R. W. Stephenson*, J. Amer. chem. Soc., 1942, **64,** 2737), *trimethyl ether*, m.p. 176°, and *triethyl ether*, m.p. 140° (*M. Krishnamurti* and *Seshadri*, Proc. Indian Acad. Sci., 1954, **39,** 144) have been synthesised thus confirming the structure advanced by *Y. Asahina* (J. pharm. Soc. Japan, 1928, **48,** A mixture of textorigenin and *pseudotectorigenin*, 5,7,4′-*trihydroxy*-8-*methoxyisoflavone*, m.p. 244–246°, has been synthesised (*Baker et al.*, Tetrahedron Letters, 1960, 6; J. chem. Soc., C, 1970, 1219). Other syntheses of textorigenin (*Farkas* and *Várady*, Magy. Kem. Foly., 1960, **66,** 446; *Várady*, Tetrahedron Letters, 1965, 4273) and textoridin (*Várady*, Acta Chim. Acad. Sci. Hung., 1966, **48,** 181) have been reported. *Tectorigenin* 4′-*methyl ether*, 5,7-*dihydroxy*-6,4′-*dimethoxyisoflavone*, m.p. 191–192°, *diacetate*, m.p. 214–215° (*Farkas*, *Várady* and *A. Gottsegen*, Magy. Kem. Foly., 1962, **68,** 238; *Várady*, Tetrahedron Letters, 1965, 4273); *tectoridin* 4′-*methyl ether*, 7-(*β*-D-*glucopyranosyloxy*)-5-*hydroxy*-6,4′-*dimethoxyisoflavone*, m.p. 229–230° (*Várady*, *ibid.*, p. 4277); *tectorigenin* 7-*methyl ether*, 5,4′-*dihydroxy*-6,7-*dimethoxyisoflavone*, m.p. 230–231°; pseudo-tectorigen 7-methyl ether (7-methylisotectorigenin), 5,4′-dihydroxy-7,8-dimethoxyiso-flavone (*Gottsegen* and *Várady*, Period. Polytech., 1964, **8,** 123; C.A., 1965, **63,** 5590); and 7-*methyltectorigenin* 4′-*glucoside*, m.p. 229–230° (*A. Szoke* and *Várady*, Acta Chim. Acad. Sci. Hung., 1968, **55,** 247). 7-Methyltectorigenin is obtained from *Dalbergia sissoo*. 5-*Hydroxy*-7,8,4′-*trimethoxyisoflavone*, m.p. 174–175°, has been isomerised to give 5-*hydroxy*-6,7,4′-*trimethoxyisoflavone*, *tectorigenin* 7,4′-*dimethyl ether*, m.p. 191–192° (*Farkas* and *Várady*, *ibid.*, 1960, **24,** 225).

Muningin, 6,4′-*dihydroxy*-5,7-*dimethoxyisoflavone*, $C_{17}H_{14}O_6$, m.p. 285° (decomp.), *diacetate*, m.p. 232–233°, *dibenzoate*, m.p. 180°, *dimethyl ether*, m.p. 176°, from muninga, *Pterocarpus angolensis* was the first of only a few naturally occurring flavo-noids with a 5-methoxy group (*King et al.*, J. chem. Soc., 1952, 96); it gives no ferric chloride reaction. It has been synthesised (*S. S. Karmaker et al.*, Proc. Indian Acad. Sci., 1955, **41A,** 192; *M. L. Dhar* and *Seshadri*, *ibid.*, 1956, **43A,** 79).

Santal, 5,3′,4′-*trihydroxy*-7-*methoxyisoflavone*, $C_{16}H_{12}O_6$, m.p. 222–223°, *triacetate*, m.p. 170°, *trimethyl ether*, m.p. 166°, is obtained from Sandalwood, *Pterocarpus santalinus*, Linn. and Barwood, *Baphia nitida*, Lodd. (*P. O'Neill* and *A. G. Perkin*, J. chem. Soc., 1918, **113,** 125, 137; *H. Raudnitz* and *G. Perlmann*, Ber., 1935, **68,** 1862). Its constitution was established by degradation (*A. Robertson et al.*, J. chem. Soc., 1949, 1571) and synthesis (*N. Narasimhachari* and *Seshadri*, Proc. Indian Acad. Sci., 1950, **32A,** 342; *Iyer*, *ibid.*, 1951, **33A,** 228). *Santal monomethyl ether*, m.p. 167–168° has been prepared and its structure established as 5,3′-dihydroxy-7-4′-dimethoxyiso-flavone (*Dhar*, *Narasimhachari* and *Seshadri*, J. sci. ind. Research India, 1956, **15B,** 285). The 7-glucoside **oroboside,** $C_{21}H_{20}O_{11}$, m.p. 220°, $[\alpha]$ −61.3° (pyridine), isolated from vetch, *Orobus tuberosus*, is hydrolysed to the aglycone **orobol,** 5,7,3′,4′-*tetrahydroxy-isoflavone*, $C_{15}H_{10}O_6$, m.p. 270°, *tetra-acetate*, m.p. 207.5° (210–212°) (*C. Charaux*, Compt. rend., 1930, **190,** 387; Bull. Soc. chim. biol., 1939, **21,** 1330) which is probably norsantol *(Robertson et al., loc. cit.)*. Oroboside has been synthesised (*Z. Luka* and *Várady*, Acta Chim. Acad. Sci. Hung., 1968, **55,** 345).

Milldurone, 6,7,2′-*trimethoxy*-4,5′-*methylenedioxyisoflavone*, $C_{19}H_{16}O_7$, m.p. 233–234°, has been obtained from the seeds of *Millettia dura* (Dunn) (*W. D. Ollis, C. A. Rhodes* and *I. O. Sutherland*, Tetrahedron, 1967, **23,** 4741) and has been synthesised (*Ollis et al.*, J. chem. Soc., C, 1970, 125; *M. Nógrádi et al.*, Ber., 1970, **103,** 999). **Lettadurone,** 6,7-*dimethoxy*-3′,4′-*methylenedioxyisoflavone*, $C_{18}H_{14}O_6$, m.p. 239–240°, also occurs in *M. dura* (Dunn) and has been synthesised.

Podospicatin, 5,7,2′-*trihydroxy*-6,5′-*dimethoxyisoflavone*, $C_{17}H_{14}O_7$, m.p. 212°, has been isolated from the heartwood of *Podocarpus spicatus* (*L. H. Briggs* and *B. F. Cain*, Tetrahedron, 1959, **6,** 143). 5-*Hydroxy*-7,8,2′,5′-*tetramethoxyisoflavone*, m.p. 174°, on isomerisation using potassium ethoxide in ethanol gives 5-*hydroxy*-6,7,2′,5′-*tetramethoxyisoflavone*, m.p. 152–153°, which on methylation with dimethyl sulphate in acetone containing potassium gives *podospicatin trimethyl ether*, m.p. 161–162° (*Farkas* and *Várady*, Acta Chim. Acad. Sci. Hung., 1963, **38,** 283). Podospicatin was the first recorded example of an isoflavone containing a 2′,5′-hydroxylation pattern.

Irigenin, 5,7,3′-*trihydroxy*-6,4′,5′-*trimethoxyisoflavone*, $C_{18}H_{16}O_8$, m.p. 185°, 7,3′-*dibenzoate*, m.p. 169°, *triacetate*, m.p. 127–128°, is obtained from the 7-glucoside, *iridin* occurring in *Iris germanica* and *I. florentian.* It yields with methyl iodide and alkali *irigenin* 7,3′-*dimethyl ether*, m.p. 166–167°, and with dimethyl sulphate and alkali the *trimethyl ether*, m.p. 163° (*Baker*, J. chem. Soc., 1928, 1022; *Baker* and *Robinson*, *ibid.*, 1925, **127,** 1981; 1929, 152). Irigenin has been synthesised along with **pseudoirigenin,** 5,7,3′-*trihydroxy*-8,4′,5′-*trimethoxyisoflavone*, m.p. 159°, *triacetate*, m.p. 147° (*Baker et al.*, Tetrahedron Letters, 1960, 6; J. chem. Soc., C, 1970, 1219). Besides irisolone and irigenin, **irisolidone,** 5,7-*dihydroxy*-6,4′-*dimethoxyisoflavone*, m.p. 195–196°, *diacetate*, m.p. 162–163°, *dimethyl ether*, m.p. 181°, has been isolated from *Iris repalensis* D. Don (*L. Prakash, A. Zaman* and *A. R. Kidwai*, J. org. Chem., 1965, **30,** 3561). The dimethyl ether is identical with tectorigenin trimethyl ether. **Irigenol,** 5,6,7,3′,4′,5′-*hexahydroxyisoflavone*, $C_{15}H_{10}O_8$, m.p. 331° (decomp.) is obtained by heating 5,7-dihydroxy-8,3′,-4′,5′-tetramethoxyisoflavone with hydrobromic acid, demethylation being accompanied by rearrangement (*Baker*, J. chem. Soc., 1928, 1022).

Caviunin, 5,7-*dihydroxy*-6,2′,4′,5′-*tetramethoxyisoflavone*, $C_{19}H_{18}O_8$, m.p. 191–193°, 7-*methyl ether*, m.p. 185.5–187.5°, *diacetate*, m.p. 198–200°, is obtained from *Dalbergia nigra* and its structure has been confirmed by synthesis (*O. R. Gottlieb* and *M. T. Magalhães*, J. org. Chem., 1961, **26,** 2449; *S. F. Dyke, Ollis* and *M. Sainsbury*, *ibid.*, p. 2453; *Farkas* and *Várady*, Ber., 1961, **94,** 2501; *Várady*, Tetrahedron Letters, 1965, 4273); *caviunin 7-methyl ether*, m.p. 187–188°, is also obtained from the same source (*Farkas* and *Várady*, Acta Chim. Acad. Sci. Hung., 1962, **33,** 183). 5,7,8,2′,4′,5′-Hexamethoxyisoflavone on treatment with aluminium chloride in nitrobenzene at 105° gives *isocaviunin 7-methyl ether*, 5-*hydroxy*-7,8,2′,4′,5′-*pentamethoxyisoflavone*, m.p. 182–183°, which on boiling with a 2% solution of potassium ethoxide in ethanol rearranges to caviunin 7-methyl ether (*Farkas* and *Várady*, Magy. Kem. Foly., 1962, **68,** 93).

Piscerythrone, 5,7,2′,4′-*tetrahydroxy*-5′-*methoxy*-3′-*phenylisoflavone*, $C_{21}H_{20}O_7$, m.p. 183.5–184.5°, *tetra-acetate*, m.p. 146–147°, *tetramethyl ether*, m.p. 174–175°, has been obtained from the roots of Jamaican Dogwood, *Piscidia erythrina*. **Piscidone,** $C_{21}H_{20}O_7$, m.p. 154–155°, has also been isolated but its structure has not been established (*C. P. Falshaw et al.*, Tetrahedron, 1966, suppl. 7,333).

Afromosin, *7-hydroxy-6,4'-dimethoxyisoflavone*, m.p. 236–237° (229°), *methyl ether*, m.p. 178–179°, has been isolated from the West African hardwood *Afromosia elata* Harms. (*T. B. H. Murray* and *C. Y. Theng*, J. chem. Soc., 1960, 1491), cabreuva or oleo pardo wood, *Myrocarpus fastigiatus* Fr. Allem., oleo vermelho, *M. balsamum* (*J. B. Harborne, Gottlieb* and *Magalhães*, J. org. Chem., 1963, **28,** 881) and *Castanospermum australe* (*Eade, Hinterberger* and *Simes*, Austral. J. Chem., 1963, **16,** 188). Formononetin is also obtained from the last species. The structure of afromosin has been established by synthesis. Methylation with dimethyl sulphate and alkali gives the methyl ether, but the use of methyl iodide results in *C*-alkylation.

Cladrastin, *7-hydroxy-6,3',4'-trimethoxyisoflavone*, $C_{18}H_{16}O_6$, m.p. 206–207°, *methyl ether*, m.p. 187–188°, and **cladrin,** *7-hydroxy-3',4'-dimethoxyisoflavone*, $C_{17}H_{14}O_5$, m.p. 257–258°, *methyl ether*, m.p. 165°, have been isolated from the heartwood of *Cladrastis lutea* (Michx. f.) K. Koch along with formononetin and afrormosin (*M. Shamma* and *L. D. Stiver*, Tetrahedron, 1969, **25,** 3887). **Fujikinetin,** *7-hydroxy-6-methoxy-3',4'-methylenedioxyisoflavone*, and its monoglucoside, **fujikinin,** have been isolated from the bark of *Cladrastis platycarpa* (*H. Imamura, Y. Hibino* and *H. Ohashi*, Mokuzai Gakkaishi, 1972, **18,** 325).

6,7,3',4'-*Dimethylenedioxyisoflavone*, $C_{17}H_{10}O_6$, m.p. 227–229°, has been isolated from the roots of *Tephrosia maxima* Aers; on boiling in benzene with anhydrous aluminium chloride it gives 6,7,3',4'-*tetrahydroxyisoflavone*, m.p. 320° (decomp.), *tetramethyl ether*, m.p. 168–170° (*A. S. Rangaswami* and *B. V. R. Sastry*, Proc. Indian Acad. Sci., 1956, **44A,** 279).

3',4'-*Methylenedioxy-7-prenylisoflavone*, m.p. 126–128°, has also been isolated from the above (*idem, ibid.*, 1963, **57A,** 135). **Purpuranin B,** $C_{21}H_{12}O_7$, m.p. 224–226°, may be assigned tentatively one of two structures, namely 5-methoxy-6,7,3',4'-dimethylenedioxy-8-prenylisoflavone or 5-methoxy-7,8,3',4'-dimethylenedioxy-6-prenylisoflavone (*N. V. Subba Rao* and *V. Sundaramurthy*, Bull. Nat. Inst. Sci., India, 1965, **31,** 83).

6,7,2',4',5'-*Pentamethoxy-*, m.p. 171–172°, 6,7,3',4'-*tetramethoxy-*, m.p. 187–188°, 6,7,3'-*trimethoxy-4',5'-methylenedioxy-*, m.p. 211–212°, and 6,7,2'-*trimethoxy-4',5'-methylenedioxy-isoflavone*, m.p. 234.5–235.5°, together with 6,7-*dimethoxy-3',4'-methylenedioxyisoflavone*, m.p. 201.5–202.5°, have been isolated from the heartwood of *Cordyla africana* and their structures established by chemical degradation, synthesis, and spectral consideration. The mass spectra of these and related isoflavones have been analysed and the presence of the 2'-methoxy group has been found to influence the fragmentation pattern profoundly (*R. V. M. Campbell, S. H. Harper* and *A. D. Kemp*, J. chem. Soc., C, 1969, 1787).

(xxv) Naturally occurring homoisoflavones

Eucomin, 5,7-*dihydroxy*-3-(4-*methoxybenzylidene*)*chroman*-4-*one*, $C_{17}H_{14}O_5$, yellow needles, m.p. 194–196°, *monoacetate*, m.p. 146–147°, *monomethyl ether*, m.p. 145–149°, *dimethyl ether*, m.p. 141–144°, and **eucomol,** 3-(4-*methoxybenzyl*)-3,5,7-*trihydroxychroman*-4-*one*, $C_{17}H_{16}O_6$, colourless hexagonal plates, m.p. 134.5–135°, $[\alpha]_D^{25}$ −32°

($CHCl_3$), *trimethyl ether*, m.p. 120–121°, have been isolated from the bulbs of *Eucomis bicolor* BAK *(Liliaceae)* (*P. Böhler* and *Ch. Tamm*, Tetrahedron Letter, 1967, 3479). Eucomin diacetate has been synthesised by condensing 5,7-dihydroxychroman-4-one with anisaldehyde in acetic anhydride; hydrolysis with ethanolic alkali affords eucomin. (±)Eucomol has also been synthesised, thus confirming its structure (*L. Farkas*, *A. Gottsegen* and *M. Nógrádi*, Tetrahedron, 1970, **26**, 2787). The biosynthesis of eucomin has been investigated (*P. M. Dewick*, Chem. Comm., 1973, 438).

Eucomin Eucomol

Punctatin, 5,7-*dihydroxy*-3-(4-*hydroxybenzylidene*)-8-*methoxychroman*-4-*one*, $C_{17}H_{14}O_6$, orange needles, m.p. 189–190°, *trimethyl ether*, m.p. 147–148°, and its 4′-*methyl ether*, $C_{18}H_{16}O_6$, yellow needles or rods, m.p. 213.5–214.5°, are constituents of the bulbs *E. punctata*. The latter, and **autumnalin, eucomnalin,** 5,7-*dihydroxy*-3-(4-*hydroxybenzylidene*)-6-*methoxychroman-4-one*, $C_{17}H_{14}O_6$, yellow rods, m.p. 244.5–247.5°, *trimethyl ether*, m.p. 115–117.5°, are obtained from the bulbs of *E. autumnalis* GRAEB *(Liliaceae)* (*W. T. L. Sidwell* and *Tamm*, Tetrahedron Letters, 1970, 475; *R. E. Finckh* and *Tamm*, Experientia, 1970, **26,** 472). Punctatin, eucomnalin, and some of their derivatives have been synthesised (*Farkas et al.*, Tetrahedron, 1971, **27,** 5049).

(xxvi) Furoisoflavones and pyranoisoflavones

The linear furoisoflavone **dehydronepseudin,** 2′,3′,4′-*trimethoxyfuro*[3,2-g]*isoflavone*, m.p. 158–160° has been obtained by dehydrogenating nepseudin (p. 288) with manganese dioxide (*L. Crombie* and *D. A. Whiting*, J. chem. Soc., 1963, 1569). Dehydronepseudin has been synthesised from 6-hydroxy-5-(2,3,4-trimethoxyphenylacetyl)-2,3-

MnO_2

Nepseudin Dehydronepseudin

dihydrobenzofuran, prepared by the Hoesch condensation of 6-hydroxy-2,3-dihydrobenzofuran and 2,3,4-trimethoxybenzyl cyanide, by treatment with ethyl orthoformate, pyridine and piperidine, followed by dehydrogenation with *N*-bromosuccinimide (*K. Fukui* and *M. Nakayama*, Experientia, 1962, **19,** 621):

$HC(OEt)_3$ C_5H_5N $C_5H_{11}N$

N.B.S.

Dehydroneotenone, *2'-methoxy-4',5'-methylenedioxyfuro*[3,2-g]-*isoflavone*, m.p. 240–241°, prepared by the dehydrogenation of neotenone (p. 288) (*Crombie* and *Whiting*, *loc. cit.*), has been synthesised (*Fukui* and *Nakayama*, Experientia, 1964, **20,** 668). **Isoelliptol,** *2',4',5'-trimethoxyfuro*[3,2-g]*isoflavone*, m.p. 190–190.5 (*Fukui et al.*, Bull. chem. Soc. Japan, 1965, **38,** 845).

A number of angular furoisoflavones have been synthesised; *furo*[2,3-h]*isoflavone* (R = H), m.p. 152–152.5° (*Fukui* and *Y. Kawase*, *ibid.*, 1959, **32,** 693); *2-methylfuro-*[2,3-h]*isoflavone* (R = Me), m.p. 187–188° (*T. Matsumoto*, *Kawase* and *M. Nanbu*, *ibid.*, 1958, **31,** 688; *L. R. Row* and *T. R. Seshadri*, Proc. Indian Acad. Sci., 1951, **34A,** 187):

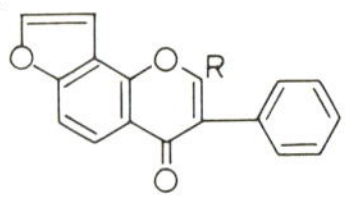

Isoderritolisoflavone (CLXXI) reacts with excess dimethylsulphoxonium methylide to yield the *cyclopentene*, CLXXII, m.p. 144–145° (*Crombie*, *J. S. Davies* and *Whiting*, J. chem. Soc., C, 1971, 304):

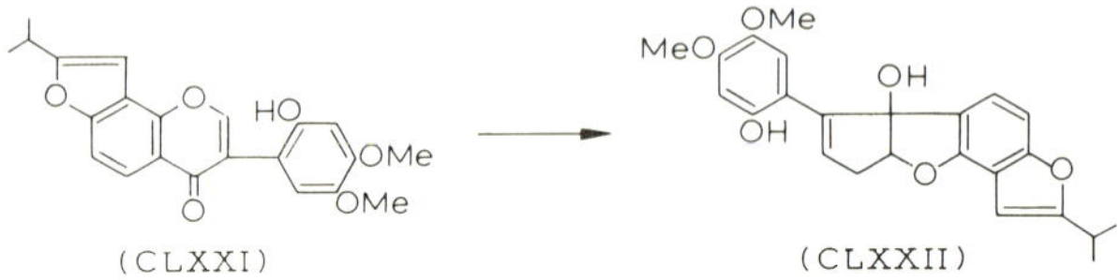

(CLXXI) (CLXXII)

Both linear and angular dimethylpyranoisoflavones have been obtained from natural sources.

Alpinumisoflavone, *5-hydroxy-7-(4-hydroxyphenyl)-8,8-dimethyl-6H-8H-benzo*[1,2-b:5,4-b']*dipyran-6-one*, *5,4'-dihydroxy-8,8-dimethyl-8H-pyrano*[3,2-g]*isoflavone*, $C_{20}H_{16}O_5$, m.p. 213–214°, *4'-methyl ether*, m.p. 136–137°, *dimethyl ether*, m.p. 119–120°, isolated from *Laburnum alpinum* J. Presl. Its structure has been confirmed by spectral data and synthesis (*B. Jackson*, *P. J. Owen* and *F. Scheinmann*, *ibid.*, p. 3389).

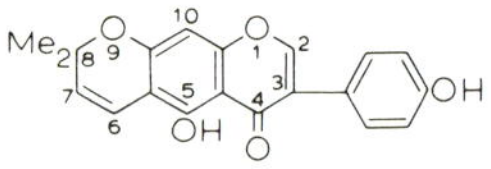

Alpinumisoflavone 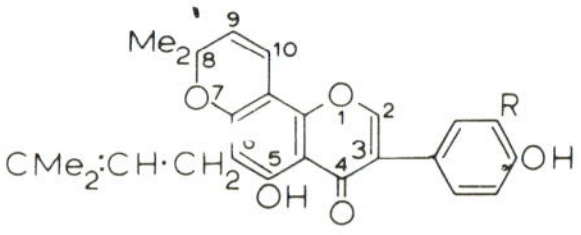 (CLXXIII)

Osajin, *5,4'-dihydroxy-8,8-dimethyl-6-prenyl-8H-pyrano*[2,3-h]*isoflavone* (CLXXIII, R = H), $C_{25}H_{24}O_5$, m.p. 189° *monoacetate*, m.p. 159°, *diacetate*, m.p. 164°, *dimethyl ether*, m.p. 118.5°, and **pomiferin** (CLXXIII, R = OH), $C_{25}H_{24}O_6$, m.p. 200.5°, *diacetate*, m.p. 134.5°, *triacetate*, m.p. 154°, the yellow pigments of the osage orange, *Maclura pomifera* (*M. L. Wolfrom et al.*, J. Amer. chem. Soc., 1946, **68,** 406). Osajin has also been isolated along with **scandenone, warangalone,** *5,4'-dihydroxy-8,8-dimethyl-10-prenyl-8H-pyrano*[3,2-g]*isoflavone*, $C_{25}H_{24}O_5$, m.p. 148.5–150.5°, *monoacetate*, m.p. 158–160° and

scandinone, 8,8-*dimethyl*-4′-*hydroxy*-5-*methoxy*-6-*prenyl*-8H-*pyrano*[2,3-h]*isoflavone*, $C_{26}H_{26}O_5$, m.p. 207–210°, *methyl ether*, m.p. 109–111°, from the roots of *Derris scandens*. The biogenesis of the above compounds has been discussed and they have been shown to be closely related to lonchocarpic acid and scandenin, pyranocoumarins obtained from the same source (*A. Pelter* and *P. Stainton*, J. chem. Soc., C, 1966, 701). The 4′-methyl ethers of osajin and scandenone have been synthesised (*A. C. Jain, P. Lal* and *Seshadri*, Tetrahedron, 1970, **26,** 1977):

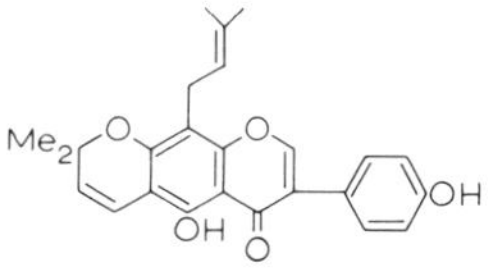

Scandenone 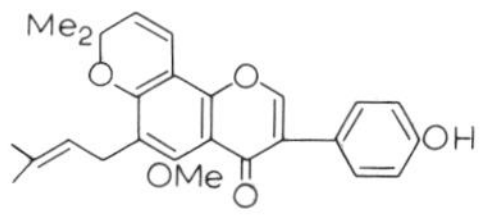 Scandinone

Toxicarol-isoflavone, 5-*hydroxy*-2′,4′,5′-*trimethoxy*-8,8-*dimethyl*-8H-*pyrano*[2,3-h]*isoflavone*, $C_{23}H_{22}O_7$, m.p. 219°, *monoacetate*, m.p. 210°, *monomethyl ether*, m.p. 178°, is obtained from *Derris malaccensis* resin; its structure has been confirmed by n.m.r. spectral data (*Harper* and *W. G. E. Underwood*, J. chem. Soc., 1965, 4203) and synthesis (*Nakayama et al.*, Experientia, 1971, **27**, 875).

Auriculatin, 5,2′,4′-*trihydroxy*-8,8-*dimethyl*-10-*prenyl*-8H-*pyrano*[3,2-g]*isoflavone*, $C_{25}H_{24}O_6$, m.p. 236–239°, *triacetate*, m.p. 123–124°, **auriculin,** *auriculatin 4′-methyl ether*, $C_{26}H_{26}O_6$, m.p. 124–125°, and **isoauriculatin,** $C_{25}H_{24}O_6$, m.p. 132–134°, *methyl ether*, m.p. 157–158°, *dimethyl ether*, m.p. 192–194°, have been isolated from *Milletia auriculata* (*M. Sharbir* and *A. Zaman*, Tetrahedron, 1970, **26**, 5041):

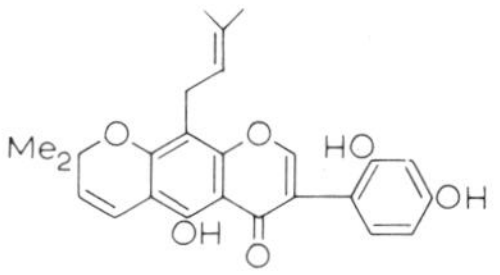

Auriculatin 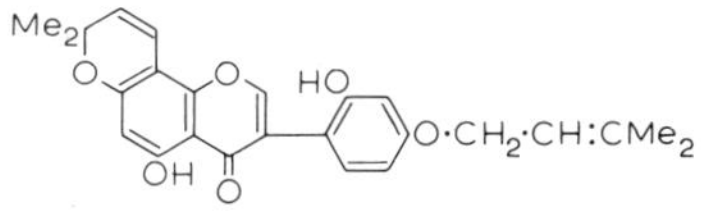 Isoauriculatin

Jamaicin, 2′-*methoxy*-8,8-*dimethyl*-4′,5′-*methylenedioxy*-8H-*pyrano*[2,3-h]*isoflavone* (CLXXIX, R = H), $C_{22}H_{18}O_6$, m.p. 193–194°, has been isolated from Jamaican Dogwood *Piscidia erythrina* L. (*J. A. Moore* and *S. Eng*, J. Amer. chem. Soc., 1956, **20,** 395; *O. A. Stamm et al.*, Helv., 1958, **41,** 2006) and **ichthynone** (CLXXIX, R = OMe), 6,2′-*dimethoxy*-8,8-*dimethyl*-4′,5′-*methylenedioxy*-8H-*pyrano*[2,3-h]*isoflavone*, $C_{23}H_{20}O_7$, m.p. 203–204°, from *P. erythrina* L. (*J. S. P. Schwarz et al.*, Tetrahedron, 1964, **20,** 1317); **lisetin,** $C_{21}H_{18}O_6$, m.p. 284–285°, *triacetate*, m.p. 254–255°, *trimethyl*

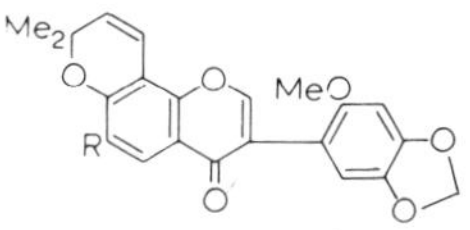

(CLXXIX) 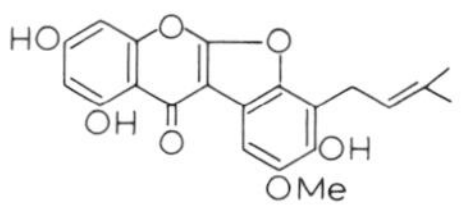 Lisetin

ether, m.p. 199–200°, has also been isolated from *P. erythrina* L. (*C. P. Falshaw et al.*, *ibid.*, 1966, suppl. 7, 333); **durmillone,** 6-*methoxy*-8,8-*dimethyl*-4′,5′-*methylenedioxy*-8H-*pyrano*[2,3-h]*isoflavone*, $C_{22}H_{18}O_6$, m.p. 171–172°, from the seeds of *Millettia dura* (Dunn) (*W. D. Ollis, C. A. Rhodes* and *I. O. Sutherland*, Tetrahedron, 1967, **23**, 4741).

Dihydrodimethylpyranoisoflavone, in which R, R^1 and R^2 may be H, or OMe have been synthesised (*T. Iwadare et al.*, J. org. Chem., 1963, **28**, 3206):

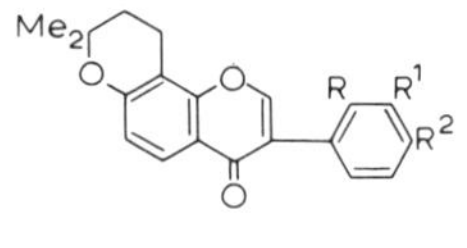

Me2 O O R R1 R2 O

4. Chroman, dihydrochromene, 3,4-dihydro-2*H*-benzo[*b*]pyran and derivatives

(*a*) *Chromans*

(*i*) *Syntheses*

Chroman (I) was first prepared by the removal of hydrogen chloride from 2-(γ-chloropropyl)phenol by gently warming the phenol with aqueous sodium hydroxide (*J. von Braun* and *A. Steindorff*, Ber., 1905, **38**, 850):

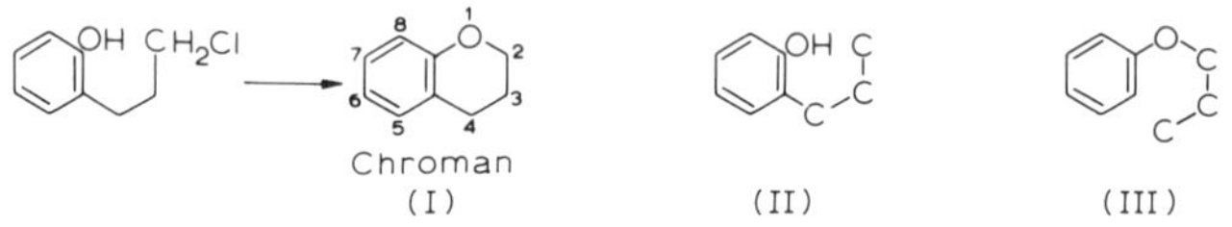

OH CH2Cl → Chroman (I) OH C C C (II) O C C C (III)

Chromans may be prepared by cyclisation of derivatives of 2-propyl phenol(II) or phenyl propyl ether (III). 2-(γ-Hydroxypropyl)phenol, formed by the reduction of coumarin with sodium amalgam, is quantitatively ring-closed to chroman when heated with ethanolic hydrogen chloride in a sealed-tube at 150° (*F. W. Semmler*, *ibid.*, 1906, **39**, 2851). The most satisfactory method of obtaining chroman appears to be the ring-closure of γ-chloro-propyl phenyl ether (IV) by means of stannic chloride (*P. Maitte*, Ann. Chim., 1954, **9**, 433). It may also be prepared by heating γ-hydroxypropyl phenyl ether with zinc chloride (*R. E. Rindfusz*, J. Amer. chem. Soc., 1919, **41**, 665):

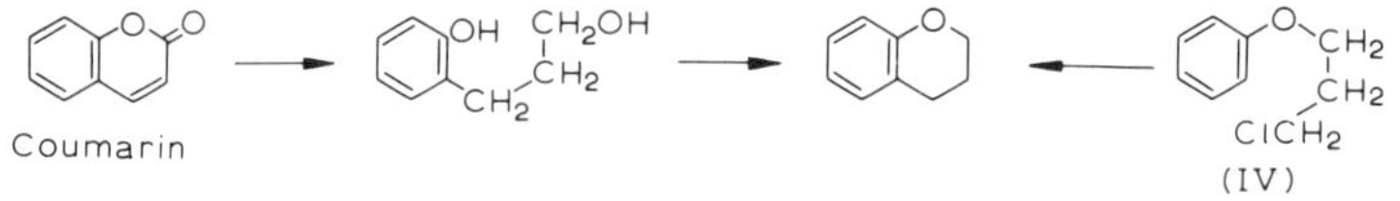

Coumarin → OH CH2OH CH2 CH2 → chroman ← O CH2 CH2 ClCH2 (IV)

Good yields of chroman, 6-methyl-, and 6-chloro-chroman have been obtained by reacting the appropriate 1,3-diphenoxypropane with aluminium chloride in boiling benzene (*L. W. Deady*, *R. D. Topson* and *J. Vaughan*, J. chem. Soc., 1963, 2094):

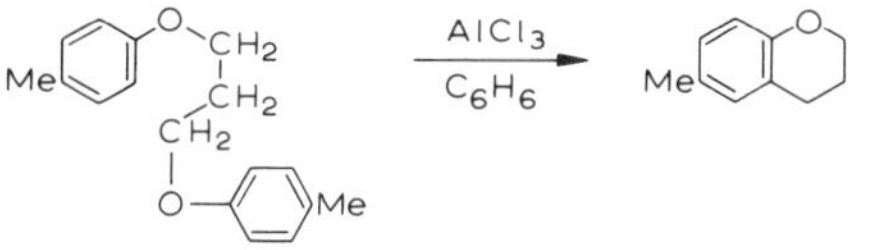

Chromans with other substituents in the benzene ring have been prepared by this method (*idem*, *ibid*., 1965, 5718).

2,2-Dialkylchromans are obtained by the action of Grignard reagents on dihydrocoumarins. Ring-fission accompanied by formation of a *tert*-alcohol first occurs, followed by ring-closure to a chroman (*A. Robertson et al.*, *ibid*., 1937, 1530; *P. M. Ruoff*, J. Amer. chem. Soc., 1940, **62,** 145):

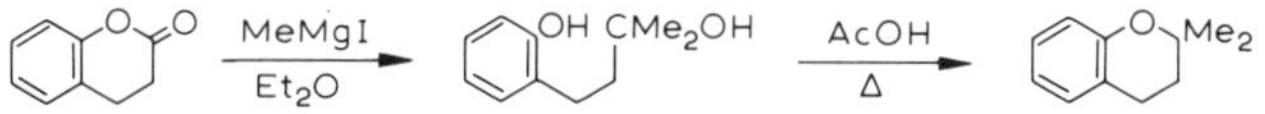

2,2-Dialkyl and 2,2-diaryl-chromans may be obtained by hydrogenation of the related chromenes using a platinum oxide catalyst (*R. Livingstone et al.*, J. chem. Soc., 1960, 602; 1964, 2978; 1970, 1758). Quinol when warmed with 3,3-dimethylally(diphenyl)phosphate gives *6-hydroxy-2,2-dimethylchroman*, m.p. 74.5–75° (*J. A. Miller* and *H. C. S. Wood*, Chem. Comm., 1965, 39):

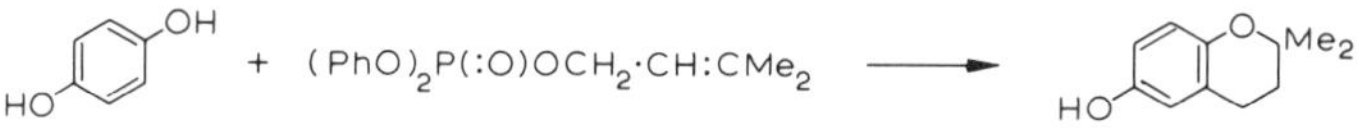

Clemmensen reduction of chromanones also yields chromans (*Robertson et al.*, *loc. cit.*, see also pp. 254, 270, and for a critical review, *G. Chatelus*, Ann. Chim., 1949, [xii], **4,** 505; *Maitte*, Ann. Chim., Fr., 1954, **9,** 431).

Chroman, colourless oil, m.p. 4.8°, b.p. 214°/742 mm, 96–98°/17 mm., d^{20} 1.0610, n_D^{20} 1.544, *picrate*, m.p. 78°, is soluble in the common organic solvents and gives a red colour with sulphuric acid. Chroman and *N*-bromosuccinimide give 4- and 6-**bromochroman** (*Maitte*, *loc. cit.*). It undergoes Friedel–Crafts acylation in the 6-position (*St. von Kostanecki et al.*, Ber., 1907, **40,** 3668; *Chatelus*, *loc. cit.*) and is stable to oxidising agents. 2-**Methyl-**, b.p. 223–226°, 100–102°/11 mm; 3-**methyl-**, b.p. 102–104°/15 mm, n_D^{20} 1.5335 (*C. D. Hurd* and *A. Hoffman*, J. org. Chem., 1940, **5**, 212); 5-**methyl-**, b.p. 71–72°/1 mm; 6-**methyl-**, b.p. 103–105°/14 mm, n_D^{15} 1.5442; 7-**methyl-**, b.p. 67–68°/1 mm, 141–143°/60 mm, n_D^{26} 1.5380; 8-**methyl-chroman**, b.p. 64–67°/2 mm, 105–108°/15 mm, n_D^{27} 1.5420 (*Deady*, *Topsom* and *Vaughan*, J. chem. Soc., 1965, 5718). 6-**Ethylchroman**, b.p. 127°/17 mm, is prepared by the Clemmensen reduction of *6-acetylchroman*, m.p. 45°, *semicarbazone*, m.p. 261°. 4-**Phenyl-**, m.p. 38.5° (*A. Greenwood* and *M. Nierenstein*, *ibid.*, 1920, **117,** 1597; *M. Sliwa*, *H. Sliwa*, *Maitte*, Compt. rend., Ser C, 1969, **268,** 263); the cyclisation of 1-phenoxy-1-phenylpropane under either Friedel–Crafts or acidic conditions leads to 4-phenyl-chroman rather than flavan as expected (*M.* and *H. Sliwa* and *Maitte*, Bull. Soc. chim. Fr., 1972, 1540). 6-**Phenyl-**, m.p. 41–42°, b.p. 151°/1 mm; 8-**phenyl-chroman**, b.p. 144–145°/1 mm, n_D^{17} 1.6223 (*Deady*, *Topsom* and *Vaughan*, *loc. cit.*); 4-(4-hydroxybenzyl)chroman derivatives (*D. M. Lynch* and *W. Cole*, J. med. Chem., 1968, **11,** 291); 5,6-**benzochroman,** 2,3-*dihydro*-1H-*naphtho*-[2,1-b]*pyran*, m.p. 40°, b.p. 132°/1 mm; 7,8-**benzochroman,** 3,4-*dihydro*-2H-*naphtho*-[1,2-b]*pyran*, b.p. 134°/1 mm, n_D^{20} 1.6373 (*Deady*, *Topsom* and *Vaughan*, *loc. cit.*); 2,2-**dimethyl-**, b.p. 98.5°/11.5 mm, 225.2–225.4°/769 mm (*L. I. Smith*, *H. E. Ungnade* and *W. W. Prichard*, J. org. Chem., 1939, **4,** 358); 2,2-**diphenyl-chroman**, m.p. 79–80° (*Livingstone*, *D. Miller* and *S. Morris*, J. chem. Soc., 1960, 602); 4-(4-*hydroxyphenyl*)-2,2,4-*trimethylchroman* (Dianin's compound), m.p. 155–156°, is obtained by the condensation of phenol and mesityl oxide in presence of anhydrous hydrogen chloride (*W. Baker et al.*, *ibid.*, 1956, 2010, 2018; *H. M. Powell*, *ibid.*, 1948, 61; *Powell* and *B. D. P. Wetters*, Chem. and Ind., 1955, 256), or $H_3PO_4 \cdot BF_3$ containing 37% BF_3 (*G. G. Kondrat'eva*, Metody Polnch. Khim. Reaktivov Prep., 1969, **20,** 199). Some substituted chromans are listed in Table 5.

TABLE 5

SUBSTITUTED CHROMANS

Substituents	*B.p. (°C)/mm Hg*	*Ref.*	*Substituents*	*M.p. (°C)*	*Ref.*
6-Bromo-	140–142/16	1	3,4-Dibromo-	127–128	3
8-Bromo-	100–104/1	1	3,4-Dichloro-	89–90	3
6-Chloro-	86–89/1	1	3,4-Dibromo-2,2--dimethyl-	81–82	4
7-Chloro-	91/1	1	3,4-Dichloro-2,2--dimethyl-	59–60	5
8-Chloro-	93/2	1			
6-Methoxy-	92/1	1	3,4-Dibromo-2,2--diphenyl-	137–138	5
7-Methoxy-	133/16	1	3,4-Dichloro-2,2--diphenyl-	99–100	5
7-Hydroxy-2,2--dimethyl-	m.p. 73°	2	3-Bromo-4-chloro-	96	6
5,7-Dihydroxy-2,2-dimethyl-	m.p. 163°	2	4-Bromo-3-chloro-	117–118	6
7-Hydroxy-2,2,5--trimethyl-	138–139/5	2			

References

1 *L. W. Deady*, *R. D. Topsom* and *J. Vaughan*, J. chem. Soc., 1965, 5718.
2 *P. R. Iyer* and *G. D. Shah*, Indian J. chem., 1968, **6,** 227.
3 *W. D. Cotterill*, *J. Cottam* and *R. Livingstone*, J. chem. Soc., C, 1970, 1006.
4 *Livingstone*, *D. Miller* and *S. Morris*, *ibid.*, 1960, 3094.
5 *Cottam*, *Livingstone* and *Morris*, *ibid.*, 1965, 5266.
6 *R. Binns et al.*, *ibid.*, Perkin II, 1974, 732.

A number of other 3,4-**dihalogenochromans** (*Livingstone et al.*, J. chem. Soc., 1960, 618; 1962, 76; 1966, 2013) and related dihalogenodihydronaphthopyrans (*Livingstone et al., ibid.*, 1967, 1472) have been synthesised. The hydrolysis of 3,4-dibromo- and 3,4-dichloro-2,2-dimethylchroman (*Binns, Cotterill* and *Livingstone, ibid.*, 1965, 5049), the methanolysis of 3,4-dihalogenochromans (*Binns et al., loc. cit.*), and the stereochemistry and reactions of some 3,4-disubstituted chromans (*Cotterill, Cottam* and *Livingstone, loc. cit.; H. Hofmann* and *G. Salbeck*, Ber., 1970, **103,** 2768) have been discussed.

The condensation of phenol with 2-methoxybutadiene gives 2-**methoxy**-2-**methylchroman**, b.p. 108–112°/14 mm, which on Birch reduction followed by hydrolysis and cyclisation affords 4,4a,5,6-tetrahydro-2-(3*H*)-naphthalenone (*L. J. Dolby* and *E. Adler*, Tetrahedron Letters, 1971, 3803):

OMe, Me — Li, liq. NH_3 / THF Bu^tOH → OMe, Me — HCl / aq. dioxane → O

2-Bromo-1-tetralone on treatment with perbenzoic acid in chloroform gives a bromolactone, V, which on hydrolysis by titration with cold alkali in the presence of hydrogen peroxide affords 2-bromo-4-(2-hydroxyphenyl)butanoic acid. The latter on warming with 2 *M*-sodium hydroxide solution yields *chroman-2-carboxylic acid* (VI), m.p. 98.5–100° (*G. Baddeley* and *J. R. Cooke*, J. chem. Soc., 1958, 2797):

O, Br → O–CO, CHBr (V) → OH, CO_2H, CHBr → CO_2H (VI)

The mechanism of hydrogenolysis of chroman-2-carboxylic acid has been discussed (*S. Mitsui et al.*, Nippon Kagaku Zasshi, 1962, **83,** 581).

(*ii*) *Chemical properties*

Chroman and 2-methylchroman on treatment with sodium in liquid ammonia probably give the corresponding 5,8-dihydrochromans (*Hurd* and *G. L. Oliver*, J. Amer. chem. Soc., 1959, **81,** 2795). Reduction of chroman with excess lithium in ethylamine–dimethylamine (1:1 by volume) gives 5,6,7,8-tetrahydrochroman (VII), which reacts with perphthalic acid in moist ether to give a *trans*-diol, VIII, cleaved with lead tetra-acetate to yield 6-oxononanolide (IX). A direct conversion of VIII to IX is obtained with an excess of 3-chloroperbenzoic acid (*I. J. Borowitz et al.*, J. org. Chem., 1966, **31,** 3032):

Li/$EtNH_2$–$MeNH_2$ → (VII) → OH, OH (VIII) → O, O (IX)

The pyrolysis of chroman over alumina at 250° gives only 2,3-dihydro-2-methylbenzo[*b*]furan in a 10% yield; at 350° the conversion is 20–25%

but 2-alkylphenols (R = Me, Et and Pr) are also obtained as by-products (*E. A. Karakhanov, N. N. Khvorostukhina*, and *E. A. Viktorova*, C.A., 1971, **74,** 53417):

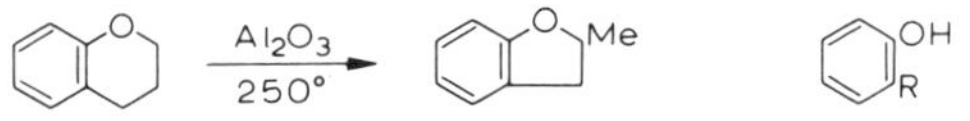

Al_2O_3, 250°

Skeletal rearrangements in chromans have been studied by the mass spectrometry of carbon-13 and deuterium-labelled compounds (*J. R. Trundell et al.*, Org. Mass. Spectrom., 1970, **3,** 753).

(b) Naturally occurring chromans

*(i) Tocopherols**

The vitamins E or tocopherols (from the Greek *tokos* and *phero* meaning "childbirth" and "bear", respectively), which are essential for reproduction, are 6-hydroxychroman derivatives with methyl and phytyl groups in the 2-position, the individual members differing only in the number and positions of methyl groups in the aromatic nucleus. In the nomenclature now adopted generally, the chroman system of the tocopherols without the methyl groups in the benzene ring is termed *tocol.* The tocopherols are therefore methylated tocols. The seven so called possible tocopherols have all been isolated from natural sources four (α, β, γ, and δ) in particular have been extensively studied, but two of them (ε and ζ) were later found not to possess a phytyl side chain:

α β γ δ

Tocopherols

R = phytyl residue $(CH_2)_3 \cdot CHMe \cdot (CH_2)_3 \cdot CHMe \cdot (CH_2)_3 \cdot CHMe_2$

The main source of the tocopherols is plant material, only small quantities being found in the animal organism. The best natural sources are vegetable oils and particularly wheat germ oil. Other sources are rice germ oil and cottonseed oil. Fish liver oils, which are excellent sources of vitamins A and D, have a poor vitamin E content. The origin of the oil is important. European wheat germ oil, for example, contains the β-compound, while the American oil gives the α-compound with small quantities of the β. Cottonseed oil, palm oil and corn oil contain mainly γ- along with a small quantity

* *W. John*, Zeit. angew. Chem., 1939, **52**, 413; *L. I. Smith*, Chem. Reviews, 1940, **27**, 287; *O. Isler et al.*, Vitamins Hormones, 1962, **20**, 389.

of β-tocopherol.

The vitamins may be efficaciously isolated from the natural oils by chromatographic adsorption (*J. Green et al.*, J. Sci. Food Agric., 1955, **6,** 274; *P. W. R. Eggitt* and *F. W. Norris*, *ibid.*, p. 689). Purification can also be effected by conversion into crystalline allophanates.

Structure. The methods used to establish the structure of α-tocopherol can be taken as typical. The phenolic nature of the vitamin is indicated by the shift of the wave-length of the absorption band when the tocopherol is acetylated.

Pyrolysis of the vitamin yields durohydroquinone (X) and heating with hydriodic acid gives pseudocumen-6-ol (XI), both fragments being obtained from the chroman nucleus of the vitamin. The presence of a chroman ring was confirmed by oxidation of the vitamin with chromic acid which yields dimethylmaleic anhydride, diacetyl, acetone, and long-chain substances of which the most important are the lactone XII, and the ketone XIII (*E. Fernholz*, J. Amer. chem. Soc., 1938, **60,** 700). As a result of these oxidations Fernholz advanced the correct formula for α-tocopherol.

That the vitamin is a chroman and not a coumaran was confirmed by the careful oxidation of α-tocopherol with silver nitrate or ferric chloride to a yellow quinone (IV) reduction of which afforded a hydroquinone. The di-4-bromobenzoate of this was stable towards chromic acid, thus showing that the hydroxyl group of the side-chain is tertiary (as seen in XIV) (*W. John et al.*, Z. physiol. Chem., 1937, **250,** 11; 1938, **252,** 208, 222).

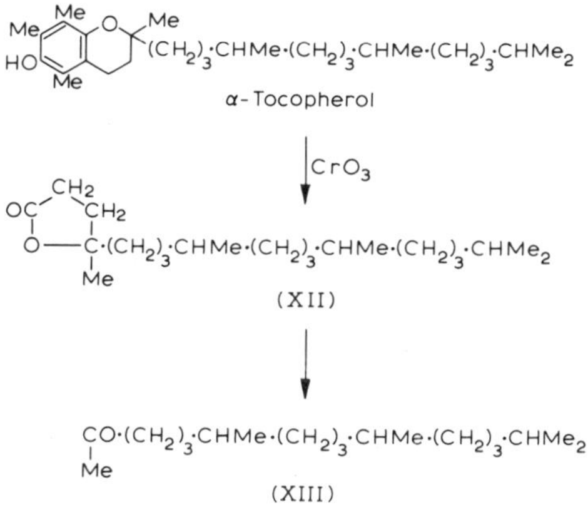

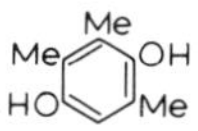

$(CH_2)_3 \cdot CHMe \cdot (CH_2)_3 \cdot CHMe \cdot (CH_2)_3 \cdot CHMe_2$

α-Tocopherol

CrO_3

(X) 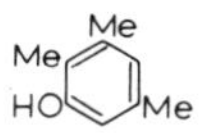 (XI)

$C \cdot (CH_2)_3 \cdot CHMe \cdot (CH_2)_3 \cdot CHMe \cdot (CH_2)_3 \cdot CHMe_2$

(XII)

$CO \cdot (CH_2)_3 \cdot CHMe \cdot (CH_2)_3 \cdot CHMe \cdot (CH_2)_3 \cdot CHMe_2$

(XIII)

$HO{-}C \cdot C_{16}H_{33}$ $\cdot CH \cdot C_{16}H_{33}$

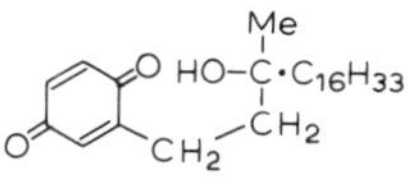

(XIV) 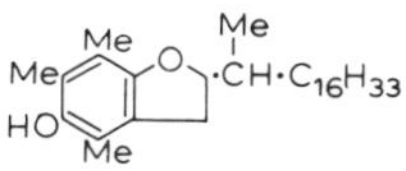 (XV)

This established the chroman structure, since the coumaran XV would have been oxidised to a quinone with a secondary alcoholic group in the side-chain, easily oxidisable by chromic acid. Further evidence against a coumaran structure was furnished by the synthesis of the coumaran analogue of α-tocopherol (XV) and its low vitamin activity (*L. I. Smith* and *G. A. Boyack*, J. Amer. chem. Soc., 1948, **70,** 2690).

The chroman structure was confirmed by synthesis. Care must, however, be exercised, since either chromans or coumarans may be formed in the cyclisation of 2-allylphenols of type XVII (*Smith et al., ibid.*, 1941, **63,** 1887). If R and R^1 are alkyl groups ring-closure occurs at the γ-carbon atom and yields a chroman XVIII, but if R and R^1 are hydrogen atoms ring-closure occurs at the β-atom and forms a coumaran XIX (*P. Karrer et al.*, Helv., 1939, **22,** 1281):

(XVI) + $BrCH_2{\cdot}CH{:}CRR^1$ → (XVII) → (XVIII) or (XIX)

It may be noted that the ring-closures follow the Markownikoff rule. In accordance with the above observations 2,3,5-trimethylhydroquinone (XVI) reacts with phytyl bromide or phytol (*Karrer, ibid.*, 1938, **21,** 520; *F. Bergel et al.*, Nature, 1938, **142,** 36; J. chem. Soc., 1938, 1382; *Smith* and *H. E. Ungnade*, J. org. Chem., 1939, **4,** 299) to give a good yield of racemic α-tocopherol. For other syntheses see *Smith* and *H. C. Miller* (J. Amer. chem. Soc., 1942, **64,** 440) and *M. E. Maurit et al.* (Doklady Akad. Nauk S.S.S.R., 1961, **140,** 1330).

Properties. The tocopherols are oils, soluble in all lipoid solvents, but insoluble in water. Their absorption of light in the u.v. (λ_{max}) region is used for quantitative estimations. The tocopherols may also be estimated quantitatively by oxidation in ethanolic solution by ferric chloride and determining the resulting ferrous chloride colorimetrically by the red colour it gives with 2,2′-bipyridyl (*A. Emmerie* and *C. Engel*, Rec. Trav. chim., 1938, **57,** 1351); and by oxidation with nitric acid to red quinones (*Smith et al.*, J. Amer. chem. Soc., 1939, **61,** 2424; J. org. Chem., 1941, **6,** 236). Tocopherols with an unsubstituted 5-position couple with diazonium salts in alkaline solution (*M. L. Quaife*, J. Amer. chem. Soc., 1944, **66,** 308; *L. Weisler et al.*, Anal. Chem., 1947, **19,** 906), and those with free 5- and 7-positions react with nitrous acid. The tocopherols are good antioxidants, their efficiency decreasing in the order $\delta > \gamma > \beta > \alpha$. The i.r. spectra have been measured (*H. Rosenkrantz* and *A. T. Milhorat*, J. biol. Chem., 1950, **187,** 83).

The structural requirements for the preferential oxidative coupling at the 5- *versus* the 7-position in tocopherols have been studied; during the investigation various methyl-substituted 6-chromanols were oxidised with 1,4-benzoquinone, which led to the characterisation of nine new dimers of the chromanols (*J. L. G. Nilsson, H. Sievertsson* and *H. Selander*, Acta Chem. Scand., 1969, **23**, 859). The oxidation of α-tocopherol in alcoholic solution affords a quinone acetal (*C. Martius* and *H. Eilingsfeld*, Biochem. Z., 1957, **328**, 507).

Treatment of 6-hydroxy-2,2,5,7,8-pentamethylchroman with aqueous alkaline potassium ferricyanide affords compound XX, (R = Me) in 73% yield (*P. Schudel et al.*, Helv., 1963, **46,** 636):

(XX)

Similarly the oxidation of (±)-α-tocopherol gives compound XX (R = $C_{16}H_{33}$) also obtained following oxidation of (±)-α-tocopherol with α,α-diphenyl-β-picrylhydrazyl in chloroform (*W. Boguth, R. Repges* and *M. Sernetz*, Ber. Bunsenges. Physik. Chem., 1966, **70,** 34; C.A., 1966, **64,** 11253). Besides the dimer, a trimer (XXI, R = Me or $C_{16}H_{33}$) has also been isolated following the oxidation of 6-hydroxy-2,2,5,7,8-pentamethylchroman or α-tocopherol with alkaline potassium ferricyanide (*W. A. Skinner* and *R. M. Parkhurst*, J. org. Chem., 1964, **29,** 3601). It has been reported that on oxidation 2-hydroxy-2,2,5,7,8-pentamethylchroman gives a mixture of five products including those already mentioned (*M. Fujimaki et al.*, Agric. Biol. Chem., 1970, **34,** 1781):

(XXI)

(XXII)

β-Tocopherol yields a dimer similar to that obtained from α-tocopherol, but γ-tocopherol affords a spiroacetal trimer XXII (*Nilsson, Sievertsson* and *Selander*, Tetrahedron Letters, 1968, 5023); for oxidation products from some other tocopherols see also *Nilsson* and *Selander* (Acta. Chem. Scand., 1970, **24,** 2885).

The Claisen rearrangements of allyl tocyl ether and allyl δ-tocophenyl ether are similar and give the 5-allyltocopherol predominantly in each case; the rearrangements have been studied kinetically (*J. Green, S. Marcinkiewicz* and *D. McHale*, J. chem. Soc., C, 1966, 1422).

α-**Tocopherol,** 5,7,8-*trimethyltocol*, $C_{29}H_{50}O_2$, $[\alpha]^{24}_{5461}$ +0.30° (EtOH), *allophanate*, m.p. 157–158°, *succinate*, m.p. 76–77°, 4-*nitrophenylcarbamate*, m.p. 129–131° (*J. G. Baxter et al.*, J. Amer. chem. Soc., 1943, **65**, 918); the fractionation into diastereoisomers of synthetic α-tocopherol by means of a complex with piperazine has been reported (*D. R. Nelan* and *C. D. Robeson*, J. Amer. chem. Soc., 1962, **84**, 3196); a number of fatty acid esters of α-tocopherol have been prepared (*P. F. G. Praill*, J. chem. Soc., 1959, 3100).

β-**Tocopherol,** 5,8-*dimethyltocol*, $C_{28}H_{48}O_2$, $[\alpha]^{25}_{5461}$ +2.9° (EtOH), *allophanate*, m.p. 138–139°, 3,5-*dinitrobenzoate*, m.p. 86–87°, 4-*nitrophenylcarbamate*, m.p. 90°, *phenylazobenzoate*, m.p. 70–71° (*Baxter et al., loc. cit.*; *Karrer* and *H. Fritzsche*, Helv., 1939, **22,** 260); it gives on pyrolysis trimethylhydroquinone (pseudocumoquinol), and is cleaved by hydriodic acid to *p*-xylenol. Oxidation yields the same lactone XII as is obtained from α-tocopherol (*O. H. Emerson*, J. Amer. chem. Soc., 1938, **60,** 1741).

γ-**Tocopherol,** 7,8-*dimethyltocol*, $C_{28}H_{48}O_2$, $[\alpha]^{25}_{5461}$ +2.2° (EtOH), m.p. −3 to −2°, *allophanate*, m.p. 148–152° (136–138°), 4-*nitrophenylcarbamate*, m.p. 119–121° (*Emerson* and *Smith, ibid.*, 1940, **62,** 1869).

δ-**Tocopherol,** 8-*methyltocol*, $C_{27}H_{46}O_2$, a dextro-rotatory yellow oil, *allophanate*, m.p. 138–139°, *phenylazobenzoate*, m.p. 41–42°, is isolated from soya bean oil and constituted about 30% of the mixed tocopherols (*M. H. Stern et al., ibid.*, 1947, **69,** 869). It has been synthesised (*Green et al.*, J. chem. Soc., 1959, 3374) and occurs with α- and γ-tocopherol in seaweed (*F. Brown*, Chem. and Ind., 1953, 174).

An extract containing α-, β-, γ-, and δ-tocopherols can be converted to predominantly α-tocopherol by treatment with paraformaldehyde and hydrogen chloride over tin dust (*Nelan*, U.S.P. 3,631,068/1971).

ε-**Tocopherol,** 6-*hydroxy*-2,5,8-*trimethyl*-2-(4,8,12-*trimethyltrideca*-3,7,11-*trienyl*)-*chroman*, 5,8-*trimethyltocotrienol* (XXIII, R=H), $C_{28}H_{42}O_2$, 4-*phenylazobenzoate*, m.p. 70–71°, isolated from wheat bran oil (*Green et al.*, J. chem. Soc., 1959, 3362; 1963, 784; Chem. and Ind., 1960, 73), was originally thought to be 5-*methyltocol*. Hydrogenation over a platinum oxide catalyst gives β-tocopherol and methylation gives ζ_1-tocopherol. ε-Tocopherol has been synthesised (*Schudel et al.*, Helv., 1963, **46,** 2517):

(XXIII)

ζ_1-**Tocopherol,** 6-*hydroxy*-2,5,7,8-*tetramethyl*-2-(4,8,12-*trimethyltrideca*-3,7,11-*trienyl*)-*chroman*, 5,7,8-*trimethyltocotrineol* (XXIII, R=Me), $C_{29}H_{44}O_2$, occurs in wheat, barley, rye and palm oil, and has been synthesised (*Green et al., loc. cit.*; *Schudel et al., loc. cit.*).

ζ_2-**Tocopherol,** 5,7-*dimethyltocol*, $C_{28}H_{48}O_2$, from rice, synthetic sample had m.p. −4°; 4-*phenylazobenzoate*, m.p. 61°, 3,5-*dinitrophenylcarbamate*, m.p. 65° (*Green et al.*, J. Sci. Food Agric., 1955, **6,** 329; J. chem. Soc., 1958, 1600; 1959, 3362).

η-**Tocopherol,** 7-*methyltocol*, $C_{27}H_{46}O_2$, *allophanate*, m.p. 148–149°, 4-*nitrophenylcarbamate*, m.p. 86–87°, 3,5-*dinitrophenylcarbamate*, m.p. 115–117°, occurs in rice, and when methylated yields ζ_2-tocopherol and α-tocopherol (*Green et al., loc. cit.*). It has been synthesised (*H. K. Pendse* and *P. Karrer*, Helv., 1958, **49,** 396).

Tocol, $C_{26}H_{44}O_2$, 3,5-*dinitrophenylcarbamate*, m.p. 97° (*idem, ibid.*, 1957, **40,** 1837; *Green et al.*, J. chem. Soc., 1958, 1850,; 1959, 3362); 5-*methyltocol*, $C_{27}H_{46}O_2$, 4-*phenylazobenzoate*, m.p. 69–70° (*D. McHale et al., ibid.*, 1959, 3358); 8-*methyltocol*, $C_{27}H_{46}O_2$, 4-*phenylazobenzoate*, b.p. 250° bath/0.001 mm (*Green et al., ibid.*, p. 3374). The condensation of toluquinol with phytol leads to a 1:2:1 mixture of 5-, 7-, and 8-methyltocol (*S. Marcinkiewicz et al., ibid.*, p. 3377). The preparation and uses of some new 6-hydroxychromans have been reviewed (*M. H. Stern*, Org. chem. Bull., 1970, **42,** 4).

(ii) Fuscin and cannabicyclol

Fuscin, $C_{15}H_{16}O_5$, orange plates, m.p. 230°, a physiologically active mould metabolite from *Oidiodendron fuscum* Rodak, forms purple salts with alkali (*S. E. Michael et al.*, Biochem. J., 1948, **43,** 528; 1951, **48,** 67). With sodium dithionite it yields the colourless *dihydrofuscin*, (XXIV), $C_{15}H_8O_5$, m.p. 206°, which on pyrolysis at 300° loses carbon dioxide to yield 7,8-*dihydroxy*-2,2,5-*trimethylchroman* (XXV, R = Me), $C_{12}H_{16}O_3$, m.p. 139°, with aqueous sodium hydroxide affords *fuscinic acid* (XXV, $R = CH_2 \cdot CO_2H$), $C_{13}H_{16}O_5$, m.p. 184°, and acetaldehyde, and when fused with potassium hydroxide yields isovaleric acid and 3,4,5-*trihydroxyphenylacetic acid*, m.p. 161° (*D. H. R. Barton* and *J. B. Hendrickson*, Chem. and Ind., 1955, 682; *A. J. Birch, ibid.*, p. 682). The assigned structure has been confirmed by i.r. spectroscopy and by synthesis (*Barton* and *Hendrickson*, J. chem. Soc., 1956, 1028):

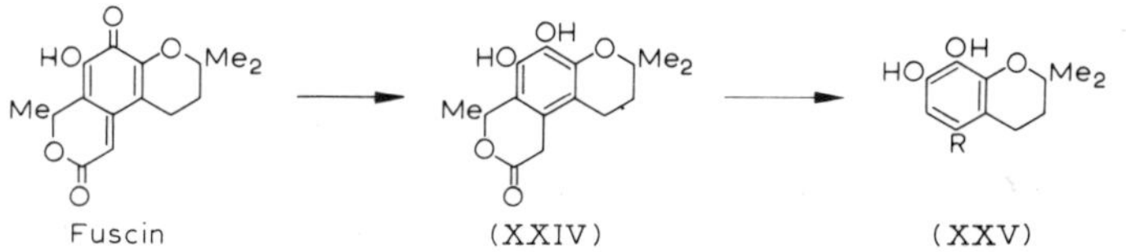

Fuscin (XXIV) (XXV)

Cannabicyclol, $C_{21}H_{30}O_2$, m.p. 146–147°, has been obtained from the resin of the female *Cannabis sativa* L. plant along with *cannabichromene*, $C_{21}H_{30}O_2$, 3,5-*dinitrophenylcarbamate*, m.p. 106–107°, and other products (*Y. Gaoni* and *R. Mechoulam*, J. Amer. chem. Soc., 1971, **93,** 217). Reasons for the proposed structure have been presented (*L. Crombie* and *R. Ponsford*, J. chem. Soc., C, 1971, 796), and it has been confirmed by X-ray studies (*M. J. Begley et al.*, Chem. Comm., 1970, 1547), and spectral data (*V. V. Kane*, Tetrahedron Letters, 1971, 4101):

Cannabicyclol Cannabichromene

(c) Phenylchromans (flavans and isoflavans), 3,4-dihydrophenyl-2H-benzo[b]pyrans

(i) Synthesis

(1) **Flavan,** 2-*phenylchroman*, 2,3-*dihydro*-2-*phenylbenzo*[b]*pyran*, colour-

less crystals, m.p. 44–45°, is obtained by dehydrating phenyl-(2-hydroxyphenylethyl)methanol with ethanolic hydrogen chloride:

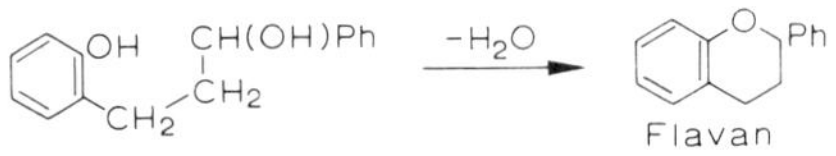

OH CH(OH)Ph CH2 CH2 $\xrightarrow{-H_2O}$ Flavan

(*2*) Flavans can be obtained in almost quantitative yield by reducing flavylium salts successively with lithium tetrahydridoaluminate and with hydrogen in the presence of Raney nickel (*W. E. Elstow* and *B. C. Platt*, Chem. and Ind., 1950, 824). Flavans have also been obtained by the Clemmensen reduction of 3-hydroxyflavanones (*M. M. Bokadia* and *B. L. Verma*, *ibid.*, 1964, 235; *C. B. Rao* and *V. Venkateswarlu*, Tetrahedron, 1964, **20**, 551; *Verma*, *P. N. Verma* and *Bokadia*, Indian J. chem., 1965, **3**, 565).

(*3*) Flavans may be obtained by the reductive desulphurisation, with Raney nickel in dioxane of the corresponding flavanone ethylene thioacetals (*E. J. Keogh et al.*, Chem. and Ind., 1961, 2100):

Flavan is unaffected by reaction with aluminium chloride and lithium tetrahydridoaluminate, but 4′-methoxyflavan gives 1-(2-*hydroxyphenyl*)-3-(4-*methoxyphenyl*)*propane*, *benzoate*, m.p. 45.5–46.5° (*B. R. Brown* and *G. A. Somerfield*, Proc. chem. Soc., 1958, 7). 4′-*Methoxyflavan*, m.p. 83–84°, isolated from the scent glands of the Canadian beaver, *Castor fiber* (*A. Gaudema* and *E. Lederer*, Compt. rend., 1964, **259**, 4167), condenses with phenol or resorcinol in ethanol in the presence of hydrogen chloride to give, after methylation, the triarylpropanes (XXVI, R = H or OMe) (*Brown* and *W. Cummings*, J. chem. Soc., 1958, 4302):

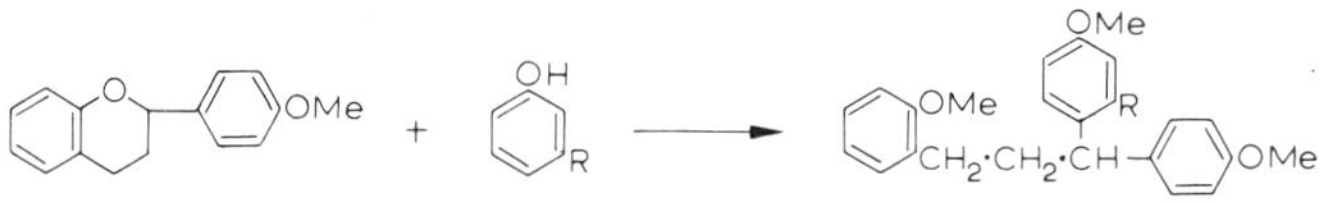

OMe + OH R → OMe OMe R $CH_2 \cdot CH_2 \cdot CH$ OMe

(XXVI)

4′-*Methoxy*-6-*methylflavan*, m.p. 55°

The absolute configurations [both (*2S*)] of (−)-7,3′,4′,5′-tetrahydroxy- and (−)-7,3′,4′-trihydroxy-flavan have been established by interconversions from flavanols and flavanones of known configuration (*D. G. Roux*, Biochem. J., 1963, **87**, 435).

(*4*) 2-Hydroxyphenylisopropenyl compounds undergo dimerisation in the presence of hydrochloric acid to give flavans, the same products being obtained by condensing phenols with acetone by means of hydrochloric acid

(*W. Baker* and *D. M. Besly*, Nature, 1939, **144,** 865; J. chem. Soc., 1940, 1105).

Thus *m*-cresol and acetone yield 2′-hydroxy-2,4,4,7,4′-pentamethylflavan (*Baker et al.*, *ibid.*, 1951, 76; 1952, 1774):

Flavans of this type form crystalline complexes with many substances (1:1 mol. ratio of compound to flavan) (*idem, ibid.*, 1957, 3060). Besides the above product, 2,2-bis(2-hydroxy-4-methylphenyl)propane and three stereoisomers of 4,4,7,4′4′,7′-hexamethyl-2,2′-spirobichroman (XXVII) have been isolated following the condensation of *m*-cresol and acetone in concentrated hydrochloric acid; in sulphuric acid XXVII is the sole compound obtained, and in acetic acid – boron trifluoride, depending on temperature, either 2′-hydroxy-2,4,4,7,4′-pentamethylflavan, or the propane derivative is obtained (*G. E. Svadkovskaya, N. E. Kologrivova* and *L. A. Kheifits*, Khim. Geterotsikl. Soedin., Sb. 1970, 191; C.A., 1972, **77,** 126369):

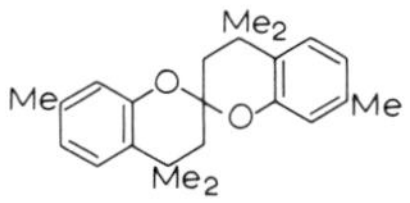

(XXVII)

The resin of the dragon's blood tree, *Dracaena Draco* (p. 79) has yielded six new natural products including (2*S*)-7-*hydroxy*-5-*methoxy*-6-*methylflavan* (XXVIII), $C_{17}H_{18}O_3$, m.p. 122–124°, $[\alpha]_D^{20}$ −9.25° (*c* 2.1 in $CHCl_3$), and (2*S*)-7-*hydroxy*-5-*methoxyflavan* (XXIX), $C_{16}H_{16}O_3$, m.p. 84–87°, $[\alpha]_D^{20}$ −6.35° (*c* 1.94 in $CHCl_3$) (*G. Cardillo et al.*, J. chem. Soc., C, 1971, 3967):

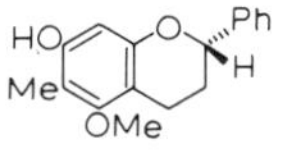

(XXVIII) 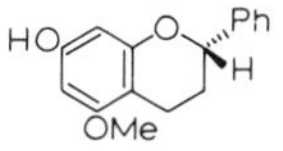 (XXIX)

(−)-4′-*Hydroxy*-7-*methoxyflavan*, m.p. 148.5–149.5°, has been isolated from *Stypandra grandis*, and (−)-4′-*hydroxy*-7-*methoxy*-8-*methylflavan*, m.p. 126–127°, from *Dianella revoluta* (*R. G. Cooke* and *J. G. Down*, Tetrahedron Letters, 1970, 1037).

5,7,4′-*Trimethoxyflavan*, m.p. 109–111°, has been isolated following the methylation of the main aqueous sodium hydroxide soluble fraction obtained from the Australian *Xanthorrhoea preissii*. On reduction with sodium and ethanol in liquid ammonia, followed by methylation, it gives 1-(2,4,6-trimethoxyphenyl)-3-(4-methoxyphenyl)propane (*A. J. Birch* and *M. Salahuddin*, Tetrahedron Letters, 1964, 2211):

The biflavone **xanthorrhone** (XXX, R = H), m.p. 193–196°, and the *hydroxyxanthorrhone* (XXX, R = OH), m.p. 190–193°, have also been obtained from *Xanthorrhoea pressii* (*Birch, C. J. Dahl* and *A. Pelter, ibid.*, 1967, 481):

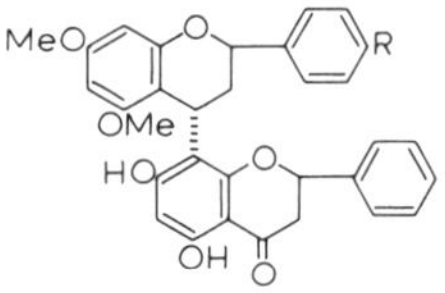

(XXX)

Equol is a derivative of isoflavan, 3-phenylchroman, 7,4′-*dihydroxyisoflavan*, m.p. 189–190.5°, *dimethyl ether*, m.p. 89°, *diacetate*, m.p. 122.5°, the first isoflavan to be obtained from natural sources. It was isolated from the urine of mares and stallions (*G. F. Marrian et al.*, Biochem. J., 1932, **26**, 1227; 1935, **29**, 1586), and its structure was established by its formation by the reduction of daidzein (p. 214) (*F. Wessely* and *F. Prillinger*, Monatsh., 1939, **72**, 197; Ber., 1939, **72**, 629). (−)-**Duartin**, 7,3′-*dihydroxy*-8,2′,4′-*trimethoxyisoflavan*, $C_{18}H_{20}O_6$, m.p. 149°, (−)-**mucronulatol**, 7,3′-*dihydroxy*-2′,4′-*dimethoxyisoflavan*, m.p. 145°, and (+)-**vestitol**, 7,2′-*dihydroxy*-4′-*methoxyisoflavan*, m.p. 156°, have been isolated from *Dalbergia variabilis* (Vogel) and five *Machaerium* species (*W. D. Ollis et al.*, Chem. Comm., 1968, 1263). Their absolute configurations and that of equol have been established (*idem, ibid.*, p. 1265). (±)7,2′,4′-*Trimethoxyisoflavan*, needles, m.p. 90°; (±)7,2′-*dimethoxy*-4′,5′-*methylenedioxyisoflavan*, needles, m.p. 120° (111–113°) (*C. A. Anirudhan, W. B. Whalley* and *M. M. E. Badran*, J. chem. Soc., C, 1966, 629); 5,7,3′,4′-*tetramethoxyisoflavan*, m.p. 134°, has been prepared by the reduction of the corresponding isoanthocyanidin (*K. Freudenberg et al.*, Ann., 1925, **446**, 87).

(d) Chromanols

The chroman-2-ols are hemiacetals in contrast to the 3- and 4-isomers which are alcohols. They are obtained by the action of 0.5 *M*-sulphuric acid on the corresponding chrom-2-enes or by decomposing the chrom-2-ene mercurichlorides with hydrochloric acid (*P. Maitte*, Ann. Chim., 1954, [xii], **9**, 464). They react both as chromanols and as open-chain aldehydes or ketones, **chroman-2-ol**, b.p. 139°/12.5 mm for instance giving a 2,4-*dinitrophenylhydrazone*, m.p. 185°. The chroman-2-ols are converted by boiling

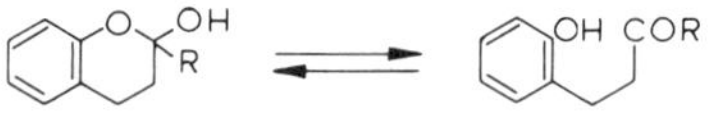

benzene and anhydrous oxalic acid or by anhydrous sodium sulphate into chrom-2-enes and by alcohols and hydrogen chloride into methyl ethers (*W. Baker* and *J. Walker*, J. chem. Soc., 1935, 646; *L. I. Smith* and *R. B. Carlin*, J. Amer. chem. Soc., 1942, **64**, 435).

2-Methylchroman-2-ol yields 2-methylchrom-2-ene (XXXI) with sodium sulphate, and 2-*methoxy-2-methylchroman* (XXXII), b.p. 107°/14 mm, with methanol and hydrogen chloride:

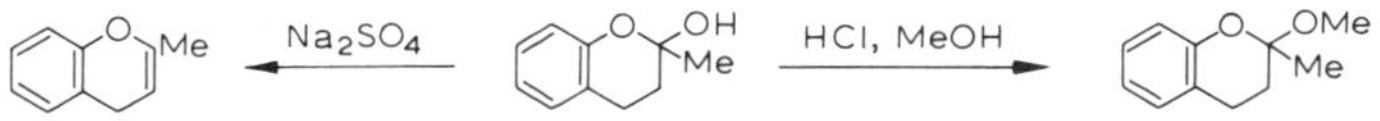

(XXXI) ← Na_2SO_4 — 2-methylchroman-2-ol — HCl, MeOH → (XXXII)

These ethers are also obtained from phenols and $\alpha\beta$-unsaturated carbonyl compounds in the presence of mineral acids (*D. W. Clayton et al.*, J. chem. Soc., 1953, 581). Thus resorcinol, acraldehyde, and ethanolic hydrogen chloride afford 2-*ethoxy*-7-*hydroxychroman*, b.p. 98° bath/0.002 mm, 3,5-*dinitrobenzoate*, m.p. 113.5°.

4,7-*Dimethylchroman-2-ol*, b.p. 138°/1.0 mm, has been prepared from 4,7-dimethylcoumarin (*W. C. Still* and *D. J. Goldsmith*, J. org. Chem., 1970, **35,** 2282). 2,4,4-*Trimethylchroman-2-ol.*, m.p. 93.5–94°, is obtained with the corresponding chromene by treating trimethylindanyl hydroperoxide with sulphuric acid. With boiling benzene oxalic acid it yields 2,4,4-trimethylchromene and when heated with dimethyl sulphate and sodium hydroxide it forms 4-(2-methoxyphenyl)-4-methylpentan-2-one, 2,4-*dinitrophenylhydrazone*, m.p. 106–107°. It has been synthesised (*Baker et al.*, J. chem. Soc., 1952, 1774; *W. Webster* and *D. P. Young*, *ibid.*, 1956, 4785).

2,4-*Diphenyl-2-methoxychroman*, m.p. 106°; 2,4-*diphenyl-2-ethoxychroman*, m.p. 122° (*R. Livingstone*, *D. Miller* and *S. Morris*, *ibid.*, 1960, 5148); in attempts to prepare the 2-ol only 1,3-diphenyl-3-(2-hydroxyphenyl)propan-1-one (XXXIII) has been obtained (*C. S. Barnes*, *J. L. Occolowitz*, and *M. I. Strong*, Tetrahedron, 1963, **19,** 839); similarly 2,3-dihydro-2,4-diphenyl-4*H*-naphtho[1,2-*b*]pyran-2-ol and 2,3-dihydro-2,4-diphenyl-

(XXXIII) (XXXIV)

4*H*-naphtho[2,3-*b*]pyran2-ol could not be obtained, but only 1,3-*diphenyl*-3-(1-*hydroxy-2-naphthyl)propan*-1-*one*, m.p. 155° (*J. Cottam* and *Livingstone*, J. chem. Soc., 1964, 5228) and 1,3-*diphenyl*-3-(3-*hydroxy*-2-*naphthyl)propan*-1-*one*, m.p. 207–208° (*idem, ibid.*, 1965, 6646); 2,3-*dihydro*-1,3-*diphenyl*-1H-*naphtho*[2,1-b]*pyran*-3-*ol* (XXXIV, R=OH), m.p. 153–154°, *methyl ether* (XXXIV, R=OMe), m.p. 106°, *ethyl ether* (XXXIV, R=OEt), m.p. 122° (*Livingstone*, *Miller* and *Morris*, *ibid.*, 1960, 5148); 2,3-*dihydro*-1,3-*bis*-(2-*methoxyphenyl*)-1H-*naphtho*[2,1-b]*pyran*-3-*ol*, m.p. 172°; 2,3-*dihydro*-1, 3-*bis*-(3-*methoxyphenyl*)-1H-*naphtho*[2,1-b]*pyran*-3-*ol*, m.p. 172°, (*Cotterill*, *Livingstone* and *M. V. Walshaw*, *ibid.*, C, 1970, 1758).

Chroman-3-ol, m.p. 79°, has been obtained by the hydroboration–oxidation of coumarin; 4,7-*dimethylchroman*-3-*ol*, m.p. 60°, b.p. 128–132°/0.5 mm (*Still* and *Goldsmith*, *loc. cit.*). The reduction of 4-phenylchroman-3-one to 4-*phenylchroman*-3-*ol*, m.p. 87°, by means of zinc dust and acetic anhydride affords a general method for the preparation of chroman-3-ols (*A. I. M. Mahil* and *M. Nierenstein*, J. Amer. chem. Soc., 1924, **46,** 2556). The treatment of certain allyl aryl ethers with thallium(III) sulphate in 2.0–2.5

M-sulphuric acid gives the corresponding chroman-3-ol as the major product (*J. R. Collier* and *A. S. Porter*, Chem. Comm., 1972, 618):

R = H, Me, MeO

2,2-*Dimethylchroman-3-ol*, m.p. 73°, *acetate*, m.p. 58–59°, is obtained by treating 2,2-dimethyl-3,4-epoxychroman, prepared from 3-bromo-2,2-dimethylchroman-4-ol, with lithium tetrahydridoaluminate (*Livingstone*, J. chem. Soc., 1962, 76):

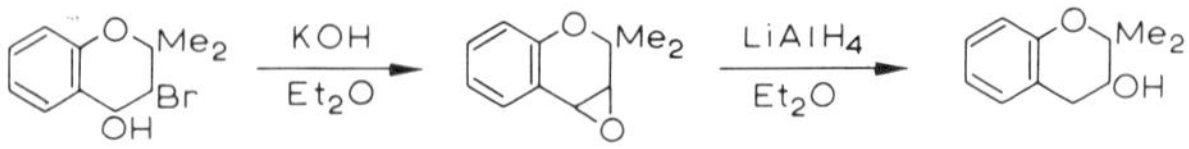

2,2-*Diphenylchroman-3-ol*, m.p. 115° (*Cottam* and *Livingstone*, *ibid.*, 1965, 5266); 6-*methoxy-*, m.p. 72°, *acetate*, m.p. 121°; 8-*methoxy-*2, 2-*dimethylchroman-*3-*ol*, m.p. 117°, *acetate*, m.p. 115° (*J. D. Hepworth* and *Livingstone*, *ibid.*, C, 1966, 2013); 2,3-*dihydro-*3,3-*dimethyl-*1H-*naphtho*[2,1-b]*pyran-*2-*ol*, m.p. 116–118°, *acetate*, 113–114°; 2,3-*di-hydro-*3,3-*diphenyl-*1H-*naphtho*[2,1-b]*pyran-*2-*ol*, m.p. 156–158°, *acetate*, m.p. 171–172° (*J. B. Abbott et al.*, *ibid.*, C, 1967, 1472).

When heated the chroman-3-ols yield chrom-3-enes.

Catalytic reduction of chroman-4-ones (*R. Mozingo* and *H. Adkins*, J. Amer. chem. Soc., 1938, **60,** 669) or reduction using lithium tetrahydridoaluminate or sodium tetrahydridoborate yields *inter alia* **chroman-4-ols** (*Maitte*, Ann. Chim., France, 1954, **9,** 431; *W. E. Parham* and *L. D. Huestis*, J. Amer. chem. Soc., 1962, **84,** 813). Some chroman-4-ols and halogenochroman-3- and -4-ols are given in Table 6.

TABLE 6

CHROMAN-4-OLS AND HALOGENOCHROMAN-3- AND -4-OLS

Derivative	*M.p. (°C)*	*Ref.*	*Derivative*	*M.p. (°C)*	*Ref.*
Chroman-4-ol	40–41.5	1	4-chloro-6-methoxy- -2,2-dimethyl-3-ol	84	6
trans-3-chloro-4-ol	106–107	2			
cis-3-chloro-4-ol	129–129.5	2	4-chloro-8-methoxy- -2,2-dimethyl-3-ol	107	6
trans-3-bromo-4-ol	107–108	2			
cis 3-bromo-4-ol	109–110	2	3-chloro-6-methoxy- -2,2-dimethyl-4-ol	73	6
2-ethyl-4-ol	78–79	3			
3-chloro-2,2-dimethyl-4-ol	77–78	4	3-bromo-6-methoxy- -2,2-dimethyl-4-ol	75–76	6
3-bromo-2,2-dimethyl-4-ol	106	5			
2,2-dimethyl-3,4-diol	57–58	5	3-chloro-7-methoxy- -2,2-dimethyl-4-ol	99	6
4-chloro-2,2-dimethyl-3-ol	83–84	4			
3-chloro-2,2-diphenyl-4-ol	115–118	4	3-chloro-8-methoxy- -2,2-dimethyl-4-ol	116	6
3-bromo-2,2-diphenyl-4-ol	140–141	4			
4-chloro-2,2-diphenyl-3-ol	143–144	4	3-bromo-8-methoxy- -2,2-dimethyl-4-ol	123–124	6

References
1 *W. E. Parham* and *L. D. Huestis*, J. Amer. chem. Soc., 1962, **84,** 813.
2 *W. D. Cotterill, J. Cottam* and *R. Livingstone*, J. chem. Soc., C, 1970, 1006.
3 *R. Mozingo* and *H. Adkins*, J. Amer. chem. Soc., 1938, **60,** 669.
4 *Cottam, Livingstone* and *S. Morris*, J. chem. Soc., 1965, 5266.
5 *Livingstone, ibid.*, 1962, 76.
6 *J. D. Hepworth* and *Livingstone, ibid.*, C, 1966, 2013.

For related halogeno derivatives of 2,3-dihydro-3,3-dimethyl- and 2,3-dihydro-3,3-diphenyl-1*H*-naphtho[2,1-*b*]pyran-1- and -2-ols see *Abbott et al.*, J. chem. Soc., C, 1967, 1472.

The rates of oxidation of chroman 4-ols with chromic acid have been determined; chromanols containing a *cis*-2-substituent had the fastest rate; the oxidation of 3-substituted chromanols also has been discussed (*S. Yamaguchi et al.*, Bull. chem. Soc. Japan, 1971, **44,** 3487).

Cycanomaclurin, $C_{15}H_{12}O_6$, m.p. 290°, $[\alpha]_D^{20}$ +215° (AcOEt), $[\alpha]_D^{20}$ +192° (H_2O); it gives a *tetra-acetate*, m.p. 136–138°, and an amorphous *trimethyl* derivative, m.p. 73–75°, is obtained in very small yield from jackwood, *Artocarpus integrifolia* (*A. G. Perkin* and *F. Cope*, J. chem. Soc., 1895, **67,** 939; *Perkin, ibid.*, 1905, **87,** 715). Cyanomaclurin with warm alkali gives a striking blue colour (hence the name); with ferric chloride a violet and with sulphuric acid a crimson. Fusion with potassium hydroxide yields phloroglucinol and *β*-resorcylic acid, and it has ben converted into morindin chloride. Chemically it is not possible to distinguish between the suggested structures XXXV (*R. Robinson* and *H. Appel, ibid.*, 1935, 752) and XXXVI for cyanomaclurin, but the n.m.r. spectra of cyanomaclurin, its acetate and trimethyl ether when compared with those of closely related catechins and derivatives (p. 238) indicated that the latter structure was correct; in D_2O (+)catechin and (−)epicatechin showed the presence of a methylene group, absent in XXXVI (*G. Chakravarty* and *T. R. Seshadri*, Tetrahedron Letters, 1962, 787):

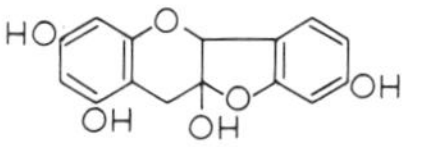

HO O OH OH OH O
(XXXV)

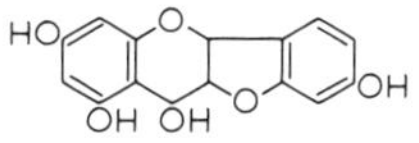

HO O OH OH O OH
(XXXVI)

On the interpretation of n.m.r. spectral data, structure XXXVII, also has been suggested for cyanomaclurin (*P. M. Nair* and *K. Venkataraman, ibid.*, 1963, 317).

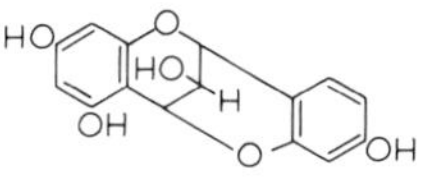

HO O HO H OH O OH
(XXXVII)

(*e*) *Flavanols: the catechins and related condensed tannins*

Important naturally occurring representatives of the flavanols (3-hydroxy-2-phenylchromans, 2,3-dihydro-3-hydroxy-2-phenyl-4*H*-benzo[*b*]pyrans) are the catechins and the related condensed tannins.

*(i) Catechins**

Several varieties of catechu are used by dyers and tanners and from these colourless, crystalline ingredients termed catechins may be isolated. The various catechins are stereoisomers of which the two primary naturally occurring compounds are (+)*catechin*, 5,7,3′,4′-*tetrahydroxyflavan-3-ol* and (−)*epicatechin*. (+)-Catechin is best obtained from Gambier catechu (cube gambier), which is obtained from *Uncaria Gambier*, a bush found in Penang and Singapore and used to dye cotton "catechu brown" (*K. Freudenberg* and *L. Purrmann*, Ber., 1923, **56,** 1185). Other catechins found in nature are the *gallocatechins* and (−)*epiatzelechin*.

The earlier investigations led to conflicting formulae for the catechins, but the formulation of catechin as a reduction product of quercetin (*A. G. Perkin* and *E. Yoshitake*, J. chem. Soc., 1902, **81,** 1160; 1905, **87,** 398) has been substantiated by later workers, particularly *Freudenberg* and his collaborators. The catechins are thus phenolic flavan-3-ols:

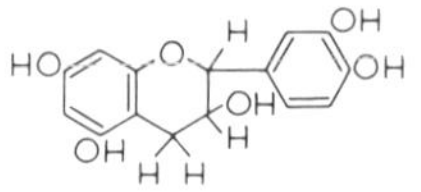

Catechin

The skeleton structure of the molecule was indicated by the fusion of catechin with alkali when phloroglucinol and protocatechuic acid were obtained, and by the oxidation of the tetramethyl ether to phloroglucinol dimethyl ether and veratric acid. More precise information came from the reduction (sodium and ethanol) of catechin tetramethyl ether followed by methylation to give a product proved to be 1-(2,4,6-trimethoxyphenyl)-3-(3,4-dimethoxyphenyl)propane (XXXVIII) (*Freudenberg*, Ber., 1920, **53,** 1416; *Freudenberg* and *E. Cohn*, *ibid.*, 1923, **56,** 2127):

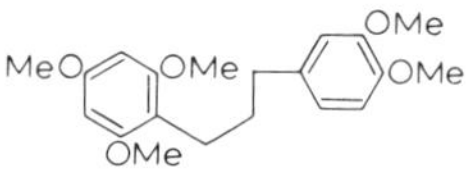

(XXXVIII) 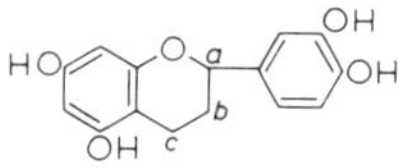 (XXXIX) 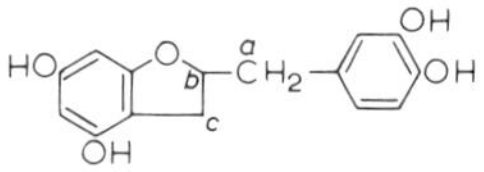 (XL)

Catechin must therefore be a hydroxy derivative of the flavan XXXIX or the coumaran XL, with a hydroxyl group at one of the positions *a*, *b* or *c*. A decision as to which of the various structures was correct proved to be difficult, and was finally reached by the brilliant researches of *Freudenberg* and his collaborators.

There must be two asymmetric centres in the catechin molecule. A single asymmetric

K. Freudenberg*, Proc. roy. Dublin Soc., 1956, **27, 153; *O. Th. Schmidt* and *W. Mayer*, Z. angew. Chem., 1956, **68,** 110.

centre only is required to account for the existence of (+), (−), and (±)catechin, but it does not account for the corresponding epicatechins which are found in nature and can be obtained from the catechins by epimerisation (p. 241). It will be recalled that in the sugar series, erythrose and its epimer threose give rise to four optically active isomers and two racemic forms. Catechin and epicatechin are similarly related and their interconversion is shown in the diagram in which the changes effected by epimerisation (E), racemisation (R), or mixing (M) are indicated (*Freudenberg* and *Purrmann*, Ann., 1942, **437,** 274):

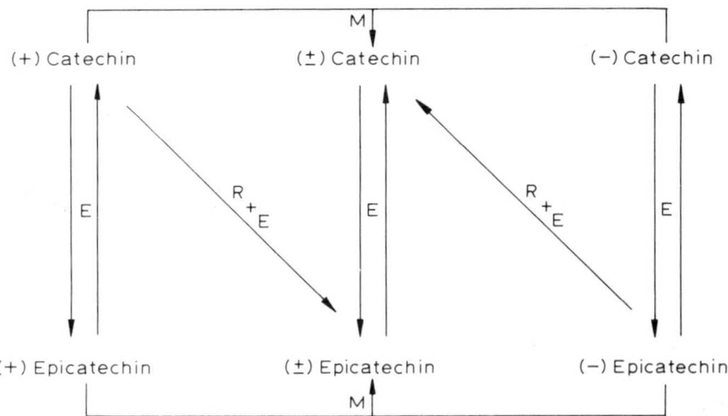

The presence of two asymmetric centres necessitates hydroxylation at a methylene group (positions *b* or *c* in XXXIX; *a* or *c* in XL). The other symmetric carbon atom results from the attachment of the catechol nucleus to the hetero ring (position *a* in XXXIX or *b* in XL).

Freudenberg's work showed that the catechins must be phenolic derivatives of flavan-3-ol (XLI) or 3-hydroxycoumaran (XLII). If the latter formula be correct it follows that tetramethylanhydroepicatechin, obtained by removal of the elements of water from

(XLI) (XLII)

tetramethylepicatechin must have formula XLIII. A compound of this structure was synthesised by Freudenberg and found to be quite different from tetramethylanhydro-epicatechin, which accordingly must have formula XLIV. Catechin and epicatechin are therefore diastereoisomeric 5,7,3′,4′-tetrahydroxyflavan-3-ols (XLI):

(XLIII) (XLIV) (XLV)

The pentahydroxyflavan structure for the catechins and epicatechins has been confirmed in a number of ways including the ring-fission of tetramethylanhydroepicatechin (XLIV) by acetic acid to yield the dihydrochalcone XLV (*Freudenberg et al.*, Ann., 1925, **442**, 309), the structure of which follows from its formation from the chalcone (*D. D. Pratt et al.*, J. chem. Soc., 1924, **125**, 199). The flavan structure is also shown by the hydrogenation of tetramethylanhydroepicatechin to the flavan (*Freudenberg et al.*, Ann., 1925, **441**, 157).

Structural proofs are given by a number of interconversions of flavonols, anthocyanidins and catechins. Considering examples of these compounds in which phloroglucinol and catechol nuclei are linked by a three-carbon chain the most highly oxidised is the flavonol quercetin and the most highly reduced is catechin with cyanidin chloride and luteolin as intermediate stages. This relationship has been demonstrated by

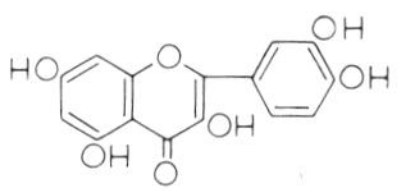

Quercetin 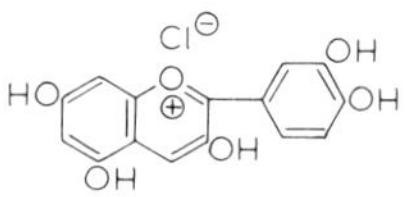 Cyanidin chloride 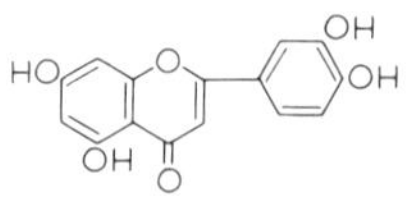 Luteolin

several reductions. Quercetin pentamethyl ether in acetic acid with a platinum catalyst absorbs three moles of hydrogen to yield pentamethyl(±)epicatechin (*Freudenberg* and *A. Kammuller*, *ibid.*, 1927, **451**, 209) and tetramethyl luteolin yields tetramethyldeoxyepicatechin identical with the flavan obtained by the hydrogenation of tetramethylanhydroepicatechin (*Freudenberg et al.*, *loc. cit.*). Cyanidin chloride with two molecules of hydrogen and platinum catalyst yields (±)epicatechin (*Freudenberg et al.*, Ann., 1925, **444**, 135) and the reverse process was effected by *Appel* and *Robinson* (J. chem. Soc., 1935, 426). Tetramethyl (+)catechin with bromine yielded bromocyanidin tetramethyl ether, which was demethylated with hydriodic acid to cyanidin iodide.

The stereochemical relationship of the diastereoisomeric catechins and epicatechins has been established. The catechins are the *trans*-compounds [$H_{(2)}$ and $H_{(3)}$ *trans*] and the epicatechins *cis*, (+)catechin and (+)epicatechin having the configuration given below (*Freudenberg*, Sci. Proc. roy, Soc. Dublin, 1956, **27**, 153; *A. J. Birch*, *J. W. Clark-Lewis* and *A. V. Robertson*, J. chem. Soc., 1957, 3586):

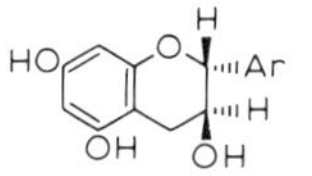

(+) Catechin (+) Epicatechin

R = 3,4-dihydroxyphenyl

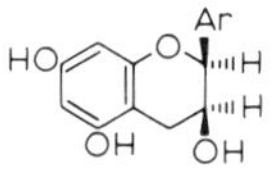

Reductive ring-opening of the heterocyclic ring in tetramethylcatechin and tetramethylepicatechin with lithium tetrahydridoaluminate and aluminium chloride is accompanied by 1,2-rearrangement and gives a phenol which is converted by methylation into 2-(3,4-dimethoxyphenyl)-3-(2,4,6-trimethoxyphenyl)propan-1-ol. Formation of the same enantiomorph of this propan-1-ol for (+)catechin and (−)epicatechin tetramethyl ether confirms the relative configurations established for these diastereoisomers

(*Clark-Lewis*, *ibid*., 1960, 2433; *K. Weinges* and *F. Nader*, Ann., 1967, **706,** 112). The above configurations have also been confirmed by the exhaustive oxonisation method (*A. Züst*, *F. Lohse* and *E. Hardegger*, Helv., 1960, **43,** 1274).

This assignment of configuration requires that the epimerisation of (+)catechin to (+)epicatechin involves inversion at the $C_{(2)}$ atom. Models show that in the catechins the heterocyclic ring is non-planar and exists in a half-chair form (see conformations given below) and there is evidence to support *W. B. Whalley's* conclusion ("Chemistry of Vegetable Tannins", Soc. of Leather Trades Chemists, London, 1956, p. 151) that the catechins have the preferred conformation 2(*a*) aryl, 3(*a*) hydroxyl, and the epicatechnins 2(*e*) aryl, 3(*a*) hydroxyl (*E. A. H. Roberts*, Chem. and Ind., 1956, 73). The n.m.r. spectrum of tetramethylcatechin appears to be incompatible with a diaxial *trans* arrangement and suggests at least a diequatorial conformation in solution (*Clark-Lewis* and *L. M. Jackman*, Proc. chem. Soc., 1961, 165; *E. J. Corey et al.*, Tetrahedron Letters, 1961, 429):

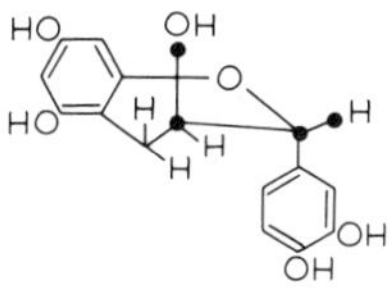

Catechin

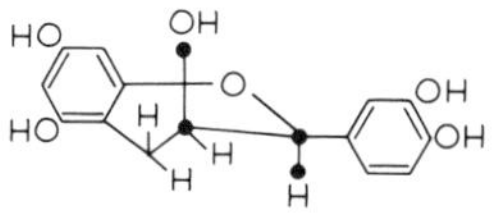

Epicatechin

These conformations account both for the presence of strong intra-molecular hydrogen bonding of the axial 3-hydroxy groups (*Roberts*, Chem. and Ind., 1955, 631, 1551) and for the chemical properties of the catechins and the epicatechins. 5,7,3′,4′-Tetramethyl (−)epicatechin 3-toluene-*p*-sulphonate with anhydrous hydrazine smoothly yields the flavene, tetramethylanhydroepicatechin (XLIV), whereas tetramethylcatechin toluene-*p*-sulphonate reacts much less readily with the same reagent and is broken down into phloroglucinol dimethyl ether (XLVI) and 3-(3,4-dimethoxyphenyl)pyrazoline(XLVII) (*Freudenberg et al.*, Ann., 1924, **436**, 286; 1925, **441,** 157):

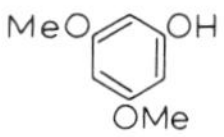

(XLVI)

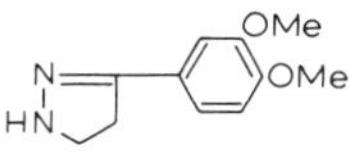

(XLVII)

The formation of the flavene is a stereospecific *trans*-elimination (E_2) of the tosyl group, which is favoured by the four co-planar participating centres (indicated by black dots in the formula). In catechin the conformation is unfavourable for this elimination and reaction occurs *via* the intermediate flav-3-ene (*Freudenberg et al.*, *ibid*., 1925, **446,** 87). It is also noteworthy that the catalytic hydrogenation of planar nuclei such as the flavylium salts and flavonols by *cis*-addition yields epicatechins.

The tosyl group can be removed from tetramethylcatechin 3-toluene-*p*-sulphonate without ring rupture by heating in quinoline. The reaction, however, is accompanied by rearrangement to an isoflavene (XLVIII) (*Freudenberg et al.*, *loc. cit*. cf. *J. J. Drumm*, Proc. roy. Irish Acad., 1923, **36B,** 41), which is oxidised by potassium permanganate in acetone or pyridine to 5,7,3′,4′-tetramethoxy-3-phenylcoumarin (XLIX) (*W. Baker*, J. chem. Soc., 1929, 1593).

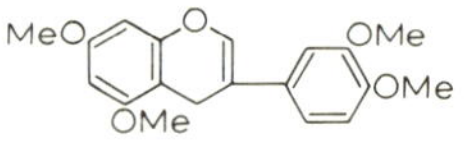

(XLVIII) 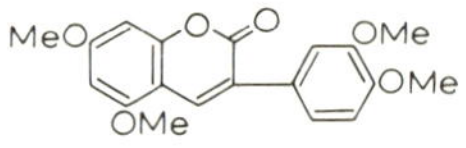 (XLIX)

(+)Tetramethylcatechin 3-toluene-*p*-sulphonate when heated with ethanolic potassium acetate suffers a Wagner–Meerwein rearrangement to (+)2-ethoxy-5,7,3′,4′-tetramethoxyisoflavan (*Clark-Lewis* and *W. Korytnyk*, *ibid.*, 1958, 2367).

A minor product from the acetolysis of tetramethyl-(+)-catechin toluene-*p*-sulphonate has been identified as (*R*, 3*S*)-3-acetoxymethyl-4,6-dimethoxy-2-(3,4-dimethoxyphenyl)coumaran (L) (*C. A. Anirudhan, D. W. Mathieson* and *Whalley*, J. chem. Soc., C, 1966, 634):

NaOAc / AcOH

(L)

Tetramethylcatechin with phosphorus pentachloride gives 2-chloro-5,7,3′,4′-tetramethoxyisoflavan (*Drumm, loc. cit.*).

The naturally occurring (+)catechin, (−)epicatechin, and (−)epiafzelechin have the same configuration at $C_{(2)}$ (*Birch et al., loc. cit.*).

(+)**Catechin,** (2*R*:3*S*)-5,7,3′,4′-*tetrahydroxyflavan*-3-*ol*, $C_{15}H_{14}O_6 \cdot 4H_2O$, m.p. 96° (*St. von Kostanecki* and *J. Tambor*, Ber., 1902, **35**, 1867); the anhydrous compound occurs in two crystalline forms, m.p. 177° and 219° (*H. L. Hergert* and *E. F. Kurth*, J. org. Chem., 1953, **18**, 521), $[\alpha]_{5780}$ +17.1° (aqueous acetic acid), *penta-acetate*, m.p. 131–132°, *pentamethyl ether*, m.p. 95°. Characteristic reactions are: colourless precipitate with lead acetate or bromine; intense red colour with warm sulphuric acid; dark green colour with ferric chloride indicative of a catechol nucleus; and a beautiful red colour with vanillin and hydrochloric acid showing the presence of a phloroglucinol nucleus; it couples readily with diazonium salts. When heated at 115° with water it is epimerised to (+)epicatechin (*Freudenberg* and *Purrmann*, Ann., 1924, **437**, 274). (+)Catechin and (+)epicatechin in dilute acid solution condense with phloroglucinol to give the same product, $C_{21}H_{18}O_8$ (*W. Mayer* and *F. Merger*, Chem. and Ind., 1959, 485; Ann., 1961, **644**, 70). **Polydin,** (+)*catechin*-7-L-*arabinoside*, m.p. 191–193°, $[\alpha]_D^{20}$ −121.6 (methanol), is obtained from *Polypodium vulgare*. Hydrolysis with 0.5% sulphuric acid gives (+)catechin and L-arabinose (*N. I. Uvarova, J. Jizba* and *V. Herout*, Coll. Czech. chem. Comm., 1967, **32**, 3075; *Weinges* and *R. Wild*, Ann., 1970, **734**, 46). (−)**Catechin,** $[\alpha]_{5780}$ −16.8° (aqueous acetone), *penta-acetate*, m.p. 131–132°, *pentamethyl ether*, m.p. 95°; (±)**catechin,** $C_{15}H_{14}O_6 \cdot 3H_2O$, m.p. 212–214°, *penta-acetate*, m.p. 164–165°, *pentamethyl ether*, m.p. 110–111° (*Freudenberg* and *Purrmann, loc. cit.*). A mechanism has been proposed and confirmed for the self-condensation of catechin (*Weinges, Y. Naya* and *F. Toribio*, Ber., 1963, **96**, 2870).

The structure of **dehydrodicatechin A,** obtained by enzymatic dehydrogenation of naturally occurring (+)catechin has been determined from the n.m.r., i.r. and u.v. spectral data for its heptamethyl ether (*Weinges et al.*, Ann., 1971, **754**, 124). Its configuration and conformation has been confirmed by X-ray analysis of its 8-bromo heptamethyl ether (*T. C. van Soest, ibid.*, p. 137):

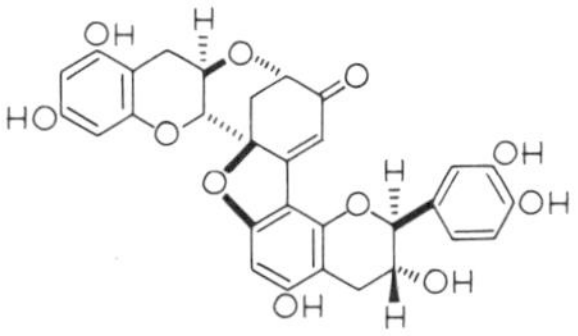

Dehydrodicatechin A

(−)**Epicatechin,** m.p. 242°, $[\alpha]_{5780}$ −60° (aqueous acetic acid), *penta-acetate*, m.p. 151–152°, *tetramethyl ether*, m.p. 153–154°, *pentamethyl ether*, m.p. 93–94°, is isolated from *Acacia catechu* (*Freudenberg* and *Purrmann, loc. cit.*), and fresh Formosan tea leaves (*Y. Oshima*, Proc. Imp. Acad., Tokyo, 1936, **12,** 189; C.A., 1937, **31,** 1902). When heated in water it affords (−)catechin. The hydrogenation of cyanidin affords epicatechin (*Freudenberg*, Sci. Proc. roy. Dublin Soc., 1956, **27,** 153).

(−)**Epigallocatechin,** (2*R*:3*R*)-5,7,3′,4′,5′-*pentahydroxyflavan*-3-*ol*, $C_{15}H_{14}O_7$, m.p. 218° (227°), $[\alpha]_D^{18}$ −67.5° (ethanol), *hexa-acetate*, m.p. 189–190°, *pentamethyl ether*, m.p. 183°, is obtained from Japanese (*M. Tsujimura*, Sci. Papers Inst. phys. chem. Res., Tokyo, 1934, **24,** 149), Formosa (*Oshima*, Bull. agri. chem. Soc. Japan, 1936, **12,** 103), Javanese (*W. B. Eijs*, Rec. Trav. chim., 1939, **58,** 805), and Ceylon green tea (*A. E. Bradfield et al.*, J. chem. Soc., 1947, 32; 1948, 2249). Its absolute configuration has been ascertained (*Birch, Clark-Lewis* and *Robertson, loc. cit.).* The pentamethyl ether is oxidised to the trimethyl ether of gallic acid and this provides an important clue to its structure. A racemic *gallocatechin*, two forms m.p.'s about 160° and 195° (decomp.), *hexa-acetate*, m.p. 158.5–159.5°, has been isolated *(Bradfield et al., loc. cit.). Epigallocatechin gallate* (3-*galloylepigallocatechin*), m.p. 215–216°, is a constituent of tea. (+)**Casuarin,** isolated from *Casuarina equisetifolia*, is regarded as (+)**gallocatechin,** (2*R*:3*R*)-5,7,3′,4′,5′-*pentahydroxyflavan*-3-*ol*, m.p. 185–188°, $[\alpha]_D$ +14.7° (aqueous acetone), *hexa-acetate*, m.p. 141–143°, $[\alpha]_D$ +31.0° (acetone), *pentamethyl ether*, m.p. 160–162°, $[\alpha]_D$ −3° (acetone), which is obtained from the bark of oak and sweet chestnut (*W. Meyer*, "The Chemistry of Vegetable Tannins", 1956, p. 127; Ann., 1958, **611,** 264). The pentamethyl ether when reduced by sodium and ethanol and then methylated with diazomethane yields 1-(2,4,6-*trimethoxyphenyl*)-3-(3,5-*dimethoxyphenyl*)*propane*, m.p. 85°, methoxy group being lost in the process.

(−)**Epiafzelechin,** (2*R*:3*R*) 5,7,4′-*trihydroxyflavan*-3-*ol*, $C_{15}H_{14}O_5$, m.p. 240–243° (decomp.) $[\alpha]_D^{19}$ −59° (5% ethanol), *tetra-acetate*, m.p. 126–127°, *trimethyl ether*, m.p. 110°, is obtained from the heartwood of *Afzelia* species (*F. E. King et al.*, J. chem. Soc., 1955, 2948; *Birch, Clark-Lewis* and *Robertson*, Chem. and Ind., 1956, 644; *loc. cit.*). It is oxidised to *p*-anisic acid and can be converted into apigenidin chloride.

(−)**Robinetinidol,** 7,3′,4′,5′-*tetrahydroxyflavan*-3-*ol*, m.p. 203–205° (207°), $[\alpha]_D$ −10.7° (aqueous acetone), *penta-acetate*, m.p. 132–133°, *tetramethyl ether*, m.p. 143–144°, is obtained from *Robinia pseudoacacia* and *Acacia mearnsil* (*D. G. Roux* and *A. E. Maihs*, Nature, 1958, **182,** 1798; Biochem. J., 1960, **74,** 44; *Roux* and *E. Paulus*, *ibid.*, 1962, **84,** 324, 416).

(−)**Fisetinidol,** 7,3′,4′-*trihydroxyflavan*-3-*ol*, m.p. 214°, *trimethyl ether*, m.p. 123°, is obtained from *Acacia mollissima* (*Roux* and *Paulus*, *ibid.*, 1961, **78,** 120).

Quebrachocatechin, 7,3′,4′-*trihydroxyflavan*-3-*ol*, $C_{15}H_{14}O_5$, has not been isolated from natural sources but (±)**epiquebrachocatechin,** m.p. 93–96°, *tetra-acetate*, m.p. 184–186°, has been synthesised (*Freudenberg* and *P. Maitland*, Ann., 1934, **510,** 193). It is converted by alkali to (±)**quebrachocatechin,** *tetra-acetate*, m.p. 144–146°, and is polymerised by boiling dilute acid to a tannin-like material which gives products identical with those obtained from quebracho tannin from the wood of *Quebracho colorado*. It is surmised that this tannin is a polymer of quebrachocatechin.

The chemistry of **brazilin** and **haematoxylin,** vegetable products related to catechin, is discussed in Chapter 22, which in the previous edition was written by the late Sir Robert Robinson. This particular piece of chemistry has been isolated so that it could be presented as a tribute to his outstanding contribution to the chemistry of organic compounds.

Hydroxylation of 5,7,3′,4′-tetramethoxyflav-2-ene (LI) yields a flavan-2,3-diol which exists almost entirely as the open-chain tautomer, a dihydro-α,2-dihydroxychalcone, 2-hydroxy-1-(3,4-dimethoxyphenyl)-3-(2,4-dimethoxy-6-hydroxyphenyl)-1-oxopropane (LII):

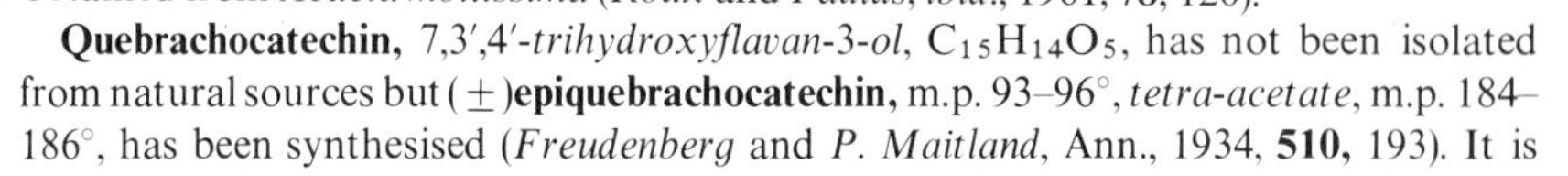

MeO OMe OMe OMe (LI) $\xrightarrow{OsO_4}$ MeO OH CO OMe OMe CHOH OMe CH_2 (LII) ⇌ MeO OH OMe OMe OMe OH

Cyclisation of 3-(2-hydroxyphenyl)-2-methoxy-1-phenylpropan-1-ol with pure P_4O_{10} in boiling benzene gives 3-methoxyflavan, 3-flavanyl methyl ether, whereas reaction with polyphosphoric acid containing P_4O_{10} yields 2-benzalcoumaran (*L. Reichel*, *P. Pritze* and *H. Gragert*, Ann., 1972, **757,** 79):

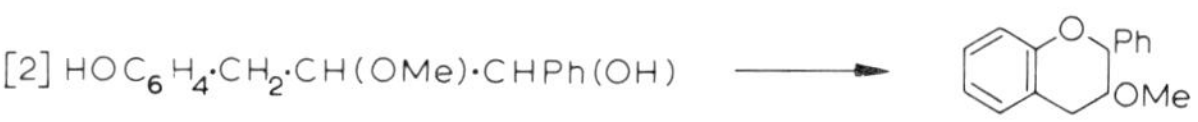

[2] $HOC_6H_4{\cdot}CH_2{\cdot}CH(OMe){\cdot}CHPh(OH)$ ⟶ Ph OMe

*(ii) Condensed tannins**

The tannins are amorphous, water-soluble metabolic products of plants and trees which have an astringent taste and form coloured precipitates with metallic salts. They are complex substances with molecular weights of at least 500 and frequently 2000 or more. They combine with proteins and this prop-

* *K. Freudenberg*, "Tannin, Cellulose and Lignin", Springer Verlag, Berlin, 1933; *O. Th. Schmidt* and *W. Mayer*, Z. angew. Chem., 1956, **68,** 103; "The Chemistry of Vegetable Tannins", Society of Leather Trades' Chemists, London 1956.

erty is the basis of the tanning process whereby skins and hides are converted into leather.

The tannins have been conveniently divided by Freudenberg into (*1*) hydrolysate tannins (see Vol. IIID, p. 227) and (*2*) condensed *tannins* or *phlobatannins* which are not hydrolysed by acids, but on the contrary are converted into complex products, the reddish-brown *phlobaphenes*. Some at least, of the condensed tannins, such as Gambier or Burma cutch, appear to be polymeric substances in which the unit is the catechin molecule. Molecular weight determinations suggest that the polymers contain about five monomer units.

The commonly used condensed tannins and their sources (see *T. White*, in "The Chemistry of Vegetable Tannins", Society of Leather Trades' Chemists, London, 1956, p. 11) are Quebracho-wood of *Schinopsis lorentzii*, Mimosa (wattle) – bark of *Acacia mollissima*, Mangrove (Borneo cutch) – bark of *Rhizophoraceae*, Spruce – bark of *Picea abies*, Hemlock – bark of *Tsuga canadensis*, Gambier – leaves of *Uncaria Gambier*, Burma cutch – wood of *Acacia catechin*, Myrtan – wood and bark of *Eucalyptus redunca*, and Oak – bark of *Quercus* species.

The tanning extracts obtained from cutch, wattle, quebracho, etc., are complex mixtures. It has been shown by countercurrent and chromatographic methods that a wattle extract, for example, contained twenty-seven different phenolic materials (*White et al.*, J. Soc. Leather Trades' Chemists, 1952, **36**, 148). The assignment of a structure or structures to the condensed tannins therefore presents a problem of great difficulty which is increased by the failure to isolate and identify breakdown products. Nevertheless substantial progress has been made in elucidating the constitution of some condensed tannins, particularly by Freudenberg and his school. It appears that the fundamental unit of these tannins is the catechin molecule.

The catechins themselves are sensitive phenolic substances with no tanning properties, which under the influence of acid, alkali, or enzymes undergo self-condensation to give products which vary according to the degree of condensation from the water-soluble tannins to the insoluble *phlobaphenes*. Thus (−)epicatechin yields *Cocoa-red*. These amorphous tannin-like products resemble the natural condensed tannins. This was strikingly shown by heating *quebrachocatechin* with dilute acid to give a tannin which closely resembled natural quebracho tannin (*K. Freudenberg* and *P. Maitland*, Ann., 1934, **510,** 193). Both gave similar analytical values and underwent condensation to phlobaphenes without loss of the elements of water. Confirmation that catechins are the building units of the condensed tannins comes from the observation that those trees which yield tanning materials also afford catechins.

For the condensation of hydroxyflavans in the presence of acids to take place (*Freudenberg* and *K. Weinges*, Ann., 1954, **590,** 140) hydroxyl groups in the 7 and 4′ positions are necessary. It is postulated that the first stage of the process involves a breakage of the pyran ring. The cleaved molecule then condenses through the $C_{(2)}$ atom with a second molecule in the reactive 6-position as shown in LIII:

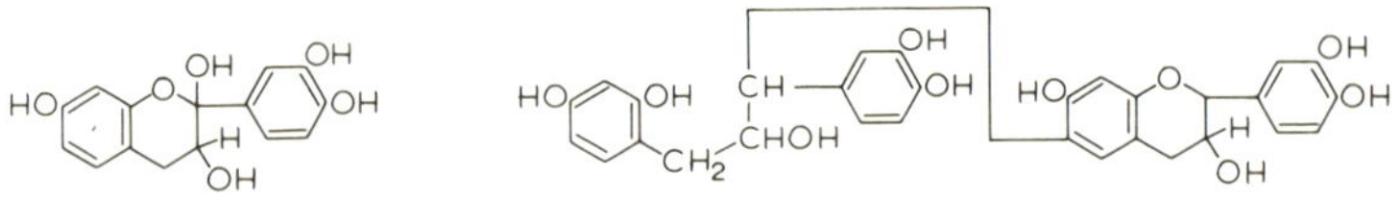

Quebrachocatechin

(LIII)

The intact pyran ring in the product is then cleaved, further condensation occurs with a third catechin molecule, and the process proceeds until a tanning material or phlobaphene is obtained. If this picture is true, no water or hydrogen is lost in the process, but the number of phenolic groups per catechin unit increases. This is borne out experimentally since the products have the same carbon and hydrogen values but higher acetyl values than the parent catechin. The observation that the methylated products give veratric acid on oxidation shows that the catechol part of the catechin molecule does not participate in the condensation. This view is taken by others (*B. R. Brown*, *W. Commings* and *G. A. Somerfield*, J. chem. Soc., 1957, 3757,) but other possibilities have been suggested for the formation of tannins (*A. Russell*, Chem. Reviews, 1935, **17,** 155; *L. R. Finch* and *D. E. White*, J. chem. Soc., 1950, 3367; *J. Jack et al., ibid.*, 1954, 3684; *Freudenberg et al.*, Ann., 1935, **518,** 37; *A. M. Stephen*, J. chem. Soc., 1949, 3082; *D. G. Roux*, J. Soc. Leather Traders' Chemists, 1950, **34,** 122; *D. E. Hathway* and *J. W. T. Seakins*, Biochem. J., 1957, 239; J. chem. Soc., 1957, 1562; *W. G. C. Forsyth* and *J. B. Roberts*, *ibid.*, 1960, 374). It is now believed that the majority of the tannins that occur in tanning woods originate by acid-catalysed intermolecular condensation of catechins and leucoanthocyanidins (hydroxyflavan-3,4-diols) without the participation of enzymes. The mechanism of the reaction and the stereoselectivity of the acid-catalysed intermolecular condensation of catechins has been described [*Weinges*, J. Polym. Sci. C, 1965 (Pub. 1968), **16,** 3625].

It had been suggested previously that there was a relationship between the leucoanthocyanidins and the condensed tannins (*G. M. Robinson* and *R. Robinson*, Biochem. J., 1933, **27,** 206; *E. C. Bate-Smith* and *T. Swain*, Chem. and Ind., 1953, 377; *Bate-Smith*, J. expt. Bot., 1953, **4,** 1; *F. E. King* and *W. Bottomley*, J. chem. Soc., 1954, 1399; *Freudenberg*, J. Polym. Sci., 1958, **29**, 433) and in the section on leucoanthocyanins (p. 93) an account is given of work which supports this viewpoint. Flavan-3,4-diols may thus be the precursors both of anthocyanidins and of tannins.

(iii) Flavan-4-ols and flavan-3,4-diols

Preparation. **Flavan-4α-ol,** 4(*a*)-*hydroxy*-2(*e*)-*phenylchroman*, m.p. 118–119°, is prepared from flavanone by conversion into the oxime, reduction to the amine and deamination; it has the 2-phenyl group equatorial and the 4-hydroxy group axial. A number of flavan-4α-ols have been produced by this method (*H. Fletcher et al.*, Acta Univ. Debrecen. Ludovico Kossuth Nom., 1962, **8,** 5; C.A., 1963, **59,** 13923). **Flavan-4β-ol**, 4(e)-*hydroxy*-2(e)-*phenylchroman*, m.p. 148°, is obtained by catalytic hydrogenation of flavone using a palladium on carbon catalyst; also by reduction of flavone or flavanone using diborane; it has the 2-phenyl and the 4-hydroxyl groups equatorial (*C. P. Lillya et al.*, Chem. and Ind., 1963, **84;** *J. W. Clark-Lewis, T. M. Spotswood* and *L. R. Williams*, Proc. chem. Soc., 1963, 20).

Some hydroxyflavanones have been reduced with sodium tetrahydridoborate in the presence of boric acid and with sodium tetrahydridoborate in 95% ethanol (*L. R. Row et al.*, Current Sci., 1963, **32,** 67), to give flavan-4β-ols. These compounds are quite stable in the presence of 3% hydrogen chloride in acetic acid, but undergo polymerisation under more rigorous conditions; the action of phosphoryl chloride–pyridine effects partial dehydration (*Row et al.*, J. Indian chem. Soc., 1963, **40,** 521). Reduction of flavanone with lithium tetrahydridoaluminate has been reported to give flavan-4β-ol (*R. Bognár* and *M. Rákosw*, Magy. Kem. Foly., 1958, **64,** 106), but it has also been reported that the reduction product is 2′-hydroxychalcone (95%) (*S. Mitsui* and *A. Kasahara*, Nippon Kagaku Zasshi, 1958, **79,** 1382). A number of flavan-4β-ols have been converted to the corresponding flavan-4α-ols by treatment with phosphorus tribromide followed by alcoholic potassium hydroxide (*M. S. Kamet*, *P. Y. Mahajan* and *A. B. Kulkarni*, Indian J. Chem., 1970, **8,** 119). Some 4-substituted flavan-4-ols have been obtained by reacting the appropriate flavanone with a Grignard reagent; in some cases with excess reagent 4,4-diarylflavans are formed (*M. A. -F. Elkaschef*, *M. H. Nosseir* and *H.-El-Din M. Mohamed*, J. chem. Soc., 1965, 494).

Chemical properties: (1) Flavan-4-ols and flavan-3,4-diols condense with phenols to yield 4-arylflavans, but the 3-hydroxyl group of the diols does not participate in further condensations to form ethers; for example, flavan-4β-ol condenses with phenol to yield, after methylation, 4-(4-methoxyphenyl)-flavan:

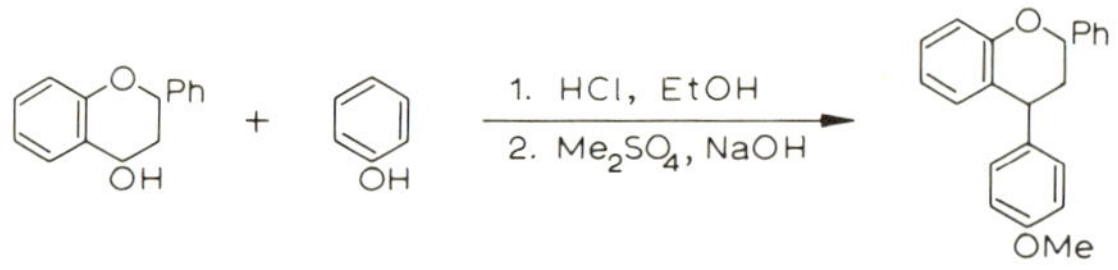

and flavan-3,4-diol with resorcinol gives, after methylation, 3-hydroxy-4-(2,4-dimethoxyphenyl)flavan:

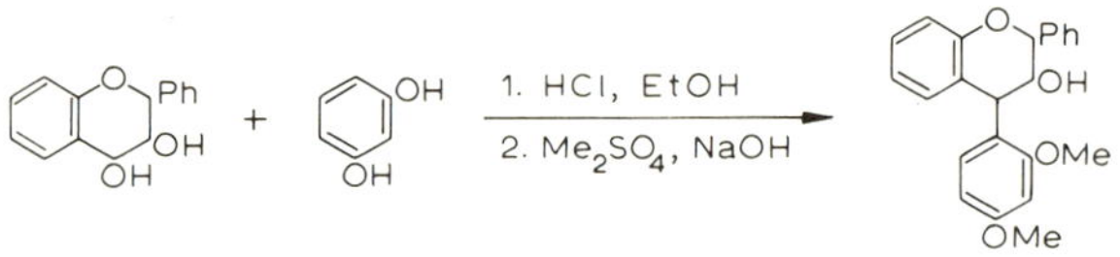

4′-Methoxyflavan-4β-ol condenses with two molecules of a phenol to afford, after methylation, 1-(2-methoxyphenyl)-1,3,3-tri(4-methoxyphenyl)propane (LIV) (*B. R. Brown, W. Cummings* and *J. Newbould*, *ibid.*, 1961, 3677):

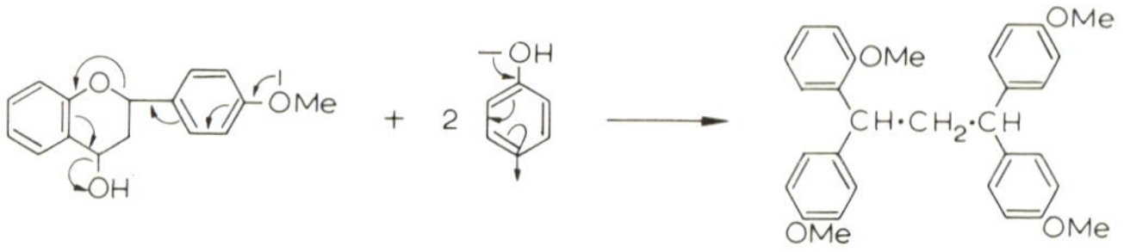

(LIV)

(*2*) Attempted *acetylation* of (±)-5,7,4′-*trimethoxyflavan*-4*β*-*ol*, m.p. 159°, with acetic anhydride and pyridine at 28° resulted in epimerisation to the 4α-ol (*H. G. Krishnamurty, T. R. Seshadri* and *D. G. Roux*, Tetrahedron Letters, 1965, 3689). *Bromination* of flavan-4β-ol gives 4α,6-*dibromoflavan*, m.p. 125°, which on treatment with silver acetate yields 4α-*acetoxy*-3-*bromoflavan*, m.p. 125–127° (*B. J. Bolger, K. G. Marathe* and *E. M. Philbin*, Chem. and Ind., 1966, 1304).

(*3*) Flavan-4-ols (α and β), unsubstituted in the 5-position are conveniently oxidised to flavanones by dimethyl sulphoxide–acetic anhydride, for example, flavan-4-ol (α and β) and 7,3′,4′-*trimethoxyflavan-4β-ol*, m.p. 136–137°, are oxidised to the corresponding flavanones, but 5,7,4′-*trimethoxyflavan-4-ol*, m.p. 158°, is recovered unchanged, indicating that the 5-methoxy group sterically hinders the oxidation (*V. K. Bhatis et al.*, Tetrahedron Letters, 1968, 3859):

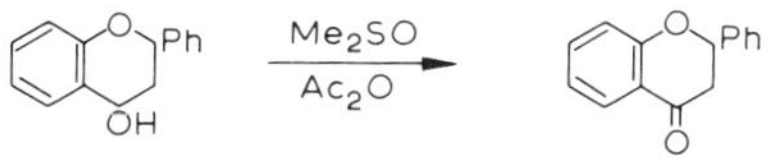

2,3-trans-**Flavan**-3,4-trans-**diol** (LV), m.p. 145°, is prepared by the catalytic reduction of dihydroflavonol, while 2,3-trans-**flavan**-3,4-cis-**diol** (LVI), m.p. 160°, is obtained through the oxime and corresponding amine (*Bognár et al.*, Tetrahedron, 1963, **19,** 391):

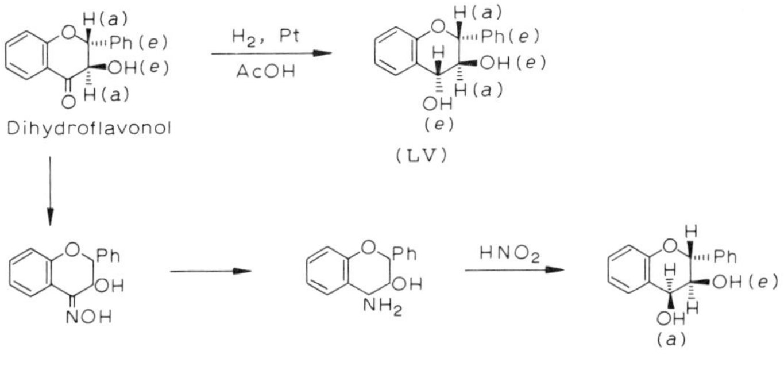

3,4-trans-*Diacetoxy*-2,3-trans-*flavan*, m.p. 90–91°; 3,4-cis-*diacetoxy*-2,3-trans-*flavan*, m.p. 97–98°; 2,3-cis-*flavan*-3,4-cis-*diol* (LVII), m.p. 111–113°, *diacetate*, m.p. 114–115°; 3,4-trans-*diacetoxy*-2,3-cis-*flavan* (LVIII), m.p. 169–170° (*M. A. Vickars, ibid.*, 1964, **20,** 2873).

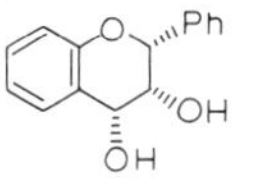

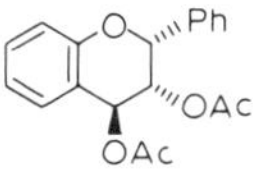

The *stereochemistry* of the four possible racemic forms (LV, LVI, LVII and LVIII) of flavan-3,4-diols have been established from their n.m.r. spectral data. Similarly the four racemates of 3′,4′-dimethoxy-6-methylflavan-3,4-diol diacetate (*Clark-Lewis, L. M. Jackman* and *Williams*, J. chem. Soc., 1962, 3858) and 7,4′-dimethoxyflavan-3,4-diol diacetate (*H. M. Saayman* and *Roux*, Biochem. J., 1965, **96,** 36) have been prepared and their geometrical configurations assigned. For n.m.r. data referring to flavan-3,4-diols see *E. J. Corey*, *Philbin* and *T. S. Wheeler*, Tetrahedron Letters, 1961, 429; *Clark-Lewis* and *Jackman*, Proc. chem. Soc., 1961, 165; *M. M. Bokadia et al.*, J. chem. Soc., 1961, 4663; *Philbin* and *Wheeler*, J. org. Chem., 1962, **27,** 4114; *Clark-Lewis*, *Spotswood* and *Williams*, Austral. J. Chem., 1963, **16,** 107). The difference in rates of oxidation using lead tetra-acetate have been used to distinguish between 2,3-*trans*-3,4-*trans*-, 2,3-*trans*-3,4-*cis*-, and 2,3-*cis*-flavan-3,4-*cis*-diols (*Clark-Lewis* and *Williams, ibid.*, p. 869). The four racemates of 4′-methoxy-6-methylflavan-3,4-diol have been synthesised (*M. D. Kashikar* and *A. B. Kulkarni*, Chem. and Ind., 1958, 1084). Film chromatography on silica and reaction with 2,2-dimethoxypropane have been shown to be diagnostic for assignment of relative stereochemistry to a number of pairs of 2,3-*trans*-flavan-3,4-diols of known stereochemistry (*Brown* and *J. A. H. MacBride*, J. chem. Soc., 1964, 3822). The stereochemistry of the 3-bromo-4′-methoxy-6-methylflavan-4-ols has been established by a consideration of their i.r. spectra and chemical properties (*C. G. Joshi* and *Kulkarni*, J. Indian chem. Soc., 1957, **34**, 753); 4′-methoxy-6-methylflavan-3,4-diols (*Kashikar* and *Kulkarni*, J. sci. ind. Research, India, 1959, **18B**, 418). The epoxides of 2′-benzyloxychalcones and 2′-(methoxy–methoxy)chalcones have been converted into *trans*-2,3-flavan-3,4-diols (*G. Piccardi* and *J. Chopin*, Bull. Soc. chim. Fr., 1971, 230). Reduction of 7-benzyloxy-3-hydroxy-2,3,4′-trimethoxyflavanone and of 3-hydroxy-2,3,7,4′-tetramethoxyflavanone with sodium tetrahydridoborate afford mainly *trans*-2-methoxyflavan-3,4-*trans*-diols. The 7-benzyloxy hemiacetal also gives a small amount of the *cis*-2-methoxyflavan-3,4-*cis*-diol and less of the corresponding *cis*-2,3-flavan-3,4-*trans*-diol (*Clark-Lewis et al.*, Austral. J. Chem., 1972, **25,** 857). Epimerisation at the $C_{(4)}$ centre of a flavan-3,4-diol has been effected by treatment of the *cis*-isomer with sodium tetrahydridoborate–boron trifluoride. Neither sodium tetrahydridoborate nor boron trifluoride, alone effected epimerisation (*Kashikar* and *Kulkarni*, Current Sci., 1961, **30,** 142). The concept that the formation of cyclic derivative from a flavan-3,4-diol cannot be regarded as proof of *cis* configuration is given support by examples of isomerisation during the formation of isopropylidene derivatives (*S. Fujise et al.*, Bull. chem. Soc. Japan, 1962, **35,** 1245).

Substitution reactions of mercaptoacetic acid with 2,3-*trans*- and 2,3-*cis*-flavan-3,4-diols and 2,3-*trans*-flavan-[4-^{2}H]-3,4-diols illustrate the mechanism of their conversion into flavanonol (dihydroflavonol) and flavanone analogues as side-reactions (*I. C. du Preez, T. G. Fourie* and *Roux*, Chem. Comm., 1971, 333).

F. E. King and *W. Bottomley* (J. chem. Soc., 1954, 1399) have shown that melacacidin, which possesses all the properties associated with leucoanthocyanins (p. 93) is a flavan-3,4-diol. All the leuco-compounds characterised since have been flavan-3,4-diols or derivatives of flavan-3,4-diols.

(iv) Isoflavan-4-ols

trans-**Isoflavan-4-ol** has been prepared in 61% yield by reacting chromene with lithium tetrachloropalladinate(II) and phenylmercury chloride in aqueous acetone. *cis*-Flavan-4-ol and isoflavene are also formed in the reaction (*M. Arai et al.*, Chem. Letters, 1972, 889); 7-methoxyisoflavan-4α-ol (*N. Inoue, S. Yamaguchi* and *S. Fujiwara*, Bull. chem. Soc. Japan, 1964, **37,** 588).

The reaction of nitrous acid on the 4-aminoisoflavan hydrochlorides (LIX; $R = R^1 = H$, and $R = OMe$, $R^1 = H$) obtained by the catalytic hydrogenation of the isoflavanone oximes, afford the corresponding *trans*-4-ols, in poor yield. The hydroboration of the isoflav-3-enes (LX, $R = R^1 = H$; $R = OMe$, $R^1 = H$; and $R = R^1 = OMe$) gives the corresponding *trans*-alcohols in good yields (*S. Yamaguchi et al.*, *ibid.*, 1968, **41,** 2073):

(LIX) (LX)

n.m.r. Spectral data has been obtained and the stereochemistry of the 2-methylisoflavan-4-ols discussed (*Inoue* and *S. Ito*, Chem. and Ind., 1965, 1382); 7-*acetoxy*-2-cis-*methylisoflavan*-trans-4-ol (LXI), m.p. 124–126°, *acetate*, m.p. 119–120°; 7-*acetoxy*-2-trans-*methylisoflavan*-trans-4-*ol* (LXII), m.p. 110–112°, *acetate*, 161–162°:

(LXI) (LXII)

(*f*) *Chromanones, dihydrobenzo*[b]*pyranones*

(*i*) *Chromanones*

Of the three chromanones the chroman-2-ones, 3,4-dihydro-2*H*-benzo-[*b*]pyran-2-ones are dihydrocoumarins (p. 103) and, as lactones, are hydrolysed by alkali to salts of β-(2-hydroxyphenyl)propionic acids. The chroman-3- and 4-ones on the other hand are ketones:

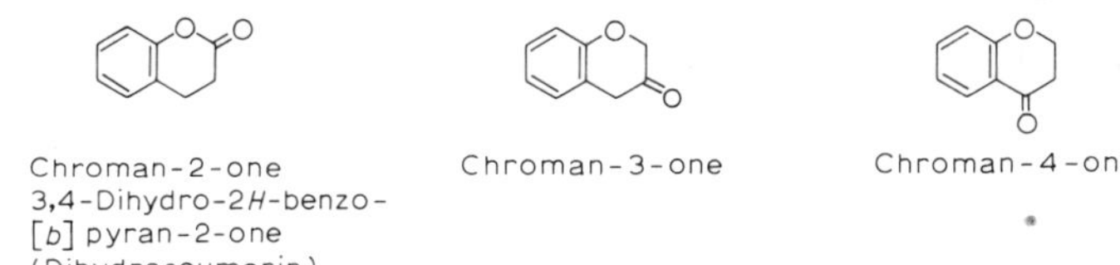

Chroman-2-one, dihydrocoumarin, m.p. 24°, b.p. 272°, 145°/13 mm, occurs in *Melilotus officinalis*, and is prepared by hydrogenating coumarin with a Raney nickel catalyst (*P. L. de Benneville* and *R. Connor*, J. Amer. chem. Soc., 1940, **62,** 283; *C. L. Palfray*, *ibid*., 1941, **63,** 3540), or by the ring-closure of melilotic acid (*P. Ramart-Lucas* and *M. J. Hock*, Bull. Soc. chim. Fr., 1935, [v], **2,** 327). With aniline or hydrazine it yields the *anilide*, m.p. 138°, or *hydrazide*, m.p. 166°, of melilotic acid (*F. M. Dean et al*., J. chem. Soc., 1950, 895).

The **chroman-3-ones** are prepared by heating 2-carboxymethylphenoxyacetic acids with acetic anhydride and sodium acetate (*J. H. Richardson et al., ibid*., 1948, 1610). They may be obtained also by the pyrolysis of 4-chlorochroman-3-ols (*J. B. Abbott et al*., J. chem. Soc., C, 1967, 1472). The cyclisation of chloromethyl 2-methoxydiphenylmethyl ketone gives 1-(2-

methoxyphenyl)indan-2-one and not 4-phenylchroman-3-one (*P. K. Grover* and *N. Anaud*, Indian J. Chem., 1969, **7**, 196).

Chroman-3-one, b.p. 104–108° (bath temp.)/4 mm, *dinitrophenylhydrazone*, m.p. 165–166°, *semicarbazone*, m.p. 188.5°, *thiosemicarbazone*, m.p. 192.5° (*P. Pfeiffer* and *E. Enders*, Ber., 1951, **84**, 247).

Ring-closure of β-phenoxypropionic acids to **chroman-4-ones** is effected by phosphorus pentoxide (*F. Krollpfeiffer* and *H. Schultze*, Ber., 1924, **57**, 206) or polyphosphoric acid (*C. D. Hurd* and *S. Hayao*, J. Amer. chem. Soc., 1954, **76**, 5065; *J. D. Loudon* and *R. K. Razdan*, J. chem. Soc., 1954, 4299). This method is of wide application since β-phenoxypropionic acids are readily accessible (*A. H. Cook* and *K. J. Reed*, *ibid.*, 1945, 920; *T. L. Gresham et al.*, J. Amer. chem. Soc., 1949, **71**, 661).

Another general method utilises halogenated, *e.g.*, 2-chloropropyl- or unsaturated phenolic ketones, *e.g.*, 1-propenyl 2-hydroxy-5-tolyl ketone which undergo ring-closure with alkali to yield 2,6-dimethylchroman-4-one (*K. von Auwers* and *Krollpfeiffer*, Ber., 1914, **47**, 2587; *Auwers et al.*, Ann., 1920, **421**, 1):

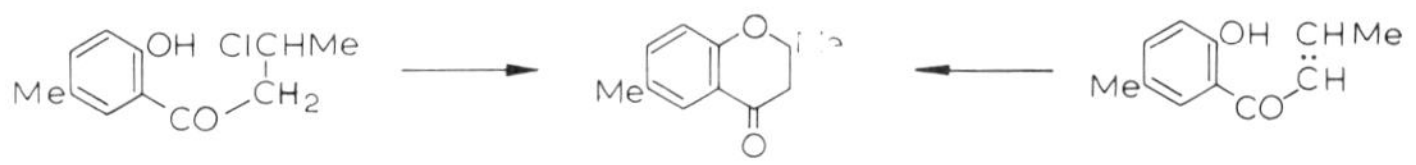

Crotonic esters of phenols in the presence of hydrogen fluoride undergo rearrangement with migration of the acyl group, the product on treatment with 1.5% sodium hydroxide undergoing ring-closure to yield a chroman-4-one (*O. Dann et al.*, *ibid.*, 1954, **587**, 16); *e.g.* *p*-tolyl crotonate yields 2,6-dimethylchroman-4-one:

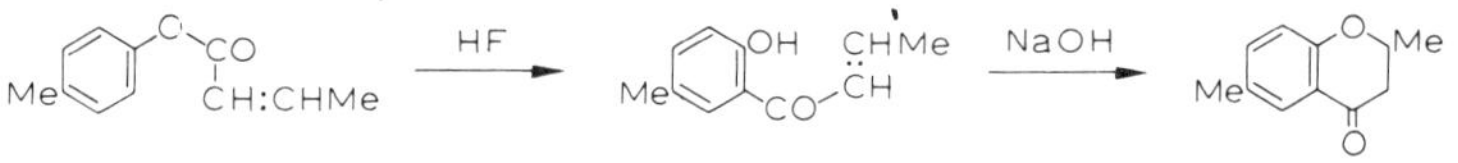

Phenols and unsaturated acids with hydrogen fluoride at 100° yield chroman-4-ones, *p*-cresol and crotonic acid thus giving an 82% yield of 2,6-dimethylchroman-4-one (see also *H. A. Offe* and *W. Barkow*, Ber., 1947, **80**, 458, 464). Similarly the condensation of resorcinol with α,β-unsaturated carboxylic acids in the presence of boron trifluoride gives the corresponding chroman-4-one (*M. Nakajima et al.*, Agric. biol. Chem., Tokyo, 1963, **27**, 700). 2,2-Dimethylchroman-4-ones may be prepared by the reaction between β,β-dimethylacrylic acid and reactive phenols using freshly fused zinc chloride and phosphoryl chloride (*P. R. Iyer* and *G. D. Shah*, Indian J. Chem.,

1968, **6,** 227). Chroman-4-ones have been prepared in a one stage process by treating α,β-unsaturated carboxylic acids with a phenol in the presence of an excess of polyphosphoric acid (*P. S. Bramwell, A. O. Fitton* and *G. R. Ramage*, B.P. 1,077,066/1967).

An aqueous solution of the sodium salt of 2-hydroxypropiophenone on treatment with an equivalent amount of formaldehyde affords 3-methylchroman-4-one (*P. Da Re* and *L. Verlicchi*, Experientia, 1960, **16,** 301):

ONa CO CH2 Me —HCHO, 50°→ O Me O

Chroman-4-one, m.p. 39°, b.p. 160°/50 mm (*W. E. Parham* and *L. D. Huestis*, J. Amer. chem. Soc., 1962, **84,** 813), *oxime*, m.p. 144–145° (*F. Mayer* and *L. van Zütphen*, Ber., 1924, **57,** 200), *semicarbazone*, m.p. 227°, 2,4-*dinitrophenylhydrazone*, m.p. 244°, and *phenylhydrazone*, m.p. 84° (*R. H. Harradence et al.*, J. Proc. roy. Soc. N.S. Wales, 1939, **72,** 273). The presence of a $CH_2 \cdot CO$ grouping is shown by the formation of a 3-*benzylidene* derivative, m.p. 147–150°; by the forming of a sodium salt of the enolic form; and by the condensation of chroman-4-one with vinyl cyanide to give 3,3-di(2-cyanomethyl)chroman-4-one (*E. M. Padfield* and *M. L. Tomlinson*, J. chem. Soc., 1950, 2272). The u.v. spectrum of chroman-4-one resembles that of 2-methoxyacetophenone (*Ramart-Lucas* and *M. van Cowenbergh*, Bull. Soc. chim. Fr., 1935, [v], **2,** 1381). Bond lengths and angles have been given for chroman-4-one (*E. M. Philbin* and *T. S. Wheeler*, Proc. chem. Soc., 1958, 167).

2-**Methyl-**, m.p. 32°, b.p. 132–134°/25 mm, 2,4-*dinitrophenylhydrazone*, m.p. 236° (*G. W. K. Cavill et al.*, J. chem. Soc., 1954, 4573; 3-**methyl-**, m.p. 68° (*G. Clerc-Bory*, *H. Pachéco* and *C. Mentzer*, Bull. Soc. chim. Fr., 1955, 1083); 6-**methyl-**, m.p. 34–36°, b.p. 141–143°/13.5 mm, *oxime*, m.p. 84–85°, 4-*nitrophenylhydrazone*, m.p. 222° (*S. G. Powell* and *N. G. Johnson*, J. Amer. chem. Soc., 1924, **46,** 2863; *D. Chakravarti* and *J. Dutta*, J. Indian chem. Soc., 1939, **16,** 639); 7-**methyl-**, b.p. 138°/13 mm, *oxime*, m.p. 98–99° (*Powell* and *Johnson, loc. cit.)*; 8-**methyl-**, b.p. 125–130°/9 mm, *semicarbazone*, m.p. 230–231° (decomp.) (*Chakravart* and *Dutta, loc. cit.)*; 2,2-**dimethyl-**, m.p. 88–89°, 2,4-*dinitrophenylhydrazone*, m.p. 222–223° (*W. Baker et al.*, J. chem. Soc., 1956, 2015; *R. Livingstone, ibid.*, 1962, 76); 2,2-**diphenyl-**, m.p. 137° (*S. Wawzonek, R. C. Nagler* and *L. J. Carlson*, J. Amer. chem. Soc., 1954, **76,** 1080; *J. Cottam et al.*, J. chem. Soc., 1965, 5261; 5266); 7-**hydroxy**-2,2-**dimethyl-**, m.p. 171°, *acetate*, m.p. 90–91°, *methyl ether*, m.p. 79°; 5,7-**dihydroxy**-2,2-**dimethyl-**, m.p. 197–198°, *diacetate*, m.p. 178°, dimethyl ether, m.p. 104–105°; 7,8-**dihydroxy**-2,2-**dimethyl-**, m.p. 142°; 7-**hydroxy**-2,2,5-**tri-methyl-**, m.p. 190°, *acetate*, m.p. 75°, *methyl ether* m.p. 88° (*P. R. Iyer* and *G. D. Shah*, Indian J. Chem., 1968, **6,** 227); 6-**methoxy**-2,2-**dimethyl-**, m.p. 75°, 2,4-dinitrophenylhydrazone, m.p. 205–206°; 8-**methoxy**-2,2-**dimethylchroman**-4-**one**, m.p. 127°, 2,4-*dinitrophenylhydrazone*, m.p. 243–244° (*J. D. Hepworth* and *Livingstone*, J. chem. Soc., C, 1966, 2013).

Chemical properties of chromanones. The chroman-4-ones with bromine in chloroform or carbon disulphide yield 3-bromochroman-4-ones, which with bases lose hydrogen bromide to give chromones (*J. Colonge* and *A. Guyot*, Bull. Soc. chim. Fr., 1958, 329); with sulphuryl chloride in chloroform they yield 3-chlorochroman-4-ones (*W. D. Cotterill*, *Cottam* and *Livingstone*, J. chem. Soc., 1970, 1006). Chroman-4-one may be converted almost quantitatively to chromone (p. 149). The Grignard reaction between 7-methoxy-2-methylchroman-4-one and vinylmagnesium bromide gives, besides the expected product, some 7-methoxy-2-methylchromone (p. 144). The 3-halogenochroman-4-ones may be obtained by oxidising the corresponding chroman-4-ols with chromic acid in acetic acid. The m.p.'s of some 3-halogenochroman-4-ones are given in Table 7.

TABLE 7

SOME 3-HALOGENOCHROMAN-4-ONES

Substituent	*M.p. (°C)*	*Ref.*	*Substituent*	*M.p. (°C)*	*Ref.*
3-Bromo	77–78	1	3-Bromo-6-methoxy-	83	5
	59.5–60.5	2	-2,2-dimethyl		
3-Chloro		3	3-Chloro-6-methoxy-	78–79	5
3-Bromo-2,2-	62–63		-2,2-dimethyl		
-dimethyl		4	3-Chloro-7-methoxy-	88	5
3-Chloro-2,2-	57–58		-2,2-dimethyl		
-dimethyl		4	3-Bromo-8-methoxy-	134	5
3-Bromo-2,2-	195		-2,2-dimethyl		
-diphenyl		4	3-Chloro-8-methoxy-	106–107	5
3-Chloro-2,2-	159–160		-2,2-dimethyl		
-diphenyl					

References

1 *J. Colonge* and *A. Guyot*, Bull. Soc. chim. Fr., 1958, 329.
2 *W. D. Cotterill*, *J. Cottam* and *R. Livingstone*, J. chem. Soc., C, 1970, 1006.
3 *Livingstone*, *ibid.*, 1962, 76.
4 *Cottam*, *Livingstone* and *S. Morris*, *ibid.*, 1965, 5266.
5 *J. D. Hepworth* and *Livingstone*, *ibid.*, C, 1966, 2013.

Reduction of chroman-2-ones gives γ-(2-hydroxyphenyl)propyl alcohols, while reduction of the chroman-4-ones yields either chroman-4-ols, chromans, or both, depending on the reducing agent and the conditions (*R. Mozingo* and *H. Adkins*, J. Amer. chem. Soc., 1938, **60,** 669; *G. B. Bachmann* and *H. A. Levine*, *ibid.*, 1947, **69,** 2341; *S. W. George* and *A. Robertson*, J. chem. Soc., 1937, 1535). The reduction products of 2-methyl- and 2-ethyl-chroman-4-one have been analysed by chromatography (*K. Hanaya* and *K. Furuse*, Nippon Kagaku Zasshi, 1968, **89,** 1002), and the stereochemistry of the reduction of 3-methylchroman-4-one has been studied using a metal hydride complex, Meerwein–Pondorf, and catalytic hydrogenation methods. The first two methods afford the *trans*-chromanol, the third yields the *cis*-isomer as the main product over Raney nickel, but over palladium the yield of the *trans*-isomer is greater than that of the *cis*-isomer (*Hanaya*, Bull. chem. Soc. Japan, 1970, **43,** 442).

3-Benzylidenechroman-4-ones formed by the condensation between chroman-4-ones and aromatic aldehydes, react with dimethyl sulphoxonium methylide to give 2′-arylchroman-3-spirocyclopropan-4-ones; stereochemical assignments have been made to the cyclopropyl ketones obtained (*P. Bennett et al.*, J. chem. Soc. Perk. I, 1972, 1554):

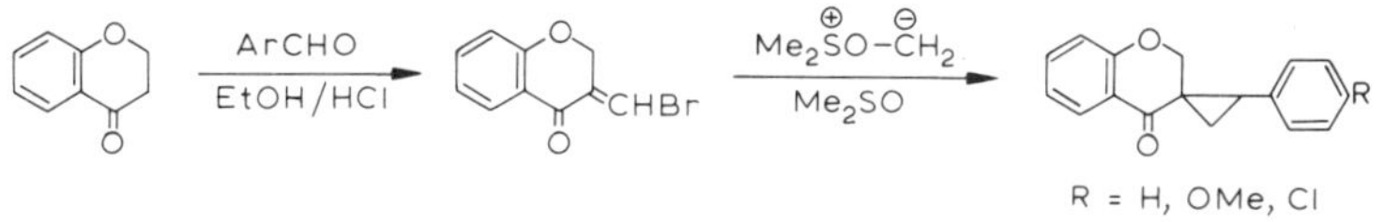

ArCHO, EtOH/HCl → CHBr; $Me_2\overset{\oplus}{S}O-\overset{\ominus}{C}H_2$, Me_2SO →

R = H, OMe, Cl

The epoxide has been formed by treating 3-benzylidenechroman-4-one with aqueous hydrogen peroxide in sodium hydroxide–methanol (*Dann* and *H. Hofmann*, Naturwiss., 1961, **48,** 162).

2′,4′,5′-Trihydroxyisoflavones have been prepared by condensing chroman-4-ones with 4,5-dimethoxy-1,2-benzoquinone (p. 210), and *via* the Schmidt rearrangement chroman-4-ones with substituents in the 5-, 6-, 7- and 8-positions have been converted into 1,4- (LXIII) and 1,5-benzoxazepinones (LXIV) (*U. T. Bhalerao* and *G. Thyagarajan*, Canad. J. chem., 1968, **46,** 3367):

(LXIII) (LXIV)

Reduction of 7-methoxy- and 7-methoxy-3-methyl-chromanone oxime with lithium tetrahydridoaluminate gives only the rearrangement product, the 2,3,4,5-tetrahydro-1,5-benzoxazepinone, but 7-methoxy-2-methyl- and 7-methoxy-2-isopropyl-chromanone oxime yields a mixture of the rearrangement product and the normal reduction product, a primary amine (*S. Ito*, Bull. chem. Soc. Japan, 1970, **43,** 1824).

Chroman-4-ones when fused with aluminium chloride yield hydroxyindanones, chroman-4-one, for example, giving 7-hydroxyindanone (*Loudon* and *Razdan*, *loc. cit.*).

3-(2-Hydroxybenzylidene)chroman-2,4-dione (LXV) and a small amount of monohydroxy biscoumarin (LXVI) are obtained on boiling 2-hydroxybenzaldehyde and 4-hydroxycoumarin in alcohol (*J. Riboulleau et al.*, Bull. Soc. Chim. Fr., 1970, 3138):

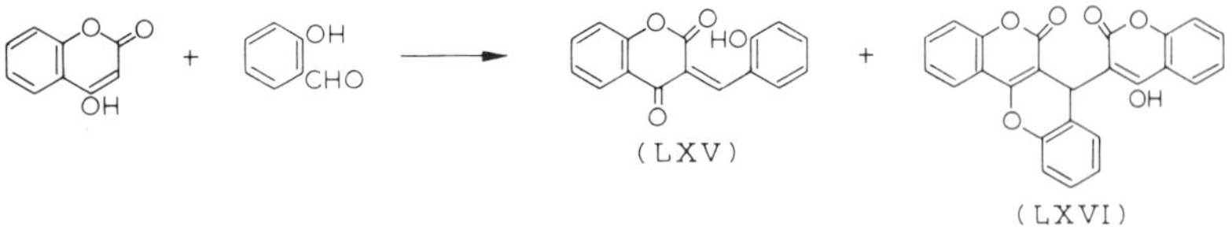

(LXV)

(LXVI)

γ-Pyronochromanones (LXVII) have been prepared by condensing 7-hydroxy-2-methylchromone, 7-hydroxyflavone and 7-hydroxy-2-methylisoflavone with acrylonitrile, hydrolysing the nitrile group, and cyclising (*J. R. Merchant, J. R. Patell* and *S. M. Thakkar*, Current Sci., 1971, **40,** 353):

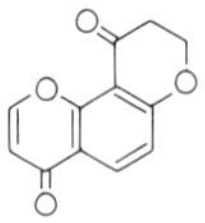

(LXVII)

Cyclohexenonyl crotonates give 5,6,7,8-tetrahydrochroman-2,5-diones (LXVIII) in alkaline solution and 5,6,7,8-tetrahydrochroman-4,5-diones (LXIX) under acid conditions (*S. Gelin* and *B. Chantegrel*, Compt. rend. Ser. C, 1971, **273,** 635):

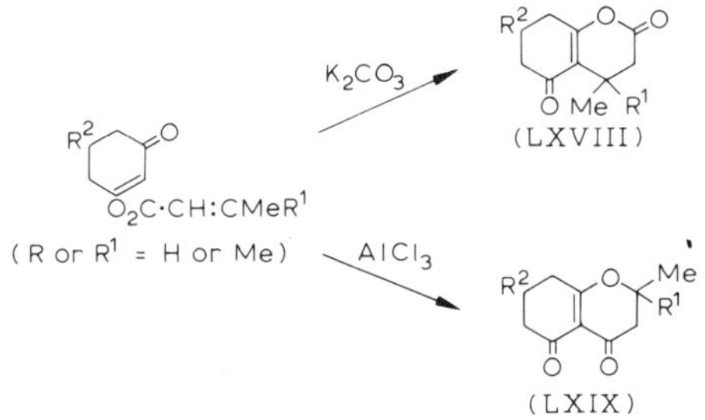

$O_2C{\cdot}CH{:}CMeR^1$

(R or R^1 = H or Me)

K_2CO_3

(LXVIII)

$AlCl_3$

(LXIX)

Rosellinic acid, $C_{11}H_{10}O_5$, m.p. 206–208°, *methyl ester*, m.p. 137–138°, *methyl ether*, m.p. 202–203°, 2,4-*dinitrophenylhydrazone*, m.p. 138°, has been isolated from the culture filtrate of *Rosellinia necatrix*. It is an inhibitor for plant growth and fusion with alkali yields 4,5-dihydroxyisophthalic acid. 8-Hydroxy-2-methylchroman-4-one-6-carboxylic acid has been proposed for its structure (*Y.-S. Chen*, Agric. biol. Chem., 1964, **28,** 431):

Rosellinic acid

alkali fusion

Papuanic acid, $C_{25}H_{36}O_6$, m.p. 73–77°, has been obtained from the bark resin of *Calophyllum papuanum* Lauterb. Its structure and absolute stereochemistry have been established by a combination of degradative, synthetic, and X-ray crystallographic methods. **Isopapuanic acid,** has also been obtained from the same source.

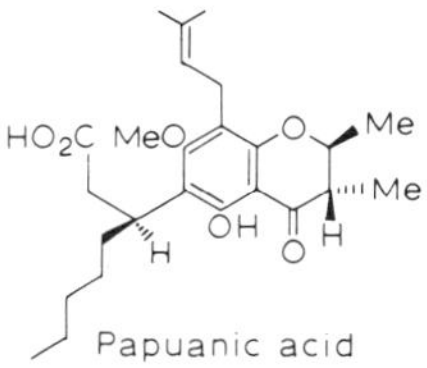

Papuanic acid 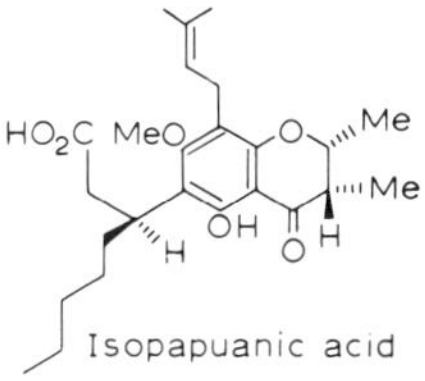 Isopapuanic acid

The 2-phenylchroman-4-ones or flavanones comprise an important group of compounds (p. 268) and in rotenone and related compounds there is a chromano[3,4-*b*] chromanone nucleus.

*(ii) Chromanochromanones, rotenone and related substances**

An arrow poison, *ipoh*, used by the natives of Malaya and a fish poison, *tuba*, employed by the Javanese contain an active principle rotenone or tubatoxin. The plants from which such fish poisons are derived belong to the botanical family *Fabaceae* or *Papilionaceae* and the genera *Derris, Tephrosia,* and *Lonchocarpus.* The derris of commerce comes from Indonesia, Malaya, and the Philippine Islands, while *Lonchocarpus*, known as *cube* or *timbo*, comes from Brazil and Peru. The toxicity of these plants is due to the presence of structurally related substances of which the best known and most thoroughly investigated is rotenone.

The value of these fish poisons rests on their toxicity when introduced into the blood stream and their comparative innocuousness when taken orally. Their commercial value, however, is based on the use of rotenone and derris root as insecticides.

Rotenone and the other constituents such as deguelin, toxicarol, etc., of the rotenone-bearing plants are chromanochromanone derivatives containing the nucleus LXX and consequently they exhibit many properties and characteristics in common. A knowledge of the structure and chemistry of rotenone accordingly leads to an understanding of the whole series (p. 264).

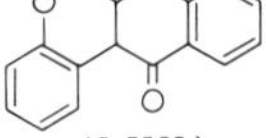

(LXX)

* *L. Feinstein* and *M. Jacobson*, Fortschrit, org. Naturstoffe, **10,** 436 (Springer Verlag, Berlin, 1953).

Rotenone, $C_{23}H_{22}O_6$, is a colourless, dimorphous substance, m.p. 163° (*A. Butenandt* and *F. Hildebrandt*, Ann., 1930, **477,** 245; *E. L. Gooden* and *C. M. Smith*, J. Amer. chem. Soc., 1935, **57,** 2616), $[\alpha]_D$ −105° (acetone) and −226° (benzene) (*H. A. Jones* and *Smith, ibid.*, 1930, **52,** 2554), and is obtained from the tropical plant *Derris elliptica.* It is one of the most powerful vegetable poisons known for fish and insects. The accepted formula was advanced independently by several groups of workers as the result of the thorough examination of the degradation products of rotenone (*F. B. LaForge* and *H. L. Haller*, *ibid.*, 1932, **54,** 810; *Butenandt* and *W. McCartney*, Ann., 1932, **494,** 17; *S. Takei et al.*, Ber., 1932, **65,** 1041; *A. Robertson*, J. chem. Soc., 1932, 1380; reviewed by *LaForge et al.*, Chem. Reviews, 1933, **12,** 181 and *Haller*, J. chem. Educ., 1942, **19,** 315). The numbering in LXXIa is that of *R. S. Cahn et al.*, (J. chem. Soc., 1938, 515), in LXXIb that based on rotexen nucleus, *e.g.*, see *L. Crombie*, *P. W. Freeman* and *D. A. Whiting* (*ibid.*, Perk. I, 1973, 1277) and in LXXIc as found in the Ring Index:

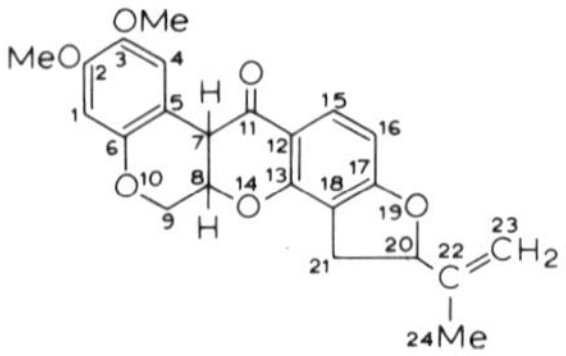

(LXXI a)

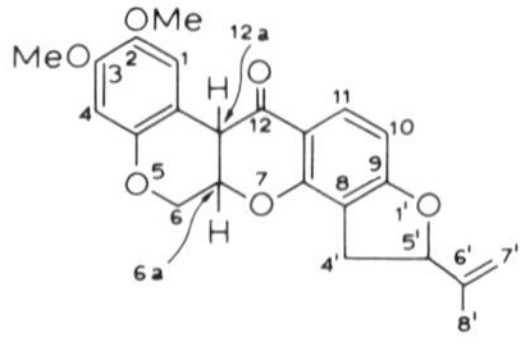

(LXXI b)

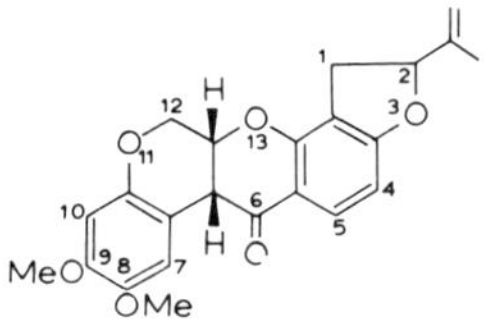

(LXXI c)

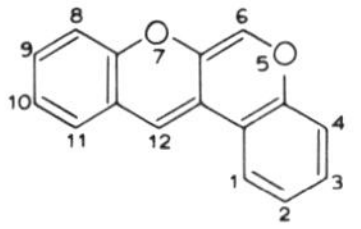

Rotoxene nucleus

Rotenone contains two methoxyl groups, an oxo group, three inactive oxygen atoms present in ether linkages, and one ethylenic bond. The carbonyl group can enolise since rotenone with sodium acetate and acetic anhydride forms an acetate which is reduced by catalytic hydrogenation to deoxyrotenone (*L. E. Smith* and *LaForge*, J. Amer. chem. Soc., 1932, **54,** 2996):

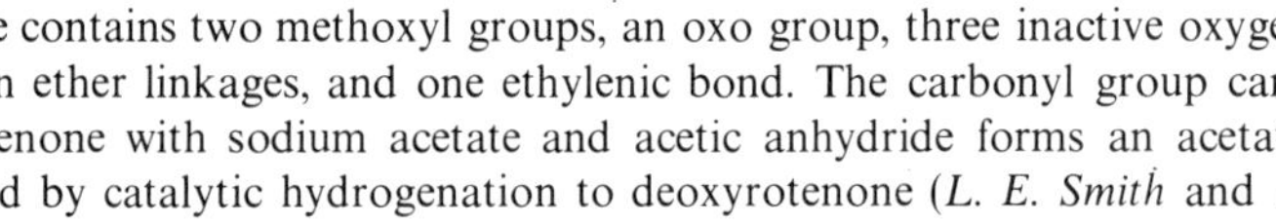

The formula shows three chiral centres at $C_{(6a)}$, $C_{(12a)}$, and $C_{(5' \text{ or } 2)}$ (LXXIb and LXXIc) and this accounts satisfactorily for the optical activity of rotenone and some of its derivatives such as dehydrorotenone and isorotenone and absence thereof in others including dehydroisorotenone.

When rotenone is cleaved by means of ethanolic potassium hydroxide (*T. Kariyone et al.*, J. pharm. Soc. Japan, 1924, **514,** 1094; 1925, **518,** 376; *Takei* and *M. Koide*, Ber.,

1929, **62,** 3030) a monobasic phenolic acid is obtained, **tubaic acid**, $C_{12}H_{12}O_4$, m.p. 128° (*Haller* and *LaForge*, J. Amer. chem. Soc., 1931, **53,** 4460). The acid is catalytically reduced to *dihydrotubaic aicd* (LXXII), m.p. 168° (*Takei et al.*, Ber., 1930, **63,** 1369). Tubaic acid can be decarboxylated to a phenol and with diazomethane yields a *methyl ether*, m.p. 78°, but with difficulty since it is a di-*ortho*-substituted phenol. When hydrogenated in the presence of a platinum oxide catalyst it affords *tetrahydrotubaic acid, 2,4-dihydroxy-3-isoamylbenzoic acid* (LXXIII), m.p. 206° (decomp.) (*Haller* and *LaForge*, *loc. cit.*), which when heated at its melting point is decarboxylated to

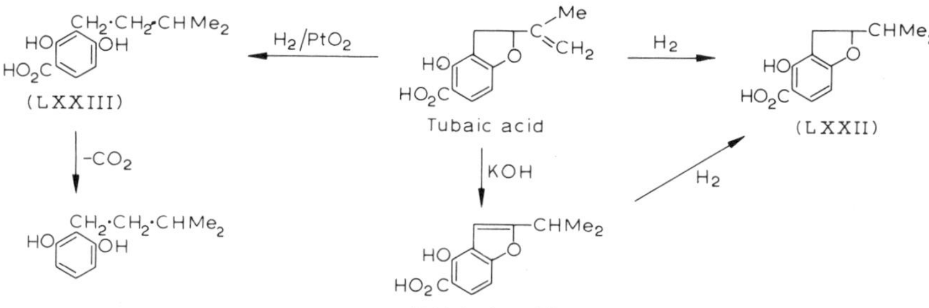

(LXXIII) Tubaic acid (LXXII) Isotubaic acid

tetrahydrotubanol, m.p. 85°, which was shown to be *2-isoamylresorcinol* by synthesis (*Haller* and *LaForge*, J. Amer. chem. Soc., 1932, **54,** 1988, 4755). A migration of the ethylenic linkage occurs when tubaic acid is fused with alkali and *isotubaic acid, rotenic acid*, m.p. 183° (rapid heating) is obtained. Due to the loss of an asymmetric carbon atom rotenic acid, unlike tubaic acid, is therefore optically inactive. This is one of the characteristic properties of rotenone and its derivatives. Rotenone and all its degradation products which contain the dihydrofuran ring with an attached unsaturated side-chain are isomerised by sulphuric acid in acetic acid, rotenone thus yielding isorotenone, m.p. 178° (*Takei et al.*, Ber., 1928, **61,** 1003; 1930, **63,** 508; *Butenandt* and *McCartney*, *loc. cit.*; *Haller*, J. Amer. chem. Soc., 1932, **54,** 2126). This accounts for the conversion of isorotenone into isotubaic acid. When hydrogenated isotubaic acid yields dihydrotubaic acid (LXXII);

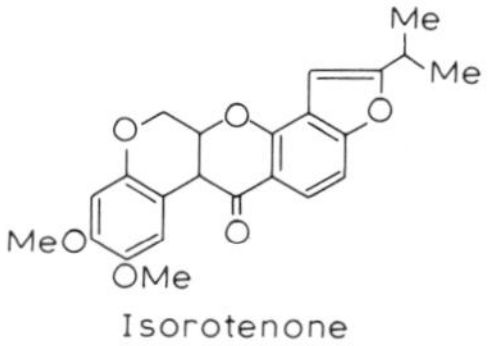

Isorotenone

A feature of rotenone and other chromanochromanones is the ease with which they undergo dehydrogenation. Rotenone with a variety of oxidising agents and particularly iodine, sodium acetate, and ethanol (see, however, below) yields *dehydrorotenone* (LXXIV), a yellow substance, m.p. ~225° (*Butenandt*, Ann., 1928, **464,** 253). This change results in the loss of chiral centres at $C_{(6a)}$ and $C_{(12a)}$, but the asymmetric atom

$C_{(2)}$ (LXXIc) remains and the compound is optically active. When dehydrorotenone, however, is isomerised to dehydroisorotenone all chirality is lost. The newly formed double bond cannot be hydrogenated catalytically (*Butenandt* and *Hildebrandt, loc. cit.*). Hydrogenation of dehydrorotenone with a platinum catalyst is accompanied by fission of the oxidic linkage to give optically inactive dehydrodihydrorotenoic acid (*LaForge* and *Smith*, J. Amer. chem. Soc., 1930, **52,** 1091; *Haller* and *LaForge, ibid.*, 1931, **53,** 3426). The presence of an isopropylidene grouping has been confirmed by n.m.r. data (*G. Büchi et al.*, J. chem. Soc., 1961, 2843):

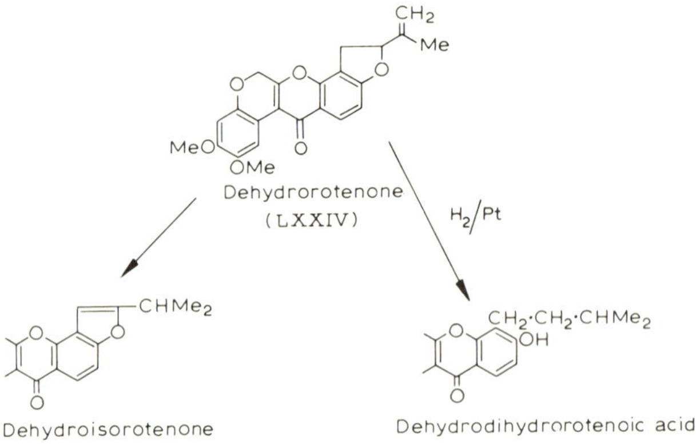

Dehydrorotenone (LXXIV)

H_2/Pt

Dehydroisorotenone

Dehydrodihydrorotenoic acid

Dehydrorotenone (LXXIV) unlike rotenone is not readily attacked by alkali, but with ethanolic potassium hydroxide both condensed oxygen rings undergo hydrolytic fission to give **derrisic acid,** m.p. 152° (*Butenandt*, Ann., 1928, **464,** 253; *Takei et al.*, Ber., 1931, **64,** 248; *Butenandt* and *Hildebrandt, loc. cit.*; formula, *La Forge* and *Haller*, J. Amer. chem. Soc., 1932, **54,** 810):

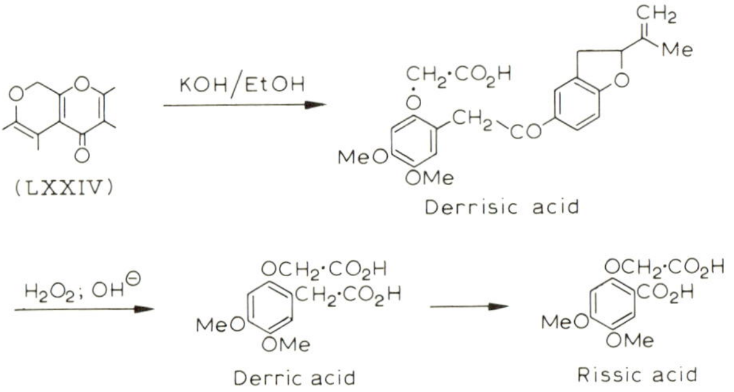

(LXXIV)

KOH/EtOH

Derrisic acid

H_2O_2; $OH^{\ominus}$

Derric acid

Rissic acid

The acid is oxidatively degraded by alkaline hydrogen peroxide to **derric acid,** m.p. 171°, *dimethyl ester*, m.p. 66°, which is oxidised by potassium permanganate to **rissic aicd,** m.p. 262°, *dimethyl ester*, m.p. 84° (*LaForge* and *Smith, ibid.*, 1930, **52,** 2878; *Takei*, Ber., 1931, **64,** 248, 1000; *Butenandt* and *McCartney, loc. cit.*). The structures of these acids were confirmed by synthesis (*Robertson*, J. chem. Soc., 1932, 1380). Derrisic acid was synthesised along with dehydrorotenone by the interaction of sodium derritol and ethyl chloroacetate (*LaForge et al.*, J. Amer. chem. Soc., 1931, **53,** 4400; *Takei et al.*, Ber., 1932, **65,** 1041).

Treatment of rotenone with iodine and sodium acetate gives as the main product not dehydrorotenone but an acetyl derivative, acetylrotenolone, which on hydrolysis yields **rotenolone-I,** m.p. 140–141° (*LaForge* and *Smith*, J. Amer. Chem. Soc., 1930, **52,** 1091, 3603). Oxidation of rotenone by air in ethanolic sodium hydroxide yields two products, rotenolone-I, and **rotenolone**-II, m.p. 220° (*Takei et al.*, Ber., 1933, **66**, 479; *M. Miyano* and *M. Matsui*, Ber., 1959, **92**, 1438).

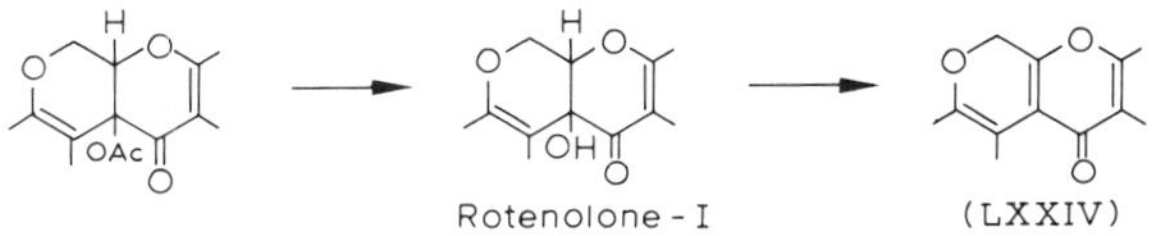

Rotenolone - I (LXXIV)

These compounds are accepted by them as individuals: but because there is an optical centre at 2 (β) (LXXIc) four diastereoisomeric hydroxyketones (LXXVa–d) must be produced instead of the two pairs of enantiomorphs obtained when isorotenone is similarly treated. By direct and indirect means these have been isolated and stereochemically identified (*Crombie* and *P. J. Godin*, J. chem. Soc., 1961, 2861):

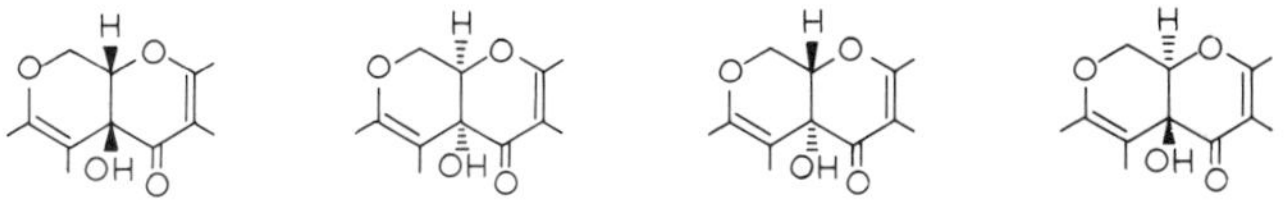

(LXXV a) (LXXV b) (LXXV c) (LXXV d)

Rotenolone-I is dehydrated by ethanolic sulphuric acid to dehydrorotenone (LXXIV) and is reduced by zinc and ethanolic potassium hydroxide to derritol and rotenol probably *via* rotenone. The same two products are obtained when rotenone is cleaved using alkali and zinc.

Derritol, $C_{21}H_{22}O_6$, yellow needles, m.p. 161°, *oxime*, m.p. 191–192°, which contains two fewer carbon atoms than rotenone, is a diphenol. It is soluble in alkali and hence can easily be separated from the alkali-insoluble rotenol. Derritol is formed by fission of the chromanone ring of rotenone followed by dehydration of the resulting secondary alcohol, LXXVI, to a chromene derivative, LXXVII. By hydrolysis and hydrogenationa two-carbon fragment (possibly acetaldehyde) is removed to give derritol *(LaForge* and *Haller*, *loc. cit.).* The structure is supported by the oxidation of derritol with permanganate to 4,5-dimethoxy-2-hydroxybenzoic acid and of derritol methyl ether to tubaic acid and 2,4,5-*trimethoxybenzoic acid*, **asaronic acid,** m.p. 143° (*Smith* and *LaForge*, J. Amer. chem. Soc., 1931, **53,** 3072). The methyl ether with alkaline hydrogen peroxide yields 2,4,5-*trimethoxyphenylacetic acid*, **homoasaronic acid,** m.p. 87° (*Takei et al.*, Ber., 1932, **65,** 279). Steric hindrance prevents the methylation of the second phenolic group of derritol.

Rotenol, $C_{23}H_{24}O_6$, is a colourless monophenol, m.p. 119°, *oxime*, m.p. 187°, which results from the reduction of the intermediate LXXVII. When fused with potassium hydroxide it yields an unexpectedly large quantity of isotubaic acid (*Haller* and *LaForge*, J. Amer. chem. Soc., 1930, **52**, 4505) and when oxidised by hydrogen peroxide it yields **netoric acid,** $C_{12}H_{14}O_5$, m.p. 91–92° (hydrated), 134° (anhyd.) (*Butenandt* and *McCartney*, *loc. cit.*), the structure of which has been confirmed by synthesis (*Robertson* and *G. L. Rusby*, J. chem. Soc., 1936, 212):

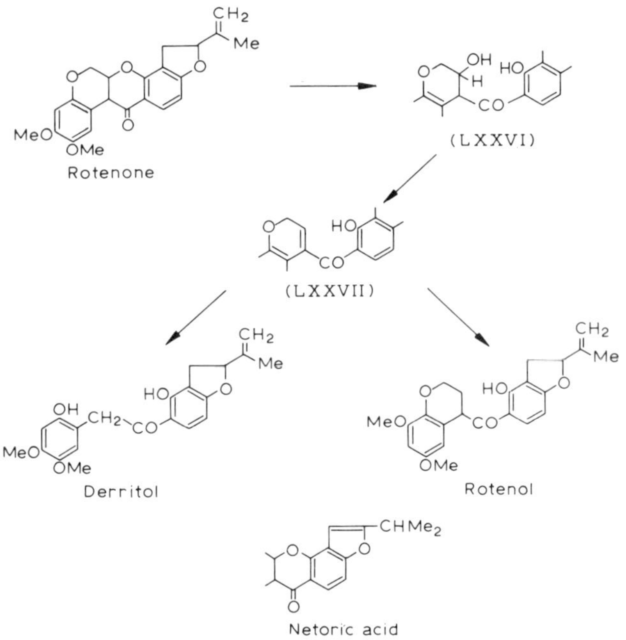

Rotenone

(LXXVI)

(LXXVII)

Derritol

Rotenol

Netoric acid

Rotenone is oxidised by chromic acid to the yellow substance **rotenonone,** m.p. 298°, a coumarin derivative. In the process the two hydrogen atoms at $C_{(6a)}$ and $C_{(12a)}$ are removed and the methylene group at $C_{(12)}$ (LXXIc) is oxidised to carbonyl. Dehydrorotenone likewise yields rotenonone when oxidised.

Rotenonone is broken down almost quantitatively by 20% ethanolic potassium hydroxide to derritol and oxalic acid probably through the oxo form LXXIX of the intermediate LXXVIII (*Takei et al.*, Ber., 1932, z65, 1041). Derritol has been converted into rotenone by interaction with diethyl oxalate in the presence of sodium acetate or the acid chloride of monoethyl oxalate (*LaForge*, J. Amer. chem. Soc., 1932, **54,** 3377). Rotenone is hydrolysed to rotenononic acid, m.p. 250°, probably through the intermediate LXXVIII. **Rotenononic acid,** is lactonised in acid solution to **β-rotenonone,** m.p. 275° (*Takei et al., loc. cit.*), which rearranges in ethanolic potassium hydroxide to rotenonone.

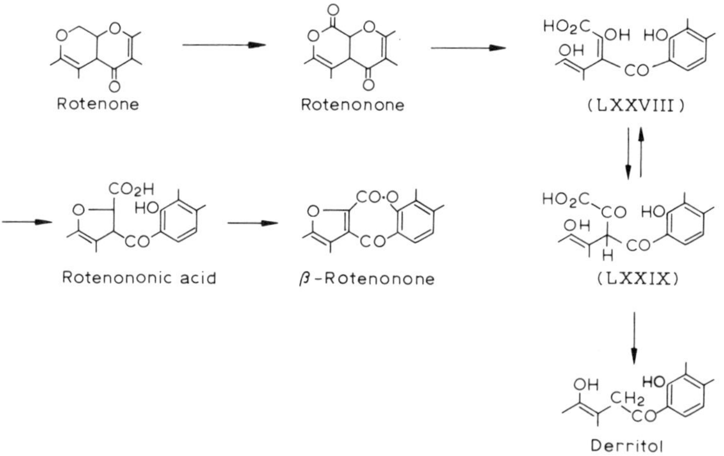

Rotenone → Rotenonone → (LXXVIII) ⇌ (LXXIX) → Derritol

→ Rotenononic acid → β-Rotenonone

Catalytic hydrogenation of rotenone gives *dihydrorotenone*, two hydrogen atoms adding to the isopropenyl group (*LaForge* and *Smith*, J. Amer. chem. Soc., 1929, **51,** 2574; *Takei et al.*, Ber., 1930, **63,** 508). It is dimorphic, m.p. 216° and 164°, the lower melting form being less stable (*LaForge* and *G. L. Keenan*, J. Amer. chem. Soc., 1931, **53,** 4450). Hydrogenation may be accompanied by fission of the hydrofuran ring especially when the reaction is carried out in alkaline solution. In neutral solution dihydrorotenone and **rotenonic acid,** m.p. 209°, are obtained (*Haller*, J. Amer. chem. Soc., 1932, **54,** 2126). Rotenonic acid with sulphuric acid in acetic acid undergoes ring-closure to give a chroman derivative, β-dihydrorotenone. (±)Rotenonic acid has been synthesised (*L. Crombie*, *P. W. Freeman* and *D. A. Whiting*, J. chem. Soc., Perk. I, 1973, 1277):

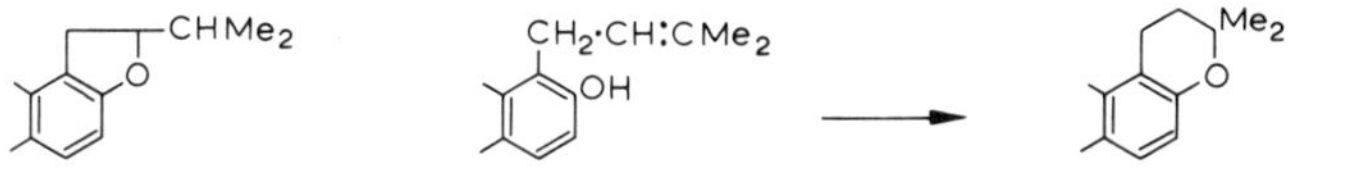

Dihydrorotenone   Rotenonic acid → β-Dihydrorotenene

In ethanol rotenone reacts with hydroxylamine to form the *α-oxime*, m.p. 237°, and in pyridine the *β-oxime*, m.p. 249° (*S. H. Harper*, J. chem. Soc., 1939, 1424). These are dimorphous forms with the higher melting isomer the more stable. The *hydrazone*, m.p. 258°, and *phenylhydrazone*, m.p. 255°, are also known. In alkaline solution, however, rotenone and hydroxylamine yield the *iso-oxime* (LXXX), m.p. 230°, which has phenolic properties (*Butenandt*, Ann., 1928, **464,** 253; 1932, **494,** 25); its structure is supported by spectral data (*Crombie et al.*, J. chem. Soc., 1961, 2876):

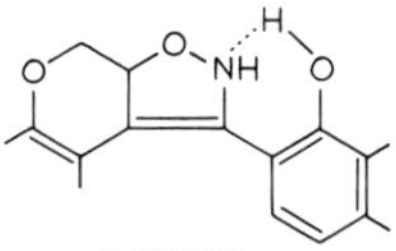

(LXXX)

An *isohydrazone*, m.p. 229°, and *isophenylhydrazone*, m.p. 203°, have also been prepared. Some aspects of the chemistry of the B/C ring system (LXXIc) of rotenone and its natural relatives, particularly the reactions in basic media have been discussed *(idem, loc. cit.)*. The absolute configuration of rotenone has been shown to be (6a*S*, 12a*S*, 2*R*) (LXXIc) *(Büchi et al., loc. cit.)*. The synthesis of rotenoids from 2′-hydroxyisoflavonoids has been developed and the intermediacy of the latter in the biosynthesis of rotenoids has been demonstrated. The synthesis involves reaction of 2′-hydroxyisoflavone with dimethylsulphoxonium methylide followed by rearrangement of the resultant vinyl coumaranone; the synthesis of isorotenone and rotenone from the corresponding 2′-hydroxyisoflavones have been described *(Crombie, Freeman* and *Whiting, loc. cit.)*:

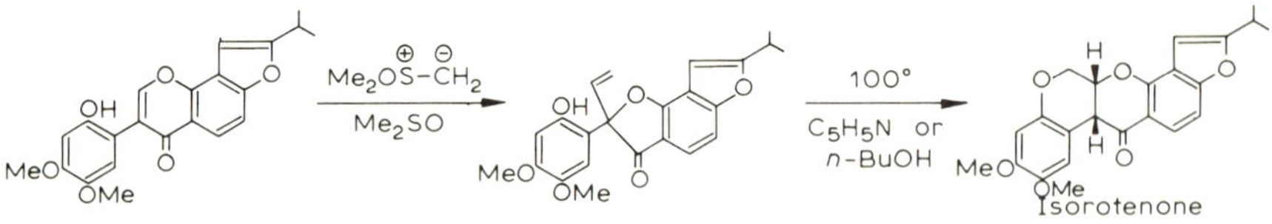

$Me_2\overset{\oplus}{S}O-\overset{\ominus}{C}H_2$, Me_2SO → ; 100°, C_5H_5N or n-BuOH →

Isorotenone

An absolute synthesis of dihydrorotenone has also been described (*Miyano* and *Matsui*, Bull. chem. Soc. Japan., 1958, **31**, 397; Ber., 1959, **92**, 1438, 2487; Bull. agric. chem. Soc. Japan, 1960, **24**, 540).

Naturally occurring substances related to rotenone (see *Haller et al.*, Chem. Reviews, 1942, **30**, 33; *Crombie*, Fortsch. Chem. org. Naturstoffe, 1963, **21**, 275) are deguelin, toxicarol, tephrosin, elliptone, malaccol and sumatrol. They give the Durham test (Ann. App. Biol., 1923, **10**, 5), an intense red colour with a drop of concentrated nitric acid which in contact with a drop of aqueous ammonia changes to a transient green or blue. The test is more sensitive in acetone (*H. A. Jones* and *C. M. Smith*, Ind. Eng. Chem. Anal., 1933, **5**, 75).

Extensive n.m.r. studies of major rotenoids have verified the *cis* B/C ring fusion of rotenone, confirmed the structure of the reduction–dehydration product of rotenone, provided considerable evidence regarding the preferred conformations of rotenoids and revealed an array of long-range couplings (*D. G. Carlon, D. Weisleder* and *W. H. Tallent*, Tetrahedron, 1973, **29**, 2741).

Deguelin, $C_{23}H_{22}O_6$, m.p. 171°, is obtained in the optically inactive form from *Cracca (Tephrosia) toxicaria, Derris elliptica,* and *Lonchocarpus nicou* (cube) (*J. J. Boam et al.*, J. Soc. chem. Ind., 1937, **56**, 91T). It is isomeric with rotenone from which it differs only in possessing a chromene ring in place of a hydrofuran ring (*E. P. Clark*, J. Amer. chem. Soc., 1932, **54**, 3000). With potassium ferricyanide it is readily

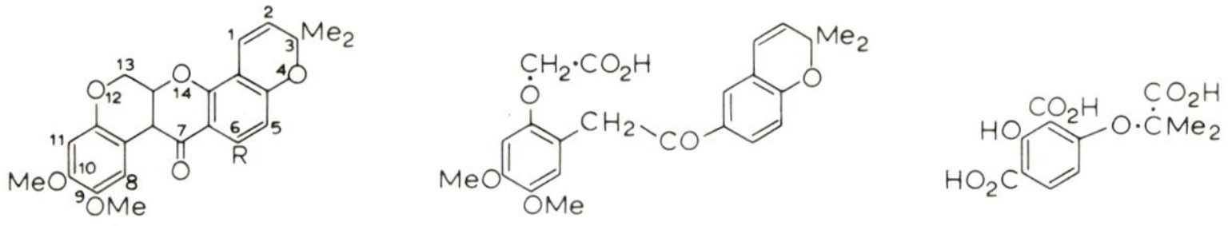

Deguelin (R = H)
α-Toxicarol (R = OH)

Deguelinic acid

Nicouic acid

dehydrogenated to *dehydrodeguelin*, $C_{23}H_{20}O_6$, m.p. 233° (cf. dehydrorotenone, LXXIV,) which with ethanolic potassium hydroxide yields *deguelinic* acid, m.p. 189°, analogous to derrisic acid (p. 260) (*Clark, ibid.*, 1931, **53**, 313). Oxidation of this acid with hydrogen peroxide gives derric acid. One moeity of the deguelin molecule is therefore identical with that of rotenone. Evidence for the 2,2-dimethylchromene nucleus in the other half of the molecule is provided by the oxidation of dehydrodeguelin with permanganate which yields, in addition to 2-hydroxy-4,5-dimethoxybenzoic acid and rissic acid, the tricarboxylic acid **nicouic acid,** m.p. 196° (*Clark, ibid.*, 1931, **53,** 2007; 1932, **54,** 300). The presence of the chromene nucleus is shown decisively by the catalytic hydrogenation of dehydrodeguelin to dehydrodihydrodeguelin (*Clark, ibid.*, 1931, **53,** 2369):

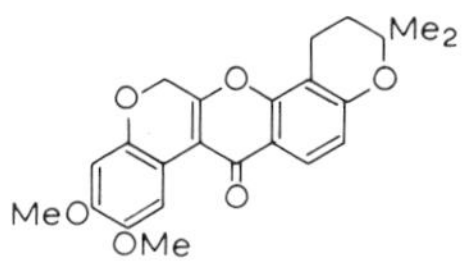

Dehydrodihydrodeguelin

Hydrogenation of deguelin gives β-dihydrorotenone.

Oxidation of racemic deguelin by air in alkaline solution gives two isomeric 7a-hydroxydeguelins [12a-hydroxy if numbered as (LXXXIb)], **tephrosin** (*cis*-B/C fused), $C_{23}H_{22}O_7$, m.p. 198°, *acetyl* deriv., m.p. 200°, and **isotephrosin** (*trans*-B/C fused), m.p. 252°; both are easily dehydrated to dehydrodeguelin (*Crombie* and *Godin, loc. cit.*). It is reported that tephrosin and isotephrosin have been obtained from *Tephrosia vogelii* and *Lonchocarpus nicou* (*Clark*, J. Amer. chem. Soc., 1931, **53,** 729), and the latter from Peruvian cube root (*Clark* and *H. V. Claborn, ibid.*, p. 4454), but only after treatment with alkali and exposure to air. This along with the lack of optical activity in the extractives suggests that they are merely artefacts. The structure of tephrosin is indicated by its oxidation using potassium permanganate to *tephrosin dicarboxylic acid* (LXXXI), $C_{23}H_{22}O_{11}$, m.p. 221° (*Clark, loc. cit.*) which when heated in diphenyl ether yields *tephrosin monocarboxylic acid* (LXXXII), $C_{19}H_{14}O_8$, m.p. 269°, and hydroxyisobutyric acid, $Me_2C(OH)\cdot CO_2H$ (*Clark*, J. Amer. chem. Soc., 1932, **54,** 3000):

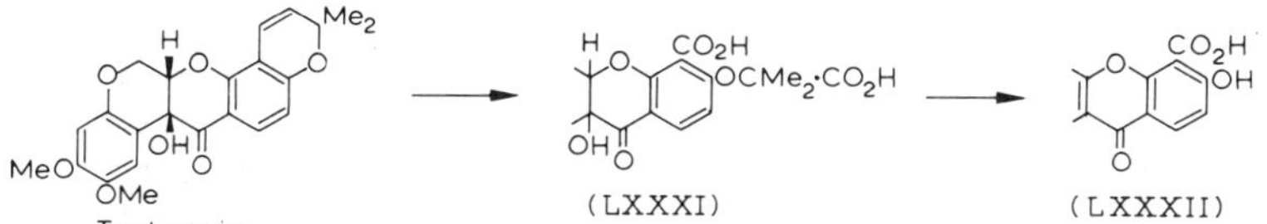

Tephrosin (LXXXI) (LXXXII)

A third isomer, *hydroxydeguelin-C*, m.p. 203°, is obtained by oxidising deguelin with alkaline hydrogen peroxide (*LaForge* and *Haller, ibid.*, 1934, **56**, 1620). Unlike tephrosin and isotephrosin it is unchanged by dehydrating agents

Deguelin has the same absolute configuration at positions 6a (7a) and 12a (13a) as rotenone (LXXIb and LXXIc) (*C. Djerrassi, W. D. Ollis* and *R. C. Russell*, J. chem. Soc., 1961, 1448). n.m.r. Data (*Carlson, Weisleder* and *Tallent, loc. cit.*). The synthesis of deguelin has been investigated (*H. Fukami et al.*, Bull. agric. chem. Soc., Japan,

1960, **24,** 327; Agric. biol. Chem., 1961, **25,** 252).

α-**Toxicarol**, 6-**hydroxydeguelin** [11-*hydroxydeguelin* if numbered as (LXXIb)], $C_{23}H_{22}O_7$, m.p. 125–127°, $[\alpha]_D^{20}$ −66° (benzene), *acetate*, m.p. 158° (*Clark*, J. Amer. chem. Soc., 1930, **52,** 2461; *Butenandt* and *G. Hilgetag*, Ann., 1933, **506,** 158; *S. W. George et al.*, J. chem. Soc., 1937, 1535), is isolated from *Tephrosia toxicaria* (*F. Tattersfield* and *J. T. Martin*, J. Soc. chem. Ind., 1937, **56,** 77T), *Derris elliptica* (*Cahn* and *Boam*, *ibid.*, 1935, **54,** 37T) and *Derris, malaccensis* (*Harper*, J. chem. Soc., 1939, 812). (±)α-*Toxicarol*, m.p. 231–232° (*Jones* and *J. W. Wood*, J. Amer. chem. Soc., 1941, **63,** 1760). α-Toxicarol differs from rotenone and deguelin in yielding with ethanolic potassium hydroxide a diphenol, **apotoxicarol**, m.p. 247°, *triacetate*, m.p. 206° (*R. G. Heyes* and *Robertson*, J. chem. Soc., 1935, 681):

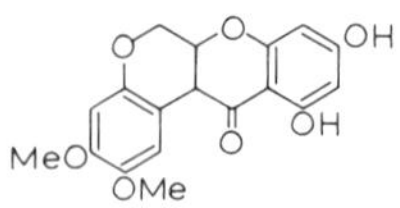

Apotoxicarol 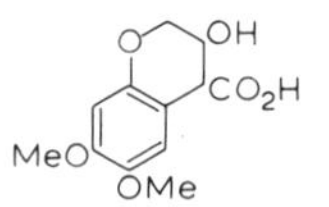 Hydroxynetoric acid 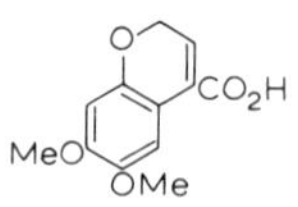 Toxicaric acid

Further oxidation with alkaline hydrogen peroxide gives **hydroxynetoric acid**, $C_{12}H_{14}O_6$, m.p. 189°, and **toxicaric acid,** $C_{12}H_{12}O_5$, m.p. 212° (decomp.), which on catalytic hydrogenation yields netoric acid (p. 262).

When toxicarol is heated with potassium carbonate in acetone it isomerises to β-**toxicarol,** the racemic substance has m.p. 169–170° (*Cahn et al.*, J. chem. Soc., 1938, **513,** 734; *Crombie et al.*, *ibid.*, 1961, 2876). β-Toxicarol is isomeric with the α-compound, but differs in the points of attachment of the dimethylpyran residue; only the former gives a positive reaction with Gibbs's reagent (*Crombie* and *R. Peace*, *ibid.*, 1961, 5445): n.m.r. data (*Carlson*, *Weisleder* and *Tallent*, *loc. cit.*):

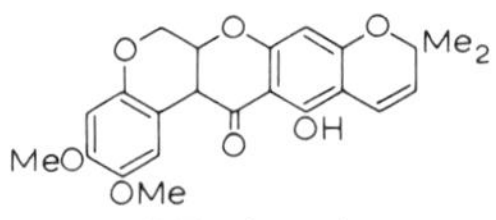

β-Toxicarol

Elliptone, $C_{20}H_{16}O_6$, has been obtained in the optically inactive form from *Derris elliptica* (*T. A. Buckley*, J. Soc. chem. Ind., 1936, **55,** 285T) and in the optically active form (*Harper*, J. chem. Soc., 1939, 1099). It has m.p. 178.5–179.5° (*Harper*, *ibid.*, 1941, 878), $[\alpha]_D^{20}$ −18° (benzene), +55° (acetone), and forms two *oximes*, α-, m.p. 222°, and β-, m.p. 236°. (±) *Elliptone* has m.p. 180–181°, and forms two *oximes* α-, m.p. 259°, and β-, m.p. 261°: n.m.r. data (*Carlson*, *Weisleder* and *Tallent*, *loc. cit.*):

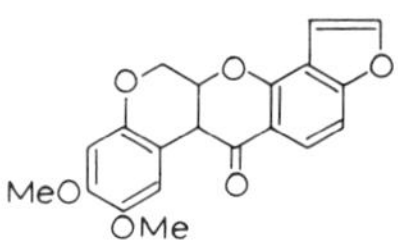

Elliptone

Malaccol, $C_{20}H_{16}O_7$, yellow prisms, m.p. 249–250° (244–246°), after partially melting at 225–227° and resolidifying (*Harper*, J. chem. Soc., 1941, 878), $[\alpha]_D^{18}$ +190° ($CHCl_3$), is obtained from *Derris malaccensis* (Kinta type) (*T. M. Meyer* and *D. R. Koolhaas*, Rec. Trav. chim., 1939, **58,** 207; *Harper*, J. chem. Soc., 1940, 309; 1941, 878). Natural (+)malaccol (α-**malaccol**) is the angular form, and when it is boiled with ethanolic sodium acetate it gives a (±) product, (±)β-malaccol, m.p. 249°, which is the linear form *(Crobie* and *Peace, loc. cit.)*:

α-Malaccol → β-Malaccol

Optical rotatory dispersion correlations with natural rotenone indicate the stereochemistry shown above for α-malaccol.

Sumatrol, $C_{23}H_{22}O_7$, m.p. 199–200° *(Harper, loc. cit.)* $[\alpha]_D^{20}$ −182° (benzene), *oxime*, m.p. 245–247°, *acetyl* deriv., m.p. 218–219°, is obtained from *Derris* species (*Cahn* and *Boam*, J. Soc. chem. Ind., 1935, **54,** 42T). It is isomeric with tephrosin and toxicarol and is phenolic giving a reddish-brown colour with ethanolic ferric chloride (*Robertson et al.*, J. chem. Soc., 1939, 1601). The stereochemistry of (−)sumatrol has been indicated by optical rotatory dispersion correlations with natural rotenone *(Crombie* and *Peace, loc. cit.)*: n.m.r. data *(Carlson, Weisleder* and *Tallent, loc. cit.)*:

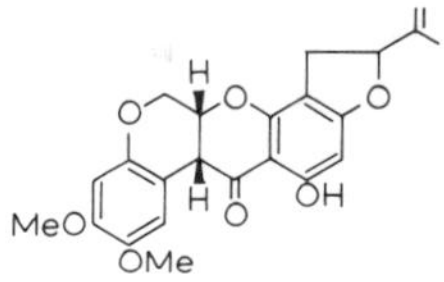

Sumatrol

Munduserone, $C_{19}H_{18}O_6$, m.p. 162°, $[\alpha]_D$ +103° ($CHCl_3$), has been isolated from *Mundulea sericea* (*N. Finch* and *Ollis*, Proc. chem. Soc., 1960, 176), (±)*munduserone*, m.p. 171–172°, and its structure confirmed by synthesis (*J. R. Herbert*, *Ollis* and *Russell*, *ibid.*, p. 177; *H. Omokawa* and *K. Yamashita*, Agric. biol. Chem., 1973, **37,** 195, 1717; *Crobie*, *Freeman* and *Whiting*, *loc. cit.*). Compounds related to munduserone have also been synthesised and their stereochemistry discussed (*D. J. Adams*, *Crombie* and *Whiting*, J. chem. Soc., C, 1966, 542):

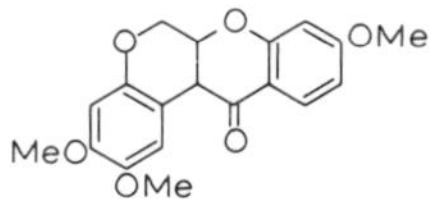

Munduserone

Pachyrrhizone, $C_{20}H_{14}O_7$, m.p. 272°, $[\alpha]_D^{20}$ +100° ($CHCl_3$), *oxime*, m.p. 279°, *phenylhydrazone*, m.p. 309° (decomp.), has been isolated from *Pachyrrhizus erosus* seeds; (±)*pachyrrhizone*, m.p. 265°; *dehydropachyrrhizone*, m.p. 259–261° (*H. Bickel* and *H.*

Schmid, Helv., 1953, **36,** 664). Its stereochemistry has been discussed (*Djerrassi, Ollis* and *Russell, loc. cit.)* and (±)pachyrrhizone has been synthesised (*M. Uchiyama* and *H. Shimotori*, Agric. biol. Chem., 1973, **37,** 1227):

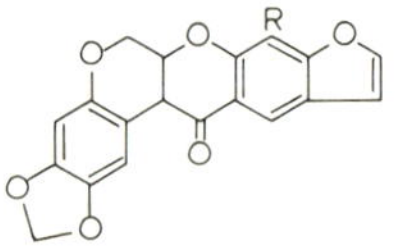

Pachyrrhizone (R = OMe)
Dolineone (R = H)

Dolineone (Dolichone), $C_{19}H_{12}O_6$, m.p. 233–235°, $[\alpha]_D^{20}$ +135° ($CHCl_3$), has been obtained from the roots of *Neorautanenia pseudopachyrrhiza*, Harms, and its structure has been determined by spectral methods (*Crombie* and *Whiting*, J. chem. Soc., 1963, 1569).

Millettone, $C_{22}H_{18}O_6$, cream needles, m.p. 180–181°, has been isolated from the seeds of *Millettia dura* (Dunn) along with (−)**millettosin,** $C_{22}H_{18}O_7$, and other known rotenoids and new isoflavones (p. 217). Millettone has also been obtained from the root of Jamaican Dogwood, *Piscidia erythrina L.* (*C. P. Falshaw et al.*, Tetrahedron, 1966, suppl. **7,** 333). Millettone gives a positive Durham test and contains no methoxy groups. Consideration of its n.m.r. spectrum led to its proposed structure and this was confirmed by chemical means. Catalytic hydrogenation of millettone gives dihydromilletone, which on treatment with iodine and anhydrous potassium acetate in boiling ethanol yields *dehydrodihydromillettone*, pale yellow needles, m.p. 234°. On similar treatment, millettone affords *dehydromillettone*, m.p. 278°. Oxidation of dehydrodihydromillettone with alkaline hydrogen peroxide gives dihydro-β-tubaic acid (LXXIII) (*Ollis, C. A. Rhodes* and *I. O. Sutherland*, Tetrahedron, 1967, **23,** 4741):

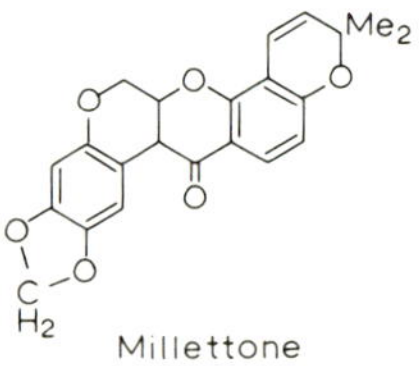

Millettone 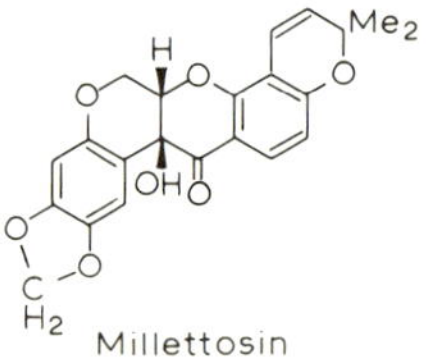   Millettosin 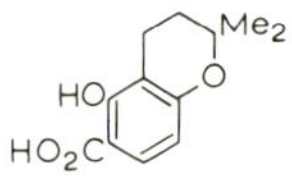   (LXXXIII)

(g) Flavanones, 2-phenylchroman-4-ones, 2,3-dihydro-2-phenyl-4H-benzo-[b]pyran-4-ones

(i) Synthesis

(*1*) Ring-closure of 2-hydroxychalcones with boiling ethanolic (*St. von Kostanecki* and *W. Szabranski*, Ber., 1904, **37,** 2634) or methanolic hydrogen chloride (*M. Vandewalle* and *A. Vandendooren*, Bull. Soc. chim. Belges, 1962, **71,** 131), alcoholic sulphuric acid (*S. C. Datta, V. V. S. Murti* and *T. R. Seshadri*, Indian J. chem., 1971, **9,** 614), dilute hydrochloric acid (*T. A. Geissman* and *R. O. Clinton*, J. Amer. chem. Soc., 1946, **68,** 697), dilute sulphuric acid (*G. Zemplén* and *R. Bognár*, Ber., 1942, **75,** 1043), or preferably

1.5% aqueous sodium hydroxide (*A. Löwenbein*, *ibid.*, 1924, **57**, 1515). By this last method flavanone is obtained in 80% yield.

2-Hydroxyacetophenone and benzaldehyde in the presence of 0.1 *M* NaOH give flavanone (*L. Reichel* and *K. Müller*, *ibid.*, 1941, **74**, 1741) whereas the same reactants with more concentrated alkali give only the chalcone. High hydroxyl ion concentration favours ring-fission of the flavanone to the chalcone, whereas low hydroxyl ion concentration favours ring-closure of the chalcone:

2′,4′-Dihydroxy-α-methyl- and 2′,4′-dihydroxy-α-ethyl-chalcone under controlled conditions may be ring-closed to give either *cis*- or *trans*-3-alkylflavanones (*M. Suzuki*, Science Repts. Tōhoku Univ., 1956, **39**, 182; C.A., 1957, **51**, 7364; *S. Fujise et al.*, Nippon Kagaku Zasshi, 1956, **77**, 109; *Suzuki et al.*, *ibid.*, 1969, **90**, 401). The cyclisation of α-methylchalcones with alkali gives in good yield the corresponding 2,3-*trans*-3-methylflavanones. Conversion of the latter compounds into their oximes and subsequent treatment with sodium bisulphite followed by dilute hydrochloric acid at 0° causes epimerisation at the 3-position and results in the formation of the 2,3-*cis*-3-methylflavanones; n.m.r. data has been reported (*D. M. X. Donnelly et al.*, Tetrahedron Letters, 1970, 1333) and the stereochemistry of the intermediate 4-hydroxyiminoflavans has been discussed (*idem*, Tetrahedron, 1972, **28**, 2545). The stereoisomers of 7-hydroxy-3-methyl- (*Suzuki*, Nippon Kagaku Zasshi, 1955, **76**, 1389; 1961, **82**, 888), 3-ethyl-7-hydroxy-, and 3-ethyl-5,7-dihydroxy-flavanone have been prepared (*Fujise*, *Suzuki* and *K. Shinoda*, C.A., 1957, **51**, 8084). 5′-*sec*- and 5′-*tert*-alkyl-2′,4′-dihydroxychalcones have reen prepared from cinnamic acid and alkylresorcinols and converted to flavanones (*S. P. Starkov* and *A. I. Panasenko*, Izv. Vyssh. Ucheb. Zaved., Khim. Khim. Tekhnol., 1972, **15**, 870). 6-Hydroxyflavanones have been synthesised from 2,5-dihydroxyacetophenone and the appropriate aldehydes (*Fujise et al.*, Nippon Kagaku Zasshi., 1954, **75**, 435); and 6-benzoyl-8-methylflavanones from 5-benzoyl-2-hydroxy-3-methylacetophenone and the appropriate aldehyde (*K. C. Amin* and *G. C. Amin*, J. Indian chem. Soc., 1961, **38**, 391). 6-Hydroxy-5-methoxy- and 5,6-dihydroxy-flavanone have been prepared (*J. Chopin et al.*, Compt. rend., 1960, **251**, 1021). Cinnamic acid condenses with resorcinol in chloroform in the presence of boron trifluoride to give the chalcone (80%), which on treatment with sodium hydroxide in aqueous ethanol affords 7-hydroxyflavanone (56%); flavanones have also been prepared from 4-alkylresorcinols (*Starkov*, *Panesenko* and *M. N. Volkotrub*, Izv. Vyssh. Ucheb. Zaved., Khim. Khim. Tekhnol, 1972, **15**, 866):

$PhCH{:}CH{\cdot}CO_2H$ / BF_3, $CHCl_3$; NaOH, H_2O, EtOH

(R = H or alkyl)

(*2*) Cinnamoyl chloride condenses with phenol ethers to give flavanones in presence of aluminium chloride (*H. Simonis*, Zeit. angew. Chem., 1926, **39,** 1461).

Most polyhydric phenols with cinnamoyl chloride in nitrobenzene in the presence of aluminium chloride yield chalcones, but phloroglucinol gives flavanones (*J. Shinoda* and *S. Sato*, J. pharm. Soc. Japan, 1928, **48,** 109, 117);

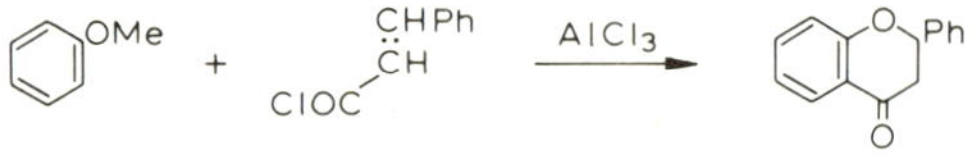

In this way phloroglucinol and *O*-ethoxycarbonyloxycinnamoyl chloride yield naringenin (p. 280).

Dihydroxybenzenes react with cinnamic acids in the presence of polyphosphoric acids to give hydroxyflavanones:

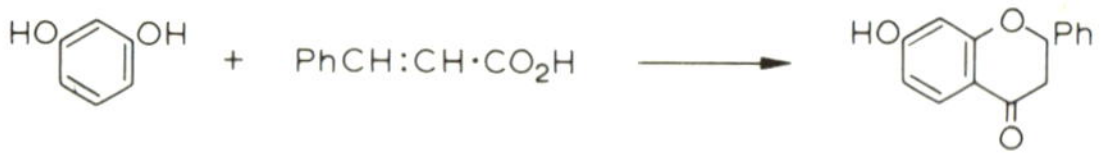

7-*Hydroxy*-, m.p. 184–185°, 7-*hydroxy*-2-*methyl*-, m.p. 182–183°, 8-*hydroxy*-, m.p. 248–250°, 8-*hydroxy*-2-*methyl*-, m.p. 239–240° (*N. Hasebe*, Nippon Kagaku Zasshi, 1961, **82,** 1728), 7-*hydroxy*-3-*methyl*-, m.p. 162–164°, 3-*ethyl*-7-*hydroxy*-*flavanone*, m.p. 168–170° (*idem, ibid.*, p. 1730).

(*3*) The hydrogenation of flavones gives flavanones (see p. 165; also *Geissman* and *Clinton*, J. Amer. chem. Soc., 1946, **68**, 697).

(*4*) 3-**Hydroxyflavanones** can be prepared by the oxidation of flavanones with alkaline hydrogen peroxide (*M. Murakami* and *T. Irie*, Proc. Imp. Acad. Japan., 1935, **11,** 229) or lead tetra-acetate (below); by the reduction of flavonols catalytically or with sodium dithionite (*J. C. Pew*, J. Amer. chem. Soc., 1948, **70,** 3031; *Geissman* and *H. Lischner*, *ibid.*, 1952, **74,** 3001) when 2-benzylcoumaran-3-one derivatives are also obtained; and from 2-hydroxychalcone dibromides (*T. Oyamada*, Ann., 1939, **538,** 44; *M. G. Marathey*, J. org. Chem., 1955, **20,** 563).

Of interest is the free radical hydroxylation of flavanones to 3-hydroxyflavanones by Fenton's reagent in acid solution (*V. B. Mahesh* and *Seshadri*, J. chem. Soc., 1955, 2503). Boron trifluoride-etherate reacts with 2′-benzyloxychalcone epoxides to give mainly 3-hydroxyflavanones if ring B is unsubstituted, but isoflavones (p. 209) are formed by rearrangement if methoxy groups are present in positions 4 or 3 and 4 (*S. C. Bhrara*, *A. C. Jain* and *Seshadri*, Tetrahedron, 1965, **21,** 963):

(R = H or OMe)

Treatment of 2′-methoxymethoxychalcone epoxides LXXXIV with acid in methanol gives inseparable *erythro–threo* mixtures LXXXV, which when treated briefly with concentrated sulphuric acid or polyphosphoric acid afford a 30–50% yield of *trans*-3-hydroxyflavanones (*J. P. Pineau* and *Chopin*, Bull. Soc. chim. Fr., 1971, 3678):

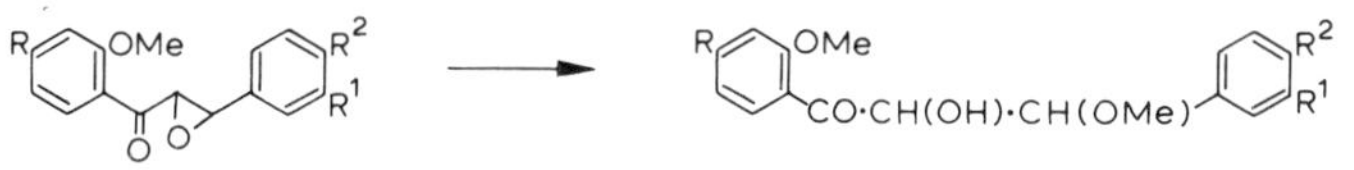

(LXXXIV)

R	R^1	R^2
H	H	OMe
H	H	OCH_2Ph
H	OMe	OMe
H	OMe	OCH_2Ph
H	OCH_2Ph	OCH_2Ph
OMe	OMe	H
OMe	OMe	OMe

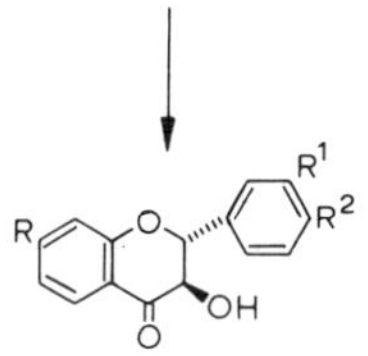

(LXXXV)

cis-3-Hydroxyflavanones are unknown (*Donnelly et al.*, Tetrahedron, 1972, **28,** 2545), but some *cis*-3-methoxy- (*Fujise, Y. Fujise* and *S. Hishida*, Nippon Kagaku Zasshi, 1963, **84,** 78; *J. W. Clark-Lewisj R. W. Jemison* and *V. Nair*, Austral. J. Chem., 1968, **21,** 3015) and *cis*-3-bromo-flavanones have been synthesised (*Bognár, M. Rákosi* and *Gy. Litkei*, Acta Chim. Hung., 1962, **34,** 253; *Clark-Lewis, T. M. Spotswood* and *L. R. Williams*, Austral. J. Chem., 1963, **16,** 107).

3-**Bromoflavanones** obtained either from acetoxychalcone dibromides or by bromination of flavanones with *N*-bromosuccinimide, have the *trans* configuration, whereas those obtained by the action of cupric bromide on the flavanones have the *cis* configuration. *cis*-3-Chloroflavanones are formed on chlorinating flavanones with sulphuryl chloride (*P. Y. Mahajan, M. S. Kamat* and *A. B. Kulkarni*, Indian J. Chem., 1970, **8,** 310). Convenient methods for the stereospecific synthesis of 3-halogeno-4′-methoxyflavanones have been described (*idem, ibid.*, 1968, **6,** 55). Sulphuryl chloride reacts with flavone and 3-chloroflavone to give 2,3,3-*trichloroflavanone*, m.p. 114–115° (*J. R. Merchant* and *D. V. Rege*, Chem. Comm., 1970, 380).

(ii) Properties

Ring fission. The flavanones are incompletely isomerised to the related chalcones by alkali, an equilibrium being established (*C. Kuroda*, J. chem. Soc., 1930, 752, 765). The reverse change occurs in ethanolic sulphuric acid. Fission of flavanones to chalcones occurs even in weakly acid solution (*Seshadri*, J. sci. ind. Research India, 1950, **9A,** 276). The chalcone butein and the related flavanone butin co-exist in the Dhak tree. 5-Hydroxyflavanones exist entirely in the flavanone form (*idem*, Proc. roy. Dublin Soc., 1956, **27,** 77; *M. Shimokoruzama*, J. Amer. chem. Soc., 1957, **79,** 4199). Thus naringenin and hesperetin are unchanged when heated with mineral acids, whereas under these conditions butin is partially changed to butein:

Butin ⇌ Butein

The equilibrium between hydroxychalcones, $-HOC_6H_4COCH{:}CHR$, where R is a phenyl or a heterocyclic group and the corresponding flavanone or chromanone has been studied (*J. Tiroflet* and *A. Corvaisier*, Compt. rend., 1961, **252,** 3818).

Oxidation of flavanones by lead tetra-acetate gives a mixture of products (*G. W. K. Cavill et al.*, J. chem. Soc., 1954, 4573; *Oyamada*, J. chem. Soc. Japan, 1943, **64,** 331, 471). Flavanone, for instance, yields flavone, 3-acetoxyflavanone and, unexpectedly, isoflavone. Phenolic oxidation of a 4(2)-hydroxychalcone[4′(2′)-hydroxyflavanone] initiates transformations to aurone, flavone, dihydroflavonol and isoflavone. The oxidation of simple model flavanones and chalcones by potassium ferricyanide has been studied; the 4′-hydroxy and 2′-hydroxy compounds rapidly give flavone or aurone in high yield, in sharp distinction to the 3′-hydroxy derivatives. The fully methylated flavanones or corresponding chalcones are not oxidised even after long periods. The results provide analogies for the proposed biosynthetic pathway (*A. Pelter, J. Bradshaw* and *R. Warren*, Phytochem., 1971, **10,** 835). For dehydrogenation of flavanones to flavones, see p. 142.

Reduction. Flavanones may be reduced to flavan-4β-ols (p. 247; also *R. Mozingo* and *H. Adkins*, J. Amer. chem. Soc., 1938, **60,** 669; *C. G. Joshi* and *Kulkarni*, Chem. and Ind., 1954, 1421; J. sci. ind. Research India, 1957, **16B,** 249; *M. D. Kashikar* and *Kulkarni*, *ibid.*, 1959, **18B,** 418; *M. Suzuki et al.*, Nippon Kagaku Zasshi, 1969, **90,** 397). The reduction of flavanone using sodium in alcohol affords mainly 2′-hydroxychalcone and a small amount of flavan-4β-ol (*S. Mitsui* and *A. Kasahara*, Nippon Kagaku Zasshi, 1958, **79,** 1382). Treatment of flavanone with lithium tetrahydrido-aluminate–aluminium chloride gives flavan in good yield (*M. M. Bokadia et al.*, J. chem. Soc., 1962, 1658).

Derivatives. Flavanone when treated with phenylhydrazine in acetic acid at room temperature gives, after standing for 12 hours, the *phenylhydrazone*, m.p. 146–147°, but when the reaction mixture is boiled 3-(2-*hydroxyphenyl*)-1,5-*diphenylpyrazoline*, m.p. 164–165°, is obtained (*P. Venturella, A. Bellino* and *S. Cusmano*, Ann. chim., Rome, 1961, **51,** 34):

$\xrightarrow[\text{AcOH, boil}]{\text{PhNH·NH}_2}$

On u.v. *irradiation* in benzene, flavanone produces 2′-hydroxychalcone (LXXXVI), 4-phenyldihydrocoumarin (LXXXVII, and salicyclic acid (*P. O. L. Mack* and *J. T. Pinhey*, Chem. Comm., 1972, 451):

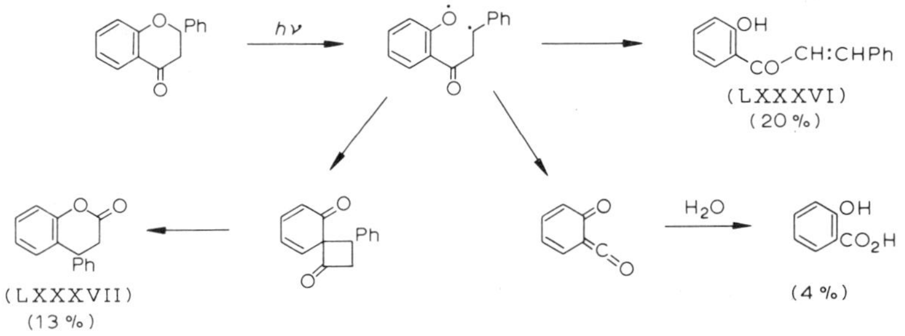

(LXXXVI) (20 %)

(LXXXVII) (13 %)

(4 %)

Irradiation in isopropanol yielded bis(4-hydroxy-2-phenyl-4-chromanyl) (LXXXVIII) and three unidentified flavopinacols (*K. Minami et al.*, Mokuzai Gakkaishi, 1972, **18,** 71).

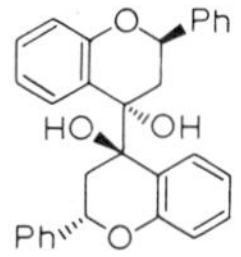

(LXXXVIII)

Flavanone, $C_{15}H_{14}O_2$, colourless needles, m.p. 75–76°, has been resolved, (−)*flavanone*, m.p. 75–76°, $[\alpha]_D^{17}$ −9.35° (benzene), (+)*flavanone*, m.p. 72–74°, $[\alpha]_D^{13.5}$ 12.4° (benzene) (*M. Kotake* and *G. Nakaminami*, Proc. Japan Acad., 1953, **29,** 56); u.v. λ_{max} 250 and 320 mμ (*S. Aronoff*, J. org. Chem., 1940, **5,** 561). Flavanone with benzaldehyde and hydrogen chloride yields the 3-*benzylidene* deriv., m.p. 103–104° (*H. Ryan* and *G. Cruess-Callighan*, Proc. roy. Irish Acad., 1929, **39B,** 124), which is oxidised by potassium permanganate to 3-*benzoyl*-3-*hydroxyflavanone*, m.p. 164–165° (*J. Alger* and *I. P. Carey*, *ibid.*, 1937, **44B,** 37).

*(h) Hydroxyflavanones**

The naturally occurring flavanones contain a hydroxyl group in the 5-position which i.r. spectroscopy shows to be chelated with the carbonyl group (*H. L. Hergert* and *E. F. Kurth*, J. Amer. chem. Soc., 1953, **75,** 1622; *B. L. Shaw* and *T. H. Simpson*, J. chem. Soc., 1955, 655). Flavanone derivatives occur in nature as glycosides; glycosides have been synthesised from

**J. E. Gowan et al.*, "The Chemistry of Vegetable Tannins", Soc. of Leather Trades' Chemists, London, 1956, p. 133.

hydroxyflavanones by the acetobromoglucose procedure (*S. Fujise* and *S. Mitsui*, Ber., 1938, **71,** 912) and by condensing hydroxyacetophenone glycosides with benzaldehyde (*G. Zemplén et al.*, *ibid.*, 1943, **76,** 1112). Eriodictyol-7-glucoside has been obtained in the chalcone form by the second method. Attempts to obtain the flavanone form, however, were unsuccessful. A number of 5,7-bis-(β-D-glucopyranosyloxy)flavanones and 7-(β-D-glucopyranosyloxy)-5-hydroxyflavanones have been synthesised (*H. Pacheco* and *A. Grouiller*, Bull. Soc. chim. Fr., 1965, 2937).

Reactions. The hydroxyflavanones dissolve in concentrated sulphuric acid to give deeply coloured solutions. With magnesium and hydrochloric acid the hydroxyflavanones, including flavanonols (3-hydroxyflavanones), give striking colours, but only the flavanonols give a positive test with zinc and acid (*J. C. Pew*, J. Amer. chem. Soc., 1948, **70,** 3031). The colour reactions of flavanones, chalcones, and aurones (see Vol. IVA, pp. 173–174) with sodium tetrahydridoborate and hydrochloric acid have been described (*H. G. Krishnamurty* and *T. R. Seshadri*, Current Sci., 1965, **34,** 681).

The *degradation* of hydroxyflavanones by alkali to polyhydric phenols and substituted benzoic or cinnamic acids gives valuable structural information. 5,7,4′-Trihydroxyflavanone, for instance, yields phloroglucinol and 4-hydroxybenzoic acid or 4-hydroxycinnamic acid according to the reagent used. 5,7-Dihydroxy-8-methoxyflavanone (dihydrowogonin) with 10% sodium hydroxide yields iretol (2,4,6-trihydroxyanisole) and cinnamic acid, but with 30% sodium hydroxide or ethanolic sodium hydroxide followed by acidification rearrangement occurs and the product is 5,7-dihydroxy-6-methoxyflavanone (*J. Chopin et al.*, Bull. Soc. chim. Fr., 1957, 192). Related isomerisations have been discussed (*Chopin*, Compt. rend., 1958, **247,** 1346; *Chopin et al.*, Bull. soc. chim. Fr., 1959, 1585; *Chopin* and *M. L. Bouillant*, Compt. rend., 1961, **252,** 2727; *M. Chadenson*, *Chopin* and *Bouillant*, Bull. Soc. chim. Fr., 1962, 1457). It is interesting that the dihydroxymethoxyflavanone when fused with potassium hydroxide loses a methoxy group to give phloroglucinol.

Rearrangements and dealkylation. 5,8-Dihydroxyflavanone rearranges with Lewis acids, bases, or heat to the 5,6-isomer. Demethylation of 5,8-*dimethoxyflavanone*, m.p. 163°, gives 5,6-*dihydroxyflavanone*, m.p. 190°, using hydrogen bromide in acetic acid or aluminium chloride in benzene, but 5-*hydroxy*-8-*methoxyflavanone*, m.p. 125–126°, is obtained using the latter in ether. This same reagent also converts 5,6-dimethoxyflavanone to 5-*hydroxy*-6-*methoxyflavanone*, m.p. 138–139° (*Chopin* and *Chadenson*, Compt. rend., 1957, **244,** 2727), and has been used to prepare a number of 5-hydroxyflavanones (*S. N. Aiyar*, *I. Dass* and *Seshadri*, Proc. Indian Acad. Sci.,

1957, **46A,** 238). The demethylation of methoxyflavanones possessing phloroglucinol units, with aluminium chloride in benzene, leads to fission and formation of hydroxyacetophenones (*R. N. Khanna et al.*, Indian J. Chem., 1965, **3,** 160). Flavanones and flavanonols containing *o*-dihydroxy groups in Ring A and 5-deoxyflavanonols form acid-labile complexes with aluminium chloride in methanol (*L. J. Porter* and *K. R. Markham*, Phytochem., 1972, **11,** 1477):

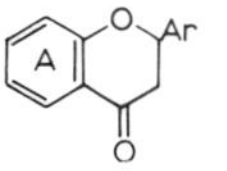

O Ar A O

The effect of aluminium chloride and sodium acetate on the u.v. spectra of flavanones has been studied (*R. M. Horowitz* and *L. Jurd*, J. org. Chem., 1961, **26,** 2446). Hydroxyflavanones with alkali and excess methyl sulphate give methoxychalcones (*N. Narasimhachari* and *Seshadri*, Proc. Indian Acad. Sci., 1948, **27A,** 223; *Chopin et al., loc. cit.).*

Autoxidation in strongly alkaline medium of the 7-glucosides of 5,7-dihydroxyflavanones substituted or not in the 3′ and/or 4′ positions by hydroxy or methoxy groups, always gives the corresponding aurones; in a neutral medium, *e.g.*, aqueous ethanol only the 3′,4′-dihydroxy derivatives are autoxidised to aurones; the 4′-methoxy and 3′,4′-dimethoxy derivatives on heating with sodium bicarbonate in methanol yield the corresponding flavones, while the 3′,4′-dihydroxy derivatives afford aurones (*G. Dellamonica* and *Chopin*, Bull. Soc. Chim. Fr., 1972, 2003).

5-Hydroxyflavanones can be oxidised to 5,6-dihydroxyflavanones by alkaline potassium persulphate (*Seshadri et al.*, J. sci. ind. Research India, 1955, **14B,** 335). By this method 5-hydroxy-7-methoxyflavanone yields 5,6-*dihydroxy*-7-*methoxyflavanone*, m.p. 248–249°, and naringenin 7,4′-dimethyl ether affords 5,6-*dihydroxy*-7,4′-*dimethoxyflavanone*, m.p. 235–237°. Nuclear methylation of naringenin using methyl iodide and potassium hydroxide in boiling methanol gives 5-*hydroxy*-7,4′-*dimethoxy*-6-*methylflavanone*, m.p. 148° (*R. N. Goel, A. C. Jain* and *Seshadri*, Proc. Indian Acad. Sci., 1958, **48A,** 180).

5,7,3′,4′-Tetramethoxyflavanone on treatment with iodine and silver acetate in boiling ethanol affords the corresponding 8-iodoflavanone, which on boiling with sodium hydroxide in aqueous ethanol yields the chalcone LXXXIX; the 8-bromoflavanone has also been prepared (*E. J. Keogh et al.*, Chem. and Ind. 1963, 412):

I_2, EtOH / AcOAg

NaOH / H_2O, EtOH

(LXXXIX)

It has been found that for the reaction between flavanones, iodine, and silver acetate to take place, the presence of either a 5-hydroxy or a 5-methoxy group is essential. A 7-hydroxy group inhibits iodination, even of 5-hydroxyflavanone derivatives. The completely methylated ethers afford the corresponding 8-iodoflavanones, whereas 5-hydroxy-7-methoxyflavanones give a mixture of 3-iodo and 8-iodo derivatives (*Jain*, *P. D. Sarpal* and *Seshadri*, Proc. Indian Acad. Sci., 1965, **42A**, 293). i.r. (*G. E. Inglètt*, J. org. Chem., 1958, **23**, 93; *L. H. Briggs* and *L. D. Colebrook*, Spectrochim. Acta, 1962, **18**, 939) and n.m.r. spectra (*Grouiller*, Bull. Soc. chim. Fr., 1966, 2405) of some flavanones have been reported.

The 3-hydroxyflavanones (flavanonols) are oxidised almost quantitatively by bismuth acetate to the corresponding flavanols (*J. M. Guider*, J. chem. Soc., 1955, 170). Thus 3,5,4′-trihydroxy-7-methoxyflavanone yields rhamnocitrin; taxifolin gives quercetin; and 3,4′-dihydroxy-5,7-dimethoxyflavanone gives 4′-hydroxy-5,7-dimethoxyflavonol. The same dehydrogenation is effected by *N*-bromosuccinimide (*R. Bógnar* and *M. Rákosi*, Chem. and Ind., 1955, 773; *B. J. Ghiya* and *M. G. Marathey*, Indian J. Chem., 1965, **3**, 420) or iodine in ethanolic sodium acetate (*V. B. Mahesh* and *Seshadri*, Proc. Indian Acad. Sci., 1955, **41**, 210). By the last named method pinobanksin gives galangin (45% yield) and taxifolin affords quercetin (65%). Other methods, which are not always satisfactory, include oxidation with alkaline hydrogen peroxide (*L. Reichel* and *J. Steudel*, Ann., 1942, **553**, 83); boiling in 2 *M*-sulphuric acid in the presence of air (*K. Freudenberg* and *L. Hartmann*, Ann., 1954, **587**, 207; *Pew*, *loc. cit.*) whereby dihydrorobinetin is converted to robinetin; and catalysed hydrogen transfer (*M. Kotake* and *T. Kubota*, Ann., 1940, **544**, 253).

Methyl 3-hemiacetals of 2-methoxy-3,4-flavandiones are obtained by oxidising flavonols with periodic acid in methanol. Solutions of these hemiacetals are an equilibrium mixture of the hemiacetal and the free dione, this being responsible for the solution's yellow colour. The hemiacetals when treated with diazomethane in ether are converted to epoxides (*M. A. Smith et al.*, J. org. Chem., 1972, **37**, 2774):

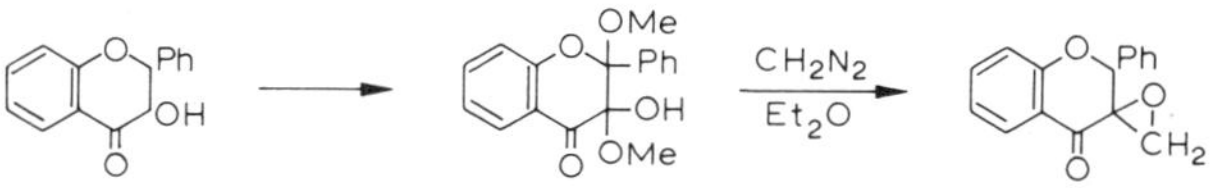

The 3-hydroxyflavanones behave as α-hydroxyketones, reducing warm Fehling's solution and forming dihydroquinoxalines with *o*-phenylenediamine. They can be isomerised with alkali to 2-benzylcoumaranones (*Kotaka* and *Kubota*, *loc. cit.*; *J. Gripenberg*, Acta Chem. Scand., 1953, **7,** 1323; *Kubota*, J. Inst. Polytechnics, Osaka City University, 1953, **4,** 253). A number of 3-hydroxyflavanones have been prepared and their rearrangement under alkali conditions studied (*Chopin*, *Chadenson* and *P. Durual*, Compt. rend., 1964, **258,** 6178). 3-Hydroxyflavanones without a free 5-hydroxyl group are slowly isomerised and converted by alcoholic mineral acids into the corresponding anthocyanidins (*Jurd*, Phytochem., 1969, **8,** 2421).

3-Hydroxyflavanones react with *p*-tosylhydrazine in the presence of acid to yield a 4-amino-3-hydroxyflavylium salt, which may be hydrolysed by alkali to the corresponding flavonol (*G. Janzsó*, *F. Kállay* and *I. Koczor*, Tetrahedron, 1966, **22,** 2909):

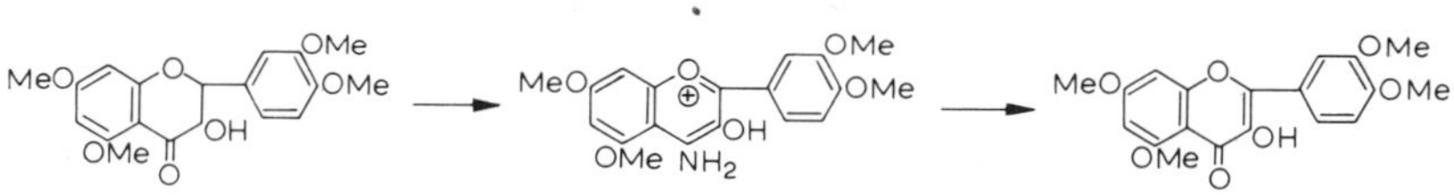

u.v. Spectral data has been used to characterise flavanonols (*F. Tomas*, *O. Carpena* and *J. Mataix*, An. Quim., 1972, **68,** 123). For optical rotatory dispersion and absolute configuration of flavanones, 3-hydroxyflavanones, and their glycosides, see *W. Gaffield* and *A. C. Waiss, Jr.*, Chem. Comm., 1968, 29.

(i) Naturally occurring hydroxyflavanones

Hydroxyflavanones and hydroxyflavanonols occur in nature particularly in the heartwood of conifers, especially of the genus *Pinus*. It is noteworthy that the hydroxyflavanones are laevorotatory while the flavanonols are dextrorotatory in most solvents (*H. Erdtman*, Proc. roy. Dublin Soc., 1956, **27,** 129).

Pincocembrin, 5,7--*dihydroxyflavanone*, $C_{15}H_{12}O_4$, m.p. 194–195° (192–193°), $[\alpha]_D^{20}$ −54.5° (methanol), $[\alpha]_D^{15}$ −45.3° (acetone), *diacetate*, m.p. 122°, *dimethyl ether*, m.p. 169° (159–160°), is obtained from *Pinus cembra*, *P. excelsa* (*G. Lindstedt*, Acta Chem. Scand., 1949, **3,** 1375), *P. virginiana* (*Lindstedt*, *ibid.*, 1949, **3,** 1381), *P. griffthii* (*V. B.*

Mahesh and *T. R. Seshadri*, J. sci. ind. Research India, 1954, **13B,** 835), cherry (*C. Mentzer* and *H. Pacheco*, Bull. Soc. Chim. biol., 1952, **34,** 956), and the leaves of *Eucalyptus sieberi* (*I. R. C. Bick*, *R. B. Brown* and *W. E. Hillis*, Austral. J. Chem., 1972, **25,** 449). Pinocembrin dimethyl ether has also been isolated from the last source along with **alpinetin,** *pinocembrin 5-methyl ether*, m.p. 223°, also obtained from *Alpina chinensis* Roscoe (*Y. Kimura*, J. pharm. Soc. Japan, 1940, **60,** 151). **Verecundin,** *pinocembrin 5-glucoside*, $C_{21}H_{22}O_9$, m.p. 135°, $[\alpha]_D^{14}$ −71.1° (aq. acetone), *7-methyl ether*, m.p. 98°, occurs in the Japanese tree *Prunus verecunda* Koehne (*M. Hasegawa* and *T. Shirato*, J. Amer. chem. Soc., 1957, **79,** 450). Pinocembrin is converted by acetic anhydride into 2,4,6-triacetoxychalcone $(AcO)_3C_6H_2 \cdot CO \cdot CH{:}CH \cdot Ph$. **Pinostrobin,** *pinocembrin 7-methyl ether*, m.p. 107–109°, has been isolated from *P. excelsa* and other trees (*Mahesh* and *Seshadri*, *loc. cit.*). *2,5-Dihydroxy-7-methoxyflavanone* (XC), $C_{16}H_{15}O_5$, m.p. 170–172°, has been isolated from *Populus nigra* and its structure confirmed by i.r., n.m.r. and mass spectral data, and synthesis (*M. Chadenson*, *M. Hauteville* and *J. Chopin*, Chem. Comm., 1972, 107).

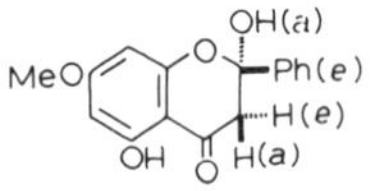

(XC)

Liquiritigenin, *7,4′-dihydroxyflavanone*, $C_{15}H_{12}O_4 \cdot H_2O$, colourless needles, m.p. 207°, *oxime*, m.p. 178°, *diacetate*, m.p. 186°, is found as the *4′-glucoside*, **liquiritin,** $C_{21}H_{22}O_9 \cdot H_2O$, m.p. 212°, in *Glycyrrhiza glabra*, Linn (*B. Puri* and *Seshadri*, J. sci. ind. Research India, 1954, **13B,** 475). The sugar residue is attached to the 4′-position since the glucoside when methylated and ring-opened affords 2-hydroxy-4-methoxyacetophenone. Liquiritigenin with 60% potassium hydroxide gives resacetophenone, and 4-hydroxybenzoic acid and its structure has been established by synthesis (*J. Shinoda* and *S. Ueeda*, Ber., 1934, **67,** 434; *D. R. Nadkarni* and *T. S. Wheeler*, J. chem. Soc., 1938, 1320). (−)Liquiritigenin has been shown to have the 2(*S*)-configuration (*H. Arakawa* and *M. Nakazaki*, Ann., 1960, **636,** 111). **Sophoranone,** *5,3′,5′-triprenylliquiritigenin, 7,4′-dihydroxy-5,3′,5′-triprenylflavanone*, $C_{30}H_{36}O_4$, m.p. 108°, $[\alpha]_D^{35}$ −13.0° (ethanol- has been obtained from the roots of *Sophora subprostrata* Chun et T. Chen. (*M. Komatsu et al.*, Chem. pharm. Bull. Tokyo, 1970, **18,** 602).

Butin, *7,3′,4′-trihydroxyflavanone*, $C_{15}H_{12}O_5$, colourless needles, m.p. 224–226°, *triacetate*, m.p. 123–125°, *tribenzoate*, m.p. 155–157°, occurs as the *7,3′-diglucoside* **butrin,** $C_{27}H_{32}O_{15} \cdot 2H_2O$, m.p. 193.5°, $[\alpha]_D^{31.5}$ −73.3° (H_2O) in the flowers of *Butea frondosa*, a tree found throughout India and Burma (*A. G. Perkin* and *J. T. Hummel*, J. chem. Soc., 1904, **85,** 1459; *J. B. Lal*, *ibid.*, 1937, 1562; *P. S. Rao* and *Seshadri*, Proc. Indian Acad. Sci., 1941, **14A,** 29). Butin when fused with potassium hydroxide yields resorcinol and protocatechuic acid, and with aqueous potassium hydroxide followed by acidification gives the orange-yellow **butein,** *3,4,2′,4′-tetrahydroxychalcone*, m.p. 213–215°. The structure was confirmed by synthesis (*A. Göschke* and *J. Tambor*, Ber., 1912, **45,** 186). Butin and butein were the first representatives of the flavanone and chalcone series to be isolated from natural sources. *Butin* 7-O-D-*glucopyranoside*, m.p. 165–167°, $[\alpha]_D^{18}$ −78°, has been obtained from *Bidens tripartita* (*A. G. Serbin*, *M. I. Borisov* and *V. T. Chernobai*, Khim. Prir. Soedin., 1972, **8,** 440).

Dihydrowogonin, *5,7-dihydroxy-8-methoxyflavanone*, $C_{16}H_{14}O_5$, yellow prisms, m.p. 151–152°, $[\alpha]_D$ −56° (ethanol), *dinitrophenylhydrazone*, m.p. 223–228° (decomp.), occurs in wild cherry wood, *Prunus avium* (*Chopin et al.*, Bull. Soc. chim. Fr., 1957, 192). Its rearrangement to *5,7-dihydroxy-6-methoxyflavanone*, **dihydro-oroxylin,** m.p. 177°, has been described (p. 274). Dihydrowogonin has been synthesised (*S. N. Aiyar, I. Dass* and *Seshadri*, Proc. Indian Acad. Sci., 1957, **46A,** 238) and the demethylation of some of its derivatives studied (*Chopin* and *Chadenson*, Compt. rend., 1958, **247,** 1625). *Dihydrowogonin 7-O-β-D-glucoside*, m.p. 214–216°, $[\alpha]_D^{29}$ −56.7° (dimethyl sulphoxide), $[\alpha]_D^{24}$ −67.8° (pyridine), *penta-acetate*, m.p. 117–119°, has been isolated from *P. cerasus* L. and its structure confirmed by synthesis (*H. Wagner et al.*, Tetrahedron Letters, 1969, 1471). 5,8-Dihydroxy-7-methoxyflavanone is methylated using excess diazomethane or dimethyl sulphate to give *5,7,8-trimethoxyflavanone*, m.p. 158–159°. However, when two moles of dimethyl sulphate are used, *5,6,7-trimethoxyflavanone*, m.p. 159–160°, is obtained. Similarly 5,8-dihydroxy-4′,7-dimethoxy- and 5,8-dihydroxyflavanone show no isomerism with excess methylating reagents; the related 5-hydroxyflavanones with other positions methylated undergo isomeric changes even when heated in acetone containing potassium carbonate (*H. G. Krishnamurty* and *Seshadri*, J. sci. ind. Research India, 1959, **18B**, 151).

Hesperetin, *5,7,3′-trihydroxy-4′-methoxyflavanone*, $C_{16}H_{14}O_6$, occurs as the *7-β-rutinoside*, **hesperidin,** $C_{28}H_{34}O_{15}$, in Citrus species (*A. Hilger*, Ber., 1876, **9,** 26) and both substances have been synthesised (*G. Zemplén* and *R. Bógnar*, Ber., 1942, **75,** 1043; 1943, **76,** 773). When cleaved with alkali they yield phloroglucinol (*W. F. Newhall* and *S. U. Ting*, J. Agric. Food Chem., 1967, **15,** 776) and isoferulic (3-hydroxy-4-methoxycinnamic) acid. On oxidation with hydrogen peroxide in sodium hydroxide solution, hesperidin affords a flavanonol, an aurone, and a flavonol (*Chopin* and *A. Durix*, Compt. rend., 1964, **258,** 4792). Hesperidin obtained from *Zanthoxylum* species of Hong Kong is hydrolysed to a mixture of (±) and (−)hesperetin and it is concluded that hesperidin is the 7-β-rutinoside of (−)hesperetin (*H. R. Arthur et al.*, J. chem. Soc., 1956, 632). (±)**Hesperetin,** m.p. 226–228°, *triacetate*, m.p. 139–141°; (−)**hesperetin,** m.p. 216–218°, $[\alpha]_D^{27}$ −37.6° (ethanol), *triacetate*, m.p. 130–132°; *hesperidin hydrate*, $C_{28}H_{34}O_{15} \cdot H_2O$, m.p. 260° (decomp.), $[\alpha]_D^{27}$ −88.2° (pyridine). (−)*Hesperetin* has been shown to have the 2(*S*)-configuration (*Arakawa* and *Nakazaki*, *loc. cit.*). Iodination and bromination of 5,7,3′,4′-tetramethoxyflavanone gives the 8-halogeno derivative. 5-Hydroxy-7,3′,4′-trimethoxyflavanone similarly gives a 3,6(or 8)-di-iodo derivative (*E. J. Keogh et al.*, J. chem. Soc., 1963, 5271). Hesperetin 7,3′-dimethyl ether may be converted to luteolin 7,3′,4′-trimethyl ether (*R. N. Goel, Mahesh* and *Seshadri*, Proc. Indian Acad. Sci., 1958, **47A,** 184). *6′-Hydroxyhesperetin*, m.p. 227–228° (decomp.), *triacetate*, m.p. 181–182°, on methylation with dimethyl sulphate disproportionation occurs with the production of *5,7,3′,4′,6′-pentamethoxyflavanone*, m.p. 175°, and a *2′-hydroxypentamethoxychalcone*, m.p. 175–176° (*L. Hörhammer et al.*, Tetrahedron, 1965, **21,** 969).

Naringenin, *5,7,4′-trihydroxyflavanone*, $C_{15}H_{12}O_5$, needles, m.p. 250–251°, *triacetate*, m.p. 126°, *oxime*, m.p. 231° (*M. K. Seikel* and *T. A. Geissman*, J. Amer. chem. Soc., 1950, **72,** 5725) occurs in the heartwood of *Ferreirea spectabilis* (*F. E. King et al.*, J. chem. Soc., 1952, 4580) and in the fruits of grape-fruit trees, *citrus decumana*, as the 7-rhamnoglucoside, **naringin,** $C_{27}H_{32}O_{14} \cdot 8H_2O$, m.p. ca. 83°, $[\alpha]_D^{19}$ −82.1° (ethanol). Naringenin gives an intense brown colour with ferric chloride and yields 4-hydroxybenzoic acid and 4-hydroxycinnamic acid by fusion with potassium hydroxide and 50% ethanolic potassium hydroxide at 160°, respectively. It has also been degraded to phloroglucinol by alkali fusion (*Newhall* and *Ting, loc. cit.*). Acetylation of naringenin with acetic anhydride and sodium acetate at 180° gives 5,7,4′-triacetoxy-4-styrylcoumarin (XCI). The formation of such a compound is dependent on the presence of a hydroxy group in the 5-position; naringenin 7,4′-dimethyl ether gives a styrylcoumarin derivative, but 7-hydroxy- or 7-methoxy-flavanone, and naringenin trimethyl ether afford the corresponding chalcone acetates (*D. K. Bhardwaj, S. Neelakantan* and *Seshadri*, Indian J. Chem., 1969, **7,** 325):

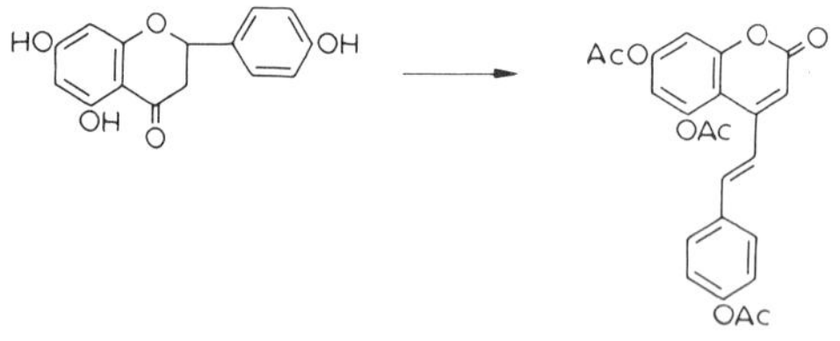

(XCI)

Naringen with silver acetate and iodine in ethanol yields 3-*acetoxynaringenin*, m.p. 228–230° (decomp.); hydrolysis gives 3-*hydroxynaringenin*, m.p. 238–240° (decomp.); naringenin 7,4′-dimethyl ether may be converted to apigenin 7,4′-dimethyl ether (*Goel, Mahesh* and *Seshadri, loc. cit.*). 5,7′-Dihydroxy-4′-methoxyflavanone on methylation with diazomethane followed by acetylation affords 5-*acetoxy*-7,4′-*dimethoxyflavanone*, m.p. 161°, which on heating with lead tetra-acetate in acetic acid at 85–90° and then with hydrochloric acid yields, 3,5-*dihydroxy*-7,4′-*dimethoxyflavanone*, m.p. 188–190°. Other related derivatives have been obtained (*S. Matsukawa, H. Kotake*, and *M. Yamaguchi*, Nippon Kagaku Zasshi, 1963, **84,** 359). Dihydrochalcones have been prepared from flavanone glycosides, for example, naringin, and compared to sucrose for sweetness (*Y-C, Lin, C-I. Ch'en* and *F-C. Ch'en*, Taiwan K'o Hsaeh, 1971, **25,** 103). Naringenin has been synthesised (*J. Shinoda* and *S. Sato*, J. pharm. Soc. Japan, 1928, **48**, 933) as has also **sakuranetin,** its 7-*methyl ether*, $C_{16}H_{14}O_5$, m.p. 153°, *oxime*, m.p. 195–196°. (+)Sakuranetin has the 2(*R*)-configuration (*Arakawa* and *Nakazaki, loc. cit.*). **Sakuranin,** *sakuranetin* 5-D-*glucoside*, $C_{22}H_{24}O_{10}$, m.p. 212°, occurs in the bark of *Prunus yedonensis* and the wood of *P. donarium* (*Hasegawa* and *Shirato*, J. Amer. chem. Soc., 1955, **77,** 3557). **Isosakuranetin,** *naringenin* 4′-*methyl ether*, m.p. 177°, *acetate*, m.p. 121°, $[\alpha]_D^{14}$ −27.1° (acetone–pyridine), which occurs in *Prunus verecunda* Koehne as the 7-D-glucoside, **isosakuranin,** $C_{22}H_{24}O_{10}$,1.5H_2O, m.p. 190°, $[\alpha]_D^{13}$ −41.4° (acetone), is hydrolysed by emulsion to the aglucone and glucose (*idem, ibid.*, 1957, **79,** 450). The (+)form has been synthesised (*Zemplén et al.*, Ber., 1942, **75,** 1432). **Poncirin,** *isosakuranetin* 7-*rhamnoglucoside*, m.p. 210°, isolated from *Poncirus trifoliata*, on hydrolysis with aqueous ethanolic hydrochloric acid gives isosakuranin (*M. Shimokoriyama, ibid.*, p. 4199; *S. Hattori, Hasegawa* and *Shimokoriyama*, J. chem. Soc. Japan, 1944, **65,** 61). **Didymine, atsinoside,** isosakuranetin 7-rutinoside, has been obtained from *Acinos thymoides* (*Wagner et al.*, Ber., 1969, **102**, 3605). **Prunin**, *naringenin 7-D-glucoside*, $C_{21}H_{22}O_{10}$, needles, m.p. 225°, has been obtained from *Prunus yedonensis* (*Hasegawa* and *Shirato*, J. Amer. chem. Soc., 1952, **74**, 6114) and synthesised (*H. Pacheco* and *A. Grouiller*, Compt. rend., 1963, **256**, 4927). **Salipurpin,** *salipurposide, naringenin 5-glucoside*, $C_{21}H_{22}O_{10}$, m.p. 227° (decomp.), is obtained from *Salix purpurea* (*C. Charaux* and *J. Rabate*, *ibid.*, 1931, **192,** 1478; *Zemplén et al.*, Ber., 1943, **76,** 386). *Naringenin 5,7-diglucoside*, m.p. 179–181°, has been obtained from the leaves of *Crataegus phenophyrus* Medicus (*Z. Kowalewski* and *K. Mrugasiewicz*, Planta Med., 1971, **19,** 311). 4′-β-D-Glucosyl-7-β-neohesperidosylnaringenin and 4′-β-D-glucosyl-7-rutinosylnaringenin have been isolated from Texas Ruby grapefruit, *Citrus paradisi* Macf. (*J. W. Mizelle, W. J. Dunlap* and *S. H. Wender*, Phytochem., 1967, **6,** 1305) and synthesised (*G. Aurnhammer et al.*, Ber., 1971, **104**, 473). Flavanone compounds reported to occur in various *Citrus* species have been reviewed (*R. F. Albach* and *C. H. Redman*, Phytochem., 1969, **8,** 127). **Selinone,** *naringenin 4′-prenyl ether*, $C_{20}H_{20}O_5$, m.p. 151–152°, $[\alpha]_D^{29}$ −50° (methanol), occurs in *Selinum vaginatum* (*Seshadri* and *M. S. Sood*, Tetrahedron Letters, 1967, 853); also reported is **archangelenone,** $C_{20}H_{20}O_5$, m.p. 148°, $[\alpha]_D$ ±0° ($CHCl_3$), obtained from *Angelica archangelica* and given the same structure (*S. C. Basa, D. Basa* and *A. Chatterjee*, Chem. and Ind., 1971, 355; Indian J. chem., 1973, **II,** 407). Naringenin 7,4′-diprenyl ether has also been obtained from the same source. **Poriol,** 6-C-*methylnaringenin*, $C_{16}H_{14}O_5$, m.p. 255–256°, has been isolated from the diseased (*Porii veirii* Murr) root-bark of Douglas fir, *Pseudotsuga menziesii* (*G. M. Barton*, Canad. J. Chem., 1967, **45**, 1020) and its structure confirmed by synthesis (*A. C. Jain, P. Lal* and *Seshadri*, Tetrahedron, 1969, **25,** 283).

Citronetin, *5,7-dihydroxy-2′-methoxyflavanone*, $C_{16}H_{14}O_5$, m.p. 224–225°, *diacetate*, m.p. 118–119°, *dimethyl ether*, m.p. 125°, *oxime*, m.p. 234–235°, occurs in the rind of *Citrus limon* as the *7-rhamnoglucoside*, **citronin,** $C_{28}H_{34}O_{14}$, m.p. 235° (*R. Yamamoto* and *Y. Oshima*, J. agric. chem. Soc., Japan, 1931, **7,** 321; *Shinoda* and *Sato*, J. pharm. Soc. Japan, 1931, **7,** 312). Citronetin has been synthesised (*N. Narasimhachari, D. Rajagopalan* and *Seshadri*, Proc. Indian Acad. Sci., 1953, **37A,** 620).

Fustin, *dihydrofisetin*, *3,7,3′,4′-tetrahydroxyflavanone*, $C_{15}H_{12}O_5$, was the first 3-hydroxyflavanone to be found in nature. (±)**Fustin,** m.p. 216–218°, *tetra-acetate*, m.p. 150–151°, is obtained from the wood of *Rhus succedanea* L., *R. Glabra* L., and *Gleditsia triacanthos* (*T. Oyamada*, Ann., 1939, **538,** 44); (+)**fustin,** m.p. 228–229°, $[\alpha]_D^{23}$ 28.3° (50% aqueous acetone), *tetra-acetate*, m.p. 116–119°, *trimethyl ether*, m.p. 138–140°, is obtained from black-wattle heartwood, *Acacia mollissima* and it possesses the (2*R*, 3*R*) configuration; (−)**fustin,** m.p. 228°, $[\alpha]_D^{25}$ −26° (50% aqueous acetone), *tetra-acetone*, m.p. 117–118°, is obtained from *Rhus cotinus*, and it possesses the (2*S*, 3*S*) configuration (*D. G. Roux* and *E. Paulus*, Biochem. J., 1960, **77,** 315; 1962, **84,** 416):

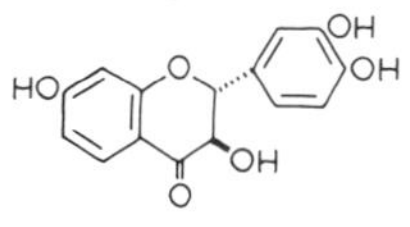

(+) Fustin

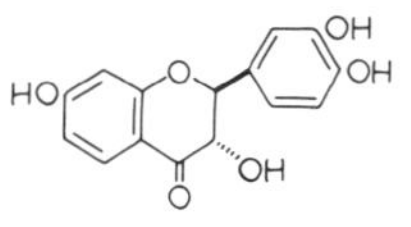

(−) Fustin

(+)*Fustin* 3-O-β-D-*glucoside*, and **lecontin**, 3-O-β-D-*glucoside* of 3,7,4′-*trihydroxyflavanone*, m.p. 125–130°, $[\alpha]_D^{25}$ −16,75° (H_2O) have been isolated from *Baptisia lecontei* (Leguminosae). o.r.d., c.d. and n.m.r. data have been presented for these and other flavanonol 3-*O*-glycosides and the relationship between absolute stereochemistry and the order and sign of Cotton effects in this class of glycosides has been established (*K. R. Markham* and *T. J. Mabry*, Tetrahedron, 1968, **24**, 823).

Carthamin, $C_{21}H_{22}O_{11}$, the red pigment of safflower, *Carthamus tinctorius* and believed to have medicinal properties, is a glucoside which when crystallised from methanol is converted into the isomeric **isocarthamin** (renamed **neocarthamin**), m.p. 228° (decomp.). There has been a great deal of discussion concerning the structure of these two compounds and it is now believed that carthamin is the chalcone and neocarthamin the 5-glucoside of 5,6,7,4′-tetrahydroxyflavanone (*Zemplén, L. Farkas* and *R. Rakusa*, Acta Chim. Acad. Sci. Hung., 1958, **14**, 471; C.A., 1959, **53**, 11357). Carthamin is hydrolysed by phosphoric acid to **carthamidin,** 5,7,8,4′-*tetrahydroxyflavanone*, $C_{15}H_{12}O_6 \cdot 2H_2O$, yellow prisms, m.p. 216–218°, giving a green colour with ethanolic ferric chloride and a blue colour with alkali, and to **isocarthamidin,** 5,6,7,4′-*tetrahydroxyflavanone*, $C_{15}H_{12}O_6$, yellow prisms, m.p. 239–240°, which gives a bluish-green colour with ferric chloride (*C. Kuroda*, J. chem. Soc., 1930, 752, 760). Both substances have been synthesised by demethylating the corresponding tetramethyl ethers by means of aluminium chloride in benzene (*Narasimhachari et al.*, Proc. Indian Acad. Sci., 1949, **29**, 404); isocarthamidin has also been obtained from *isocarthamidin* 5,7-*dimethyl ether*, m.p. 198° (*Zemplén, Farkas* and *Rakusa, loc. cit.*) and by use of the Dakin hydroxylation reaction (*V. K. Ahluwalia et al.*, J. Indian chem. Soc., 1965, 42, 279).

Eriodictyol, 5,7,3′,4′-*tetrahydroxyflavanone*, $C_{15}H_{12}O_6$, colourless needles, m.p. 262–265° (decomp.) (267°), depending on rate of heating, 7,3′,4′-*trimethyl ether*, m.p. 136° (*T. A. Geissman*, J. Amer. chem. Soc., 1940, **62**, 3258), is found in *Eriodictyon Californicum*. *Eriodictyol* 5,3′-*diglucoside*, m.p. 194–196°, is found in the leaves of *Crataegus phenophyrum* Medicus (*Z. Kowalewski* and *M. Krugasiewicz*, Planta. Med., 1971, **19**, 311).

Plathymenin, 6,7,3′,4′-*tetrahydroxyflavanone*, $C_{15}H_{12}O_6$, m.p. 229° (decomp.), *tetraacetate*, m.p. 149–151°, has been isolated from the heartwood of *Plathymenia reticulata* (*F. E. King et al.*, J. chem. Soc., 1953, 1055). With alkali it yields **neoplathymenin,** 2,4,5,3′,4′-*pentahydroxychalcone*, m.p. 232°. There is no naturally occurring flavone comparable with plathymenin

Ponkanetin, m.p. 152°, isolated from the peel of *Citrus poonensis* and *C. deliciosa* and reported to be 5,6,7,8,4′-pentamethoxyflavanone (*N. Ichikawa* and *T. Yamashita*, J. chem. Soc. Japan, 1941, **62**, 1006), has been shown to be 5,6,7,8,4′-pentamethoxyflavone (*A. Bellino, P. Venturella* and *S. Cusmano*, Ann. Chim., Rome, 1962, **52**, 795); 5,6,7,8,4′-*pentamethoxyflavanone*, m.p. 108–109° (*J. M. Sehgal et al.*, Proc. Indian Acad. Sci., 1955, **42A**, 252).

C-Methylflavanones are frequently met with and the following flavanones containing a *C*-methylated phloroglucinol nucleus have been isolated from certain pine species. **Strobopinin,** 5,7-*dihydroxy*-6-*methylflavanone*, $C_{16}H_{14}O_4$, yellow prisms, m.p. 225–227°, $[\alpha]_D^{20}$ −61° (methanol), obtained from *Pinus strobus* and *Alnus sieboldiana* (*Erdtman*, Svensk. kem. Tids., 1944, **56**, 2; C.A., 1946, **40**, 1309; *Y. Asakawa*, Bull. chem.

Soc. Japan, 1971, **44,** 2761); **cryptostrobin,** 5,7-*dihydroxy*-8-*methylflavanone*, $C_{16}H_{14}O_4$, m.p. 202–203°, $[\alpha]_D^{20}$ $-33°$ (methanol), obtained from *Pinus strobus* (*S. Matsuura*, Pharm. Bull., Tokyo, 1957, **5,** 195). Strobopinin and cryptostrobin have been synthesised (*S. C. Bharara et al.*, Indian J. chem., 1964, **2,** 399); **strobobanksin,** 3,5,7-*trihydroxy*-6-*methylflavanone*, $C_{16}H_{14}O_5$, m.p. 177–178°, $[\alpha]_D^{20}$ $+17°$ (methanol) (*Lindstedt* and *A. Misiorny*, Acta Chem. Scand., 1951, **5,** 1). **Obtusifolin,** $C_{24}H_{22}O_7$, m.p. 202–204°, α_D $+91.2°$, *triacetate*, m.p. 164–170°, has been isolated from *Gnaphalium obtusifolium*. On methylation with diazomethane it yields a *monomethyl ether*, m.p. 232–236°, a *dimethyl ether*, m.p. 168–171°, and a *trimethyl ether*, m.p. ~60°. Spectral data have been correlated on the basis of a structure previously established by X-ray analysis (*R. Hänsel, D. Ohlendorf* and *A. Pelter*, Z. Naturforsch., 1970, **25b,** 989; *W. Hoppe et al.*, Tetrahedron Letters, 1970, 3643; Acta Cryst., 1971, **27B,** 718):

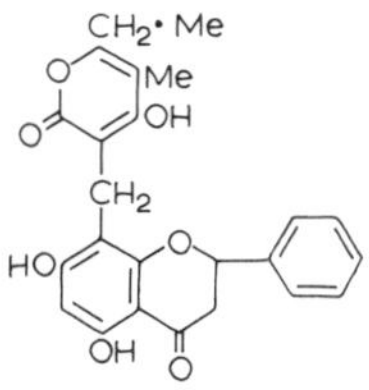

Obtusifolin

The absolute configuration of obtusifolin has been determined (*D. W. Engel, K. Zechmeister* and *Hoppe*, Tetrahedron Letters, 1972, 1323).

Matteucinol, 5,7-*dihydroxy*-4′-*methoxy*-6,8-*dimethylflavanone*, $C_{18}H_{18}O_5$, (±), m.p. 172.5–173°, 7-*methyl ether*, m.p. 102.5°, (−), m.p. 175.5–176°, $[\alpha]_D^{18}$ $-30°$ (acetone), occurs naturally as the (−)isomer in *Rhododendron simsii* (*Arthur* and *W. H. Hui*, J. chem. Soc., 1954, 2782) and the fern *Matteucia orientalis* (*T. Munesada*, J. pharm. Soc. Japan, 1924, **505,** 185).

Farrerol, 5,7,4′-*trihydroxy*-6,8-*dimethylflavanone*, 4′-*demethylmatteucinol*, $C_{17}H_{16}O_5$, m.p. 207–208°, $[\alpha]_D$ $-16°$ (acetone) is obtained from the leaves of *Rhododendron farrerae* (*Arthur*, J. chem. Soc., 1955, 3740) and *Cryptomium falcatum* (*Y. Kishimoto*, Pharm. Bull. Japan, 1956, **4,** 2; *Arthur* and *Kishimoto*, Chem. and Ind., 1956, 738). It gives a green colour with ferric chloride, an intense vermilion with aqueous sodium hydroxide, and bright red with magnesium, hydrochloric acid and methanol. The *racemic* modification, m.p. 223–224°, is obtained by deacetylation of the (±)*triacetate*, m.p. 192°. **Cyrtominetin,** 5,7,3′,4′-*tetrahydroxy*-6,8-*dimethylflavanone*, $C_{17}H_{16}O_6$, has been isolated from *Cyrtomium* species and its *racemate*, m.p. 237–238° synthesised, *tetra-acetate*, m.p. 173–174° (*Kishimoto*, Pharm. Bull. Japan, 1956, **4,** 24). **Protofarrerol,** $C_{17}H_{18}O_6$, m.p. 210°, $[\alpha]_{589}^{25}$ 201°, has been isolated from *Leptorumohra miqueliana* and its u.v. spectrum is similar to that of 5,7-*dihydroxy*-6,8-*dimethylchroman*-4-*one*, m.p. 222°. On dehydration it gives farrerol, and treatment with acetic anhydride and a drop of sulphuric acid affords triacetylfarrerol (*T. Noro et al.*, Yakugaku Zasshi, 1969, **89,** 851):

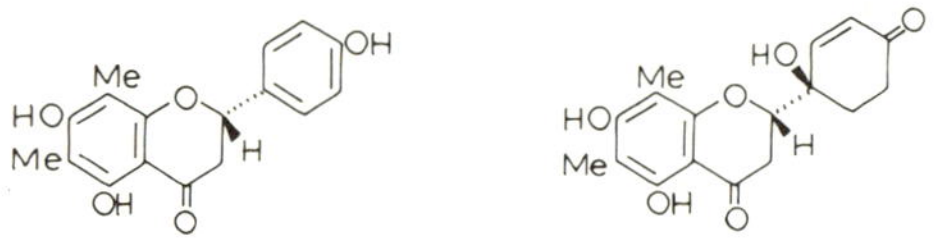

Farrerol Protofarrerol

Farrerol and protofarrerol possess the 2(*S*)-configuration (*S. Fukushima et al., ibid.*, p. 1272).

Pinobanksin, *dihydrogalangin, 3,5,7-trihydroxyflavanone*, $C_{15}H_{12}O_5$, yellow prisms, m.p. 177–178°, $[\alpha]_D^{20}$ +14° (methanol), is isolated from *Pinus strobus* or *P. banksiana* (*Erdtman*, Svensk. Kem. Tidskr., 1944, **56,** 26, 95; *G. Lindstedt*, Acta Chem. Scand., 1949, **3,** 755, 759; 1950, **4,** 772). **Alpinone,** *pinobanskin 7-methyl ether* (*J. Gripenberg et al., ibid.*, 1956, **10,** 393), *3,5-dihydroxy-7-methoxyflavanone*, $C_{16}H_{14}O_5$, m.p. 186–187°, $[\alpha]_D^{20}$ +91° (pyridine), occurs in *Alpinia japonica* Miq (*Y. Kimura* and *M. Hoshi*, J. pharm. Soc. Japan, 1937, **57,** 147). Its structure has been established by synthesis (*H. Kotake, Jr. et al.*, Chem. and Ind., 1954, 1562). **Aromadendrin, katsuranin,** *dihydrokaempferol, 3-hydroxynaringenin, 3,5,7,4′-tetrahydroxyflavanone*, $C_{15}H_{12}O_6 \cdot 2.5H_2O$, m.p. 247–249°, [224–225°, 237–241° (decomp.)], $[\alpha]_D^{20}$ +45° (aqueous acetone), $[\alpha]_D^{25}$ +26.5° (methanol) is obtained from katsura, *Cercidiphyllum japonicum* (*H. Uota et al.*, J. agric. chem. Soc. Japan, 1943, **19,** 467), *Larix decidua*, *Nothofagus dombeyi*, coigue and black cherry wood (*J. C. Pew*, J. Amer. chem. Soc., 1948, **70,** 3031). **Deodarin,** *8-methyl-3,5,6,3′,4′-pentahydroxyflavanone*, $C_{16}H_{14}O_7 \cdot \frac{1}{2}H_2O$, m.p. 248–251° (decomp.), $[\alpha]_D^{25}$ +28.5 (acetone), *penta-acetate*, m.p. 145–147°, *pentamethyl ether*, m.p. 150–151°, has been obtained from the stem bark of *Cedrus deodara* (*D. Adinarayana* and *Seshadri*, Tetrahedron, 1965, **21,** 3727). 8-*Methyl*-3,5,7,3′,4′-pentahydroxyflavanone has also been isolated from the above source (*K. Raghunathan, S. Rangaswami* and *Seshadri*, Current Sci., 1971, **40,** 464).

Taxifolin, *dihydroquercetin, 3,5,7,3′,4′-pentahydroxyflavanone*, $C_{15}H_{12}O_7$, colourless needles, m.p. 240–242° (decomp.), $[\alpha]_D^{20}$ +13° (ethanol), is extracted from Douglas fir heartwood (*Pew, loc. cit.*). The *trans*-configuration has been proved for (+)dihydroquercetin and its racemate, and the absolute configuration of the two asymmetric centres in (+)dihydroquercetin have been shown to be identical with those in (+)catechin (*J. W. Clark-Lewis* and *W. Korytnyk*, Chem. and Ind., 1957, 1418; J. chem. Soc., 1958, 2367). For *acetates* of taxifolin and aromadendrin see *H. V. Brewerton* (New Zealand J. Sci. Technol, 1957, **38B,** 697; C.A., 1958, **52,** 4566). The acetate derivatives of taxifolin, eriodictyol, astilbin, and sakuranetin have been prepared and their chemistry studied (*H. Aft*, J. org. Chem., 1958, **26,** 1958). The reactions between related partially acetylated polyhydroxyflavanones, and *N*-bromo- and *N*-iodo-succinimide have been investigated as a synthetic method for the preparation of halogenated flavanone derivatives (*idem, ibid.*, 1965, **30,** 897). One of the major products obtained from the reaction between taxifolin tetramethyl ether and sodium acetate is the corresponding chalcone acetate and not the enol-acetate; with potassium acetate the chalcone acetate is only formed as a minor component, the major product is the pseudobase acetate, XCII (*V. Krishnamoorthy* and *Seshadri*, Indian J. Chem., 1968, **6,** 469):

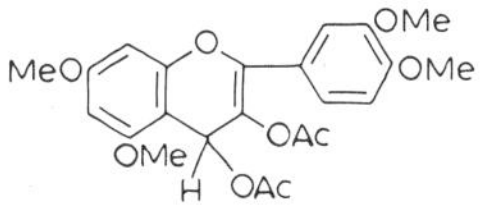

(XCII)

The rearrangement of taxifolin tetramethyl ether with alkali has been discussed (*C. Enebäck* and *Gripenberg*, J. org. Chem., 1957, **22,** 220). **Astilbin,** *dihydroquercetin* 3-L-*rhamnoside*, $C_{21}H_{22}O_{11}$, m.p. 180° (decomp.), *tetramethyl ether*, m.p. 226–227°, has been isolated from rhizomes of *Astilbe odontophylla* and *thunbergi* and from *Litsea glauca* (*K. Hayashi* and *K. Ouchi*, C.A., 1953, **47,** 7493; 1954, **48,** 13841). On heating in aqueous pyridine astilbin yields *neoastilbin*, $C_{21}H_{22}O_{11} \cdot 4H_2O$. m.p. 176–177°, $[\alpha]_D^{19}$ $-152°$, and *isoastilbin*, $C_{21}H_{22}O_{11}$, m.p. 278–280° (decomp.), $[\alpha]_D^{20}$ $-276°$ (*T. Tominaga*, Yakugaku Zasshi, 1958, **78,** 1077; 1960, **80,** 1202). **Phellamurin,** $C_{26}H_{32}O_{12}$, colourless needles, m.p. 205°, is isolated from the Japanese tree *Phellodendron amurense* Ruprecht, and is the 7-glucoside of 3,5,7,4′-tetrahydroxy-8-(γ-hydroxyisovaleryl)-flavanone (*Hasegawa* and *Shirato*, J. Amer. chem. Soc., 1953, **75,** 5507). It occurs along with the corresponding flavonol glucoside **amurensin,** yellow needles, m.p. 290°. **Phellodendroside,** 3-*glucoside* of 3,5,7,4′-*tetrahydroxy*-8-(γ-*hydroxyisovaleryl*)*flavanone*, $C_{26}H_{32}O_{12}$, m.p. 154–156°, is obtained from the leaves of *Phellodendron japonicum* (*T. Bodalski* and *E. Lamer*, Diss. Pharm., 1963, **15,** 319).

Dihydrorobinetin, 3,7,3′,4′,5′-*pentahydroxyflavanone*, $C_{15}H_{12}O_7$, m.p. 246° [226–228° (decomp.)], $[\alpha]_D$ $+29°$ (50% aqueous acetone,) occurs in the wood of *Robinia pseudacacia* (*K. Freundenberg* and *L. Hartmann*, Ann., 1954, **587,** 207; *Roux* and *Paulus*, Biochem. J., 1962, **84,** 416). The 7,3′,4′,5′-*tetramethyl ether*, m.p. 167–168°, has been synthesised (*V. R. Shah et al.*, Chem. and Ind., 1955, 1062). The alkaline benzilic rearrangement which occurs when the tetramethyl ether is treated with sodium hydroxide solution has been investigated (*D. Molho* and *Chadenson*, Bull. Soc. chim. Fr., 1959, 453).

Dihydromorin, 3,5,7,2′,4′-*pentahydroxyflavanone*, $C_{15}H_{12}O_7$, m.p. 228° (anhyd.), *pentaacetate*, m.p. 190–193°, is obtained from E. African mulberry, *Marus lactea* (*W. R. Carruthers et al.*, J. chem. Soc., 1957, 4440). **Ampelopsin,** *dihydromuricetin*, 3,5,7,3′,4′,5′-*hexahydroxyflavanone*, $C_{15}H_{12}O_8 \cdot 2.5H_2O$, m.p. 245–246°, *hexa-acetate*, m.p. 175–176°, occurs in the Chinese plant drug, *Ampelopsis meliaefolia* kudo (*M. Kotake* and *T. Kubota*, Ann., 1940, **544,** 253). **Citromitin,** 5,6,7,8,3′,4′-*hexamethoxyflavanone*, $C_{21}H_{24}O_8 \cdot \frac{1}{2}H_2O$, m.p. 134–136° and 5-O-*desmethylcitromitin*, 5-*hydroxy*-6,7,8,3′,4′-*pentamethoxyflavanone*, $C_{20}H_{22}O_8$, m.p. 146–147°, have been isolated from the peels of *Citrus mitis* Blanco. (*G. P. Sastry* and *L. R. Row*, Tetrahedron, 1961, **15,** 111).

(i) Biflavanones

Three biflavanones have been isolated from the heartwood of *Garcinia buchananii* Baker (Guttiferae) and *G. eugeniifola* Wall, **GB-1** (XCIII, $R = R^2 = H$, $R^1 = OH$), **GB-1a** (XCIII, $R = R^1 = R^2 = H$), and **GB-2** (XCIII, $R = H$, $R^1 = R^2 = OH$). The isolation of a fourth biflavanone has also been reported (XCIII, $R = R^1 = H$, $R^2 = OH$). Their structures have been assigned largely on the basis of u.v., i.r., n.m.r. and mass spectral evidence and by degradation with alkali, which, in the case of GB-1, and GB-1a, gives phloroglucinol and 4-hydroxybenzoic acid. The structures have been proposed with the reservation that the flavanone units may be alternatively 3,6-linked between rings X and Y (*B. Jackson et al.*, Tetrahedron Letters, 1967, 3049; *Jackson, H. D. Locksley* and *F. Scheinmann*, Chem. Comm., 1968, 1125):

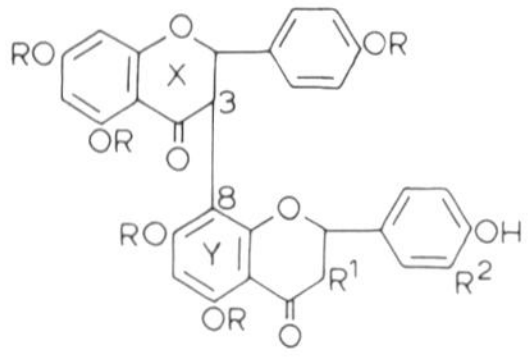

(XCIII)

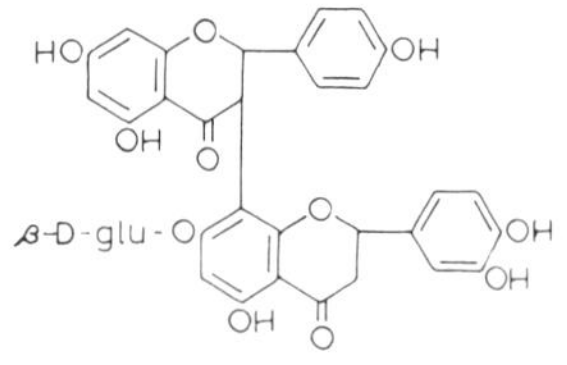

Xanthochymusside

Alternative views on the structures of the above compounds have been given (*A. Pelter*, Tetrahedron Letters, 1967, 1767; 1968, 897). **Xanthochymusside,** $C_{36}H_{32}O_{16} \cdot 2H_2O$, a colourless, amorphous powder, m.p. 219°, has been obtained from the fresh leaves of *G. xanthochymus* Hook. (*M. Konoshima, Y. Ikeshiro* and *S. Migahara, ibid.*, 1970, 4203).

(j) Isoflavanones, 3-*phenylchroman*-4-*ones*, 2,3-*dihydro*-3-*phenyl*-4H-*benzo*[b]*pyran*-4-*ones*

The isoflavanones probably occur more widely in nature than had been imagined. They are prepared by the reduction of isoflavones either catalytically or by alkaline sodium metabisulphite (*N. Narasimhachari* and *T. R. Seshadri*, Proc. Indian Acad. Sci., 1952, **35A**, 202). Isoflavanones on reduction with sodium tetrahydridoborate in 95% ethanol afford isoflavan-4-ols in 80–85% yields (*L. R. Row, A. S. R. Anjaneyula* and *C. S. Krishna*, Current Sci., 1963, **32**, 67).

Ferreirin, 5,7,2'-*trihydroxy*-4'-*methoxyisoflavanone*, $C_{16}H_{14}O_6$, m.p. 210–213°, 7,2'-*dimethyl ether*, m.p. 119–120°, *trimethyl ether*, m.p. 163°, and **homoferreirin,** 5,7-*dihydroxy*-2',4'-*dimethoxyisoflavanone*, $C_{17}H_{16}O_6$, m.p. 168–169°, *diacetate*, m.p. 132–133°, *dimethyl ether*, m.p. 163°, are present in the heartwood of *Ferreirea spectabilis* (*F. E. King* and *K. G. Neill*, J. chem. Soc., 1952, 4752; *Neill, ibid.*, 1953, 3454). Ferreirin has been synthesised (*L. Farkas et al.*, J. chem. Soc., C, 1971, 1994). **Padmakastein,** $C_{16}H_{14}O_5$, m.p. 236–238°, and its 4'-*glucoside*, *padmakastin*, m.p. 225–227°, have been isolated from the bark of *Prunus puddum* (*Narasimhachari* and *Seshadri*, *loc. cit.*).

It was believed that padmakastin was 5,4'-*dihydroxy-7-methoxyisoflavanone, dihydroprunetin, 4'-methyl ether*, m.p. 131–132°, 5,4'-*dimethyl ether*, m.p. 146–147°, but it is reported that the compound prepared by hydrogenation of the corresponding isoflavone is not identical and the same is observed in the case of padmakastin (*A. Malhotra, V. V. S. Murti* and *Seshadri*, Current Sci., 1968, **37,** 668; *Farkas et al.*, Tetrahedron, 1969, **25,** 1013).

Sophorol, 7,2'-*dihydroxy-4',5'-methylenedioxyisoflavanone*, $C_{16}H_{12}O_6$, pale yellow needles, m.p. 215° (180–181°), $[\alpha]_D^{15}$ +9.5° (acetone), −13.6° (ethanol), *dimethyl ether*, m.p. 136–138°, *oxime*, m.p. 195° (decomp.), has been isolated from *Sophora japonica*, L. (*H. Suginome*, J. org. Chem., 1959, **24,** 1655; Tetrahedron Letters, 1960, No. 19, 16; Bull. chem. Soc. Japan, 1966, **39**, 1525) and synthesised (*Farkas, A. Gottsegen* and *M. Nógrádi*, Chem. Comm., 1972, 825). Sophorol is unique in that it is the first example of an optically active isoflavanone. Sophorol on warming with dilute sulphuric acid is dehydrated to give *anhydrosophorol* (XCIV), $C_{16}H_{10}O_5$, colourless needles, m.p. 225–226°:

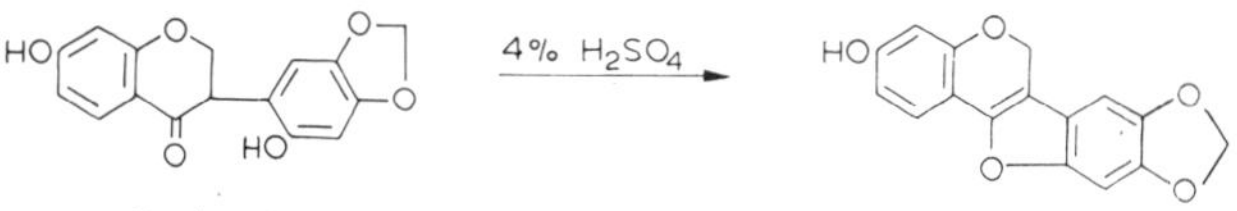

Sophorol —(4% H_2SO_4)→ (XCIV)

o.r.d. Curves have been used in the absolute and relative stereochemical correlation of isoflavanone, sophorol, and naturally-occurring isoflavan derivatives (*Suginone, ibid.*, pp. 409 and 1544).

Dalbergioidin, 5,7,2',4'-*tetrahydroxyisoflavanone*, $C_{15}H_{12}O_6$, m.p. 231–233° (decomp.), and its 7-*methyl ether*, **Ougenin,** $C_{16}H_{14}O_6$, have been isolated along with homoferreirin from the heartwood of *Ougeinia dalbergioides* Linn. Ougenin dimethyl ether has been synthesised and biogenetic relations discussed (*S. Balakrishna et al.*, Proc. roy. Soc. A, 1962, **268**, 1). Dalbergioidin and ougenin have also been synthesised (*A. C. Jain, P. Lal* and *Seshadri*, Indian J. Chem., 1969, **7,** 61; *Farkas et al.*, J. chem. Soc., C, 1971, 1994).

6,7-*Dimethoxy-3',4'-methylenedioxyisoflavanone*, $C_{18}H_{16}O_6$, m.p. 201.5–202.5°, has been isolated along with a number of isoflavones from the heartwood of *Cordyla africana* and its structure confirmed by synthesis (*R. V. M. Campbell, S. H. Harper* and *A. D. Kemp*, J. chem. Soc., C, 1969, 1787).

Isoflavanones have been obtained by the reaction between methylene iodide and *o*-hydroxydeoxybenzoins in acetone in the presence of potassium carbonate (*S. K. Aggarwal, S. K. Grover* and *Seshadri*, Indian J. chem., 1969, **7,** 1059):

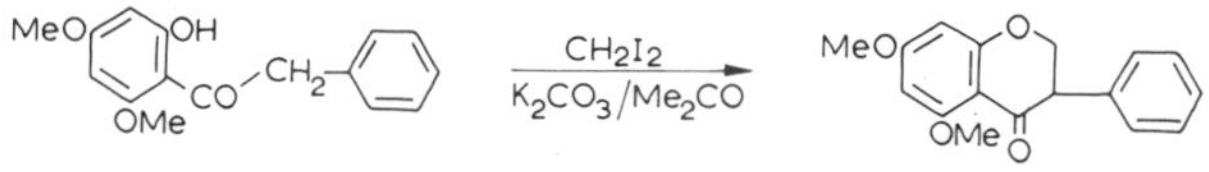

MeO, OH, OMe, CO–CH_2–Ph —(CH_2I_2, K_2CO_3/Me_2CO)→ MeO, OMe isoflavanone

While the use of ethyl formate in isoflavone synthesis leads to the isoflavone itself, except in special cases, the use of methyl formate affords 2-hydroxyisoflavanones (*Narasimhachari, D. Rajagopalan* and *Seshadri*, J. sci. ind. Research India, 1953, **12B,**

287). *2-Hydroxy-5,7-dimethoxy-8-methylisoflavanone*, m.p. 191–192° (*Farkas, J. Varady* and *Gottsegen*, Tetrahedron, 1964, **20**, 351). Methods of synthesis of isoflavanones have been discussed (*Murti, P. V. Raman* and *Seshadri*, Symp. Syn. Heterocycl. Compounds. Physiol. Interest Hyderabad, India, 1964, 64).

Structurally related to the isoflavanones are **pterocarpin,** $C_{17}H_{14}O_5$ (XCV, methylenedioxy group at R^1 and R^2), m.p. 164.5°, $[\alpha]_{5461}^{20.5}$ −207.5° ($CHCl_3$), *dinitrophenylhydrazone*, m.p. 305° (decomp.) and **homopterocarpin,** $C_{17}H_{16}O_4$ (XCV, R^1=OMe, R^2=H), m.p. 87°, $[\alpha]_{5461}^{20.5}$ −236.6° ($CHCl_3$), which are found in red sandalwood *Pterocarpus santalinus*, Linn., and barwood *Baphia nitida*, Lodd (*A. McGookin et al.*, J. chem. Soc., 1940, 787; *A. Robertson* and *W. B. Whalley*, *ibid.*, 1954, 1440):

(XCV)

The structure proposed for pterocarpin is supported by n.m.r. spectral data (*J. B. Bredenberg* and *J. N. Schoolery*, Tetrahedron Letters, 1961, 285). **Nepseudin,** 2′,3′,4′-*trimethoxyfuro*[3,2-g]*isoflavone*, $C_{20}H_{18}O_{61}$, m.p. 115–116°, and **neotenone,** 2′-*methoxy-4′,5′-methylenedioxyfuro*[3,2-g]*isoflavanone*, $C_{19}H_{14}O_6$, m.p. 180–180.5°, have been obtained from *Neorautanenia pseudopachyrrhiza* (*L. Crombie* and *D. Whiting*, *ibid.*, 1963, 1569).

5. 1*H*-Benzo[*c*]pyran, isobenzopyran, 3,4-benzopyran and its derivatives

This group of substances attracted little attention until about twenty years ago, isochromene, 1*H*-benzo[*c*]pyran (I) being obtained in 1958, after the preparation of isochroman in 1954 (p. 297). There are about twenty natural products, which can be looked upon as derivatives of isochromene. Isocoumarin has been more thoroughly studied and some naturally occurring substances such as brevifolinecarboxylic acid (Vol. IIID, p. 240), eleutherin (see Vol. IIIG), and mellein (p. 303) belong to,this class of compound.

Isochromene

(I)

Isochromene is obtained (*1*) by reacting isocoumarin (II) with lithium tetrahydridoaluminate in ether, and ring-closing the resulting carbaldehyde (III) by boiling with acetic anhydride and sodium acetate (*J. N. Chatterjea*, Ber., 1958, **91**, 2636):

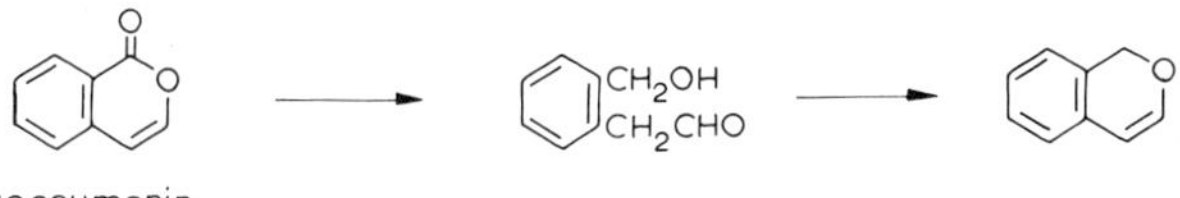

Isocoumarin
(II)

(III)

Isochromene, leaflets, m.p. 21.5°, b.p. 140°/115 mm, n_D^{24} 1.5818, *picrate*, orange-red needles, m.p. 85°, is unstable polymerising at room temperature during two days; on hydrogenation it affords isochroman; 6,7-*dimethoxyisochromene*, prisms, m.p. 64°, which changes within a few days to a viscous mass, *picrate*, m.p. 102–103°; 3-*phenylisochromene*, leaflets, m.p. 124–125°, which changes within a few days to a viscous mass.

(*2*) Irradiation of indene oxide (IV) in benzene or *n*-hexane affords isochromene and indan-2-one (V) in nearly equivalent amounts (*H. Kristinsson, R. A. Mateer* and *G. W. Griffin*, Chem. Comm., 1966, 415):

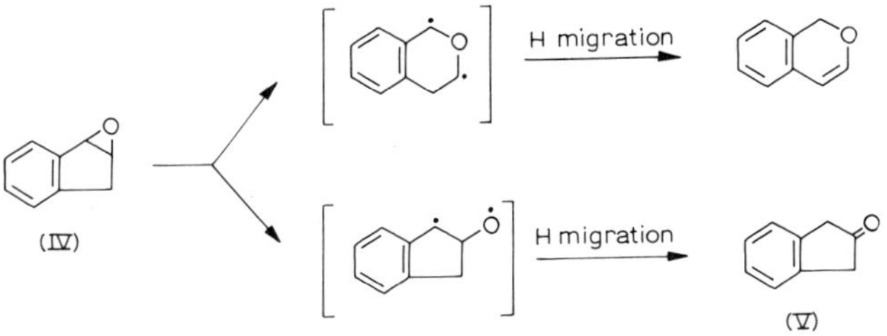

H migration

(IV)

H migration

(V)

(*3*) 3-Methylisochromene has been prepared by heating a dilute solution of 1-acetylbenzocyclobutene in decane; similarly 3-isopropylisochromene and 3-phenylisochromene have been obtained (*R. Hug, H. J. Hansen* and *H. Schmid*, Helv., 1972, **55,** 10):

Purpurogenone, $C_{14}H_{12}O_5$, dark-red prisms, m.p. 310° (decomp.), *triacetate*, m.p. 226°, a metabolic product of *Penicillium purpurogenum* contains an isochromene nucleus (*J. C. Roberts* and *C. W. H. Warren*, J. chem. Soc., 1955, 2992):

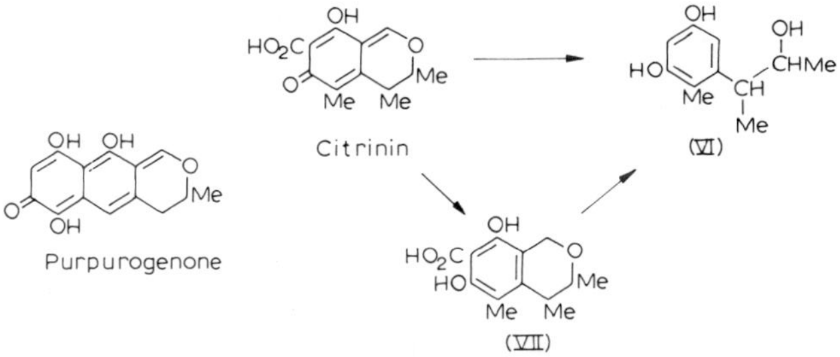

Citrinin

(VI)

Purpurogenone

(VII)

It bears a structural resemblance to (−)**citrinin,** $C_{13}H_{14}O_5$, m.p. 178–179°, the yellow antibiotic metabolite of *Penicillium citrinum* first isolated by *A. C. Hetherington* and *H. Raistrick* (Phil. Trans., 1931, **220B,** 269). The assigned structure is in harmony with the breakdown of citrinin by acid or alkali to the dihydric phenol VI, which has been obtained both in racemic and (−)forms (*J. P. Brown et al.*, J. chem. Soc., 1949, 859). Hydrogenation of citrinin results in aromatisation of part of the molecule and the formation of *dihydrocitrinin* (VII), m.p. 171°, which can also be transformed into the phenol VI (*Brown et al., ibid.*, 1949, 867). The (−)form of VI when fused with potassium hydroxide yields 5-ethyl-4-methylresorcinol. (±)Citrinin has been synthesised from the (+)phenol VI and the natural (−)form, obtained by resolution with brucine (*D. H. Johnson et al., ibid.*, 1950, 2971).

The absolute configuration of citrinin (VIII) has been determined (*R. K. Hill* and *L. A. Gardella*, J. org. Chem., 1964, **29,** 766; *D. W. Mathieson* and *W. B. Whalley*, J. chem. Soc., 1964, 4640):

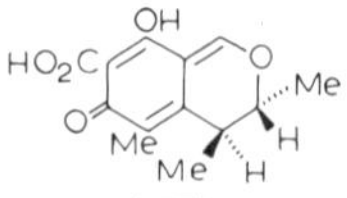

(VIII)

A single-crystal X-ray diffraction study of citrinin has shown that it is a 1,4-benzoquinone methide in which intramolecular repulsion between the methyl groups has resulted in a marked distortion of the molecule (*O. R. Rodig, M. Shiro* and *Q. Fernando*, Chem. Comm., 1971, 1553). **Antimycin,** isolated from cultures of *Penicillium expansum* is identical to (−)citrin (*G. Haese*, Arch. Pharm., 1963, **296,** 227). **Ascochitine,** $C_{15}H_{16}O_5$, m.p. 196–198°, $[\alpha]_D^{25}$ −86° ($CHCl_3$), a yellow pigment isolated from culture filtrates of *Ascochyta pisi* Lib. and obtained from *Ascochyta fabae* Speg (*I. Iwai* and *H. Mishima*, Chem. and Ind., 1965, 186). (±)Ascochitine has been synthesised (*M. N. Galbraith* and *Whalley*, J. chem. Soc., C, 1971, 3557):

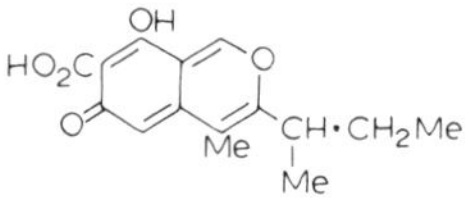

Ascochitine

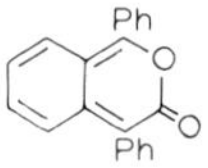

(IX)

1,4-*Diphenylbenzo*[c]*pyran*-3-*one* (IX), m.p. 182–183°, is prepared by dehydration–rearrangement of 1,3-dihydroxy-1,3-diphenylindan-2-one and by dehydration of 2-benzoylphenyl(phenyl)acetic acid (*J. M. Holland* and *D. W. Jones, ibid.*, 1970, 530).

(a) Isocoumarins, 1H-benzo[c]pyran-1-ones

(i) Synthesis

Isocoumarin itself has been described in Vol. IIIE (p. 234) as the anhydride of *o*-β-carboxyphenylvinyl alcohol and its preparation and chemistry has been reviewed (*R. D. Barry*, Chem. Reviews, 1964, **64,** 229).

(*1*) Isocoumarins are often prepared by the Dieckmann method (*W. Dieckmann* and *W. Meiser*, Ber., 1908, **41,** 3253; critical summary, *H. W. Johnston et al.*, J. org. Chem., 1948, **13,** 477). A homophthalic ester, *e.g.* X, is condensed with ethyl formate and the product is cyclised by heating with hydrochloric acid. Hydrolysis of the resultant ester with hydrochloric acid followed by decarboxylation with copper gives a 75% yield of isocoumarin:

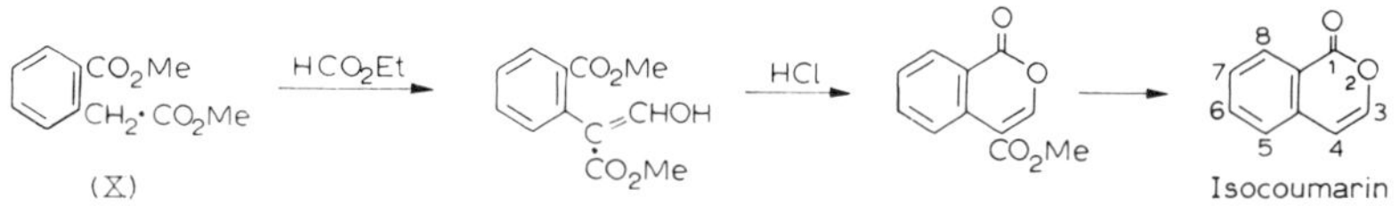

CO_2Me $CH_2 \cdot CO_2Me$ (X) $\xrightarrow{HCO_2Et}$ CO_2Me $C{=}CHOH$ CO_2Me $\xrightarrow{HCl}$ CO_2Me $\longrightarrow$ Isocoumarin

Boron trifluoride can be used for the second stage (*A. Kamal et al.*, J. chem. Soc., 1950, 3375).

(*2*) Treatment of homophthalic anhydride with ethyl chlorocarbonate in the presence of *N,N*-diethylaniline gives the "enol" ester XI, whereas in pyridine the *C*-aryl compound XII is obtained; at room temperature, in the presence of pyridine and a catalytic quantity of homophthalic anhydride XI rearranges by the Claisen–House rearrangement to XII (*J. Schnekenburger*, Arch. Pharm., 1965, **298,** 405):

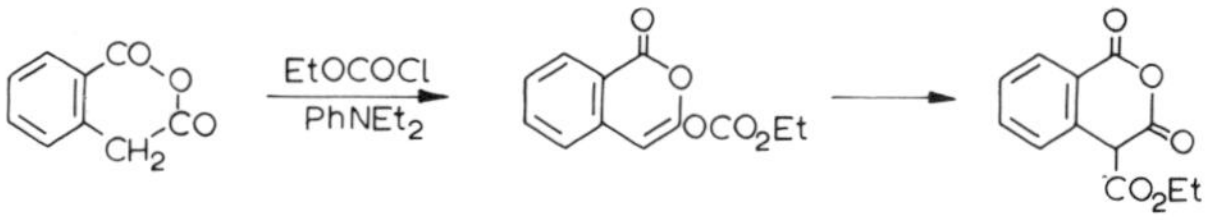

$\xrightarrow[PhNEt_2]{EtOCOCl}$ OCO_2Et (XI) $\longrightarrow$ CO_2Et (XII)

(*3*) Homophthalic anhydride on heating with potassium acetate at 180–190°, undergoes self-acylation, decarboxylation, and intramolecular cyclisation to give 3-(2-carboxybenzyl)isocoumarin (*D. Muceniece* and *V. Oskaja*, Uch. Zap. Latv. Univ., 1970, **117,** 146; C.A., 1972, **77,** 34253):

2 (homophthalic anhydride) $\longrightarrow$ HO_2C, CH_2

Autocondensation of 4-nitro- and 4-methoxy-homophthalic anhydrides in acetic acid in the presence of potassium acetate yield 3-(2-carboxy-4-nitrobenzyl)-7-nitroisocoumarin (XIII; R = NO_2) and 3-(-2-carboxy-4-methoxybenzyl)-7-methoxyisocoumarin (XIII, R = OMe) (*Muceniece*, *S. Berzina* and *Oskaja*, Latv. PSR Zinat. Akad. Vestis, Kim. Ser., 1972, 103):

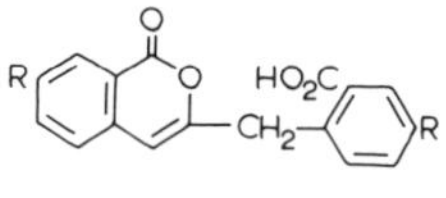

R, HO_2C, CH_2, R

(XIII)

(*4*) Isocoumarins may be obtained by the cyclisation of 2-carboxylbenzyl ketones; usually boiling the ketone with traces of mineral acid will bring about lactone formation. 3-Phenylisocoumarin has been obtained by using thionyl chloride as the cyclisation agent (*S. Siegel*, *S. K. Colburn* and *D. R. Levering*, J. Amer. chem. Soc., 1951, **73**, 3163):

2-Carboxy-4,5-dimethoxyphenylacetaldehyde has been cyclised using orthophosphoric acid to give 6,7-dimethoxyisocoumarin (*N. K. Bose* and *D. N. Chaudhury*, J. Indian, chem. Soc., 1965, **42**, 211).

(5) Halogenated methyl ketones condense with the alkali metal salts of phthaldehydic acids to give high yields of esters XIV, which cyclise when heated with piperidine, yielding 3-substituted isocoumarins (*S. I. Kanevskaya*, *I. N. Kovsharova* and *L. I. Lenevich*, C.A., 1955, **49**, 5477; *Kanevskaya* and *M. N. Shemyakin*, J. pr. Chem., 1932, **132**, 341):

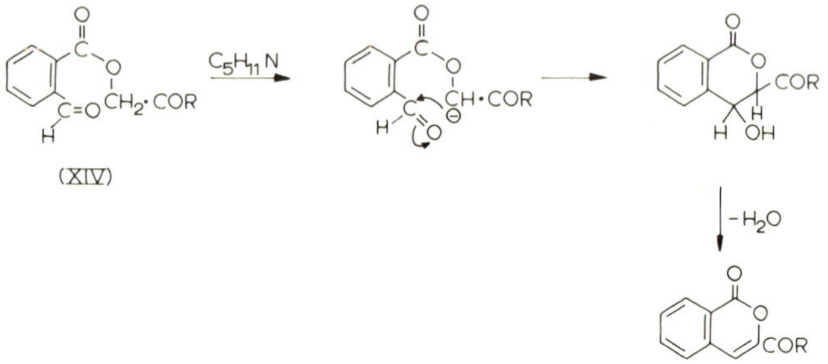

(XIV)

(*6*) Isocoumarin is obtained in small yield by boiling nitromethylenephthalide (Vol. IIIE, p. 233) with hydriodic acid (*S. Gabriel*, Ber., 1903, **36**, 570):

Other phthalide derivatives such as benzylidene- or alkylidene-phthalides can be converted by a series of reactions into isocoumarins, *p*-anisylidenephthalide, for instance, yielding 3-(4-methoxyphenyl)isocoumarin (*A. Horeau* and *J. Jacques*, Bull. Soc. chim. Fr., 1948, 53).

(*7*) Some 3-benzoyloxybutan-2-ones on treatment with sulphuric acid, boron trifluoride or titanium tetrachloride, give isocoumarin,s for example, 3-(3-methoxybenzoyloxy)butan-2-one yields 7-methoxy-3,4-dimethylisocoumarin (*S. Yamaguchi*, *S. Kuroda* and *Y. Kawase*, Nippon Kagaku Zasshi, 1971, **92,** 95):

(*8*) Dihydroisocoumarins may be converted into isocoumarins by treatment with *N*-bromosuccinimide, followed by triethylamine (*J. N. Srivastava* and *Chaudhury*, J. org. Chem., 1962, **27,** 4337; *N. S. Narasimhan* and *B. H. Bhide*, Tetrahedron, 1971, **27,** 6171).

(*9*) Isocoumarins may be prepared *via* the cyclisation of phenylpropionic acids to indanones, which on treatment with isoamyl nitrite give the corresponding hydroxyiminoindanones; conversion to the (2-carboxyphenyl)-acetic acid and cyclisation using acetic anhydride yields the isocoumarin (*J.-Y. Lin*, *S. Yoshida* and *N. Takahashi*, Agric. Biol. Chem., 1972, **36,** 506):

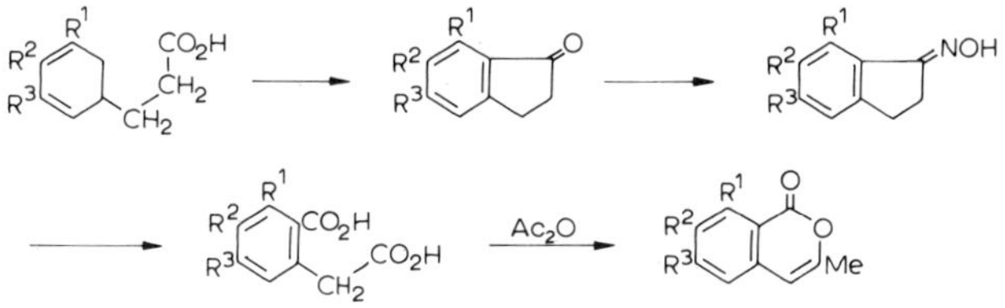

$R^1 = R^2 = R^3 = H$ or OMe
$R^1 = OH$, $R^2 = R^3 = OMe$
$R^1 = R^3 = OMe$, $R^2 = H$
$R^1 = OH$, $R^2 = H$, $R^3 = OMe$

(*10*) 4-Phenylisocoumarin-3-carboxylic acids have been prepared by condensing 2-benzoylbenzoic acids with diethyl bromomalonate for <7 h; when heating is prolonged to 12 h, 3-phenylphthalides are obtained (*J. N. Chatterjea*, *H. C. Jha* and *A. K. Chattopadhyay*, Tetrahedron Letters, 1972, 3409):

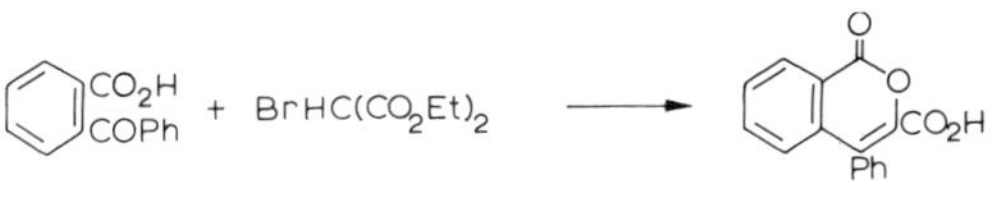

(ii) Properties and reactions

The isocoumarins are unsaturated lactones and are stable to acid and neutral reagents, but are hydrolysed by alkali. In this way 3-methylisocoumarin yields methyl 2-carboxybenzyl ketone. The isocoumarins differ from the coumarins in being odourless and in reacting with ammonia

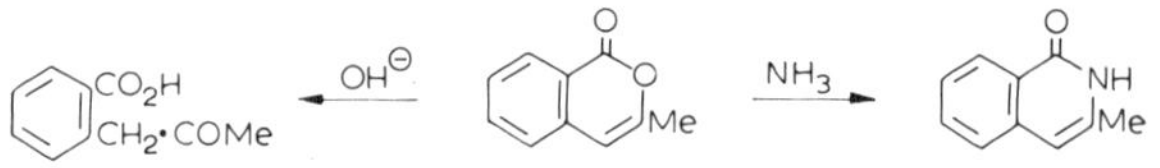

to form isocarbostyrils, isoquinolones (*E. Bamberger* and *W. Frew*, Ber., 1894, **27,** 198; *H. E. Ungnade et al.*, J. org. Chem., 1945, **10,** 533). Heating XV with 2% ammonia gives the corresponding isoquinolone *(Lon, Yoshida* and *Takahashi, loc. cit.)*:

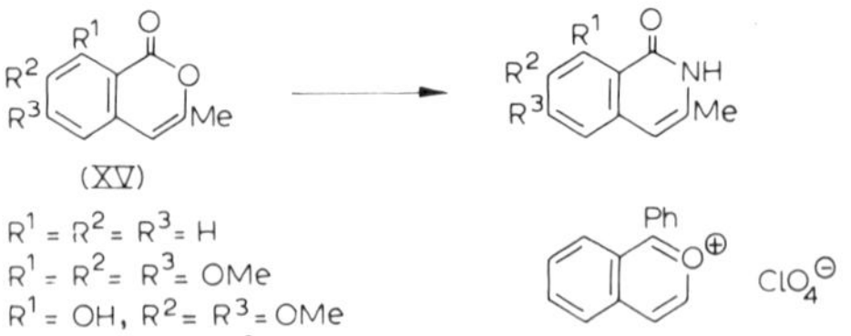

(XV)

$R^1 = R^2 = R^3 = H$
$R^1 = R^2 = R^3 = OMe$
$R^1 = OH$, $R^2 = R^3 = OMe$
$R^1 = R^3 = OMe$, $R^2 = H$
$R^1 = OH$, $R^2 = H$, $R^3 = OMe$

(XVI)

The reaction of ammonia, methylamine, or hydrazine hydrate with a large number of 3-arylisocoumarins has been investigated with a view to the preparation of derivatives of quinoline and 5*H*-2,3-benzodiazepine (*A. Rose* and *N. P. Buu-Hoï*; J. chem. Soc., C, 1968, 2205).

Isocoumarins with Grignard reagents and perchloric acid yield isobenzopyrylium salts, isocoumarin and phenylmagnesium bromide, for instance, giving 1-*phenylisobenzopyrylium perchlorate* (XVI), orange crystals, m.p. 210–211° (*R. L. Shriner et al.*, J. org. Chem., 1949, **14,** 204). It is reported that 3-phenylisocoumarins on reaction with Grignard reagents afford isochromenes (XVII) (*S. Gulgule* and *R. N. Usgaonkar*, J. Indian chem. Soc., 1971, **48,** 707):

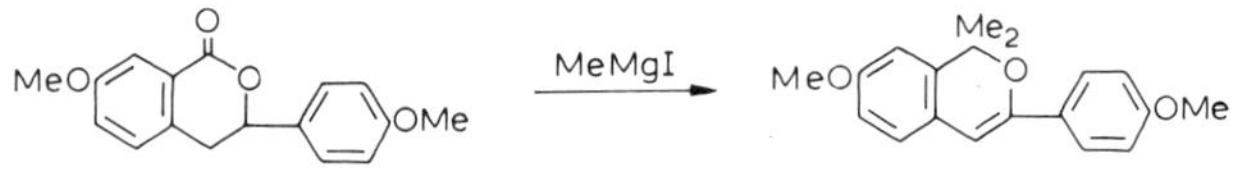

(XVII)

3-(2-Methoxycarbonylbenzyl)isocoumarin (XVIII) on reduction with lithium tetrahydridoaluminate yields the *hemiacetal* XIX, m.p. 95–96°, as the main product; this on treatment in methanol with a little perchloric or sulphuric acid gives an 86% yield of *spiroacetal* XX, m.p. 205–206° (*J. Knabe* and *K. Schaller*, Arch. Pharm., 1968, **301**, 457):

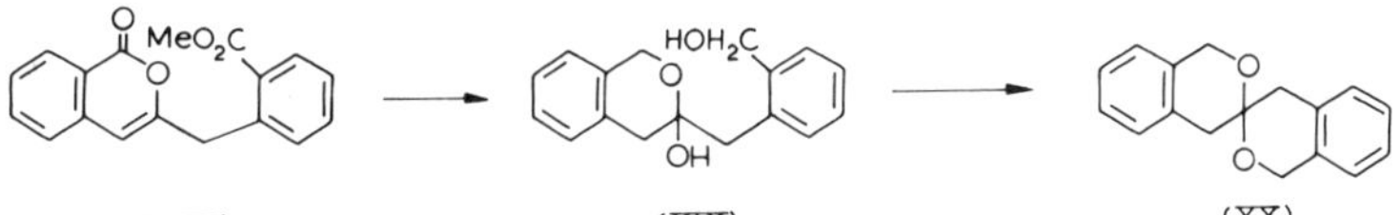

(XVIII) (XIX) (XX)

Other reactions of XVIII have been discussed. Isocoumarins possess characteristic absorption spectra.

Isocoumarin, $C_9H_6O_2$, m.p. 45–46°, b.p. 285–286°/719 mm, u.v. λ_{max} 228 (main band), 239 (4.22), 253 (3.86), 261 (3.87) and 318 mμ (3.58) (log ε values in brackets), prepared by decarboxylation of the 3- or 4-carboxylic acid, forms the 3,4-dibromide with bromine (*Bamberger* and *Frew, loc. cit.).*

3-*Methyl*-, m.p. 73–74° (78°) (*H. Nogami*, J. pharm. Soc. Japan, 1941, **61,** 46; *R. B. Tirodkar* and *Usgaonkar*, J. Indian chem. Soc., 1969, **46,** 935; Indian J. Chem., 1970, **8,** 123); 4-methyl-isocoumarins (*H. K. Desai* and *Usgaonkar*, J. Indian chem. Soc., 1964, **41,** 821, 1019); 3-*ethyl*-, m.p. 76–77° *(idem, loc. cit.)*; 3-*phenyl*-, m.p. 90° (115–116°) (*G. Wang* and *U. Walbe*, Ber., 1938, **71**, 1448; *S. Siegel et al.*, J. Amer. chem. Soc., 1951, **73,** 3163; *M. Renson* and *L. Christiaens*, Bull. Soc. chim. Belges, 1962, **71,** 394; *P. Kaiser* and *J. Schnekenburger*, Z. Naturforsch. B, 1970, **25,** 1190), 3,4-*diphenyl-isocoumarin*, m.p. 101° (*C. F. H. Allen* and *J. W. Gates Jr.*, J. Amer. chem. Soc., 1943, **65,** 1230). **Isocoumarin-3-carboxylic acid,** m.p. 245–246°; *isocoumarin-4-carboxylic acid*, m.p. 250–252° *(Johnson et al., loc. cit.)*; 5-**methoxy**-, m.p. 107–108°, 6-*methoxy*-, m.p. 96°, 8-*methoxy-isocoumarin*, m.p. 114° *(Narasimhan* and *Bhide, loc. cit.)*; 3,4-dimethylisocoumarins (*R. B. Telang* and *Usgaonkar*, *ibid.*, 1968, **45,** 473); 3-ethylisocoumarin-4-carboxylic acid *(Tirodkar* and *Usgaonkar, loc. cit.)*; 7-*methoxy-3-methyl*-, m.p. 93–94°, 3-*ethyl-7-methoxy-isocoumarin*, m.p. 45–46°, 7-*methoxy-3-methylisocoumarin-4-carboxylic acid*, m.p. 210–211° (*idem*, Current Sci., 1968, **37,** 164); 5,6-*dimethylisocoumarin*, m.p. 136° (*Srivastava* and *Chaudhury*, J. Indian chem. Soc., 1963, **40,** 865); 6,7-*dimethoxyisocoumarin*, m.p. 122–123° (*Bose* and *Chaudhury*, *loc. cit.*); 3-alkyl-6,7-dimethoxy-isocoumarins (*Tirodkar* and *Usgaonkar*, J. Indian chem. Soc., 1971, **48,** 192); 7,8-*dimethoxyisocoumarin*, m.p. 149–150° (*P. K. Banerjee* and *Chaudhury*, *ibid.*, 1964, **41,** 221). 10-*Methoxy*-5H, 7H-*benzo*[b]*pyrano*[4,3-b]*benzopyran*-5,7-*dione* (XXI), m.p. 279–281° has been synthesised (*G. Wurm* and *H. Loth*, Arch. Pharm., 1970, **303,** 413):

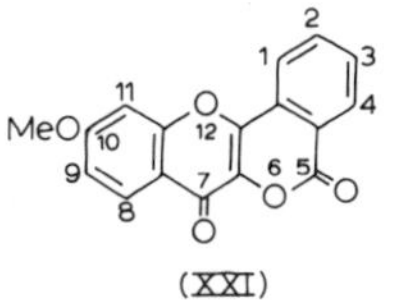

(XXI)

Some isocoumarin derivatives are found in nature. **Bergenin,** $C_{14}H_{16}O_9 \cdot H_2O$, m.p. 138–139° (anhydrous), resolidifying and melting at 230° (238°), $[\alpha]_D^{24}$ −40.7° (ethanol), *tetra-acetate*, m.p. 208–209° (192.5–193.5°,) is obtained from the roots of *Saxifraga siberica* and also from *Astilbe thunbergii* Miguel, *Corylopsis spicata*, *Mallotus japonica*, *Shorea leprosula* and *Humiria balsamifera* (*B. M. Dean* and *J. Walker*, Chem. and Ind.,

1958, 1696). It reduces ammoniacal silver nitrate and Fehling's solution, and gives a violet colour with ferric chloride.

Bergenin

It can be converted into 5,6,7-trimethoxyisocoumarin-3-carboxylic acid (see below) (*S. Fujise et al.*, Bull. chem. Soc., Japan, 1959, **32,** 97). Its structure has been confirmed by synthesis (*Miss J. Evelyn Hay* and *L. J. Hayes*, J. chem. Soc., 1958, 2231; *T. Posternak* and *K. Durr*, Helv., 1958, **41,** 1159). **Vakerin,** obtained from *Caesalpinia digyna*, appears to be identical with bergenin (*W. R. Carruthers et al.*, Chem. and Ind., 1957, 76).

Galloflavin, $C_{12}H_6O_8$, m.p. 290°, *tetra-acetate*, m.p. 230–233°, first prepared by the aeration of an alkaline solution of gallic acid (*R. Bohn* and *C. Graebe*, Ber., 1887, **20,** 2327). With diazomethane it yields a *tetramethyl ether*, m.p. 236–237°, and is converted almost quantitatively by 10% potassium hydroxide under nitrogen into **isogalloflavin,** $C_{12}H_6O_8 \cdot H_2O$, colourless needles, m.p. 275–280°, *triacetate*, m.p. 223° (decomp.), *tetramethyl ether*, m.p. 233–234° (*R. D. Haworth et al.*, J. chem. Soc., 1952, 1583; 1955, 833). The structure of galloflavin was established by its conversion into 5,6,7-trimethoxyisocoumarin-3-acetic acid, identical with a sample prepared from the corresponding 3-carboxylic acid.

Galloflavin

Isogalloflavin (R = CO_2H)
(XXIV) (R = *n*-Pr)

(XXII)

(XXIII)

The structure of isogalloflavin was indicated by methylation and conversion into the isocoumarin derivative XXII and was confirmed by conversion into 2-*n*-propyl-7,8,9-trimethoxy-5*H*-benzo[*c*]pyrano[4,3-*b*]pyran-5-one (XXIV), which with aqueous methanolic potassium hydroxide yielded 3-acetyl-4,5,6-trimethoxyphthalide (XXIII, R = Me), 4,5,6-trimethoxyphthalide-3-carboxylic acid (XXIII, R = OH), methyl *n*-propyl ketone, and *n*-butyric acid (*J. Grimshaw* and *Haworth*, *ibid.*, 1956, 418.

Reticulol, $C_{11}H_{10}O_5$, m.p. 193–193.5°, is produced by *Streptomyces rubrireticuli*; its structure has been confirmed by chemical transformations and spectral data

Reticulol

Canescin-A

Canescin-B

(XXV)

(*L. A. Mitscher*, *W. W. Andres* and *W. McCrae*, Experientia, 1964, **20,** 258). Reticulol dimethyl ether has been synthesised (*Chatterjea et al.*, J. Indian chem. Soc., 1972, **49,** 791). 8-**Hydroxy**-6-**methoxy**-3-**methylisocoumarin** (XXV), $C_{11}H_{10}O_4$, m.p. 129°, has been isolated from *Streptomyces mobaraensis* (*M. A. W. Eaton* and *D. W. Hutchinson*, Tetrahedron Letters, 1971, 1337). **Canescin,** $C_{15}H_{14}O_7$, m.p. 200–202°, $[\alpha]_D^{23}$ +17.8, the mould metabolite from *Aspergillus malignus*; its structure has been assigned on the basis of chemical evidence and spectral data (*A. J. Birch et al.*, *ibid.*, 1965, 29). Canescin has been shown to consist of a mixture of two stereoisomers **canescin-A** and **canescin-B** (*idem*, Austral. J. Chem., 1969, **22,** 1933), and its biosynthesis has been investigated (*idem*, Tetrahedron Letters, 1969, 1519). **Artemidin,** 3-(*but*-1-*enyl*)*isocoumarin*, $C_{13}H_{12}O_2$, m.p. 48°, obtained from *Anthemis fuscata* Brot. (*F. Bohlmann* and *C. Zdero*, Ber., 1970, **103,** 2856) and *Artemisia dracunculus* (*A. Mallabaev*, *I. M. Saitbaeva* and *G. P. Sidyakin*, Khim. Prir. Soedin., 1970, **6,** 467, 531):

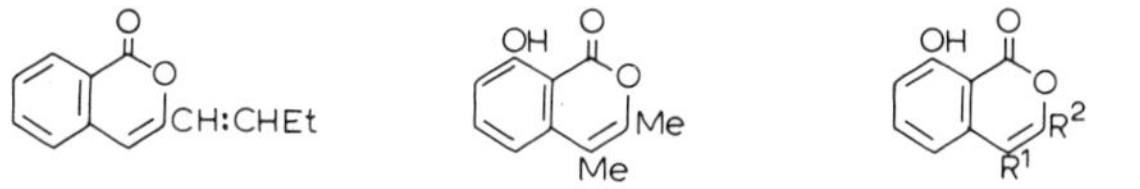

CH:CHEt

Artemidin

OH Me Me

Oospolactone

OH R^2 R^1

Oosponol (R^1 = CO·CH_2OH, R^2 = H)
Oospoglycol (R^1 = CHOH·CH_2OH, R^2 = H)

Oospolactone, 3,4-*dimethyl*-8-*hydroxyisocoumarin*, $C_{11}H_{10}O_3$, colourless needles, m.p. 129°, *acetate*, m.p. 145°, is a metabolic product of *Oospora astringenes* (*K. Nitta et al.*, Agric. biol. Chem. Tokyo, 1963, **27,** 813). **Oosponol**, and **oospoglycol** also isolated as metabolities of the above fungus have been synthesised (*M. Uemura* and *T. Sakan*, Chem. Comm., 1971, 921).

6. Isochroman, 3,4-dihydro-1*H*-benzo[*c*]pyran and derivatives

(*a*) *Isochromans*

(i) Synthesis

Isochroman, 3,4-*dihydro*-1H-*benzo*[c]*pyran* (I), b.p. 91°/13 mm, n_D^{20} 1.5422, is prepared by the interaction of β-phenylethanol, trioxymethylene, and hydrochloric acid at about 0° (*P. Maitte*, Ann. Chim., 1954, [xii], **9,** 431) or by the ring-closure of chloromethyl β-phenylethyl ether (*J. Colonge* and *P. Boisde*, Bull. Soc. chim. Fr., 1956, 1337). The distillation of homophthalyl alcohol (II) at 70°/3 mm with a drop of sulphuric acid gives a nearly quantitative yield of isochroman and water (*J. L. Warnell* and *R. L. Shriner*, J. Amer. chem. Soc., 1957, **79,** 3165):

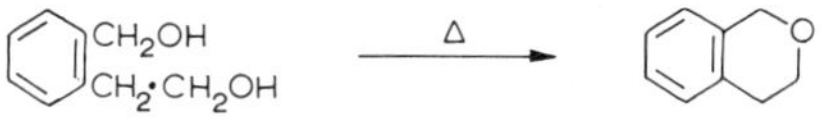

CH_2OH / CH_2·CH_2OH (II) $\xrightarrow{\Delta}$ (I)

It is oxidised by selenium dioxide to *isochromanone* (III) (see p. 300) and unlike chroman does not form a picrate. With hydrogen at 200°/150 atm in the presence of Raney nickel it sluggishly yields *o*-tolylethanol and 5,6,7,8-*tetrahydroisochroman*, b.p. 73°/12.5 mm.

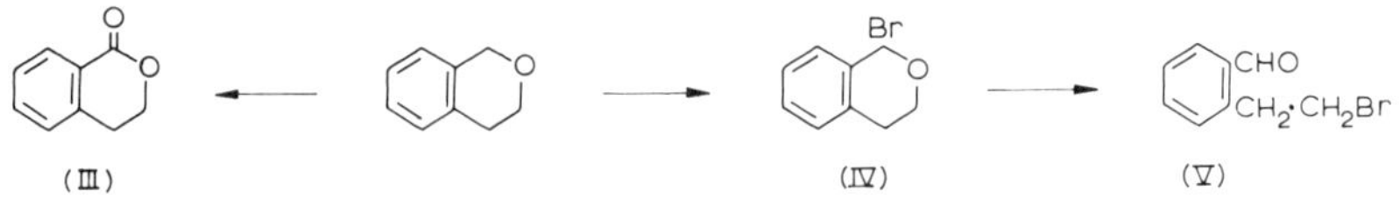

(III) ← isochroman → (IV) → (V) CHO / $CH_2 \cdot CH_2Br$

Isochroman with bromine yields 1-bromoisochroman (IV), which when distilled in a vacuum rearranges to 2-(2-bromoethyl)benzaldehyde (V) (*A. Rieche* and *E. Schmitz*, Ber., 1956, **89**, 1254). It undergoes ring-fission with hydrogen bromide and acetic acid to give 2-bromomethylphenylethyl bromide.

2-Dimethylaminomethylbenzyllithium and some related amino-organolithium reagents have been prepared by metalation of the appropriate 2-methyl- or 2-benzylbenzyldimethylamines with *n*-butyllithium, and condensed with benzaldehyde and certain ketones to form the corresponding amino alcohols. Several of these products have been cyclised through their methiodides to give isochromans (*R. L. Vaulx*, *F. N. Jones* and *C. R. Hauser*, J. org. Chem., 1964, **29**, 1387):

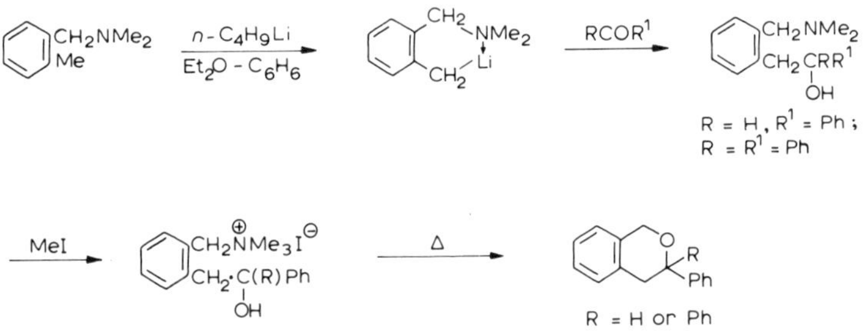

CH_2NMe_2, Me $\xrightarrow[Et_2O - C_6H_6]{n\text{-}C_4H_9Li}$ CH_2–NMe_2 → Li, CH_2 $\xrightarrow{RCOR^1}$ CH_2NMe_2, CH_2CRR^1(OH)

$R = H, R^1 = Ph$;
$R = R^1 = Ph$

$\xrightarrow{MeI}$ $CH_2\overset{\oplus}{N}Me_3I^{\ominus}$, $CH_2 \cdot C(R)Ph$(OH) $\xrightarrow{\Delta}$ (isochroman with R, Ph)

R = H or Ph

(±)Glyceraldehyde reacts with benzene in liquid hydrogen fluoride to give 1-(1,2-dihydroxyethyl)-3-(hydroxymethyl)-4-phenylisochroman (*F. Micheel* and *Z. H. Schleifstein*, Tetrahedron Letters, 1970, 1613):

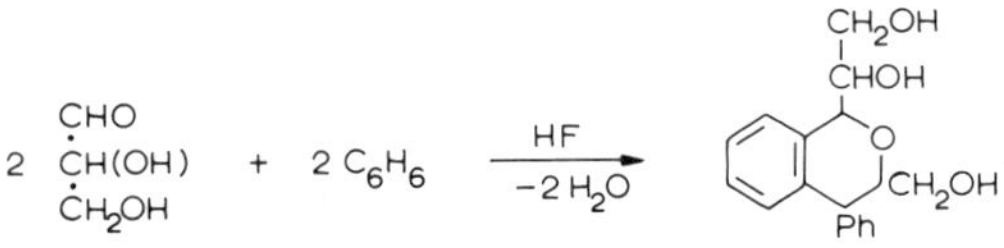

$$2\ CHO \cdot CH(OH) \cdot CH_2OH + 2\ C_6H_6 \xrightarrow[-2H_2O]{HF}$$ (isochroman with CH_2OH–CHOH, CH_2OH, Ph)

The 3-phenylisochroman (VII) has been prepared *via* the deoxybenzoin (VI) (*W. Wiegrebe et al.*, Tetrahedron, 1972, **28**, 2849):

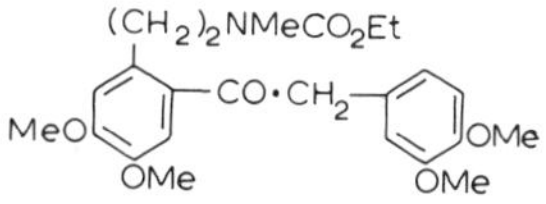

(VI) 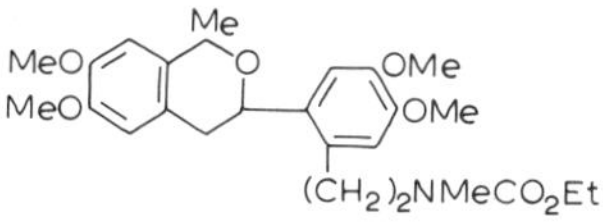 (VII)

Some 1,1-dialkyl-3-arylisochromans have been prepared by reacting the corresponding 3-arylisocoumarin with the appropriate Grignard reagent and then catalytically reducing the resulting isochromene (*S. Gulgule* and *R. N. Usgaonkar*, J. Indian chem. Soc., 1971, **48,** 707).

(ii) Reactions

When 4,4-dimethyl-1-phenylisochroman (VIII) is boiled with a mixture of hydriodic acid and phosphorus for 20 h, it affords a mixture of 10,11-dihydro-10,10-dimethyl-5*H*-dibenzo[*a,d*]cycloheptene (IX) and 9,10-dihydro-9-isopropylanthracene (X) (2:1), and anthracene (*C. van der Stelt et al.*, Rec. Trav. chim. Pays-Bas, 1967, **86,** 1316):

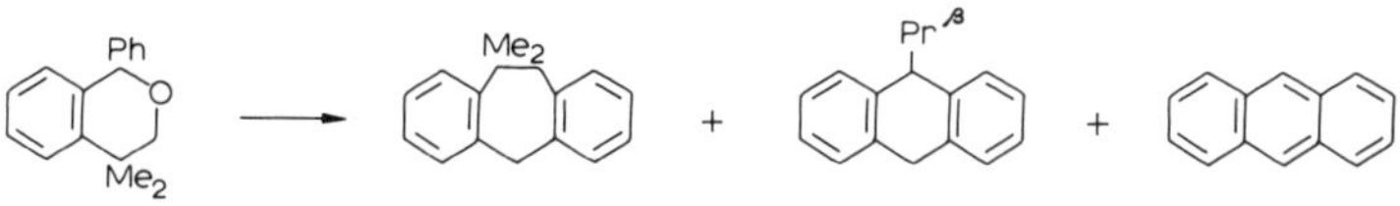

(VIII) (IX) (X)

The mono- and di-substituted isochromans are attacked at $C_{(1)}$ on halogenation or oxidation with selenium dioxide, nitric acid, or hydrogen peroxide; when $C_{(1)}$ is blocked, halide attack occurs at $C_{(4)}$. The exception, 1-phenylisochroman, undergoes ring-opening on oxidation to give *o*-benzylphenylacetic acid and treatment with *N*-bromosuccinimide and sodium ethoxide yields *o*-vinylbenzophenone (*J. Thibault*, Ann. chim., Paris, 1971, **6,** 263). *trans*- and *cis*-4-Methoxycarbonyl-6,7-methylenedioxy-3-phenylisochromans have been synthesised (*M. Palamareva* and *M. Khaimova*, Ber., 1971, **104,** 1400). The isochromans readily undergo autoxidation (*Rieche* and *Schmitz*, *ibid.*, 1957, **90,** 1082).

The reduction products obtained from some isochromenochromones with lithium tetrahydridoaluminate, sodium bis(2-methoxyethoxy)aluminium hydride, and by catalytic hydrogenation have been described (*J. W. Clark-Lewis* and *M. M. Mahandru*, Austral. J. Chem., 1971, **24,** 563).

When di(isochroman-1-yl) ether is heated with concentrated sulphuric acid it undergoes disproportionation with the formation of 1 mole of isochroman and 1 mole of isochroman-1-one (*Rieche* and *Schmitz*, Ber., 1957, **90,** 531):

Isochromone

5,6-*Dimethoxy-*, m.p. 62°, 5-*methoxy-7-methyl-*, b.p. 106–108°/0.5 mm, n_D^{25} 1.5447, 6,7-*dimethoxy-*, m.p. 82–83°, and 6,7-*methylenedioxy-isochroman*, m.p. 87–88° (*J. N. Srivastava* and *D. N. Chaudhury*, J. org. Chem., 1962, **27**, 4337); 5,7-*dimethoxyisochroman*, m.p. 61° (*idem*, J. Indian chem. Soc., 1963, **40**, 865); 3-*phenylisochroman*, m.p. 114–115° (76–77.5°) (*S. Siegel* and *S. Coburn*, J. Amer. chem. Soc., 1951, **73**, 5494); 3,3-*diphenyl-*, m.p. 120–121°, and 6,8-*dimethyl-3-phenyl-isochroman*, m.p. 84.5–86° (*Vaulx, Jones* and *Hauser, loc. cit.*).

Fusarubin, *oxyjavanicin*, $C_{15}H_{14}O_7$, red prisms, m.p. 220°, *dimethyl ether*, m.p. 220° (190°) is the pigment from *Fusarium solani* and was believed to possess structure XI (*H. W. Ruelius* and *A. Gauhe*, Ann., 1950, **569**, 28; **570**, 121; *W. B. Whalley*, Chem. and Ind., 1958, 131). It has now been assigned structure XII, indicating that it is a derivative of *javanicin*, 5,8-dihydroxy-6-methoxy-2-methyl-3-(2-oxopropyl)-1,4-naphthoquinone (*E. Hardegger et al.*, Helv., 1964, **47**, 2027):

(XI) (XII)

(b) Isochromanones

Isochroman-1-one, 3,4-*dihydroisocoumarin*, $C_9H_8O_2$, b.p. 176°/20 mm, 165°/16 mm, $d_4^{18.5}$ 1.203, $n_D^{18.5}$ 1.5664, is prepared by heating 2-(*β-hydroxyethyl*)*benzoic acid*, m.p. 87° (*R. Wegler* and *W. Frank*, Ber., 1937, **70**, 1279; *Warnell* and *Shriner*, *loc. cit.*) or by the oxidation of isochroman with selenium dioxide (*Maitte*, Ann. Chim., 1954, [xii], **9**, 473). 3-*Methylisochroman-1-one*, m.p. 30° (53°) (*R. B. Tirodkar* and *Usgaonkar*, J. Indian chem. Soc., 1969, **46**, 935); 5,6-*dimethoxy-*, m.p. 72–73° (*Srivastava* and *Chaudhury*, *loc. cit.*); 5,7-*dimethoxy-*, m.p. 99–100° (*idem*, J. Indian chem. Soc., 1963, **40**, 865); 6,7-*dimethoxy-*, m.p. 140–141°; 7-*methyl-5-methoxy-*, m.p. 80°; 6,7-*methylenedioxy-isochroman-1-one*, m.p. 127° (*idem*, J. org. Chem., 1962, **27**, 4337). Some isochroman-1-ones have been prepared from the chlorides of *o*-(α,α-dimethylacetonyl)- and *o*-(α,α-dimethylphenacyl)-benzoic acids; 4,4-*dimethyl-3-methylene-*, b.p. 160°/under 15 mm, 3-*methoxy-3,4,4-trimethyl-*, b.p. 165–170°/15 mm, 3,3,4,4-*tetramethyl-*, b.p. 140°/1 mm, 3-*methoxy-4,4-dimethyl-3-phenyl-*, m.p. 174°, 3,4,4-*trimethyl-3-phenyl-*, m.p. 124–125°, 3-*acetoxy-4,4-dimethyl-3-pl enyl-isochroman-1-one*, m.p. 204°; also recorded is the i.r. data of the above compounds (*M. Renson* and *L. Christiaens*, Bull. Soc. chim. Belges, 1962, **71**, 394). Benzaldehyde-2-carboxylic acid and ethyl diazoacetate condense together in ether at room temperature to give 3-*ethoxycarbonyl-4-hydroxyisochroman-1-one*, m.p. 98–99° (*J. Aknin* and *D. Molho*, Bull. Soc. chim. Fr., 1967, 1813):

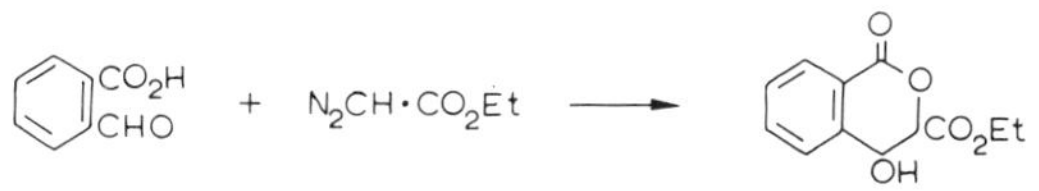

The dehydration of 4-hydroxyisochroman-1-ones to isocoumarins is affected by the nature of the 3-substituent (*S. K. P. Sinha*, J. Indian chem. Soc., 1971, **48,** 369). The relationship between the chemical structure and sweet taste of some isochroman-1-ones has been studied (*M. Yamato et al.*, Yakugaku Zasshi, 1972, **92,** 367).

Homophthalic anhydride condenses with piperonal, 3,4-methylenedioxybenzaldehyde, to give *trans*-3-(3,4-methylenedioxyphenyl)isochroman-1-one-4-carboxylic acid (*S. F. Dyke, M. Sainsbury* and *B. J. Moon*, J. chem. Soc., C, 1971, 3935):

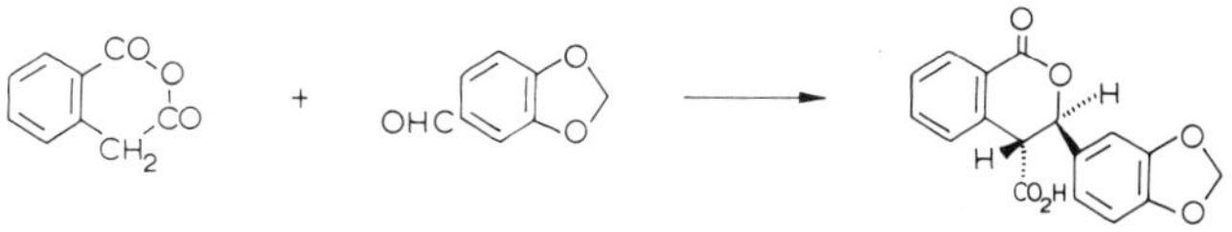

Isochroman-3-one (XIII), m.p. 82–83°, is prepared by the oxidation of indan-2-one by persulphuric acid (*G. A. Swan*, J. chem. Soc., 1949, 1720) or by the hydrolysis of 2-methoxymethylacetonitrile (*F. G. Mann* and *F. H. C. Stewart*, *ibid.*, 1954, 2819);

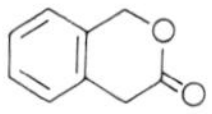

(XIII) 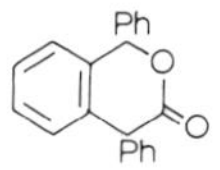 (XIV)

1,4-*diphenylisochroman-3-one*, (XIV), m.p. 166°, is prepared by heating *o*-benzoylbenzilic acid with hydriodic acid (*R. Weiss* and *K. Bloch*, Monatsh., 1933, **63,** 39).

Isochroman-4-ones have been prepared by treating a mixture of trioxymethylene and an aroylmethanol with anhydrous hydrogen chloride; **isochroman-4-one** (XV), b.p. 99°/14 mm (118–120°/12 mm), d_{25} 1.098, n_D^{25} 1.5314 (1.5670), *semicarbazone*, m.p. 222°, 2,4-*dinitrophenylhydrazone*, m.p. 181°; 7-*methylisochroman-4-one*, b.p. 120°/16 mm, *semicarbazone*, m.p. 228°, 2,4-*dinitrophenylhydrazone*, m.p. 196° (*Colonge, G. Descotes* and *M. Fournier*, Compt. rend., 1964, **258,** 4083). Isochroman-4-one has also been prepared by the cyclisation of 2-carboxybenzyloxyacetic acid in acetic anhydride containing potassium acetate (*C. Normant-Chetnay, Thibault* and *Maitte*, Compt. rend. C, 1968, **267,** 547), by the Dieckmann cyclisation of [2]$EtO_2C \cdot C_6H_4 \cdot CH_2OCH_2 \cdot CO_2Et$, and by the Perkin cyclisation of [2]$HO_2C \cdot C_6H_4 \cdot CH_2OCH_2 \cdot CO_2H$ (*Thibault*, Ann. Chim., Paris, 1971, **6,** 381). Reduction of isochroman-4-one using lithium tetrahydridoaluminate affords *isochroman-4-ol* (XVI), m.p. 68°, which on dehydration with potassium hydrogen sulphate yields isochromene (*Normant-Chefnay, Thibault* and *Maitte, loc. cit.*). Treatment of isochroman-4-one with methylmagnesium iodide or phenylmagnesium bromide gives the corresponding 4-methyl- or 4-phenyl-isochroman-4-ol (XVII, R = Me or Ph), which on dehydration afford the corresponding isochromenes (*Thibault, loc. cit.*).

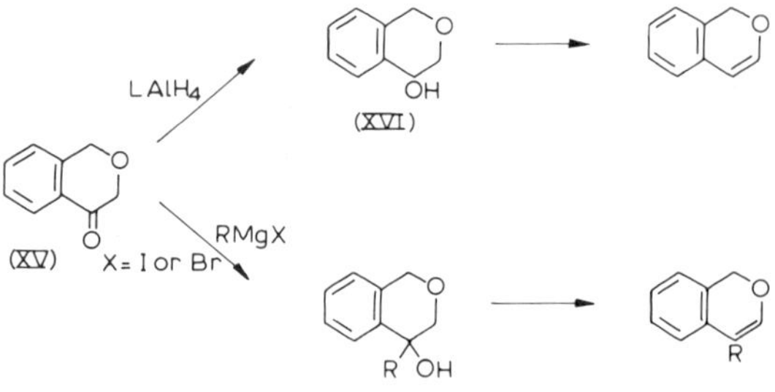

(XV) (XVI) (XVII)

1,1-Dialkylisochromans on treatment with chromium trioxide, selenium dioxide, or hydrogen peroxide and vanadium pentoxide give 1,1-dialkylisochroman-4-ones; n.m.r. spectral data are recorded (*Thibault* and *Maitte*, Bull. Soc. chim. Fr., 1969, 915).

4-Methoxyhomophthalic acid reacts with acetic anhydride in the presence of pyridine at room temperature to give mainly **4-acetyl-7-methoxyisochroman-1,3-dione** (XVIII), m.p. 165–166°, which on treatment with 80% sulphuric acid is converted quantitatively to 7-methoxy-3-methylisocoumarin-4-carboxylic acid (XIX) (*Tirodkar* and *Usgaonkar*, *loc. cit.*):

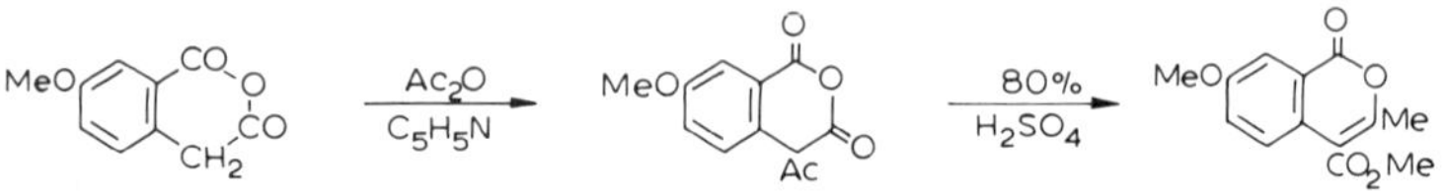

(XVIII) (XIX)

The reaction of phthalic acid monomethyl ester chloride with diazomethane gives the diazoketone XX, which on warming with 3% sulphuric acid affords **isochroman-1,4-dione** (XXI). The reaction occurs *via* the hydroxyketone, which splits out methanol. The 1,4-dione XXI, on boiling with 20% hydrochloric acid, yields naphthacene-5,6,11,12-tetrone (XXII) (*L. Reichel* and *W. Hampel*, Z. Chem., 1963, **3,** 28):

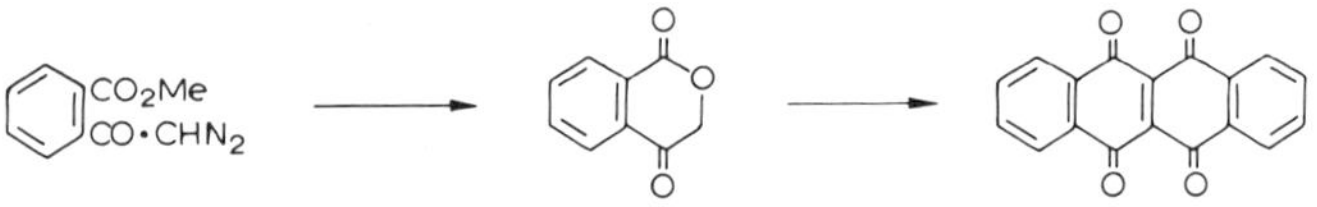

(XX) (XXI) (XXII)

When formaldehyde is added to a solution of isochroman-1, 4-dione in water it gives 3,3′-*bis-(isochroman*-1,4-*dionyl)methane* (XXIII), m.p. 235° (*N. P. Buu-Hoï, P. Jacquignon* and *M. Mangane*, Rec. Trav. chim., 1965, **84,** 334):

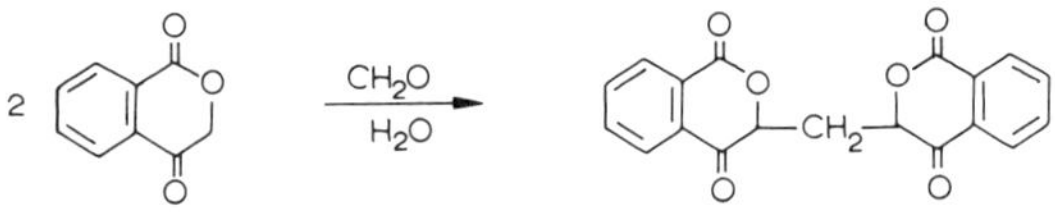

(XXIII)

Mellein, $C_{10}H_{10}O_3$, m.p. 58°, $[\alpha]_D^{12}$ −108.15° ($CHCl_3$), *acetate*, m.p. 126–127°, obtained from the fungus *Aspergillus melleus* grown on a medium containing sucrose (*E. Nishikawa*, J. agric. Chem. Soc. Japan, 1933, **9**, 772) or maltose (*H. S. Burton*, Nature, 1950, **165**, 274) is identical with **ochracin,** (−)-3,4-dihydro-8-hydroxy-3-methylisocoumarin, 8-hydroxy-3-methylisochroman-1-one, a metabolic product of *A. ochraceus* (*T. Yabuta* and *Y. Sumiki*, J. agric. chem. Soc. Japan, 1934, **10,** 703). (±)Mellein has been synthesised from 4-bromo-7-methoxyindanone (*M. Matsui, K. Mori* and

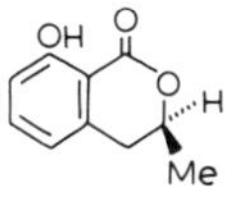

Mellein
Ochracin

S. Arasaki, Agric. biol. Chem., Tokyo, 1964, **28,** 896). 2-Methoxy-*N*-methylbenzamide in tetrahydrofuran on treatment with a 5 molar excess of *n*-butyllithium in ether, followed by reaction with propylene oxide gives the alcohol XXIV, which on hydrolysis gives (±)*mellein methyl ether*, m.p. 68°, converted by treatment with hydrobromic acid to (±)*mellein*, m.p. 38–39° (*N. S. Narasimhan* and *B. H. Bide*, Chem. Comm., 1970, 1552):

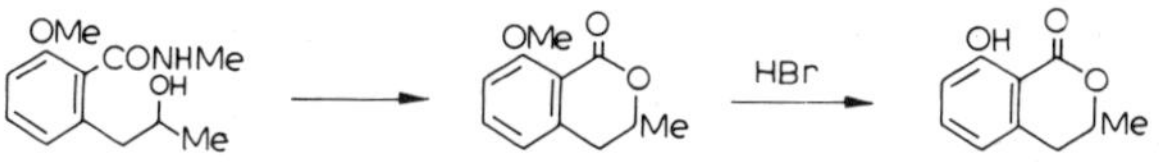

(XXIV)

The (±)methyl ether has also been synthesised by a longer route starting with 3-methoxy-2-nitrobenzoic acid (*J. Blair* and *G. T. Newbold*, J. chem. Soc., 1955, 2871). The absolute configuration of mellein has been determined by circular dichroism and it has been shown to be the (3*R*)-isochroman-1-one derivative (*H. Arakawa*, Bull. chem. Soc. Japan, 1968, **41,** 2451). (±)5-*Methylmellein* has been synthesised (*J. N. Chatterjea et al.*, J. Indian chem. Soc., 1972, **49,** 791). (−)-8-**Hydroxy-6-methoxy-3-methylisochroman-1-one,** (−)-3,4-*dihydro-8-hydroxy-6-methoxy-3-methylisocoumarin*, $C_{11}H_{12}O_4$, m.p. 75–76° $[\alpha]_D^{24}$ −56° (aq. MeOH) occurs in carrots (*E. Sondheimer*, J. Amer. chem. Soc., 1957, **79,** 5036).

Erythrocentaurin, $C_{10}H_8O_3$, m.p. 140–141°, *semicarbazone*, m.p. 219–220°, hydrolysis product of swertiamorin (*T. Kubota* and *Y. Tomita*, Chem. and Ind., 1958, 230; Tetrahedron Letters, 1961, **176,** 223) has been synthesised (*E. Wenkert, D. B. R. Johnston* and *K. G. Dave*, J. org. Chem., 1964, **29,** 2534):

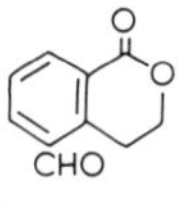

Erythrocentaurin

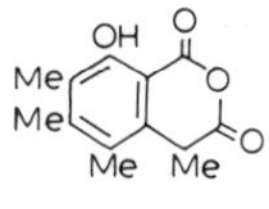

Sclerin

Sclerin, $C_{13}H_{14}O_4$, m.p. 123°, *monoacetate*, m.p. 156°, *methyl ether*, m.p. 104–105°, is a metabolite of *Sclerotinia libertiana* (*T. Tokoroyama, T. Kamikawa* and *Kubota*, Tetrahedron, 1968, **24,** 2345). Its structure has been confirmed by synthesis (*Tokoroyama et al., ibid.*, 1969, **25,** 1047) and its biosynthesis has been discussed (*Tokoroyama* and *Kubota*, J. chem. Soc., C, 1971, 2703).

The structure and stereochemistry of the hexahydroisochromanone (XXV), obtained from the carvenone–benzaldehyde condensation has been discussed (*C. A. R. Baxter, G. C. Forward* and *D. A. Whiting, ibid.*, 1968, 1162, 1169):

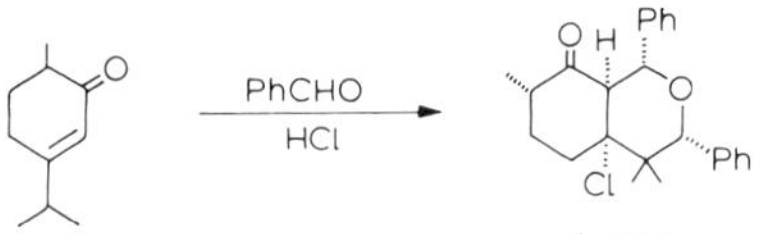

PhCHO / HCl → (XXV)

7. Xanthene, 2,3-5,6-dibenzopyran, dibenzo[*a,e*]pyran and its derivatives

Xanthene, the parent compound of a number of naturally* occurring substances, and some synthetic dyes, is dibenzo[*a,e*]pyran. From it are derived xanthydrol (I), xanthylium salts (II) and the carbonyl compound xanthone (III). The I.U.P.A.C. numbering is as shown.

Xanthene derivatives have long been known and include euxanthic acid and the related euxanthone.

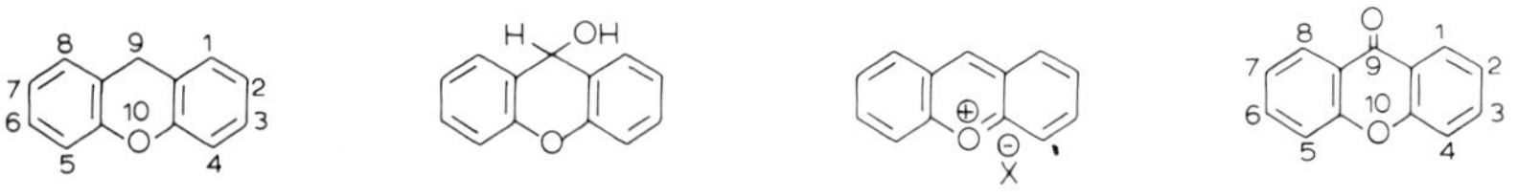

Xanthene (I) (II) (III)

(a) Xanthene, alkyl- and aryl-xanthenes

(i) Xanthene

Xanthene was first prepared by the distillation of euxanthone with zinc dust. The methods of preparation generally give poor yields. It is formed by the dehydration of 2,2′-dihydroxydiphenylmethane by heat (*C. A. Buehler et al.*, J. org. Chem., 1943, **8,** 316): along with xanthone by the pyrolysis of xanthydrol (*A. Schönberg* and *A. Mustafa*, J. chem. Soc., 1944, 305); and by Wolff–Kishner reduction of xanthone (p. 316).

* "Xanthones, Xanthenes, Xanthydrols and Xanthylium Salts", by *S. Wawzonek* in Heterocyclic compounds, Ed. *R. C. Elderfield*, (Wiley and Sons, Inc., New York, 1951), Vol. II, p. 419.

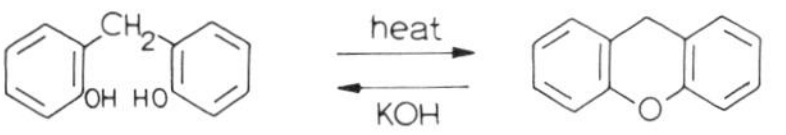

The *structure* follows from its synthesis: its relationship to xanthone into which it is converted by oxidation with chromic acid or potassium permanganate; and its degradation to 2,2′-dihydroxydiphenylmethane by fusion with potassium hydroxide (*O. Kruber* and *H. Lauenstein*, Ber., 1941, **74**, 1693). The anthracene-like framework of xanthene is reflected in the stability of xanthene towards hydrogen iodide at 180–190°, and its ready oxidation to xanthone. The dipole moment, 1.28 D, is similar to that of diphenyl ether, 1.13 D (*E. Bergmann* and *A. Weizmann*, Trans. Faraday Soc., 1936, **32,** 1318). For the conformational analysis of xanthene see *Z. Aizenshtat et al.*, Israel J. chem., 1972, **10,** 753.

Xanthene has also been obtained by reducing xanthone with diphenylsilane at 260° (*H. Gilman* and *J. Diehl*, J. org. Chem., 1961, **26,** 4817):

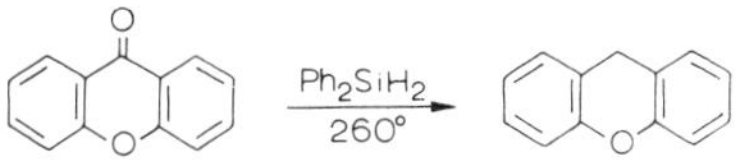

and by the reduction of xanthydrols by controlled-potential electrolysis on a mercury cathode in dimethylformamide (*P. H. Given* and *M. O. Peover*, Nature, 1959, **184,** Suppl. No. 14, 1064). High yields of pure xanthene may be prepared by heating an intermediate phenol–formaldehyde condensation product in the presence of a phosphorus pentoxide catalyst (*J. A. Crowder*, *E. E. Gilbert* and *H. R. Nychka*, U.S. P. 2,783,246/1957).

The 3-methyl-1,4-naphthoquinon-2-ylmethyl anion (IV) adds to trimethyl-1,4-benzoquinone giving, by 1,4-cycloaddition, the xanthene derivative V but not the isomer VI. (*F. M. Dean* and *L. E. Houghton*, J. chem. Soc., C, 1970, 722):

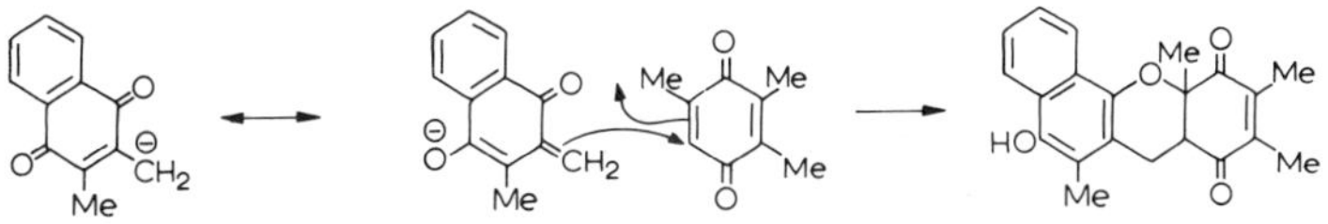

(IV)

(V)

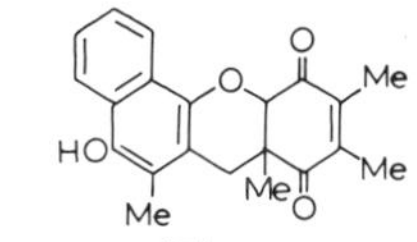

(VI)

Xanthene crystallises in colourless plates, m.p. 99.15–100.5°, b.p. 315°. It dissolves with a yellow colour and intense yellowish-green fluorescence in concentrated sulphuric acid. Many xanthene derivatives such as fluorescein and eosin are strongly fluorescent. Xanthene with nitric acid yields xanthone or 2,7-dinitroxanthone and with sulphuric acid yields a mixture of sulphonic acids (*A. von Baeyer*, Ann., 1910, **373,** 138). With benzoyl chloride in the presence of aluminium it gives 2-*benzoylxanthene*, m.p. 148°, which is reduced to 2-*benzylxanthene* m.p. 93–94°, and oxidised by nitric acid to 2-*benzoylxanthone*, m.p. 146–147°, the structure of which was established by synthesis (*J. Heller* and *St. von Kostanecki*, Ber., 1908, **41,** 1324). Xanthene is oxidised by ozone or lead tetra-acetate to xanthydrol and xanthone (*C. W. K. Cavill et al.*, J. chem. Soc., 1949, 1567), and is autoxidised to xanthone in dimethyl sulphoxide-*tert*-butyl alcohol containing potassium *tert*-butoxide (*G. A. Russell et al.*, J. Amer. chem. Soc., 1962, **84,** 2652). The mechanism of the acid-catalysed autoxidation of xanthene to xanthylium ion and hydrogen peroxide in mixtures of perchloric and acetic acids at 20° has been studied (*A. P. ter Borg, H. R. Gersmann* and *A. F. Bickel*, Rec. Trav. chim., 1966, **85,** 899).

Xanthene in sunlight condenses with ketones and yields with xanthone, for example, 9-*xanthylxanthydrol* (VII), m.p. 194° (decomp.), which loses water easily to give **bixanthylene** (VIII), $C_{26}H_{16}O_2$, m.p. 315°, when heated with acetyl chloride (*Schönberg* and *Mustafa*, J. chem. Soc., 1944, 67). Bixanthylene is also prepared by the reduction of xanthone by zinc and hydrochloric acid (*G. Gurgenjanz* and *Kostanecki*, Ber., 1895,

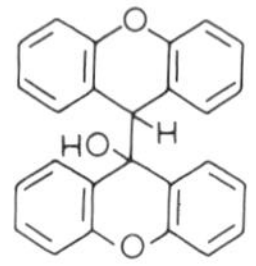

(VII) 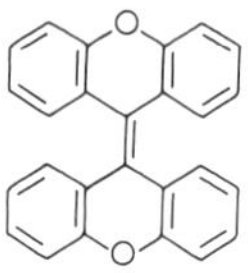 (VIII) 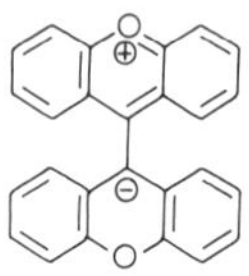 (IX) 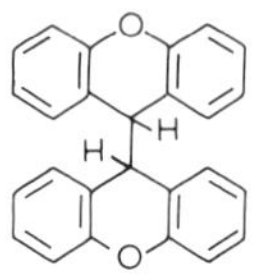 (X)

28, 2310) or by treating xanthione with copper (*Schönberg*, Ber., 1924, **57,** 2133). The mild oxidation of xanthene (*e.g.* by benzoquinone) in sunlight affords 9,9′-**bixanthyl** (X), m.p. 209°, which is also obtained by the reduction of xanthione by zinc and hydrochloric acid (*Schönberg* and *Mustafa*, *loc. cit.*). Bixanthylene is thermochromic, *i.e.* it is colourless at low temperatures, but bluish-green when fused (*Schönberg et al.*, J. chem. Soc., 1946, 442; *Mustafa* and *M. E. E. Sobhy*, J. Amer. chem. Soc., 1955, **77,** 5124). This property Schönberg ascribes to temperature-dependent resonance between the colourless ethylene structure VIII and the coloured betaine structure IX, but this has not won general acceptance, principally because of the independence of the thermochromic absorption spectrum on the acidity of the solvent. A diradical theory has been advanced to account for the green colour (*Schönberg et al.*, J. Amer. chem. Soc., 1953, **75,** 337). Molecules of bixanthylene in the yellow or blue-green crystal modification show the same untwisted conformation. This does not agree with the view that thermochromism is caused by "conformation twisting"; some implications have been discussed (*J. F. D. Mills* and *S. C. Nyburg*, J. chem. Soc., 1963, **308,** 927). Later it was recognised that the phenomenon is related to the mutal interference of the four hydrogen atoms in the immediate vicinity of the double bond; substitution of at least two of these four hydrogen atoms destroys the thermochromism of the system (*E. D. Bergmann et al.*, Tetrahedron, 1966, Suppl. **7,** 349). Photosensitised dehydrogenation of bixanthylene and the constitution of its thermochromic form have been studied (*Schönberg* and *K. Junghans*, Ber., 1965, **98,** 2539). Other properties not usually associated with the ethylenic linkage are the formation of xanthione and 9,9-dichloroxanthene when bixanthylene is heated with sulphur and thionyl chloride, respectively (*Schönberg* and *W. Asker*, J. chem. Soc., 1942, 272; *Mustafa* and *Sobhy*, *loc. cit.*). Treatment of bixanthylene in boiling xylene with Raney nickel for 8–9 h affords xanthene (53%) and unchanged starting material (*Schönberg*, *K. H. Brosowski* and *E. Singer*, Ber., 1962, **95,** 2984).

Xanthene cyclic ethers (XI and XII) are quantitatively cleaved by lithium tetrahydridoaluminate in boiling ether to give xanthene (*N. Latif* and *N. Mishriky*, J. org. Chem., 1962, **27,** 846):

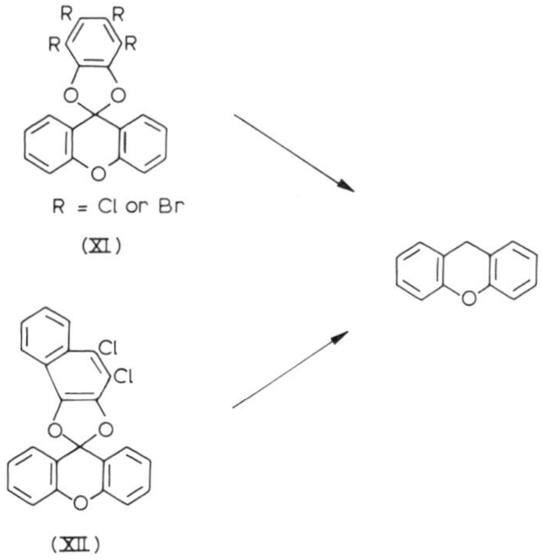

R = Cl or Br

(XI)

(XII)

(ii) Alkyl- and aryl- xanthenes

Alkyl- and aryl-xanthenes are found in coal-tar distillates (*F. Ruszig*, Z. angew Chem., 1919, **32,** 37; *J. Marcusson*, *ibid.*, p. 385) and include α-pyrocresol, 3,6-*dimethylxanthene*, m.p. 196° (*F. Zmerzlikar*, Monatsh., 1910, **31,** 897). Those xanthenes unsubstituted in the 9-position are oxidised by chromic anhydride and acetic acid to the corresponding xanthones or xanthonecarboxylic acids depending on the temperature. The alkyl groups are sometimes very resistant to oxidation (*A. Eckert* and *F. Seidel*, J. pr. Chem., 1921, [ii], **102,** 338). The m.p.'s of some alkyl- and aryl-xanthenes and of the corresponding xanthones and xanthydrols are given in Table 8.

TABLE 8

XANTHENES, XANTHYDROLS AND XANTHONES

Substituents	*Xanthene m.p. °C*	*Ref.*	*Xanthydrol m.p. °C*	*Ref.*	*Xanthone m.p. °C*	*Ref.*
2-Methyl-					125.5	1,2
3-Methyl-	121	3			98	3
4-Methyl-					126	1
9-Methyl-			110–111	4		
1,8-Dimethyl-	215(200)	5	132	19	172	5
2,4-Dimethyl-					152	1
2,7-Dimethyl-	167–168	6–8			143	6,9
3,6-Dimethyl-	200–201	6,10			166–167	10
4,5-Dimethyl-	No m.p. reported	11			172	11,12
9-Benzyl-	69–70(80)	13,14	152	13		
9-Benzylidene-	116	13,14				
3-Phenyl-					141–141.5	16
9-Phenyl-	145	14–16	157–158	See text		
9,9-Diphenyl-	200	9,15				
2-Methyl-9--phenyl-	145	18				
3,9-Diphenyl-	146.5–147	17	128.5–129	17		
3,6-Diphenyl-					193.5–194.5	17

References

1 *F. Ullmann* and *M. Zlokasoff*, Ber., 1905, **38,** 2114.
2 *J. Meisenheimer*, J. pr. Chem., 1928, [ii], **119,** 360.
3 *W. Borsche* and *A. Geyer*, Ber., 1914, **47,** 1157.
4 *J. B. Conant et al.*, J. Amer. chem. Soc., 1926, **48,** 1747.
5 *G. Illari* and *L. Orsolini*, Gazz, 1948, **78,** 813.
6 *T. Sengoku*, J. pharm. Soc. Japan, 1933, **53,** 177, 962; C.A., 1935, **29,** 5444.
7 *A. T. Carpenter* and *R. F. Hunter*, J. chem. Soc., 1954, 2731.
8 *J. Zehenter*, Monatsh., 1961, **37,** 587.
9 *A. Schönberg* and *W. Asker*, J. chem. Soc., 1946, 609.
10 *J. Postowsky* and *B. Lugowkin*, J. pr. Chem., 1929, [ii], **122,** 141.
11 *R. Möhlau*, Ber., 1916, **49,** 168.
12 *A. Lespagnol* and *J. Dupas*, Bull. Soc. chim. Fr., 1937, [v], **4**, 541.
13 *E. Bergmann* and *S. Fujise*, Ann., 1930, **480,** 196.
14 *H. E. Ungnade et al.*, J. Amer. chem. Soc., 1953, **75,** 3333.
15 *Ullmann* and *G. Engi*, Ber., 1904, **37,** 2371.
16 *M. H. Hubacher*, J. Amer. chem. Soc., 1943, **65,** 1655.
17 *C. S. Schoepfle* and *J. H. Truesdale*, J. Amer. chem. Soc., 1937, **59,** 372.
18 *H. Berlitzer*, Monatsh., 1915, **36,** 207.
19 *Lespagnol et al.*, Bull. Soc. chim. Fr., 1939, [v], **6,** 1625.

Xanthene-9-carboxylic acid, xanthanoic acid, m.p. 223–224°, *methyl ester*, m.p. 85–86° (*J. B. Conant* and *B. S. Garvey*, J. Amer. chem. Soc., 1927, **49,** 2085; *A. Wander, A.-G.Swiss* P. 306, **540,** 1955), has been obtained also by the carbonation of 9-xanthyllithium (*R. A. Heacock, R. L. Wain* and *F. Wightman*, Ann. Appl. Biol., 1958, **46,** 352).

A number of 9-xanthenyl and 9-thioxanthenyl esters, ethers and amines have been synthesised (*L. Capuano, R. Zander* and *A. Bolourtschi*, Ber., 1971, **104**, 3750).

3,6-*Bisdimethylaminoxanthene*, m.p. 116°, is the leuco base of Pyronine G. It is obtained by condensing formaldehyde (1 mole) with *m*-dimethylaminophenol (2 moles) and cyclising the product in sulphuric acid.

The isomerisation of 14*H*-dibenzo[*a,j*]xanthene has been investigated (*J. Suszko* and *W. Mitura*, Bull. acad. polon. sci., Sér. sci., Chim., géol. et géograph., 1959, **7**, 391; C.A., 1960, **54**, 21072). 2-Hydroxy-1-naphthaldehyde in the presence of hydrogen chloride affords dibenzo[*a,j*]xanthylium chloride (XIII) (*K. G. Dzhaparidze* and *I. A. Mzhavanadze*, Soobshch. Akad. Nauk Gruz. SSR, 1971, **63**, 325):

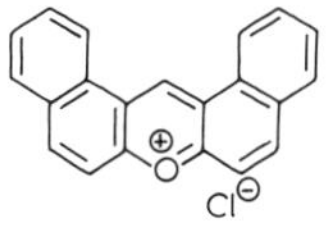

(XIII) 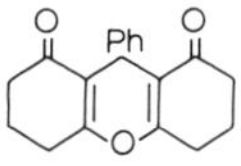 (XIV)

Benzaldehyde condenses with cyclohexane-1,3-dione in acetic anhydride to give 9-*phenyl*-1,2,3,4,5,6,7,8-*octahydroxanthene*-1,8-*dione* (XIV), m.p. 255° (*G. Swoboda* and *P. Schuster*, Monatsh., 1964, **95**, 398).

Cyclohexene-1,3-diones condense with 2-hydroxybenzyl alcohol on heating with hexamethylphosphoramide at 185° to give 3,4-dihydro-1(2*H*)-xanthenones (*P. Yates, D. J. Bichan* and *J. E. McCloskey*, Chem. Comm., 1972, 839):

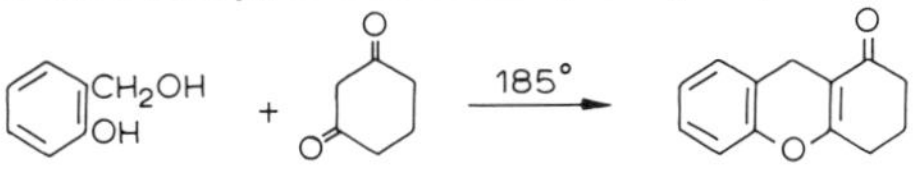

CH_2OH, OH + → 185°

The diketone XV on treatment with triphenylmethyl perchlorate in acetonitrile affords an 80–90% yield of 1,2,3,4,5,6,7,8-*octahydroxanthylium perchlorate* (XVI), m.p. 174–175° (*A. T. Balaban* and *N. S. Barbulescu*, Rev. Roumaine Chim., 1966, **11**, 109):

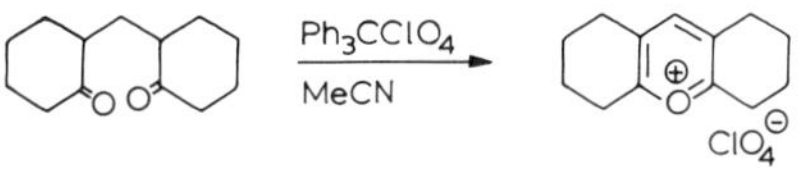

Ph_3CClO_4 / MeCN; $ClO_4^{\ominus}$

(XV) (XVI)

The heartwood of the red sandalwood *Pterocarpus santalinus*, contains a number of red pigments, the two major ones being **santalin A** and **santalin B**. On methylation both of them give the same "*santalin permethyl ether*" (XVII) $C_{38}H_{36}O_{10}$, orange-yellow needles, m.p. 155–156°, resolidifies and melts at 229–230°.

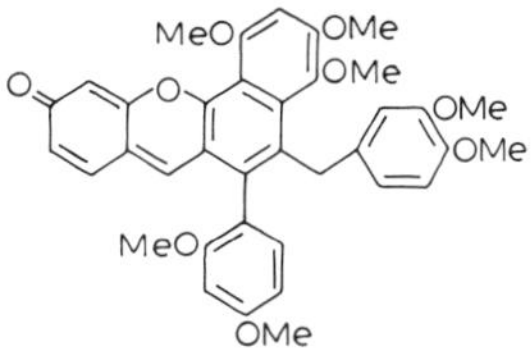

(XVII)

(b) Xanthydrols and xanthene colouring matters

Xanthydrol (9-hydroxyxanthene) and 9-phenylxanthydrol resemble the di- and tri-phenylmethanols in their basicity and in the mobility of the hydroxyl group, which is readily replaced by acid anions. The products are salt-like (compare benzopyrylium salts) and, though the simple chloride and bromide are colourless in the solid state, they become coloured in strong acid solution and the perchlorates and certain double salts are coloured. 9-Phenylxanthydryl chloride with molecular silver in benzene gives a deep red solution of the free radical 9-phenylxanthyl (*W. Schlenk*, Ann., 1912, **394,** 188). The colour bases of the pyronine and rhodamine dyes are derivatives of 3,6-diaminoxanthydrols. The cation of the xanthylium salts can be represented as a resonance hybrid of the oxonium and carbonium forms XVIIIa and XVIIIb.

(XVIII a) ⟷ (XVIII b)

The xanthylium salts are hydrolysed by water to the xanthydrols (*W. Kropp* and *H. Decker*, Ber., 1909, **42,** 578). With alcohol in the cold they form xanthydryl ethers but at higher temperatures are reduced to xanthenes by the alcohol (*M. Gomberg* and *L. H. Cone*, Ann., 1909, **370,** 142). They are also reduced to xanthenes by zinc and hydrochloric acid (*W. Dilthey* and *H. Stephan*, J. pr. Chem., 1939, [ii], **152,** 114) and by phosphorus and hydriodic acid (*Decker* and *T. von Fellenberg*, Ann., 1907, **356,** 281). They do not react with ammonia to give acridines. Of the substituted xanthydrols (XIX, R = Ph, C_6H_{11}, or Et) only the 9-phenyl compound gives condensation products with unsaturated amides and urea (*A. Lespagnol, D. Bar* and *Ch. Mizon-Capron*, Bull. Soc. pharm. Lille, 1967, 111):

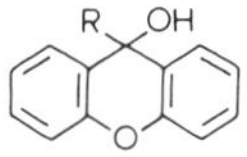

(XIX)

Xanthydrol, $C_{13}H_{10}O_2$, m.p. about 123°, is obtained by the reduction of xanthone using sodium amalgam in 95% ethanol (*A. F. Holleman*, Org. Synth. Coll. Vol. 1, 1941, p. 539). When heated with dilute hydrochloric acid it undergoes disproportionation to xanthone and xanthene. It is readily dehydrated to *dixanthyl ether*, m.p. 200°. It reacts with carbon disulphide to form dixanthyl sulphide and with malonic acid to give 9-xanthylmalonic acid. It reacts with ureas to form crystalline compounds useful for quantitative determination in minute amounts (*N. N. Ivanov*, Biochem. Z., 1923, **135,** 6); 8-*dixanthylurea*, m.p. 260° (decomp.). Xanthydrol is oxidised by warm air or by manganese dioxide in an organic solvent (*M. Z. Barakat et al.*, J. chem. Soc., 1956, 4685) to xanthone and is reduced by stannous chloride and hydrochloric acid to bixanthyl. The rate of this oxidation of xanthydrol has been studied (*E. F. Pratt* and *J. F. van de Castle*, J. org. Chem., 1961, **26,** 2973).

9-**Phenylxanthydrol,** m.p. 157–158°, is obtained from xanthone and phenylmagnesium bromide, and is also formed from diphenyl ether and benzaldehyde in the presence of aluminium chloride (*H. E. Ungnade et al.*, J. Amer. chem. Soc., 1953, **75,** 3333). When phenylxanthylium perchlorate is nitrated the product is 9-(3-nitrophenyl)xanthylium perchlorate (91% yield), from which it is argued that the cation has the carbonium form XVIIIb (*R. L. Shriner* and *C. N. Wolf*, *ibid.*, 1951, **73,** 891; cf. the nitration of triphenylmethyl sulphate, *idem*, J. org. Chem., 1950, **15,** 367). Oxidation of 9-phenylxanthylium perchlorate (XX) with hydrogen peroxide in acetic acid gives 2-(2-hydroxyphenoxy)benzophenone probably through the intermediate peroxide (*Dilthey* and *F. Quint*, Ber., 1931, **64,** 2082; *Dilthey* and *F. Dahm*, J. pr. Chem., 1934, [ii], **141,** 61):

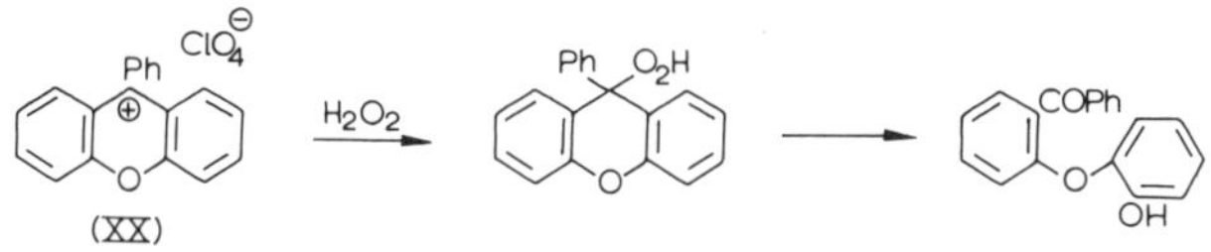

(i) Fluorones, 3H-*xanthen*-3-*one and fluorimes*

3-Hydroxyxanthydrols lose water spontaneously to pass into fluorones or isoxanthones; similarly 3-aminoxanthydrols pass into fluorimes. 9-Phenylfluorone (XXI) is the analogue of fuchsone (Vol. IIIF, p. 136) and 9-phenylfluorime (XXII) that of fuchsonimine (Vol. IIIF, p. 126); these compounds are, respectively, the parents of the fluorescein and rhodamine dyes. The fluorones have no ketonic properties. They are reduced by sodium amalgam or by catalytic hydrogenation to xanthenes (*J. Liebschütz* and *F. Wenzel*, Monatsh., 1901, **25,** 319). When they are fused with alkali the pyran ring is broken, 6-hydroxy-9-phenylfluorone, 6-hydroxy-9-phenyl-3*H*-xanthen-3-one (XXIII), for example, yielding resorcinol and 2,4-dihydroxybenzophenone (*F. G. Pope*, J. chem. Soc., 1914, **105,** 251):

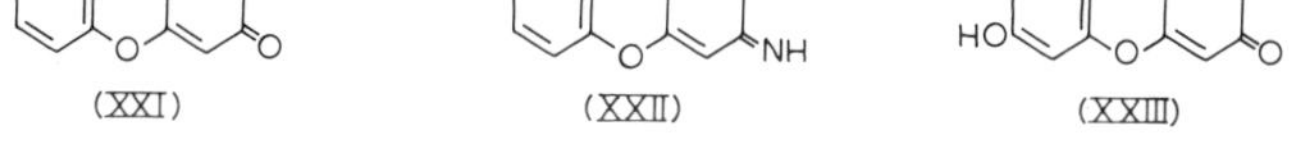

9-**Phenylfluorone,** 9-phenyl-3*H*-xanthen-3-one (XXI), red needles, m.p. 207°, is formed by oxidation of 3-hydroxy-9-phenylxanthene (*Pope* and *Howard*, *ibid.*, 1910, **97,** 1023) or by demethylation of the 3-methoxy compound obtained from 3-methoxyxanthone and phenylmagnesium bromide (*M. Gomberg* and *C. J. West*, J. Amer. chem. Soc.,

1912, **34,** 1529).

The condensation products of benzotrichloride with mono- and poly-hydric phenols (two moles) were in the past called benzeins, whether the products were derivatives of triphenylmethanol (*i.e.* benzaurins, Vol. IIIF, p. 135) or of xanthydrol. 6-**Hydroxy-9-phenylfluorone** (resorcinolbenzein) (XXIII), brick red crystals, m.p. 333°, is formed by condensing one mole of benzotrichloride with two moles of resorcinol and treating the product with water or by heating benzoic acid with resorcinol in presence of zinc chloride. Instead of benzoic acid or its chloride, benzaldehyde or benzyl alcohol can be used, the condensation being effected in sulphuric acid which acts also as oxidising agent; in this manner a large number of fluorone compounds have been prepared using different aldehydes or alcohols and different phenols. The parent 6-hydroxy-fluorone has been prepared by condensing 2,4-dihydroxybenzyl alcohol with resorcinol in presence of zinc chloride and oxidising the resulting 3,6-dihydroxyxanthene with air in alkaline solution (*R. W. Sen* and *N. N. Sinha*, J. Amer. chem. Soc., 1923, **45,** 2984; *Sen* and *N. N. Sarkar*, *ibid.*, 1925, **47,** 1079). By condensing resorcinol with chloroform or carbon tetrachloride, 6-*hydroxy*-9-(2,4-*dihydroxyphenyl*)*fluorone*, orange-yellow, m.p. above 290°, has been obtained (*Sen et al.*, J. Indian chem. Soc., 1925, **1,** 303). 6-Hydroxy-9-methylfluorone has been obtained by condensing 2,4-dihydroxyaceto-phenone with resorcinol in the presence of zinc chloride.

Hydroxyxanthydrols not convertible into fluorones are obtained, for example, by condensing benzaldehyde with hydroquinone (two moles) and oxidising the 2,7-di-hydroxyxanthene so obtained with ferric chloride to give 2,7-dihydroxy-9-phenyl-xanthydrol (*F. Kehrmann*, Ann., 1910, **372**, 301).

6-**Amino-9-phenylfluorone,** dark red needles, m.p. 305°, is obtained as its acetyl derivative, together with 3,6-*diacetamido*-9-*phenylxanthydrol*, m.p. 248°, from benzo-trichloride and *N*-acetyl-3-aminophenol. Hydrolysis of the diacetamido compound gives 6-amino-9-phenylfluorene, forming a red hydrochloride (*Kehrmann, loc. cit.*).

(*ii*) *Xanthene dyes**

The xanthene dyes are close analogues of the di- and tri-arylmethane colouring matters, from which they differ in having a 2,2′-oxygen bridge between two benzenoid nuclei. The high intensity of colour (high extinction coefficient) is similar in the two series (*A. B. Peterson* and *W. C. Holmes*, J. phys. Chem., 1932, **36,** 633). The oxygen bridge confers greater stability, shown strikingly in Pyronine G which is a stable dye whereas salts of Michler's hydrol, which are strictly analogous, are far too unstable for use as dyes. The oxygen bridge has a hypsochromic effect, xanthene dyes absorbing at shorter wave-lengths than corresponding di- and tri-aryl-methane dyes.

This phenomenon may be related to the fact that, for example, 3,6-

* See "Triarylmethane and Xanthene Dyes," *J. F. Thorpe*, Dict. appl. Chem., 4th Edn., Longmans, Green and Co., London 1955, Vol. XI, p. 715.

diaminoxanthene dyes may be represented not only, like their analogues, as ammonium or as carbonium salts but also as oxonium salts and are to be regarded as resonance hybrids embracing all three forms:

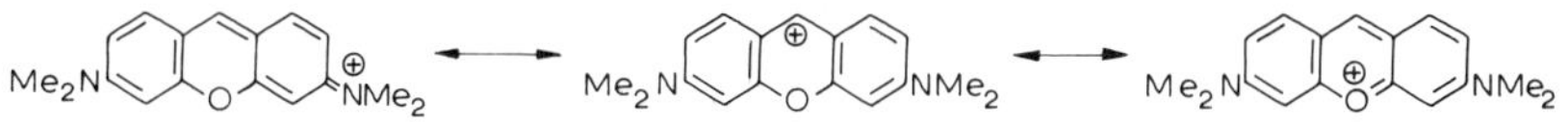

Pyronine G

The principal types of xanthene dyes are:

(1) pyronines, derived from 3,6-diaminoxanthydrol;

(2) rosamines and rhodamines, derived from 3,6-diamino-9-phenylxanthydrol;

(3) fluoresceins, derived from 3,6-dihydroxy-9-phenylxanthydrol.

The rhodamines and fluoresceins have the 9-(2-carboxyphenyl) group, and are derivatives of 9-(2-carboxyphenyl)xanthydrol. The last compound exists as the lactone, fluoran.

Fluoran, m.p. 173–175°, is formed as a by-product in the preparation of phenolphthalein (Vol. IIIF, p. 141) from phthalic anhydride and phenol; the corresponding *phenylimino* derivative has m.p. 242°. It is strongly fluorescent in concentrated sulphuric acid. Fluoran can be reduced to hydrofluoranic acid, 9-(2-carboxyphenyl)xanthene, m.p. 226–228°. When it is fused with lime, xanthone is formed. 3-*Hydroxyfluoran*, m.p. 181°, is formed in the condensation of 2,4-dihydroxybenzoylbenzoic acid with phenol. By condensing phthalic anhydride with *p*-cresol, 2,7-*dimethylfluoran*, m.p. 246°, is obtained (*M. Copisarow*, J. chem. Soc., 1920, **117**, 215). This compound can be cyclised to an anthraquinone derivative (compare coerulein, p. 316) which, be reduction of the oxo group forms dimethylcoeroxene which has been used as a fluorescent dye for oils (see Dict. appl. Chem., Vol. XI, p. 725):

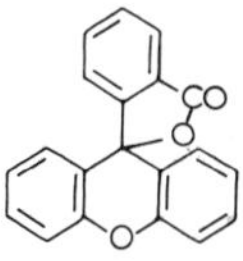

Fluoran

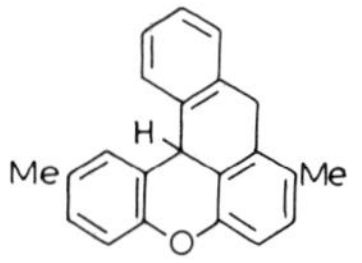

Dimethylcoeroxene

(*1*) *Pyronines.* The pyronines are yellowish red, fluorescent basic dyes, derivatives of 3,6-diaminoxanthydrol. The most important is **Pyronine G,** the bisdimethylamino compound (XXIV), prepared by oxidising 3,6-bisdimethylaminoxanthene (p. 309). **Pyronine B** is the bisdiethylamino compound.

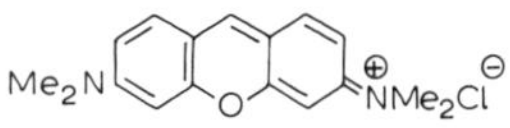

(XXIV)

(*2*) *Rosamine and rhodamines.* The rosamines, phenylpyronines with no carboxyl group, are unimportant; far more important are the rhodamines, prepared from phthalic anhydride. The basic rhodamine dyes are brilliant pink to violet colouring matters of reasonable fastness on mordanted cotton and also in the form of complex salts as pigments. Certain sulphonated dyes are of use for wool dyeing especially for yellowish red shades.

Methods of preparation (*a*) Phthalic anhydride is condensed with two moles of an *N*-alkyl- or *N*-aryl- substituted 2-aminophenol. The reaction can be carried out in two stages, the intermediate benzoylbenzoic acid derivative from one mole of 3-aminophenol being condensed with another mole of the same or a second 3-aminophenol.

(*b*) 3,6-Dichlorofluoran (fluorescein dichloride) is reacted with a primary alkyl- or aryl-amine, amino groups displacing the chloro groups.

(*c*) An aromatic aldehyde is condensed with a 3-aminophenol (two moles) to form a 4,4′-diamino-2,2′-dihydroxydiphenylmethane, which is cyclised to a xanthene in concentrated sulphuric acid and oxidised with a mild oxidising agent such as ferric chloride.

(*d*) From suitable diaminoxanthones by methods similar to those used to obtain triphenylmethane dyes from 4,4′-diaminobenzophenones (see, *e.g.*, *E. H. Rodd et al.*, B.P., 314, 825; 320, 345, 1929).

Rhodamine B, C.I. Basic Violet 10, is manufactured by heating phthalic anhydride with 3-diethylaminophenol without any condensing agent. It is marketed as the chloride. It is more soluble than the dye from *m*-dimethylaminophenol. The solubility of these dyes is improved by esterifying the carboxyl group, and **Rhodamine 3B** is the ethyl ester of **Rhodamine B.** A more modern dye is **Rhodamine 6G,** C.I. Basic Red 1, from phthalic anhydride and 3-ethylamino-4-cresol, the condensation product being esterified:

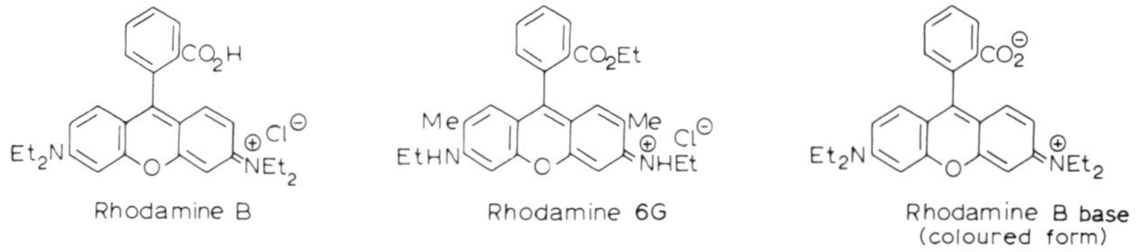

Rhodamine B | Rhodamine 6G | Rhodamine B base (coloured form)

A series of esters of some Rhodamine dyes, from methyl to octyl and benzyl, has been described (*H. E. Fierz-David* and *J. P. Rufener*, Helv., 1934, **17,** 1452). The free base of Rhodamine B has been described in two forms, a colourless form considered to be the lactonic form in non-polar solvents and a coloured form (as formulated) in polar solvents (*E. Noelting* and *K. Dziewonski*, Ber., 1905, **38,** 3516). For the structure in acid solution see *R. W. Ramette* and *E. B. Sandell* (J. Amer. chem. Soc., 1956, **78,** 4872).

A typical sulphonated rhodamine is prepared by condensing 3,6-dichlorofluoran with two moles of *o*-toluidine and sulphonating the product. The resulting dye, as the sodium salt, is **Fast Acid Violet A2R,** C.I. Acid Violet 9.

(*3*) *Fluorescein and related compounds.* Fluorescein, formed by the condensation of phthalic anhydride (one mole) with resorcinol (two moles) is the analogue of phenolphthalein (Vol. IIIF, p. 141). Its formation proceeds when the reactants are heated together, 2-(2,4-dihydroxybenzoyl)benzoic acid being first formed and condensing with a second molecule of resorcinol to give the phthalein XXV which loses water spontaneously to give fluorescein (XXVIa or XXVIb).

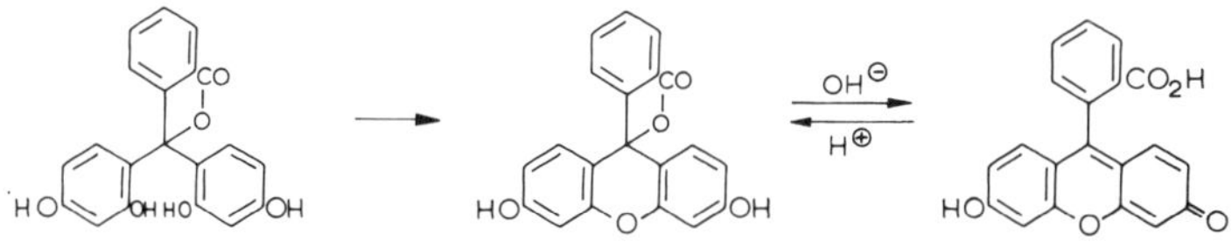

(XXV) (XXVIa) Fluorescein (XXVIb)

In the manufacture of fluorescein a condensing agent, zinc chloride, is used. Suitably oriented trihydroxybenzenes, *e.g.*, pyrogallol and 1,2,4-trihydroxybenzene, condense similarly with phthalic anhydride to give hydroxylated fluoresceins. When fluorescein is fused with potassium hydroxide it is broken down to 2-(2,4-dihydroxybenzoyl)-benzoic acid and resorcinol. By the action of phosphorus pentabromide it is converted into a tribromofluoran which on reduction gives hydrofluoranic acid (p. 313). By these and other reactions the structure is confirmed.

Fluorescein, C.I. Acid Yellow 73, discovered by A. von Baeyer in 1871, is known in a yellow, amorphous form and a red crystalline form, m.p. (sealed tube) 314–316° (decomp.). The yellow form is obtained when an alkaline solution of fluorescein is acidified cold with acetic acid, the red form by acidifying at the boil with mineral acid or by heating the yellow form. The lactoid structure, 3,6-**dihydroxyfluoran** (XXVIa) has been assigned to the yellow form and the quinonoid structure (XXVIb) to the red (*W. R. Orndorff* and *A. J. Hemmer*, J. Amer. chem. Soc., 1927, **49,** 1272). Fluorescein is very sparingly soluble in water but dissolves readily in alkalis. The alkaline solution is red, becoming yellow on dilution and showing a beautiful green fluorescence which is still detectable at a concentration of one part in 40,000,000. The quinonoid structure XXVIb is assumed to be present in alkaline solution and the u.v. spectrum is said to give support to this supposition (*Orndorff et al., ibid.*, 1928, **50,** 819; cf. *V. Zanker* and *W. Peter*, Ber., 1958, **91,** 572). The i.r. spectrum of solid films of fluorescein has been measured by *M. Davies* and *R. L. Jones* (J. chem. Soc., 1954, 120) who conclude that the solid must have the lactoid structure (XXVIa). However, the *dimethyl ether* of XXVIa, m.p. 198°, is known and is colourless and it is strange therefore that fluorescein itself is coloured. The *methyl ether methyl ester* corresponding with XXVIb is orange-red, m.p. 208°. The latter compound is formed by further methylation of *fluorescein methyl ester* (XXVIb, with carboxyl group methylated), m.p. 281–282° (269–270°), obtained by esterification of fluorescein with methyl alcohol in sulphuric acid.

The disodium salt of fluorescein is **Uranine A** of commerce. Fluorescein dyes wool and silk yellow, but the dyeings are fugitive.

Fluorescin, m.p. 125–127°, is the leuco compound of fluorescein. (*H. von Liebig*, J. pr. Chem., 1913, **88,** 42).

By esterification of the dimethyl ether of fluorescein, the trimethyl derivative is obtained, *i.e.* methyl 3,6-dimethoxy-9-phenylxanthydrol-2′-carboxylate, which is a fairly strong base, forming salts which dye tannined cotton, although no quinonoid structure is possible (*F. Kehrmann*, Ann., 1910, **372**, 295).

Fluorescein dichloride, 3,6-*dichlorofluoran* (XXVIa, with Cl in place of both OH groups), m.p. 250–253°, is obtained by the action of phosphorus pentachloride or phosphoryl chloride on fluorescein or by condensing 3-chlorophenol with phthalic anhydride in presence of zinc chloride or sulphuric acid. From it rhodamines may be prepared.

Technically more important colouring matters are the halogeno and nitro derivaties of fluorescein. Bromination of fluorescein gives 4,5-dibromofluorescein (*R. M. Harris et al.*, J. chem. Soc., 1936, 1838; *R. B. Sandin et al.*, J. Amer. chem. Soc., 1939, **61,** 2919) and **Eosin G,** (or **A**), C.I. Acid Orange 11, 2,4,5,7-tetrabromofluorescein, which is used extensively in making red inks. The stable form of the latter is colourless (*Orndorff*, *ibid.*, 1914, **36,** 680). The ethyl ester of Eosin G is manufactured as **Eosin S.** The various brands of **Erythrosine** are mono- to tetra-iodofluoresceins; **Rose Bengal B** is made by iodinating tetrachlorofluorescein, obtained from tetrachlorophthalic anhydride and resorcinol.

Nitration of fluorescein gives 4,5-dinitrofluorescein (*J. T. Hewitt* and *A. W. G. Woodfore*, J. chem. Soc., 1902, **81,** 893) which has been used as a cosmetic pigment; dinitrodibromofluorescein is manufactured as **Eosine BN.**

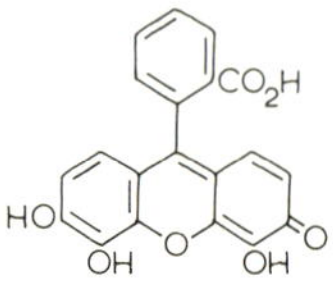

CO_2H HO O OH OH O

Gallein

CO HO O OH OH O

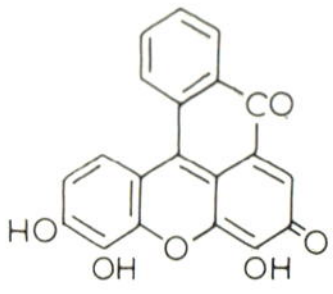

Coerulein

Pyrogallophthalein, gallein, is obtained from phthalic anhydride and pyrogallol or gallic acid. It dissolves in alkali with a red colour turning blue with excess of alkali and is important as a mordant dye and because it is cyclised by concentrated sulphuric acid to an anthracene derivative, **coerulein,** a blue to green mordant dye used in calico printing, especially in the form of its sodium bisulphite derivative known as **Coerulein S.**

(c) Xanthones

Xanthone is the analogue of 4-pyrone and chromone.

Method of synthesis

(*1*) One of the best is the thermal decomposition of phenyl salicylate into phenol, carbon dioxide and xanthone (*A. F. Holleman*, Org. Synth., **7,** 84; Coll. Vol. 1, p. 537):

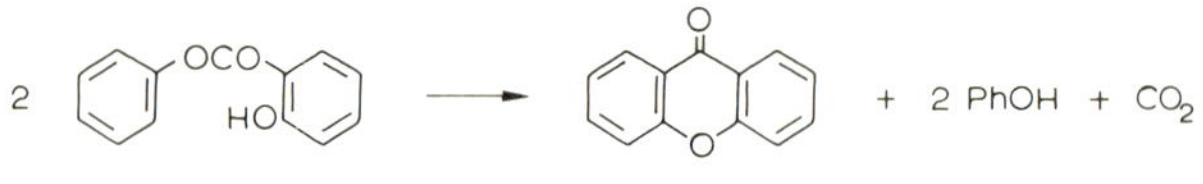

2 OCO HO ⟶ O O + 2 PhOH + CO_2

Xanthone

(*2*) By the ring-closure of the phenyl ether of salicylic acid, *e.g.* by sulphuric acid, or by the elimination of hydrogen chloride from the acid chloride by distillation (*G. Lock* and *F. H. Kempter*, Monatsh., 1935, **67,** 24).

Many substituted xanthones can be synthesised in this way. 2-Chlorobenzoic acid condenses with 3-nitrophenol to give 2-(3-nitrophenoxy)benzoic acid, which readily undergoes ring-closure to form mainly 1-nitroxanthone (XXVII) (*S. N. Dhar*, J. chem. Soc., 1920, 1061; *A. A. Goldberg* and *H. A. Walker*, *ibid.*, 1953, 1348):

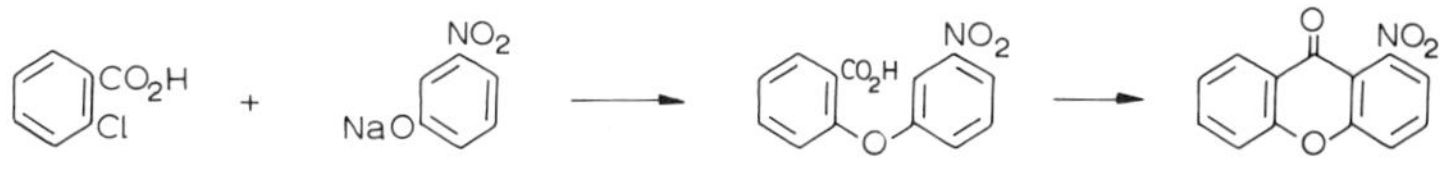

(XXVII)

2-Aryloxybenzoic acids prepared from 3-substituted phenols can give rise to 1- or 3-substituted xanthones. In the above example the product is mainly 1-nitroxanthone, whereas 3-acetamidophenol yields 3-acetamidoxanthone (*M. Julie*, Bull. Soc. chim. Fr., 1952, 546; *Goldberg* and *Walker*, *loc. cit.*).

(*3*) 2,2′Dihydroxybenzophenone undergoes ring-closure with acids to give xanthone. The corresponding ketimine may be used, 2,4,6,2′-tetrahydroxydiphenylketimine, for instance, with alkali yielding 1,3-dihydroxyxanthone and ammonia (*H. Nishikawa* and *R. Robinson*, J. chem. Soc., 1922, **121,** 839).

(*4*) Oxidation of xanthydrol with potassium persulphate gives xanthone in a 45% yield (*Z. Horii et al.*, J. pharm. Soc. Japan, 1956, **76,** 1101). When xanthydryl chloride is treated with silver nitrate in ether at room temperature, xanthone may be crystallised from the solution following filtration (*G. W. H. Cheeseman*, J. chem. Soc., 1959, 458). The oxidation of xanthylium perchlorate using manganese dioxide in either boiling chloroform or acetonitrile affords xanthone in high yield (*I. Degani* and *R. Fochi*, Ann. Chim., Rome, 1968, **58,** 251).

The *structure* of xanthone follows from its synthesis and its breakdown to 2,2′-dihydroxybenzophenone by heating with potassium hydroxide in cyclohexanol (*H. Hoyer*, Ber., 1953, **86,** 1027), or to 2-phenoxybenzoic acid by potassium *tert*-butoxide (*G. A. Swan*, J. chem. Soc., 1948, 1408). The unexpectedly large dipole moment of xanthone, $\mu = 3.07$ D (*E. Bergmann* and *A. Weizmann*, Trans. Faraday Soc., 1936, **32,** 1318) or $\mu = 3.11$ D (*C. G.* and *R. J. W. Le Fèvre*, J. chem. Soc., 1937, 200; *R. J. W. Le Fèvre*, *ibid.*, B, 1971, 82), suggests that the ketonic form is in resonance with the dipolar ionic form:

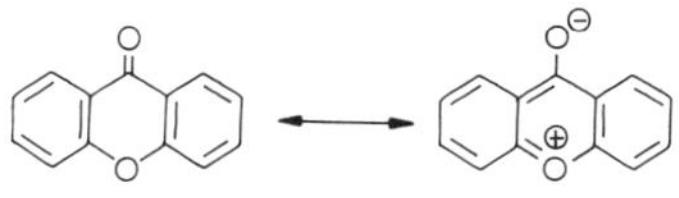

The inertness of xanthone towards the usual carbonyl reagents is in agreement with this hypothesis. A basicity scale for carbonyl compounds including xanthone has been discussed (*E. M. Arnett, R. P. Quirk* and *J. W. Larsen*, J. Amer. chem. Soc., 1970, **92,** 3977).

The xanthones like many other ketones are colourless in the pure state, but give coloured solutions with sulphuric acid (halochromism), which are strongly fluorescent.

Xanthone forms colourless needles, m.p. 173–174°, b.p. 351°, and sublimes easily. It gives a brilliant blue fluorescence in sulphuric acid, and is a weak base giving a stable crystalline perchlorate. Salt formation with mineral acids is shown by the change in colour and absorption spectrum when xanthone is dissolved in these acids. Although it has been stated that, with the exception of hydrazine, xanthone does not react with carbonyl reagents (*A. M. von den Knesebeck* and *F. Ullman*, Ber., 1922, **55,** 306) under suitable conditions the *oxime*, m.p. 161°, *phenylhydrazone*, m.p. 150°, and 2,4-*dinitrophenylhydrazone*, m.p. 278°, may be obtained (*N. Campbell et al.*, J. chem. Soc., 1957, 1922). Xanthone is reduced to xanthydrol by zinc dust and ethanolic potassium hydroxide, sodium amalgam and 95% ethanol (*Goldberg* and *A. H. Wragg*, *ibid*., 1957, 4823), aluminium isopropoxide (*H. Lund*, Ber., 1937, **70,** 1520), or lithium tetrahydridoaluminate in boiling ether (*R. C. Shah et al.*, J. sci. ind. Res. India, 1954, **13B,** 186). Reduction to xanthene is achieved by the Wolff–Kishner method (*N. D. Zuong* and *N. P. Buu-Hoï*, J. chem. Soc., 1952, 3741) or lithium tetrahydridoaluminate in boiling benzene–ether mixture (*A. Mustafa et al.*, *ibid*., 1952, 1343; J. Amer. chem. Soc., 1955, **77,** 5121). The polarographic reduction of xanthone is a one-electron reversible process (*W. E. Whitman* and *L. E. Wiles*, J. chem. Soc., 1956, 3016). Catalytic hydrogenation of xanthone in ethanol using Raney nickel or palladium on charcoal at 80° and 100 atmospheres gives octahydroxanthone (*G. Gatti* and *C. Morandi*, J. heterocycl. Chem., 1968, **5,** 145):

Reduction with triphenylmethylmagnesium bromide gives the pinacol (*W. E. Bachmann*, J. Amer. chem. Soc., 1931, **53,** 2758). With other Grignard reagents xanthone yields 9-substituted xanthydrols giving with phenylmagnesium bromide, for example, 9-phenylxanthydrol. u.v. Spectra also show the presence of the carbonyl group (*R. C. Gibbs et al.*, *ibid*., 1930, **52,** 4985; *L. C. Anderson* and *C. M. Gooding*, *ibid*., 1935, **57,** 999; *R. A. Morton* and *W. T. Earlam*, J. chem. Soc., 1941, 159).

Fusion of xanthone with phosphorus pentasulphide affords the sulphur analogue **xanthione,** a dark red crystalline substance, m.p. 156°, which reacts much more readily than xanthone with hydroxylamine and phenylhydrazine (*C. Graebe* and *P. Röder*, Ber., 1899, **32,** 1688):

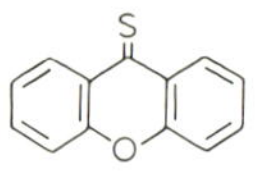

Xanthione

Xanthione reacts with diphenyldiazomethane to give the diphenylethylene sulphide derivative, which on treatment with copper bronze produces 9-(diphenylmethylene)-xanthene (*A. Schönberg et al.*, J. Amer. chem. Soc., 1957, **79,** 6020):

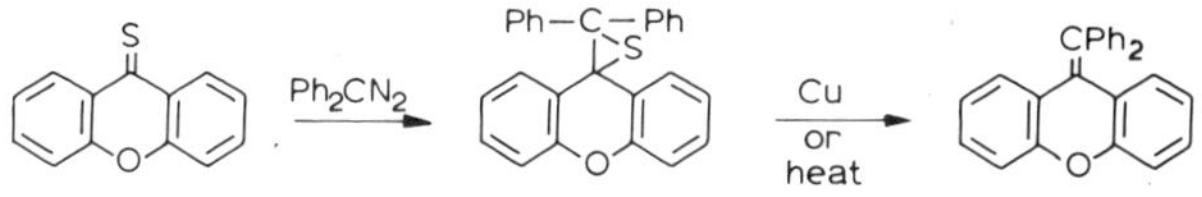

Electrophilic substitution of xanthone occurs most readily in the 2- and 7-positions and less readily in the 4- and 6- positions. The crystal and molecular structure of xanthone has been discussed (*S. C. Biswas* and *R. K. Sen*, Indian J. pure appl. Phys., 1969, **7,** 408). The cleavage of xanthone and a number of its derivatives with the reagent formed by adding water (3 equiv.) to potassium *tert*-butoxide (10 equiv.) in 1,2-dimethoxyethane has been studied (*D. G. Davies, M. Dereberg* and *P. Hodge*, J. chem. Soc., C, 1971, 455).

3,6-*Bisdimethylaminoxanthone*, m.p. 240–242°, can be obtained in high yield by boiling 3,6-*bisdimethylaminoxanthione*, m.p. 308–309°, with 15% hydrochloric acid. The latter is readily obtained by heating the bisdimethylaminoxanthene with sulphur in xylene solution (*E. H. Rodd* and *H. H. Stocks*, B.P., 314, 826 1929).

2-Styrylchromone (XXVIII) reacts with 1,2-bis(4-chlorobenzoyl)ethylene in boiling anisole to give the xanthone derivative XXIX (*G. Aziz*, J. org. Chem., 1962, **27,** 2954):

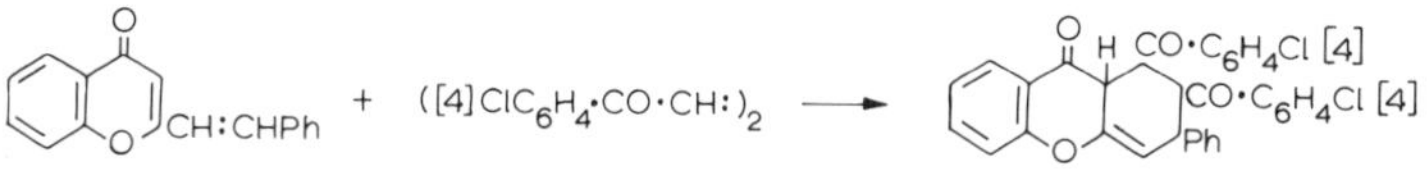

(i) *Halogenoxanthones*

Bromination of xanthone in sulphuric acid gives 2-*bromoxanthone*, m.p. 150°, and in acetic acid with a trace of iodine, 2,7-*dibromoxanthone*, m.p. 212°, identical with the synthetic product (*S. Dahr*, J. chem. Soc., 1916, **109,** 744; 1920, **117,** 1067; *O. Diels et al.*, Ber., 1905, **38,** 1486; 1906, **39,** 2358). Bromination in acetic acid or nitrobenzene with one molecule of bromine and a trace of iodine is stated to give 3-*bromoxanthone*, m.p. 133° (*Dhar, loc. cit.*), but this must be accepted with reserve since further bromination of the compound gives 2,7-dibromoxanthone. Further bromination of xanthone in a sealed tube gives *tetrabromoxanthone*, m.p. 298°, and *hexabromoxanthone*, m.p. 308°. Most monohalogenoxanthones have been obtained synthetically. The xanthones XXXI, R = H, Cl, NO_2; R^1 = Me, OH) have been prepared by cyclising the corresponding halogenomethoxybenzophenone (XXX, X = Cl, Br; R^1 = Me, MeO) in the presence of pyridine hydrochloride (*R. Royer, J. P. Lechartier* and *P. Demerseman*, Bull. Soc. chim. Fr., 1972, 2948):

1-*Chloro-*, m.p. 136–137° (*A. A. Goldberg* and *A. H. Wragg*, J. chem. Soc., 1958, 4234); 2-*chloro-*, m.p. 165° (171°) (*Dhar, loc. cit.*; *F. Ullmann*, Ann., 1907, **355,** 366; *Ullmann* and *C. Wagner, ibid.*, 1910, **371,** 388); 3-*chloro-*, m.p. 130° (*M. Gomberg* and *L. H. Cone*, Ann., 1909, **370,** 142; *Goldberg* and *Wragg, loc. cit.*); 4-*chloro-*, m.p. 130° (*Dhar, loc. cit.; Ullmann, loc. cit.; Ullmann* and *Wagner, loc. cit.*); 2-*bromo-*, m.p. 150° (*Dhar, loc. cit.*); 3-*bromo-*, m.p. 126° (*Gomberg* and *Cone, loc. cit.*); 4-*bromo-*, m.p. 140° (*Dhar, loc. cit.*); 2-*iodo-*, m.p. 156° (*R. Q. Brewster* and *F. Strain*, J. Amer. chem. Soc., 1934, **56,** 117); and 3-*iodo-xanthone*, m.p. 173° (*A. C. Sircar* and *S. C. Dutt*, J. Indian chem. Soc., 1934, **11,** 877).

Halogen in the 1-position is readily replaced by, for example, amino groups, resembling that of 1-halogenoanthraquinones. 1-Chloro-4-methylxanthone with methylamine yields 1-methylamino-4-methylxanthone (*H. Mauss*, Ber., 1948, **81,** 19). Reduction of 2-chloro-, 4-chloro-, 2-bromo-, and 4-bromo-xanthone by lithium tetrahydridoaluminate yields xanthene (*A. Mustafa et al.*, J. Amer. chem. Soc., 1955, **77,** 5121); reduction of 2,7-dibromoxanthone with aluminium isopropoxide gives the propyl ether of 2,7-dibromoxanthydrol (*A. Lespagnol* and *J. Bertrand*, Bull. Soc. chim. Fr., 1943, [v], **10,** 50).

(ii) Nitro- and amino-xanthones

Xanthone is nitrated fairly readily in the 2- and 7-positions. Cold fuming nitric acid gives the benzene-soluble 2,4,7-trinitroxanthone and the benzene-insoluble 2,7-dinitroxanthone (*C. G.* and *R. J. W. Le Fèvre*, J. chem. Soc., 1937, 200; *Goldberg* and *H. A. Walker, ibid.*, 1953, 1348). The latter compound is also obtained in 90% yield by nitration of xanthone with nitric and sulphuric acids (*A. Baeyer*, Ann., 1910, **372,** 80). By careful nitration of xanthone in sulphuric acid with fuming nitric acid (1 mole) it is possible to obtain 2-nitroxanthone (*Goldberg* and *Walker, loc. cit.*). Further nitration

TABLE 9

NITRO- AND AMINO-XANTHONES[a]

	M.p. (°C)		*M.p. (°C)*
1-Nitro	202	1-Amino	148
2-Nitro	204	2-Amino	212
3-Nitro	174	3-Amino	230
4-Nitro	190	4-Amino	202
1,5-Dinitro	304	1,5-Diamino	266
1,6-Dinitro	250–252	1,6-Diamino	254–256
1,7-Dinitro	264	1,7-Diamino	276
2,5-Dinitro	218	2,5-Diamino	258–260
2,6-Dinitro	274	2,6-Diamino	320
2,7-Dinitro	262–264	2,7-Diamino	276
3,5-Dinitro	234	3,5-Diamino	268–270
		3,6-Diamino	324–326
2,4,7-Trinitro	200–264	2,4,7-Triamino	252–254

[a] *Goldber* and *Walker, loc. cit.; M. Julia*, Bull. Soc. chim. Fr., 1952, 546.

of 1-, 2-, 3-, and 4-nitro-xanthone gives 1,7-, 2,7-, 2,6-, and 2,5-dinitro-xanthone, respectively.

The cyclisation of 2-(3-nitrophenoxy)benzoic acid gives mainly 1-nitro- with about 10% of 3-nitro-xanthone. Reduction of the nitroxanthones with stannous chloride and hydrochloric acid gives the corresponding aminoxanthones, which can be diazotised. Phenyl ethers with nitro groups in the *ortho*- and *para*-positions are readily split by warm piperidine; similarly 2,7-dinitroxanthone, for example, yields 5,5′-dinitro-2-hydroxy-2′-piperidinobenzophenone (*R. J. W. Le Fèvre*, J. chem. Soc., 1928, 3249).

Table 9 gives m.p.'s, of nitro- and amino-xanthones.

Certain derivatives of 1-amino-4-methylxanthone, such as Miracil A, 1-(β-diethylaminoethyl)amino-4-methylxanthone, have been found to be effective against Bilharzia in animals (*H. Mauss*, Ber., 1948, **81,** 19).

3,6-*Bisdimethylaminoxanthone*, (p. 319), is formed by oxidation of Pyronine G in alkaline solution.

(iii) Hydroxyxanthones

Hydroxyxanthones are prepared by heating hydroxysalicylic acids with phenols in the presence of acetic anhydride or preferably a mixture of phosphoryl chloride and zinc chloride (*P. K. Grover et al.*, J. chem. Soc., 1955, 3982; *S. R. Dalal* and *R. C. Shah*, Chem. and Ind., 1957, 140). 2,3,4-Trihydroxybenzoic acid and phloroglucinol thus yield 1,3,5,6-tetrahydroxy-xanthone:

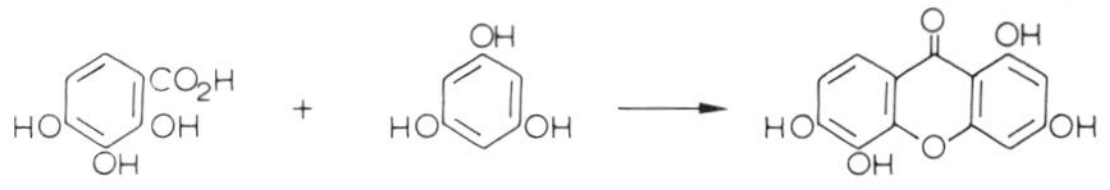

Hydroxybenzophenones obtained by this synthesis can be cyclised to hydroxyxanthones by heating at 200° (*R. Meyer* and *A. Conzetti*, Ber., 1897, **30,** 969) or with water in an autoclave at about 200° (*E. R. Watson* and *J. M. Dutta*, J. Soc. chem. Ind., 1911, **30,** 196). Thus 2,2′,3,4-tetrahydroxybenzophenone (from pyrogallol and salicylic acid) yields 3,4-dihydroxyxanthone. When 2-hydroxy-3,4-dimethoxybenzoic acid is reacted with phloroglucinol in the presence of zinc chloride and phosphoryl chloride 6-*methoxy*-1,3,5-*trihydroxyxanthone*, m.p. 290–292°, is obtained and not the expected, 1,3-dihydroxy-5,6-dimethoxyxanthone (*E. D. Burling* and *F. Scheinmann*, Chem. and Ind., 1962, 1756).

2-*C*- and 4-*C*-methylxanthones with different oxygenation patterns have been synthesised by nuclear methylation of 1,3-dihydroxyxanthones to the corresponding 1,3-dihydroxy-2-methylxanthones:

and condensation of 2-methyl- and 4-methyl-3,5-dimethoxyphenols with substituted salicylic acids (*A. C. Jain, V. K. Khanna* and *T. R. Seshadri*, Tetrahedron, 1969, **25,** 275):

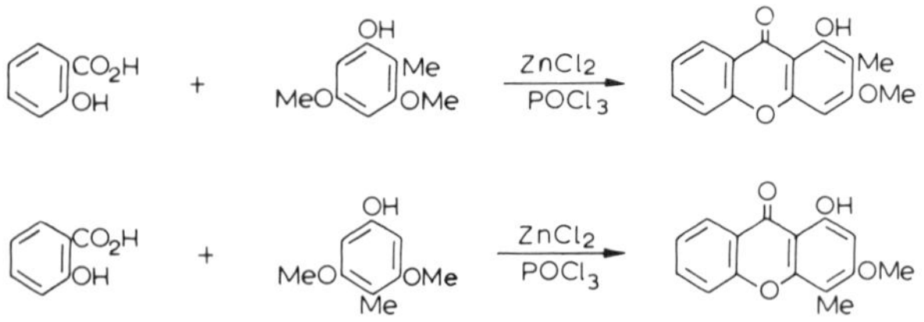

Nuclear alkylation of 1,3-dihydroxyxanthone gives mainly 2-alkyl-1,3-dihydroxyxanthone (*Jain, Khanna* and *Seshadri*, Indian J. Chem., 1969, **7,** 1182).

Since the chelated OH group in the $C_{(1)}$ position of polyhydroxyxanthones forms boroacetate complex compounds from which the chelated OH can be readily regenerated, this method may be used to produce intermediates for the selective alkylation of compounds like 1,3-dihydroxyxanthone to 3-*hydroxy*-1-*methoxyxanthone*, m.p. 318° (*G. S. Puranik* and *S. Rajagopal*, Monatsh., 1963, **94,** 410; Rec. Trav. Chim., 1965, **16,** 1014).

1,5-Dihydroxy-3-methoxyxanthone has been prepared by condensing 2,3-dihydroxybenzoic acid with phloroglucinol dimethyl ether. Selective benzylation of 1,5-dihydroxy-3-methoxyxanthone followed by methylation and subsequent debenzylation affords 1,3-dimethoxy-5-hydroxyxanthone (*B. R. Samant* and *A. B. Kulkarni*, Indian J. Chem., 1969, **7,** 463).

A number of polyhydroxyxanthones have been prepared by the Elbs–Seshadri persulphate oxidation of related hydroxyxanthones; 1,3- and 1,6-dihydroxy-3-methylxanthones were partially methylated before being treated with potassium persulphate (*V. V. Kane, Kulkarni* and *Shah*, J. sci. ind. Research India, 1959, **18B,** 75).

The photo-Fries rearrangement of 4-methoxyphenyl 2,3-dimethoxybenzoate (XXXII) is advantageously employed in the preparation of 2-hydroxy-2′,3′,5-trimethoxybenzophenone which, after demethylation and cyclodehydration affords 2,5-*dihydroxyxanthone* (XXXIII), yellow needles, m.p. 303–305° (*R. A. Finnegan* and *K. E. Merkel*, J. org. Chem., 1972, **37,** 2986):

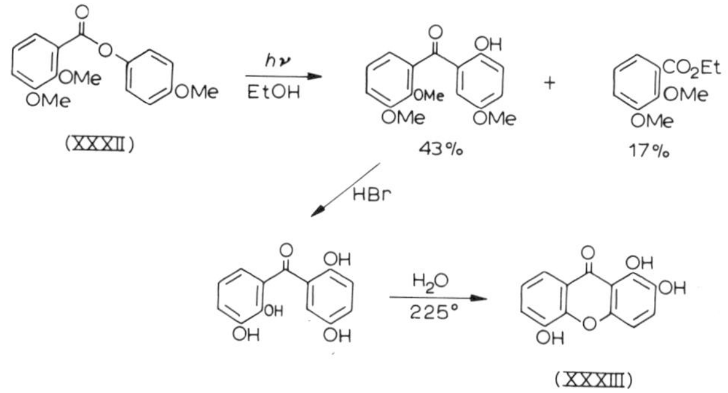

(XXXII) —hν, EtOH→ 43% + 17%; HBr; H_2O, 225° → (XXXIII)

Similarly starting with 2-methoxyphenyl 2,3-dimethoxybenzoate 4,5-*dihydroxyxanthone*, m.p. 273–274°, may be obtained.

The oxidation of 2,3′-dihydroxybenzophenone at pH 13, with excess aqueous potassium ferricyanide gives a mixture of 2-hydroxyxanthone, and 4-hydroxyxanthone (*J. E. Atkinson* and *J. R. Lewis*, Chem. Comm., 1967, 803):

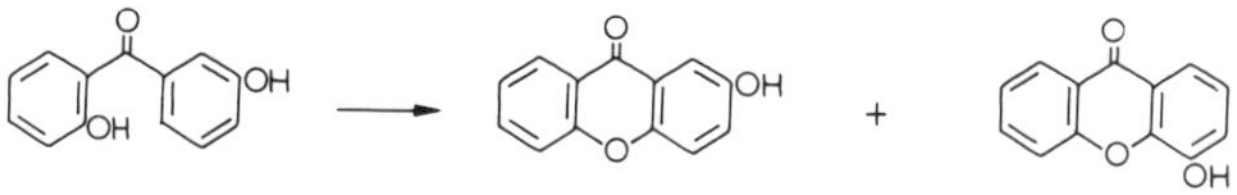

The oxidation of certain hydroxymethoxybenzophenones with dichlorodicyanobenzoquinone to give xanthones has been discussed; the best interpretation of the reaction is *via* phenoxonium-ion intermediates (*J. W. A. Findlay*, *P. Gupta* and *Lewis*, J. chem. Soc., C, 1969, 2761).

The Wessely–Moser type of rearrangement is encountered in the xanthone series, 1,4-dihydroxy-7-methoxyxanthone, for example, giving 1,2,7-trihydroxyxanthone when heated with hydriodic acid under pressure (*E. M. Philbin et al.*, J. chem. Soc., 1956, 4455; cf. *K. V. Rao* and *Seshadri*, Proc. Indian Acad. Sci., 1947, **26A,** 288). Boron trichloride readily demethylates methoxy groups that stand *ortho* to a carbonyl group in xanthones (*F. M. Dean et al.*, Tetrahedron Letters, 1966, 4153).

For certain cases, if the 1-hydroxyxanthone is hydroxylated in the 4-position by a modified Elbs persulphate oxidation, then the product is susceptible to mild oxidation, yielding benzenoid nucleus A, with its attendant groups intact, in the form of an identifiable salicylic acid XXXIV (*J. C. Roberts*, J. chem. Soc., 1960, 785):

1 % aq. NaOH
3 % H_2O_2
2-18 days

(XXXIV)

Monohydroxyxanthones, esters, and ethers are listed in Table 10.

TABLE 10

MONOHYDROXYXANTHONES

	M.p. (°C)	*Acetate m.p. (°C)*	*Benzoate m.p. (°C)*	*Methyl ether m.p. (°C)*	*Ref.*
1-Hydroxy	147	167	206.5	138	1–3
2-Hydroxy	242	161	151	138	3–7
3-Hydroxy	246	157–158	147	135	3,5,8–10
4-Hydroxy	242	145	172	173	3,4,6,11

References
1 *J. Tamber*, Ber., 1910, **43,** 1882.
2 *F. Ullmann* and *L. Panchard*, Ann., 1906, **350,** 108.
3 *E. König* and *St. von Kostanecki*, Ber., 1894, **27,** 1994.
4 *Ullmann* and *M. Zlokasoff*, *ibid.*, 1905, **38,** 2111.
5 *Ullmann* and *W. Denzler*, *ibid.*, 1906, **39,** 4332.
6 *Kostanecki* and *R. Rutishausen*, *ibid.*, 1892, **25,** 1648.
7 *Ullmann* and *H. Kipper*, *ibid.*, 1905, **38,** 2120.
8 *H. Atkinson* and *I. M. Heilbron*, J. chem. Soc., 1926, 2688.
9 *Ullmann* and *C. Wagner*, Ann., 1907, **356,** 359.
10 *A. Baeyer*, *ibid.*, 1910, **372,** 80.
11 *O. Dimroth*, *ibid.*, 1925, **446,** 114.

*(iv) Naturally occurring hydroxyxanthones**

Euxanthone, 1,7-*dihydroxyxanthone*, $C_{13}H_8O_4$, yellow needles, m.p. 240°, occurs in the wood of *Platonia insignis* Mart and as the glucuronate in **euxanthic acid** from which it can be prepared by hydrolysis (*A. Robertson* and *R. B. Waters*, J. chem. Soc., 1929, 2239). Euxanthone with methyl iodide and potassium hydroxide gives 1-*hydroxy*-7-*methoxyxanthone*, m.p. 130.5°. The 1-hydroxyl group participates in the formation of a

* *J. C. Roberts*, Chem. Reviews, 1961, **61,** 591.

boroacetate XXXV on treatment with boroacetic anhydride, the 7-hydroxyl group being at the same time acetylated. Decomposition of the product with warm water yields 7-*acetoxyeuxanthone*, m.p. 160°, which when methylated with diazomethane in ether and nitrobenzene and then hydrolysed affords 7-*hydroxy*-1-*methoxyxanthone*, m.p. 235°:

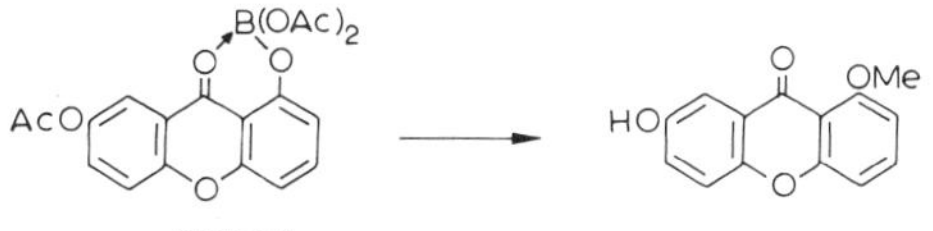

(XXXV)

The structure of euxanthone was confirmed by synthesis (*F. Ullmann* and *L. Panchaud*, Ann., 1906, **350,** 108; *H. D. Locksley, I. Moore* and *F. Scheinmann*, J. chem. Soc., C, 1966, 430). Euxanthic acid is 1-hydroxy-7-β-glucuronoxyxanthone, $C_{19}H_{10}O_{10}$ (*J. Herzig* and *R. Stanger*, Monatsh., 1914, **35,** 47; *Robertson* and *Waters*, J. chem. Soc., 1931, 1709). **Indian Yellow** or **Piuri,** a dye prepared in India from the urine of cows fed on mango leaves, contains euxanthic acid. Euxanthone along with 2- and 4-hydroxyxanthone and 1,5-*dihydroxyxanthone*, m.p. 260–261°, have been isolated from the seed of *Mammea americana L.* (*R. A. Finnegan* and *J. K. Patel*, J. chem. Soc. Perk. I, 1972, 1896).

Gentisin, 1,7-*dihydroxy*-3-*methoxyxanthone*, $C_{14}H_{10}O_5$, yellow needles, m.p. 266–267°, *diacetate*, m.p. 196°, *dibenzoate*, m.p. 192°, occurs as the glycoside **gentiin,** $C_{25}H_{28}O_{14}$, yellow needles, m.p. 274°, in the colouring matter of gentian root, *Gentiana lutea*, Linn. (*J. Shinoda*, J. chem. Soc., 1927, 1983). Gentisin is prepared by methylating **gentisein,** 1,3,7-*trihydroxyxanthone*, $C_{13}H_8O_5$, m.p. 318°, and has been synthesised (*St. von Kostaneckw* and *J. Tambor*, Monatsh., 1894, **15,** 1; *N. Anand* and *K. Venkataraman*, Proc. Indian Acad. Sci., 1947, **25A,** 438). **Isogentisin,** 1,3-*dihydroxy*-7-*methoxyxanthone*, $C_{14}H_{10}O_5$, yellow crystals, m.p. 241°, *diacetate*, m.p. 211–212°, also occurs in gentian roots and has been synthesised (*Shinoda, loc. cit.; P. K. Grover et al.*, J. chem. Soc., 1955, 3982). **Gentioside,** $C_{25}H_{28}O_{14}$, m.p. 267° (decomp.), $[\alpha]_D$ $-107°$, (60% ethanol), is a 3-glycoside of isogentisin obtained from the above source (*L. Canonica* and *F. Pellizzoni*, Gazz., 1955, **85,** 1007).

Ravenelin, 1,4,8-*trihydroxy*-3-*methylxanthone*, $C_{14}H_{10}O_5$, m.p. 267–268°, *triacetate*, m.p. 204–205°, *tribenzoate*, m.p. 255°, *trimethyl ether*, m.p. 178–179°, is a yellow metabolic product of *Helminthosporium Ravenelii* and *H. turcicum*. Its structure follows from its degradation with sodamide in boiling xylene to urea and 3-methyl-2,5,3′-trihydroxydiphenyl ether (*H. Raistrick, R. Robinson* and *D. E. White*, Biochem. J., 1936, **30,** 1303) and its synthesis by the oxidation of 1-hydroxy-8-methoxy-3-methylxanthone with alkaline potassium persulphate and subsequent demethylation (*V. K. Ahluwalia* and *T. R. Seshadri*, Proc. Indian Acad. Sci., 1956, **44,** 1; see also *R. P. Mull* and *F. F. Nord*, Arch. Biochem., 1944, **4,** 419):

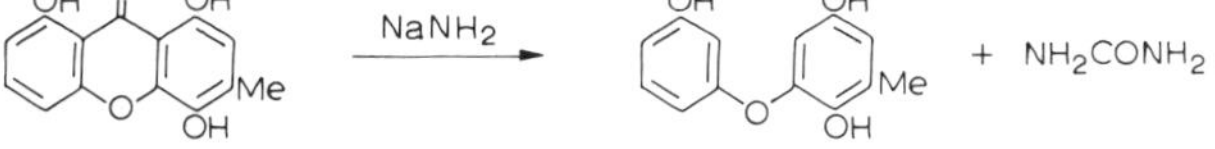

$$\text{Ravenelin} \xrightarrow{NaNH_2} \text{3-methyl-2,5,3'-trihydroxydiphenyl ether} + NH_2CONH_2$$

Ravenelin

Decussatin, $C_{16}H_{14}O_6$, yellow needles, m.p. 149–150°, *acetate*, m.p. 167°, and **swertinin,** $C_{15}H_{12}O_6$, yellow needles, m.p. 217°, are isolated from the Indian plant *Swertia decussata* (*S. R. Dalal et al.*, J. Indian chem. Soc., 1953, **30,** 457, 463). Chemical evidence and u.v. spectra suggest that decussatin is 1-hydroxy-2,6,8-trimethoxyxanthone and swertinin is 1,2-dihydroxy-6,8-dimethoxyxanthone (*R. C. Shah et al.*, J. sci. ind. Research Indian, 1954, **13B,** 175). Swertinin on methylation yields decussatin. A revised structure has been proposed for swertinin, namely 1,3-dihydroxy-7,8-dimethoxyxanthone (*P. Rivaille et al.*, Phytochem., 1969, **8,** 1533). **Swerchirin,** 1,8-*dihydroxy*-3,5-*dimethoxyxanthone*, $C_{15}H_{12}O_6$, yellow needles, m.p. 185–186°, *dimethyl ether*, m.p. 210–211°, is obtained from the Indian medicinal plant *Swertia chirata* and is identical with methyl bellidifolin (*Dalal* and *Shah*, Chem. and Ind., 1956, 664; 1956, 140; *K. R. Markham*, Tetrahedron, 1964, **20,** 991). It has also been isolated from the roots of *Frasera caroliniensis* and *F. albicaulis* (*G. H. Stout* and *W. J. Balkenhol*, Tetrahedron 1969, **25,** 1947; *Stout, E. N. Christensen* and *Balkenhol, ibid.*, p. 1961). **Swertianolin,** $C_{20}H_{20}O_{11} \cdot \frac{1}{2}H_2O$, m.p. 226°; $[\alpha]_D^{18}$ −112.3°, a constituent of *Swertia tosaensis* Makino and *S. japonica* is the 1-glucoside of **swertianol,** 1,3,8-*trihydroxy*-5-*methoxyxanthone*, $C_{14}H_{10}O_6$, yellow needles, m.p. 267°, *triacetate*, m.p. 238–240° (*T. Nakaoki* and *Y. Hida*, J. pharm. Soc. Japan, 1943, **63,** 554). **Bellidifolin** 1,5,8-*trihydroxy*-3-*methoxyxanthone*, $C_{14}H_{10}O_6$, m.p. 270–271°, and **isobellidifolin,** 1,3,7-*trihydroxy*-5-*methoxyxanthone*, $C_{14}H_{10}O_6$, m.p. 263–264°, *acetate*, m.p. 225°, which is identical with swertianol have been obtained from *Gentiana bellidifolia*. Bellidifolin has been synthesised (*Markham*, Tetrahedron, 1965, **21,** 1449). **Corymbiferin,** renamed 4,5-*di*-O-*methylcorymbin*, 1,3,8-*trihydroxy*-4,5-*dimethoxyxanthone*, $C_{15}H_{12}O_7$, fine yellow needles, m.p. 268°, has been isolated along with 4,7-*di*-O-*methylbellidin*, 1,3,8-*trihydroxy*-4,7-*dimethoxyxanthone*, yellow needles, m.p. 220–221°, from the roots of *Gentiana bellidifolia* (*Markham, ibid.*, p. 3687).

Mangiferin, euxanthogen, 2-*β*-D-*glucopyranosyl*-1,3,6,7-*tetrahydroxyxanthone*, $C_{19}H_{18}O_{11}$, m.p. 269° (decomp.), $[\alpha]_D^{18}$ +33.6 (pyridine), *dimethyl ether*, m.p. 276°, *hepta-acetate*, m.p. 150°, obtained from the mango tree *Mangifera indica* and the Malayan timber betis. Treatment with hydriodic acid in phenol gives 1,3,6,7-tetrahydroxyxanthone (*B. J. Hawthorne et al.*, Recent Progr. Chem. nat. synth. Colouring Matters and Related Fields, 1962, 331). n.m.r. Spectral data and chemical degradations confirm the assigned structure (*L. J. Haynes* and *D. R. Taylor*, J. chem. Soc., C, 1966, 1685):

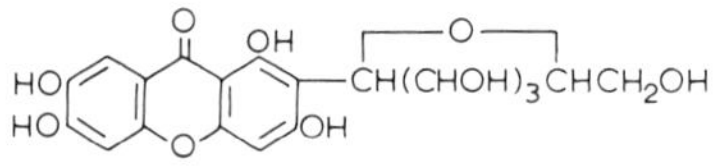

Mangiferin

Oxidation of mangiferin tetramethyl ether and 2-allyl-1,3,6,7-tetramethoxyxanthone with potassium permanganate yield 1,3,6,7-tetramethoxyxanthone-2-carboxylic acid (*V. K. Bhatia, J. D. Ramanathan* and *Seshadri*, Tetrahedron, 1967, **23,** 1363).

Gentiakochianine, 1,7,8-*trihydroxy*-3-*methoxyxanthone* and 1-*hydroxy*-3,7,8-*trimethoxyxanthone*, have been obtained from *Gentiana kochiana* leaves. Gentiakochianine is hydrolysed to yield 3,8-dihydroxy-1,7-dimethoxyxanthone. n.m.r.-Spectral data for the acetates derived from the above compounds have been reported (*M. Guyot, J. Massicot* and *Rivaille*, Compt. rend. Ser. C, 1968, **267,** 423). 1,7-Dihydroxy-3,8-dimethoxy- (free and as the primeveroside), and 3,8-dihydroxy-1,7-dimethoxyxanthone (as the primeveroside) have also been isolated from *G. kochiana.* 1,7,8-Trihydroxy-3-methoxyxanthone has also been obtained from *Swertia decussata* and 1,8-dihydroxy-3,7-dimethoxyxanthone has been isolated from *S. perennis (Rivaille et al., loc. cit.).*

The root of *Frasera caroliniensis* Walt. contains 1-*hydroxy*-2,3,4,7-*tetramethoxyxanthone*, m.p. 117–118°, 1-*hydroxy*-2,3,4,5-*tetramethoxyxanthone*, m.p. 155–156°, 1-*hydroxy*-2,3,7-*trimethoxyxanthone*, m.p. 177–177.5°, 1-*hydroxy*-2,3,5-*trimethoxyxanthone*, m.p. 189–190°, swerchirin, a swerchirin glycoside, 1,3-*dihydroxy*-4,5-*dimethoxyxanthone*, m.p. 274–275°, and a compound that yields 1,2,3,5,8-*pentamethoxyxanthone*, m.p. 151–154°, on methylation. The new xanthones have been synthesised by a general method from suitable benzophenone precursors by elimination of a methoxyl group under strongly basic conditions. The method has also been used to prepare 1,2,3,4,8-*pentamethoxy*-, m.p. 112–113°, and 1,3,4,8-*tetramethoxy-xanthone*, m.p. 191–192° *(Sout* and *Balkenhol, loc. cit.).* 7-*Hydroxy*-1,2,3,4-*tetramethoxy*-, m.p. 197–198°, 1,2,3,4,6,7-*hexamethoxy*-, m.p. 156–157°, 2,3,6,7-*dimethylenedioxy*-1-*methoxy*-, m.p. 250–252°, and 6,7-*methylenedioxy*-1,2,3-*trimethoxy-xanthone*, m.p. 177°, have been isolated from *Polygala macradenia* Gray *(Polygalaceae).* Evidence for their structures are based on spectroscopic properties of the parent compounds and some of their simple derivatives (*D. L. Dreyer*, Tetrahedron, 1969, **25,** 4415).

Lichexanthone, 1-*hydroxy*-3,6-*dimethoxy*-8-*methylxanthone*, $C_{16}H_{14}O_5$, m.p. 189–190°, *acetate*, m.p. 195–196°, occurs in the lichens *Parmelia quercina* Wainio and *P. formosana* Zahlbr. (*Y. Asahina* and *N. Nogami*, Bull. chem Soc. Japan, 1942, **17,** 202; C.A., 1947, **41,** 4488) and has been synthesised (*K. Aghoramaurthy* and *Seshadri*, J. sci. ind. Research India, 1953, **12B,** 350; *Grover, G. D.* and *R. C. Shah, ibid.*, 1956, **15B,** 629.).

Polygalaxanthone A, 1,2,4-*trimethoxy*-6,7-*methylenedioxyxanthone*, $C_{17}H_{14}O_7$, m.p. 170°, and **polygalaxanthone B,** 1,2,3,4,7-*pentamethoxyxanthone*, $C_{18}H_{18}O_7$, m.p. 122–123°, have been isolated from the roots of *Polygala paenea* (*J. Moron, J. Polonsky* and *H. Pourrat*, Bull. Soc. chim. Fr., 1967, 130). Polygalaxanthone B has been synthesised by double hydroxylation of 1,3-dihydroxy-7-methoxyxanthone, followed by methylation. **Isopolygalaxanthone A,** 1,3,4-*trimethoxy*-6,7-*methylenedioxyxanthone*, m.p. 215°, has been obtained from 1,3-dihydroxy-6,7-methylenedioxyxanthone by hydroxylation with alkaline persulphate followed by methylation (*A. C. Jain, V. K. Khanna* and *Seshadri*, Indian J. Chem., 1970, **8,** 667). 1,2,3-Trimethoxy-6,7-methylene-$C_{17}H_{14}O_7$, m.p. 177°, has been obtained from *Polygala macradenia* Gray. (*Dreyer*, Tetrahedron, 1969, **25,** 4415).

Celebixanthone, 3,4,8-*trihydroxy*-2-*methoxy*-1-*prenylxanthone*, $C_{19}H_{18}O_6$, m.p. 219–220°, *triacetate*, m.p. 164–165°, has been obtained from the bark of *Cratoxylon celebicum.* Its structure has been determined by X-ray methods without recourse to a heavy-atom derivative (*Stout et al.*, Tetrahedron Letters, 1962, 541; Tetrahedron, 1963, **19,** 667):

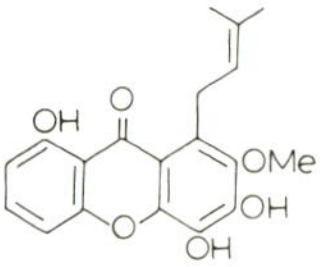

Celebixanthone

One of the xanthones isolated from *Frasera albicaulis* Dougl. ex Gries b. is *2-hydroxy-1,3,4,7-tetramethoxyxanthone*, $C_{17}H_{16}O_7$, m.p. 144.5–145.5°, by X-ray crystallographic analysis (*Stout, T. Shun Lin* and *I. Singh, ibid.*, 1969, **25,** 1975).

Jacareubin, 5,9,10-*trihydroxy*-2,2-*dimethyl*-2H,6H-*pyrano*[3,2-b]*xanthen*-6-*one*, $C_{18}H_{14}O_6$, bright yellow prisms, m.p. 256–257° (decomp.), *triacetate*, m.p. 212–213°, *trimethyl ether*, m.p. 182–183°, isolated from the heartwood of *Calophyllum brasiliense* (Guttiferea) (*F. E. King et al.*, J. chem. Soc., 1953, 3932; 1957, 563). Dihydrojacareubin has been synthesised and a general method described for the synthesis of 2,2-dimethylchromanones (*H. B. Bhat* and *Venkataraman*, Tetrahedron, 1963, **19,** 77). **Osajaxanthone,** 5,8-*dihydroxy*-2,2-*dimethyl*-2H,6H-*pyrano*[3,2-b]*xanthen*-6-*one*, $C_{18}H_{14}O_5$, m.p. 265°, is the pigment from osage orange, *Maclura pomifera* Raf. (*M. L. Wolfrom et al.*, Tetrahedron Letters, 1963, 749):

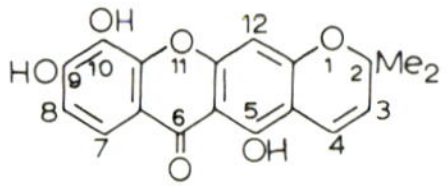

Jacareubin 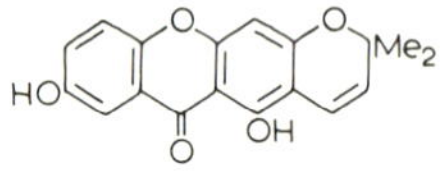 Osajaxanthone

Jacareubin and osajaxanthone have been obtained along with 6-**dehydroxyjacareubin,** yellow plates, m.p. 211–213°; **guanandin,** lustrous yellow plates, m.p. 206–208°, **isoguanandin,** yellow plates, m.p. 175–176°, and **dehydrocycloguanandin,** 167–169°, from Brazilian *Kielmeyera* and *Calophyllum* species (*O. R. Gottlieb et al.*, Tetrahedron, 1968, **24,** 1601).

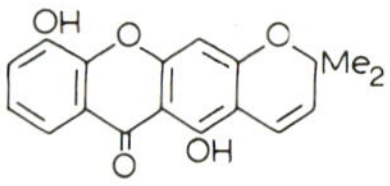

6-Desoxyjacareubin 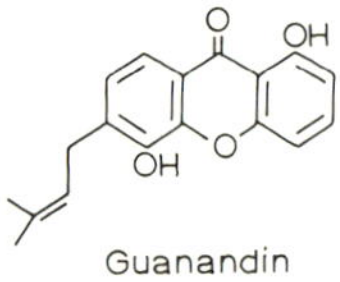 Guanandin

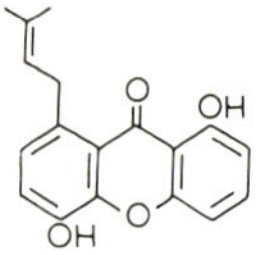

Isoguanandin 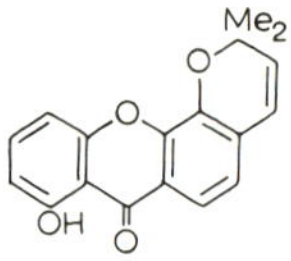 Dehydrocycloguanandin

Unambiguous syntheses have been described of 6-deoxyjacareubin and ether derivatives of the related metabolite 2-prenyl-1,3,5-trihydroxyxanthose (XXXVI). Also related pyrano-, furo- and (3,3-, and 1,1-dimethylallyl)-xanthones derived from 1,3,5- and 1,3,7-trihydroxyxanthones have been prepared. The nuclear Overhauser effect can differentiate a linear pyranoxanthone from its angular isomer (*Locksley, A. J. Quillinan* and *F. Scheinmann*, J. chem. Soc., C, 1971, 3804).

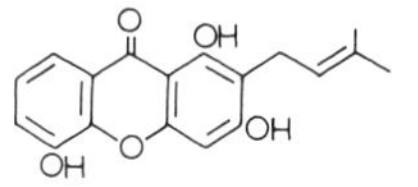

(XXXVI)

The synthesis of 3,4-dihydroxy-2-methoxyxanthone (a metabolite of *Kielmeyera corymbosa*), 2,3,4-trimethoxyxanthone, 1,2,3-trimethoxyxanthone, and various methyl ethers of constituents of the *Kielmeyera* species from 2,3-dimethoxyxanthone have been described (*Gottlieb et al.*, An. Acad. Brasil. Cienc., 1970, **42**(Suppl.), 117).

2-*Prenyl*-1,3,5,6-*tetrahydroxyxanthone*, m.p. 255–257°, 2-*prenyl*-1,3,7-*trihydroxyxanthone*, yellow needles, m.p. 227–229°, and osajaxanthone have been obtained from the heartwood of *Calophyllum canum* Hook f. Jacareubin, present in small quantity has been isolated as its 5,7-*dimethyl ether*, yellow needles, m.p. 192–194° (*I. Carpenter, Locksley* and *Scheinmann*, J. chem. Soc., C, 1969, 486). Jacareubin (*W. M. Bandaranayake, L. Crombie* and *D. A. Whiting, ibid.*, 1971, 811; *P. Helboe* and *P. Arends*, Arch. Pharm. Chemi, Sci. Ed., 1973, **1**, 549) and osajaxanthone (*Jain, Khanna* and *Seshadri*, Tetrahedron, 1969, **25**, 2787) have been synthesised.

Prenylation of 1,3-dihydroxy-5-methoxyanthone with prenyl bromide in the presence of methanolic methoxide yields in the ratio of 2:1, 2,4-di-*CC*-prenyl and 2-*C*-prenyl derivatives which are 5-methyl ethers of natural xanthones isolated from *Garcinia mangostana* and *Calophyllum scriblitifolium*, respectively. The structures of these natural compounds are supported by syntheses of their derivatives. Oxidative cyclisation of the synthetic 2-*C*-prenyl compound with 2,3-dichloro-5,6-dicyano-1,4-benzoquinone (DDQ), yields the 5-methyl ether of 6-desoxyjacareubin (*S. M. Anand* and *Jain, ibid.*, 1972, **28**, 987).

Buchanaxanthone, 1,6-*dihydroxy*-5-*methoxyxanthone*, $C_{14}H_{10}O_5$, pale yellow crystals, m.p. 243–246°, 1,5,6-*trihydroxyxanthone*, m.p. 287–292° (decomp.), and 1,5-*dihydroxyxanthone*, m.p. 266–267°, have been obtained from the heartwood of *Garcinia buchananii* Baker. The structure of buchanaxanthone is supported by u.v. and n.m.r. spectral data and by synthesis (*B. Jackson et al.*, J. chem. Soc., C, 1968, 2579). Euxanthone, buchanaxanthone, 1,5,6-trihydroxyxanthone, 6-desoxyjacareubin, nacareubin, and 2-prenyl-1,3,5,6-tetrahydroxyxanthone have been isolated from the heartwood of *Calophyllum fragrans* Ridley. By methylation of the crude extracts, 1,2-dimethoxy-8-hydroxyxanthone has been isolated and 1,3,5,6- and 1,3,6,7-tetramethoxyxanthones were detected by t.l.c. (*Locksley* and *I. G. Murray, ibid.*, 1969, 1567).

Macluraxanthone, 5,9,10-*trihydroxy*-2,2-*dimethyl*-12-(1,1-*dimethylallyl*)-2H,6H-*pyran*[3,2-b]*xanthen*-6-*one*, $C_{23}H_{22}O_6$, diamorphous, m.p. 181–182° and 204–205°, *dimethyl ether*, m.p. 161–163°, *trimethyl* ether, m.p. 98°, isolated from the root bark of the osage orange, *Maclura pomifera* Raf. (*Wolfrom et al., loc. cit.)*. Also isolated from the same source are osajaxanthone and alvaxanthone. Macluraxanthone was believed to be the first case of a natural phenolic compound substituted with an isoprenoid unit in the form of a 1,1-dimethylallyl group. The biosynthetic implications

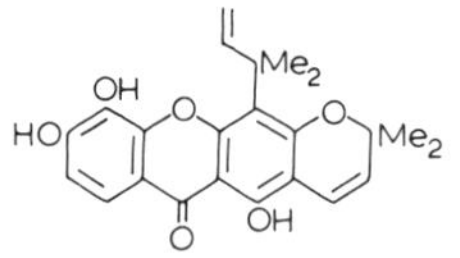

Macluraxanthone

of this have been discussed and a possible biosynthetic route to macluraxanthone has been proposed (p. 334) (*Wolfrom et al.*, J. org. Chem., 1964, **29**, 692).

Alvaxanthone, 1,3,5,6-tetrahydroxy-2-[2-(2-*methyl*-3-*butenyl*)]-8-*prenylxanthone*, $C_{23}H_{24}O_6$, m.p. 155–156°, *triacetate*, m.p. 170–171°, has been obtained from the root bark of the osage orange, *Maclura pomifera* Raf. (*Wolfrom, F. Komitsky*, Jr., and *P. M. Mundell, ibid.*, 1965, **30,** 1088):

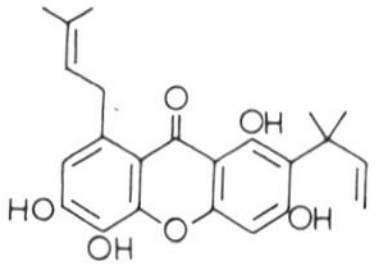

Alvaxanthone

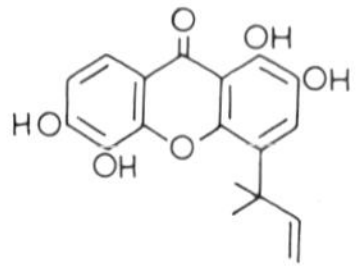

Symphoxanthone

Symphoxanthone, 1,2,5,6-*tetrahydroxy*-4-[2-(2-*methyl*-3-*butenyl*)]*xanthone*, $C_{18}H_{16}O_6$, yellow needles, m.p. 222–224°, and **globuxanthone,** 1,2,5-*trihydroxy*-4-[2′-(2″-*methyl*-3″-*butenyl*)]*xanthone*, $C_{18}H_{16}O_5$, orange needles, m.p. 162–163°, along with four other polyhydroxyxanthones namely, euxanthone, 1,5,6-trihydroxy-, 1,3,5,6-tetrahydroxy- and 1,3,6,7-tetrahydroxy-xanthone have been isolated from the heartwood of *Symphonia globulifera* L. (*Locksley, I. Moore* and *Scheinmann*, (J. chem. Soc., C, 1966, 2186).

Mangostin, 1,3,6-*trihydroxy*-7-*methoxy*-2,8-*diprenylxanthone*, $C_{24}H_{26}O_6$, m.p. 181.6–182.6°, *dimethyl ether*, m.p. 123.3–123.8°, is a constituent of the latex from the mangosteen tree *Garcinia mangostana* (*P. Yates* and *Stout*, J. Amer. chem. Soc., 1958, **80,** 1691). Acid-catalysed cyclisation of mangostin gives products in which one of the two isopentenyl side chains forms dihydropyran rings. The structure assigned to mangostin can thus be confirmed (*Yates* and *Bhat*, Canad. J. Chem., 1970, **48,** 680). *β*-**Mangostin,** 1,6-*dihydroxy*-3,7-*dimethoxy*-2,8-*diprenylxanthone*, $C_{25}H_{28}O_6$, is a minor colouring matter obtained from the above source (*Yates* and *Bhat, ibid.*, 1968, **46,** 3770):

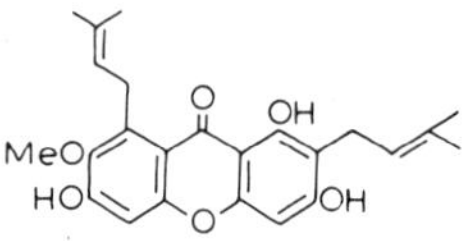

Mangostin

Scriblitifolic acid, $C_{19}H_{18}O_6$, yellow needles, m.p. 164–167°, obtained from the heartwood of *Calophyllum scriblitifolium* Henderson and Wyatt-Smith. This metabolite was the first 1,5-dihydroxyxanthone derivative to be isolated from Guttiferae, although oxy-substituents in these positions were common features of more fully hydroxylated xanthones found in this and in other plant families. The second novel feature is the presence of the 2-methylbutanoic acid side chain at $C_{(6)}$, which is probably derived by stepwise modification of a 3,3-dimethylallyl (prenyl) group (*Jackson, Locksley* and *Scheinmann*, J. chem. Soc., C, 1967, 785).

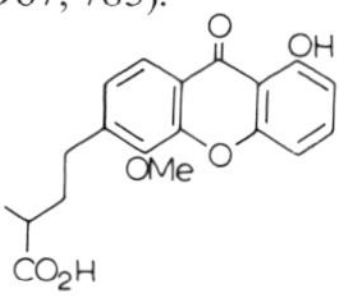

Scriblitifolic acid

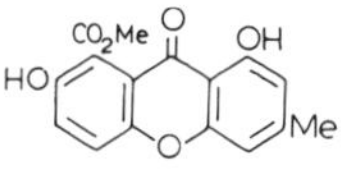

Pinselin

Pinselin, cassiollin, 1,7-*dihydroxy*-8-*methoxycarbonyl*-3-*methylxanthone*, $C_{16}H_{12}O_6$, yellow needles, m.p. 227–229°, *diacetate*, m.p. 211–212°, *dimethyl ether*, m.p. 222–223°, is a constituent of *Cassia occidentalis* Linn and a metabolite of *Pencicillium amarum*. Previously cassiollin had been wrongly reported as 1,7-dihydroxy-5-methoxycarbonyl-3-methylxanthone (*C. E. Moppett*, Chem. Comm., 1971, 423). **Cassiaxanthone,** 8-*hydroxyxanthone*-1,3-*dicarboxylic acid*, $C_{15}H_8O_7$, m.p. 330–333°, *acetate*, m.p. 215–216°, *dimethyl ether*, m.p. 183–184°, is a constituent of *Cassia* species (*M. S. R. Nair, T. C. McMorris* and *M. Anchel*, Phytochem., 1970, **9,** 1153) and has been synthesised (*T. Arunachalam, Anchel* and *Nair*, J. org. Chem., 1972, **37,** 1262). Other hydroxyxanthonedicarboxylic acids have been synthesised. 1,6-Dihydroxy-3-methylxanthone-8-carboxylic acid has been obtained as a degradation product of sulochrin; decarboxylation by subliming at 300° affords 1,6-*dihydroxy*-3-*methylxanthone*, *diacetate*, m.p. 168°, (*M. Itabashi, H. Nishikawa* and *S. Sagisaka*, Tohoku J. Agric. Res., 1955, **5,** 277; C.A., 1957, **51,** 1394).

Thuringione, 2,4,7-*trichloro*-1,6-*dihydroxy*-3-*methoxy*-8-*methylxanthone*, $C_{15}H_9Cl_3O_5$, yellow needles, m.p. 278–279°, *dimethyl ether*, m.p. 202–203°, has been obtained from *Lecidea carpathica* (*S. Huneck* and *J. Santesson*, Z. Naturforsch., 1969, **24B,** 756).

Aphloiol, $C_{14}H_{14}O_8$, yellow needles, m.p. 304–305°, $[\alpha]_D^{25}$ 35.6° (pyridine), has been isolated from the dry leaves of *Aphloia madagascariensis* and *A. theaeformis*. Its structure has been proposed on the basis of chemical evidence and physico-chemical data (*S. Adjangba*, Bull. Soc. chim. Fr., 1964, 376; *C. Mentzer* and *Adjangba*, Probl. Evolyutsionnoi i Tekhn. Biokhim., Akad. Nauk, S.S.S.R., Inst. Biokhim., 1964, 287; C.A., 1964, **61,** 10648). Other workers believe that aphloiol is identical with mangriferin, and is probably a 3-glucoside of 1,3,6,7-tetrahydroxyxanthone (*R. R. Paris* and *S. Etchepare*,

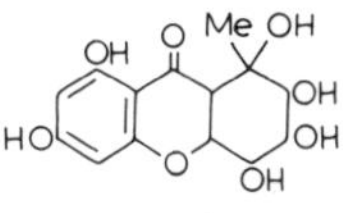

Aphloiol

Compt. rend., 1964, **258,** 5277; *D. Billet et al.*, Bull. Soc. chim. Fr., 1965, 3006.

Secalonic acid A, $C_{32}H_{30}O_{14}$, m.p. 246–247° (decomp.), $[\alpha]_D^{20}$ —76° ($CHCl_3$), $[\alpha]_D^{20}$ −177° (pyridine), and **secalonic acid B,** $C_{32}H_{30}O_{14}$, m.p. 254–256° (decomp.), $[\alpha]_D^{20}$ +196° ($CHCl_3$), $[\alpha]_D^{20}$ +194° (pyridine) have been obtained from ergot, *Claviceps purpurea.* The diastereoisomers A and B may be separated by chromatography (*B. Franck* and *E. M. Gottschalk*, Angew. Chem., 1964, **76**, 438). **Chrysergonic acid** is a 2:1 mixture of secalonic acids A and B.

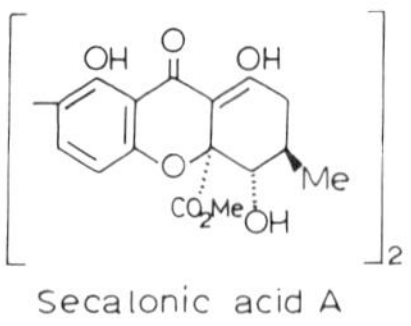

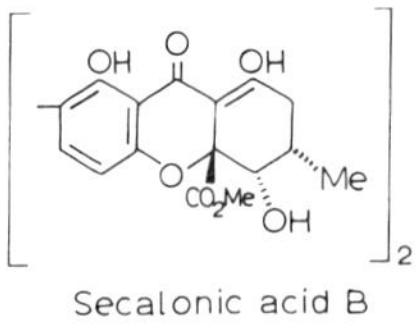

Secalonic acid A

Secalonic acid B

The biphenyl linkage in secalonic acids A and B is 2,2′- (*J. W. Hooper et al.*, J. chem. Soc., C, 1971, 3580) and not 4,4′- as was previously thought (*Franck* and *Gottschalk*, *loc. cit.*). Other secalonic acids have been obtained; **secalonic acid C** from ergot (*Franck et al.*, Ber., 1966, **99**, 3842; *Hooper et al., loc. cit.*); **secalonic acid D,** m.p. 253–255°, $[\alpha]_D^{20}$ +82° ($CHCl_3$), $[\alpha]_D^{20}$ +192° (pyridine), a toxic metabolite from *Penicillium oxalicum* (*P. S. Steyn*, Tetrahedron, 1970, **26,** 51; *Hooper et al., loc. cit.*); **secalonic acid E,** m.p. 206–208°, $[\alpha]_D^{22}$ −212° (pyridine), from *Phoma terrestris* (*C. C. Howard, R. A. W. Johnstone* and *I. D. Entwistle*, Chem. Comm., 1973, 464). The m.s. of secalonic acids A, C, D, E (B) show features closely related to the stereochemistry at the six chiral centres in these compounds. Analysis of the spectra suggest that the stereochemistry of any new secalonic acid could be defined largely by its m.s. (*Howard* and *Johnstone*, J. chem. Soc. Perk. I, 1973, 2033, 2440).

Ergoflavin, $C_{30}H_{26}O_{14}$, yellow needles, m.p. 350° (decomp.), $[\alpha]_D^{11}$ +37.5° (*c.* 1.236 in acetone), *hexa-acetate*, m.p. 248–249° (decomp.), *tetramethyl ether*, m.p. 282° (decomp.), obtained from ergot, *Claviceps purpurea*; it gives an intense green colour with ferric chloride:

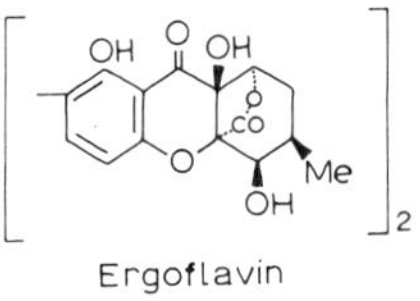

Ergoflavin

An extensive examination of the chemistry and spectral properties of ergoflavin and its numerous derivatives in conjunction with an X-ray analysis has enabled the constitution of ergoflavin to be defined. Ozonolysis of ergoflavin disrupts one half of the symmetrical molecule to give hemiergoflavin-2-carboxylic acid (XXXVII), which can be decarboxylated to hemiergoflavin (XXXVIII) (*J. W. ApSimon et al.*, J. chem. Soc., 1965, 4130):

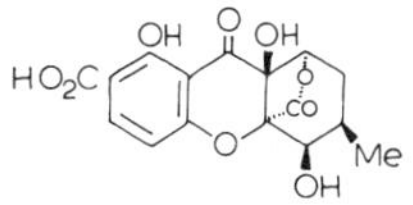

(XXXVII) 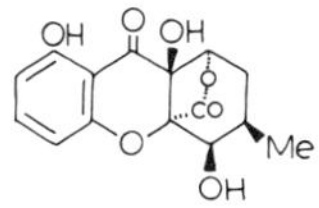 (XXXVIII)

X-Ray crystal-structure analysis of tetra-*O*-methylergoflavin di-4-iodobenzoate has defined the constitution and absolute stereochemistry of ergoflavin (*A. T. McPhail et al., ibid.*, B, 1966, 18). A partial synthesis of ergoflavin has been described (*Hooper et al.*, Chem. Comm., 1971, 111).

Ergochrysin A, $C_{31}H_{28}O_{14}$, m.p. 285° (decomp.), $[\alpha]_D^{20}$ −65.4° (pyridine) has been obtained from ergot (*W. Bergmann*, Ber., 1932, **65,** 1489; *ApSimon et al.*, J. chem. Soc., 1965, 4144) and shown to possess a linear structure, whereas its base-catalysed transformation product, **isoergochrysin A,** yellow prisms, m.p. 255–259° (decomp.), $[\alpha]_D^{20}$ +58.6° (pyridine), has the corresponding angular structure (*Hooper et al.*, J. chem. Soc., C, 1971, 3580).

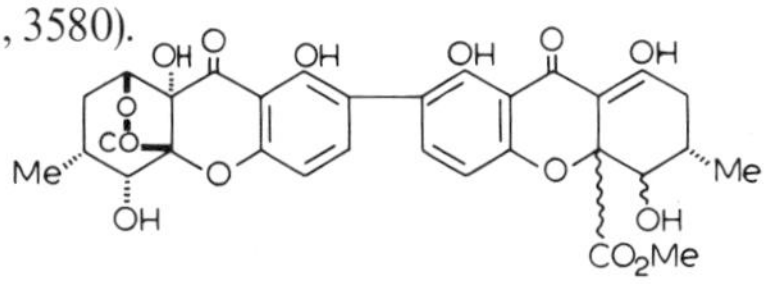

Ergochrysin A

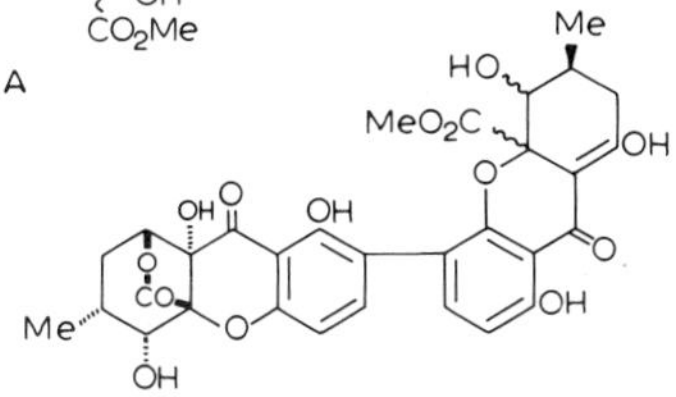

Isoergochrysin A

Some furo[3,2-*b*]xanthones, *e.g.* 5*H*-furo[3,2-*b*]xanthen-5-one (XXXIX), have been synthesised (*Y. S. Agasimundin* and *S. Rajagopal*, J. org. Chem., 1971, **36,** 845). Claisen rearrangement of three 1-hydroxy-3-prenyloxyxanthones at 200° for 3 h yields the corresponding deprenylated xanthones and linear and angular furoxanthones, in

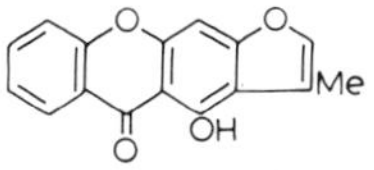

(XXXIX)

contrast to previous results (*S. M. Anand* and *A. C. Jain*, Chem. Comm., 1972, 1026):

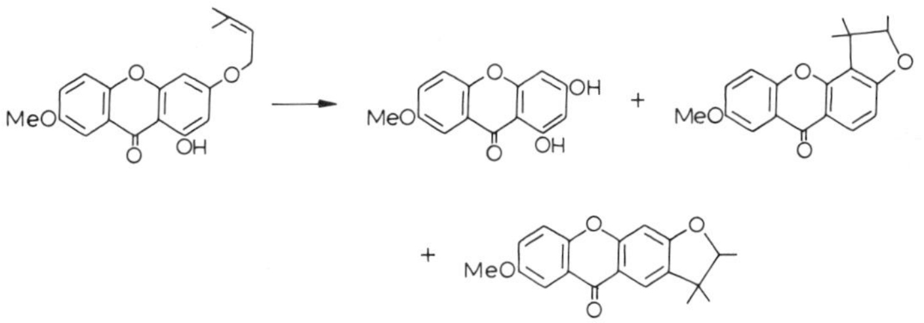

The above reaction has been investigated in relationship to possible routes for the synthesis of macluraxanthone (p. 329).

Some 1- and 2-phenyl-6*H*-furo[2,3-*c*]xanthen-6-ones, *e.g.* XL have been prepared by the cyclisation of simple and substituted 3-phenylacyloxyxanthones using polyphosphoric acid, under different experimental conditions (*C. S. Angadiyavar* and *Rajagopal*, Indian J. chem., 1969, **7**, 1088).

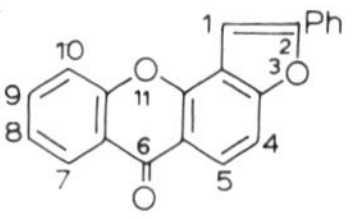

(XL)

8. 6*H*-Dibenzo[*b,d*]pyran, 3,4-benzochromene and its derivatives

6*H*-**Dibenzo**[*b,d*]**pyran-6-one,** 3,4-**benzocoumarin,** 6-**dibenzopyranone** (I), $C_{13}H_8O_2$, needles, m.p. 92.5°, is prepared from diazotised phenyl 2-aminobenzoate by various methods (*R. A. Heacock* and *D. H. Hey*, J. chem. Soc., 1954, 2481) or by treating 2′-methoxydiphenyl-2-carboxylic acid with thionyl chloride (*H. G. Rule* and *E. Bretscher*, *ibid*., 1927, 925). When distilled with zinc dust it yields biphenyl, 2-methylbiphenyl, and fluorene and as a lactone is converted by alkali and dimethyl sulphate into 2′-methoxybiphenyl-2-carboxylic acid:

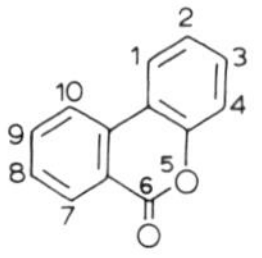

(I) 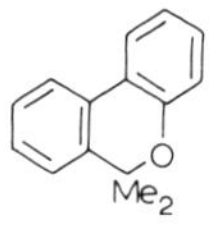 (II)

3-*Hydroxy*-6H-*dibenzo*[b,d]*pyran*-6-*one*, m.p. 233° (*R. Ghosh et al*., *ibid*., 1940, 1393).

Dibenzo[*b,d*]pyran-6-one and methylmagnesium chloride afford 6,6-*dimethyldibenzo*[b,d]*pyran* (II), b.p. 186–187°/27 mm (*R. S. Cahn*, *ibid*., 1933, 1400), 2,6,6-*trimethyldibenzo*[b,d]*pyran*, m.p. 57°, b.p. 201°/30 mm, with hydrochloric acid yields 2-*hydroxy*-5-*methylbiphenyl*, m.p. 110–111°.

Intramolecular cyclisation of *o*-phenylphenoxymethyl radicals, formed by oxidation of *o*-phenylphenoxyacetic acid (III) with persulphate, yields cyclohexadienyl radicals which are either oxidised to a dibenzopyran IV and a related bis-acetal VI, or dimerise to give V (*P. S. Dewar*, *A. R. Forrester* and *R. H. Thomson*, Chem. Comm., 1970, 850):

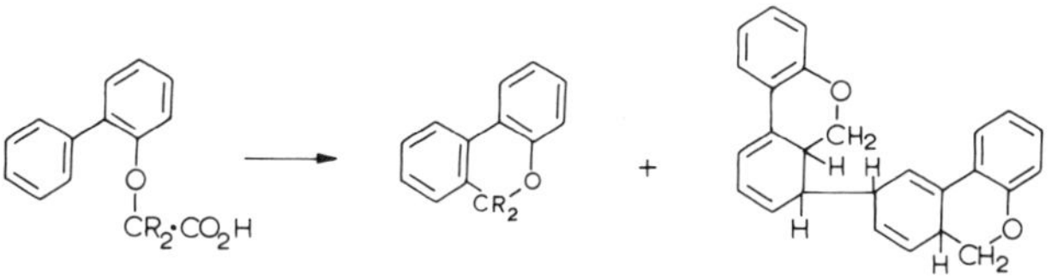

(III) R = H; R = Me → (IV) 60 %; 82 % + (V)

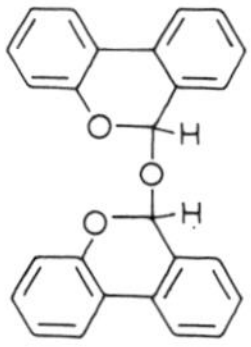

(VI)

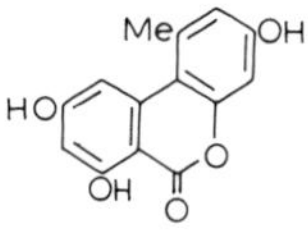

Alternariol

The bis-acetal VI derived from III, R = H, is isolated most easily when the reaction is terminated after 15 min.

Alternariol, 3,7,9-*trihydroxy*-1-*methyl-dibenzo*[b,d]*pyran*-6-*one*, $C_{14}H_{10}O_5$, m.p. 350° (decomp.), *monomethyl ether*, m.p. 267° (decomp.), diacetate, m.p. 162–163°, *trimethyl ether*, m.p. 162.5–164°, occurs with the monomethyl ether in myceluine or *Alternaria tenuis* and was the first dibenzo-2-pyrone of fungal origin to be isolated (*H. Raistrick et al.*, Biochem. J., 1953, **55,** 421). It yields fluorene when distilled with zinc dust and has been synthesised. Biogenetic-type syntheses of alternariol and lichexanthone have been investigated (*J. V. Hay* and *T. M. Harris*, Chem. Comm., 1972, 953).

Autumnariol, 3,7-*dihydroxy*-1-*methyldibenzo*[b,d]*pyran*-6-*one*, $C_{14}H_{10}O_4$, sublimes 240–258°, *diacetate*, m.p. 206–208°, *dimethyl ether*, m.p. 151–151.5°, and **autumnariniol,** 3,7-*dihydroxy*-4-*methoxy*-1-*methyldibenzo*[b,d]*pyran*-6-*one*, $C_{15}H_{12}O_5$, m.p. 176–179°, *diacetate*, m.p. 174–176.5°, *dimethyl ether*, m.p. 207.5–208.5°, have been isolated from the bulbs of *Eucomis autumnalis* Graeb. (*W. T. L. Sidwell, H. Fritz* and *Ch. Tamm*, Helv., 1971, **54,** 207):

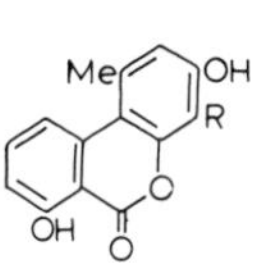

Autumnariol (R = H)
Autumnariniol (R = OMe)

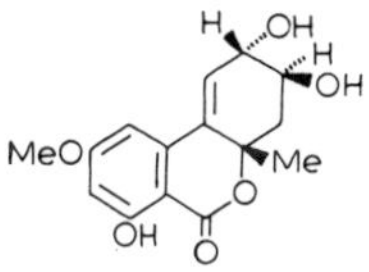

Altenuene

(±)**Altenuene,** 2,3,7-*trihydroxy*-9-*methoxy*-4a-*methyl*-2,3,4,4a-*tetrahydrodibenzo*[b,d]-*pyran*-6-*one*, $C_{15}H_{16}O_6$, m.p. 190–191°, is a toxic fungal metabolite from *Alternaria tenuis* (*R. W. Pero et al.*, Biochim. Biophys. Acta, 1971, **230,** 170). Single-crystal X-ray analysis has established unequivocally that (±)altenuene is correctly represented by the given formula and its mirror image (*D. Harvan* and *Pero*, Chem. Comm., 1973, 682).

Cannabinol, 3-n-*amyl*-1-*hydroxy*-6,6,9-*trimethyldibenzo*[b,d]*pyran*, $C_{21}H_{26}O_2$, colourless crystals, m.p. 76–77°, b.p. 263–264°/20 mm, *acetate*, m.p. 75–76°, *methyl ether*, m.p. 66°, 4-*nitrobenzoate*, m.p. 165–166°, does not react with diazomethane and is insoluble in aqueous sodium hydroxide. Cannabinol is isolated along with **cannabidiol,** $C_{21}H_{30}O_2$, b.p. 160–180°/10^{-3} mm, $[\alpha]_D^{18}$ −126.6° (ethanol), *bis*-3,5-*dinitrobenzoate*, m.p. 106–107° (*R. Adams et al.*, J. Amer. chem. Soc., 1940, **62,** 196; *A. Jacob* and *A. R. Todd*, J. chem. Soc., 1940, 649), from the resinous secretion from hemp, *Cannabis sativa* L., *Cannabis Indica*, etc., known by different names such as Hashish (Egypt), bhang (Indian), and marihuana (America) and for centuries used as an intoxicating drug. Neither cannabinol nor cannabidiol is the active principle; (for a general account of the hemp drugs see *Todd*, Endeavour, 1943, **2,** 69):

Cannabinol Cannabidiol

The structure of cannabinol was formulated, except for the exact positions of the hydroxyl and amyl groups, by *R. S. Cahn* (J. chem. Soc., 1932, 1342; 1933, 1400) and was established by synthesis (*Adams et al.*, J. Amer. chem. Soc., 1940, **62,** 2204; *Todd et al.*, J. chem. Soc., 1940, 1121, 1393; *T. Petrzilka, W. Haefliger* and *C. Sikemeier*, Helv., 1969, **52,** 1102). Isomerisation of cannabidiol gives a mixture of tetrahydrocannabinols which on dehydrogenation yield cannabinol. Cannabidiol is an olivetol (5-*n*-amylresorcinol) derivative since it is cleaved by pyridine hydrochloride to this compound and both substances have similar u.v. absorption spectra (*Adams et al.*, J. Amer. chem. Soc., 1940, **62,** 732, 735).

From Westheimer and extended Hückel molecular orbital calculations, structures and energies have been obtained for λ-Δ^9- and $\Delta^{6a(10a)}$-tetrahydro-, and hexahydrocannabinol, and cannabinol. Conclusions reached on the basis of these studies, concerning the conformation of the pyran ring, the preferred orientation of the phenolic O–H bond, and ring C conformational preferences in the first three compounds, and Δ^8-tetrahydrocannabinol, are in substantial agreement with p.m.r. observations resulting from nuclear Overhauser effect and solvent effect studies (*R. A. Archer et al.*, *ibid.*, 1970, **92,** 5200).

9. Naphthopyrans, benzochromenes and some of their derivatives

The naphthopyrans include 2*H*-naphtho[1,2-*b*]pyran (I), 3*H*-naphtho[2,1-*b*]pyran (II), 2*H*-naphtho[2,3-*b*]pyran (III), naphtho[1,8-*bc*]pyran (IV), and 1*H*,3*H*-naphtho[1,8-*cd*]pyran (V):

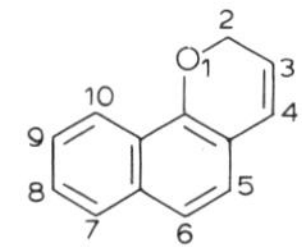

2H-Naphtho[1,2-b]pyran,
7,8-benzochromene
(I)

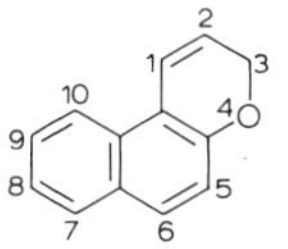

3H-Naphtho[2,1-b]pyran,
5,6-benzochromene
(II)

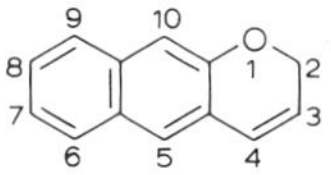

2H-Naphtho[2,3-b]pyran,
6,7-benzochromene
(III)

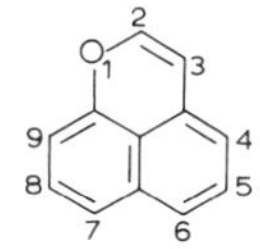

Naphtho[1,8-bc]pyran,
1-oxaphenalene
(IV)

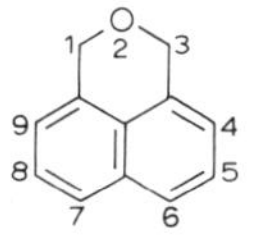

1H,3H-Naphtho[1,8-cd]pyran,
perinaphthopyran, 2-oxaperinaphthan
(V)

The preparation and properties of some of the derivatives of these compounds have been described previously under the headings which deal with the derivatives of the related benzopyrans. This is because their method of preparation and properties are very similar, for example dialkyl- and diarylnaphthopyrans (pp. 55, 59). The u.v. spectra of a number of naphtho-4-pyrones, benzochromones and related compounds have been determined and compared with those of natural benzochromones. The spectra of linear compounds have been distinguished from those of angular compounds by their characteristic absorption bands (*S. Fukushima, Y. Akahori* and *A. Ueno*, Chem. pharm. Bull., Tokyo, 1964, **12,** 316).

(a) Some derivatives of 2H- *and* 4H-*naphtho*[1,2-b]*pyran, 7,8-benzochrom-3-ene and 7,8-benzochrom-2-ene*

4*H*-**Naphtho**[1,2-*b*]**pyran-4-one,** *7,8-benzochromone* (VI, R = H), $C_{13}H_8O_2$, colourless needles, m.p. 125°, intense blue fluorescence in concentrated sulphuric acid, is obtained by heating the *2-carboxylic acid* (VI, R = CO_2H), m.p. 277–278° (*St. von Kostanecki* and *G. Froemsdorff*, Ber., 1902, **35,** 859). *2-Methyl-*4H-*naphtho*[1,2-b]*pyran-4-one* (VI, R = Me), m.p. 179° (*A. S. Bhullar* and *K. Venkataraman*, J. chem. Soc., 1931, 1165). 4*H*-Naphtho[1,2-*b*]pyran-4-one and its 2-propyl derivative are dimerised by sodium ethoxide (*A. Schönberg* and *E. Singer*, Ber., 1963, **96,** 1529, 3062). 2-**Phenyl**-4H-**naphtho**-[1,2-b]**pyran-4-one,** *2-phenyl-7,8-benzochromone, 7,8-benzoflavone, α-naphthaflavone* (VI, R = Ph), yellow plates, m.p. 155–156°, green fluorescence in sulphuric acid, is obtained from 1-naphthol and benzoylacetic ester and by other methods (*F. E. Smith, ibid.*, 1946, 542; *Bhullar* and *Venkataraman, loc. cit.*; *D. Pillon* and *J. Massicot*, Bull. Soc. chim. Fr., 1954, 26):

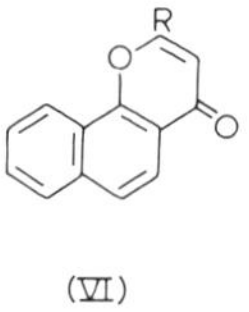

(VI) 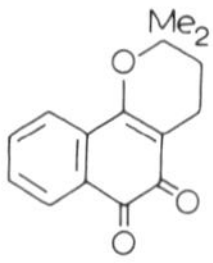 (VII) 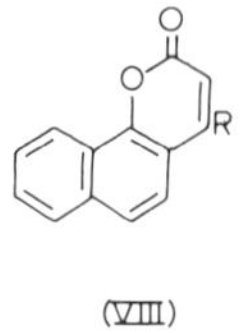 (VIII)

2,3-*Dihydro-2,2-diphenyl-*, m.p. 117°, 2,3-*dihydro-2,4-diphenyl-4H-naphtho*[1,2-b]*pyran*, m.p. 128–130° (*J. Cottam* and *R. Livingstone*, J. chem. Soc., 1964, 5228). 2,3-*Dihydro-4H-naphtho*[1,2-b]*pyran-4-one*, *7,8-benzochroman-4-one*, $C_{13}H_{10}O_2$, m.p. 104.5°, *oxime*, m.p. 144–144.5°, *semicarbazone*, m.p. 259–260° (decomp.), is prepared by the hydrogenation of the chromone, 4*H*-naphtho[1,2-*b*]pyran-4-one (Pt black) (*P. Pfeiffer* and *J. Grimmer*, Ber., 1917, **50,** 924), and by the cyclisation under acid conditions of β-(1-naphthoxy)propionitrile (*J. Colonge* and *A. Guyot*, Bull. Soc. chim. Fr., 1958, 325). 2,3-**Dihydro**-2,2-**dimethyl**-4*H*-**naphtho**[1,2-*b*]**pyran**-4-**one,** *2,2-dimethyl-7,8-benzochroman-4-one*, b.p. 136–140°/0.8 mm, has been prepared by condensing 1-naphthol with β,β-dimethylacrylic acid in the presence of antimony trichloride, zinc chloride or polyphosphoric acid (*A. S. R. Anjaneyulu et al.*, Current Sci., 1968, **37,** 513). 2,2-*Dimethyl-2H-naphtho*[1,2-b]*pyran*, b.p. 172–176°/15 mm, *picrate*, red needles, m.p. 123–125°, on boiling with 2,4-dinitrophenylhydrazine in ethanolic sulphuric acid gives the 2,4-*dinitrophenylhydrazone* of 2,3-*dihydro-2,2-dimethyl-4H-naphtho*[1,2-b]*pyran*-4-one, m.p. 279–280° (*Livingstone* and *R. B. Watson*, J. chem. Soc., 1957, 1509).

β-**Lapachone** (VII), $C_{15}H_{14}O_3$, orange needles, m.p. 155–156°, *monoxime*, m.p. 168.5–169.5°, *phenylhydrazone*, m.p. 188–189°, is prepared by treating lapachol or α-lapachone (p. 342) with sulphuric acid (*S. C. Hooker*, J. Amer. chem. Soc., 1936, **58,** 1168).

2*H*-**Naphtho**[1,2-*b*]**pyran-2-one,** *7,8-benzocoumarin*, *α-naphthacoumarin* (VIII, R = H), $C_{13}H_8O_2$, pale yellow needles, m.p. 141°, blue fluorescence in sulphuric acid, is prepared from 1-naphthol by the Pechmann reaction (*K. Bartsch*, Ber., 1903, **36,** 1966; *C. F. Koelsch* and *P. T. Masley*, J. Amer. chem. Soc., 1953, **75,** 3596). It is converted by sodium sulphite and 20% potassium hydroxide into β-(1-hydroxy-2-naphthyl)acrylic acid (*B. B. Dey* and *K. K. Row*, J. chem. Soc., 1924, **125,** 563). 2-**Oxo**-2*H*-**naphtho**-[1,2-*b*]**pyran-4-acetic acid**, *7,8-benzocoumarin-4-acetic acid* (VIII, $R = CH_2 \cdot CO_2H$), yellow crystals, m.p. 212–213°, is obtained from 1-naphthol, acetonedicarboxylic acid and sulphuric acid (*Dey*, *ibid.*, 1915, **107,** 1606). When heated it yields 4-**methyl**-2*H*-**naphtho**[1,2-*b*]**pyran-2-one** (VIII, R = Me), m.p. 167°, also obtained from 1-naphthol and acetoacetic ester (*Dey* and *A. K. Lakshminarayanan*, J. Indian chem. Soc., 1932, **9,** 149), and by the dehydrogenation of the product obtained by the cyclisation of compound IX using polyphosphoric acid (*P. N. Chakrabortty*, Indian, J. chem., 1971, **9,** 87):

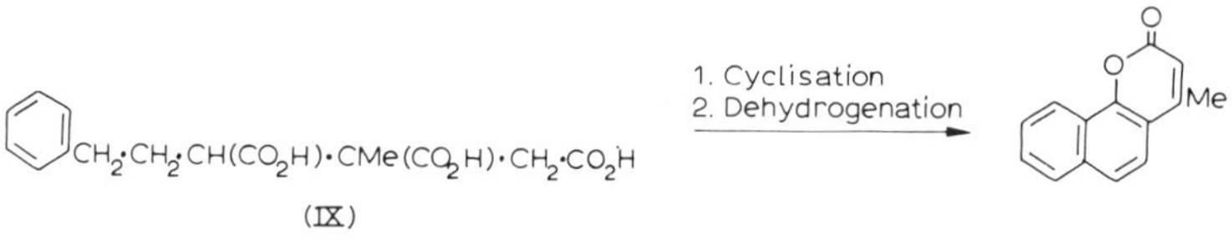

$C_6H_5 \cdot CH_2 \cdot CH_2 \cdot CH(CO_2H) \cdot CMe(CO_2H) \cdot CH_2 \cdot CO_2H$

(IX)

1. Cyclisation
2. Dehydrogenation

1-Naphthol condenses with methyl acrylate in the presence of aluminium chloride and hydrochloric acid to give 3,4-dihydro-2*H*-naphtho[1,2-*b*]pyran-2-one, which may be dehydrogenated to give 2*H*-naphtho[1,2-*b*]pyran-2-one; similarly, 3-methyl-2*H*-naphtho[1,2-*b*]pyran-2-one may be obtained from methyl 2-methylacrylate (*A. K. Das Gupta et al.*, J. Indian chem. Soc., 1968, **45**, 960).

3,4-**Dihydro**-2*H*-**naphtho**[1,2-*b*]**pyran,** 7,8-*benzochroman* (p. 224) $C_{13}H_{12}O$, b.p. 140–142°/5 mm, is prepared from 2-(3-hydroxypropyl)-1-naphthol (*P. L. de Benneville* and *R. Connor*, J. Amer. chem. Soc., 1940, **62**, 3067); 2,3-*dihydro-2,2-dimethyl-6-methoxy-naphtho*[1,2-b]*pyran, dihydrolapachenole* (p. 64), m.p. 78° (*Das Gupta, R. M. Chatterje* and *T. P. Bhowmic*, Tetrahedron, 1969, **25**, 4207). 2*H*-Naphtho[1,2-*b*]pyran-2-one is reduced by hydrogen and Raney nickel to the *dodecahydro* derivative, *decahydro*-7,8-*benzochroman*, $C_{13}H_{22}O$, b.p. 111–113°/5 mm.

Some derivatives of naphtho[1,2-*b*]pyran (*Cottam* and *Livingstone*, *loc. cit.*), and 4*H*-anthra[1,2-*b*]pyran have been prepared (*N. H. Shah* and *S. Sethna*, J. Indian chem. Soc., 1960, **37**, 699).

(b) Some derivatives of 1H- *and* 3H-*naphtho*[2,1-b]*pyran*, 5,6-*benzochrom-2-ene and* 5,6-*benzochrom-3-ene*

3-**Methyl**-1*H*-**naphtho**[2,1-*b*]**pyran**, 2-*methyl*-5,6-*benzochrom-2-ene*, m.p. 44–45°, b.p. 190°/16 mm (*W. S. Rapson* and *R. Robinson*, J. chem. Soc., 1935, 1533). 1*H*-**Naphtho**-[2,1-*b*]**pyran-1-one**, 5,6-*benzochromone* (X), $C_{13}H_8O_2$, pale yellow needles, m.p. 103°, blue fluorescence in concentrated sulphuric acid, is prepared by the interaction of 1-acetyl-2-naphthol and ethyl formate in ethanolic sulphuric acid (*B. K. Menon* and *Venkataraman*, *ibid.*, 1931, 2591); it has also been prepared by dehydrogenating 2,3-dihydro-1*H*-naphtho[2,1-*b*]pyran-1-one using triphenylmethyl perchlorate (*W. Bonthrone* and *D. H. Reid*, Chem. and Ind., 1960, 1192); some reactions of 3-*methyl*-1H-*naphtho*[2,1-b]*pyran*-1-*one*, m.p. 168°, have been described (*A. Sammour, T. Zimaity* and *S. Kamel*, J. pr. Chem., 1972, **314**, 271); 3-**phenyl**-1*H*-**naphtho**[2,1-*b*]**pyran-1-one,** 2-*phenyl*-5,6-*benzochromone*, 5,6-*benzoflavone*, m.p. 164° (*J. Tamber, et al.*, Helv., 1926, **9**, 463):

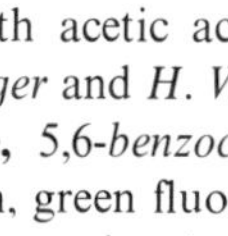

(X) 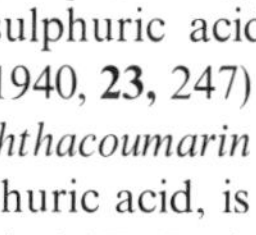 (XI)

3*H*-**Naphtho**[2,1-*b*]**pyran,** m.p. 41–42° (*I. Iwai* and *J. Ide*, Chem. pharm. Bull., Tokyo, 1962, **10**, 926; cf. p. 59). 3,3-Diphenyl-3*H*-naphtho[2,1-*b*]pyran (p. 55) gives a coloured solution with acetic acid containing sulphuric acid, but the preparation reported by *R. Wizinger* and *H. Wenning* (Helv., 1940, **23**, 247) is incorrect. 3*H*-**Naphtho**[2,1-*b*]**pyran-3-one,** 5,6-*benzocoumarin*, *β-naphthacoumarin* (XI), $C_{13}H_8O_2$, m.p. 118°, b.p. 240°/12 mm, green fluorescence in sulphuric acid, is obtained either by the Pechmann–Duisberg reaction (p. 98) on 2-naphthol (*T. Boehm* and *E. Profft*, Arch. Pharm., 1931, **269**, 25), or the Perkin reaction (p. 97) on 2-hydroxy-1-naphthaldehyde (*G. Kauffmann*, Ber., 1883, **16**, 685). It is broken down by aqueous alkali in a sealed tube to β-(2-hydroxy-1-naphthyl)acrylic acid (*Kauffmann, ibid.*, p. 693). 3*H*-Naphtho[2,1-*b*]-pyran-3-one and its 2-methyl derivative may be obtained by the dehydrogenation of the dihydrocoumarins obtained by condensing 2-naphthol with methyl acrylate and methyl 2-methylacrylate, respectively (*Das Gupta et al., loc. cit.*). The reaction between *N*,*N*-dialkyl-α-ethoxycarbonyl-α-alkylacetamide and 2-naphthol under Vilsmeier–Haack conditions affords 3*H*-naphtho[2,1-*b*]pyran-3-ones (XII) (*A. Ermili, G. Roma* and *F. Braguzzi*, Ann. Chim., Rome, 1972, **62,** 458).

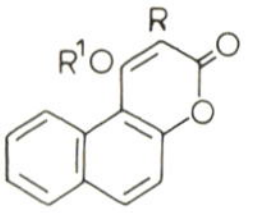

(XII)

R = H, 2-naphthyl

R^1 = H, Me, Et

2,3-**Dihydro**-1*H*-**naphtho**[2,1-*b*]**pyran,** 5,6-*benzochroman, benzo*[f]*chroman*, $C_{13}H_{12}O$, m.p. 41–42° (*R. E. Rindfusz et al.*, J. Amer. chem. Soc., 1920, **42,** 164; *C. O. Guss, ibid.*, 1951, **73,** 608); 3-*methyl*-2,3-*dihydro*-1H-*naphtho*[2,1-b]*pyran*, 2-*methyl*-5,6-*benzochroman*, m.p. 90–91° (*F. J. McQuillin* and *R. Robinson*, J. chem. Soc., 1941, 586); 2,3-*dihydro*-1,3-*diphenyl*-, m.p. 192–193°, 2,3-*dihydro*-3,3-*diphenyl*-1H-*naphtho*[2,1-b]*pyran*, m.p. 162° (*Livingstone, D. Miller* and *S. Morris, ibid.*, 1960, 5148). 2-(3-Methylbut-2-enyl-oxy)naphthalene on heating at 200° *in vacuo* gives 2,3-dihydro-3,3-dimethyl-1*H*-naphtho[2,1-*b*]pyran in 74% yield. If the reaction is carried out in quinoline solution 3,3-dimethyl-3*H*-naphtho[2,1-*b*]pyran is obtained (*D. R. Buckle* and *E. S. Waight*, Chem. Comm., 1969, 922). The Wittig reaction between 1,2-naphthoquinone and $Me_2C{:}CH \cdot CH{:}PPh_3$ affords a mixture of 2,2-dimethyl-2*H*-naphtho[1,2-*b*]pyran and 3,3-dimethyl-3*H*-naphtho[2,1-*b*]pyran (*G. Cardillo, L. Merlini* and *S. Servi*, Ann. Chim., Rome, 1970, **60,** 564).

2,3-**Dihydro**-1*H*-**naphtho**[2,1-*b*]**pyran**-1-**one,** 5,6-*benzochroman*-4-*one, benzo*[f]*chroman*-4-*one* (XIII), $C_{13}H_{10}O_2$, m.p. 50–51°, 2,4-*dinitrophenylhydrazone*, m.p. 312°, is prepared from 2-naphthol and acrylonitrile followed by hydrolysis and ring-closure of the resulting acid by sulphuric acid (*G. B. Bachman* and *H. A. Levine*, J. Amer. chem. Soc., 1947, **69,** 2341).

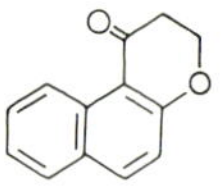

(XIII) 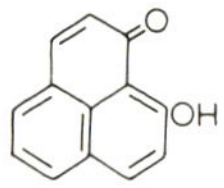 (XIV) 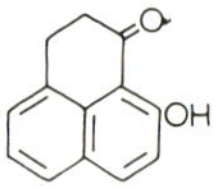 (XV)

The dihydronaphthopyranone with aluminium chloride at 190° yields 9-*hydroxyperinaphthen*-1-*one*, (XIV), yellow plates, m.p. 199°, and the 2,3-dihydroderivative (XV) (*J. D. Loudon* and *R. K. Razdan*, J. chem. Soc., 1954, 4299).

1,2-**Dihydro**-3*H*-**naphtho**[2,1-*b*]**pyran**-3-**one,** 3,4-*dihydro*-5,6-*benzocoumarin*, m.p. 69–70°, b.p. 210–220°/5 mm (*A. F. Hardmann*, J. Amer. chem. Soc., 1948, **70,** 2219). On treatment with aluminium chloride it gives a product which when heated with palladium-charcoal in trichlorobenzene yields 4-*hydroxyperinaphthen*-1-*one*, orange needles, m.p. 266–267°, *methyl ether*, m.p. 53° (*Loudon* and *Razdan*, *loc. cit.*). Some 2,3-dihydro-1*H*-naphtho[2,1-*b*]pyran-1-ones and 1,2-dihydro-3*H*-naphtho[2,1-*b*]pyran-3-ones have been synthesised (*Anjaneyulu et al.*, *loc. cit.*).

3-**Phenylnaphtho**[2,1-*b*]**pyrylium chloride,** 2-*phenyl*-5,6-*benzopyrylium chloride* (XVI), orange needles, m.p. 133–134°, *ferrichloride*, m.p. 188°, is obtained by passing hydrogen chloride into a solution of 2-naphthol and ethynyl phenyl ketone in acetic acid (*A. W. Johnson* and *R. R. Melhuish*, J. chem. Soc., 1947, 346). 2,3-Dihydro-1,3-diphenyl-3-methoxy-1*H*-naphtho[2,1-*b*]pyran on boiling with ferric chloride in acetic acid and

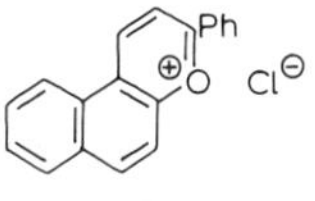

Ph Cl⊖

(XVI)

acetic anhydride gives 1,3-*diphenylnaphtho*[2,1-*b*]*pyrylium ferrichloride*, m.p. 167–168° (*Livingstone*, *Miller* and *Morris*, *loc. cit.*).

Some derivatives of naphtho[2,1-*b*]pyran (*idem*, *loc. cit.*; *J. B. Abbott et al.*, J. chem. Soc., C, 1967, 1472; *W. D. Cotterill*, *Livingstone* and *M. V. Walshaw*, *ibid.*, 1970, 1758; *E. Bradley et al.*, *ibid.*, 1971, 3028; *Cotterill et al.*, J. heterocycl. Chem., 1974, **11,** 283), and of 1*H*- and 3*H*-anthra[2,1-*b*]pyran derivatives have been prepared (*Shah* and *Sethna*, J. org. Chem., 1959, **24,** 1783).

(*c*) *Some derivatives of* 2H- *and* 4H-*naphtho*[2,3-b]*pyran*, 6,7-*benzochrom*-3-*ene and* 6,7-*benzochrom*-2-*ene*

Few representatives of these linear substances are known 2-**Methyl**-4*H*-**naphtho** [2,3-*b*]-**pyran-4-one,** 2-*methyl*-6,7-*benzochromone* (XVII, R = Me), $C_{14}H_{10}O_2$, pale yellow needles, m.p. 134–135°, *benzylidene* derivative, m.p. 168–169°, is prepared by heating 3-methoxy-2-naphthoylacetone with hydriodic acid and acetic acid (*S. Wawzonek* and *H. A. Ready*, J. org. Chem., 1952, **17,** 1419; *H. Schmid* and *H. Seiler*, Helv., 1952, **35,** 1990); 4-*oxo*-4H-*naphtho*[2,3-b]*pyran*-2-*carboxylic acid*, 6,7-*benzochromone*-2-*carboxylic acid* (XVII, R = CO_2H), $C_{14}H_8O_4$, m.p. 273° (decomp.) (*Schmid* and *Seiler*, *loc. cit.*), 10-*hydroxy*-2-*methyl*-4H-*naphtho*[2,3-b]*pyran*-4-*one*, 8-*hydroxy*-2-*methyl*-6,7-*benzochromone*, m.p. 272°; 5,10-*dihydroxy*-2-*methyl*-4H-*naphtho*[2,3-b]*pyran*-4-*one*, m.p. 250°

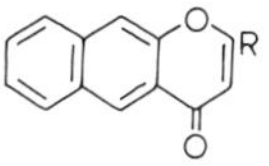

(XVII)

(decomp.), may be isomerised to give 5,6-dihydroxy-2-methyl-4*H*-naphtho[1,2-*b*]-pyran-4-one, 5,6-dihydroxy-2-methyl-7,8-benzochromone (*Fukushima* and *Akahori*, Chem. pharm. Bull., Tokyo, 1964, **12,** 307).

Rubrofusarin, 5,6-*dihydroxy*-8-*methoxy*-2-*methyl*-4H-*naphtho*[2,3-b]*pyran*-4-*one*, $C_{15}H_{12}O_5$, orange-red needles, m.p. 210–211°, *monoacetate*, m.p. 211°, *diacetate*, m.p. 260°, *methyl ether*, m.p. 203–204°, *dimethyl ether*, m.p. 187–188°, a pigment obtained

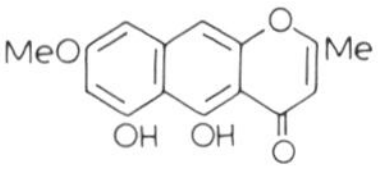

Rubrofusarin

from *Fusarium graminearum* (*R. P. Mull* and *F. F. Nord*, Arch. Biochem., 1944, **4**, 419). Originally it was believed to be a xanthone derivative, but from zinc dust distillation experiments and alkaline degradation studies it was shown to be a naphthalene derivative (*H. Tanka et al.*, Tetrahedron Letters, 1961, **151**; Agric. biol. chem., Tokyo, 1962, **26,** 767; 1963, **27,** 48, 249). X-Ray analysis results supported the naphtho[2,3-*b*]pyran-4-one structure (*G. H. Stout, D. L. Dreyer* and *L. H. Jensen*, Chem. and Ind., 1961, 289; *Sout* and *Jensen*, Acta Cryst., 1962, **15,** 1061). Demethylation of flavasperone (p. 154) on boiling with hydriodic acid and acetic anhydride is accompanied by a molecular rearrangement to give nor-rubrofusarin (XVIII) (*B. W. Bycroft, T. A. Dobson* and *J. C. Roberts*, J. chem. Soc., 1962, 40):

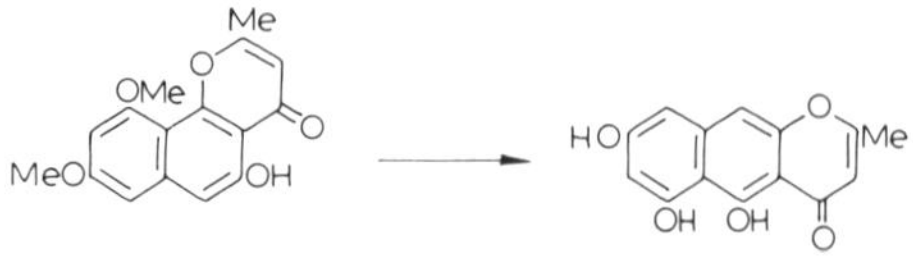

(XVIII)

2-**Phenyl-4*H*-naphtho**[2,3-*b*]**pyran**-4-**one**, 6,7-*benzoflavone* (XVII, R = Ph), m.p. 171–172° (*V. V. Virkar* and *T. S. Wheeler*, *ibid.*, 1939, 1681); 6,7-*cyclohexanoflavone*, m.p. 170–171°, blue fluorescence in sulphuric acid (*M. P. O'Farrell*, *ibid.*, 1955, 3986).

2*H*-**Naphtho**[2,3-*b*]**pyran**-2-**one**, 6,7-*benzocoumarin*, $C_{13}H_8O_2$, m.p. 163–164°, is prepared from 3-formyl-2-naphthol by the Perkin synthesis (*Boehm* and *Profft*, *loc. cit.*). **Lambertellin,** 9-*hydroxy*-3-*methylnaphtho*[2,3-b]*pyran*-2,5,10-*trione*, $C_{14}H_8O_5$, red-gold spangles, m.p. 253–254°, a metabolite of two *Lambertella* fungi (*E. Widmer et al.*, Helv., 1965, **48,** 550). It has been synthesised (*P. M. Brown et al.*, J. chem. Soc., C, 1970, 109):

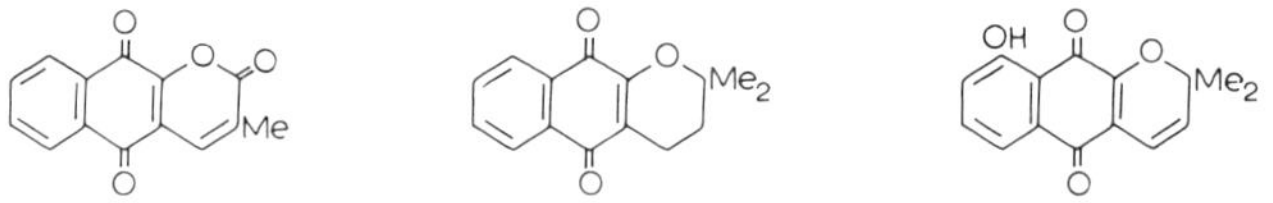

Lambertellin (XIX) α-Caryopterone

Lapachol, 2-hydroxy-3-(3-methyl-2-butenyl)-1,4-naphthoquinone, is converted by hydrochloric acid into α-**lapachone**(XIX), $C_{15}H_{14}O_3$, yellow crystals, m.p. 116–117°, *monoxime*, m.p. 204° (decomp.), 2,4-*dinitrophenylhydrazone*, m.p. 277–278° (*Hooker*, J. Amer. chem. Soc., 1936, **58,** 1168). α-**Caryopterone,** 3,4-*dehydro*-9-*hydroxy*-α-*lapachone*, $C_{15}H_{12}O_4$, needles, m.p. 143.5–145.5° (decomp.), has been isolated from *Caryopteris clandonensis* Bunge (*T. Matsumoto, C. Mayer* and *C. H. Eugster*, Helv., 1969, **52,** 808).

Some derivatives of naphtho[2,3-*b*]pyran have been prepared (*Cottam* and *Livingstone*, J. chem. Soc., 1965, 6646).

(*d*) *Some derivatives of naphtho*[1,8-bc]*pyran and* 1H,3H-*naphtho*[1,8-cd]*pyran, perinaphthopyran*

Naphtho[1,8-*bc*]**pyran,** 1-*oxaphenalene* (XXI), $C_{12}H_8O$, m.p. 44–54°, is obtained by treating 3-acetoxy-1-oxaphenalene in methanol with a solution of potassium tetrahydridoborate in dilute aqueous sodium hydroxide (*S. O'Brien* and *D. C. C. Smith*, *ibid.*, 1963, 2907):

O—CO; NaOH, $ClCH_2 \cdot CO_2H$ → $HO_2C \cdot CH_2O$ CO_2H; Ac_2O, NaOAc → (XX) (OAc) → (XXI)

1-Oxaphenalene may also be obtained by the following route (*N. S. Narasimhan* and *R. S. Mali*, Tetrahedron Letters, 1973, 843):

OMe Li; 1. PhN(Me)CHO 2. $Ph_3\overset{\oplus}{P}CH_2OMe\ Cl^{\ominus}$, $KOBu^t$ → OMe CH:CHOMe; C_5H_5N, HCl →

The preparation of 2,3-**dihydronaphtho**[1,8-*bc*]**pyran**-3-**one,** 1-*oxaphenalen*-3-(2H)-*one* (XXII), m.p. 104°, from 1-oxaphenalen-3-yl acetate and from ethyl 3-oxo-2,3-dihydro-1-oxaphenalene-2-carboxylate obtained by the cyclisation of methyl 8-ethoxycarbonylmethoxy-1-naphthoate has been described (*D. Alderson* and *Livingstone*, J. chem. Soc., C, 1967, 1782):

(XXII) Xanthorrhoein

Xanthorrhoein, 2,3-*dihydro*-5-*methoxy*-2-*methyl*-1-*oxaphenalene*, m.p. 68–69° $[\alpha]_D^{13}$ +55.6° (benzene), has been obtained from *Xanthorrhoea preissi* and *X. reflexa* (*H. H. Finlayson*, J. chem. Soc., 1926, 2763), and its structure confirmed by chemical reaction, u.v. and n.m.r. spectral data (*A. J. Birch*, *M. Salahud-Din* and *Smith*, Tetrahedron Letters, 1964, 1623).

The vinyl alcohol XXIII when boiled with *p*-toluenesulphonic acid in benzene, furnishes, after chromatography over acid-washed alumina, 2,3,3a,4,5,6-*hexahydro*-7-*methyl*-1-*oxaphenalene* (XXIV) (25%), m.p. 79–80°, and compound XXV (10–11%) (*S. Swaminathan et al.*, J. org. Chem., 1966, **31**, 656):

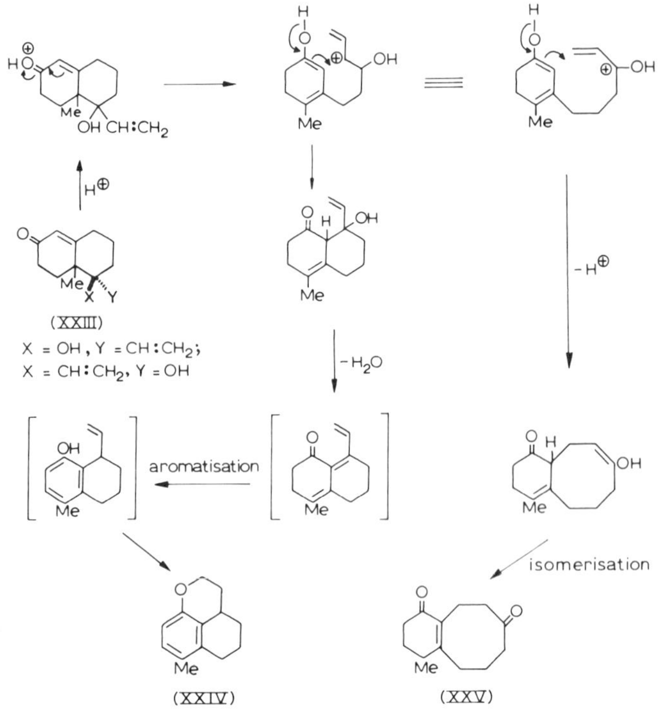

(XXIII)
X = OH, Y = CH:CH₂;
X = CH:CH₂, Y = OH

aromatisation

isomerisation

(XXIV) (XXV)

Biflorin, *7,8-dihydro-6,9-dimethyl-2(4-methyl-pent-3-enyl)-1-oxaphenalen-7,8-dione*, $C_{20}H_{20}O_3$, m.p. 97°, a bluish red antibiotic obtained from the roots of *Capraria biflora* (*J. Comin et al.*, Helv., 1963, **46**, 409):

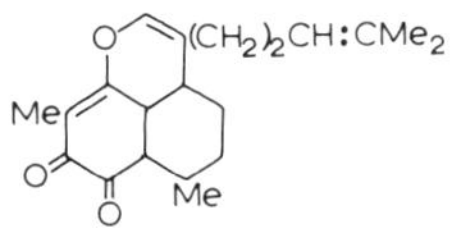

Biflorin

It may be regarded as an *o*-quinone of the diterpene series.

1*H*,3*H*-**Naphtho**[1,8-*cd*]**pyran,** *perinaphthopyran, 2-oxaperinaphthan*, (V; p. 337) m.p. 80–81°, *picrate*, m.p. 173–175° (*J. Cason* and *J. D. Wordie*, J. org. Chem., 1950, **15**, 608) is obtained by the action of water on 1,8-bisbromomethylnaphthalene. 1,3-**Diphenyl**-1*H*,3*H*-**naphtho**[1,8-*cd*]**pyran** (XXVI), m.p. 201.5–202°, is obtained by heating 1,8-bis(phenylhydroxymethyl)naphthalene with *p*-toluene sulphonic acid and 90% formic acid on a steam-bath for two days, or by treating it with phosphorus tribromide in benzene containing a drop of pyridine (*R. L. Letsinger* and *P. T. Lansbury*, J. Amer. chem. Soc., 1959, **81**, 935). Photolysis of *cis*-1,3-diphenyl-1*H*,3*H*-naphtho[1,8-*cd*]pyran under acid conditions gives 1,2-diphenylacenaphthylene (XXVII); this unusual photo-

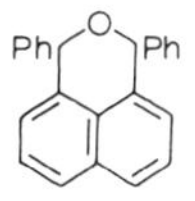

(XXVI)  (XXVII)

dehydration does not occur in the absence of acid (*A. G. Schultz* and *R. H. Schlessinger*, Chem. Comm., 1970, 1044). Naphthalic anhydride on reduction with $(Me_2CH \cdot CH_2)_2$-AlH in toluene at 5–10° gives 1*H*,3*H*-naphtho [1,8-*cd*]pyran-1-one (XXVIII) (*J. W. Burnham* and *E. J. Eisenbraun*, Org. Prep. Proced. Int., 1972, **4**, 35):

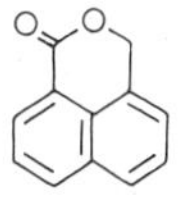

(XXVIII)

10. Benzo- and dibenzo-xanthene derivatives

The benzoxanthones are readily prepared by heating the naphthyl ethers of salicylic acid with sulphuric acid or by heating a mixture of salicylic acid and certain hydroxynaphthoic acids with acetic anhydride.

12*H*-**Benzo**[*a*]**xanthen**-12-**one,** 1,2-*benzoxanthone*, (I), colourless needles, m.p. 145°, is obtained by heating salicylic acid and either 2-hydroxy-1-naphthoic acid or 2-hydroxy-3-naphthoic acid with acetic anhydride (*W. Dilthey et al.*, J. pr. Chem., 1934, [ii], **141**, 65). In both cases the naphtholcarboxylic acid undergoes decarboxylation. The benzoxanthone is also obtained by ring-closure of the 2-naphthyl ether of salicylic

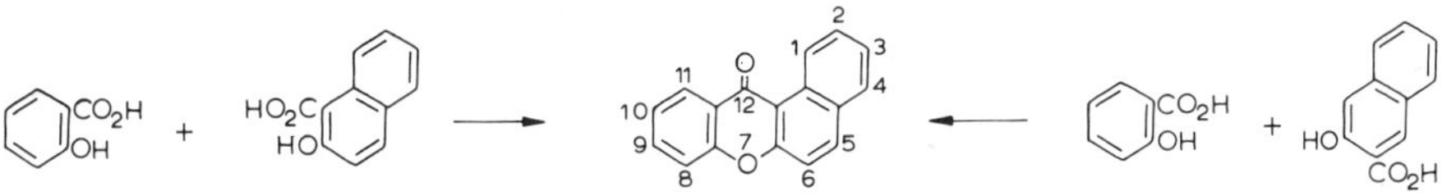

(I)

acid with sulphuric acid (*F. Ullman* and *M. Zlokasoff*, Ber., 1905, **38**, 2111; *W. Knapp*, J. pr. Chem., 1936, [ii], **146**, 113). Zine dust distillation of the xanthone yields 12H-**benzo**[a]**xanthene,** 1,2-*benzoxanthene*, m.p. 80°.

12-**Benzo**[b]**xanthen**-12-**one,** 2,3-*benzoxanthone*, (II), pale yellow needles, m.p. 201–202°, is prepared by cyclising 2-phenoxy-3-naphthoic acid (*Dilthey* and *F. Quint*, *ibid.*, 1934, [ii], **141,** 306) or the 2-naphthyl ether of salicylic acid (*Knapp, loc. cit.)*:

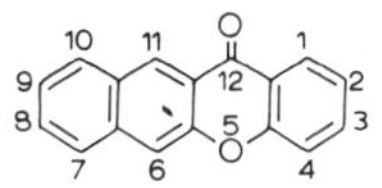

(II) 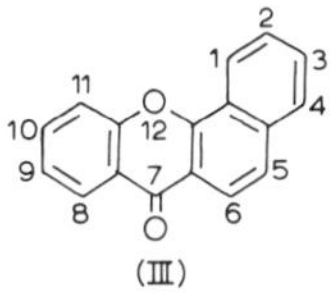 (III)

1-*Hydroxy-3-methyl-12H-benzo*[b]*xanthen-12-one*, 1-*hydroxy-3-methyl-6,7-benzoxanthone*, m.p. >300°, has been converted to a benzofuranoxanthone (*A. N. Goud* and *S. Rajagopal*, Proc. Indian Acad. Sci., Sect. A, 1969, **69**, 129). 7*H*-**Benzo**[c]**xanthen-7-one,** 3,4-*benzoxanthone* (III), m.p. 155° (161°), is similarly obtained from the 1-naphthyl ether of salicyclic acid (*Knapp, loc. cit.*), or by the distillation of 1-hydroxy-2-naphthoic acid, salicylic acid, and acetic acid (*St. von Kostanecki*, Ber., 1892, **25**, 1643). 7*H*-**Benzo**[c]**xanthene,** 3,4-*benzoxanthene*, m.p. 90–91°, *picrate*, m.p. 125–126°, is obtained by heating 2-(2-methoxybenzylidene)-1-tetralone with fused potassium hydrogen sulphate and potassium sulphate (*F. G. Baddar* and *M. Gindy*, J. chem. Soc., 1951, 64). It is oxidised to 7*H*-benzo[*c*]xanthen-7-one by potassium permanganate or by exposure of its benzene solution to sunlight.

14H-**Dibenzo**[a,j]**xanthene,** 1,2;7,8-*benzoxanthene* (IV), m.p. 203–204°, is readily prepared by passing hydrogen chloride into a benzene solution of 2-naphthol and formaldehyde (see also *R. N. Sen* and *N. N. Sarkar*, J. Amer. chem. Soc., 1925, **47**, 1089):

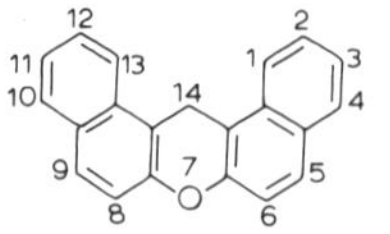

(IV) 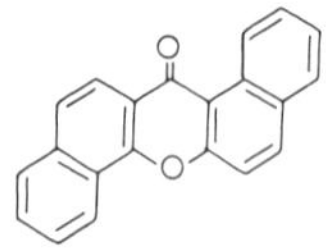 (V) 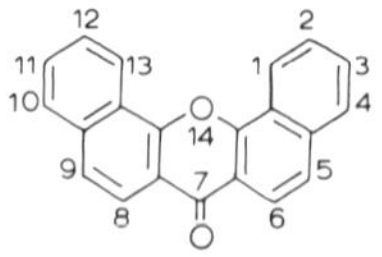 (VI)

Distilling 2-hydroxy-1-naphthoic acid with acetic anhydride gives 14*H*-**benzo**[a,j]-**xanthen-14-one**, 1,2;7,8-*dibenzoxanthone*, m.p. 194° (*Kostanecki, loc. cit.*; for proof of structure see *Dilthey* and *H. Stephan*, J. pr. Chem., 1939, [ii], **152,** 114). Distillation of 2-hydroxy-3-naphthoic acid with phosphoryl chloride or acetic anhydride affords 14*H*-**dibenzo**[*a,h*]**xanthen-14-one,** 1,2;6,7-*dibenzoxanthone* (V), m.p. 241° (*Dilthey* and *Stephan, loc. cit.*). 7*H*-**Dibenzo**[*c,h*]**xanthen-7-one**, 3,4;5,6-*dibenzoxanthone* (VI), m.p. 297°, is obtained by distilling 1-hydroxy-2-naphthoic acid with oxalic acid and sulphuric acid (*E. Clar*, Ber., 1929, **62**, 350). The linear compound, 13*H*-dibenzo[*b,i*]-xanthen-13-one, 2,3;6,7-dibenzoxanthone, appears to be unknown (*Dilthey et al.*, J. pr.

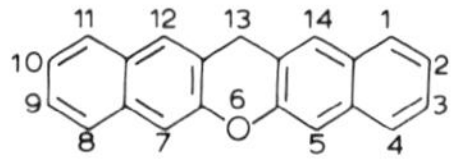

(VII)

Chem., 1934, [ii], **141**, 65; cf. *G. R. Clemo* and *R. Spence*, J. chem. Soc., 1928, 2819). Some novel results concerning the formation of dibenzoxanthones and dibenzodisalicylides from *o*-hydroxynaphthoic acids and their derivatives have been discussed. The discussion includes brief comments on the mechanism of formation of dibenzoxanthones (*M. Kamel* and *H. Shoeb*, Tetrahedron, 1966, **22**, 1539).

Chapter 21

Six-membered Ring Compounds with one Hetero Atom: Sulphur, Selenium, Tellurium, Silicon, Germanium, Tin, Lead or Iodine

R. LIVINGSTONE

1. Thiopyran derivatives

(a) Thiopyrans, thiopyrylium salts, and thiopyrones

The thiopyrans, thiapyrans*, thiins are the sulphur analogues of pyrans, the oxygen of the latter being replaced by sulphur. They have not the importance of the pyrans and are almost entirely of synthetic origin, though tetrahydrothiopyran (thiacyclohexane, thiane) and thia-adamantane, which has a thiacyclohexane ring, and a few related compounds have been found in petroleum (*S. F. Birch et al.*, J. Inst. Petroleum, 1953, **39,** 185; Ind. Eng. Chem., 1955, **47,** 240).

(i) Thiopyrans

2*H*-**Thiopyran,** *α-thiopyran*, 2H-*thiin*, (II), b.p. 32–34°/12 mm, n_D^{20} 1.5708, has been prepared by cyclising the product I obtained by treating ethyl vinyl sulphide with lithium in liquid ammonia and then adding 3-bromopropyne (*L. Brandsma* and *P. J. W. Schuijl*, Rec. Trav. chim., 1969, **88,** 30):

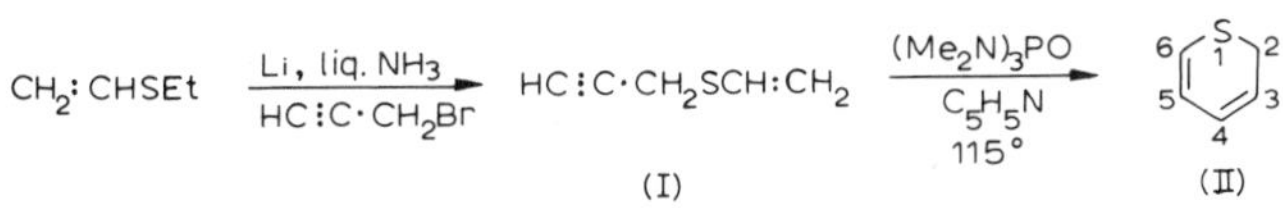

$$CH_2{:}CHSEt \xrightarrow[HC{:}C\cdot CH_2Br]{Li,\ liq.\ NH_3} \underset{(I)}{HC{:}C\cdot CH_2SCH{:}CH_2} \xrightarrow[C_5H_5N,\ 115°]{(Me_2N)_3PO} (II)$$

X-ray diffraction has shown the compound obtained by treating 2,3,5,6-tetrabromothiopyran 1,1-dioxide with zinc to be 2*H*-thiopyran 1,1-dioxide (*E. Boelema, C. J. Visser* and *A. Vos, ibid.*, 1967, **86,** 1275).

* The name “thiapyran”, which is sometimes used for this compound (and many of its derivatives), is not in accordance with, and therefore not recommended in, the I.U.P.A.C. Rules on the Nomenclature of Organic Chemistry (see Rule C-501). The I.U.P.A.C. systematic name is “thiin’.

3-*Phenyl*-2H-*thiopyran* 1,1-*dioxide*, m.p. 100–101° (*G. Pagani*, Gazz., 1967, **97**, 1518) reacts with diazomethane in ether to give the isopyrazoline III, readily isomerised by boiling methanol to the pyrazoline IV (*S. Bradamante et al.*, Tetrahedron Letters, 1969, 2971):

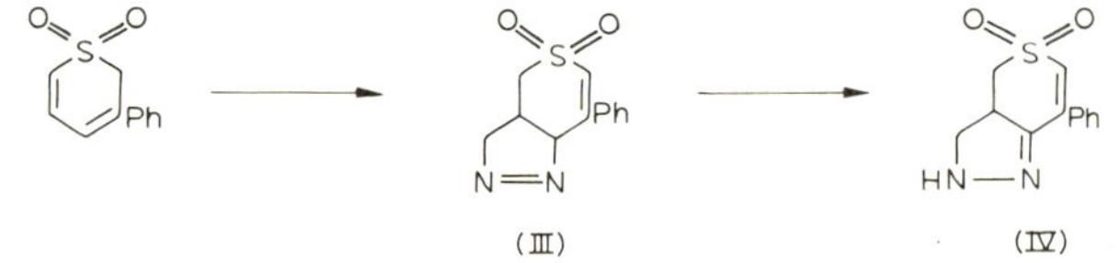

(III) (IV)

Alkyl- or aryl- substituted β-chlorovinylaldehydes react with sodium sulphide to give substituted β-mercaptoacroleines (V) and symmetrically substituted bis(β-formylvinyl) sulphides (VI). Reactions of compounds of type V with β-chlorovinylaldehydes containing a methyl or a methylene group in the β-position result in unsymmetrical *bis*(β-formylvinyl) sulphides (VII), which undergo, under conditions of the aldol reaction, alkaline-catalysed cyclisation to 2-formylmethylene-2*H*-thiopyrans. Similarly the *bis*(β-formylvinyl) sulphides (VI, $R^2 = R^3$), which have methyl or methylene groups in the α-position also give 2-formylmethylene-2*H*-thiopyrans (*W. Weissenfels* and *M. Pulst*, Tetrahedron, 1972, **28**, 5197):

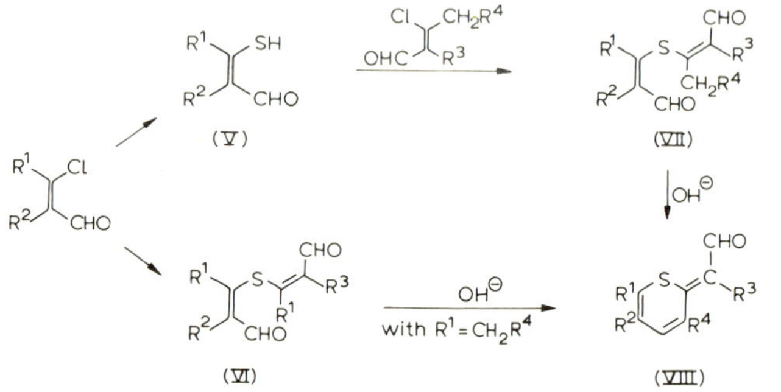

(V) (VI) (VII) (VIII)

$OH^{\ominus}$ with $R^1 = CH_2R^4$

The addition of amines to *bis*(2-propynyl) sulphone leads to mixtures of 2*H*-thiopyran 1,1-dioxides, substituted at $C_{(3)}$ or $C_{(5)}$ with the respective amino residue (*L. Skattebøl, B. Boulette* and *S. Solomon*, J. org. Chem., 1968, **33**, 548):

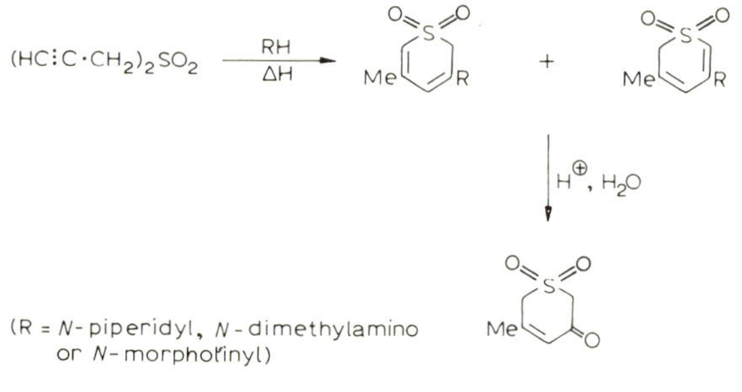

$(HC{:}C{\cdot}CH_2)_2SO_2 \xrightarrow[\Delta H]{RH}$

$H^{\oplus}$, H_2O

(R = *N*-piperidyl, *N*-dimethylamino or *N*-morpholinyl)

Hydrolysis of these cyclic dienamine mixtures gives a single ketone, *viz.*, 2,6-dihydro-3-methyl-5-oxo-5*H*-thiopyran 1,1-dioxide.

Compounds IX, obtained by the nucleophilic addition of 2-propynethiol to cyanoalkynes, when heated in silicone oil afford 5-cyano-2*H*-thiopyrans (*R. A. Van der Welle* and *Brandsma*, Rec. Trav. chim., 1973, **92,** 667):

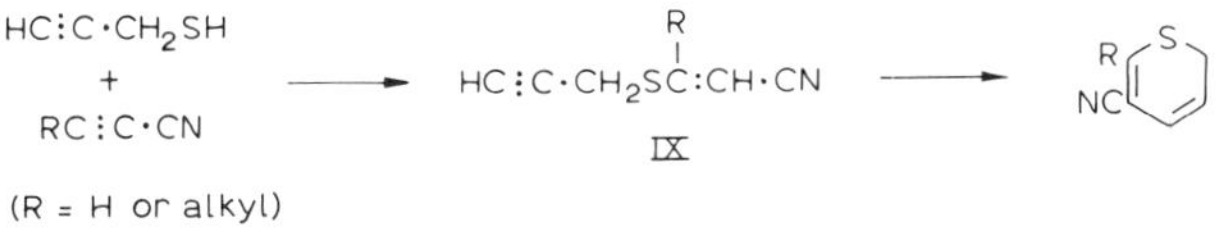

$HC\vdots C\cdot CH_2SH$ + $RC\vdots C\cdot CN$ ⟶ $HC\vdots C\cdot CH_2SC(R)\vdots CH\cdot CN$ (IX) ⟶ 5-cyano-2*H*-thiopyran

(R = H or alkyl)

Treatment of 2-benzyl-2,4,6-triphenyl-2*H*-thiopyran with 1 mol. equivalent of *N*-bromosuccinimide in the presence of benzoyl peroxide gives the 3-bromo derivative; with three equivalents of *N*-bromosuccinimide a dibromo derivative is obtained (*U. Eisner* and *T. Krishnamurthy*, J. org. Chem., 1972, **37,** 150):

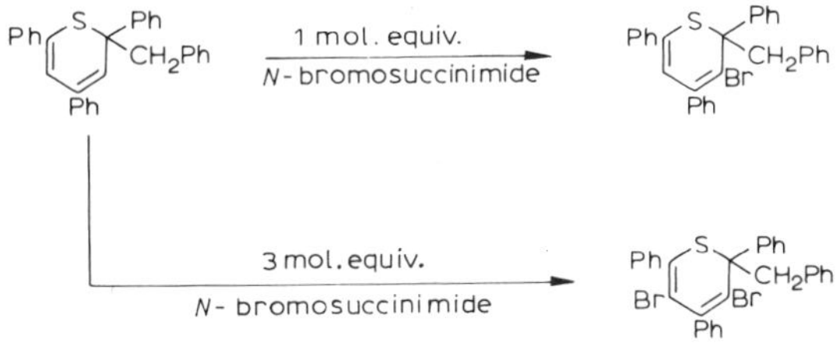

1 mol. equiv. *N*-bromosuccinimide

3 mol. equiv. *N*-bromosuccinimide

The 2,4-disubstituted 2*H*-thiopyran X is obtained by treating retinal at low temperature in amine solvents (*e.g.* aniline, pyridine, etc.) with hydrogen sulphide (*A. J. Chechak, M. H. Stern* and *C. D. Robeson, ibid.*, 1964, **29,** 187). Desulphurisation of X with amalgamated zinc in pyridine gives a good yield of *β*-carotene:

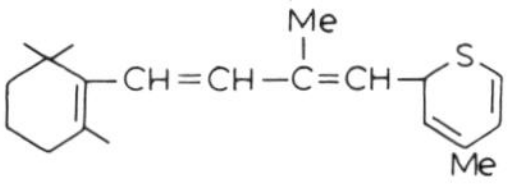

(X)

The structure of 3-benzyl-2,6-diphenyl-2*H*-thiopyran-5-carbaldehyde obtained by treating benzaldehyde with sodium polysulphide in boiling aqueous ethanol, has been solved by an X-ray diffraction study (*M.-U. Hague* and *C. N. Caughlan, ibid.*, 1967, **32**, 3017).

1-Methyl-3,5-diphenylthiabenzene 1-oxide (XI) on treatment with 1 mol. equivalent of trichlorosilane in boiling benzene for 9 h gives a 3:2 mixture of 2*H*- and 4*H*-3,5-diphenylthiopyran, from which the 2H-*isomer*, m.p. 104.6–106.1°, can be isolated by chromatography (Florisil).

(ii) Thiopyrylium salts and thiopyranyl 1,1-dioxide anions

Reaction of 3,5-diphenyl-2*H*-thiopyran with methyl iodide–silver tetrafluoroborate, MeI–$AgBF_4$, yields 1-*methyl*-3,5-*diphenyl*-2H-*thiinium tetrafluoroborate*, 1-*methyl*-3,5-*diphenyl*-2H-*thiopyrylium tetrafluoroborate* (XII), m.p. 144.5–145.3° (decomp.), which with base under a variety of conditions effects the generation of 1-methyl-3,5-diphenylthiabenzene (XIII) (*A. G. Hortmann* and *R. L. Harris*, J. Amer. Chem. Soc., 1970, **92,** 1803):

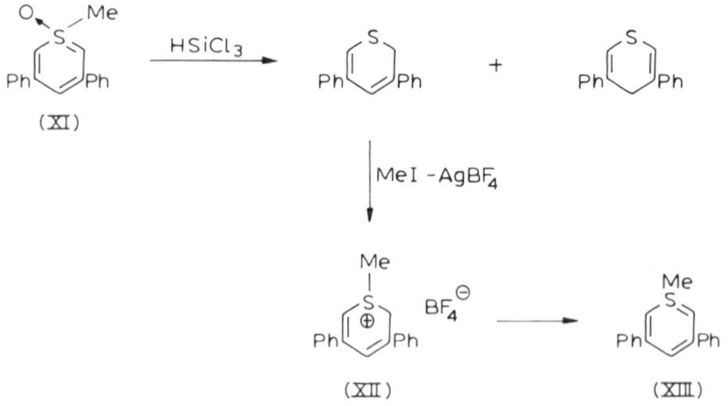

(XI) —$HSiCl_3$→ [3,5-diphenyl-2H-thiopyran] + [3,5-diphenyl-4H-thiopyran]; MeI-$AgBF_4$ → (XII) $BF_4^{\ominus}$ → (XIII)

3-Phenyl- and 6-methyl-3-phenyl-2*H*-thiopyran 1,1-dioxide have been condensed with 4-halogeno-*N*-methylpyridinium iodides in the presence of potassium carbonate to give thiopyranylidenedihydropyridine *S,S*-dioxides and the conjugation of the sulphonyl group has been studied (*G. Pagani*, J. chem. Soc. Perk. II, 1973, 1184).

Thiopyranyl 1,1-dioxide anions, thiinyl 1,1-**dioxide** anions (XIV) have been obtained by treating the appropriate thiopyran with sodium dimsyl in dimethyl sulphoxide (*Bradamante*, *A. Mangia* and *Pagani*, Tetrahedron Letters, 1970, 3381):

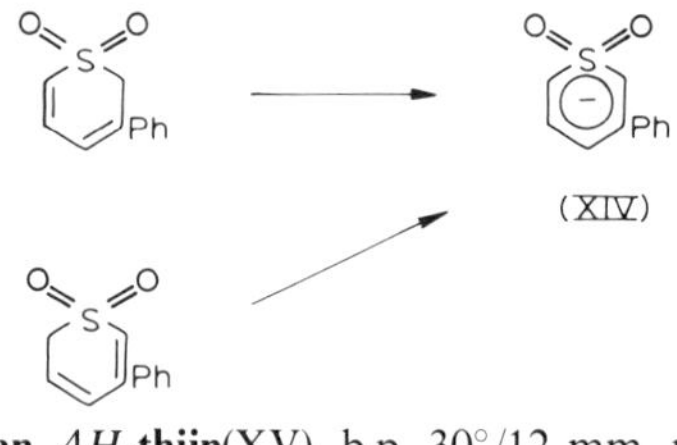

(XIV)

4*H*-**Thiopyran,** α-**thiopyran**, 4*H*-**thiin**(XV), b.p. 30°/12 mm, m.p. −28°, n_D^{20} 1.5623, is obtained in 20% yield from glutaraldehyde and hydrogen sulphide in methylene dichloride at 120°/60 mm (*J. Strating et al.*, Angew. Chem., 1962, **74,** 456; Org. Prep. Proced., 1969, **1,** 21):

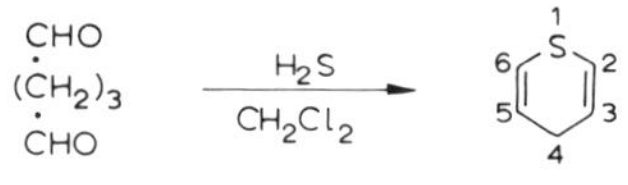

CHO·$(CH_2)_3$·CHO —H_2S, CH_2Cl_2→ (XV)

It decomposes rapidly in air at room temperature, but can be stored in the solid state. 4*H*-Thiopyran reacts with dichlorocarbene to form simple and double addition products (*K. Dimroth, W. Kinzebach*, and *M. Soyka*, Ber., 1966, **99**, 2351).

3-**Methyl-4*H*-thiopyran,** volatile oil, b.p. 134°, d^{19} 0.994, having a smell similar to that of xylene, has been obtained by heating α-methylglutaric acid with phosphorus trisulphide (*K. Krekeler*, Ber., 1886, **19**, 3266). It resembles thiophene, giving a colour with isatin and sulphuric acid, with phenanthraquinone, and with acetyl chloride and aluminium chloride forming an *acetyl* compound, b.p. 233–235°, *oxime*, m.p. 68°. It is, however, oxidised by potassium permanganate to acetic acid and oxalic acid. 4-*Benzyl*-2,4,6-*triphenyl*-4H-*thiopyran*, m.p. 116–117° (*Eisner* and *Krishnamurthy, loc. cit.*). In acid solution 3,5-dimethyl-2,6-diphenyl-4*H*-thiopyran isomerises to the corresponding 2*H*-thiopyran (*V. G. Kharchenko et al.*, Zhur. Org. Khim., 1969, **5**, 1711).

1,5-Diketones have been converted into 2*H*- and 4*H*-thiopyrans and thiopyrylium salts by treatment in acetic acid with hydrogen sulphide (*idem*, Khim. Geterotsikl. Soedin., 1970, 900):

$(PhCO{\cdot}CH_2)_2CHPh$ —(1. AcOH, H_2S, 1h at 15°; 2. AcOH, H_2S-HCl, 3h at 25-35°)→ 2,4,6-triphenylthiopyrylium (45%) + 2,4,6-triphenyl-4*H*-thiopyran (20%)

$PhCO{\cdot}CH_2{\cdot}CHPh{\cdot}CHMe{\cdot}COPh$ —(AcOH, H_2S-HCl)→ (95%)

The 4*H*-compounds are isomerised to 2*H*-thiopyrans in acetic acid containing hydrogen chloride. A number of substituted thiopyrans have been prepared by reacting the appropriate 1,5-diketone with phosphorus pentasulphide in pyridine at 90–115°. Oxidation of some of the thiopyrans with hydrogen peroxide–acetic acid affords the corresponding sulphones (*idem, ibid.*, Sb. 3, 1971, 73).

2,6-Diphenyltetrahydrothiopyran when boiled with triphenylmethane in acetic acid containing 70% perchloric acid yields 2,6-*diphenylthiopyrylium perchlorate* m.p. 186–188° (decomp.); 4-*methyl*-2,6-*diphenylthiopyrylium perchlorate*, m.p. 231–232° (decomp.) (*R. Wizinger* and *H. J. Angliker*, Helv., 1966, **49**, 2046):

—(Ph_3CH, AcOH, 70% $HClO_4$)→ $ClO_4^{\ominus}$

The above thiopyrylium salts have been converted to thiopyrylocyanines.

trans-trans-1,4-Diphenyl-1,3-butadiene and thiophosgene on boiling in benzene afford 2-*chloro*-3,6-*diphenylthiopyrylium chloride*, m.p. 142–144° (*G. Laban* and *R. Mayer*, Z. Chem., 1967, **7**, 227). 1,5-Dioxo-1,2,4,5-tetraphenylpentane and phosphorus pentasulphide give the thiopyrylium salts XVI (X=Cl, ClO_4) after treatment with the appropriate acid. The thiopyrylium chloride (XVI, X=Cl) on reduction with lithium tetrahydridoaluminate yields 2,3,5,6-tetraphenyl-4*H*-thiopyran; the corresponding perchlorate reacts with a Grignard reagent to give the related 4-substituted thiopyran (*Kharchenko* and *V. I. Kleimenova*, Khim. Geterotskl. Soedin., 1971, 613):

PhCO·CHPh·CH_2·CHPh·COPh —(1. P_4S_{10}; 2. HX)→ (XVI) (X = Cl, ClO_4)

(XVI) —(HCl, $LiAlH_4$)→ 2,3,5,6-tetraphenyl-4*H*-thiopyran

(XVI) —($HClO_4$, RMgI (R = Et, $PhCH_2$))→ 4-R-2,3,5,6-tetraphenyl-4*H*-thiopyran

When 4,6-diphenyl-2*H*-thiopyran-2-one or -2-thione is treated with oxalyl halides, the acylation products fragment and yield reactive 4,6-diphenyl-2-halogenothiopyrylium salts (*J. Faust*, *G. Speier* and *Mayer*, J. pr. Chem., 1969, **311**, 61). Some thiopyrylium trifluoroacetates have been prepared (*Kharchenko et al.*, Khim. Geterotsikl. Soedin., 1972, 916). Disproportionation occurs to yield thiopyrylium salts and thiacyclohexanes, tetrahydrothiopyrans (XVII) when 4*H*-thiopyrans are treated with perchloric, hydrochloric or trifluoroacetic acid (*idem*, Zhur. org. Khim., 1971, **7**, 422):

4*H*-thiopyran —($HClO_4$, HCl or $CF_3·CO_2H$)→ thiopyrylium $X^⊖$ + (XVII)

(X = ClO_4, Cl, $CF_3·CO_2$)

Phenyllithium reacts with 2,4,6-triphenylthiopyrylium ion (XVIII) to give 1,2,4,6-tetraphenylthiabenzene (XIX), whereas reaction with methyl, ethyl, and butyl Grignard reagents gives a mixture from which 2- or 4-alkyltriphenylthiopyrans are isolated (*G. Suld* and *C. C. Price*, J. Amer. chem. Soc., 1962, **84**, 2090):

(XVIII) (X = ClO_4, BF_4, I, I_3) —PhLi→ (XIX)

(XVIII) —MeMgI→ [1-Me-2,4,6-triphenylthiabenzene] —(25°, 20 min)→ 2-Me-2,4,6-triphenyl-2*H*-thiopyran (m.p. 119°) + 4-Me-2,4,6-triphenyl-4*H*-thiopyran (m.p. 97°)

2,4,6-Triphenylthiopyrylium perchlorate with lithium tetrahydridoaluminate yields 2,4,6-triphenyl-4*H*-thiopyran:

It couples with phenylethynyl-lithium at carbon to give a mixture of 2- and 4-*phenylethynyl-2,4,6-triphenylthiopyran* rather than at sulphur to give the thiabenzene (*Price et al.*, J. org. Chem., 1971, **36,** 791):

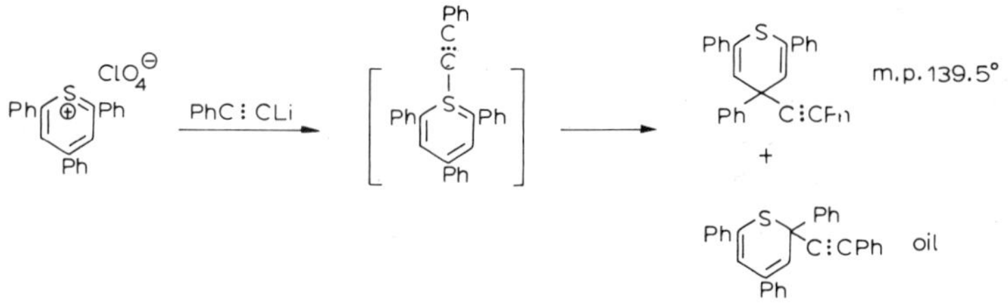

U.v. and n.m.r. spectra of thiopyrylium chloride show that at pH < 6 the compound is not hydrolysed; at pH > 6 the thiopyrylium ion, 4-mercapto-1,3-butadiene-1-carbaldehyde, HSCH:CH·CH:CH·CHO, and its sulphide ion $\overset{\ominus}{S}$CH:CH·CH:CH·CHO coexist; and at pH ⩾ 11 only the mercaptocarbaldehyde is formed. Thiopyrylium iodide treated with methanol and sodium hydrogen carbonate gives 2-*methoxy*-2H-*thiopyran*, b.p. 70–71°/14 mm (*I. Degani, R. Fochi* and *C. Vincenzi*, Gazz., 1967, **97,** 397). The n.m.r. spectrum of thiopyrylium tetrafluoroborate has been discussed (*Dimroth, Kinzbach* and *Soyka, loc. cit.*).

The relative stabilities of pyrylium, thiopyrylium, selenopyrylium, their benzo and dibenzo derivatives have been compared using liquid sulphur dioxide, acetonitrile, or nitromethane and u.v. and n.m.r. spectroscopy to determine the concentration of the final products. The S-containing compounds are more stable than the O or Se compounds, but for the same hetero atom the dibenzo cations are more stable than the benzo cations which are more stable than unsubstituted cations (*Degani, Fochi* and *Vincenzi*, Boll. Sci. Fac. Chim. Ind. Bologna, 1965, **23,** 21).

The **dithiopyrylium dication,** in which two thiopyrylium rings are directly connected by a covalent C–C bond has been synthesised. Thiopyrylium iodide is first reacted with zinc in acetonitrile at 0° under a nitrogen atmosphere. After removal of the unreacted zinc, the filtrate is treated at 0° with triphenylmethyl tetrafluoroborate to give 4,4′-dithiopyrylium bistetrafluoroborate. The diiodide and diperchlorate have also been prepared (*Z. Yoshida et al.*, Tetrahedron Letters, 1971, 3999):

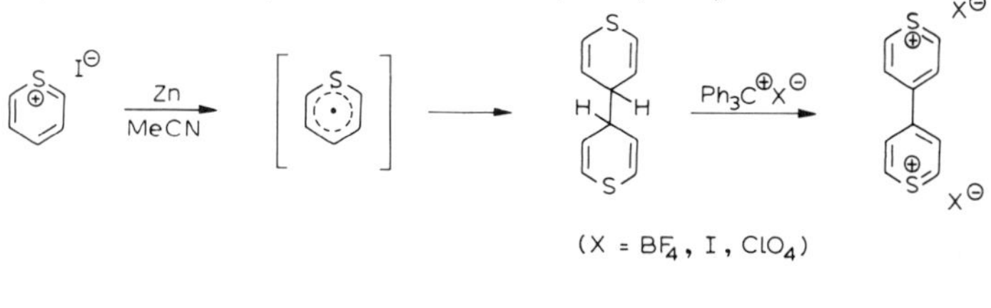

(X = BF_4, I, ClO_4)

(iii) Thiopyranones

2*H*-**Thiopyran**-2-**one,** 2*H*-**thiin**-2-**one,** (XX), m.p. 18–20°, b.p. 105–106°/ 3 mm, *picrate*, m.p. 118°, is obtained by treating 2*H*-thiopyran-2-thione in a little water with excess of a saturated aqueous solution of mercuric chloride (*Mayer*, Ber., 1957, **90,** 2362):

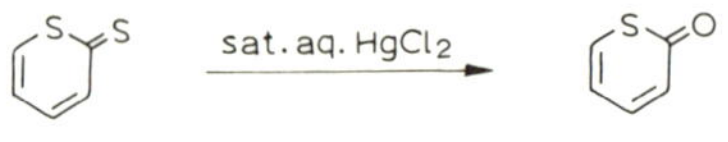

(XX)

Alkylidene- and aralkylidene-malononitriles, *e.g.*, XXI, of the type $R(R^1CH_2)C{:}C(CN)X$ ($X = CN$ or CO_2Et) with carbon disulphide at room temperature yield **thiopyran-2-thiones,** which are converted by treatment with mineral acid in the presence of an alcohol to 6-(alkylthio)thiopyran-2-ones (*K. Gewald*, J. pr. Chem., 1966, **31,** 205):

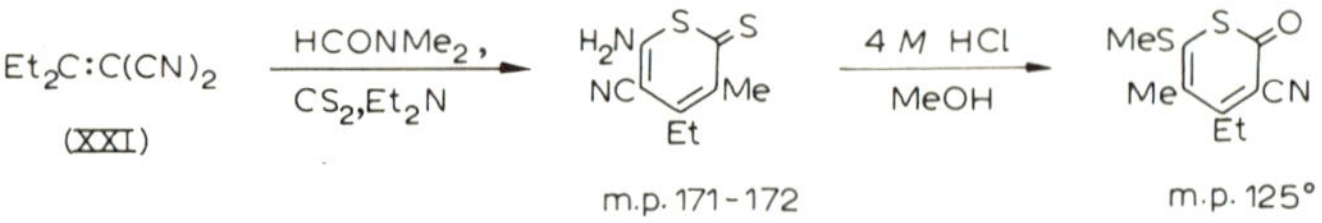

3,6-*Diphenyl*-2H-*thiopyran*-2-*one*, m.p. 183.5–184°, may be obtained starting with *trans-trans*-1,4-diphenyl-1,3-butadiene and thiophosgene in boiling benzene (*Laban* and *Mayer*, *loc. cit.*).

Thioamide vinylogues of the type $ArCSCH{:}CHNRR^1$, on treatment with phenylacetyl chloride give 2*H*-thiopyran-2-ones (*J. C. Meslin* and *H. Quiniou*, Compt. rend. Ser. C, 1971, **273,** 148):

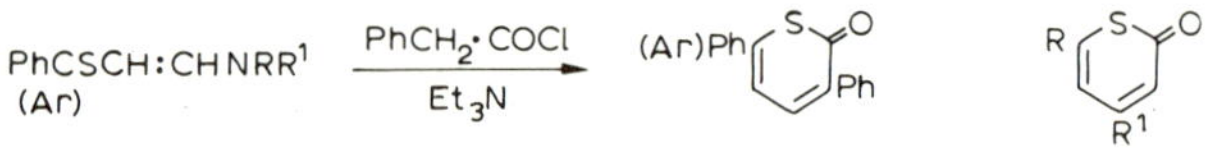

(XXII)

The 2*H*-thiopyranones XXII (R = Ph, [4]BrC_6H_4, [4]MeC_6H_4, [4]$MeOC_6H_4$; R^1 = H, Ph) have been prepared by the cycloaddition of ketone to aminothioacrylophenones $R^2R^1C{:}CHCSR$ (R^1 = secondary amino) (*idem*, Bull. Soc. chim. Fr., 1972, 2517).

Substituted 2*H*-thiopyran-2-ones have been synthesised by use of the Knoevenagel reaction (*M. Weissenfels* and *S. Illing*, Z. Chem., 1973, **13,** 130):

4H-**Thiopyran-4-ones,** *4H*-**thiin-4-ones,** are prepared by the dehydrogenation of tetrahydrothiopyrones with phosphorus pentachloride (*F. Arndt* and *N. Bekir*, Ber., 1930, **63,** 2393). They resemble the 4-pyrones and are colourless crystalline substances, soluble in water, almost odourless. They form salts with mineral acids, and the carbonyl and olefinic groups are inert, suggesting that the carbonyl group is highly polarised. This is confirmed by the high dipole moments of the thiopyran-4-ones (*Arndt et al.*, J. chem. Soc., 1935, 602) and the lack in thiopyran-4-one of the absorption band in the 1660 cm^{-1} region characteristic of a conjugated carbonyl group (*D. S. Tarbell* and *P. Hoffman*, J. Amer. chem. Soc., 1954, **76,** 2451). The *4H*-thiopyran-4-ones with phosphorus pentasulphide yield *4H*-thiopyran-4-thiones (p. 358) (*Arndt et al.*, Ber., 1925, **58**, 1644).

4H-*Thiopyran-4-one*, 4H-*thiin-4-one*, (XXIII) m.p. 109.5–111.5°, *hydrochloride*, m.p. 125–135°, μ 3.96D (*M. Rolla et al.*, Ann. chim., Rome, 1952, **42**, 673; C.A., 1953, **47**, 11934; *Arndt et al., loc. cit.; Mayer, loc. cit.*), absorbs at 246 mμ (log ε 4.08) and 260 mμ (log ε 3.72) (*P. Franzosini et al.*, Ann. chim., Rome, 1955, **45**, 128; C.A., 1955, **49**, 13779). It has been indicated that thiopyran-4-one, while resembling 4-pyrone in some ways is more stable, the structure XXIV being of considerable importance. No evidence for "aromatic"-type chemical properties could be found (*P. L. Pauson, G. R. Proctor* and *W. J. Rodger*, J. chem. Soc., 1965, 3037).

(XXIII) (XXIV)

Oxidation of thiopyran-4-one with hydrogen peroxide yields the 1,1-*dioxide* (S-*dioxide, sulphone*), yellow crystals, m.p. 173–174°, *oxime*, m.p. 196°, *semicarbazone*, m.p. 237–239° (*G. Traverso*, Ber., 1958, **91**, 1224) also obtained by the dehydrogenation of tetrahydrothiopyran-4-one 1,1-dioxide, penthianone sulphone (LIII, p. 364) (*E. A. Fehnel* and *M. Carmack*, J. Amer. chem. Soc., 1948, **70,** 1813). Deuterium substitution in the *4H*-thiopyran-4-one 1,1-dioxide system has been investigated (*L. A. Paquette* and *L. D. Wise*, *ibid.*, 1968, **90,** 807). Treatment of thiopyran-4-one in chloroform at −40° with chlorine in chloroform gives a quantitative yield of *thiopyrylium chloride*, m.p. 90–92°; similarly with bromine at −35°, 2,3,5,6-*tetrabromotetrahydrothiopyran*, m.p. 149–150°, 1-*oxide*, m.p. 157–160°, 1,1-*dioxide*, m.p. 185–186° is obtained (*E. Molenaar* and *Strating*, Tetrahedron Letters, 1965, 2941).

2,6-**Dimethyl-4*H*-thiopyran-4-one,** m.p. 102.6–103.4°, μ 4.30D (*Rolla et al., loc. cit.*); 2,6-**diphenyl-4*H*-thiopyran-4-one,** m.p. 132°, μ 4.41D (*idem*, Ann. chim., Rome, 1954, **44,** 430; C.A., 1955, **49,** 19; *Arndt et al., loc. cit.*) is oxidised by hydrogen peroxide to the 1,1-*dioxide*, yellow needles, m.p. 144–145°, μ 0.093D, *oxime*, m.p. 197–198° (*Arndt et al.*, Ber., 1925, **58,** 1633); 3,5-**diphenyl-4*H*-thiopyran-4-one,** m.p. 167° (*A. Schönberg* and *W. Asker*, J. chem. Soc., 1946, 604).

2,6-Dimethylthiopyran-4-one reacts with malononitrile, ethyl cyanoacetate and rhodanine, but not with diethyl malonate, acetylacetone, cyclopentadiene or nitromethane, *e.g.*, malononitrile affords 4-(dicyanomethylene)-2,6-dimethyl-4*H*-thiopyran (*M. Ohta* and *H. Kato*, Bull. chem. Soc. Japan, 1959, **32**, 707).

The major product from the reaction of 2,6-dimethylthio-3,5-diphenyl-4*H*-thiopyran-4-one with hydroxide ion is the hydroxythiopyran-4-one (XXV), which on treatment with diazomethane gives the isomeric enol ethers XXVI and XXVIa. These conclusions have been supported by spectral and chemical evidence (*H. J. Teague* and *W. P. Tucker*, J. org. Chem., 1970, **35**, 1968):

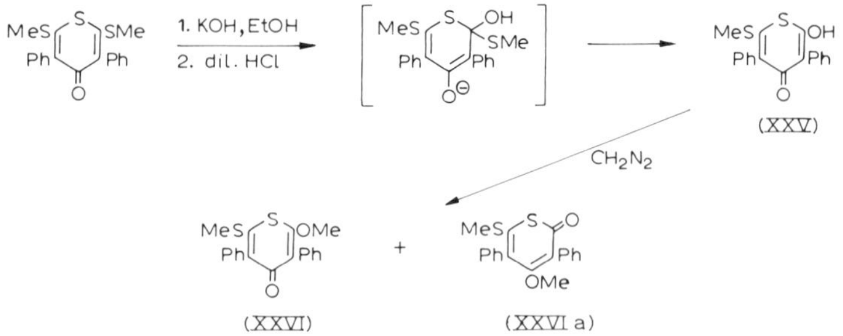

1. KOH, EtOH
2. dil. HCl

CH_2N_2

(XXV)

(XXVI) + (XXVI a)

Dimercaptothiopyran-4-ones, *e.g.*, 2,6-dimercaptothiopyran-4-one and some of its derivatives may be used in colorimetry, amperometry, and gravimetry, since they are oxidised at a rotating platinum microelectrode and form precipitates or soluble coloured compounds with heavy metals, *e.g.*, bismuth and copper (*Yu. I. Usatenko* and *O. M. Arishkevich*, Dopovidi Akad. Nauk Ukr. R.S.R., 1962, 504).

The electronic spectra and chemical reactivities of pyran-4-one, thiopyran-4-one and related compounds have been studied and the results discussed in terms of the MO–LCAO treatment (*R. Zahradnik*, *C. Parkanyi* and *J. Koutecky*, Coll. Czech. chem. Comm., 1962, **27**, 1242). The basicities of some pyran-4-ones and thiopyran-4-ones in sulphuric acid have been determined spectrophotometrically and pK_a values obtained (*A. I. Tolmachev*, *L. M. Shulezhko* and *A. A. Kisilenko*, Zhur. obshcheĭ Khim., 1968, **38**, 118).

2*H*-**Thiopyran**-2-**thione**, 4*H*-**thiin**-2-**thione**, (XXVII; R = H), deep wine-red needles, m.p. 53–54°, orange in water, deep yellow to red-orange in organic solvents, is obtained by heating tetrahydrothiopyran, thiacyclohexane with sulphur (*Mayer*, *loc. cit.*). It gives immediately with Grote reagent (*I. W. Grote*, J. biol. Chem., 1931, **93**, 25) a stable deep blue colour and with aqueous mercuric chloride a yellow precipitate; it reacts with hydroxylamine and phenylhydrazine with the elimination of hydrogen sulphide; it is easily converted to thiopyran-4-one (p. 355):

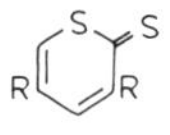

(XXVII)

Carbon disulphide reacts with enamines $RCH{:}CHNR^1_2$ or dienamines $R^1_2NCH{:}CRCH{:}CHR$ at room temperature to give 2*H*-thiopyran-2-thiones (XXVII, R = Me, Et, pentyl, Ph); several 4,6-disubstituted derivatives have been prepared (*Mayer* and *Gewald*, Angew, Chem. intern. Edn., 1967, **6**, 294; *Mayer*, *Laban* and *M. Wirth*, Ann., 1967, **703**, 140). 3,6-*Diphenyl*-2H-*thiopyran-2-thione*, m.p. 122–123°, (*Laban* and *Mayer*, *loc. cit.*). A number of substituted 2*H*-thiopyran-2-thiones have been obtained by reacting $RCOCR^1{:}CR^2Cl$ (R = H, Me, Ph, [4]ClC_6H_4; R^1 = H; R^2 = H, Me, Ph) with $R^3CH_2 \cdot C(S)SH$ (R^3 = H, Me, Ph) (*S. Scheithauer* and *Mayer*, Z. Chem., 1969, **9**, 59).

Substituted 3-acylthiopyran-2-thiones, XXX, are obtained when 1,2-dithiolylium acid sulphates and perchlorates (XXVIII, R and R^2 = H or aryl; R^1 = H, Ph or Me) are treated with acyldithioacetate esters (XXIX, R^3 = Me or aryl); mixtures of the 4,6-isomers are obtained when R and R^2 are different (*Quiniou*, Compt. rend. Ser. C, 1969, **269**, 1059):

(XXVIII) + $R^3C(OH){:}CH \cdot CS_2Me$ (XXIX) ⟶ (XXX)

3-Acyl-6-aryl-2*H*-thiopyran-2-thiones may be prepared from β-oxodithiocarboxylic acids and α-acetylenic ketones or β-chlorovinylcarbaldehydes (*J. P. Pradere et al.*, *ibid.*, p. 929).

4,6-*Diamino*-3,5-*dicyano*-2H-*thiopyran-2-thione*, yellow needles turning brownish near 300°, is easily obtained in good yield from malononitrile and carbon disulphide in aqueous alkali (*T. Takeshima et al.*, J. org. Chem., 1970, **35**, 2438):

$NC \cdot CH_2 \cdot CN$ $\xrightarrow[\text{room temp., 6h}]{CS_2,\ \text{aq. } NH_4OH}$ (4,6-diamino-3,5-dicyano-2H-thiopyran-2-thione)

3,6-Diarylthiopyran-2-ones on treatment with tetraphosphorus decasulphide, P_4S_{10}, give the corresponding **thiopyran-2-thiones** (*Meslin* and *Quiniou*, *loc. cit.*):

(Ar, Ph thiopyran-2-one) $\xrightarrow{P_4S_{10}}$ (Ar, Ph thiopyran-2-thione)

The thiopyran-2-thione XXXI with *N*-bromosuccinimide yields compound XXXII and by pyrolysis compound XXXIII (*Schönberg* and *R. v Ardenne*, Ber., 1966, **99**, 3316):

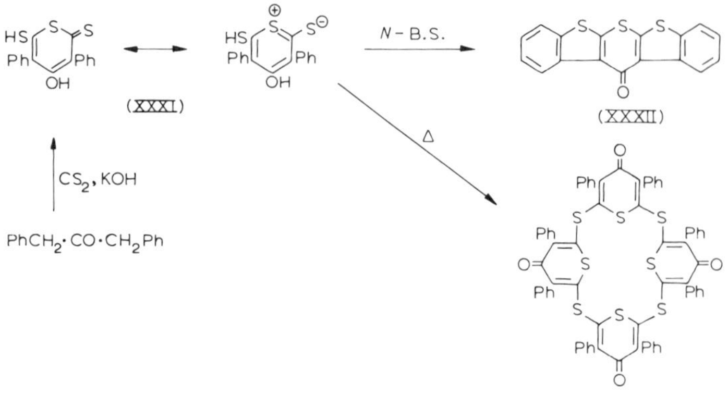

The m.s. fragmentation of compound XXXIII has been discussed.

Treatment of 4,6-diphenyl-2*H*-thiopyran-2-thione with diphenyldiazomethane yields 4,6-diphenyl-2-diphenylmethylene-2*H*-thiopyran (*idem*, *ibid*., 1968, **101**, 346):

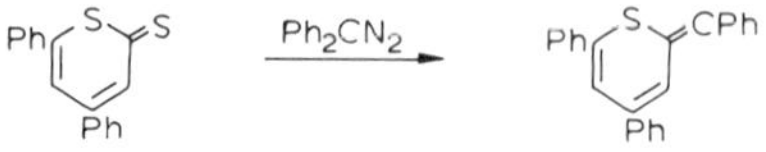

4*H*-**Thiopyran-4-thione,** 4*H*-**thiin-4-thione,** (XXXIV), m.p. 48°, b.p. 104–105°/4 mm, *oxime*, m.p. 126–127°, may be obtained by boiling 4*H*-thiopyran-4-one in benzene with powdered tetraphosphorus decasulphide (*Traverso, loc. cit.*). On treatment with aqueous sodium carbonate it gives thiopyran-4-one (*Mayer, loc. cit.*). 2,6-*Diphenyl*-4H-*thiopyran-4-thione*, m.p. 129–130°, *oxime*, m.p. 197–198° (*Traverso* and *M. Sanesi*, Ann. chim., Rome, 1953, **43,** 795; 1955, **45,** 695).

2,6-Diphenyl-4*H*-thiopyran-4-thione and 2,6-bis(methylthio)-3,5-diphenyl-4*H*-thiopyran-4-thione with diphenyldiazomethane give the corresponding 4-diphenylmethylene derivatives (*vide supra*):

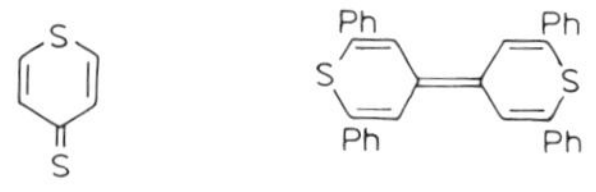

Compounds with a thioketone structure frequently dimerise with the elimination of sulphur. In the case of 2,6-diphenyl-4*H*-thiopyran-4-thiones the thermochemical dimersation with the elimination of sulphur gives 2,2′,6,6′-tetraphenyl-4,4′-di(thiopyranylidene) (XXXV), also obtained by the photolysis of the parent compound (*N. Ishibe, M. Sunami* and *M. Odani*, Tetrahedron, 1973, **29,** 2005).

(b) Di- and tetra-hydro-thiopyrans and -thiopyranones

(i) Dihydrothiopyrans, dihydrothiins, dihydrothiapyrans

Dihydrothiopyrans are prepared by heating the corresponding dihydropyrans with hydrogen sulphide and a catalyst. 3,4-Dihydro-2*H*-pyran, for instance, with hydrogen sulphide and alumina at 425° yields **3,4-dihydro-2*H*-thiopyran** (XXXVI, R = H), b.p. 142.5°, 66°/57 mm, 40–41°/10 mm, n_D^{20} 1.5330, also obtained when a mixture of tetrahydrothiopyran 1-oxide, pre-

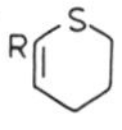

(XXXVI)

pared by the oxidation of tetrahydrothiopyran with 30% hydrogen peroxide in absence of solvent, and benzoic anhydride in dry benzene is boiled for 14 h (*W. E. Parham et al.*, J. org. Chem., 1964, **29**, 2211):

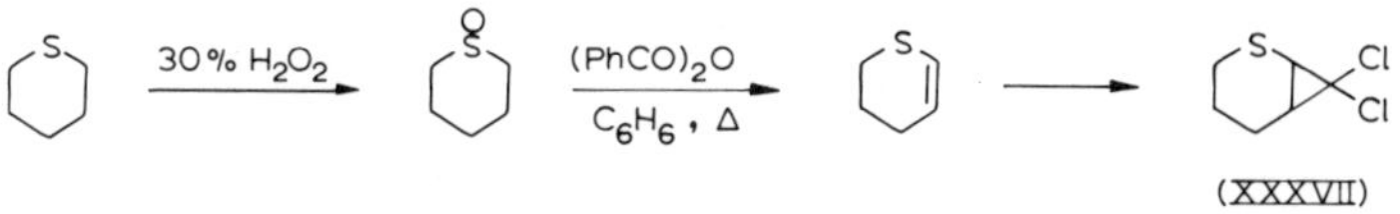

30% H_2O_2 → ; $(PhCO)_2O$, C_6H_6, Δ → ; →

(XXXVII)

3,4-Dihydro-2*H*-thiopyran may be prepared by the pyrolysis of the crude ester obtained by reacting tetrahydrothiopyran with *tert*-butyl perbenzoate (*L. A. Cohen* and *J. A. Steele*, *ibid*., 1966, **31**, 2333); 3,4-*dihydro*-6-*methyl*-2H-*thiopyran* (XXXVI, R = Me), b.p. 105–107°, 67–68°/30 mm, n_D^{20} 1.4457.

3,4-Dihydro-2*H*-thiopyran reacts with methyl iodide to yield mainly trimethylsulphonium iodide (*R. F. Naylor*, J. chem. Soc., 1949, 2749); with alcohols, under acid catalysis, by addition to the double; and with ethyl trichloroacetate and sodium methoxide to give 7,7-dichloro-2-thiabicyclo-[4.1.0]heptane (XXXVII) *(Parham et al., loc. cit.)*. The preparation of adducts with cholestanol, cyclopentanol, and 5′-*O*-acetylthymidine has been described. The protective grouping can be removed readily by reaction with silver ion at neutral pH. Implications of the technique for the synthesis of ribonucleotides have been discussed *(Cohen* and *Steele, loc. cit.)*. α-Benzoyl-δ-valerothiollactone (2-benzoyl-4-methyl-4-butanethiolide) on treatment with hydrogen sulphide in ethanolic solution in the presence of hydrogen chloride at room temperature affords 5-*ethoxycarbonyl*-3,4-*dihydro*-6-*phenyl*-2H-*thiopyran* (*F. Duus* and *S.-O. Lawesson*, Tetrahedron, 1971, **27**, 387); α-thioacyl-δ-thiolides rearrange almost quantitatively in acid alcoholic solution to give 5-alkoxycarbonyl-6-alkyl-3,4-dihydro-2*H*-thiopyrans:

EtOH, $H^{\oplus}$; 40 h, boil →

These reactions proceed with the evolution of hydrogen sulphide and the yields of the rearrangement products are practically quantitative.

5,6-**Dihydro-2*H*-thiopyran** (XXXVIII, b.p. 35–36°/12 mm, 75°/58 mm, n_D^{20} 1.5328, $n_D^{24.5}$ 1.5305, is obtained by heating tetrahydrothiopyran-4-ol with copper sulphate and pumice at 190° or with anhydrous magnesium sulphate in a nitrogen atmosphere at 200° for 4 h (*Parham et al., loc. cit.*):

(XXXVIII)

Reaction of 5,6-dihydro-2*H*-thiopyran with ethyl trichloroacetate and sodium methoxide gives a mixture, supported by spectral data, of 4-chloromethyl-3,4-dihydro- and 2-chloromethyl-5,6-dihydro-2*H*-thiopyran, which on desulphurisation with Raney nickel afford 3-methylpentane and *n*-hexane:

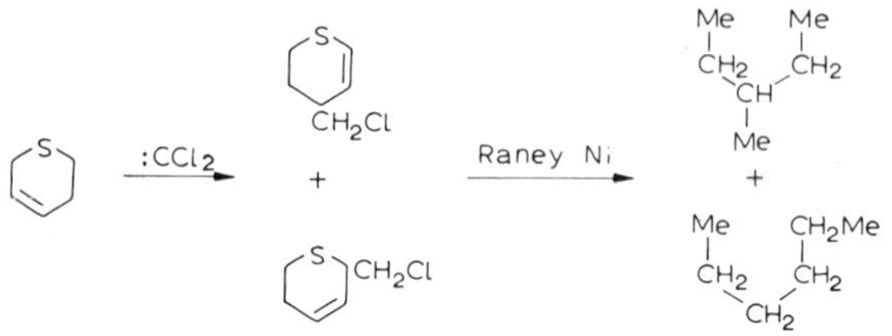

5,6-*Dihydro*-4-*methyl*-2H-*thiopyran*, b.p. 54°/11 mm, n_D^{20} 1.5239, with methyl iodide at 24° gives the *methylsulphonium iodide*, m.p. 142.5°, but at 100° ring-cleavage occurs giving *dimethyl*-(5-*iodo*-3-*methylpent*-3-*enyl*) *sulphonium iodide*, red-brown needles, m.p. 200° (decomp.) (*Naylor, loc. cit.*):

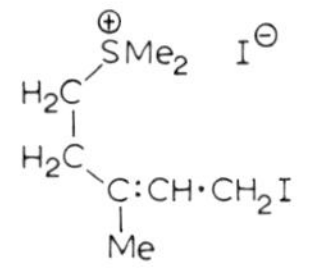

The reaction between mercaptoacetone and ω-bromoacetophenone in chloroform solution in the presence of triethylamine affords acetonyl phenacyl sulphide, which is oxidised to the corresponding sulphone with potassium permanganate in water–benzene. This sulphone on heating in acetic acid in the presence of sodium acetate undergoes cyclisation to 5,6-*dihydro*-3-*phenyl*-2H-*thiopyran*-5-*one* 1,1-*dioxide* (XXXIX), m.p. 158°; reduction with sodium tetrahydridoborate in diluted ethanol solution yields 5,6-*dihydro*-3-*phenyl*-2H-*thiopyran*-5-*ol* 1,1-*dioxide* (XL), oil, *acetyl* derivative, m.p. 144°, which on dehydration with 85% phosphoric acid at 100° gives 3-*phenyl*-2H-*thiopyran* 1,1-*dioxide* (XLI), m.p. 99–100° (*S. Rossi* and *G. Pagani*, Tetrahedron Letters, 1966, 2129):

$MeCO \cdot CH_2SH$ + $PhCO \cdot CH_2Br$ ⟶ $PhCO \cdot CH_2SCH_2 \cdot COMe$ ⟶ $PhCO \cdot CH_2SO_2CH_2 \cdot COMe$

(XLI) ⟵ (XL) ⟵ (XXXIX)

Catalytic hydrogenation using palladium on charcoal in methanol converts XLI to 3-*phenyltetrahydrothiopyran* 1,1-*dioxide*, m.p. 163°. The reactions of 5,6-dihydro-3-phenyl-2*H*-thiopyran-5-one 1,1-dioxide also have been studied and discussed (*Pagani*, Gazz., 1967, **97,** 1518).

5-Substituted 5,6-dihydro-2*H*-thiopyran 1,1-dioxides (XLII, R = an amino, alkylthio, or arylthio group) have been prepared, for example, 5-*butylamino*-5,6-*dihydro*-2H-*thiopyran* 1,1-*dioxide hydrochloride* (XLII, R = NHBu) m.p. 170–172° (*E. Molenaar* and *J. Strating*, Rec. Trav. chim., 1967, **86,** 436):

$\xrightarrow[\text{boil, 30 min}]{BuNH_2}$ (XLII) (XLIII)

6-*Cyano*-5,6-*dihydro*-6-*methylthio*-2H-*thiopyran* (XLIII), colourless crystals, m.p. 32–33°, is obtained in 75% yield by adding cyanodithioformate in methylene dichloride to an excess of liquid buta-1,3-diene at −78°, and allowing the mixture to stand at room temperature for 0.5 h in a pressure bottle (*D. M. Vyas* and *G. W. Hay*, Chem. Comm., 1971, 1411). The autoxidation (*L. Bateman, J. I. Cunneen* and *J. Ford*, J. chem. Soc., 1957, 1539), and catalytic hydrogenation of 5,6-dihydro-2*H*-thiopyran (*Bateman* and *F. W. Shipley*, *ibid.*, 1958, 2888) have been investigated, as well as the hydrogenation of 3,4-dihydro-2*H*-thiopyran.

Allylthioacetyl chloride on treatment with aluminium chloride in 1,1,2,2-tetrachloroethane at 50–55° cyclises to give 3,4-dihydro-2*H*-thiopyran-3-one (XLIV) and 5,6-dihydro-2*H*-thiopyran-5-one (XLV) (*K. Sato, S. Inoue* and *K. Kondo*, J. org. Chem., 1971, **36,** 2077):

$CH_2{:}CH \cdot CH_2SCH_2 \cdot COCl \xrightarrow[TCE]{AlCl_3}$ (XLIV) 53% + (XLV) 47%

Under similar conditions 3-methyl-, 2-methyl-, and 3,3-dimethylallyl-thioacetyl chlorides, yield 4-methyl- and 3-methyl-5,6-dihydro-2*H*-thiopyran-5-one, and 4-isopropylidenetetrahydrothiophen-3-one, respectively. The substituent effects on the directionality of cyclisation have been discussed.

(ii) Tetrahydrothiopyrans, thianes, tetrahydrothiapyrans

The saturated substances, tetrahydrothiopyrans or 1-thiacyclohexanes, occur as homologues in crude Mexican petroleum or Middle East oil distillates. They are separated from mixtures by standard procedures such as adsorption on silica (*D. Haresnape et al.*, Ind. Eng. Chem., 1949, **41,** 2691); by the formation of mercurichlorides with mercuric chloride followed by regeneration of the heterocyclic compound by means of hydrogen sulphide, hydrochloric acid or ammonia (*O. L. Polly et al., ibid.*, 1942, **34,** 755); or by making use of the solubility of water of the mercuriacetates (*S. F. Birch* and *D. T. McAllan*, J. Inst. Petroleum, 1951, **37,** 443). In the last instance addition of hydrochloric acid precipitates the mercurichlorides from which the sulphur compounds can be recovered.

The tetrahydrothiopyrans are characterised by means of the m.p.'s of their 1,1-dioxides (sulphones) and mercurichlorides. With Raney nickel they undergo hydrogenolysis, tetrahydrothiopyran, for instance, yielding *n*-pentane (91.8%) and cyclopentane (7.8%) (*Birch* and *R. A. Dean*, Ann., 1954, **585,** 234).

Tetrahydrothiopyran, thiane, 1-thiacyclohexane, pentamethylene sulphide (XLVI, R=H), $C_5H_{10}S$, b.p. 141°, m.p. 19.07°, d_4^{15} 0.9889, n_D^{20} 1.5067, *mercuri-iodide*, m.p. 72–74°, *mercurichloride*, m.p. 138–139° (*E. H. Thierry*, J. chem. Soc., 1925, **127,** 2756), is prepared from pentamethylene dibromide and sodium sulphide in ethanol (*R. F. Naylor, ibid.*, 1947, 1106; *Bateman, Cunneen* and *Ford, ibid.*, 1957, 1539), or from 1,5-diiodopentane and potassium sulphide (*J. von Braun* and *A. Trümpter*, Ber., 1910, **43,** 545). It gives a golden yellow colour with tetranitromethane in ethanol (*A. K. Macbeth*, J. chem. Soc., 1915, **107,** 1824), yields the *methylsulphonium iodide*, subliming at 192°, 1-*oxide*, a *sulphoxide*, yellow liquid and *tetrahydrothiopyran* 1,1-*dioxide*, a *sulphone* (XLVII) m.p. 98.5–99°; i.r. spectrum (*N. Sheppard*, Trans. Faraday Soc., 1950, **46,** 429):

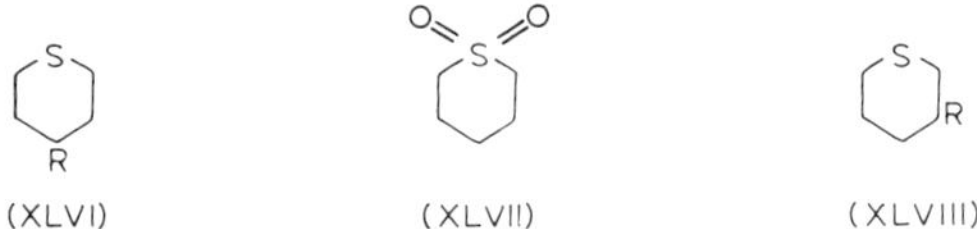

The autoxidation of tetrahydrothiopyran has been discussed (*Bateman, Cunneen* and *Ford, loc. cit.*), and on heating in nitromethane at 100° it gives a 31% yield of 2-methylthiophene (*E. A. Viktorova et al.*, Khim. Geterotsikl. Soedin., 1973, 141). A crude estimate has been made of the entropy changes accompanying ring closure in the formation of thia-alkanes, including tetrahydrothiopyran (*F. A. Cotton* and *F. E. Harris*, J. phys. Chem., 1956, **60,** 1451). An X-ray study has been made of tetrahydrothiopyran (*S. Kondo*, Bull. chem. Soc. Japan, 1956, **29,** 999). Its physical properties and u.v. data have been reported (*G. H. Jeffrey*, *R. Parker* and *A. I. Vogel*, J. chem. Soc., 1961, 570) and the latter related to ring size (*R. E. Davis*, J. org. Chem., 1958,

23, 1380). 2-**Methyl-**, b.p. 55.0°/26 mm, m.p. −58.14°, d^{20} 0.9428 g/ml, n_D^{20} 1.4905, 1,1-*dioxide*, m.p. 65–66°, *mercurichloride*, m.p. 102°; 3-*methyl-*, b.p. 157.9°, m.p. −60.17°, d^{20} 0.9473 g/ml, 1,1-*dioxide*, m.p. 83°, *mercurichloride*, m.p. 136°; and 4-**methyl-tetrahydrothiopyran**, b.p. 54.0°/22 mm, m.p. −28.11°, d^{20} 0.9471 g/ml, n_D^{20} 1.4923, 1,1-*dioxide*, m.p. 121.5°, *mercurichloride*, m.p. 135–136° (*E. V. Whitehead et al.*, J. Amer. chem. Soc., 1951, **73,** 3632). These methyl compounds have been isolated from a brand of kerosine.

The passage of 3-alkyltetrahydropyrans over 7.5% ThO_2–Al_2O_3 at 300° in a rapid stream of hydrogen sulphide affords 3-alkyltetrahydrothiopyrans (XLVIII, R = Me, Et, Pr, Bu or Bu^{β}) (*Yu. K. Yur'ev* and *O. M. Revenko*, Zhur. obshcheĭ Khim., 1962, **32,** 1822).

The methylation of tetrahydrothiopyran with diazomethane under u.v. radiation gives a mixture containing 2-methyl- (41%), and 3- and 4-methyl-tetrahydrothiopyran (59%) (*G. N. Gordadze*, Zhur. org. Khim., 1971, **7,** 1999).

The reduction of 2-benzoyltetrahydrothiophene with a zinc/mercury couple in hydrochloric acid yields 2-phenyltetrahydrothiopyran (*N. A. Nesmeyanov* and *V. A. Kalyavin*, *ibid.*, 1971, **7,** 608):

S, Bz —Zn/Hg, HCl→ S, Ph

2,6-**Diphenyltetrahydrothiopyran** is prepared by the Clemmensen reduction of *cis-* and *trans-*diphenyltetrahydrothiopyran-4-one and is obtained in the cis-, m.p. 98°, and trans-*form*, m.p. 142° (*F. Arndt* and *E. Schauder*, Ber., 1930, **63,** 313); 3,5-*dimorpholinotetrahydrothiopyran* 1,1-*dioxide*, m.p. 184–185° (*E. Molenaar* and *J. Strating*, Rec. Trav. chim., 1967, **86,** 436).

Conformation studies of 2-alkoxy- and 2-alkylthiotetrahydrothiopyrans XLIX show that when R = OMe, OBu, OH the compounds exist as ~80% axial conformer, whereas when R = SPr, SBu, SPh, they exist as 33–50% axial conformer, owing to repulsion between the S atoms (*N. S. Zefirov et al.*, Zhur. org. Khim., 1971, **7,** 594):

(XLIX)

chloride to give a *S*-ylide, which is then attacked by $Cl^{\ominus}$ at the carbon atom (*S. Bory*
by (dichloroiodo)benzene in pyridine has been examined; the mechanism involves initial addition of $Cl^{\oplus}$ to the S atom to give a species which eliminates hydrogen

(L) (LI)

chloride to give a *S*-ylide, which is then attacked by $Cl^{\ominus}$ at the carbon atom (*S. Bory et al.*, Compt. rend. Ser. C, 1973, **276,** 1323). It has been shown that the proportions of *cis-* and *trans-*sulphoxides obtained by the oxidation of 4-substituted tetrahydrothiopyrans vary over a wide range with change in the oxidising agent (*C. R. Johnson* and *D. McCants, Jr.*, J. Amer. chem. Soc., 1965, **87,** 1109).

Ethyl tetrahydrothiopyran-4-carboxylate (XLVI, R=CO_2Et), b.p. 118–120°/15 mm, yielding the *acid* (XLVI, R=CO_2H), m.p. 111.5–112.5° and *amide* (XLVI, R=$CONH_2$), m.p. 184.5° (*V. Prelog* and *E. Cěrkovnikov*, Ann., 1939, **537,** 214) and *tetrahydrothiopyran-4-acetic acid* (XLVI, R=$CH_2 \cdot CO_2H$), m.p. 169–171°, *ethyl ester*, b.p. 137–143°/10 mm, repulsive odour (*Prelog* and *D. Kohlbach*, Ber., 1939, **72,** 672) have been synthesised.

Tetrahydrothiopyran-4-one, *penthian-3-one, tetrahydrothio-4-pyrone* (LII), C_5H_8OS, m.p. 65–66°, *oxime*, m.p. 85–86°, *semicarbazone*, m.p. 151°, 2,4-*dinitrophenylhydrazone*, m.p. 186°, 2,6-*dibenzylidene* deriv., yellow m.p. 149–151°, is prepared by the reduction of 4*H*-thiopyran-4-one with zinc dust and acetic acid, or by treating *ethyl 4-oxotetrahydrothiopyran-3-carboxylate*, m.p. 59°, obtained by a Dieckmann condensation of diethyl *β,β*-thiodipropionate, with hot dilute hydrochloric acid or cold dilute alkali (*G. M. Bennett* and *L. V. D. Scorah*, J. chem. Soc., 1927, 194):

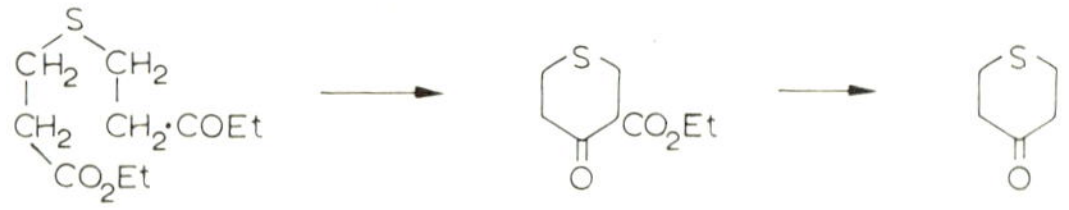

(LII)

It forms the *methylsulphonium iodide*, pale yellow crystals, m.p. 112–113°, and is reduced by aluminium isopropoxide and isopropanol to *tetrahydrothiopyran*-4-ol, m.p. 53°, b.p. 70°/0.05 mm, also prepared by the action of hydrogen sulphide on tetrahydropyran-4-ol at 400° (*Naylor*, *ibid.*, 1949, 2749).

The 1,1-*dioxide* (LIII), m.p. 65–66°, prepared by oxidation of tetrahydrothiopyran-4-one by hydrogen peroxide, undergoes Clemmensen reduction to give tetrahydrothiopyran 1,1-dioxide (LIV) and 2*H*-5,6-*dihydrothiopyran* 1,1-*dioxide* (LV), m.p. 75–76°, which when boiled with methanolic sodium methoxide yields 3-*methoxytetrahydrothiopyran* 1,1-*dioxide* (LVI), m.p. 66–66.5° (*E. A. Fehnel* and *P. A. Lackey*, J. Amer. chem. Soc., 1951, **73,** 2473). The 1,1-dioxide LIII is reduced by hydrogen and Raney nickel to 4-*hydroxytetrahydrothiopyran* 1,1-*dioxide* (LVII), m.p. 138–139.5°, which with dimethyl sulphate affords 4-*methoxytetrahydrothiopyran* 1,1-*dioxide*, m.p. 61–63°, and forms an *acetate* (LVIII) m.p. 124–125°, which on pyrolysis at 330° gives LV:

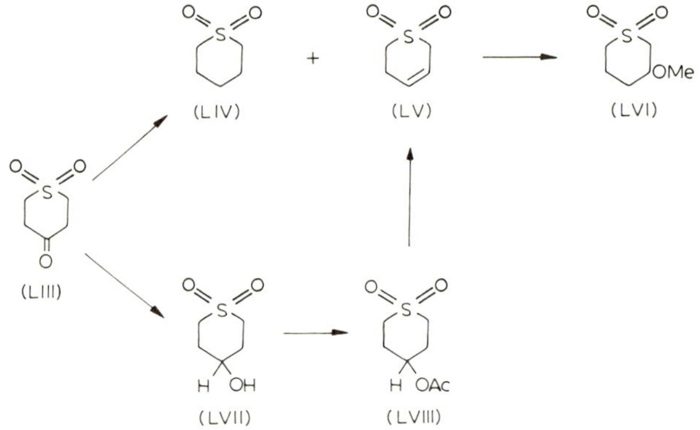

(LIV) (LV) (LVI)

(LIII)

(LVII) (LVIII)

Irradiation of tetrahydrothiopyran-4-one in *tert*-butyl alcohol gives thiacyclobutan-2-one (LIX) and *tert*-butyl-4-thiahexanoate (LX) (*P. Y. Johnson* and *G. A. Berchtold*, J. org. Chem., 1970, **35**, 584):

27%
(LIX)

18%
(LX)

Tetrahydrothiopyran-4-ones may be prepared by the action of hydrogen sulphide on the appropriate unsaturated ketones, *e.g.*, alkenyl β-(dialkylamino)alkyl ketones, in alcohol in the presence of anhydrous sodium acetate (*B. V. Unkovskii et al.*, Russ, P., 239,968 1969).

2,2-**Dimethyl**-6-**phenyltetrahydrothiopyran**-4-**one,** unpleasant smelling crystals, m.p. 42°, b.p. 175–176° (*F. Arndt* and *J. Pusch*, Ber., 1925, **58,** 1648); 2,6-**diphenyltetrahydrothiopyran**-4-**one,** prepared by action of hydrogen sulphide on dibenzylidene-acetone in the presence of sodium acetate, is obtained in two *forms* cis, m.p. 113–114°, μ1.64 D, *phenylhydrazone*, m.p. 155°, *semicarbazone*, m.p. 206–207°, 1,1-*dioxide*, m.p. 235° (forms a *compound*, m.p. 134°, with hydrogen peroxide) and trans, m.p. 87–88°, μ1.62 D, *phenylhydrazone*, m.p. 142°, 1,1-*dioxide*, m.p. 96° (*compound*, m.p. 123°, with hydrogen peroxide (*Arndt et al.*, *ibid.*, 1925, **58,** 1633).

2,5-Dimethyltetrahydrothiopyran-4-one reacts with phenylmagnesium bromide to give 2,5-*dimethyl*-4-*phenyltetrahydrothiopyran*-4-*ol*, b.p. 123–124°/2 mm, also obtained in greater yield by using phenyllithium; 2,2-*dimethyl*-4-*phenyltetrahydrothiopyran*-4-*ol*, b.p. 119–120°/2 mm; 2,2,4-*trimethyltetrahydrothiopyran*-4-*ol*, m.p. 82–83° (*I. N. Nazarov* and *E. T. Golovin*, Zhur. obshcheĭ Khim., 1956, **26,** 477).

The Mannich reaction between 3-methyltetrahydrothiopyran-4-one, dimethylamine, and formaldehyde gives 3-[(dimethylamino)methyl]-3-methyltetrahydrothiopyran-4-one (LXI). Reduction of LXI with lithium tetrahydridoaluminate yields a mixture of the tetrahydrothiopyranols LXII and LXIII with the former predominating; with aluminium *tert*-butoxide in isopropranol the same mixture is obtained with the latter as the major product. Treatment of LXI with phenyllithium affords the corresponding 4-phenyltetrahydrothiopyran-4-ol LXIV (*Golovin*, *B. M. Glukhov* and *Unkovskii*, Zhur. org. Khim., 1973, **9**, 619):

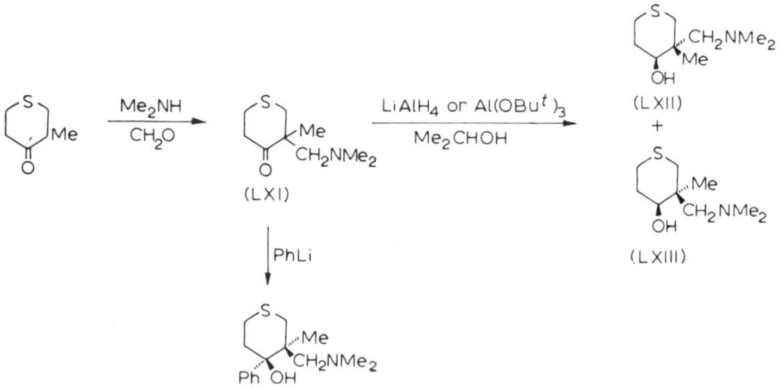

(LXI) (LXII) (LXIII) (LXIV)

A similar Mannich reaction with tetrahydrothiopyran-4-one gives 3-(dimethylamino)methyltetrahydrothiopyran-4-one (LXV) and 3,5-bis(dimethylamino)methyltetrahydrothiopyran-4-one (LXVI) (*Golovin et al., ibid.*, p. 614):

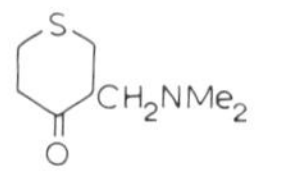

(LXV) 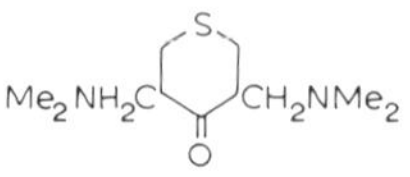 (LXVI)

Ketones of the type $RCH_2 \cdot CO \cdot CH_2R$ yield tautomeric substances when treated with carbon disulphide and sodium hydroxide (*H. Apitsch*, Ber., 1905, **38,** 2888). Thus diethyl ketone affords 2,6-*dimercapto*-3,5-*dimethyl*-4H-*thiopyran*-4-*one* or its tautomeric form 2,6-*dithiono*-3,5-*dimethyltetrahydrothiopyran*-4-*one*, orange prisms, m.p. 157°:

Evidence for the mercapto form is given by the formation of a disodium salt which with methyl iodide yields the *bismethylthio* derivative m.p. 123°. The 3-*methyl*-, orange red prisms, m.p. 144.5–145°, 3-*phenyl*-, 135°, and 3,5-*diphenyl* compound, red prisms, m.p. 165°, have also been prepared.

Tetrahydrothiopyran-3-**one** (LXVII), b.p. 82–83°/4 mm, gives an *oxime*, m.p. 77–77.5°, which is an approximately 50:50 mixture of *syn* and *anti*, m.p. 96°, isomers. Treatment of the *anti*-oxime with either concentrated sulphuric or polyphosphoric acid regenerates the parent ketone. This is unexpected since tetrahydrothiopyran-4-one oxime has been reported to rearrange normally under these Beckmann conditions (*R. K. Hill* and *D. A. Cullison*, J. Amer. chem. Soc., 1973, **95,** 2923):

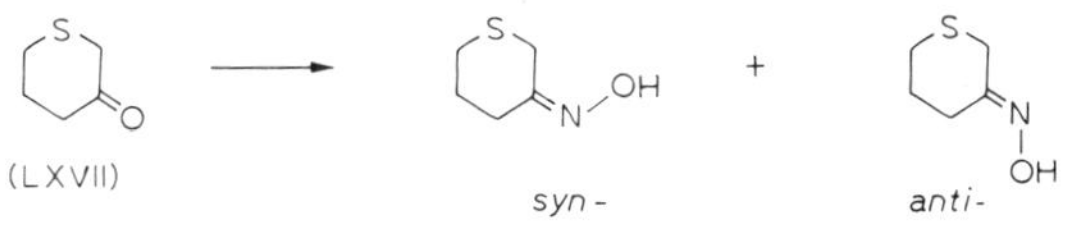

(LXVII) syn- anti-

2. Benzo[b]thiopyrans, benzo[b]thiins, thiochromenes, 5,6-benzothiapyrans and derivatives, and some naphthothiopyrans and derivatives

(a) Benzo[b]thiopyrans and benzo[b]thiopyranones and related naphtho-derivatives

(i) Benzo[b]thiopyrans, benzo[b]thiins, thiochromenes

The 2*H*-benzo[*b*]thiopyrans, benzo[*b*]thiins, 2*H*-thiochromenes are obtained from the thiochroman-4-ols by distillation with phosphorus pentoxide, or by heating with freshly fused potassium hydrogen sulphate and then distilling (*F. Krollpfeiffer et al.*, Ber., 1925, **58**, 1654; *W. E. Parham* and *R. Koncos*, J. Amer. chem. Soc., 1961, **83,** 4034).

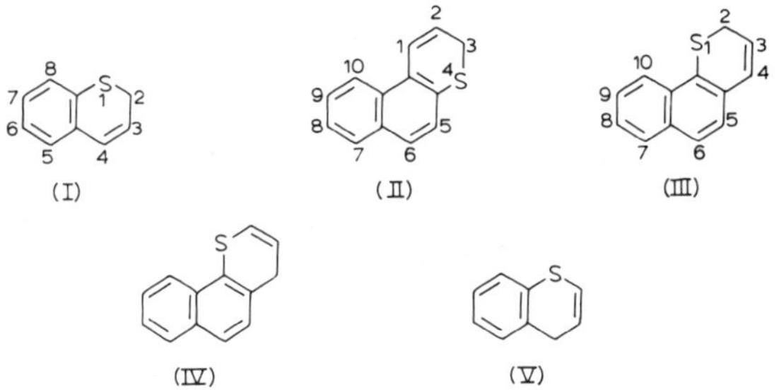

(I) (II) (III)

(IV) (V)

2*H*-**Benzo[*b*]thiopyran,** 2*H*-**benzo[*b*]thiin, thiochrom**-3-**ene** (I), b.p. 86–87°/0.85 mm, 95°/1.4 mm, n_D^{25} 1.6438; dehydration of 4-hydroxythiochroman 1,1-dioxide on heating with 85% phosphoric acid at 140°, gives 93% of 2H-*benzo*[b]*thiopyran*1,1-*dioxide*, m.p. 72° (*S. Rossi* and *G. Pagani*, Tetrahedron Letters, 1966, 2129). 4-*Methyl*-, b.p. 138°/12 mm, deep-blue colour in sulphuric acid; 4,6-*dimethyl*-, b.p. 145–146°/16 mm; 4,6,8-*trimethyl*-, b.p. 155–157°/12 mm; 4-*ethyl*-7-*methyl*-, b.p. 158–160°/12 mm; 6-*methyl*-4-*phenyl*-2H-*benzo*[b]-*thiopyran*, m.p. 47–48°, b.p. 211°/12 mm (*Krollpfeiffer et al., loc. cit.*). 3*H*-**Naphtho**[2,1-*b*]**thiopyran** (II), m.p. 90–92°, is obtained by the dehydrobromination of 1-bromo-2,3-dihydro-1*H*-naphtho[2,1-*b*]thiopyran or the dehydration of 2,3-dihydro-1*H*-naphtho[2,1-*b*]thiopyran-1-ol with anhydrous copper sulphate in boiling benzene or on boiling with acetic acid. Attempts to prepare 3*H*-naphtho-[2,1-*b*]thiopyran or dehydrating the above alcohol with potassium hydrogen sulphate or polyphosphoric acid gives 2,3-dihydro-1*H*-naphtho[2,1-*b*]pyran, which presumably arises through an intermolecular hydride shift reaction (*W. D. Cotterill et al.*, J. chem. Soc. Perk. I., 1972, 787).

Dehydration of 3,4-dihydro-2*H*-naphtho[1,2-*b*]thiopyran-4-ol in boiling benzene containing anhydrous copper sulphate yields a mixture (4:1) of 2*H*-naphtho[1,2-*b*]-thiopyran (III) and 4*H*-naphtho[1,2-*b*]thiopyran (IV), which cannot be separated by distillation.

4*H*-Benzo[*b*]thiopyran, thiochrom-2-ene, 4*H*-benzo[*b*]thiin (V), b.p. 64°/0.7 mm, n_D^{25} 1.6242, is prepared by heating a mixture of thiochroman 1-oxide and acetic anhydride at 90° for 5 h (*Parham* and *Koncos, loc. cit.*).

The reaction of 2*H*-benzo[*b*]thiopyran with dichlorocarbene gives 2-chloromethyl-2*H*-benzothiopyran (VI) and 4-dichloromethyl-4*H*-benzothiopyran (VII). None of the expected cyclopropane derivative is obtained. The reaction of dichlorocarbene with 4*H*-benzo[*b*]thiopyran affords only the cyclopropyl adduct VIII, which on thermal decomposition in hot quinoline yields 2-chloronaphthalene (IX) (*Parham* and *Koncos, loc. cit.*):

:CCl$_2$ → (VI) CH$_2$Cl + (VII) CH$_2$Cl

→ (VIII) Cl, Cl → (IX) Cl

Acid catalysed disproportionation of benzo[*b*]thiopyrans gives 3,4-dihydro-2*H*-benzo-[*b*]thiopyrans, thiochromans (XI) and thiobenzopyrylium salts, thianaphthalenium salts (XII). The reaction involves an intermolecular transfer of hydride from the 2-position of one molecule of thiochromene to the 4-position of its conjugated acid X:

HClO$_4$ → [(X)] → (XI) + (XII) $ClO_4^{\ominus}$

Since the hydride addition can occur from either face of the planar carbonium ion X, in cases when the thiochromene is disubstituted in the 3,4-positions, *cis/trans* isomerism in the resulting thiochroman would occur depending on the steric course of hydride attack. In the case of 3,4-dimethyl-2*H*-benzo[*b*]thiopyran (XIII) stereoselective intermolecular hydride transfer occurs to give 85% *cis*- XV and 15% *trans*-3,4-dihydro-3,4-dimethylbenzo[*b*]thiopyran (XVI) along with the benzothiopyrylium salt XVII; the intermediacy of a bridge sulphonium ion XIV with a predictable steric configuration has been proposed (*B. D. Tilak, R. B. Mitra* and *Z. Muljiani*, Tetrahedron, 1969, **25**, 1939):

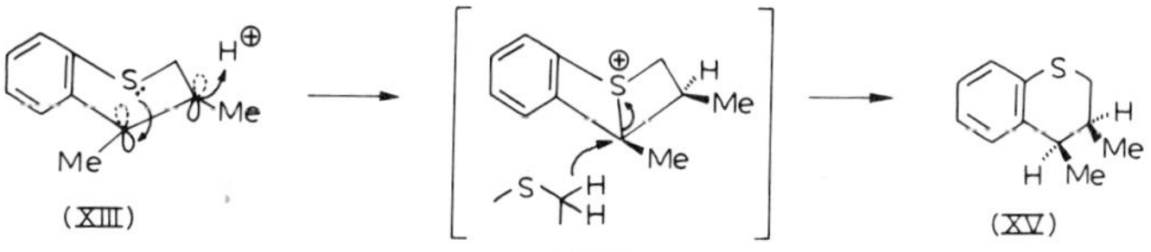

(XIII) (XIV) (XV)

(XVI) (XVII)

2-Aryl-2*H*- and -4*H*-benzo[*b*]thiopyrans are aromatised by triphenylmethyl perchlorate to give 2-arylbenzothiopyrylium perchlorates (XVIII) (*I. Degani, R. Fochi* and *G. Spunta*, Ann. Chim., Rome, 1971, **61,** 793):

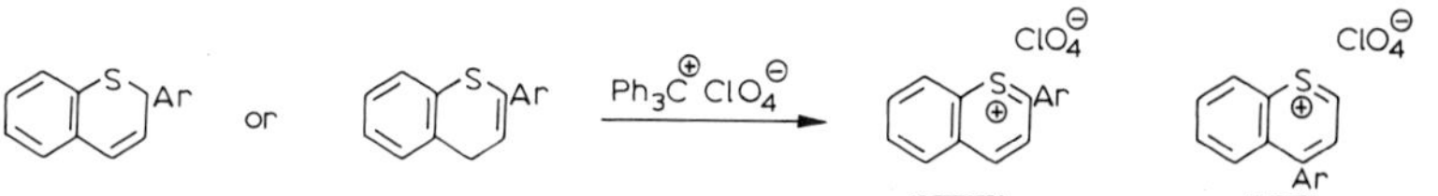

$Ph_3C^{\oplus}\ ClO_4^{\ominus}$

(XVIII) (XIX)

The 4-aryl compounds XIX are prepared by the same method. 3-Phenylbenzothiopyrylium perchlorate has been obtained by the aromatisation of a mixture of *cis*- and *trans*-3-phenylthiochroman-3-ols.

2,2,4-Trimethyl-2*H*-benzo[*b*]thiopyran on treating with polyphosphoric acid at 100° undergoes a ring contraction to give 2-isopropylbenzo[*b*]thiophene (XX) (*D. D. MacNicol* and *J. J. McKendrick*, Tetrahedron Letters, 1973, 2593). A plausible mechanism for the ring contraction has been suggested:

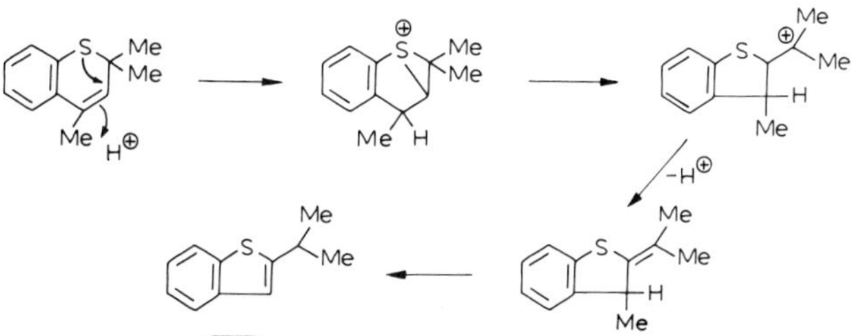

(XX)

2- and 4-Substituted benzothiopyrans, thiochromenes, are converted by dehydrogenation with sulphuryl chloride or tetrachloro-1,2-benzoquinone in the presence of strong acids into the corresponding benzothiopyrylium salts, thianaphthalinium salts (*A. Leuttringhaus, N. Engelhard* and *A. Kolb*, Ann., 1962, **654**, 189):

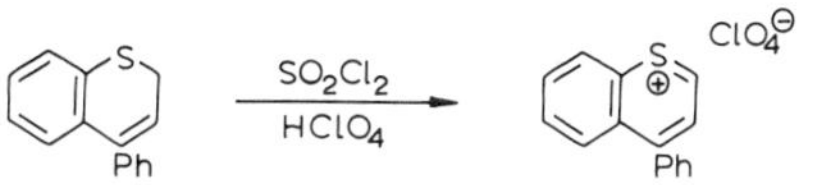

$\xrightarrow[HClO_4]{SO_2Cl_2}$

4-**Phenyl-**, m.p. 138–139° (decomp.); 2-**phenyl-**, m.p. 163° (decomp.); 2,4-**diphenyl-benzothiopyrylium perchlorate**, m.p. 228–229°. A number of 4- and 2,4-disubstituted benzothiopyrylium perchlorates have been obtained by cyclodehydration of appropriate alkyl or aryl β-arylthioethyl ketones on reaction with perchloric acid alone in the presence of triphenylmethyl chloride (*Tilak* and *G. T. Panse* Indian J. Chem., 1969, **7,** 191); naphthothiopyrylium salts (*Tilak* and *S. L. Jindal*, *ibid.*, p. 737).

6- and 7-Derivatives of benzothiopyrylium perchlorate (XXI, R = Me, MeO, MeS, Cl and Br) on hydrolysis are converted into the pseudo-base XXII. The reactions have been studied spectrophotometrically and the cationic system appears to be stabilised by the presence of Me, MeO and MeS groups. The stabilisation can be considered as a consequence of the presence of a reaction centre constituted by the whole cationic structure, and not by a given position, as it is on the symmetrical distribution of charges on the benzene nucleus (*I. Degani*, *R. Fochi* and *G. Spunta*, Gazz., 1967, **97,** 388):

(XXI) (XXII)

Oxidation of the hydrolysis products of benzothiopyrylium perchlorates with manganese dioxide leads to 2*H*-benzothiopyran-2-ones, thiocoumarins (*idem*, Boll. Sci. Fac. Chim. Ind. Bologna, 1965, **23,** 151; *A. Ruwet*, *J. Meessen* and *M. Renson*, Bull. Soc. chim. Belg., 1969, **78,** 459).

Thiochromanones are dehydrogenated with triphenylmethyl perchlorate in acetic acid to give the corresponding benzothiopyrylium perchlorates (XXIV), which on treatment with aqueous sodium hydrogen carbonate solution are converted to thiochromones, for instance, 6-methylthiochroman-4-one (XXIII) gives an 80% yield of 6-methylthiochromone (XXV) (*A. Schönberg* and *G. Shütz*, Ber., 1960, **93,** 1466):

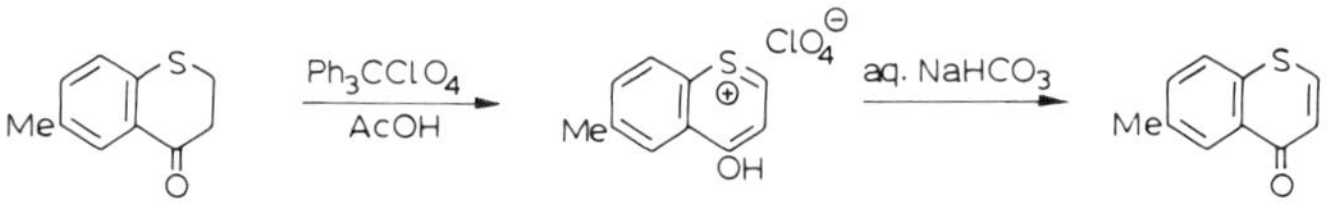

Ph_3CClO_4 / AcOH; $ClO_4^{\ominus}$; aq. $NaHCO_3$

(XXIII) (XXIV) (XXV)

(ii) 2H-Benzothiopyran-2-ones, 2H-benzothiin-2-ones, thiocoumarins

2*H*-**Benzo[*b*]thiopyran-2-one,** 2*H*-**benzo[*b*]thiin-2-one, thiocoumarin,** C_9H_6OS, (XXVI), m.p. 80–80.5°, *picrate*, m.p. 148° (*A. Kent et al.*, J. chem. Soc., 1939, 1858), having an odour similar to that of coumarin, has insecticidal properties (*L. E. Smith et al.*, J. econ. Entomol., 1936, **29,** 1027; C.A., 1937, **31,** 1147). It is prepared by heating 2-mercaptocinnamic acid with acetic anhydride or phosphorus pentoxide (*Ch. Chmelewsky* and *P. Friedländer*, Ber., 1913, **46,** 1903; *H. Simonis* and *A. Elias*, *ibid.*, 1916, **49,** 763); an improved yield is obtained by effecting the cyclisation with polyphosphoric acid. Application of the Perkin reaction to 2-mercaptobenzaldehyde gives thiocoumarin in very low yield (*W. D. Cotterill et al.*, J. chem. Soc. Perk. I, 1972, 816):

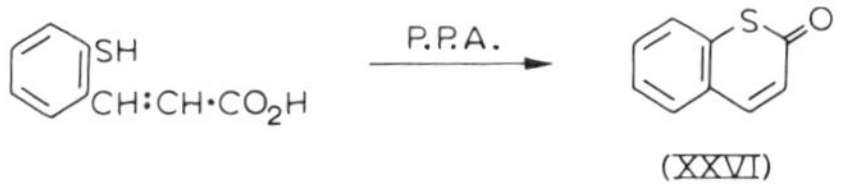

(XXVI)

It dissolves in alkali from which it is regenerated by acid and with alkali and dimethyl sulphate followed by acidification affords an impure product, possibly 2-methylthiocinnamic acid. Phosphorus pentasulphide converts thiocoumarin into **thionothiocoumarin, 2*H*-benzo[*b*]thiopyran-2-thione** (XXVI, S for O), red prisms or needles, m.p. 104°, which with phenylhydrazine affords a product m.p. 140°, claimed to be *thiocoumarin phenylhydrazone. U.v.* thiocoumarin (*A. Mangini* and *D. D. M. Casoni*, Atti accad. sci. ist. Bologna, Classe sci. fis. Rend., 1958, **5,** 20); *i.r.* some thiocoumarins and benzo[*b*]thiopyran-2-thiones (*Ruwet* and *Renson*, Bull. Soc. chim. Belg., 1970, **79,** 89).

Thiocoumarin reacts with phenylmagnesium bromide to give *thioflavone* (XXVII), m.p. 123–124.5°, and 2-*phenyl*-4H-*thiochromene* (XXVIII), m.p. 60–61°, and with bromine to yield 3,4-*dibromo*-3,4-*dihydrothiocoumarin* (XXIX), m.p. 91–93°. Treatment of 3-*bromothiocoumarin* (XXX), m.p. 129–131°, with alkali gives benzo[*b*]thiophene-2-carboxylic acid (*Cotterill et al., loc. cit.*):

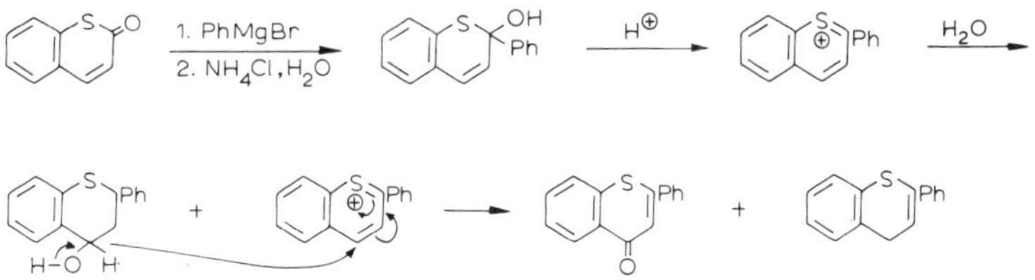

(XXVII) (XXVIII)

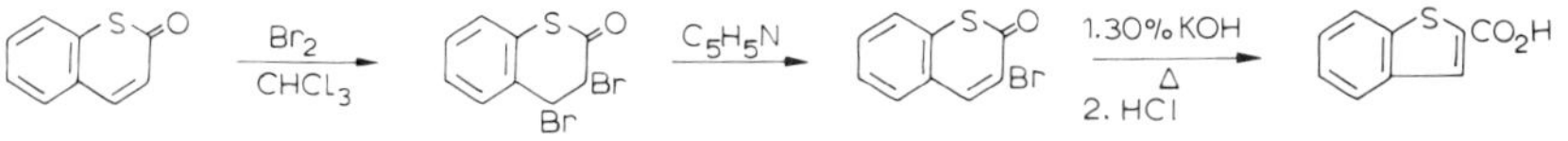

(XXIX) (XXX)

6-*Methyl*-, m.p. 86–87°; 6-*methoxy*-, m.p. 129–130°; 6-*chloro*-, m.p. 155–160°; 6-*bromo*-, m.p. 129–130°; 7-*methyl*-, m.p. 61–62°; 7-*methoxy*-, m.p. 106–107°; 7-*chloro*-, m.p. 135–136°; 7-*bromo-thiocoumarin*, m.p. 122–123° (*Degani, Fochi* and *Spunta*, Boll. Sci. Fac. Chim. Ind. Bologna, 1968, **26,** 31). Thiocoumarins may be prepared by the action of phosphoryl chloride on the appropriate 2-methylthiocinnamic acid, or its chloride, or by the action of aluminium chloride on the acid chloride in carbon disulphide or benzene (*Ruwet* and *Renson*, Bull. Soc. chim. Belg., 1968, **77,** 465; 1969, **78,** 449).

A number of 4-**hydroxythiocoumarins** have been obtained by the condensation of the appropriate arenethiol with malonic acid or substituted malonic acid in the presence of phosphoryl chloride and subsequent cyclisation of the resulting dithiophenylmalonic ester with aluminium chloride [*P. S. Jamkhandi* and *S. Rajagopal*, Symp. Syn. Heterocycl. Compounds Physiol. Interest, Hyderabad, India, 1964, (Pub. 1966) 83; C.A., 1968, **69,** 10329; Monatsh., 1968, **99,** 1390], or by treating arenethiols with substituted malonic acids in polyphosphoric acid (*Ruwet*, *C. Draguet* and *Renson*, Bull. Soc. chim. Belg., 1970, **79,** 639).

The reaction between 4-hydroxythiocoumarins and aliphatic and aromatic aldehydes has been investigated; 4-hydroxythiocoumarin on treatment with formaldehyde in water affords 3,3′-*methylenebis*(4-*hydroxythiocoumarin*) m.p. 306° (*Jamakhandi* and *Rajagopal*, Monatsh., 1966, **97,** 1732):

3-(α-Acetonylbenzyl)-4-hydroxythiocoumarins have been obtained by the Michael condensation of appropriate 6-substituted 4-hydroxythiocoumarins with *para*-substituted benzalacetones (*idem*, Indian J. Chem., 1972, **10,** 1114). 4-Hydroxythiocoumarin-3-carbaldehyde, 3-formyl-4-hydroxythiocoumarin, has been reported (*idem*, *ibid*., 1973, **11,** 708).

3,4-**Dihydroxy**-2*H*-**benzo**[*b*]**thiopyran**-2-**one**, 3,4-**dihydroxythiocoumarin**, red crystals, m.p. 120–121°, is tautomeric and reacts as the thiocoumarin XXXI, or 2,3-dihydroxy-4*H*-benzo[*b*]thiopyran-4-one, 2,3-dihydroxythiochromone (XXXII). With dimethyl sulphate it yields 4-*hydroxy*-3-*methoxythiocoumarin*, m.p. 125–126° (anhydrous), which on treatment with diazomethane affords 3,4-*dimethoxythiocoumarin*, m.p. 52–53°, whereas methylation with diazomethane yields 3-*hydroxy*-2-*methoxythiochromone*, m.p. 157°, and 2,3-*dimethoxythiochromone*, m.p. 120° (*F. Arndt* and *B. Eistert*, Ber., 1935, **68,** 1572):

(XXXI) (XXXII)

(iii) 4H-*Benzo*[b]*thiopyran*-4-*ones*, 4H-*benzo*[b]*thiin*-4-*ones*, *thiochromones and* 2-*phenyl*-4H-*benzo*[b]*thiopyran*-4-*ones*, *thioflavones*

The thiochromones are prepared by dehydrogenating thiochroman-4-ones with phosphorus pentachloride; by the action of pyridine or alkali on 3-bromothiochromanones (*F. Krollpfeiffer et al*., Ber., 1925, **58,** 1654); by condensing arenethiols with β-oxocarboxylic acid esters (*F. Bossert*, Ann., 1964, **680,** 40) or α-cyanoketones in polyphosphoric acid (*idem*, Tetrahedron Letters, 1968, 4377); by the cyclisation of 3-arylthioacryloyl chlorides using polyphosphoric acid or stannic chloride in chloroform (*N. Lozac'h, L. Legrand* and *N. Bignebat*, Bull. Soc. chim. Fr., 1964, 3247); and by reacting thiochromanone in acetonitrile with triphenylmethyl perchlorate and then treating the resulting thiochromone perchlorate with sodium hydrogen carbonate (*Degani, Fochi* and *Spunta*, Boll. Sci. Fac. Chim. Ind. Bologna, 1968, **26,** 3). They are oxidised by hydrogen peroxide to 1,1-dioxides (sulphones) and are more resistant to alkali than the chromones, long heating with sodium ethoxide being required, for instance, to break down thioflavone to 2-mercaptoacetophenone, 2-mercaptobenzoic acid, acetophenone and other products. Thiochromones condense with compounds containing active methyl or methylene groups (*A. I. Tolmachev* and *V. P. Sribnaya*, Zhur. obshcheĭ Khim., 1962, **32,** 383). The dipole moment of thiochromone has been measured and related to its i.r. spectrum and dielectric constant (*Yu. P. Borovikov et al.*, Teor. Eksp. Khim., 1973, **9,** 232; C.A., 1973, **79,** 41758).

4*H*-**Benzo**[*b*]**thiopyran-4-one**, 4*H*-**benzo**[*b*]**thiin-4-one**, **thiochromone** (XXXIII), m.p. 79°, 1,1-*dioxide*, m.p. 144°, blue fluorescence in sulphuric acid, gives a *dibromide*, m.p. 130–135°, with bromine (*F. Arndt et al.*, Ber., 1925, **58,** 1620). 2,3-*Dimethylthiochromone*, m.p. 108–109°, is prepared by condensing methyl acetoacetate with thiophenol in the presence of phosphorus pentoxide (*H. Simonis* and *A. Elias*, *ibid.*, 1916, **49,** 768). With sodium hydroxide it yields 2-mercaptobenzoic acid and ethyl methyl ketone.

(XXXIII)

Table 1 gives the melting points of some thiochromone derivatives.

TABLE 1

SOME THIOCHROMONE DERIVATIVES

	M.p. (°C)	*Ref.*		*M.p. (°C)*	*Ref.*
2-Methyl	105	1	6-Methoxy	110–111	2
3-Methyl	105	1	7-Methoxy	111–115	2
6-Methyl	69–70	2	6-Chloro	137–138	2
7-Methyl	81–83	2	7-Chloro	132–134	2
2,6-Dimethyl	122	1	6-Nitro	189–190	2
2,8-Dimethyl	125	1	7-Nitro	190–191	2
2,5,8-Trimethyl	89–90	1			

References
1 *F. Bossert*, Ann., 1964, **680,** 40.
2 *I. Degani, R. Fochi* and *G. Spunta*, Boll. Sci. Fac. Chim. Ind. Bologna, 1968, **26,** 3.

Ethyl trichloroacetate reacts with 3-methoxybenzo[*b*]thiophene and alcohol-free sodium methoxide in ether at $-10°$ to give 3-*chlorothiochromone*, m.p. 149° (*D. G. Hawthorne* and *Q. N. Porter*, Austral. J. Chem., 1966, **19**, 1751):

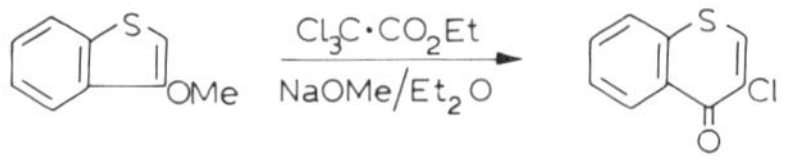

$Cl_3C \cdot CO_2Et$ / $NaOMe/Et_2O$

Arenethiols react with potassium hydrogen acetylenedicarboxylate to give (arylthio)-fumaric acids which are cyclised by sulphuric acid to thiochromone-2-carboxylic acids (*R. Hazard*, B.P. 1,291,865/1972):

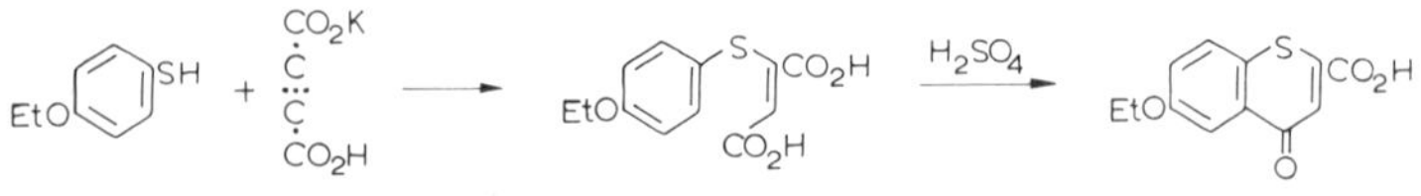

H_2SO_4

Several 2,3-cycloalkenothiochromones have been synthesised (*G. V. Boyd, D. Hewson* and *R. A. Newberry*, J. chem. Soc. C, 1969, 935).

The u.v. spectra in ethanol and sulphuric acid have been obtained, and the pK_B (*Degani, Fochi* and *Spunta, loc. cit.*) and $-pK_a$ determined for a number of thiochromones (*Tolmachev, L. M. Schulezhko* and *A. A. Kisilenko*, Zhur. obshcheĭ Khim., 1967, **37**, 367).

2-Phenyl-4*H*-benzo[*b*]thiopyran-4-one, 2-phenyl-4*H*-benzo[*b*]thiin-4-one, thioflavone, $C_{15}H_{10}OS$, (XXXIV), m.p. 129–130° (123–124.5°), yellow colour in sulphuric acid but no fluorescence, is prepared by the interaction of *β*-phenylthiocinnamic acid, phosphorus pentachloride and aluminium chloride (*S. Ruhemann*, Ber., 1913, **46**, 2197), by dehydrogenating thioflavanone, or from 3-bromothioflavanone (*Arndt et al., loc. cit.*; see also p. 371).

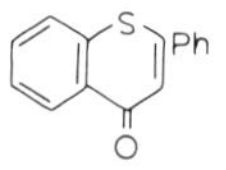

(XXXIV) 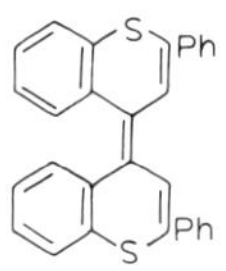 (XXXV)

It yields the 1,1-*dioxide*, yellow needles, m.p. 132–133°, and with phosphorus pentasulphide affords **4-thionothioflavone,** brown needles, m.p. 112–113° (*Arndt et al.*, Ber., 1925, **58**, 1644). When heated with thionyl chloride thioflavone yields 4,4′-**bithioflavylidene,** $C_{30}H_{20}S_2$, (XXXV), yellow crystals, m.p. 285° (*A. Schönberg* and *W. Asker*,

J. chem. Soc., 1942, 272), which yields 4-thionothioflavone when heated with sulphur. 6-*Methyl-*, m.p. 153–154°; 8-*methyl*, yellow needles, m.p. 124–125°; 5,8-*dimethyl-*, m.p. 133–134°; 6,8-*dimethyl-*, yellow needles m.p. 152–153°; 8-*hydroxy-*, yellow needles, m.p. 292°, yellow-red colour with ferric chloride, *methyl ether*, m.p. 129–130°; 6-*methoxy-thioflavone*, m.p. 155–156° (*Ruhemann*, Ber., 1913, **46,** 3384).

Addition of methyl thiosalicylate to sodium methyl sulphinylmethide in dimethyl sulphoxide–benzene solution gives 2-mercapto-ω-(methylsulphinyl)acetophenone (XXXVI), which is unstable. It rapidly eliminates methanesulphinic acid, CH_3SOH, and forms thioindoxyl, (XXXVII), the thio-analogue of indoxyl, which spontaneously oxidises to give thioindigo. In spite of the fast rate of thioindigo formation, the reaction can be used for the preparation of thioflavones XXXVIII, provided that the arenecarbaldehyde is added shortly after the formation of the methylsulphinyl ketone (*M. von Strandtmann et al.*, J. heterocycl. Chem., 1972, **9,** 171):

SH CO₂Me → ⊖CH₂S(→O)Me → [SH CO·CH₂S(→O)Me] (XXXVI) → ArCHO → (XXXVIII)

Ar = 3,4-dimethoxyphenyl (OMe, OMe), m.p.163-164°

Ar = 4-pyridyl (N), m.p. 140-141.5°

[(XXXVII)] → [O] → Thioindigo

The reaction between 4,4-dichloro-2-phenylthiochromene and di(4-methoxyphenyl)thiocarbonyl yields ~90% of 2-**phenyl**-4*H*-**benzo**[*b*]**thiopyran**-4-**thione**, 1,4-**dithioflavone** (*Schönberg* and *E. Frese*, Tetrahedron Letters, 1968, 697):

Ph, Cl₂ → [4]-MeOC₆H₄)₂CS → Ph, S

For u.v. spectra of thioflavones and related compounds see *R. E. David*, *R. Bognar* and *M. Rakosi*, Acta Phys. Chim. Debrecina, 1964, **10,** 97.

Some **thioisoflavones,** 3-aryl-4*H*-benzo[*b*]thiopyran-4-ones, *e.g.*, (XL) 3-phenyl-4*H*-benzo[*b*]thiin-4-one, have been obtained by heating the corresponding thioisoflavanones XXXIX with selenium dioxide in *n*-amyl alcohol. Heating the thioisoflavones with hydrogen peroxide in acetic acid gives the corresponding 1-oxide and 1,1-dioxide (*G. F. Katekar* and *R. M. Thomson*, Austral. J. Chem., 1972, **25,** 647):

(XXXIX) —SeO_2, n-$C_5H_{11}OH$→ (XL) (XLI)

6-Chlorothioisoflavone 1,1-dioxide on treatment with lithium tetrahydridoaluminate affords 6-chlorothioisoflav-2-en-4-ol (XLI).

Thiochromono[3,2-*b*]**benzofuran**(XLII) has been synthesised by the following route (*Bossert*, Tetrahedron Letters, 1968, 4375):

P.P.A. 90 - 100° $-H_2O$; P.P.A. 110 - 120° - EtOH

R = H or Cl

(XLII)

(*b*) *2,3-Dihydro-*4H*-benzo*[b]*thiopyrans, 2,3-dihydro-*4H*-benzo*[b]*thiins, thiochromans and 2,3-dihydro-*4H*-benzo*[b]*thiopyran-4-ones, 2,3-dihydro-*4H*-benzo*[b]*thiin-4-ones, thiochroman-4-ones*

(*i*) *2,3-Dihydro-*4H*-benzo*[b]*thiopyrans*

2,3-**Dihydro-4*H*-benzo**[*b*]**thiopyran,** 2,3-**dihydro-4*H*-benzo**[*b*]**thiin, thiochroman, 1-thiatetralin,** $C_9H_{10}S$, (XLIII) yellow oil, b.p. 254° (decomp.), 128–130°/15 mm, 124–125°/10 mm, 76–78°/1.2 mm, d^{20} 1.249, n_D^{20} 1.6145, steam volatile, occurs in certain petroleums (*S. F. Birch*, J. Inst. Petroleum, 1953, **39,** 185). It forms a *mercurichloride*, m.p. 89–90.5°, an oily *methiodide* and a deliquescent 1-*monoxide (sulphoxide)*, m.p. 31–34°. Thiochroman is prepared by treating diazotised 2-amino-(3-chloropropyl)benzene with potassium xanthate (*J. von Braun*, Ber., 1910, **43,** 3225), in poor yield by reducing the sulphoxide with lithium tetrahydridoaluminate (*F. G. Bordwell* and *W. H. McKellin*, J. Amer. chem. Soc., 1951, **73,** 2251), and by Clemmensen reduction of thiochroman-4-one (*F. Krollpfeiffer et al.*, Ber., 1925, **58,** 1654; *Birch et al.*, J. Inst. Petroleum, 1954, **50,** 76).

When equimolar quantities of allyl phenyl sulphide and quinoline are boiled for 2–4 h the sulphide is converted to 2-methyl-2,3-dihydrobenzothiophene (XLIV) and thiochroman (XLIII) (*C. Y. Meyers, C. Rinaldi* and *L. Bonoli*, J. org. Chem., 1963, **28**, 2440):

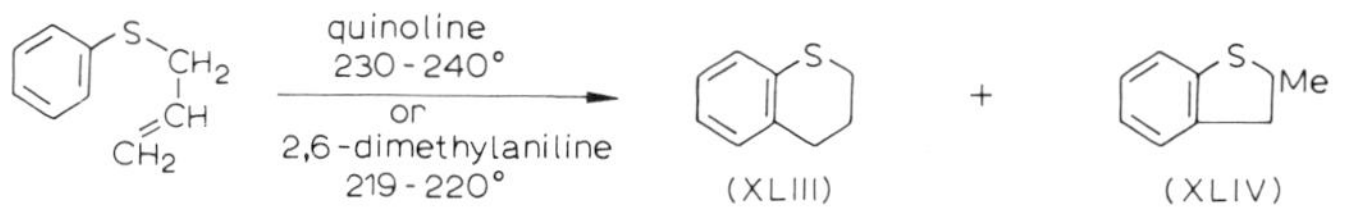

quinoline 230 - 240° or 2,6-dimethylaniline 219 - 220°

(XLIII) 33-37% (30%) + (XLIV) 40-47% (25%)

The thiochroman and 2-methyl-2,3-dihydrobenzothiophene are not interconvertible under these reaction conditions. Crotyl phenyl sulphide does not undergo the thio-Claisen reaction under any conditions so far tried, but crotyl *m*-tolyl sulphide apparently benefits greatly from the nuclear methyl substituent and is readily transformed to the thio-Claisen product. Appropriate methyl substitution in the side-chain has a similar beneficial effect; *β*-methylallyl phenyl sulphide undergoes the thio-Claisen reaction without use of high-boiling amine solvent shown to be indispensable in all previous cases studied. Proposed mechanisms for the above rearrangement have been discussed (*H. Kwart* and *E. R. Evans*, *ibid.*, 1966, **31**, 413).

Thiochroman 1,1-**dioxide,** m.p. 87.5–88°, prepared by oxidation of thiochroman by potassium permanganate or hydrogen peroxide, can be synthesised by ring-closure of 3-phenyl-1-propanesulphonyl chloride with aluminium chloride (*W. E. Truce* and *J. P. Milionis*, J. Amer. chem. Soc., 1952, **74,** 974):

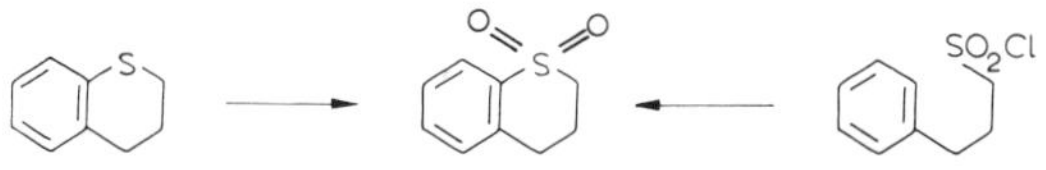

Under cracking conditions sulphur reacts with thiochroman to produce 4H-*benzo*[b]-*thiopyran*-4-*thione, dithiochromone* (XLV), m.p. 92–93°, and 2H-*benzo*[b]*thiopyran*-2-*thione, dithiocoumarin* (XLVI), m.p. 104–105° (*R. Mayer* and *H. Damme*, Z. Chem., 1965, **5,** 150):

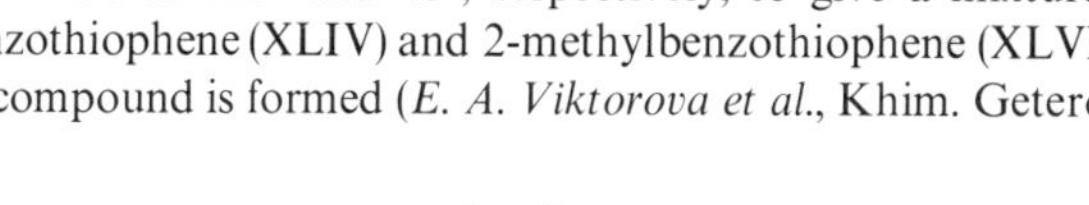

(XLV) + (XLVI)

The later compound is usually the main product; the former being formed in small amounts in special conditions. Thiochroman is dehydrogenated in nitromethane or methylene dichloride at 100° and 40°, respectively, to give a mixture of 2-methyl-2,3-dihydrobenzothiophene (XLIV) and 2-methylbenzothiophene (XLVII); in benzene only the latter compound is formed (*E. A. Viktorova et al.*, Khim. Geterotsikl. Soedin., 1973, 141):

(XLVII)

The thermal fragmentation of thiochroman and related compounds has been investigated and compared with the fragmentation under electron impact (*A. G. Loudon, A. Maccoll* and *S. K. Wong*, J. chem. Soc. B, 1970, 1733). For the i.r. spectrum see *Birch et al., loc. cit.* and for n.m.r. spectra of thiochromans and related compounds see *M. M. Dhingra et al.*, Tetrahedron Letters, 1963, 497.

2-**Methyl-**, b.p. 125–127° (bath temp.)/9 mm, n_D^{20} 1.5878, 1,1-*dioxide*, m.p. 80°; 3-**methyl-**, b.p. 132–134° (bath temp.)/19 mm, n_D^{20} 1.6020; 4-**methyl-**, b.p. 134–136° (bath temp.)/14 mm, n_D^{20} 1.6008 (*J. C. Petropoulos et al.*, J. Amer. chem. Soc., 1953, **75,** 1130); 6-*methyl-*, b.p. 154°/13.5 mm, 137°/12 mm, n_D^{20} 1.6004; 6-**ethyl-,** b.p. 163–164°/23 mm, n_D^{23} 1.5845 (*P. Cagniant* and *A. Deluzarche*, Compt. rend., 1946, **223,** 1012), and 6,8-**dimethyl-thiochroman,** b.p. 146–147°/12 mm, 1,1-*dioxide*, m.p. 101–102°. 2,2-Dimethylthiochroman has been obtained in ≥82% yield by the catalytic hydrogenation of 2,2-dimethylthiochroman-4-one in sulphuric acid–acetic acid over palladium/carbon; in the absence of sulphuric acid a mixture of products is formed (*J. B. Campbell et al.*, Ann. N.Y. Acad. Sci., 1970, **172** (Art 9), 261):

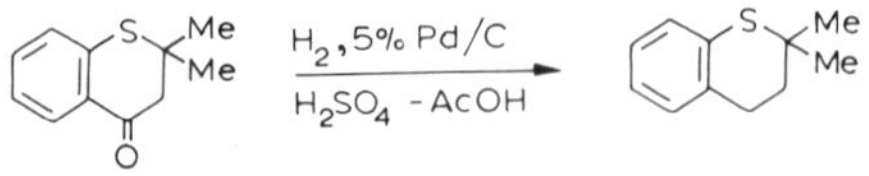

$$\xrightarrow[H_2SO_4\ -\ AcOH]{H_2,\ 5\%\ Pd/C}$$

2,2-*Dimethylthiochroman* 1-*oxide*, m.p. 60–61°, on boiling with freshly distilled acetic anhydride gives 2-acetoxymethyl-2-methylthiochroman in 80% yield (*R. B. Morin, D. O. Spry*, and *R. A. Mueller*, Tetrahedron Letters, 1969, 849):

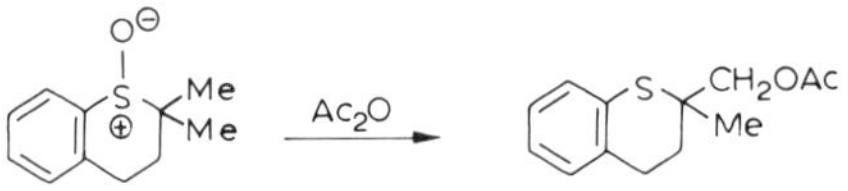

$$\xrightarrow{Ac_2O}$$

6-Methylthiochroman on acetylation yields 8-*acetyl*-6-*methylthiochroman*, m.p. 33°, b.p. 203°/17.5 mm, *semicarbazone*, m.p. 194°, *oxime*, m.p. 136°, 2,4-*dinitrophenylhydrazone*, m.p. 216°, which is reduced to 8-*ethyl*-6-*methylthiochroman*, b.p. 165°/13.5 mm, $n_D^{18.6}$ 1.5915. Similarly 8-*methylthiochroman*, b.p. 149°/14 mm, n_D^{20} 1.6033, affords 6-*acetyl*-8-*methylthiochroman*, m.p. 68°, b.p. 198°/16 mm, *semicarbazone*, m.p. 207°, *oxime*, m.p. 156°, 2,4-*dinitrophenylhydrazone*, m.p. 235°, and 6-*ethyl*-8-*methylthiochroman*, b.p. 165.5°/13 mm, $n_D^{22.8}$ 1.5843 (*Cagniant* and *Mme. P. Cagniant*, Compt. rend., 1961, **253,** 1702). Treatment of 6-methylthiochroman 1-oxide with acetic anhydride gives 2-acetoxy-6-methylthiochroman and 6-methyl-4*H*-benzothiopyran (*E. N. Karaulova et al.*, Neftekhim., 1972, **12,** 104: C.A., 1972, **76,** 153506). 2,3-*Dihydro*-1H-*naphtho*[2,1-b]*thiopyran*, 5,6-*benzothiochroman*, m.p. 92.5–93.5° (*W. D. Cotterill et al.*, J. chem. Soc. Perk. I, 1972, 787); 3,4-*dihydro*-2H-*naphtho*[2,3-b]*thiopyran*, 6,7-*benzothiochroman*, m.p. 167° (*Cagniant* and *Cagniant*, Bull. Soc. chim. Fr., 1961, 1560). U.v. spectra of 3,4-dihydro-2*H*-naphtho[1,2-*b*]pyran and its 6-methyl-, and 6-ethyl derivatives, and of 2,3-dihydro-1*H*-naphtho[2,1-*b*]pyran and its 8-ethyl derivative have been reported (*idem, ibid.*, 1966, 228).

The hydrolysis of *trans*-3,4-dibromothiochroman (XLVIII), *trans*-1,2-dibromo-2,3-dihydro-1*H*-naphtho[2,1-*b*]thiopyran (XLIX), and *trans*-3,4-dibromo-3,4-dihydro-2*H*-

naphtho[1,2-*b*]thiopyran (L) yields 2-hydroxymethylbenzo[*b*]thiophene, 2-hydroxymethylnaphtho[2,1-*b*]thiophene, and 2-hydroxymethylnaphtho[1,2-*b*]thiophene, respectively. Reduction of 3-halogenothiochroman-4-ones and related bromodihydronaphthothiopyranones with sodium tetrahydridoborate gives *cis*-3-halogenothiochroman-4-ols and the corresponding *cis*-bromodihydronaphthothiopyranols. The ^{1}H n.m.r. spectra of these compounds show that the hetero-ring adopts the 'sofa' conformation with both substituents axial in the *trans*-isomers and that the most stable conformation of the *cis*-isomer is that in which the substituent at the benzylic position is pseudo-axial (*Cotterill et al., loc. cit.*):

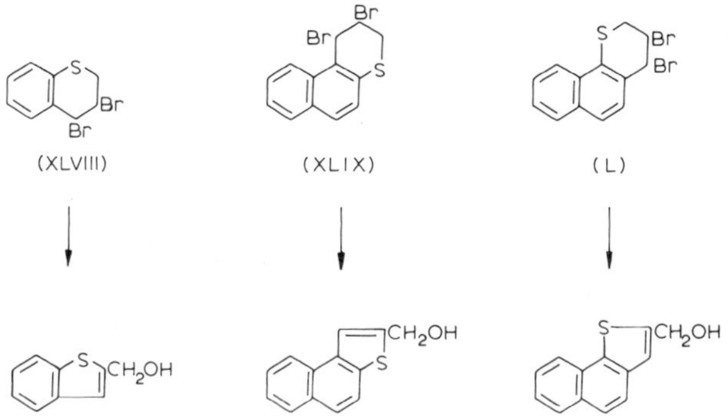

(XLVIII) (XLIX) (L)

3-Bromothiochroman-4-ol on heating in aqueous dioxane gives 2-bromomethylbenzo-[*b*]thiophene (*H. Hofmann* and *G. Salbeck*, Angew. Chem. intern. Edn., 1969, **8,** 456).

trans-1-**Thiadecalin,** *trans*-**decahydrobenzothiopyran,** b.p. 80–82°/2.0 mm, n_D^{20} 1.5210, m.p. −23.5° (approx.) (*Birch*, J. Inst. Petroleum, 1953, **39,** 185).

(ii) 2,3-*Dihydro*-4H-*benzo*[b]*thiopyran-4-ones, thiochroman-4-ones*

The thiochroman-4-ones are prepared by the ring-closure of 2-(phenylmercapto)propionic acids by sulphuric acid (*F. Arndt*, Ber., 1923, **56,** 1278; *F. Krollpfeiffer* and *H. Schultze*, *ibid*., p. 1819) or polyphosphoric acid (*C. D. Hurd* and *S. Hayao*, J. Amer. chem. Soc., 1954, **76,** 5056).

They give intensely coloured solutions or solids with sulphuric acid and perchloric acid, 6-methylthiochroman-4-one, for instance, yielding an orange-red perchlorate (*Arndt* and *J. Pusch*, Ber., 1925, **58,** 1648). These are sulphonium salts which when shaken with water yield the parent thiochromanones.

The thiochromanones with bromine yield 3-bromo- (*W. D. Cotterill et al.*, J. chem. Soc. Perk. I, 1972, 787), and 3,3-dibromo derivatives. The former with alkali undergo ring-fission, 3-bromo-6-methylthiochromanone (LI) yielding 2-mercapto-5-methylacetophenone and formic acid and 3-bromo-2,6-dimethylthiochromanone (LII) giving 2-mercapto-5-methylbenzoic acid (*Krollpfeiffer et al.*, Ber., 1925, **58,** 1654):

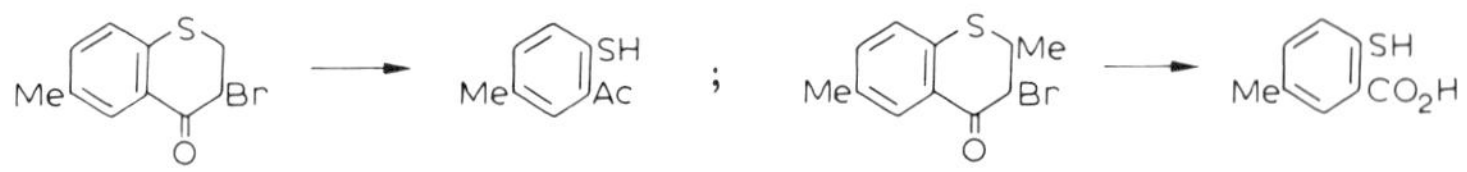

(LI) (LII)

The dibromo derivatives, on the other hand, with alkali yield benzo[*b*]thiophene-carbaldehydes, 3,3-dibromothiochroman-4-one (LIII) affording 3-hydroxybenzo[*b*]-thiophene-2-carbaldehyde (LIV). 3-Bromothiochromones are intermediates in this reaction:

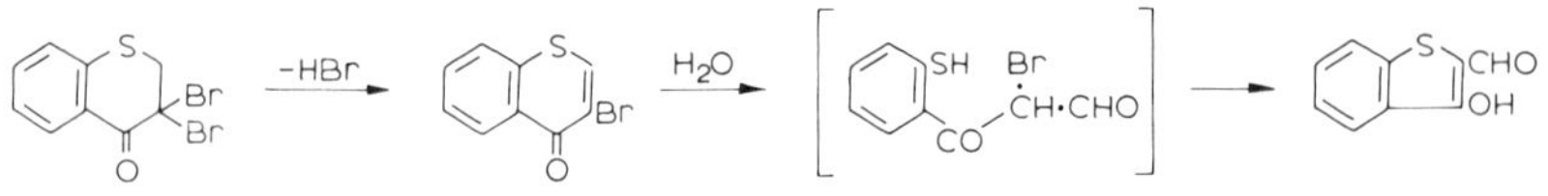

(LIII) (LIV)

Dibromides can also be formed, which with water yield sulphoxides:

Reaction of 2,x-(R,R[1]-disubstituted)-3-bromothiochroman-4-ones (LV, R = H or Me; R[1] = H, 6-Me, or 7-MeO) with formaldehyde gives 2,x-(R,R[1]-disubstituted)-3,α-epoxy-3-methylthiochroman-4-ones (*Hofmann* and *Salbeck*, *ibid*., 1970, **103**, 2084):

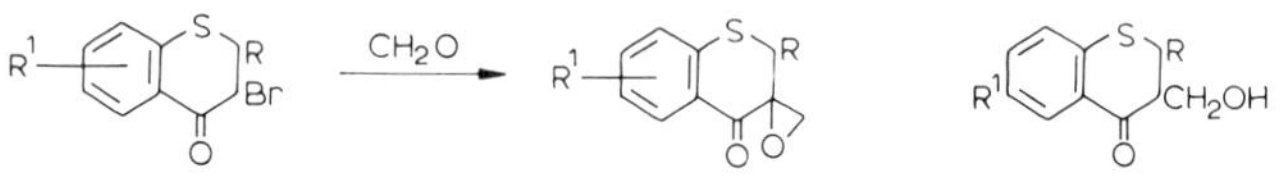

(LV) (LVI)

Similarly, 2,6-(R,R[1]-disubstituted)-3-hydroxymethylthiochroman-4-ones (LVI, R = H or Me; R[1] = H or Cl) are obtained from the related bromo compound LV.

3-Bromothiochroman-4-ones with anhydrous sodium acetate in acetic acid yield the so-called "dithiachromanols", *e.g.*, 6,6′-**dimethyl**-3,3′-**dithiachromanol** (LVII), m.p. 151–152°. These with bromine form dibromides which with pyridine undergo dehydrohalogenation to dithiochromones, *e.g.*, 6,6′-**dimethyl**-3,3′-**dithiochromone** (LVIII), brown-red needles when crystallised from nitrobenzene (*Krollpfeiffer et al., loc. cit.*):

(LVII) (LVIII)

The thiochroman-4-ones react normally with Grignard reagents, phenylhydrazine, etc., and with 4-nitrosodimethylaniline yield 3-phenylimino derivatives (*Arndt, loc. cit.*). Thiochroman-4-ones may be converted to benzothiopyrylium perchlorates (p. 369) and then to thiochromones.

Ring contraction occurs on boiling 2,2′-dimethylthiochroman-4-one 1-oxide with freshly distilled acetic anhydride (*Morin, Spry*, and *Mueller, loc. cit.*):

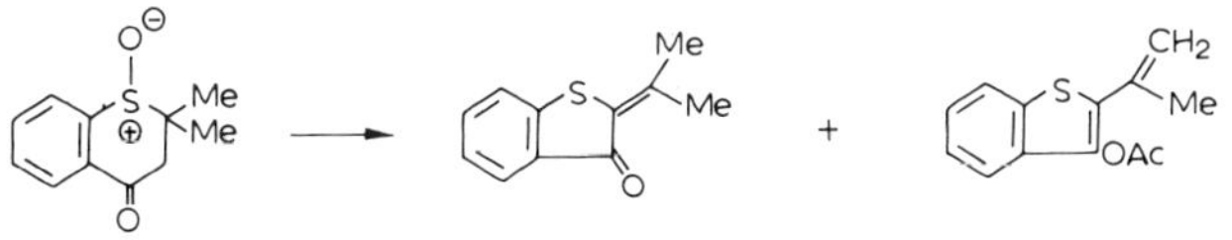

Irradiation of 3,3-dimethylthiochroman-4-one 1-oxide in benzene solution, using a Hanovia 115W medium pressure mercury arc with a Vycor filter, gives a mixture (~50%) containing isobutyrophenone (7%), 2-benzoylisobutyraldehyde (7%), 2-benzoylisobutyraldehyde (31%), and benzoic acid (12%) (*I. W. J. Still* and *M. T. Thomas*, Tetrahedron Letters, 1970, 4225):

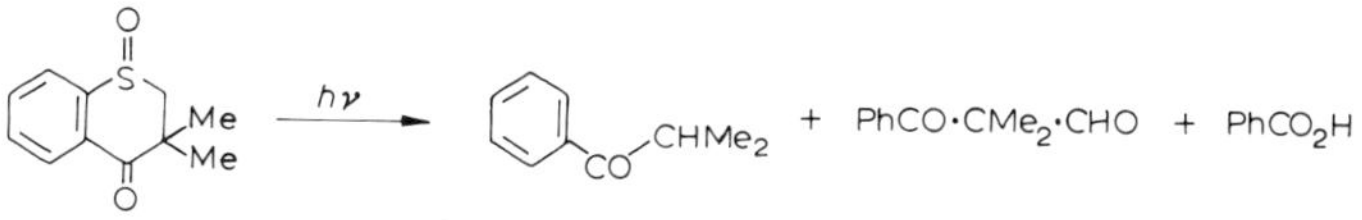

7-Methoxythiochroman-4-one on treatment with dimethylsulphoxonium methylide in dimethyl sulphoxide gives a 1:1 mixture of the benzo[b]thiophene LIX and cyclopropyl 2-mercapto-4-methoxyphenyl ketone (LX) (*W. N. Speckamp et al., ibid.*, p. 2743):

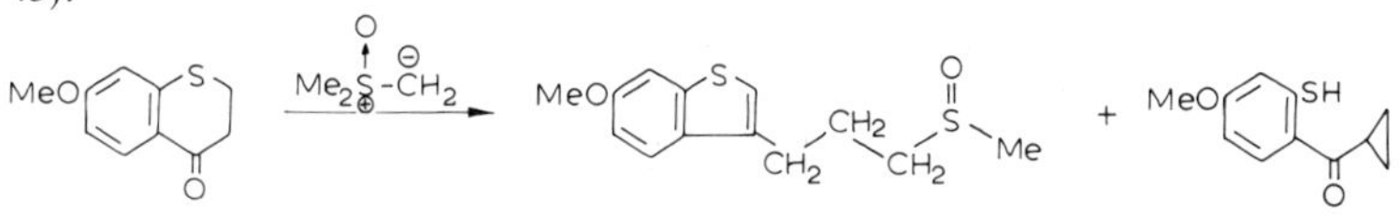

(LIX) (LX)

Related thiochromanones have been treated in a similar manner. The Wolff–Kishner–Huang reduction of 6-*tert*-butylthiochroman-4-one yields 6-*tert*-butylthiochroman (*N. Bellinger, D. Cagniant* and *P. Cagniant*, Tetrahedron Letters, 1971, 49).

2,3-**Dihydro-4*H*-benzo[*b*]thiopyran-4-one, thiochroman-4-one,** C_9H_8OS, m.p. 29–30°, b.p. 154°/12 mm, n_D^{20} 1.6397, *semicarbazone*, m.p. 219–220°, 2,4-*dinitrophenylhydrazone*, orange-red needles, m.p. 237–238°, *oxime*, m.p. 98–100° (*G. M. Bennett* and *W. B. Waddington*, J. chem. Soc., 1931 1692; *W. E. Parham* and *R. Koncos*, J. Amer. chem. Soc., 1961, **83,** 4034), orange-red colour in sulphuric acid, does not form a cyanohydrin, with hydrogen peroxide gives a 1,1-*dioxide*, m.p. 131–132°, and with 4-nitrosodimethylaniline and alkali yields the 4-*dimethylaminophenylimino* derivative, brownish-red crystals, m.p. 142°, which is broken down by alkali to 3-*hydroxythiochromone* or its tautomer *thiochroman*-3,4-*dione*, yellow needles, m.p. 172° (*Arndt, loc. cit.*). Beckmann rearrangement of thiochroman-4-one oxime in polyphosphoric acid gives a 94% yield of 2,3,4,5-*tetrahydro*-1,5-*benzothiazepin-4-one* (LXI), m.p. 216–217°. Hydrogenation of thiochromanone oxime over Raney nickel yields 4-aminothiochroman (LXII) (*V. A. Zagorevskii* and *N. V. Dudykina*, Zhur. obshcheĭ Khim., 1963, **33**, 322):

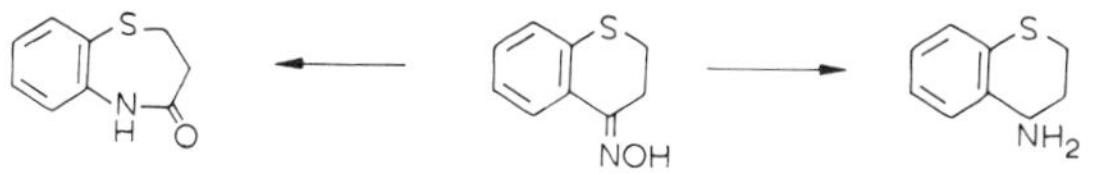

(LXI) (LXII)

2,3-**Dihydro-2-methyl-4*H*-benzo[*b*]thiopyran-4-one, 2-methylthiochroman-4-one,** m.p. 18–19°, b.p. 146–147°/9 mm, n_D^{20} 1.6125, *semicarbazone*, m.p. 167–168° *(Petropoulos et al., loc. cit.)*; 3-*methyl*-, m.p. 41–42°, b.p. 145–146° (bath temp.)/10 mm; 6-*methyl*-, m.p. 41–42°, 1-*oxide*, m.p. 61–62°, 1,1-*dioxide*, m.p. 163°; 2,6-*dimethyl*-, m.p. 64–65°, b.p. 179°/20 mm, *semicarbazone*, m.p. 205–206°, 1-*oxide*, m.p. 97–98°; 8-*methyl*-, m.p. 65–66°; 6-*chloro*-, m.p. 67–69°; 6-*methoxy-thiochroman*-4-*one*, m.p. 29–30°, b.p. 185–186°/12 mm, *semicarbazone*, m.p. 221° *(Krollpfeiffer et al., loc. cit.)*; 3,4-**dihydro-2*H*-naphtho[1,2-*b*]thiopyran-4-one, 7,8-benzothiochroman-4-one** LXIII, m.p. 108–109° *(Krollpfeiffer* and *Schultze, loc. cit.; Cotterill et al., loc. cit.)*:

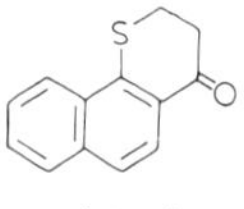

(LXIII) 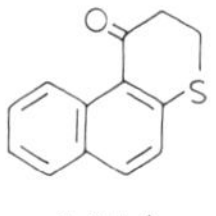 (LXIV)

2,3-*dihydro*-1H-*naphtho*[2,1-b]*thiopyran*-1-*one*, 5,6-*benzothiochroman*-4-*one* (LXIV), m.p. 68–69° *(Krollpfeiffer* and *Schultze, loc. cit.)*.

Some *tert*-butyl derivatives of thiochroman-4-one have been synthesised in order to examine the relationship between their odour and chemical constitution (*N. P. Buu-Hoï et al.*, Bull. Soc. chim. Fr., 1966, 1210) and some 5,8-disubstituted thiochroman-4-ones have been prepared (*N. Cagnoli, A. Martani* and *C. Rossi*, Ann. Chim., Rome, 1966, **56,** 1075). The preparation of several γ-oxo sulphones related to thiochroman-4-one 1,1-dioxide has been carried out by oxidation of the corresponding thiochromanone with hydrogen peroxide. U.v. irradiation of these oxo sulphones in methanol leads in most cases to the corresponding pinacols LXV by bimolecular reduction of the carbonyl group. The photochemical reduction seems to be relatively insensitive to the substitution pattern, although the lack of reaction in two cases, *viz.* 5-methylthiochromanone and methyl thiochromanone-8-carboxylate, is attributed to steric effects of neighbouring groups (*Still* and *Thomas*, J. org. Chem., 1968, **33,** 2730):

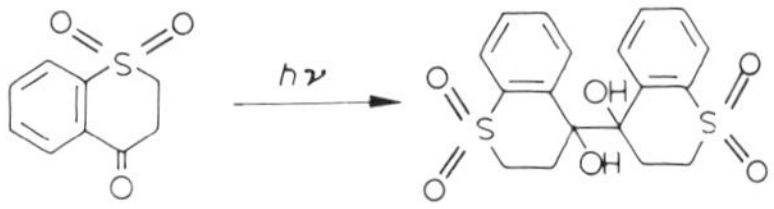

hν

(LXV)

The crystal structure of thiochroman-4-one 1,1-dioxide has been determined (*L. Preuss et al.*, Acta Crystallogr. B, 1971, **27,** 920).

The thiochroman-4-ones react with Grignard reagents to give *thiochroman*-4-*ols* and in this way 4-*methyl*-, m.p. 109–110°, 4,6-*dimethyl*-, m.p. 119–120°, 4,6,8-*trimethyl*-, m.p. 46–49° 4-*ethyl*-6-*methyl*-, m.p. 52°, b.p. 159–160°/12 mm, and 6-*methyl*-4-*phenyl-thiochroman*-4-*ol*, m.p. 112–113°, have been obtained *(Krollpfeiffer et al., loc. cit.)*. Reduction of thiochroman-4-one with lithium tetrahydridoaluminate gives *thiochroman*-4-*ol*, m.p. 68–69° *(Parham* and *Koncos, loc. cit.)*; 3,4-*dihydro*-2H-*naphtho*-[1,2-b]*thiopyran*-4-*ol*, m.p. 94–95°; 2,3-*dihydro*-1H-*naphtho*[2,1-b]*thiopyran*-1-*ol*, m.p. 134–135° (128–130°) *(Cotterill et al., loc. cit.)*. 4-Methylthiochroman-4-ols on treatment with polyphosphoric acid yield thiochromans and benzothiopyrylium salts (thianaphthalenium salts) or thiacyanine dyes (*B. D. Tilak et al.*, Tetrahedron, 1966, **22,** 7).

Cyclisation of methyl 2-mercaptobenzoate by acrylonitrile in dioxane containing sodium methoxide gives an 82% yield of 3-cyanothiochroman-4-one (LXVI), which on hydrolysis affords 4-oxothiochroman-3-carboxylic acid (LXVII) (*A. P. Momsenko*, Zhur. org. Khim., 1973, **9,** 775):

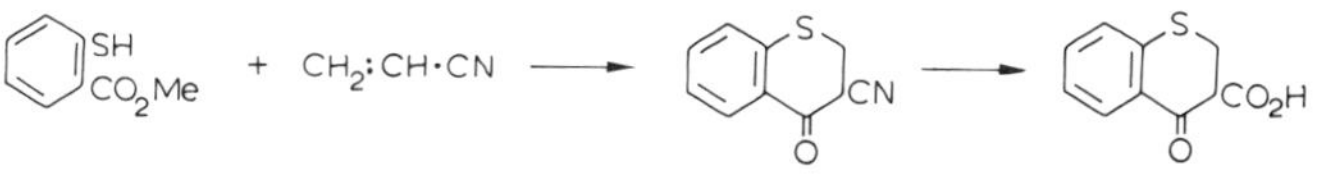

SH, CO_2Me + CH_2:CH·CN ⟶ (LXVI) ⟶ (LXVII)

3-Cyano-3-methylthiochroman-4-one and 3-methoxycarbonyl-3-methylthiochroman-4-one have been synthesised during attempts to synthesise thia-steroid analogues (*T. Moriwake*, J. med. Chem., 1966, **9,** 163).

2,3-**Dihydro**-2-**phenyl**-4*H*-**benzo**[*b*]**thiopyran**-4-**one,** 2-**phenylthiochroman**-4-**one, thioflavanone,** (LXVIII), m.p. 55–56°, 3-*benzylidene* deriv., m.p. 132–133°, is obtained by heating *β*-phenylmercaptocinnamic acid; with bromine in acetic acid it affords 3-*bromo*-, m.p. 131°, and 3,3-*dibromo-thioflavanone*, oil. It gives a blue colour in sulphuric acid (*F. Arndt*, Ber., 1923, **56,** 1269). 6-**Methylthioflavanone,** m.p. 96°, *phenylhydrazone*, m.p. 206°, 1-*oxide*, m.p. 177–178° (impure), 1,1-*dioxide*, m.p. 177–178° resolidifying to m.p. 191–192°, 3-*benzylidene* deriv., m.p. 108–109°.

The reduction of thioflavanone, 6-methyl-, and 4′-chloro-thioflavanone with lithium tetrahydridoaluminate or sodium tetrahydridoborate gives the corresponding 2,4-*cis*-thioflavan-4-ols LXIX. Deamination of 2,4-*cis*-4-aminoflavans LXX with nitrous acid yields the 2,4-*trans*-thioflavan-4-ols LXXI (*G. F. Katekar*, Austral. J. Chem., 1966, **19,** 1251):

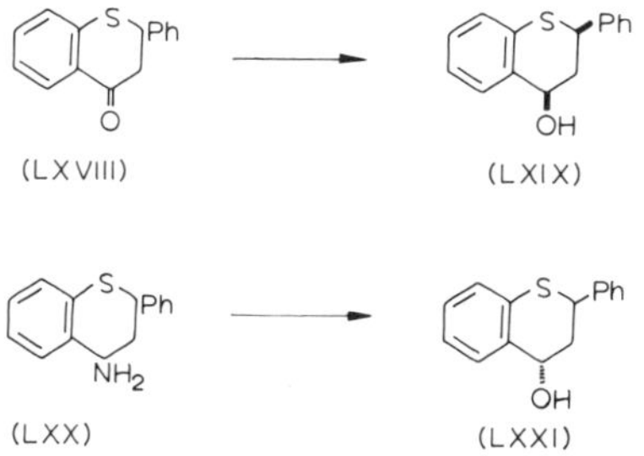

(LXVIII) ⟶ (LXIX)

(LXX) ⟶ (LXXI)

For u.v. spectra of thioflavanone and derivatives, thioflavones and *β*-4-hydroxythioflavan see *D. E. Rakosi et al.*, Acta phys. Chem. Debrecina, 1970, 163.

3-Arylthio-2-phenylpropanoic acids (LXXIII) prepared by heating atropic acid (LXXII) and the appropriate arenethiol with hydrogen chloride in a sealed tube at 100°, on treatment with concentrated sulphuric acid gives the corresponding **thioisoflavanone,** 2,3-dihydro-3-phenyl-4*H*-benzo[*b*]thiopyran-4-one (LXXIV) (*Katekar, R. M. Thomson*, Austral. J. Chem., 1972, **25,** 647):

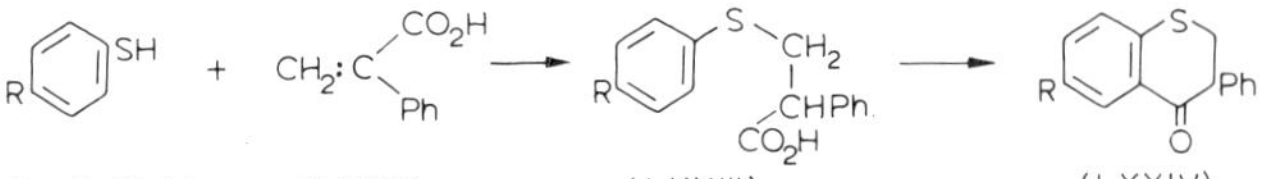

R = H,Cl,Me (LXXII) (LXXIII) (LXXIV)

3. 1*H*-Benzo[*c*]thiopyran, isothiochromene, 3,4-benzothiopyran and its derivatives

(a) 1H-*Benzo*[c]*thiopyran, isothiochromene and* 1H-*benzo*[c]*thiopyran*-1-*ones, isothiocoumarins*

Isothiochromene, 1*H*-benzo[*c*]thiopyran (II), C_9H_8S, b.p. 124°/13 mm, an unstable substance with an odour of naphthalene, is obtained by heating isothiochroman-4-ol (I) with potassium hydrogen sulphate (*J. von Braun* and *K. Weissbach*, Ber., 1929, **62,** 2416):

(I) (II)

Methyl phthalaldehydate (III) condenses with rhodanine (IV) to give 5-o-*methoxycarbonylbenzylidenerhodanine* (V), yellow needles, m.p. 215–216°, which with hot sodium hydroxide yields **isothiocoumarin-3-carboxylic acid** (VI), $C_{10}H_6O_3S$, m.p. 261–263°, *methyl ester*, m.p. 138–139°, which when heated at 330° for 10 min affords **isothiocoumarin** (VII), C_9H_6OS, m.p. 78–79° (*D. J. Dijksman* and *G. T. Newbold*, J. chem. Soc., 1951, 1213). With methanolic ammonia it yields 1,2-*dihydro*-1-*oxoisoquinoline*, m.p. 209°, but does not reduce Fehling's solution. Isothiocoumarin or its 3-carboxylic acid with Raney nickel yields not the expected *o*-formylcinnamic acid but indan-1-one (VIII) (*Dijksman* and *Newbold*, *ibid*., 1952, 13):

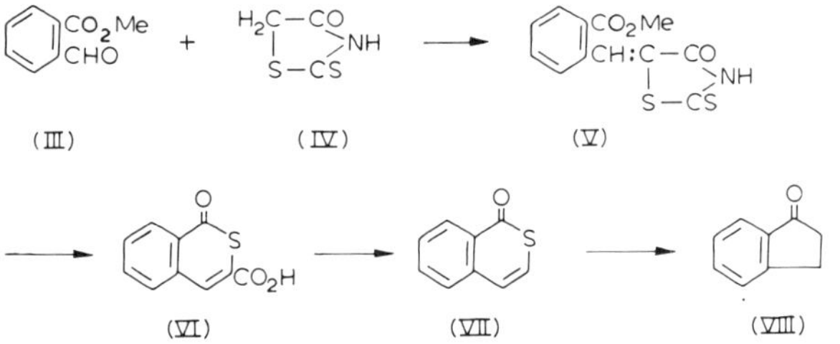

(III) (IV) (V)

(VI) (VII) (VIII)

4-*Methylisothiocoumarin*, colourless needles by sublimation, m.p. 75–76°.

3-Arylthioisocoumarins react with hydrazine to give the 1,2-dihydro-5*H*-benzo-2,3-diazepinones IX, which are converted to X on treatment with tetraphosphorus decasulphide in pyridine (*L. Legrand* and *N. Lozac'h*, Bull. Soc. chim. Fr., 1970, 2237):

N_2H_4 → (IX) $\xrightarrow[C_5H_5N]{P_4S_{10}}$ (X)

The reaction between secondary aliphatic amines and 3-aryl-benzo[*c*]pyran-1-thiones, 3-aryl-1-thioisocoumarins (XI), afford dialkylthioamides, XII, which on reaction with tetraphosphorus decasulphide yield 3-**arylbenzo[*c*]thiopyran-1-thiones** (XIII) (*idem*, *ibid*., p. 2227):

(XI) $\xrightarrow{R_2NH}$ $CSNR_2$ / $CH_2{\cdot}COAr$, R = alkyl (XII) $\xrightarrow{P_4S_{10}}$ (XIII)

1,2-**Diphenyl-2-thianaphthalene** (XIV) has been obtained as a purple amorphous solid by the route indicated below and is readily interconvertible to XV (*C. C. Price* and *D. H. Follweiler*, J. org. Chem., 1969, **34,** 3202):

$ClO_4^{\ominus}$ $\xrightarrow{PhMgCl}$ (Ph, H) $\xrightarrow[HClO_4]{SO_2Cl_2}$ (Ph, $ClO_4^{\ominus}$) $\xrightarrow{PhLi}$ (XIV) (Ph, SPh)

(XV) (Ph, H, $\oplus$S–Ph) $\underset{base}{\overset{H^{\oplus}}{\rightleftarrows}}$ (XIV)

For attempted synthesis of 2-thianaphthalene derivatives see *C. Kissel*, *R. J. Holland*, and *M. C. Caserio*, *ibid*., 1972, **37,** 2720.

(*b*) *3,4-Dihydro-1H-benzo[c]thiopyran, isothiochroman and its derivatives*

3,4-**Dihydro-1*H*-benzo[*c*]thiopyran, isothiochroman, 2-thiatetralin,** (XVII), $C_9H_{10}S$, b.p. 115°/7 mm, *mercurichloride*, m.p. 203.3–204.3°, 2-*oxide* (S-*monoxide*), m.p. 92–92.5°, *methiodide*, m.p. 124° (decomp.), is prepared in small yield by the Clemmensen reduction (*J. von Braun* and *K. Weissbach*, Ber., 1929, **62,** 2416), and in 89% yield by the Wolff–Kishner–Minlon reduction of isothiochroman-4-one (XVI) (*P. Cagniant* and *Mme P. Cagniant*, Bull. Soc. chim. Fr., 1959, 1998):

(XVI) (XVII)

or by heating isochroman with anhydrous hydrogen bromide and treating the resulting 2-(2-bromoethyl)benzyl bromide with sodium sulphide nonahydrate (*von Braun* and *F. Zobel*, Ber., 1923, **56,** 2142; *S. F. Birch et al.*, J. Inst. Petroleum, 1954, **40,** 76).

It gives a lemon yellow colour in sulphuric acid, and with chloramine-T gives the p-*toluenesulphonylimine*, m.p. 165°. 1,1-Disubstituted derivatives of isothiochroman (*H. Boehme, R. Priesner* and *B. Unterhalt*, Arch. Pharm., 1966, **299,** 931; *Boehme* and *K. Lindenberg*, *ibid.*, 1968, **301,** 584). The thermal fragmentation of isothiochroman and related compounds has been investigated and compared with the fragmentation under electron impact (*A. G. Loudon, A. Maccoll* and *S. K. Wong*, J. chem. Soc. B, 1970, 1733).

Stereoselective syntheses for *cis*-1,4-, and *cis*-1,3-dimethylisothiochromans and their 2,2-dioxides have been described. In the 1,4- and 1,3-dimethyl dioxides, the *cis*- and *trans*-forms, respectively, are the more thermodynamically stable stereoisomers. A preferred boat conformation is inferred for these compounds, and has been demonstrated for the *cis*-1,4-dimethyl dioxide by X-ray analysis. 1-Methyl- and 1,3-dimethyl-isothiochroman-4-ones (XVIII) intermediate in the synthesis are unstable to light, rearranging on irradiation to thiochroman-3-ones (XX) (*D. A. Pulman* and *D. A. Whiting*, J. chem. Soc. Perk. I, 1973, 410):

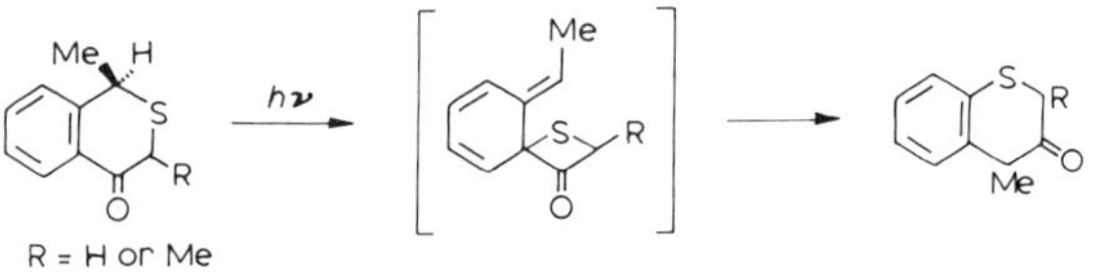

R = H or Me

(XVIII) (XIX) (XX)

Similar rearrangements of isothiochromanones substituted only in the aromatic ring have been reported and are believed to implicate intermediates such as XIX (*W. C. Lumma* and *G. A. Berchtold*, J. org. Chem., 1969, **34,** 1566).

S-Benzylthioacetyl chloride with aluminium chloride yields **isothiochroman-4-one** (XXI), m.p. 64°, faint but characteristic odour, *oxime*, m.p. 134–135°, *semicarbazone*, m.p. 236°, 4-*nitrophenylhydrazone*, red needles, m.p. 214°, which with diazonium salts gives azo-compounds but does not react with nitrous acid. Two molecules of isothiochroman-4-one condense with one of 4-nitrosodimethylaniline to give *compound* XXII, orange red crystals, m.p. 176° (*R. Lesser*, Ber., 1923, **56,** 1642).

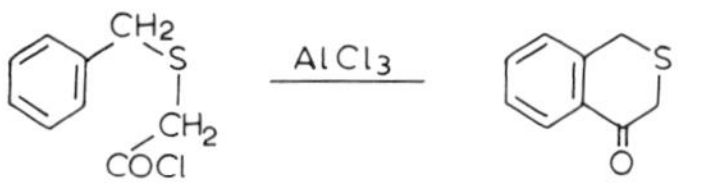

(XXI)

Isothiochroman-4-one undergoes rearrangement when reduced by the Clemmensen method to give 1-*methyl-2-thiaindane*, 1,4-*dihydro*-1-*methylbenzo*[c]*thiophene* (XXIII), m.p. 115–117° (*von Braun* and *Weissbach*, Ber., 1929, **62,** 2416; *N. J. Leonard* and *J. Figueras*, Jr., J. Amer. chem. Soc., 1952, **74**, 917), and with sodium amalgam yields **isothiochroman-4-ol** (I), b.p. 138–140°/0.1 mm, 170–175°/12 mm.

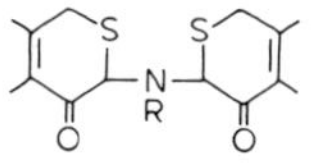

R = $C_6H_4NMe_2$

(XXII) (XXIII)

Isochroman-1-one (XXIV) reacts with tetraphosphorus decasulphide to give isothiochroman-1-thione (XXV), which with diazomethane gives *cis*- and *trans*-dispiro(isothiochroman-1,4′-[1,3]dithiolane-5′,1″-isothiochroman) (XXVI) (*M. Ebel* and *Legrand*, Bull. Soc. chim. Fr., 1971, 176):

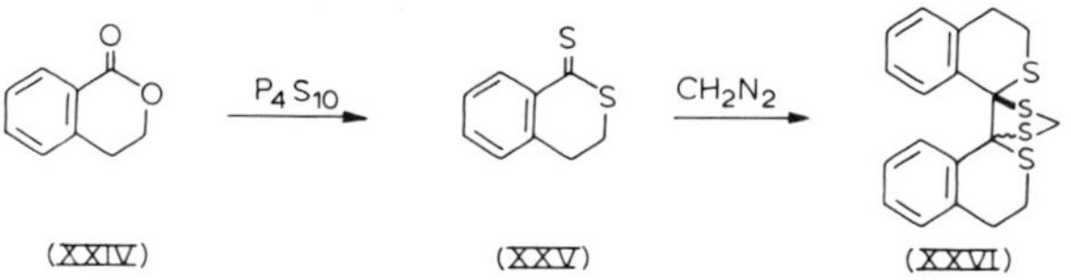

(XXIV) (XXV) (XXVI)

The **hexahydroisothiochromans, 2-thiadecalins,** $C_9H_{16}S$, are prepared by the ring-fusion of *cis*- and *trans*-hexahydroisochroman by hydrogen bromide and interaction of the resulting 1-bromomethyl-2-β-bromoethylcyclohexane with sodium sulphide (*S. F. Birch et al.*, J. org. Chem., 1954, **19,** 1449).

cis-2-*Thiadecalin*, b.p. 246°/760 mm, 122.5–123.25°/20 mm, d^{20} 1.0350g/ml, n_D^{20} 1.53346, 2-*oxide*, S-*monoxide*, m.p. 104.5–105.5°, prepared by oxidation of the sulphide by *tert*-butylhydroperoxide, 2,2-*dioxide*, S-*dioxide*, m.p. 86.0–87.4°, *methiodide*, m.p. 159.6–161.6° (sealed tube), *mercurichloride*, m.p. 148.5–151.5°. trans-2-*Thiadecalin*, b.p. 115.5–116.5°/20 mm, d^{20} 1.0058g/ml, n_D^{20} 1.52173, 2-*oxide*, m.p. 69–70°, 2,2-*dioxide*, m.p. 91–92.8°, *methiodide*, m.p. 173.5–174.5° (sealed tube), *mercurichloride*, m.p. 149.8–150.4°. Refluxing the 2-thiadecalins in ethanol with Raney nickel affords cis-2-*ethyl-methylcyclohexane*, b.p. 155.97°, n_D^{20} 1.4432 and trans-2-*ethyl-methylcyclohexane*, b.p. 151.69°, n_D^{20} 1.4382. The i.r. spectra have been reported (*Birch et al., loc. cit.*). 2-Methyl-1-thiadecalin (*E. N. Karaulova, V. Sh. Shaikhrazieva* and *G. D. Gal'pern*, Khim. Geterotsikl. Soedin., 1967, 51; Neftekhimiya, 1967, **7**, 812; C.A., 1968, **68,** 49404).

4. Dibenzothiopyran, thioxanthene and derivatives

The sulphur analogue of xanthene is **thioxanthene** (I), $C_{13}H_{10}S$, m.p. 128°, b.p. 340°/730 mm, μ 1.44D (*A. Weizmann*, Trans. Faraday Soc., 1940, **36,** 978), yellow solution in sulphuric acid with a faint fluorescence, prepared by the pyrolysis of phenyl *o*-tolyl sulphide or preferably by reducing thioxanthone (V) by phosphorus and iodine (*C. Graebe* and *O. Schultess*, Ann., 1891, **263,** 12), lithium tetrahydridoaluminate (*A. Mustafa* and *M. K. Hilmy*, J. chem. Soc., 1952, 1343), sodium tetrahydridoborate in 95% ethanol (*A. L. Ternay, Jr., D. W. Chasar* and *M. Sax*, J. org. Chem., 1967, **32,** 2465), the Wolff–Kishner method (*E. A. Fehnel*, J. Amer. chem. Soc., 1949, **71,** 1063), sodium in ethanol (*R. A. Heacock, R. L. Wain* and *F. Wightman*, Ann. appl. Biol., 1958, **46**, 352), diphenylsilane at 260° (*H. Gilman* and *J. W. Diehl*, J. org. Chem., 1961, **26**, 4817), or diborane in tetrahydrofuran at 0° when a quantitative yield is obtained (*W. J. Wechter*, *ibid.*, 1963, **28,** 2935). Substituted thioxanthenes are obtained by the ring-closure of *o*-phenylmercaptobenzaldehydes with sulphuric acid, the resulting thioxanthylium salts (*e.g.*, III) when treated with water undergoing disproportionation to give equal quantities of thioxanthene and thioxanthone (*J. D. Loudon et al.*, J. chem. Soc., 1941, 747; *E. D. Amstutz* and *C. R. Neumoyer*, J. Amer. chem. Soc., 1947, **69**, 1925). Thus 4-nitro-2-*p*-tolylthiobenzaldehyde affords 2-*methyl-6-nitrothioxanthene*, m.p. 155°, and -*thioxanthone*, m.p. 276°.

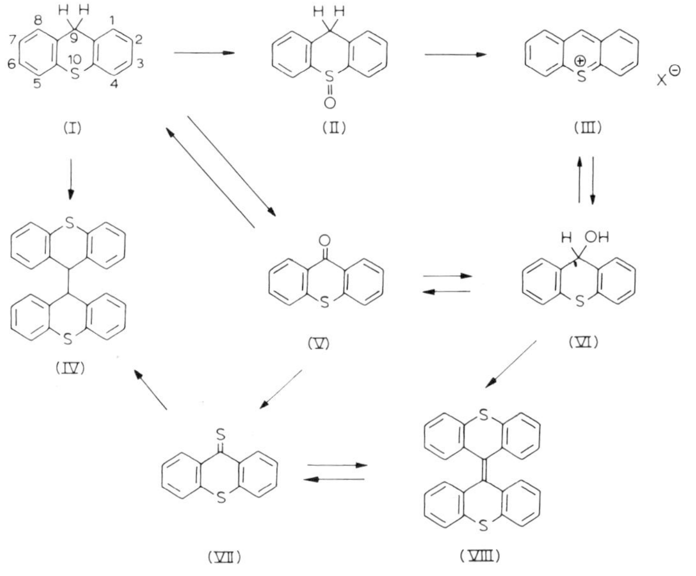

(I) (II) (III) (IV) (V) (VI) (VII) (VIII)

Thioxanthene with hydrogen peroxide according to the conditions yields thioxanthone (V), the 10,10-*dioxide (sulphone)*, m.p. 172°, or the 10-*oxide (sulphoxide)* (II), m.p. 109° (*T. P. Hilditch* and *S. Smiles*, J. chem. Soc., 1910, **99,** 145) and in the presence of oxygen and sunlight gives 9,9′-**bithioxanthyl,** (IV), $C_{26}H_{18}S_2$, m.p. 325° (*A. Schönberg* and *Mustafa, ibid.*, 1945, 657), which is also obtained by reducing thionothioxanthone, xanthene-9-thione (VII) with zinc and hydrochloric acid (*Schönberg*, Ber., 1925, **58,** 1793), or thioxanthylium perchlorate with one of several reducing reagents, such as zinc, cobaltocene, dithionite, and potassium and phosphonium iodide (*C. C. Price, M. Siskin* and *C. K. Miao*, J. org. Chem., 1971, **36,** 794). The reaction of the 10-oxide (II) with a 33% solution of hydrobromic acid affords a mixture of thioxanthene and thioxanthone, whereas no reaction occurs with the 10,10-dioxide (*Gilman* and *Diehl, ibid.*, 1961, **26,** 2132).

The reaction between thioxanthene and bromine in carbon tetrachloride, even in total absence of light, results in the quantitative precipitation of thioxanthylium perbromide (III, X = Br_3), which is hydrolysed to a 1:3 mixture of thioxanthene and thioxanthone with 5% sodium hydroxide solution (*E. K. Fields* and *S. Meyerson, ibid.*, 1965, **30,** 937). Xanthene reacts with silver perchlorate–diiodomethane to give the *S*-alkylated compound, 10-iodomethylthioxanthylium perchlorate (IX) (*R. M. Acheson* and *J. K. Stubbs*, J. chem. Soc. Perk. I, 1972, 899):

$AgClO_4$, CH_2I_2 → CH_2I $ClO_4^{\ominus}$

(IX)

Methods have been examined for the preparation of 10-alkyl- and 10-aryl-9,9-dimethylthioxanthylium salts (*M. Hori et al.*, Yakugaku Zasshi, 1973, **93,** 476).

Dehydrogenation of thioxanthene with a stoichiometric amount of triphenylmethyl perchlorate in acetic acid yields thioxanthylium perchlorate (III, X = ClO_4) (*W. Bonthrone* and *D. H. Reid*, Chem. and Ind., 1960, 1192). A radical mechanism has been proposed for the reaction between thioxanthylium salts and organolithium or organomagnesium reagents (*Hori et al.*, Chem. Letters, 1973, 391).

The crystal structure of thioxanthene (*J. A. Gillean et al.*, Acta Crystallogr. B, 1973, **29,** 2296) and 9-isobutylthioxanthene (*S. S. C. Chu, ibid.*, p. 1690) have been established; conformational analysis of thioxanthene (*Z. Aizenshtat*, Israel J. Chem., 1972, **10,** 753); crystal structure and conformation of *cis*-9-methylthioxanthene 10-oxide (*J. J. Jackobs* and *M. Sundaralingam*, Acta Crystallogr. B, 1969, **25,** 2487); stereochemistry and reactivity of 9-substituted thioxanthenes and their oxides (*Chasar*, Diss. Abstr. Int. B, 1969, **30,** 116).

2,7,9-*Trimethylthioxanthene*, m.p. 85–86°; 9-*ethyl*-2,7-*dimethylthioxanthene*, m.p. 76.5–77.5°; 9-*benzylidene*-2,7-*dimethylthioxanthene*, m.p. 109.5–110.5°; 9-*benzyl*-2,7-*dimethylthioxanthene*, m.p. 121–122°; and 2,7-*dimethyl*-9-*phenylthioxanthene*, m.p. 156–157° (*F. Krollpfeiffer* and *A. Wissner*, Ann., 1951, **572,** 195). Nitration of 9-phenylthioxanthylium perchlorate with a mixed acid gives a dinitrated product (*Hori et al.*, Chem. pharm. Bull., 1973, **21,** 1272). **Perhydrothioxanthene,** m.p. 70–71.5 (*V. G.*

Kharchenko et al., Khim. Geterotsikl. Soedin., 1967, 468; 1971, 82).

Thioxanthene 10,10-dioxide, 2*H*-benzo[*b*]thiopyran 1,1-dioxide (p. 367), 1*H*-benzo-[*c*]thiopyran 2,2-dioxide, and dibenzo[*bd*]thiopyran 5,5-dioxide are all methylated at the active methylene group (*S. Bradamante* and *G. Pagani*, J. chem. Soc. Perk. I, 1973, 163).

Thioxanthydrol, thioxanthen-9-ol (VI), $C_{13}H_{10}OS$, m.p. 104–105°, *benzoate*, m.p. 124–126°, oxidised by air in light, is prepared by reducing thioxanthone with zinc and sodium ethoxide (*A. Werner*, Ber., 1901, **34**, 3310), potassium and ethanol (*F. Mayer*, *ibid.*, 1909, **42**, 1135), or sodium amalgam (*H. F. Oehlschlaeger* and *I. R. MacGregor*, J. Amer. chem. Soc., 1950, **72**, 5332).

Thioxanthydrol on warming in acetic anhydride containing a trace of 72% perchloric acid precipitates thioxanthene containing traces of thioxanthone; disproportionation can also be achieved with hydrogen chloride and sulphuric acid but not with Lewis acids, *e.g.*, aluminium chloride (*G. E. Ivanov* and *V. A. Izmail'skii*, Zhur. org. Khim., 1967, **3**, 1142). It is oxidised by potassium permanganate to thioxanthone and yields, when boiled in acetic acid, *dithioxanthyl ether*, m.p. 314–316° (*Hilditch* and *Smiles*, *loc. cit.*). Thioxanth-9-yl ethers have been prepared from the thioxanth-9-yl acetate or *p*-toluenesulphonate (*Ivanov* and *Izmail'skii*, Zhur. org. Khim., 1968, **4**, 904). Thioxanthydrol on pyrolysis yields thioxanthone (V), thioxanthene (I), and bithioxanthylidene (VIII) (*Schönberg* and *A. Mustafa*, J. chem. Soc., 1944, 305). *α-Thioxanthydrol* 10-*oxide*, *α-thioxanthen*-9-*ol* 10-*oxide* (X), m.p. 218–218.5°, is prepared by oxidising thioxanthydrol with hydrogen peroxide (30%) in acetone; *β-thioxanthydrol* 10-*oxide*, *β-thioxanthen*-9-*ol* 10-*oxide* (XI), m.p. 205–206° is isolated following the treatment of thioxanthydrol in acetone with *m*-chloroperbenzoic acid with a mixture of the α- and β-forms (~60% of mixture) is obtained. The X-ray analysis of β-thioxanthydrol 10-oxide has revealed it to have the *trans* configuration (*Ternay*, *Chaser* and *Sax*, *loc. cit.*); an n.m.r. investigation of the conformational preferences of the isomeric thioxanthydrol 10-oxides and related compounds has been carried out (*Ternay* and *Chasar*, J. org. Chem., 1968, **33**, 2237); and the behaviour of 10-oxides in trifluoroacetic acid and its anhydride has been studied (*idem*, *ibid.*, p. 3641). *Thioxanthydrol* 10,10-*dioxide*, pale yellow needles, m.p. 184–185°, is obtained by reducing thioxanthone 10,10-dioxide with zinc and acetic acid (*Fehnel*, *loc. cit.*).

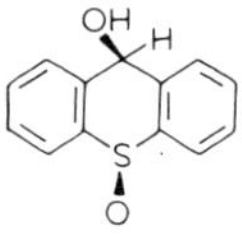

OH H S O (X)

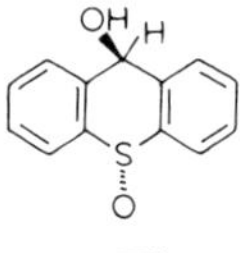

OH H S O (XI)

9-Substituted thioxanthydrols are prepared by the interaction of thioxanthone and Grignard reagents, phenylmagnesium bromide, for example, yielding 9-**phenylthioxan-**

thydrol, m.p. 105–106° (*H. Bünzly et al.*, Ber., 1904, **37,** 2937) which is reduced by sodium and absolute ethanol to 9-phenylthioxanthene, m.p. 96° (*Schönberg* and *Mustafa, loc. cit.*) and with hydrochloric acid gives the dark red *salt* $(C_{19}H_{13}S)Cl$, HCl, converted by air to the colourless *chloride* $C_{19}H_{13}SCl$, m.p. 114–115° (*M. Gomberg* and *L. H. Cone*, Ann., 1910, **376,** 202). Thioxanthone and methylmagnesium bromide yield 9-**methylthioxanthydrol,** isolated in the form of its *ferrichloride*, m.p. 200° and its *methyl ether*, m.p. 98–99° (*H. Decker*, Ber., 1905, **38,** 2510). 9-**Benzylthioxanthydrol,** m.p. 134°, when heated or treated with acetyl chloride affords 9-*benzylidenethioxanthene*, m.p. 121–122° (*E. D. Bergmann et al.*, Bull. Soc. chim. Fr., 1952, 262). 1-Chlorothioxanthydrol has been prepared by reducing the corresponding thioxanthone with sodium amalgam and ethanol (*I. Okabayashi, F. Miyoshi* and *M. Arimoto*, Yakagaku Zasshi, 1972, **92,** 1386).

Thioxanthydrol is a pseudo-base and with mineral acids forms thioxanthylium salts, giving, for example, with hydrogen chloride at 0° *thioxanthylium chloride* (III, X = Cl), brick-red needles, which when kept in a vacuum afford the colourless *chloride*, m.p. 112–113° and with ferric chloride gives the *ferrichloride*, $C_{13}H_9S^{\oplus} \cdot FeCl_4^{\ominus}$, bright red needles, m.p. 193–194°. The thioxanthylium salts are hydrolysed to thioxanthydrols.

Thioxanthone, thioxanthen-9-one (V), yellow crystals, m.p. 211°, yellow solution in sulphuric acid with green fluorescence, is obtained by the condensation of 2-mercaptobenzoic acid and benzene in the presence of sulphuric acid (*E. G. Davis* and *Smiles*, J. chem. Soc., 1910, **97,** 1290), by the action of thionyl chloride on bithioxanthylidene (VIII) (*Schönberg* and *W. Asker*, *ibid.*, 1942, 725), and by the interaction of diphenyl sulphide and carbonyl chloride with aluminium chloride as catalyst (*H. H. Szmant et al.*, J. org. Chem., 1953, **18,** 745). Diphenyl sulphide reacts with dichloromethyl methyl ether in methylene chloride and stannic chloride to give 26% of thioxanthone and a large amount of a complex of diphenyl sulphide, stannic chloride and dichloromethyl methyl ether (*R. Oda* and *K. Yamamoto*, Nippon Kagaku Zasshi, 1963, **84,** 348).

Thionanthone is broken down by fusion with alkali to 2-phenylthiobenzoic acid and is oxidised by hydrogen peroxide or potassium persulphate to the 10,10-*dioxide (sulphone)*, m.p. 187° (*F. Ullmann* and *O. von Glenck*, Ber., 1916, **49,** 2509). Thioxanthone is converted to the 10-*oxide*, m.p. 202–204°, by ceric ammonium nitrate in water–acetonitrile at room temperature (*T.-L. Ho*, and *C. M. Wong*, Synthesis, 1972, 561), or by treatment with iodisobenzene diacetate in boiling acetic acid. (*J. P. A. Castrillón* and *Szmant*, J. org. Chem., 1967, **32,** 976). The 10-oxide is reduced by hydrochloric acid–dioxane without the expected chlorination, and in this novel way it is possible to prepare the 2,4-dinitrophenylhydrazone of thioxanthone which cannot be prepared directly. It does not react with hydroxylamine or phenylhydrazine, perhaps because of the dipolar structure indicated by its high dipole moment, μ 5.4D (*Weizmann, loc. cit.*) and i.r. spectrum (*Bergmann* and *S. Pinchas*, J. Chim. phys., 1952, **49,** 537).

Thioxanthone may be reduced by a number of reagents to give thioxanthydrol (p. 390) or thioxanthene (p. 388). The oxime of thioxanthone, or its 10,10-dioxide on heating with polyphosphoric acid gives the corresponding lactam (XII) (*P. Catsoulacos*, Chim. Ther., 1970, **5**, 401):

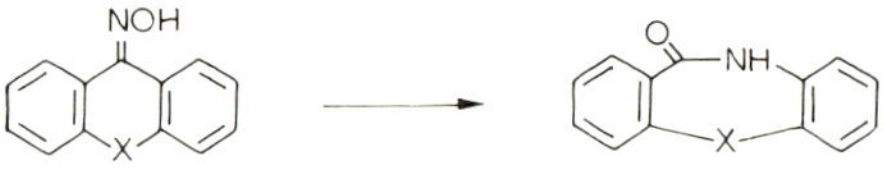

(X = S or SO_2) (XII)

Thioxanthone 10,10-dioxide oxime acetate on treatment with excess of diborane at room temperature, is converted into 9-aminothioxanthene 10,10-dioxide (*idem*, J. heterocycl. Chem., 1967, **4,** 645).

Thioxanthone in liquid ammonia reacts with alkaline salts of acetylene to give 9-*ethynylthioxanthydrol* (XIII), m.p. 98–99° (*W. Reid* and *J. Schönherr*, Angew. Chem., 1958, **70,** 271):

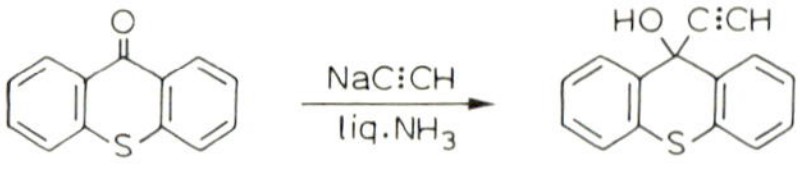

NaC:CH / liq.NH_3

(XIII)

The *ion-pair complex of phosphorus pentachloride with thioxanthone*, $C_{13}H_8OS$ $PCl_4^{\oplus} - PCl_6^{\ominus}$, m.p. 231–232° (decomp.), has been prepared (*R. C. Duty* and *R. E. Hallstein*, J. org. Chem., 1970, **35,** 4226).

1-**Chlorothioxanthone** has been prepared by the cyclisation of 2-chloro-6-(phenylthio)benzoic acid and by reacting benzene with 2-chloro-6-mercaptobenzoic acid in sulphuric acid. A number of derivatives of 1-chlorothioxanthone have been obtained (*Okabayashi, Miyoshi* and *Arimoto, loc. cit.*). The chlorine in 1-chlorothioxanthone derivatives is reactive, 1-chloro-4-methylthioxanthone, for instance, reacting with aniline to give 1-*anilino*-4-*methylthioxanthone*, orange needles, m.p. 127°, and with hydrazine hydrate to give 2H-*pyrazolothioxanthene* (XIV), m.p. 251° (decomp.) (*Ullmann* and *von Glenck*, Ber., 1916, **49,** 2495):

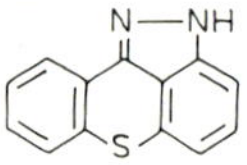

(XIV)

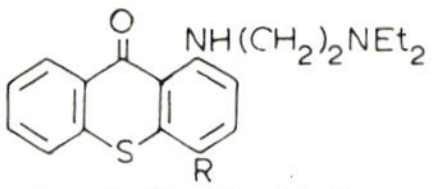

Miracil-D (R = Me)
Hycanthone (R = CH_2OH)

Miracil-D, *lucanthone* B.P., *nilodin*, 1-(2-*diethylaminoethylamino*)-4-*methylthioxanthone*, $C_{20}H_{24}ON_2S$, yellow crystals, m.p. 64–65°, *hydrochloride*, yellow needles, m.p. 196–197°, *methiodide*, orange needles, m.p. 237° (decomp.), is synthesised by the interaction of 1-*chloro*-4-*methylthioxanthone*, m.p. 143–145°, and diethylaminoethylamine (*T. M. Sharp*, J. chem. Soc., 1951, 2961) and is used clinically in the treatment of human schistosomiasis (*H. Mauss*, Ber., 1948, **81,** 19).

Hycanthone, 1-(2-*diethylaminoethylamino*)-4-(*hydroxymethyl*)*thioxanthone*, m.p. 101–102.5°, a schistosomicidal agent has been synthesised (*G. M. Laidlaw et al.*, J. org. Chem., 1973, **38,** 1743; *J. W. Schulenberg*, U.S. P. 3,711,513, 1973). 2-Chlorothioxanthone reacts with oxalyl chloride to give 2,9,9-trichlorothioxanthene (*Schönberg* and *E. Frese*, Ber., 1968, **101,** 694). Bromic acid and potassium bromate oxidise 1,4-dihydroxythioxanthone at room temperature to the corresponding 1,4-quinone (*A. G. Bayer*, B.P. 786,925, 1957). A number of compounds have been synthesised based on thioxanthone-2-carbaldehyde (*G. Vasiliu* and *N. Rasanu*, Rev. Chim., Bucharest, 1969, **20,** 545).

Thioxanthone 10,10-dioxides XV ($R=R^1=H$; $R=R^1=CO_2H$; $R=Cl$, $R^1=H$; $R=Me$, $R^1=H$; $R=MeO$, $R^1=H$; $R=NO_2$, $R^1=H$) have been synthesised and boiled with 2% sodium hydroxide in 65% aqueous dioxane. For these systems, facile hydroxide displacement of the sulphone linkage is found to occur exclusively on the more electrophilic ring to give novel 2′-hydroxybenzophenone-2-sulphinic acids. Further, in alkaline media, these sulphinic acids generally undergo unique intramolecular cyclisation to xanthones illustrated by XVI. The following mechanism has been suggested for the reaction (*O. F. Bennett et al.*, J. org. Chem., 1972, **37,** 1356):

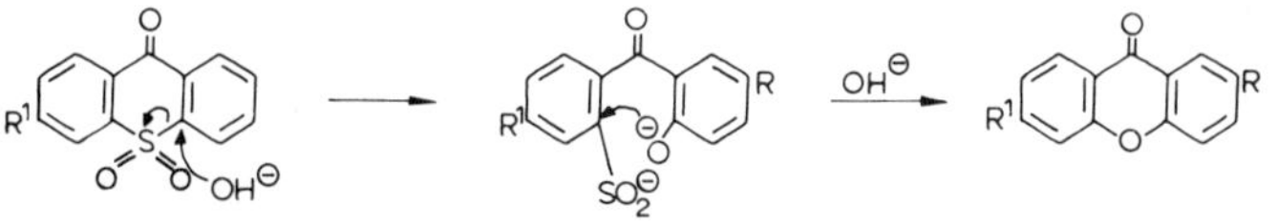

(XV) (XVI)

Thioxanthene-9-thione, *thionothioxanthone*, *thioxanthione*, (VII, p. 388), m.p. 168° (176°), is prepared by the interaction of thioxanthone and thionyl chloride (*F. G. Mann* and *J. H. Turnbull*, J. chem. Soc., 1951, 757), phosphorus pentasulphide (*F. Arndt* and *L. Lorenz*, Ber., 1930, **63,** 3129), or oxalyl chloride followed by heating the products with thioacetic acid (*Schönberg et al.*, *ibid.*, 1928, **61,** 1382) and by the action of sulphur on bithioxanthylidene (VIII) (*Schönberg* and *Asker*, J. chem. Soc., 1942, 272). It has a dipole moment of 5.2D (*Weizmann*, *loc. cit.*), with hydroxylamine yields *thioxanthone oxime*, m.p. 195–196°, and when heated with copper in xylene yields **bithioxanthylidene** (VIII), $C_{26}H_{16}S_2$, m.p. 365° (decomp.), feebly thermochromic, with a blue fluorescence in u.v. light (*Schönberg* and *Asker*, *loc. cit.*). Bithioxanthylidene is also prepared by reducing thioxanthone with zinc and acetic acid (*F. Mayer*, Ber., 1909, **42,** 2232) by treating 9,9-diazothioxanthene with triphenylphosphine at 60–80° (*M. Sidky*, *M. R. Mahran* and *L. S. Boulos*, J. Indian chem. Soc., 1972, **49,** 985), and is oxidised by hydrogen peroxide to the *disulphone*, m.p. >380° (decomp.) (*Bergmann et al.*, *loc. cit.*). 2-Chlorothioxanthene-9-thione (*Schönberg* and *Frese*, *loc. cit.*). *Thioxanthene-9-thione S-oxide* (XVII), m.p. 140–142° (decomp.), *thioxanthene-9-thione 10,S-dioxide* (XVIII), m.p. 198–200° (decomp.) (*B. Zwanenburg*, *L. Thijs* and *J. Strating*, Rec. Trav. chim., 1967, **86,** 577):

(XVII) (XVIII)

Some substituted thioxanthones and some derivatives are listed in Table 2.

TABLE 2

SUBSTITUTED THIOXANTHONES

Substituents	*M.p. (°C)*	*Derivatives*	*M.p. (°C)*	*Ref.*
2-Methyl	123	10,10-*dioxide*	199	1
3-Methyl	98			2
1,3-Dimethyl	127			3
1,4-Dimethyl	112			3
2,7-Dimethyl	128–129			4
3-Bromo	165			5
1-Nitro	237			6
2-Nitro	224–225 (227–229)	10,10-*dioxide*	254–255	7,8
3-Nitro	252			7
4-Nitro	215	10,10-*dioxide*	240	7
2,4-Dinitro	225–226			1
2,7-Dinitro	271.9–273.4	10,10-*dioxide*	302–307.5	8
1-Amino	249–250	*acetyl* deriv.	273	6
2-Amino	227–228	*acetyl* deriv.	236–237	7,9
3-Amino	252–254	*acetyl* deriv.	267	7
4-Amino	202	*acetyl* deriv.	233–234	6
2,7-Diamino	200–221	*diacetyl* deriv.	345.5–349.6	8
		10,10-*dioxide*	291–293	
2-Hydroxy-5-methyl	211			9
2-Hydroxy-7-methyl	217			9
2-Acetyl	179–180	10,10-*dioxide*	235–237	10
2,6-Dicarboxy		10,10-*dioxide*	426–429 (decomp.)	11

References

1 *F. Mayer*, Ber., 1910, **43,** 594.
2 *W. Borsche* and *A. Geyer*, *ibid.*, 1914, **47,** 1158.
3 *E. G. Marsden* and *S. Smiles*, J. chem. Soc., 1911, **99,** 1353.
4 *F. Krollpfeiffer* and *A. Wissner*, Ann., 1951, **572,** 195.
5 *M. Gomberg* and *L. H. Cone*, *ibid.*, 1910, **376,** 205.
6 *Mayer*, Ber. 1909, **42,** 3065.
7 *F. G. Mann* and *J. H. Turnbull*, J. chem. Soc., 1951, 757.
8 *E. M. Amstutz* and *C. R. Neumoyer*, J. Amer. Chem. Soc., 1947, **69,** 1925.
9 *H. Christopher* and *Smiles*, J. chem. Soc., 1911, **99,** 2046.
10 *N. Rasanu*, Rev. Chim., Bucharest, 1969, **20,** 659.
11 *O. F. Bennett* and *P. Gauvin*, J. org. Chem., 1969, **34,** 4165.

Cyclisation by the procedure of Smiles of 2-mercaptobenzoic acids with mono-substituted benzene derivatives in concentrated sulphuric acid leads to the formation of thioxanthones in which the substituent is in the 2-position (*Gilman* and *Diehl*, J. org. Chem., 1959, **24,** 1914). A number of thioxanthone derivatives have been synthesised starting with 2-bromomethylthioxanthone (*Vasiliu*, *Rasanu* and *O. Major*, Rev. Chim., Bucharest, 1968, **19,** 561). For i.r. spectra of some hydroxyanthones see

Mustafa et al., Tetrahedron, 1963, **19**, 1335.

Hydrogenation of 5,7-dimethyl-*trans*-4a,9a-dihydrothioxanthene 10,10-dioxide over Raney nickel gives 5,7-dimethyl-*trans*-1,2,3,4,4a,9a-**hexahydrothioxanthene** 10,10-**dioxide** (*V. N. Drovd* and *L. I. Zefirova, ibid.*, 1968, **4,** 852).

2,2′-Alkylidenebiscyclohexanones XIX react with hydrogen sulphide in polar solvents, such as ethanol or acetic acid in the presence of hydrogen chloride at 20° to give octahydrothioxanthenes XX (*V. G. Kharchenko, T. I. Krupina* and *S. K. Klimenko*, Zhur. org. Khim., 1966, **2,** 1899):

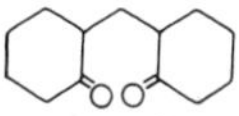

(XIX) 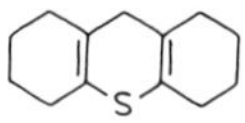 (XX)

1,2,3,4,5,6,7,8-**Octahydrothioxanthene,** m.p. 38–39.5°, *perchlorate*, m.p. 120–122°; 9-*methyl* analogue, m.p. 61–62°, 10,10-*dioxide*, m.p. 138.5–140°. The action of perchloric acid or hydrogen chloride/ferric chloride on *symm-9-methyloctahydrothioxanthene* gives, by disproportionation, 9-*methylperhydrothioxanthene*, m.p. 127–129°, and the corresponding perchlorate; *perhydrothioxanthene*, m.p. 70–71.5° (*idem, ibid.*, 1967, **3,** 1344; Khim. Geterotsikl. Soedin., 1971, 76; *Kharchenko* and *A. A. Rassudova, ibid.*, 1973, 196). Octahydrothioxanthenes have been obtained by heating 2,2′-alkylidene-bicyclohexanones with tetraphosphorus decasulphide in pyridine (*Kharchenko et al., ibid.*, 1972, 1196); reacting methylenedicyclohexanones in acetic acid with hydrogen sulphide in the presence of hydroquinone (*Kharchenko, Krupina* and *Rassudova, ibid.*, 1969, 226); and from Grignard reagents and the appropriate salt XXI (*Kharchenko et al., ibid.*, 1970, 338):

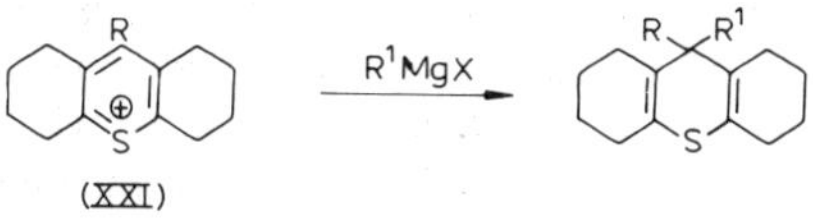

R^1MgX

(XXI)

R = H, Me

R^1 = Pr, Ph, CH_2Ph, [4]MeC_6H_4

Methylene- and ethylidene-2,2′-bicyclohexanones react with hydrogen sulphide and hydrogen halide (HX; X=Cl, Br, I) to give a mixture of the corresponding *symm*-octahydrothioxanthylium halide and related perhydrothioxanthene (*Kharchenko, N. M. Yartseva* and *Rassudova*, Zhur. org. Khim., 1970, **6,** 1513).

Dibenzo[*bd*]thiopyrylium perchlorate, 9-**thiaphenanthrenium perchlorate** (XXII) has been synthesised (*S. L. Jindal* and *B. D. Tilak*, Indian J. Chem., 1969, **7,** 637).

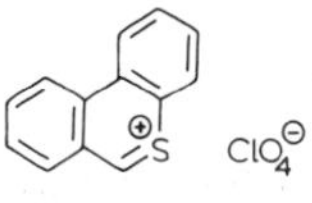

$ClO_4^{\ominus}$

(XXII)

5. Bridged ring sulphur compounds and related compounds

(a) Thiabicycloalkanes

(i) Thiabicyclo[3.2.1]octanes

cis-3-**Thiabicyclo**[3.2.1]**octane** (II), $C_7H_{12}S$, m.p. 174–175°, readily sublimes, *mercurichloride*, m.p. 189.5–190.5° (sealed tube); the 3,3-*dioxide*, S-*dioxide*, m.p. 229–230°, has a camphor-like odour and is prepared by the action of sodium sulphide on 1,3-bistosyloxymethylcyclopentane (I) (*S. F. Birch* and *R. A. Dean*, Ann., 1954, **585,** 234):

$CH_2-CH-CH_2OTos$ / CH_2 / $CH_2-CH-CH_2OTos$ (I) ⟶ $^7CH_2-\overset{1}{C}H-\overset{2}{C}H_2$ / $^8CH_2$, $S\,3$ / $_6CH_2-\underset{5}{C}H-\underset{4}{C}H_2$ (II)

It gives an unstable sulphoxide with *tert*-butyl hydroperoxide, and with Raney nickel undergoes hydrogenolysis to yield cis-1,3-*dimethylcyclopentane*, b.p. 90.6–90.8°, n_D^{20} 1.40903 and *bicyclo*[2.2.1]*heptane*, m.p. 87.5–88° (Vol. II C, p. 89):

2-**Thiabicyclo**[3.2.1]**octane** (III), m.p. 165–166° (sealed tube), b.p. 197°/774 mm, *mercurichloride*, m.p. 193–194° (decomp.), 2,2-*dioxide*, S-*dioxide*, m.p. 257° (decomp.), *methiodide*, 210–211° (decomp.) (sealed tube); 6-thiabicyclo[3.2.1]octane (IV), m.p. 172.5–174°, b.p. 197°/769 mm, *mercurichloride*, 226° (decomp.), 6,6-*dioxide*, S-*dioxide*, m.p. 236–237°, *methiodide*, m.p. 162.5–163.5° (decomp.) (sealed tube); 8-**thiabicyclo**[3.2.1]**octane,** (V), m.p. 176.5–178.5° (sealed tube), b.p. 194.5°/769 mm, 8-*oxide*, S-*monoxide*, m.p. 269° (decomp.), 8,8-*dioxide*, S-*dioxide*, m.p. 281.5–282°, *mercurichloride*, m.p. 185–187° (decomp.) [226° (decomp.)], *methiodide*, m.p. 242.5–243° (decomp.) (sealed tube); and 2-**thiabicyclo**[2.2.2]**octane** (VI), m.p. 210–212° (155–157°), 2,2-*dioxide*, S-*dioxide*, m.p. ~310°, *methiodide*, m.p. 225–226° (decomp.) (sealed tube) have been synthesised and their physical properties recorded (*Birch et al.*, J. org. Chem., 1957, **22,** 1590; *J.-M. Surzur et al.*, Tetrahedron Letters, 1971, 2035). 2-**Thiabicyclo**[2.2.2]-**oct-5-ene,** m.p. 142–143°, has been synthesised by 1,4 addition of thiophosgene to 1,3-cyclohexadiene giving 3,3-dichloro-2-thiabicyclo[2.2.2]oct-5-ene followed by reduction with lithium tetrahydridoaluminate (*H. J. Reich* and *J. E. Trend*, J. org. Chem., 1973, **38,** 2637):

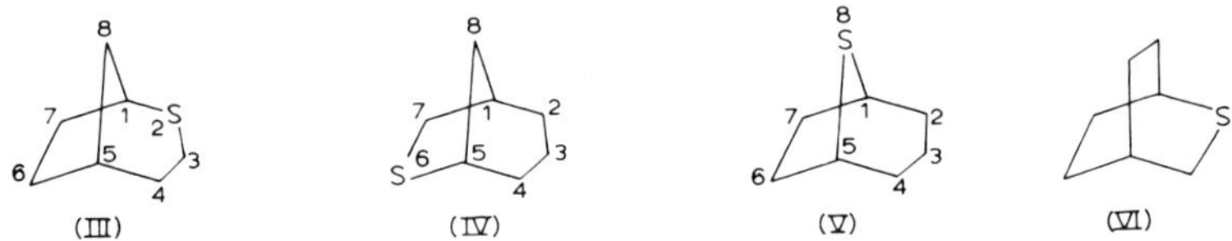

(III) (IV) (V) (VI)

8-Thiabicyclo[3.2.1]octane occurs in Iranian kerosine and with Raney nickel gives cycloheptane (*Birch et al.*, Ind. Eng. Chem., 1955, **47,** 240). In this kerosine 2-thiabicyclo[2.2.2]octane or 6-thiabicyclo[3.2.1]octane probably occur since methylcyclohexane is obtained among the desulphurisation products.

endo-4-*Bromo-6-thiabicyclo*[3,2.1]*octane* (VIII), b.p. 82°, and 6-*thiabicyclo*[3.2.1]-*oct-3-ene* (IX), b.p. 197–200°, n_D^{20} 1.5601, have been synthesised starting from 3-cyclohexenylmethyl[4]bromobenzene-sulphonate (VII) (*C. R. Johnson* and *F. L. Billman*, J. org. Chem., 1971, **36,** 855):

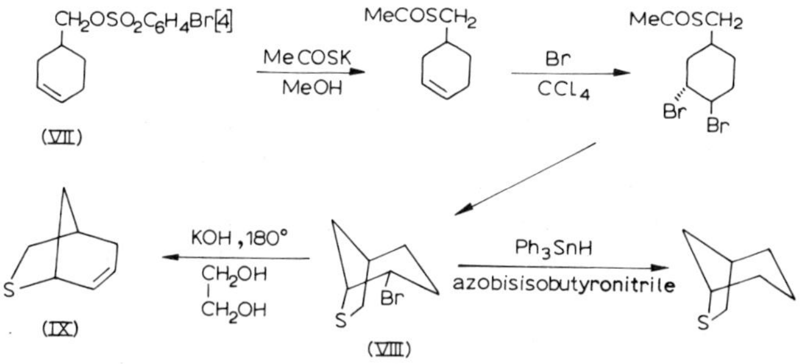

$CH_2OSO_2C_6H_4Br[4]$ (VII) $\xrightarrow[MeOH]{MeCOSK}$ $MeCOSCH_2$-cyclohexene $\xrightarrow[CCl_4]{Br}$ $MeCOSCH_2$-dibromocyclohexane → (VIII) $\xrightarrow[\text{azobisisobutyronitrile}]{Ph_3SnH}$ 6-thiabicyclo[3.2.1]octane

(VIII) $\xrightarrow[CH_2OH-CH_2OH]{KOH, 180°}$ (IX)

8-**Thiabicyclo**[3.2.1]**octan**-3-**one** has been prepared in excellent yield from tropinone methiodide and aqueous sodium sulphide.

Reduction of the ketone with sodium tetrahydridoborate leads to a mixture of stereoisomeric thia-alcohols, which on chromatography give the pure *endo*-isomer X, (33%), m.p. 238–239°, and pure exo-*isomer* XI (58%), m.p. 158–159° (*R. E. Ireland* and *H. A. Smith*, Chem. and Ind., 1959, 1252); other reducing agents, lithium tetrahydridoaluminate, aluminium isopropoxide and *tert*-butoxide afford a mixture of the same isomers (*V. Horak* and *J. Zavada*, Coll. Czech. chem. Comm., 1962, **27,** 1224):

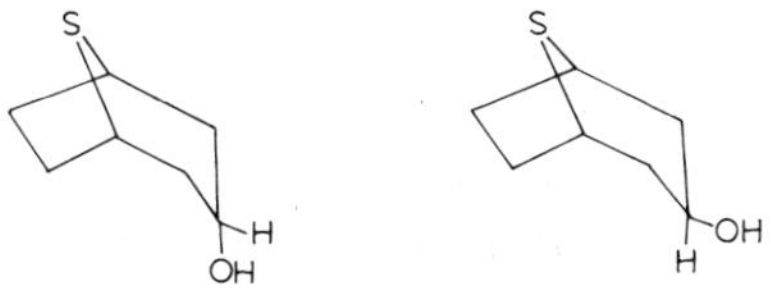

(X) (XI)

(ii) Thiabicyclo[3.3.0]nonane

Thermal extrusion of sulphur dioxide from 9-thiabicyclo[3.3.1]nonane 9,9-dioxide (XII) affords bicyclo[3.3.0]octane (XIII), as the major product. Photochemical extrusion of sulphur from 9-**thiabicyclo**[3.3.1]**nonane** (XIV) in trivalent organophosphorus solvents, *e.g.*, iso-octyl phosphite, gives an equimolar mixture of bicyclo[3.3.0]octane and cyclo-octene (XV) (*E. J. Corey* and *E. Block*, J. org. Chem., 1969, **34,** 1233).

710°

(XII) (XIII)

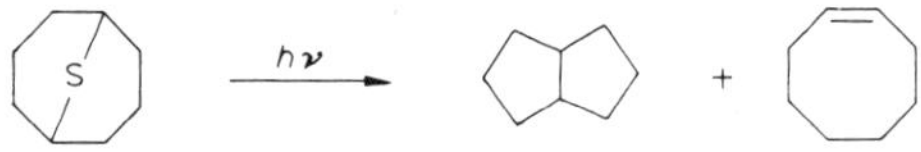

hν

(XIV) (XV)

The u.v. irradiation of 6-hydroxy-9-thiabicyclo[3.3.1]nonan-2-one (*C. Ganter* and *J. F. Moser*, Helv., 1969, **52,** 725), and 9-thiabicyclo[3.3.1]non-6-en-2-one (*A. Padwa* and *A. Battisti*, J. Amer. chem. Soc., 1971, **93,** 1304) have been studied; 2,6-dichloro-9-thiabicyclo[3.3.1]nonane is obtained by the transannular addition of sulphur dichloride to *cis, cis*-1,5-cyclo-octadiene (*F. Lautenschlaeger*, Canad. J. Chem., 1966, **44**, 2813).

(iii) Thiabicyclotridecenes

cis-13-**Thiabicyclo**[8.2.1]-5-**tridecenes**(XVI) or *trans*-13-**thiabicyclo**[7.3.1]-4-**tridecenes** (XVII) reacts with hydrogen chloride to give the perhydro-9b-thiaphenalenium halide XVIII; compound XVII with chlorine or bromine affords XVIII (*N. N. Novitskaya et al.*, Izvest. Akad. Nauk S.S.S.R., Ser. Khim., 1972, 2817):

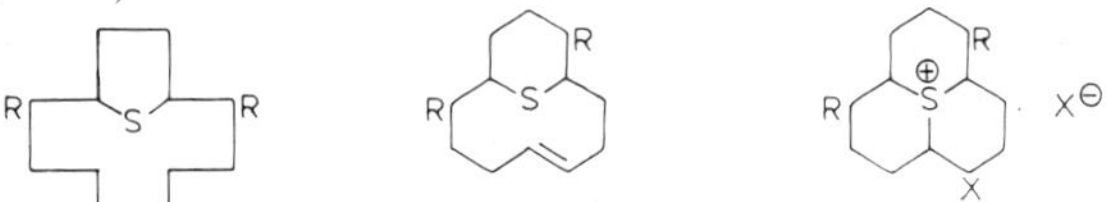

R = H, Cl, Br, OAc, OH   X = Cl, Br

(XVI) (XVII) (XVIII)

(iv) Thia-adamantane

2-**Thia-adamantane** (XIX), $C_9H_{14}S$, m.p. 320° (sealed tube), b.p. 227–228°/760 mm, *mercurichloride*, m.p. 186–188°, S-*dioxide*, sublimes at 340° (decomp.), is obtained from petroleum distillates and with Raney nickel yields bicyclo[3.3.1]nonane. Its similarity to adamantane (Vol. II C, p. 129) is shown by X-ray examination and i.r. spectra *(Birch et al., loc. cit.)*:

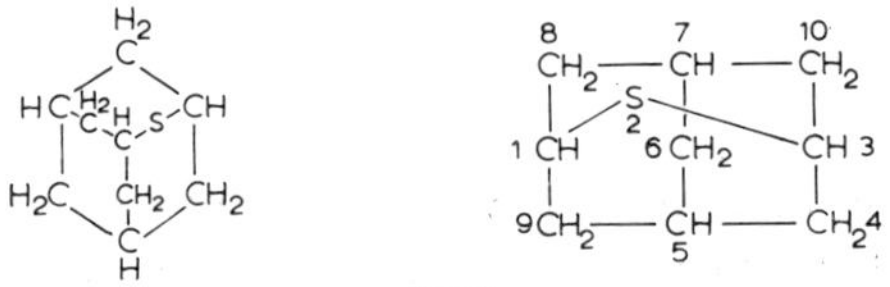

(XIX)

2-Thia-adamantane has been synthesised from bicyclo[3.3.1]nonane-2,6-dione and pyrrolidine or morpholine, which give 2,6-dipyrrolidino- or 2,6-dimorpholino-bicyclo[3.3.1]nona-2,7-diene. Either of the enamines with sulphur dichloride give 2-thia-adamantane-4,8-dione, which on Wolff–Kishner reduction yields 2-thia-adamantane (*H. Stetter* and *H. Held*, Angew. Chem., 1961, **73**, 114). For the synthesis of 2-thia-adamantane and derivatives see *Stetter* and *Held*, Ber., 1962, **95**, 1687, and for 4,8-dialkyl-2-thia-adamantane-4,8-diols, *S. Landa* and *J. Janku*, Coll. Czech. chem. Comm., 1969, **34**, 2014.

2-**Thia-adamantane 2-oxide** takes up deuterium in positions α to the sulphonyl group on deuteration with D_2O–K_2CO_3 or Bu^tOD–BuOK; 2-thia-adamantane is not deuterated under these conditions, confirming that the existence of the α-carbanion of the 2-oxide is due to Coulomb forces (*Janku* and *J. Mitera*, Z. Chem., 1970, **10,** 224). The heat of combustion of thia-adamantane has been measured (*J. L. Lacina*, *W. D. Good* and *J. P. McCullough*, J. phys. Chem., 1961, **65**, 1026).

(b) Perinaphthothiopyrans, thiaphenalenes

The naphthothiopyrans and some of their derivatives have been included in the sections containing the related benzothiopyrans. The peri compounds, for instance, naphtho[1,8-*bc*]thiopyran, 1-thiaphenalene (XXII) are given below.

Naphtho[1,8-*bc*]**thiopyran,** 1-**thiaphenalene** (XXII), orange cubes, m.p. 120–122°, obtained from 3-acetoxy-1-thiaphenalene (XX), which on combined hydrolysis and reduction using sodium tetrahydridoborate affords 2,3-dihydro-1-thiaphenalen-3-ol (XXI, R = OH). Phosphorus pentachloride converts the alcohol into the chloro compound (XXI, R = Cl), from which naphtho[1,8-*bc*]thiopyran is obtained by warm ethanolic potassium hydroxide (*S. O'Brien* and *D. C. Smith*, J. chem. Soc., 1963, 2907):

S OAc (XX) → S R (XXI) → S (XXII)

2,3,9-*Trimethylnaphtho*[1,8-bc]*thiopyran*, m.p. 76–77° (*J. F. Muller* and *P. Cagniant*, Compt. rend. Ser. C, 1967, **264,** 455). Naphtho[1,8-*bc*]thiopyran, its 3-acetoxy-derivative, and 2,3-dihydro-3-hydroxynaphtho[1,8-*b*]thiopyran upon nitration give mixtures of the corresponding 6- and 7-nitro-derivatives in the respective proportions 10:1, 10:1 and 1:8 (*C. C. Cook* and *F. K. Sutcliffe*, J. chem. Soc. C, 1968, 957), 6-*Methyl*-1,2,3,4-*tetrahydrobenzo*[e]*naphtho*[1,8-bc]*thiopyran* (XXIII), m.p. 103°, *picrate*, m.p. 192° (*Muller* and *Cagniant*, Compt. rend. Ser. C, 1968, **266,** 1072):

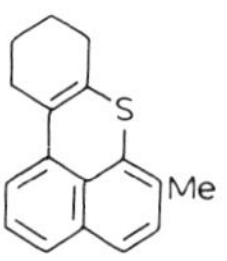

(XXIII) 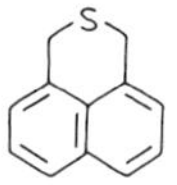 (XXIV)

1*H*,3*H*-**Naphtho**[1,8-*cd*]**thiopyran,** 1,3-*dihydro-2-thiaphenalene* (XXIV), m.p. 96–97°, 2-*oxide* (XXV), m.p. 236–241° (decomp.), obtained by the reaction of 1,8-bis(hydroxymethyl)naphthalene with phosphorus pentasulphide in carbon disulphide. Dehydration of the 2-oxide (XXV) in boiling acetic anhydride in the presence of *N*-phenylmaleimide affords a single adduct XXVII, m.p. 267–270°. The trapping of 2-thiaphenalene (XXVI) by appropriate dienophiles offers a simple new route to pleiadene derivatives for example, benzo[*l*] pleiadene-8,13-quinone (XXVIII) (*M. P. Cave, N. M. Pollack* and *D. A. Repella*, J. Amer. chem. Soc., 1967, **89**, 3640).

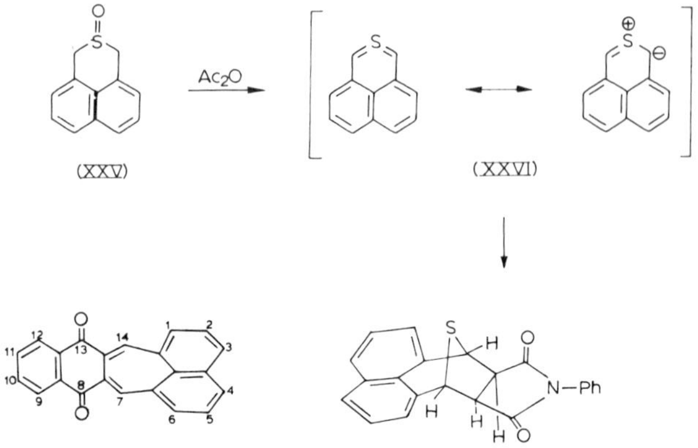

(XXV) (XXVI)

(XXVIII) (XXVII)

1,3-*Diphenyl*-1H,3H-*naphtho*[1,8-cd]*thiopyran*, 1,3-*dihydro*-1,3-*diphenylthiaphenalene*, m.p. 168°, cis-2-*oxide*, m.p. 247° (decomp.), trans-2-*oxide*, m.p. 250° (geometry indicated by n.m.r. spectrum) (*R. H. Schlessinger* and *A. G. Schultz*, *ibid.*, 1968, **90**, 1676). It has been found that sensitized irradiation of the 2-oxide (XXIX) gives 1-benzyl-8-benzoylnaphthalene (XXXI) by a unique photodesulphurisation process. This reaction proceeds from triplet-state sulphoxide to sulphine XXX, which then undergoes singlet-state collapse into ketone XXXI accompanied by loss of sulphur (*Schultz* and *Schlessinger*, Chem. Comm., 1970, 1051):

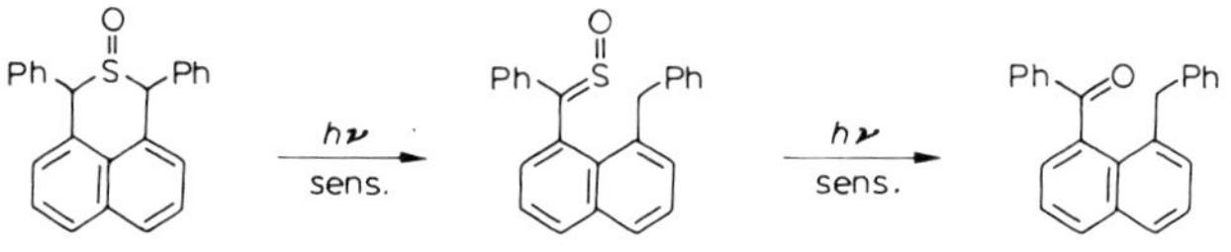

(XXIX) (XXX) (XXXI)

The crystal structure of 1*H*,3*H*-naphtho(1,8-*cd*)thiopyran has been established (*B. M. Lunden*, Acta Crystallogr. B, 1973, **29**, 1219).

Related highly reactive heterocycles **acenaphtho**[5,6-*cd*]**thiopyran** (XXXII) (*Schlessinger* and *I. S. Ponticello*, J. Amer. chem. Soc., 1967, **89**, 3641), and **acenaphthylo-**[5,6-*cd*]**thiopyran** (XXXIII) (*idem*, Tetrahedron Letters, 1967, 4057; Chem. Comm., 1969, 1214) have also been generated and their properties studied.

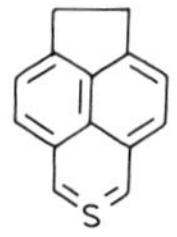

(XXXII) 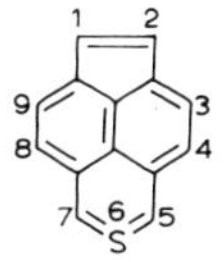 (XXXIII) 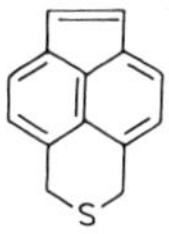 (XXXIV)

5H,7H-*Dihydroacenaphthylo*[5,6-cd]*thiopyran*, (XXXIV), m.p. 172°, 6-*oxide*, *sulphoxide*, m.p. 263° (decomp.), is prepared by treating 5,6-bis(chloromethyl)acenaphthylene with either tetraphosphorus decasulphide in carbon disulphide or anhydrous sodium sulphide in ethanol. The oxide is obtained by treating the thiopyran with sodium periodate in aqueous methanol.

(c) Cerothiene derivatives

Cerothiene, naphtho[3,2,1-*kl*]**thioxanthene** (XXXV, R = H), is known in the form of its 9-*phenyl* derivative (XXXV, R = Ph), m.p. 177–179° (*R. Painco*, Compt. rend., 1952, **234**, 852) and the oxo derivative (XL). 1-Phenylmercaptoanthraquinone (XXXVI) when heated with sulphuric acid yields the sulphonium salt XXXVII, which with water gives the tertiary *methanol base* XXXVIII, yellow prisms, m.p. 227° (decomp.), *ferrichloride*, $C_{20}H_{11}O\overset{\oplus}{S}FeCl_4^{\ominus}$, violet-black crystals, m.p. 227° (decomp.) and reduction affords 9-**hydroxycerothiene** (XXXIX) or the oxo form XL, which has a yellow fluorescence in organic solvents (*H. Decker*, Ann., 1906, **348**, 238):

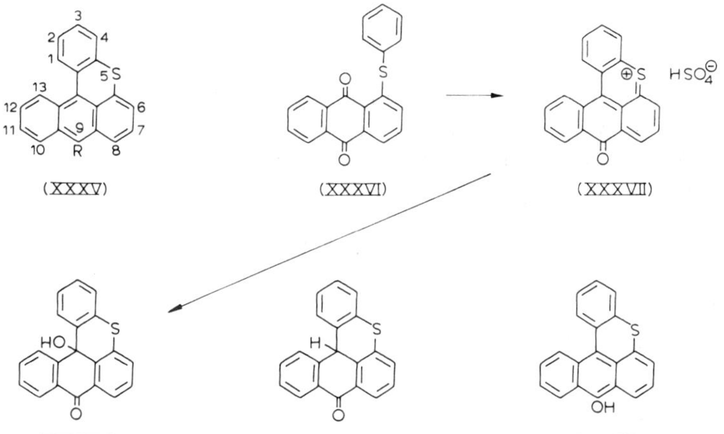

(XXXV) (XXXVI) (XXXVII)

(XXXVIII) (XL) (XXXIX)

2-(2-Carboxybenzoyl)-4,4′-ditolyl sulphide when heated with sulphuric acid yields 2,7-**dimethylthiofluoran** (XLI), m.p. 220–221° (228–230°), red colour in sulphuric acid (*R. Weiss* and *W. Knapp*, Monatsh., 1928, **50,** 392; *F. Krollpfeiffer* and *A. Wissner*, Ann., 1951, **572,** 195). The fluoran with sodium amalgam and ethanol yields 2,7-*dimethyl*-9-(2-*carboxyphenyl*)*thioxanthene*, m.p. 192–193°, which is cyclised by phosphorus pentoxide in benzene to 2,8-*dimethyl*-9-*hydroxycerothiene* (XLII), m.p. 188–189°.

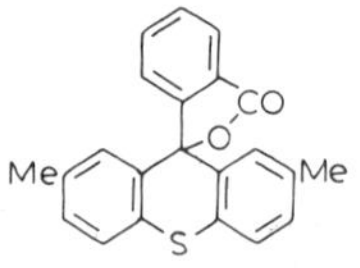

(XLI) 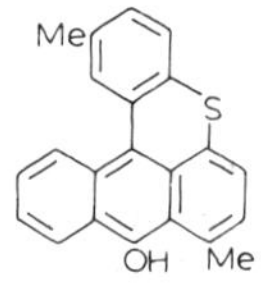 (XLII)

The product obtained by the interaction of benzenethiol potassium hydroxide, and 1,5-dinitroanthraquinone in ethanol is cyclised by a mixture of phosphoric and sulphuric acids to a sulphonium salt, which can be isolated as the *diferrichloride* $(C_{26}H_{14}S_2)^{2\oplus}2FeCl_4^{\ominus}$ m.p. 258–260°, yields the colourless *pseudo-base* XLIII, m.p. 248°, when treated with water, and when reduced in acid solution gives the red cerdithiene 8,16-**dithiadibenzo**[a,j]-**perylene,** 3,9-**dithia**(1,2;7,8)**dibenzoperylene** (XLIV), $C_{26}H_{14}S_2$, (*Decker* and *T. von Fellenberg*, *ibid*., 1907, **356,** 337):

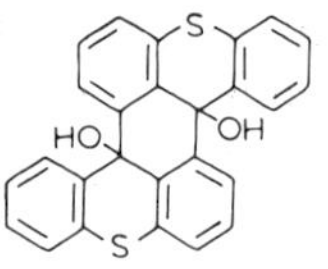

(XLIII) 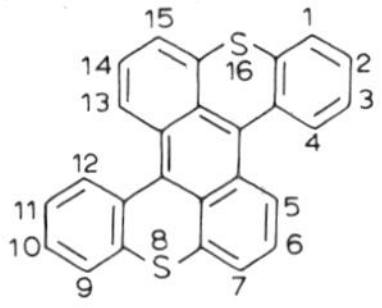 (XLIV)

Thiofluorenone reacts with anthracene, cyclopentadiene, and tetrachloro-1,2-benzoquinone at room temperature to give the Diels–Alder adducts (XLV, XLVI and XLVII), respectively (*A. Schönberg* and *B. König*, Ber., 1968, **101,** 725):

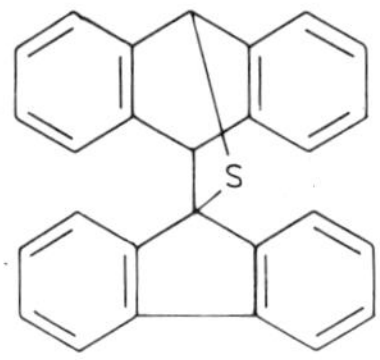

(XLV) 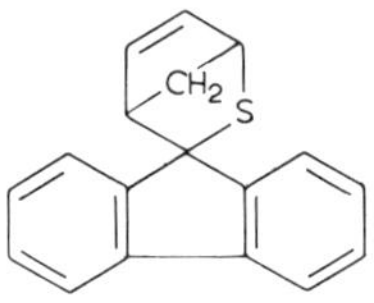 (XLVI) 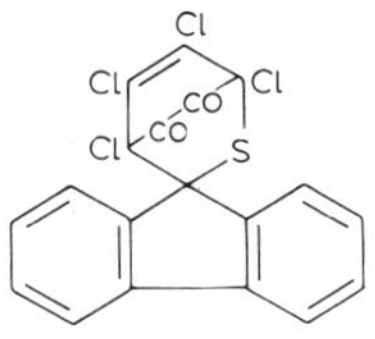 (XLVII)

6. Selenopyrans and related compounds

In recent years interest has been shown in the chemistry of organoselenium compounds* particularly in the fields of biochemistry, physiology, toxicology and botany.

(a) Selenopyrans and derivatives

4*H*-**Selenopyran** (I), an unstable liquid, b.p. 55–60°/20 mm, is prepared by treating glutaraldehyde in methylene dichloride at −10° to 0° with dry hydrogen chloride and hydrogen selenide and then with diethylaniline at 130–170° (*I. Degani, R. Fochi* and *C. Vincenzi*, Gazz., 1964, **94,** 203):

H_2Se, HCl, -10 to 0°; PhNEt$_2$, 130-170°

(I)

4-**Methyl**-4*H*-**selenopyran** (*Degani* and *Vincenzi*, Boll. sci. Fac. Chim. ind. Bologna, 1967, **25,** 51): n.m.r. spectra of selenopyrans (*Degani et al., ibid.*, 1965, **23,** 131; 1967, **25,** 61). **Selenopyrylium perchlorate,** *seleninium perchlorate*, explodes at 228°, the *bromide* is unstable and *bis(selenopyrylium)platinum hexachloride*, decomposes at 210° (*Degani, Fochi* and *Vincenzi, loc. cit.;* Boll. sci. Fac. chim. ind. Bologna, 1965, **23,** 21); 4-*methyl*-4H-*selenopyrylium perchlorate* (II) m.p. 90.5–91.5°, (*chloroplatinate*, decomposes at 170°) is obtained by treating 4-methylselenopyran with triphenylmethyl perchlorate (*Degani* and *Vincenzi, loc. cit.*):

Ph_3CClO_4 → $ClO_4^{\ominus}$

(II)

Selenopyrylium salts (III, R = Ph, [4]MeC$_6$H$_4$, [4]MeOC$_6$H$_4$, R^1 = H, Me, Ph) have been prepared by cyclisation of 1,5-diketones of the types RCO·CH$_2$·CHR^1·CH$_2$·COR with hydrogen chloride – hydrogen selenide, followed by treatment with perchloric acid in acetic acid (*M. A. Kudinova, S. V. Krivun* and *A. I. Tolmachev*, Khim. geterotsikl. Soedin., 1973, 857):

$ClO_4^{\ominus}$

(III)

For u.v. (*Degani* and *Vincenzi*, Boll. sci. Fac. Chim. ind. Bologna, 1965, **23,** 249; *J. Fabian, A. Mehlhorn* and *R. Zahradnik*, Theoret. Chim. Acta, 1968, **12,** 247), and

* "Organic Selenium Compounds", *D. L. Klayman* and *W. H. H. Günther*, Wiley-Interscience, John Wiley and Sons, Inc., 1973.

n.m.r. spectra of selenopyrylium salts see *Degani et al., loc. cit.*

2,3,5,6-*Tetraphenyl*-4H-*selenopyran*-4-*one* (V) m.p. 288–292°, is prepared by the photolysis of the di-iron carbonyl complex IV in the presence of potassium polyselenide (*K. W. Hubel* and *E. H. Braye*, U.S. P., 3,280,017/1966):

$$\text{(IV)} \xrightarrow[h\nu]{K_2Se_x} \text{(V)}$$

Selenane, *tetrahydroselenopyran* (VII), b.p. 158°/759 mm, n_D^{18} 1.5475, d_4^{20} 1.399, *mercurichloride*, m.p. 175–176°, 1,1-*dichloride*, m.p. 103°, 1,1-*dibromide*, m.p. 117–118° (a number of other derivatives are reported), is prepared (*1*) by the reaction between 1,5-dibromopentane (VI) and sodium selenide in ethanol (*G. T. Morgan* and *F. H. Burstall*, J. chem. Soc., 1929, 2197):

$$\text{(VI)} \longrightarrow \text{(VII)}$$

(*2*) by the addition of 1,5-dibromopentane to a mixture obtained by boiling sodium formaldehydesulphoxylate, powdered selenium, sodium hydroxide and water (*J. D. McCullough* and *A. Lefohn*, Inorg. Chem., 1966, **5**, 150); and (*3*) by passing oxane and hydrogen selenide over alumina at 400° (*Yu. K. Yur'ev*, J. gen. Chem. U.S.S.R., 1946, **16,** 851). The dissociation of the iodine complex of selenane ($C_5H_{10}SeI_2$) has been studied spectrophotometrically (*McCullough* and *A. Brunner*, Inorg. Chem., 1967, **6,** 1251) and the structures of the complexes between selenane and bromine or iodine have been determined in solution (*J. B. Lambert et al.*, J. Amer. chem. Soc., 1972, **94,** 8172). The axial preference of the 1-oxide, the proton in 1-protonated selenane, and the methyl group in 1-methylselenanium iodide has been demonstrated by n.m.r. spectroscopy (*Lambert*, *C. E. Mixan* and *D. H. Johnson*, Tetrahedron Letters, 1972, 4335; J. Amer. chem. Soc., 1973, **95,** 4634).

2-*Methylselenane*, 2-*methyltetrahydroselenopyran*, b.p. 90°/65 mm, 169–171°/764 mm, n_D^{18} 1.5205, d_4^{22} 1.287, *mercurichloride*, m.p. 112° (*Morgan* and *Burstall*, J. chem. Soc., 1931, 173); *selenane*-2,6-*dicarboxylic acid*, *tetrahydroselenopyran*-2,6-*dicarboxylic acid*, meso-cis-, m.p. 185–186.5°, trans-(±)-, m.p. 174–175°, trans(+)-, m.p. 146.5–148.5°, $[\alpha]_D^{25}$ 112° (acetone), 139.2° (ethanol), trans-(−)-, m.p. 146.5–148.5°, $[\alpha]_D^{25}$ −112° (acetone) (*A. Fredga* and *K. Styrman*, Arkiv. Kemi., 1959, **14,** 461); *diethyl* 2,6-*dimethylselenan*-4-*one*-3, 5-*dicarboxylate*, *diethyl* 1-*selena*-2,6-*dimethylcyclohexan*-4-*one*-3, 5-*dicarboxylate* (VIII), m.p. 129°, is obtained in ~10% yield from the reaction between acetaldehyde, diethyl 3-oxoglutarate and hydrogen selenide in the presence of 1,4-diazabicyclo[2.2.2]octane in ethanol at —10°; n.m.r. spectroscopy shows that the Me groups are *cis* to each other as are the CO_2Et groups and that the chair form is the preferred conformation with substituents at $C_{(2)}$, $C_{(3)}$, $C_{(5)}$ and $C_{(6)}$ equatorial (*W. Hänsel* and *R. Haller*, Naturwiss., 1968, **55,** 83; Arch. Pharm., 1969, **302,** 147):

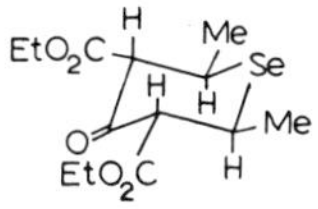

(VIII) 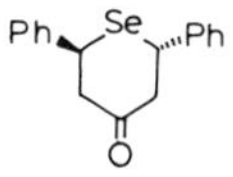 (IX)

trans-2,6-*Diphenylselenan*-4-*one* (IX) has been prepared in 82% yield by the reaction between distyryl ether, $(PhCH{:}CH)_2O$, and hydrogen selenide in the presence of 1,4-diazabicyclo[2.2.2]octane. Its configuration has been established by chemical methods and n.m.r. spectra (*W. Ziriakus*, *Hänsel* and *Haller*, *ibid.*, 1971, **304,** 681).

(b) Selenopyrans containing fused rings

(i) Benzoselenopyrans and derivatives

2*H*-**Benzo**[*b*]**selenopyran, selenochrom**-3-**ene** (X), b.p. 71°/0.01 mm, is obtained by the dehydration of selenochroman-4-ol using potassium hydrogen sulphate (*Degani*, *Fochi* and *Vincenzi*, Gazz., 1964, **94,** 451) alumina, or phosphorus pentoxide in boiling xylene (*M. Renson*, Bull. Soc. chim. Belge., 1964, **73,** 483):

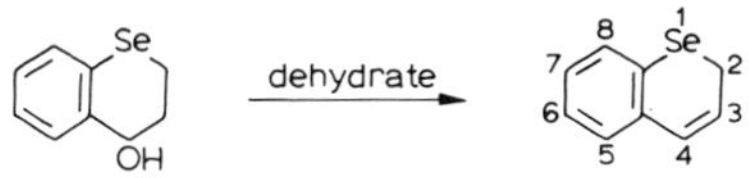

dehydrate

(X)

Selenochromenes are photochromic; the effect of the selenium atom on the coloured isomer absorption band has been determined by comparison with chromene, thiochromene and dihydroquinoline (*B. S. Luk'yanov et al.*, Tetrahedron Letters, 1973, 2007). Oxidation of selenochromenes using chromic oxide–pyridine gives small amounts of *o,o'*-biscinnamaldehyde diselenide (XI) (*A. Ruwet*, *J. Meessen* and *Renson*, Bull. Soc. chim. Belge., 1969, **78,** 459):

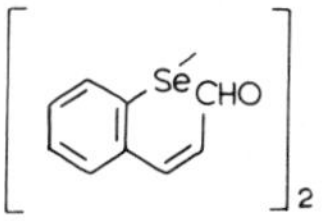

(XI)

Some selenochrom-3-enes are listed in Table 3.

TABLE 3

SELENOCHROM-3-ENES

Substituents	*M.p. (°C)*	*B.p. (°/mm)*	*Ref.*
2-Me		107–108/1.5	1
3-Me		118–119/2	1
6-Me		135–140/1.5	1
7-Me		118/2	1
8-Me		135–140/0.8	1
4-Ph	85		2
2-Me-4-Ph		135–138/1.5	1
3-Me-4-Ph	68–70		1
6-Me-4-Ph	75–76		1
7-Me-4-Ph	58–59	165/3	1
8-Me-4-Ph	79	175–180/1.5	1

References
1 *J. M. Danze* and *M. Renson*, Bull. Soc. chim. Belge., 1966, **75**, 169.
2 *Renson*, *ibid.*, 1964, **73**, 483.

Treatment of selenochrom-3-ene in acetonitrile with triphenylmethyl perchlorate yields **benzoselenopyrylium perchlorate,** *selenochromylium perchlorate*, *benzoseleninium perchlorate* (XII), decomp. 132° *(Degani*, *Fochi* and *Vincenzi*, *loc. cit.)*:

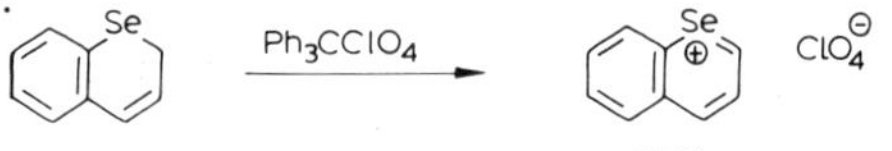

Ph_3CClO_4 → $ClO_4^{\ominus}$

(XII)

The selenochrom-3-enes listed in Table 3 afford benzoselenopyrylium perchlorates which decompose at around 120°; some selenochromones have been converted by a series of reactions to the corresponding benzoselenopyryliums *(Danze* and *Renson*, *loc. cit.)*. The relative stability of benzoselenopyrylium and related derivatives have been compared in liquid sulphur dioxide, acetonitrile and nitromethane (*Degani*, *Fochi*, and *Vincenzi*, Boll. sci. Fac. Chim. ind. Bologna, 1965, **23**, 21), and the hydrolysis of their perchlorates at various pH's has been studied (*Degani*, *Fochi* and *G. Spunta*, *ibid.*, p. 151). Benzoselenopyrylium perchlorate is oxidised by manganese dioxide in either boiling chloroform or acetonitrile to give benzo[*b*]selenophene-2-carbaldehyde in high yield (*Degani* and *Fochi*, Ann. Chim., Rome, 1968, **58**, 251; *Ruwet*, *Meessen* and *Renson*, *loc. cit.)*. Nucleophilic attack by, *e.g.*, benzeneselenol and benzenethiol on benzoselenopyrylium perchlorate occurs at $C_{(2)}$ to give the selenochromene XIII (*A. Tadino*, *L. Christiaens* and *Renson*, Bull. Soc. roy. Sci. Liege, 1973, **42**, 146):

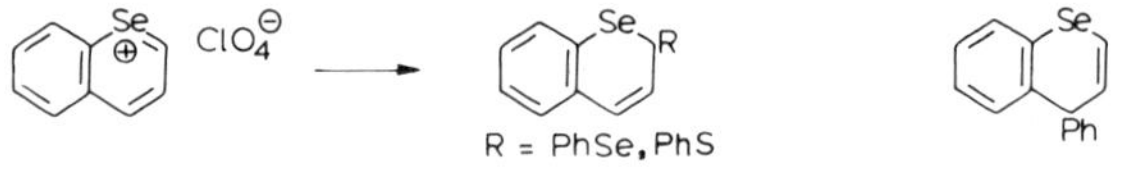

$ClO_4^{\ominus}$ →

R = PhSe, PhS

(XIII) (XIV)

Treatment of benzoselenopyrylium perchlorate with phenylmagnesium bromide gives 2-phenyl-2*H*-selenochromene (XIII, R = Ph) and 4-phenyl-4*H*-selenochromene (XIV). 2- or 4-Arylbenzoselenopyrylium cations react with nucleophiles only in the unsubstituted reactive position, independent of the nature of the nucleophile. Electron-donor or electron-withdrawing groups on the aryl substituent do not effect the reaction (*idem, ibid.*, p. 129).

4-Methyl-2-phenylbenzoselenopyrylium perchlorate, 4-methylselenoflavylium perchlorate (*A. I. Tolmachev* and *M. A. Kudinova*, Khim. geterotsikl. Soedin., 1971, **7,** 1177).

Some 2*H*-benzoselenopyran-2-ones, selenocoumarins, have been prepared by the cyclisation of 2-(methylseleno)cinnamic acids using either phosphoryl chloride or aluminium chloride in carbon disulphide, or by adding a selenocinnamic acid chloride to pyridine with cooling and then boiling the mixture (*Ruwet* and *Renson*, Bull. Soc. chim. Belge., 1968, **77**, 465; 1969, **78**, 449); **3-methylselenocoumarin** (XV), m.p. 80°.

Se O Me

(XV)

4-Hydroxyselenocoumarin (XVI, R = H), m.p. 202°, has been prepared from malonyl dichloride and benzeneselenol (*E. Ziegler* and *E. Nölken*, Monatsh., 1958, **89,** 737):

$CH_2(COCl)_2$ + 2 PhSeH $\xrightarrow{\Delta}$ $CH_2(COSePh)_2$ $\xrightarrow[\text{2. dil. HCl}]{\text{1. } AlCl_3\text{, 170°}}$ (XVI)

Se O R OH

(XVI)

Some 4-hydroxyselenocoumarins (XVI, R = Et, Pr, Bu, Ph) have been prepared by treating selenophenols with a substituted malonic acid in polyphosphoric acid (*Ruwet, C. Draguet* and *Renson*, Bull. Soc. chim. Belge., 1970, **79,** 539). For forms of 4-hydroxyselenocoumarin see *N. P. Buu-Hoï et al.*, J. Heterocycl. Chem., 1969, **6,** 825.

4*H*-Benzoselenopyran-4-one, selenochromone (XVII) m.p. 92°, is obtained by oxidising selenochroman-4-one with chloranil in boiling xylene (*Renson, loc. cit.*), or by treating selenochroman-4-one with triphenylmethyl perchlorate and then with sodium hydrogen carbonate (*Degani, Fochi* and *Vincenzi*, Gazz., 1964. **94,** 451):

Se O (XVII) Se Ph O (XVIII) Se Ph O (XIX)

A number of selenochromones and selenoflavones, 2-phenylselenochromones, 2-phenyl-4*H*-benzoselenopyran-4-ones, have been prepared by reacting the appropriate areneselenol with an equivalent amount of an acylacetic ester in 85% phosphoric acid (*F. Bossert*, Angew. Chem., 1965, **77,** 913; *Ruwet* and *Renson*, Bull. Soc. chim. Belge., 1966, **75,** 260); 2-*methyl-*, m.p. 99–100°; 2,3-*dimethyl-*, m.p. 117°; 2,6-*dimethyl-*, m.p. 98°; 2,7-*dimethyl-*, m.p. 89–90°; 2,8-*dimethyl-*, m.p. 112–113°; 2,3,6-*trimethyl-*, m.p. 111°; 2,3,7-*trimethyl*-, m.p. 125°; and 2,3,8-*trimethyl-selenochromone*, m.p. 115°; 2-**phenylselenochromone, selenoflavone** (XVIII), m.p. 133°; 6-*methyl*-, m.p. 138° 7-*methyl*-, m.p. 126°; 8-*methyl-selenoflavone*, m.p. 103°; 3-**phenylselenochromone, seleno-isoflavone** (XIX) m.p. 133°. For i.r. spectra of selenocoumarins and selenochromones see *Ruwet* and *Renson*, *ibid*., 1970, **79**, 89, and for n.m.r. spectra of selenochromones see *Bossert*, *loc. cit*.

2,3-**Dihydro**-4*H*-**benzo**[*b*]**selenopyran, selenochroman** (XX), b.p. 108–110°/0.7 mm, obtained by Clemmensen reduction of selenochroman-4-one (*Renson*, *ibid*., 1964, **73,** 483). One of the products, m.p. 47–48°, formed in 25% yield when phenyl 2-propenyl selenide, $PhSeCH_2 \cdot CH{=}CH_2$, is boiled with an equimolar amount of quinoline in dry nitrogen is probably selenochroman (*G. A. Chmutova*, Sb. Aspir. Rab., Kazan. Gos. Univ. Khim., Goel., 1967, 70; C.A., 1969, **70,** 87444):

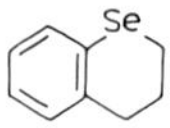

(XX)

A number of selenochromans have been prepared by the reductive cyclisation of the appropriate β-arylselenopropionic acid using Clemmensen, Wolff–Kishner–Minlon methods or sodium tetrahydridoborate (*N. Bellinger et al*., Bull. Soc. chim. Fr., 1971, 2689).

2,3-**Dihydro**-4*H*-**benzo**[*b*]**selenopyran**-4-**one, selenochroman**-4-**one,** (XXI), m.p. 38°, b.p. 114°/0.8 mm, 2,4-*dinitrophenylhydrazone*, m.p. 235° *oxime*, m.p. 83°, is prepared by cyclisation of β-phenylselenopropionic acid with polyphosphoric acid (*Renson*, *loc. cit.*):

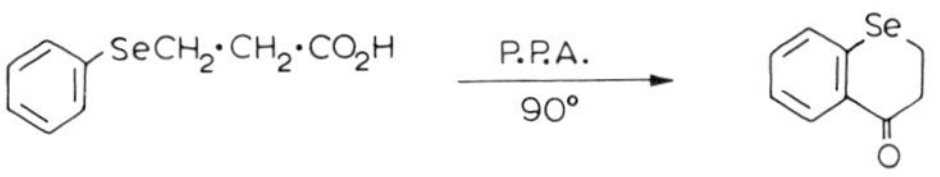

$$C_6H_5SeCH_2 \cdot CH_2 \cdot CO_2H \xrightarrow[90°]{P.P.A.} \text{(XXI)}$$

(XXI)

Condensation with benzaldehyde yields 3-*benzylidene-selenochroman*-4-*one*, m.p. 101°. For forms of selenochroman-4-one see *Buu-Hoï*, *loc. cit*..

The Wolff–Kishner–Minlon reduction of selenochroman-4-one gives selenochroman and 2,3-dihydro-2-methylbenzo[*b*]selenophene (*Bellinger*, *D. Cagniant* and *P. Cagniant*, Tetrahedron Letters, 1971, 49).

Selenochroman-4-*ol*, m.p. 90°, is obtained by reducing selenochroman-4-one with sodium tetrahydridoborate (*Degani*, *Fochi* and *Vincenzi*, *loc. cit.*), or with lithium tetrahydridoaluminate. On treatment with phenylmagnesium bromide selenochroman-4-one affords 4-*phenylselenochroman*-4-*ol*, m.p. 89°; 4-*aminoselenochroman hydrochloride*, m.p. 151° (*Renson*, *loc. cit.*).

2,3-**Dihydro**-2-**phenyl**-4*H*-**benzo**[*b*]**selenopyran**-4-**one, selenoflavanone** (XXII), needles, m.p. 72°, *dibromide*, yellow leaflets, m.p. 117–118°, prepared by boiling 2-cinnamoyl phenyl methyl selenide, [2]$PhCH{:}CH \cdot CO \cdot C_6H_4SeMe$, in a saturated solution of hydrogen bromide in acetic acid (*J. Gosselck* and *E. Wolters*, Ber., 1962, **95,** 1237):

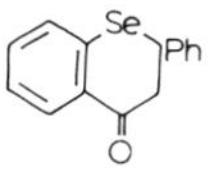

(XXII) 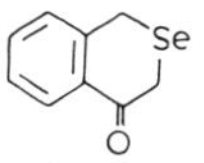 (XXIII) 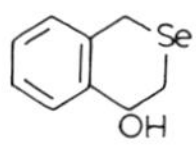 (XXIV)

3,4-**Dihydro**-1*H*-**benzo**[*c*]**selenopyran-4-one, isoselenochroman-4-one** (XXIII), m.p. 67–68°, 2,4-*dinitrophenylhydrazone*, m.p. 212°, is prepared by boiling 2-carboxybenzylselenoacetic acid, [2]$HO_2C \cdot C_6H_4 \cdot CH_2SeCH_2 \cdot CO_2H$ and potassium acetate in acetic anhydride and hydrolysing the resulting product in the cold with 10% alcoholic potassium hydroxide. Isoselenochroman-4-one on reduction with sodium tetrahydridoborate in methanol yields *isoselenochroman-4-ol* (XXIV), m.p. 103°, which on distilling from potassium hydrogen sulphate at 1 mm gives isoselenochrom-3-ene (XXV):

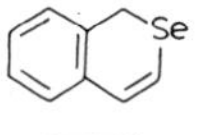

(XXV) 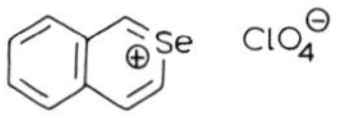 (XXVI)

Treatment is isoselenochrom-3-ene in ether with sulphuryl chloride and then 70% perchloric acid gives *benzo*[c]*selenopyrylium perchlorate* (XXVI), m.p. 101° (*Renson* and *P. Pirson*, Bull. Soc. chim. Belge., 1966, **75**, 456).

(ii) Dibenzoselenopyran, selenoxanthene and derivatives

Selenoxanthene (XXVII), needles, m.p. 145°, obtained in nearly 100% yield by heating selenoxanthone and red phosphorus slowly to 180° with fuming hydriodic acid in a sealed tube;

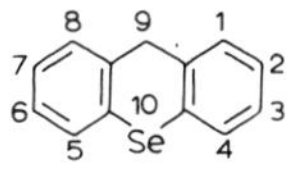

(XXVII)

it sublimes readily without decomposition (*B. R. Muth*, Ber., 1960, **93**, 283). 3,6-*Bis(dimethylamino)selenoxanthene* (XXIX), m.p. 161–162°, has been prepared by the reduction of 3,6-*bis(dimethylamino)selenoxanthylium chloride–zinc chloride double salt* Oxidation of the zinc chloride double salt with alkaline permanganate gives the corresponding selenoxanthone XXX (*R. H. Nealey* and *J. S. Driscoll*, J. heterocycl. Chem., 1966, **3**, 228):

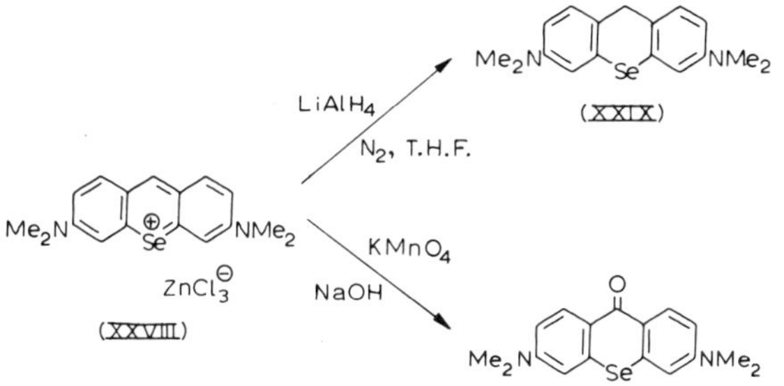

$LiAlH_4$, N_2, T.H.F. → (XXIX)

(XXVIII) $ZnCl_3^{\ominus}$

$KMnO_4$, NaOH → (XXX)

9-(3-Dimethylaminopropyl)selenoxanthone possesses antihistamine and antireserpine activity and is thus a potential antiallergic and antidepressant drug (*M. Protiva et al.*, Czech. Pat., 143, 236 1972). 9-Phenylselenoxanthyl and its analogues have been studied by e.s.r. techniques, and the odd-electron distribution in the radicals determined (*K. Maruyama et al.*, Bull. chem. Soc. Japan, 1970, **43**, 152).

Selenoxanthydrols of type XXXII, obtained by reducing the appropriate selenoxanthone XXXI with sodium tetrahydridoborate, give on treatment with 70% perchloric acid at $-40°$ the corresponding selenoxanthylium perchlorate XXXIII. Reduction of the salt XXXIII with lithium tetrahydridoaluminate yields the selenoxanthene XXXIV and treatment with a Grignard reagent gives the related 9-substituted selenoxanthene (*M. Renson* and *L. Christiaens*, Bull. Soc. chim. Belge., 1970, **79**, 511):

R = H, Me (XXXI) → (XXXII) —70% $HClO_4$, −40°→ $ClO_4^{\ominus}$ (XXXIII)

(XXXI) —R^1MgX→ R^1 = Me, Ph (XXXV) —70% $HClO_4$, −40°→ $ClO_4^{\ominus}$ (XXXVI)

(XXXIII) —$LiAlH_4$→ (XXXIV)

Treatment of the selenoxanthone XXXI with methyl- or phenyl-magnesium halide affords the 9-methyl- and 9-phenyl-selenoxanthydrol XXXV, respectively, which in turn could be converted into the corresponding selenoxanthylium perchlorate XXXVI. Reduction of the latter with lithium tetrahydridoaluminate gives the 9-substituted selenoxanthene, whereas treatment with a Grignard reagent gives the 9,9-disubstituted selenoxanthene.

9-Methylselenoxanthylium perchlorate (XXXVII) on treatment with excess of phenyllithium gives 13% of 9,9′-dimethylbiselenoxanthyl (XXXVIII) and 21% of 9-methyl-10-phenyl-10-selenanthracene (XXXIX, R = Me). With 9-phenylselenoxanthylium perchlorate only one product, 9,10-diphenyl-10-selenanthracene (XXXIX, R = Ph) is obtained in 28% yield (*M. Hori et al.*, Chem. Letters, 1973, 391):

$ClO_4^{\ominus}$ (XXXVII) —PhLi→ (XXXVIII) + (XXXIX)

Selenoxanthen-9-one, selenoxanthone (XLI) m.p. 189.5–190.5°, is prepared by cyclising 2-(phenylseleno) benzoic acid (XL) with sulphuric acid at 95° (*K. Šinderlár et al.*, Coll. Czech. chem. Comm., 1969, **34,** 3792):

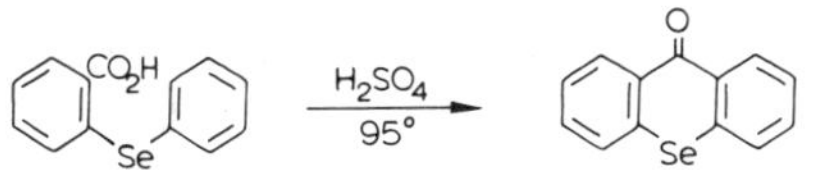

H_2SO_4, 95°

(XL) (XLI)

2-*Chloroselenoxanthone* m.p. 156°, and a number of derivatives have been reported. Selenoxanthone may be obtained in 88% and 99% yield by boiling selenoxanthylium perchlorate and manganese dioxide in chloroform and acetonitrile, respectively (*T. Degani* and *R. Fochi*, Ann. Chem., Rome, 1968, **58,** 251): for u.v. and pK_B data see *Degani, Fochi* and *G. Spunta*, Boll. sci. Fac. Chim. ind. Bologna, 1968, **26,** 3.

E.s.r. studies of the electronic distribution on monomeric metal ketyls of xanthone, thioxanthone and selenoxanthone, obtained by the reduction of the ketones with alkali or alkali earth metals, indicates that the covalency of the bond between the carbonyl O atom and the metal increases in the order xanthone, thioxanthone and selenoxanthone ketyl (*Maruyama et al.*, Rev. Phys. Chem. Japan, 1969, **39,** 117). For the structure of metal ketyls derived from selenoxanthone see *idem, ibid.*, p. 123. 3,6-*Bis(dimethylamino)-selenoxanthone*, m.p. 261° (*Nealey* and *Driscoll, loc. cit.*).

6*H*-**Dibenzo**[*b,d*]**selenopyran** (XLII), m.p. 98–99, *dibenzo*[b,d]*selenopyrylium perchlorate* (XLIII) m.p. 85°, decomposes rapidly in air (*Degani, Fochi* and *Spunta*, Boll. sci. Fac. Chim. ind. Bologna, 1965, **23,** 151).

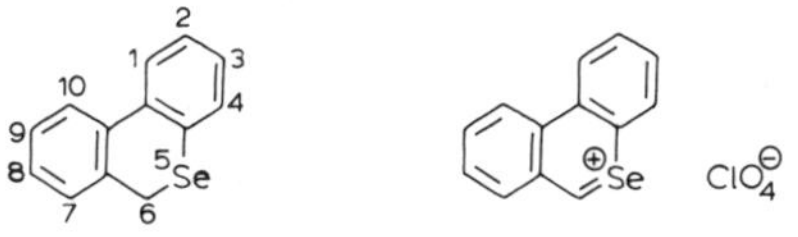

(XLII) (XLIII)

(iii) Naphthoselenopyrans

Treatment of 3-(2′-naphthylseleno)proprionic acid (XLIV) with polyphosphoric acid, or of its chloride with stannic chloride gives 2,3-dihydro-1*H*-naphtho[2,1-*b*]selenopyran-1-one (XLV):

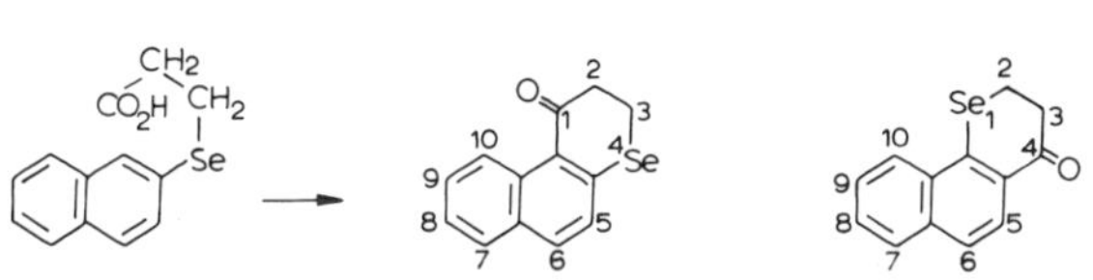

(XLIV) (XLV) (XLVI)

2,3-Dihydro-4*H*-naphtho[1,2-*b*]selenopyran-4-one (XLVI) is prepared in a similar manner. Reduction of the ketones XLV and XLVI with sodium tetrahydridoborate yields 2,3 dihydro-1*H*-naphtho[2,1-*b*]selenopyran-1-ol and 2,3-dihydro-4*H*-naphtho-[1,2-*b*]selenopyran-4-ol, which on dehydration by distillation from polyphosphoric acid afford 3*H*-naphtho[2,1-*b*]selenopyran (XLVII) and 2*H*-naphtho[1,2-*b*]seleno-pyran (XLVIII), respectively. Wolff–Kishner–Minlon reduction of ketones XLV and XLVI yield the corresponding dihydronaphthoselenopyrans. The u.v., i.r. and n.m.r. spectra of the above compounds have been discussed (*N. Bellinger* and *D. Cagniant*, Bull. Soc. chim. Fr., 1971, 2699):

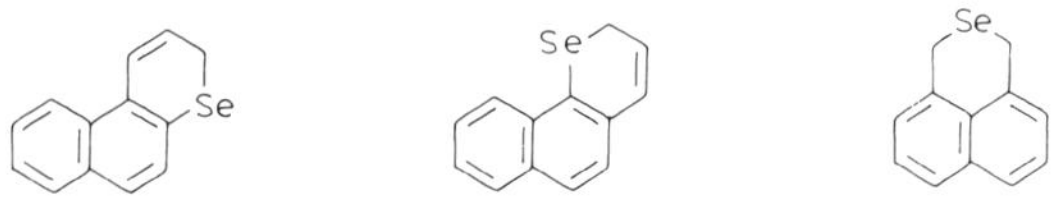

(XLVII) (XLVIII) (XLIX)

1*H*,3*H*-Naphtho[1,8-*cd*]selenopyran (XLIX) has been prepared and its n.m.r. spectra have been studied (*A. Biezais-Zirnis* and *A. Fredga*, Acta Chem. Scand., 1971, **25**, 1171).

7. Tellurane, tetrahydrotelluropyran derivatives and related compounds

Amorphous tellurium reacts with pentamethylene diiodide at 150° to give **tellurane 1,1-diiodide,** *tetrahydrotelluropyran* 1,1-*diiodide, cyclotelluropentane diiodide, telluracyclohexane diiodide, tetrahydro*-2H-*tellurin* 1,1-*diiodide* (I), m.p. 135–136°; the 1,1-*dibromide*, m.p. 106–107°, is prepared by treating with bromine the product obtained from a solution of sodium telluride, formed from finely powdered crystalline tellurium, sodium formaldehyde-sulphoxylate and sodium hydroxide in water and pentamethylene dichloride; the 1,1-*dichloride*, m.p. 105–106°, is prepared as described above but chlorine used in place of bromine (*W. V. Farrer* and *J. M. Gulland*, J. chem. Soc., 1945, 11).

Tellurane, *cyclotelluropentane, tetrahydrotelluropyran, tetrahydro*-2H-*tellurin* (II), is a lemon-yellow oil, b.p. 82–83°/12 mm, 44–45°/1–2 mm, having an extremely unpleasant alliaceous odour; it oxidises rapidly in air to a colourless solid. The 1,1-*dioxide*, is obtained as a white, amorphous powder on addition of hydrogen peroxide to tellurane dissolved in warm methanol (*G. T. Morgan* and *H. Burgess*, *ibid*., 1928, 321):

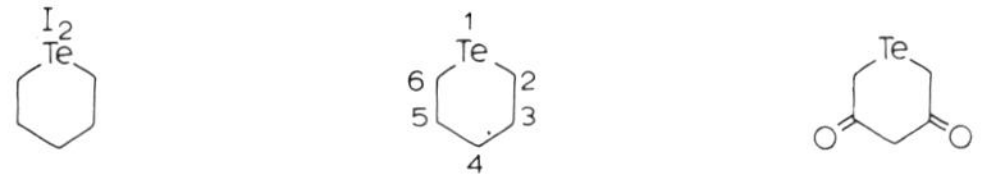

(I) (II) (III)

Tellurane-3,5-dione 1,1-dichloride, 1,1-*dichlorotelluracyclohexane*-3,5-*dione*, is a silver-white crystalline solid, darkening at 146° and melting with decomposition at 162°. Its yellow reduction product **tellurane-3,5-dione, telluracyclohexane-3,5-dione** (III), has been obtained following recrystallisation from water and drying in a vacuum. Its structure is supported by its n.m.r. spectrum (*D. H. Dewar et al., ibid.*, 1964, 688); crystal and molecular structure (*G. L. Parks*, Diss. Abstr. B, 1969, **29,** 3450).

Telluroflavanone, 2-phenyltellurochroman-4-one, 2,3-dihydro-4*H*-benzotelluropyran-4-one (V) have been prepared from the dioxolane IV by the following route (*J. L. Piette* and *M. Renson*, Bull. Soc. chim. Belge., 1971, **80,** 669):

(IV) —(1. Li; 2. Te and MeI)→ [TeMe₂]⊕ dioxolane —(1. C_5H_5N; 2. H_2SO_4)→ o-(TeMe)C₆H₄COMe —(PhCHO)→ o-(TeMe)C₆H₄CO·CH:CHPh —(HBr, AcOH)→ o-(TeBr)C₆H₄CO·CH:CHPh (70%) + (V) (20%)

The conformational analysis of pentamethylene heterocycles containing one hetero-atom O, S, Se, and Te is discussed by *J. B. Lambert* and *S. I. Featherman* (Chem. Reviews, 1975, **75,** 611).

8. Compounds containing an element from Group 4

(a) Silicon compounds

(i) Silacyclohexanes and their derivatives

The earlier method of preparing silacyclohexanes (I) involved adding silicon tetrachloride to pentane-1,5-dimagnesium dibromide in ether giving 1,1-**dichloro-1-silacyclohexane** (II), b.p. 169.5–170.5°/764 mm, which was then alkylated by a Grignard reagent (*A. Bygdén*, Ber., 1915, **48,** 1236; *G. Grüttner* and *M. Wiernik, ibid.*; p. 1474):

(I)

$$BrMg(CH_2)_5MgBr + SiCl_4 \longrightarrow$$ (II)

R. West (J. Amer. chem. Soc., 1954, **76,** 6012, 6015) introduced one or two methyl groups directly by using methyltrichloro- or dimethyldichlorosilane in place of silicon tetrachloride and found that the yield of cyclic product falls off with decrease in the number of chlorine atoms in the reactant. A 23% yield of 1,1-dichlorosilacyclohexane was obtained by passing 1,5-dichloropentane over a 9:1 mixture of silicon and copper at 360–370° (*N. S. Nametkin et al.*, Doklady. Akad. Nauk SSSR, 1966, **171,** 1345).

1,1-Disubstituted-silacyclohexanes have been prepared by the reaction of 1,1-dichlorosilacyclohexane with an alkyl- or aryl-lithium compound. Although attempts to effect cyclisation of 1,5-dilithiopentane and a dialkyl- or diaryl-dichlorosilane were originally unsuccessful (*C. Tamborski* and *H. Rosenberg*, J. org. Chem., 1960, **25,** 246), **1,1-dimethylsilacyclohexane** (I, $R = R^1 = Me$) was obtained by this method (*R. Fessenden* and *M. D. Coon*, *ibid.*, 1961, **26,** 2530). 1-Chloromethyl-1-methylsilacyclopentane with aluminium chloride at room temperature undergoes a vigorous reaction to yield over 80% of **1-chloro-1-methylsilacyclohexane** (II, $R = Me$, $R^1 = Cl$); its structure has been confirmed spectroscopically (*V. M. Vdovin et al.*, Izvest. Akad. Nauk S.S.S.R., Otdel. Khim. Nauk, 1962, 1127; 1963, 274, *Nametkin et al.*, Izvest. Nauk S.S.S.R., Ser. Khim. 1967, 2530). On treatment with methanolic sodium methoxide solution, 1-chloromethyl-1-methylsilacyclopentane gives 36% of **1-methyl-1-methylsilacyclohexane** (*idem, ibid.*, 1968, 836). Dimethyl(4-pentenyl)silane, $HSiMe_2(CH_2)_3CH{:}CH_2$ with hexachloroplatinic(IV) acid at high dilution yields 13% of 1,1-dimethylsilacyclohexane and 75% of 1,1-dimethylsilacyclopentane (*K. I. Kobrakov*, *T. I. Chernysheva* and *Nametkin*, Doklady Akad. Nauk S.S.S.R., 1971, **198**, 1340).

The hydrosilation reaction has been utilised for the synthesis of silicon heterocyclic compounds. With an appropriately substituted silane, the ring-closure reaction results in a silacyclopentane rather than the expected silacyclohexane, for example, 5-(dimethylsilyl)hex-1-ene, upon treatment with hexachloroplatinic(IV) acid, gives equal amounts of *cis*- and *trans*-1,1,2,5-tetramethylsilacyclopentane (*Fessenden* and *W. D. Kray*, J. org. Chem., 1973, **38,** 87). Some 3-substituted derivatives of 1,1-dimethylsilacyclohexane have been prepared (*Yu. N. Ogibin, G. I. Nikishin* and *A. D. Petrov*, Izvest. Akad. Nauk S.S.S.R., Ser. Khim., 1963, 1816).

1,1-Dichlorosilacyclohexane is hydrolysed by water and is reduced by lithium tetrahydridoaluminate to the parent compound, **silacyclohexane** (I, $R = R^1 = H$), b.p. 102°; also obtained in 9.5% yield by passing the silane over alumina at 375° (*Yu. K. Yur'ev* and *N. V. Makarov*, Doklady Akad. Nauk S.S.S.R., 1959, **128,** 121). Attempts to dehydrogenate this compound with palladium at 500° resulted in "cracking" and failed to produce an aromatic silane (silabenzene). Cyclic silanes with hydrogen bonded to silicon are subject to alkaline hydrolysis, as are the corresponding straight chain compounds:

$$R_3SiH + H_2O \xrightarrow{OH^{\ominus}} R_3SiOH + H_2$$

The properties of some compounds of type I (and II) are collected in Table 4.

TABLE 4

1-SILACYCLOHEXANE AND ITS 1-SUBSTITUTED DERIVATIVES (TYPE I)

R	*R*[1]	*B.p.(°C)*	n_D^{25}	d_4^{25}	*Ref.*
Cl	Cl	107	1.4679		1
H	H	102	1.4533	0.818	1
H	Cl	143	1.467	1.018	1
H	Me	118	1.4462	0.809	1
Cl	Me	167(157)	1.466	1.01	1
Me	Me	133	1.4380	0.798	1
Et	Et	184	1.4565	0.8346	2
Pr^n	Pr^n	78/3.5 mm	1.4590	0.8338	2
Bu^n	Bu^n	52/0.10 mm	1.4614	0.8365	2
Also $R = R^1 = C_5H_{11}{}^n$ to $C_{18}H_{37}{}^n$					
Ph	Ph	122/0.12 mm	1.5820	1.0319	2
OMe	OMe	171			1
OH	OH	m.p. 130–132			1
Me	OMe	173–175	n_D^{20} 1.4355		3

References
1 *R. West*, J. Amer. chem. Soc., 1954, **76,** 6012, 6015.
2 *C. Tamborski* and *H. Rosenberg*, J. org. Chem., 1960, **25,** 246.
3 *N. S. Nametkin et al.*, Izvest. Akad. Nauk S.S.S.R., Ser. Khim., 1968, 836.

The i.r. spectra of 1,1-dichlorosilacyclohexane (*G. D. Oshesky* and *F. F. Bentley*, J. Amer. chem. Soc., 1957, **79,** 2057), 1,1-diphenylsilacyclohexane and related compounds have been reported (*F. J. Bajer* and *H. W. Post*, J. org. Chem., 1962, **27,** 1422). Conformationally stable 1-methyl-4-*tert*-butyl-1-silacyclohexanes have been prepared and the *cis* and *trans* structures assigned from n.m.r. data; the stereochemistry of insertion of dimethylsilylenes into the silicon–hydrogen bond has also been studied (*H. Sakurai* and *M. Murakami*, J. Amer. chem. Soc., 1972, **94,** 5080).

Concentrated sulphuric acid causes a rupture of the C–Si bond in cyclic 5–6 membered silanes; the 5-membered ring is somewhat more stable (*A. F. Plate, N. A. Belikova* and *Yu. P. Egorov*, Izvest. Akad. Nauk S.S.S.R., Otdel. Khim. Nauk, 1956, 1085). *West* has shown that the relative reactivity of the silanes varies with ring size and decreases in the order 5 > 7 > straight chain > 6-membered rings.

The monochlorination of 1,1-dimethylsilacyclohexane results primarily in the formation of the chloromethyl derivative. The addition of free radical catalysts does not substantially alter the extent of methyl chlorination, but does result in increased attack on the ring carbon β to the silicon and decreased attack on the ring carbon α to the silicon (*Fessenden* and *F. J. Freenor*, J. org. Chem., 1961, **26,** 2003). The reaction of 1,1-dichloro- and 1,1-dimethyl-silacyclohexane with sulphuryl chloride has been studied; the chlorination of 1,1-dichlorosilacyclohexane causes little opening of the ring which is extensive during the chlorination of 1,1-dimethylsilacyclohexane. The distributions of isomers obtained from these cyclic silanes have been compared with those obtained by chlorination of *n*-alkylchlorosilanes. Chlorine atoms on silicon disfavour free-radical chlorination on the carbon atoms directly attached to such a silicon both in cyclic and linear silanes (*R. A. Benkeser, S. D. Smith* and *J. L. Noe, ibid.*, 1968, **33,** 597).

When a solution of 1,1-dimethylsilacyclohexane and phenyl(bromodichloromethyl)-mercury in benzene is boiled, insertion of CCl_2 into the βC–H linkage occurs to give as sole product 3-**dichloromethyl**-1,1-**dimethylsilacyclohexane** (III) in 68% yield (*D. Seyforth* and *S. S. Washburne*, J. Organomet. Chem., 1966, **5**, 389):

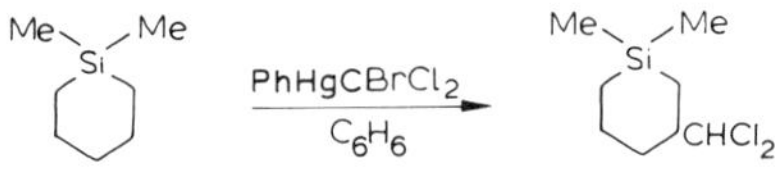

$PhHgCBrCl_2$, C_6H_6

(III)

The relationship between chair and twist forms of 1,1-dimethylsilacyclohexane has been studied by n.m.r. spectroscopy (*F. R. Jensen* and *C. H. Bushweller*, Tetrahedron Letters, 1968, 2825), and by using variable temperature p.m.r. spectroscopy and total p.m.r. line shape analysis the activation parameters for chair–chair interconversion have been determined (*Bushweller, J. W. O'Neil* and *H. S. Bilofsky*, Tetrahedron, 1971, **27,** 3065). The Raman and i.r. spectra of some 1,1-disubstituted derivatives of sila-cyclohexane have been compared (*G. Fogarasi, F. Torok* and *M. Vdovin*, Acta Chim. Acad. Sci. Hung., 1967, **54,** 277).

Slow addition of diethyl 3-(iodomethyldimethylsilyl)propylmalonate, ICH_2SiMe_2-$(CH_2)_3CH(CO_2Et)_2$, to potassium carbonate suspended in boiling ethyl methyl ketone yields 3,3-**bisethoxycarbonyl**-1,1-**dimethylsilacyclohexane,** which when heated with 50% potassium hydroxide yields 1,1-**dimethylsilacyclohexane**-3-**carboxylic acid** (*Ogibin*, U.S.S.R. Pat., 289,092/1970; *Ogibin* and *Nikishin*, Zhur. obshcheĭ Khim., 1971, **41,** 1277).

Diallyldimethylsilane has been polymerised with Ziegler-type catalysts to give polymers containing 6-membered cyclic recurring units (*C. S. Marvel* and *R. G. Woolford*, J. org. Chem., 1960, **25,** 1641):

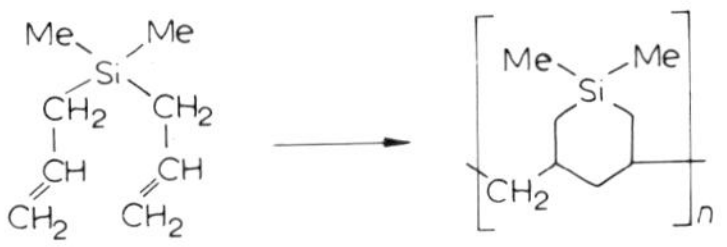

4,4-Dimethyl-4-silacyclohexanone (IV), b.p. 73–74°/11 mm, n_D^{20} 1.4635, d^{20} 0.9285 g/ml, *semicarbazone*, m.p. 191–192° (decomp.); *4,4-dimethyl-4-silacyclohexanol* (V), b.p. 70–71/5 mm, n_D^{31} 1.4648 (*Benkeser* and *E. W. Bennett*, J. Amer. chem. Soc., 1958, **80,** 5414; *W. P. Weber et al.*, 1971, **36,** 4060).

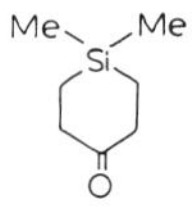

(IV) 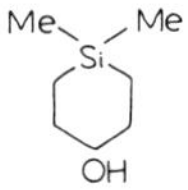 (V)

The m.s. of 4,4-dimethylsilacyclohexanone has been discussed *(Weber et al., loc. cit.)*. Photolysis of diphenylsilacyclohexanone in diethyl fumarate yields the cyclopropane resulting from trapping of a siloxycarbene by the alkene (*A. G. Brook*, and *H. W. Kucera*, Canad. J. Chem., 1971, **49,** 1618).

(ii) Silacyclohexenes and silacyclohexadienes

4,4-**Dimethylsilacyclohexadien-1-one** has been prepared by oxidation of 4,4-dimethylsilacyclohexan-1-one with selenium dioxide in boiling *tert*-butanol (*Weber* and *R. Laine*, Tetrahedron Letters, 1970, 4169); 4,4-*diphenyl-4-silacyclohexanone*, m.p. 93–94°; 4,4-*diphenylsilacyclohexadien-1-one*, m.p. 67–68°; 4,4-*dimethyl-4-silacyclohex-2-enone*; has also been obtained, the effect of the silyl centre on n→π^* transition in their u.v. spectra has been discussed (*R. A. Felix* and *Weber*, J. org. Chem., 1972, **37**, 2323).

Ring closure to form 1,1-**dichloro-1-silacyclohex-2-ene** (VII) proceeds smoothly when the olefin mixture containing either 75 or 85% of *cis*-5-chloro-1-(trichlorosilyl)pent-1-ene (VI) is treated with powdered magnesium in ether, the yield of cyclic product is increased from 44 to 59% in the latter instance (*Benkeser* and *R. F. Cunico*, J. organomet. Chem., 1965, **4,** 284):

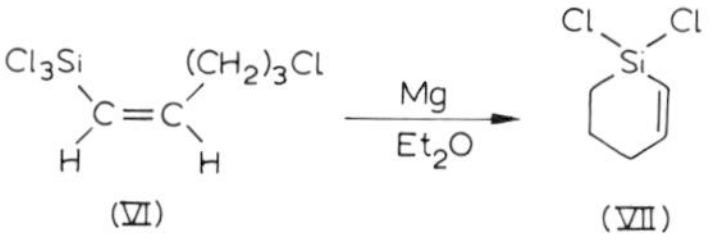

Cl_3Si(H)C=C(H)$(CH_2)_3Cl$ (VI) —Mg, Et_2O→ (VII)

Treatment of 1,1-dichloro-1-silacyclohex-2-ene with phenyllithium affords 1,1-**diphenyl-1-silacyclohex-2-ene** (VIII) and with *N*-bromosuccinimide it gives 4-**bromo**-1,1-**dichloro-1-silacyclohex-2-ene** (IX), which on pyrolysis yields a mixture of 1,1-**dichloro-1-silacyclohexadiene** (X), along with minor amounts of 1,1-dichloro-1-silacyclohex-2-ene and 1,1-**dichloro-1-silacyclohex-3-ene** (XI):

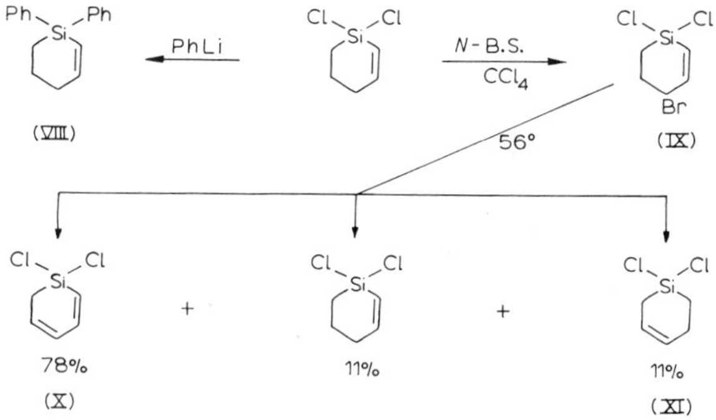

(VIII) ←PhLi— 1,1-dichloro-1-silacyclohex-2-ene —N-B.S., CCl_4→ (IX)

(IX) —56°→ (X) 78% + 1,1-dichloro-1-silacyclohex-2-ene 11% + (XI) 11%

Hydroboration of 1,1-diphenylsilacyclohex-2-ene gives predominantly the product where boron is attached to the carbon adjacent to silicon; hydroboration followed by oxidation appears to give exclusively the α-methanol 1,1-*diphenyl-silacyclohexan-2-ol* (XII), b.p. 160–164°/35 mm, n_D^{24} 1.5914. While oxidation of the alcohol with chromium trioxide gives mainly cleavage products, oxidation with dicyclohexylcarbodiimide in dimethyl sulphoxide yields 1,1-*diphenylsilacyclohexan-2-one* (XIII), m.p. 63–65° (*A. G. Brook* and *J. B. Pierce*, J. org. Chem., 1965, **30,** 2566):

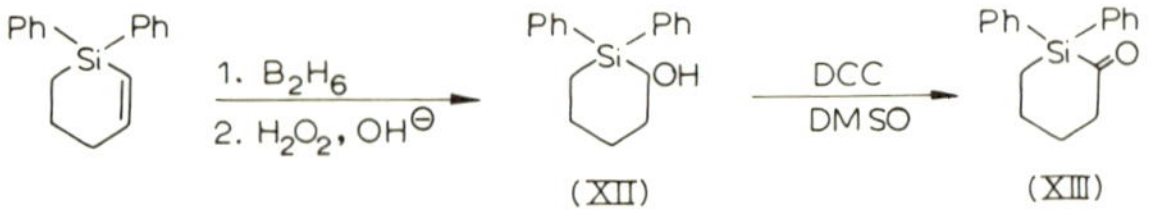

1. B_2H_6
2. H_2O_2, $OH^{\ominus}$
DCC
DMSO
(XII) (XIII)

2,2-Dimethyl-7,7-dibromo-2-silanorcarane (XV) obtained by reacting 1,1-dimethylsilacyclohex-2-ene (XIV) with phenyl(tribromomethyl)mercury in benzene, on treatment with tri-*n*-butyltin hydride affords either *endo*- (XVI) and *exo*-7-bromo-2,2-dimethyl-2-silanorcarane (XVII) or 2,2-dimethyl-2-silanorcarane (XVIII) depending on the temperature of the reaction (*E. Rosenberg* and *J. J. Zuckerman*, J. organomet. Chem., 1971, 33, 321):

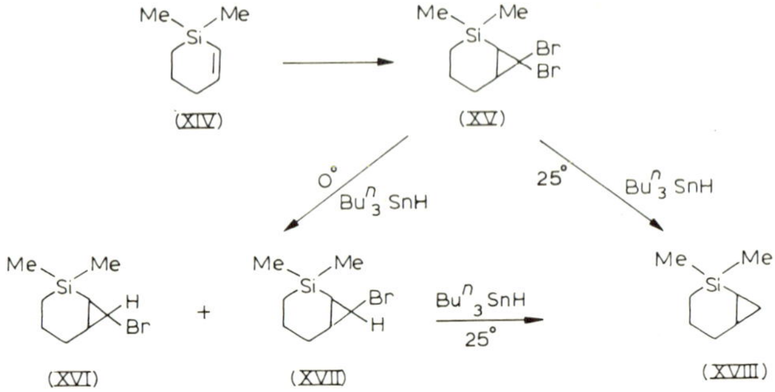

(XIV) (XV)
0° Bu^n_3SnH
25° Bu^n_3SnH
(XVI) + (XVII) Bu^n_3SnH 25° (XVIII)

The solvolysis of 4-tosyl-1,1-dimethylsilacyclohexane in glacial acetic acid at 70° gives 1,1-dimethylsilacyclohex-3-ene (13%) 1,3-bis(4-pentenyldimethyl)disiloxane (7%), 2-acetoxy-1,1-dimethylsilacyclohex-2-ene (53%), and 4-acetoxy-1,1-dimethylsilacyclohexane (27%) (*Washburne* and *R. R. Chawla*, *ibid*., 1971, **31**, C20).

Chloro(chloromethyl)dimethylsilane, $Me_2SiClCH_2Cl$, reacts with 1,4-dilithiotetraphenylbutadiene, LiCPh:CPh·CPh:CPhLi, in tetrahydrofuran to give a **silacyclohexadiene** (XIX), which reacts with butyllithium in dimethoxyethane to yield a red anion XX. Treatment of the latter with deuterium oxide, chlorotrimethylsilane and (dichloroiodo)benzene affords the respective deuterio-, trimethylsilyl- (XXI, R = D, Me_3Si) and 1,1-dimethyl-2,3,4,5-tetraphenyl-1-silacyclohexa-2,4-dien-1-yl (*G. Märkl* and *P. L. Merz*, Tetrahedron Letters, 1971, 1303):

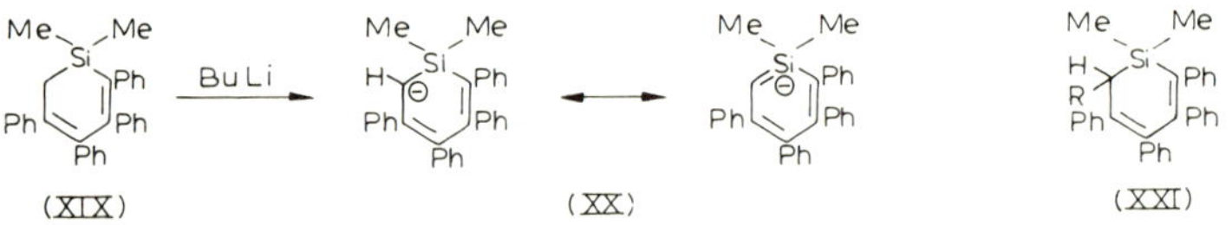

BuLi
(XIX) (XX) (XXI)

(iii) Spirocyclosilanes

Spirocyclosilanes are obtained when 1,1-dichloro-1-silacyclohexane (II) is treated with the "double Grignard" or dilithio-derivative of 1,4- or 1,5-dibromoalkanes giving 5-**silaspiro**[4,5]**decane** (10-*silaspirodicyclodecane*) (XXII), b.p. 203°, n_D^{25} 1.4860, d_4^{25} 0.899, and 6-**silaspiro**[5,5]**undecane** (11-*silaspirodicycloundecane*) (XXIII), b.p. 227°, n_D^{25} 1.4869, d_4^{25} 0.900 (*West, loc. cit.*):

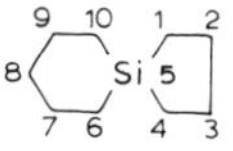

(XXII)

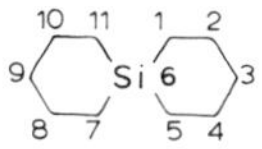

(XXIII)

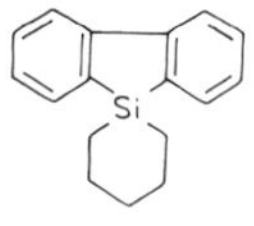

(XXIV)

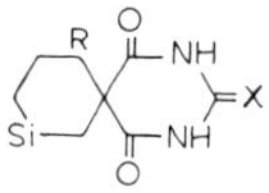

(XXV)

Spiro[**dibenzosilole**-5,1′-**silacyclohexane**] (XXIV), m.p. 144–145.5° (*H. Gilman* and *R. D. Gorsich*, J. Amer. chem. Soc., 1958, **80,** 1883). Spirobarbiturates of the general structure (XXV, X = O or S) have been synthesised (*Fessenden et al.*, J. med. Chem., 1964, **7,** 695).

(iv) Polycyclic silanes with six-membered rings

1,1-**Diphenyl**-1,2,3,4-**tetrahydro**-1-**silanaphthalene** (2,3-*benzo*-1,1-*diphenyl*-1-*silacyclohex-2-ene*) (XXVII), m.p. 78–79.5°, is obtained by reacting the di-Grignard reagent of 3-(2-bromophenyl)propyl bromide (XXVI) with dichlorodiphenylsilane (*Gilman* and *O. L. Marrs*, J. org. Chem., 1965, **30,** 325):

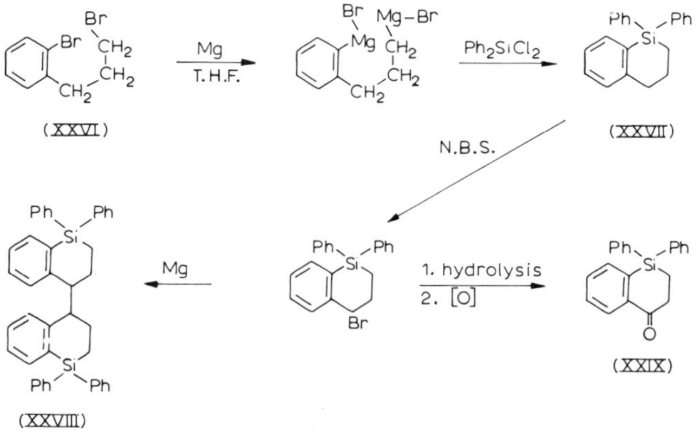

(XXVI) —Mg, T.H.F.→ —Ph_2SiCl_2→ (XXVII) —N.B.S.→ —1. hydrolysis, 2. [O]→ (XXIX); —Mg→ (XXVIII)

1,1-Diphenyl-1,2,3,4-tetrahydro-1-silanaphthalene readily reacts with *N*-bromosuccinimide to give the 4-bromo derivative, which with magnesium affords 4,4′-**bi**-

(1,1-**diphenyl**-1,2,3,4-**tetrahydro**-1-**silanaphthyl**), 4,4′-**bi**(2,3-**benzo**-1,1-**diphenyl**-1-**silacyclohex**-2-**ene**) (XVIII) and on hydrolysis and oxidation with silver acetate yields 1,1-*diphenyl*-4-*oxo*-1,2,3,4-*tetrahydro*-1-*silanaphthalene*, 2,3-*benzo*-1,1-*diphenyl*-1-*silacyclohex*-2-*en*-4-*one* (XXIX), m.p. 111–112° (127°), 2,4-*dinitrophenylhydrazone*, m.p. 228–230° (240° decomp.).

9,10-**Dihydro**-9-**silanthracenes** of type XXXI ($R = Cl$, Me, Ph, etc.; $R^1 = Cl$, Me) have been prepared by reacting the Grignard reagent XXX with the appropriate dichlorosilane, $RR'SiCl_2$:

(XXX) (XXXI)

H^1-n.m.r. spectra confirms the structure of these compounds (*P. Jutzi*, Ber., 1971, **104,** 1455).

For further studies in silicon heteroaromatic systems see *R. F. Grossman* Diss. Abs., 1963, **23,** 2693; for the synthesis and chemistry of some cyclic silicon systems see *J. C. Noe, ibid.*, 1965, **26,** 707; *F. H. Pinkerton, ibid.*, B, 1972, **32,** 5114); for ^{29}Si chemical shifts of silacyclohexanes see *R. L. Scholl, G. E. Maciel*, and *W. K. Musker*, J. Amer. chem. Soc., 1972, **94,** 6376).

(b) Germanium, tin, and lead compounds

(i) Germanium compounds

Treatment of pentane-1,5-dimagnesium dibromide under nitrogen with germanium tetrachloride in ether gives crude 1,1-dichlorogermacyclohexane, which on treatment with lithium tetrahydridoaluminate yields 20% of **germacyclohexane** XXXII, b.p. 119–120°, n_D^{20} 1.4835, d_{20} 1.1846, on distillation:

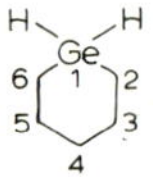

(XXXII) 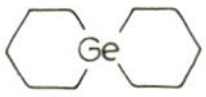 (XXXIII)

The residue, b.p. 124°/30 mm, contains 6-*germaspiro*[5,5]*undecane* (XXXIII) and germane.

1,1-*Diiodo*-, b.p. 108°/2 mm, n_D^{20} 1.6610, d^{20} 2.3698; 1,1-*dibromo*-, b.p. 115°/17 mm, n_D^{20} 1.5659; 1,1-*dichloro*-, b.p. 99°/24 mm, 208–209°, n_D^{20} 1.5107, d^{20} 1.4675; 1-*iodo*-, b.p. 132°/58 mm, n_D^{20} 1.5835, d^{20} 1.8681; 1,1-*diethyl*-, b.p. 116–117°/72 mm, n_D^{20} 1.4742, d^{20} 1.0663; 1,1-*diphenyl*-, b.p. 150°/1 mm, n_D^{20} 1.5932, d^{20} 1.2061; 1-*bromo*-1-*phenyl*-, b.p. 125°/0.8 mm, n_D^{20} 1.5828, d^{20} 1.5045; 1-*phenyl-germacyclohexane*, b.p. 128°/14 mm, n_D^{20} 1.5550, d^{20} 1.989 (*P. Mazerolles*, Bull. Soc. chim. Fr., 1962, 1907). 1,1-**Diphenylgermacyclohexane** has been prepared by the action of phenyllithium on 1,1-dichlorogermacyclohexane (*F. J. Bajer* and *H. W. Post*, J. org. Chem., 1962, **27,**

1422). For spiro compounds containing germanium see *Bajer* and *Post*, J. organomet. Chem., 1968, **11,** 187.

Dichlorocarbene derived from phenyl(bromodichloromethyl)mercury becomes inserted into the βC–H bond of 1,1-diethylgermacyclohexane to give 3-dichloromethyl-1,1-diethyl-1-germacyclohexane (XXXIV), which on thermolysis affords 1-chloro-2-(3-diethylchlorogermylpropyl)cyclopropane (XXXV) (*D. Seyferth et al.*, J. organomet. Chem., 1971, **29,** 371; *J. V. Scibelli*, Diss. Abs. B, 1973, **33,** 11, 252):

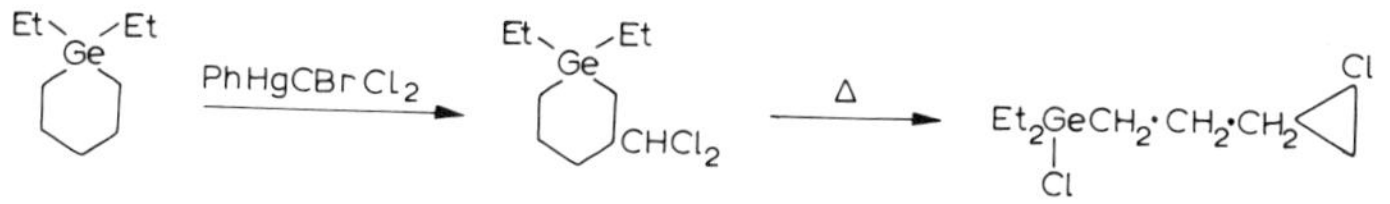

PhHgCBrCl₂ → (XXXIV) —Δ→ $Et_2Ge(Cl)CH_2{\cdot}CH_2{\cdot}CH_2$-cyclopropyl(Cl) (XXXV)

The frequency of Ge–H valence oscillations in some germacyclohexanes, $(CH_2)_5GeHX$ (X = alkyl, Ph, H, I) have been reported and discussed (*R. Mathia* and *Mazerolles*, Bull. Soc. chim. Fr., 1962, 1913). For the synthesis and i.r. spectra of heterocyclic compounds of germanium see *Bajer*, Diss. Abs., 1964, **25,** 1557).

The addition of dihalogenocarbenes to 1-germacyclopent-3-enes to give 1-**germacyclohexa-2,4-dienes** (*e.g.*, XXXVI, XXXVII and XXXVIII) has been investigated (*Seyferth et al.*, J. Amer. chem. Soc., 1970, **92,** 657):

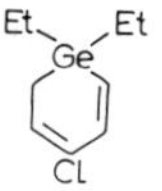

(XXXVI) 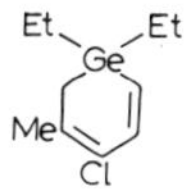 (XXXVII) 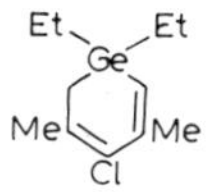 (XXXVIII)

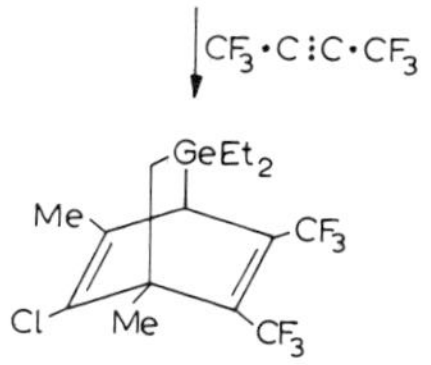

$\downarrow CF_3{\cdot}C{:}C{\cdot}CF_3$

(XXXIX)

Germacyclohexadiene (XXXVIII) reacts in a sealed tube with a twofold excess of perfluoro-2-butyne to provide quantitative conversion (by n.m.r.) to the expected Diels–Alder adduct XXXIX, the pyrolysis of which has been studied (*T. J. Barton, E. A. Kline* and *P. M. Garvey*, *ibid.*, 1973, **95,** 3078).

9,10-**Dihydro-9-germanthracenes** (XL, R = Cl, Me, Ph, etc.; R^1 = Cl, Me) have been obtained *(Jutzi, loc. cit.)*.

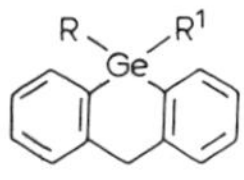

(XL)

(ii) Tin compounds

1,1-**Dimethyl-1-stannacyclohexane** (XLI, R = Me), is obtained by interaction of dimethyltin di-iodide with pentane-1,5-dimagnesium dibromide and is a colourless liquid with a terpene-like odour, b.p. 64°/16mm, n_D 1.5024, d_4^{23} 1.3357, and the corresponding *diethyl* derivative (XLI, R = Et) has b.p. 95°/14 mm, n_D 1.5003, d_4^{20} 1.2693 (*R. Schwarz* and *W. Reinhardt*, Ber., 1932, **65,** 1743). These compounds are stable in the absence of air, but, like the open chain tin alkyls, show a pronounced tendency to resinify by atmospheric oxidation. Bromine in ethyl acetate at 0° immediately causes ring fission to dimethyl- or diethyl-5-bromopentyltin bromide:

(XLI) $\xrightarrow{Br_2}$ R = Me, Et

1,1-**Diphenyl-1-stannacyclohexane, cyclopentamethylenediphenylstannane** (XLI, R = Ph), b.p. 138–140°/0.10 mm, n_D^{26} 1.6007, is obtained from diphenyltin dichloride and 1,5-dilithiopentane. On treatment with potassium bromide–potassium bromate in carbon tetrachloride or anhydrous hydrogen bromide, it gives **1,1-dibromostannacyclohexane**; and with iodine in carbon tetrachloride, 1-*iodo*-1-*phenylstannacyclohexane, cyclopentamethylenephenyltin iodide*, b.p. 219°/0.45–0.85 mm (wax bath temperature) (*Bajer* and *Post*, J. org. Chem., 1962, **27,** 1422).

tris[3,3-Dimethylpentenyl(dialuminium and dibutyltin)dichloride] give 1,1-*dibutyl*-4,4-*dimethylstannacyclohexane* (XLII), b.p. 71–77°/0.0001 mm, n_D^{20} 1.4948; the former and stannic chloride afford 3,3,9,9-**tetramethyl-6-stannaspiro**[5,5]**undecane** (XLIII),

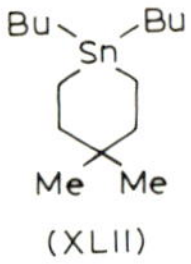

(XLII) 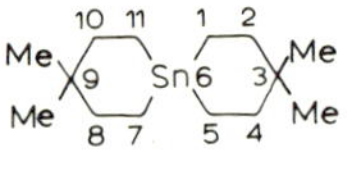 (XLIII)

b.p. 75–82°/0.0001 mm, n_D^{20} 1.5182 (*R. Polster*, G.P., 1,150,388; 1,156,807 1963); 1-chloro-1-vinylstannacyclohexane (*H. E. Ramsden*, U.S. P., 2,965,661 1960); 1,1-*di*-n-*dibutyl*-, b.p. 55°/0.04 mm, n_D^{25} 1.4958; 1,1-*dineopentyl-stannacyclohexane*, b.p. 61°/0.32 mm, n_D^{25} 1.4954 (*H. Zimmer et al.*, Tetrahedron Letters, 1968, 1615).

6-**Stannaspiro[5,5]undecane** (XLIV) undergoes ring cleavage when treated with bromine, hydrogen bromide, or iodine. Similarly ring cleavage occurs when either 1,1-dimethyl- or 1,1-diethyl-stannacyclohexane (XLV) is reacted with bromine (*Bajer* and *Post*, J. organomet. Chem., 1968, **11,** 187):

1,1-Dimethylstannacyclohexane reacts with phenyl(bromodichloromethyl)mercury to give 3-*dichloromethyl*-1,1-*dimethylstannacyclohexane*, n_D^{25} 1.5330 (*D. Seyferth* and *Ṡ. S. Washburne*, J. organometal. Chem., 1966, **5,** 389). For synthesis and i.r. spectra of heterocyclic compounds of tin see *Bajer*, Diss. Abs., 1964, **25,** 1557 and for heterocyclic compounds of tin *O. F. Bennel*, *ibid.*, 1960, **21,** 1370.

Stannohydration of diethynylmethane (XLVI) with dibutylstannane in boiling heptane followed by heating at 200° gives 1,4-dihydro-1,1-dibutylstannabenzene (XLVII), which on treatment with phenylboron dibromide yields dibutyltin dibromide and 1-**phenyl**-1,4-**dihydroborabenzene** (XLVIII). The latter could be deprotonated using *tert*-butyllithium in pentane–tetrahydrofuran to afford the 1-phenylborabenzene anion (XLIX). Treating the above solution with excess acetic acid gies 1,4-pentadiene, *cis*-1,3-pentadiene, and benzene (*A. J. Ashe* and *P. Shu*, J. Amer. chem. Soc., 1971, **93,** 1804):

1,4-Dihydro-1,1-dibutylstannabenzene and antimony trichloride react exothermally in tetrahydrofuran to give dibutyltin dichloride and 1,4-*dihydro*-1-*chlorostibabenzene* (L), m.p. 115–117° (decomp.), which in "tetraglyme" with 1,5-diazabicyclo[4.3.0]non-5-ene produces **stibabenzene** (antimonin) (LI), an extremely labile compound which rapidly polymerises to an intractable brown tar at $-80°$ (*Ashe, ibid.*, p. 6690). 1,4-*Dihydro*-1-*chlorobismabenzene*, yellow crystals, m.p. 144–145° (decomp.), has been prepared in a similar manner, but attempts to isolate **bismabenzene** failed. However, spectral evidence has been obtained for the formation of an adduct LII from bismabenzene and hexafluorobutyne (*Ashe* and *M. D. Gordon, ibid.*, 1972, **94**, 7596).

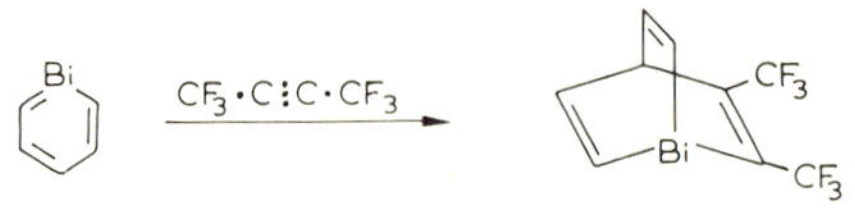

$CF_3 \cdot C{:}C \cdot CF_3$

(LII)

9,10-Dihydro-9-stannanthracenes (LIII, R=Cl, Me, Ph, etc.; R^1=Cl) have been obtained (*Jutzi, loc. cit.*).

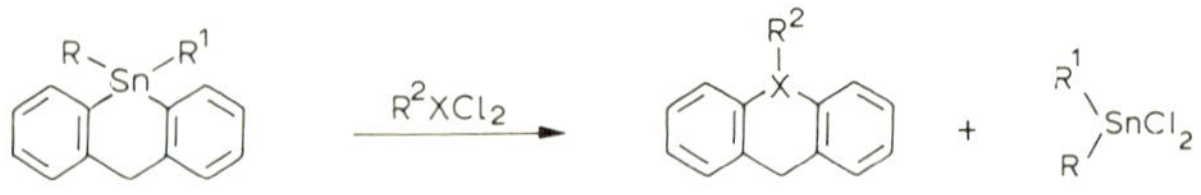

R^2XCl_2

(LIII) (LIV)

R^2 = Ph, Me, Cl, Br
X = P, As, Sb

The reaction of 9,10-dihydro-9,9-dimethyl-9-stannanthracene (LIII, R=R^1=Me) with the appropriate halides affords dihydroanthracenes LIV having elements of the fifth main group as substituents in position 9 (*Jutzi* and *K. Deuchert*, Angew. Chem. internat. Edn., 1969, **8**, 991).

(iii) Lead compounds

1,1-**Diethyl**-1-**plumbacyclohexane** (LV, R=Et), b.p. 111°/13.5 mm, n_D^{20} 1.5484, d_4^{20} 1.6866, obtained in the same way as the related tin compound (*G. Grüttner* and *E. Krause*, Ber., 1916, **49**, 2666), has an odour of lime blossom:

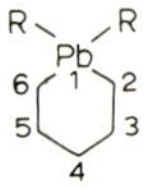

(LV)

It reduces silver nitrate immediately and though stable in the absence of air for long periods, resinifies readily and inflames when exposed to air on a filter paper. As with tin compounds, the ring system is immediately opened by bromine. 1,1-**Diphenyl**-1-**plumbacyclohexane** (LV, R=Ph) has been prepared by reacting diphenyllead dichloride with 1,5-dilithiopentane (*Bajer* and *Post*, J. org. Chem., 1962, **27**, 1422). 1,1-*Dipentylplumbacyclohexane* (LV, R=C_5H_{11}), b.p. 86–87°/0.05 mm (*E. C. Juenge* and *E. Jack*, J. organomet. Chem., 1970, **21**, 359).

9. Compounds containing iodine

Compounds having a six-membered ring containing iodine have been prepared by *R. Sandin et al.*, (J. Amer. chem. Soc., 1956, **78,** 3819) by the following method:

giving 10*H*-**dibenzo[*b,e*]iodinium chloride** (9-**iodonia**-10*H*-**anthracene chloride**) (I), m.p. 244–245° (decomp.).

From the chloride, the corresponding *bromide*, m.p. 222–224° (decomp.), and *iodide*, m.p. 184.5–185.5° (decomp.), are obtained by precipitating with the appropriate potassium halide and recrystallising the product from water.

3,7-**Dinitro**-10*H*-**dibenzo[*b,e*]iodininium bisulphate** (2,7-**dinitro**-9-**iodonia**-10*H*-**anthracene bisulphate**) (II) is prepared (*W. K. Hwang*, Sci. Sinica, 1957, **6**, 123; C.A., 1957, **51**, 16476) by a reaction very similar to that used by *I. Masson* and *E. Race* (J. chem. Soc., 1937, 1718) to produce diphenyliodonium salts (cf. Vol. IIIA, p. 267).

4,4′-Dinitrodiphenylmethane is stirred into a solution of sodium iodate and iodine in concentrated sulphuric acid at room temperature, and, when homogeneous, the mixture is poured on to ice. The precipitated *bisulphate*, m.p. 217.5° (decomp.), is purified by crystallisation from 85% formic acid. Other salts, *chloride*, m.p. 211.5°, *bromide*, m.p. 217.5°, *iodide*, m.p. 164°, and *thiocyanate*, m.p. 167–167.5° (all with decomposition), are obtained by precipitation reactions carried out in aqueous formic acid. The structure of the cation is established by degradation to the expected diphenylmethane derivative. 3,7-*Bis(dimethylamino)*-10H-*dibenz*[b,e]*iodininium iodide dihydriodide*, m.p. 179–181° (decomp.), *iodide*, m.p. 195–195.5° (decomp.) (*Hwang*, Hua Hsüeh Hsüeh Pao, 1957, **23**, 438; C.A., 1958, **52**, 16356); 3,7-*dinitro*-10-(4-*nitrophenyl*)-10H-*dibenz*[b,e]*iodininium bisulphate*, m.p. 237°; 3,7-*dicyano*-10H-*dibenz*[b,e]*iodininium bisulphate*, m.p. 236° (*Hwang*, *C.-H. Wang* and *S.-Y. Chen*, K'O Hsüeh T'ung Pao, 1957, 49; C.A., 1959, **53**, 4280); a number of 3,7-dialkylamino- derivatives have been synthesised (*Huang* and *C.-C. Chang*, *ibid.*, 1963, 47; C.A., 1964, **60**, 9280); for a review of the chemistry of 10*H*-dibenz[*b,e*]iodininium salts see *Chang*, Hua Hsüeh Tung Pao, 1966, 100; C.A., 1966, **65,** 8872.

It has been reported that 1,5-diiodopentane yields some six-membered ring iodonium ion of type III, along with the rearranged ion IV upon reaction with appropriate proportions of antimony pentafluoride in sulphur dioxide. A procedure has been developed for isolating and recrystallising iodonium ions formed from diiodides in this way:

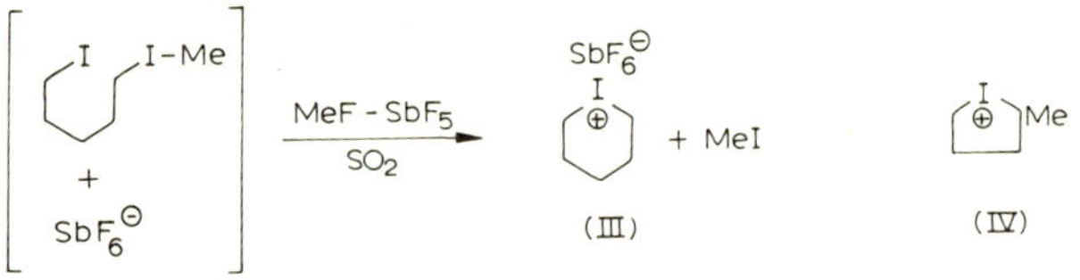

I I–Me + $SbF_6^{\ominus}$ $\xrightarrow[SO_2]{MeF - SbF_5}$ $SbF_6^{\ominus}$ (III) + MeI (IV) Me

However, the bromonium ion can be prepared only in the presence of the 2-methyltetramethylene isomer (*P. E. Peterson, B. R. Bonazza* and *P. M. Henricks*, J. Amer. chem. Soc., 1973, **95**, 2222).

Chapter 22

Brazilin and Haematoxylin *

SIR ROBERT ROBINSON

Brazilin and haematoxylin constitute a small group related to the catechins. Their study brought to light a series of molecular rearrangements of interest comparable with those encountered among camphor or morphine derivatives.

The sources of these substances are hard-woods of species of *Caesalpinia* which are disintegrated and then extracted with hot water. These extracts may be concentrated and then used directly for commercial purposes. Pernambucu redwood, *C. crista*, is the richest in brazilin content; the true Brazilwood, *C. braziliensis*, contains about half as much. Some kind of redwood was employed in India before the discovery of America and was imported into Europe from the East Indies for use as a dye. Kimichi mentioned the name bresil (braza = fiery red) in 1190, so that brazilin is not so named because it was found in Brazil. The converse is true; the Spaniards named the country Brazil because the already known dye-wood was abundant there. Logwood extract (campeachy wood) is obtained from *Haematoxylon campechianum* (Central America and Caribbean). Though not as important as formerly it is still used, especially for dying silk and furs. The actual adjective dye is haematein (q.v.) which is a polygenetic mordant dye. The most favoured shades are grey to black obtained with chromium, and especially iron, mordants.

1. The brazilin group

The wood extracts deposit crusts on standing and from these *Chevreul* (Ann. chim., 1808, [i], 66, 225) first obtained brazilin by recrystallisation from aqueous alcohol. The formula, $C_{16}H_{14}O_5$, was proposed by *C. Liebermann* and *O. Burg* (Ber., 1876, 9, 1883). On dry distillation, brazilin affords resorci-

* No significant new material on the chemistry of these two substances has been published since 1960. Chapter IX of the original edition is reproduced here as a tribute to Sir Robert Robinson's memory, and his long connection with *Rodd's* Chemistry of Carbon Compounds as Chairman of the Advisory Board from its inception over 25 years ago, particularly as it describes in his own words one of his many elegant contributions to organic chemistry.

nol in not inconsiderable yield and this process, due to *Kopp* (Ber., 1873, 6, 446), was acutally used for many years for the preparation of resorcinol. The action of nitric acid on brazilin gave rise to styphnic acid (*F. Reim*, Ber., 1871, 4, 332; 1879, 12, 1392).

Liebermann and *Burg* (*loc. cit.*; also *K. Buchka* and *A. Erck*, Ber., 1885, 18, 1139) found that brazilin gives a tetra-acetyl derivative when heated with acetic anhydride and this was confirmed by *G. Dralle* (Ber., 1884, 17, 370). The latter chemist showed that brazilin gave a trimethyl ether, insoluble in aqueous alkali, which still contains a hydroxyl group (acetyl derivative); and a tetramethyl ether under very vigorous conditions. Brazilin was thus shown to be a trihydric phenol also containing an alcoholic hydroxyl group.

Brazilin forms colourless needles (with 1·5 H_2O) or prisms (with 1·0 H_2O), m.p. circ. 250° (optical rotation, see p. 445). Derivatives are: *triacetate*, needles, m.p. 105–6°; *tetra-acetate*, needles, m.p. 149–151°; *bromotetra-acetate*, m.p. 203–4°; *dibromotetra-acetate* m.p. 185°; *trimethyl ether*, prisms, m.p. 138–9°; *trimethyl ether acetate*, m.p. 171–3°; *tetramethyl ether*, m.p. 137–9°; *bromotetramethyl ether*, prisms, m.p. 180–181°.

The first indication of a part of the molecular structure was given by *C. G. Schall* and *G. Dralle* (Ber., 1888, 21, 3017; 1889, 22, 1559; 1892, 25, 19) who oxidized brazilin in alkaline solution by means of atmospheric oxygen. The first product is brazilein (see p. 435) but the red colour due to this substance disappeared and β-resorcylic acid and a substance $C_9H_6O_4$, needles m.p. 271°, were then isolated. The latter formed a diacetyl derivative and a dimethyl ether $C_9H_4O_2(OMe)_2$, which could be oxidized by means of permanganate to 2-hydroxy 4-methoxybenzoic acid. *Schall* (Ber., 1894, 27, 528), suggested that the substance, $C_9H_6O_4$, was a dihydroxychromone (I) and *W. Feuerstein* and *St. von Kostanecki* (Ber., 1899, 32, 1025) confirmed this view, since the dimethyl ether (II) could be hydrolysed into formic acid and the fisetol dimethyl ether (III) already known as a degradation product of methylated fisetin (*J. Herzig*, Monatsh., 1891, 12, 187).

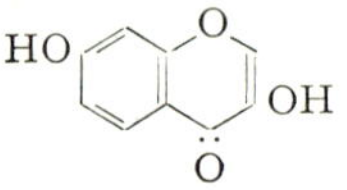

HO, O, OH, O (I) — MeO, O, OMe, O (II) — MeO, OH, $CO\cdot CH_2\cdot OMe$ (III)

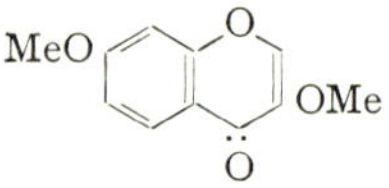

The synthesis of (II) was effected in 1924 (*P. Pfeiffer* and *J. Oberlin*, Ber., 1924, 57, 208) from 7-methoxychromanone by way of its 3-oximino-derivative. This was hydrolysed by means of sulphuric acid in aqueous acetic acid and the resulting mono-methyl ether of (I) was methylated.

Oxidation of O-trimethylbrazilin by means of permanganate

The investigations of *W. H. Perkin*, jun. in this field followed the model work of Goldschmidt on papaverine and his own studies of berberine. The oxidizing agent attacked various parts of the molecules and a variety of products was obtained. The information from the structures of the products, which sometimes had overlapping sections, was combined.

It happened in this case that one of the fragments obtained in very small yield, namely brazilic acid, was found to be of exceptional importance.

The investigation was started as long ago as 1883 and accounts were given in 1899 and 1900 (Proc. chem. Soc., 1899, 15, 27, 45, 241; 1900; 16, 106, 107). The products were (1) **brazilinic acid,** $C_{19}H_{18}O_9$, a substance obtained from trimethylbrazilin without loss of carbon atoms, (2) an acid, $C_{10}H_{10}O_6$, m.p. 174°, (3) an acid, $C_{10}H_{10}O_6$, m.p. 195° and (4) **brazilic acid,** $C_{12}H_{12}O_6$. The two acids, $C_{10}H_{10}O_6$, were identified as 2-carboxy-5-methoxyphenoxyacetic acid (IV) and metahemipinic acid (V) respectively.

MeO — $\overset{*}{O}CH_2CO_2H$, CO_2H (IV)   MeO, MeO — CO_2H, CO_2H (V)

On heating, (IV) lost a carboxyl group and furnished *m*-methoxyphenoxyacetic acid, which was synthesized. Later (*Perkin* and *R. Robinson*, J. chem., Soc., 1908, 93, 504) (IV) was also synthesized from 2-hydroxy-4-methoxybenzaldehyde by introduction of the acetic residue and oxidation of the resulting aldehyde-acid. The isolation of metahemipinic acid led Perkin to suggest that the constitution of brazilin should be (VI).

HO — O — OH, OH, OH (VI)

The study of brazilic acid (*Perkin*, *ibid.*, 1902, 81, 221) soon excluded one of the alternatives considered and the indane moiety of the formula received support from the later isolation of 4:6-dimethoxyhomophthalic acid (VII) in small yield from among the permanganate oxidation products (*ibid.*, 1902, 81, 1028; *Perkin* and *Robinson*, *ibid.*, 1907, 91, 1082).

The yield of brazilic acid was only 0.7 %; it furnished an oxime and a semicarbazone, and when reduced with sodium amalgam and the solution acidified it gave the lactone, $C_{12}H_{12}O_5$, of dihydrobrazilic acid. On very gentle heating of brazilic acid with concentrated sulphuric acid, the solution acquired intense violet fluorescence, and was found to contain anhydrobrazilic acid (VIII). The constitution of this substance was deduced from its decomposition on boiling with barium hydroxide solution into β-(6-hydroxy-4-methoxybenzoyl) propionic acid (IX) and formic acid.

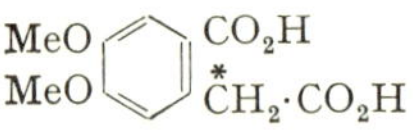

MeO MeO CO_2H $\overset{*}{C}H_2 \cdot CO_2H$ (VII) MeO O $CH_2 \cdot CO_2H$ O (VIII) MeO OH $CO \cdot CH_2 \cdot CH_2 \cdot CO_2H$ (IX)

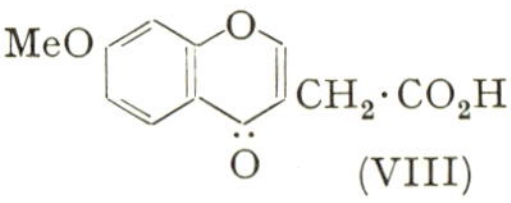

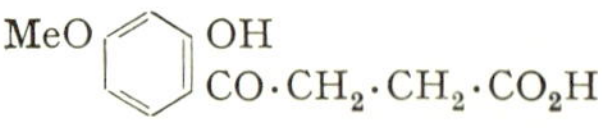

The methyl ether (IX) was synthesized. Later the acid (IX) was again obtained (*idem, ibid.*, 1908, 93, 502, 509) and its ester condensed with ethyl formate with the help of sodium and anhydrobrazilic acid isolated. This was the first example of a synthesis of a 3-substituted chromone, later applied by a number of workers to the synthesis of *iso*flavones. An alternative to (X) for brazilic acid can hardly be entertained in view of the simultaneous formation of (IV) in the oxidation process.

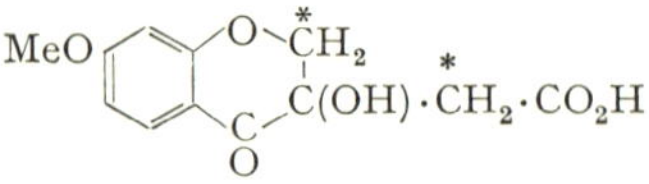

MeO O $\overset{*}{C}H_2$ C $C(OH) \cdot \overset{*}{C}H_2 \cdot CO_2H$ O

(X) Brazilic acid

In 1904 *Werner* and *Pfeiffer* (Chem. Zeitschr., 1904, 3, 421) suggested that brazilin has the structure (XI) but advanced no more experimental evidence in favour of this view and combined it with fallacious interpretations of other aspects of the chemistry of brazilin derivatives. Nevertheless the suggestion was an acute one and was eventually proved to be correct as the result of synthetic studies by Perkin and Robinson (Proc. chem. Soc., 1906, 22, 160; J. chem. Soc., 1907, 91, 1073). Apart from these an argument that gradually acquired force was that the CH_2 groups marked with an asterisk in (IV) and (X), were not likely to be produced from $-CH^{*}\langle^{C:}_{C:}$ groups in the course of a permanganate oxidation. One of these CH_2 groups also survives in dimethoxyhomophthalic acid (VII).

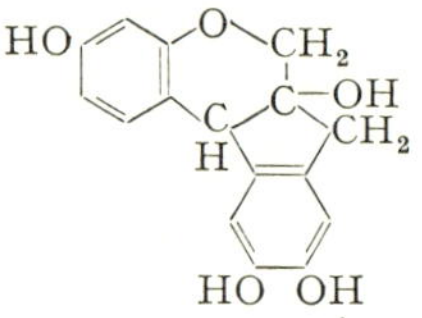

HO O CH_2 C–OH C H CH_2 HO OH

(XI) Brazilin

MeO $O \cdot CH_2 \cdot CO_2H$ CO CO_2H MeO OMe

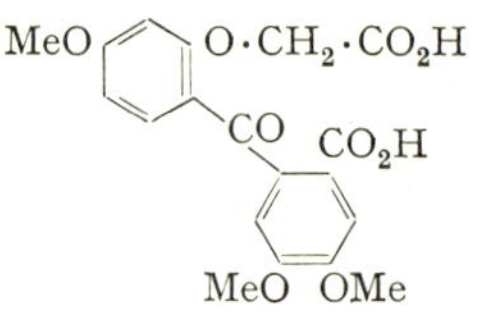

(XII) Brazilinic acid

Structure (XI), but not (VI), contains the two methylene groups which occur in the oxidation products.

Brazilinic acid, $C_{19}H_{18}O_9$, is a keto-dicarboxylic acid reducible to a dihydro-derivative which formed a characteristic lactone. On the basis of (VI) it would be an aldehyde, hardly likely to be produced in a permanganate oxidation at 90°, but the skeleton (XI) would be opened to (XII). The lactone of dihydrobrazilinic acid would clearly be (XIV).

A small yield of brazilinic acid was obtained by a Friedel-Crafts condensation of ethyl *m*-methoxyphenoxyacetate and metahemipinic anhydride. It was better to condense resorcinol dimethyl ether with metahemipinic anhydride under moderately vigorous conditions with hydrolysis of the *o*-MeO group and reduce to the lactone (XIII). Introduction of the acetic residue then afforded the desired lactone (XIV).

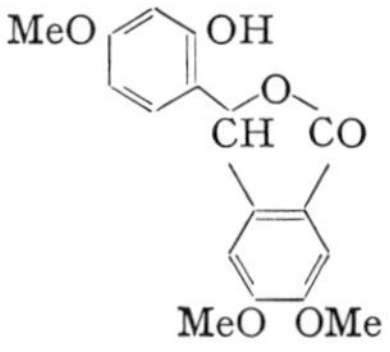

MeO OH O CH CO MeO OMe

(XIII)

MeO O·CH_2·CO_2H O CH CO MeO OMe

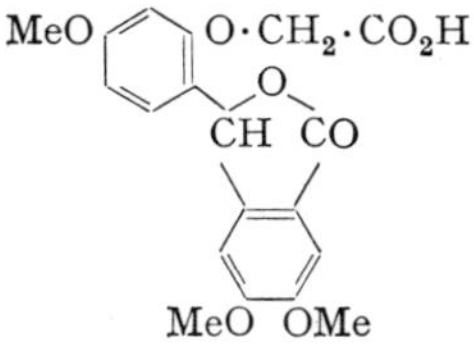

(XIV)

The evidence thus available was consistent with the structure (XI) for brazilin but there was nothing to exclude (XV) which was the only possible alternative.

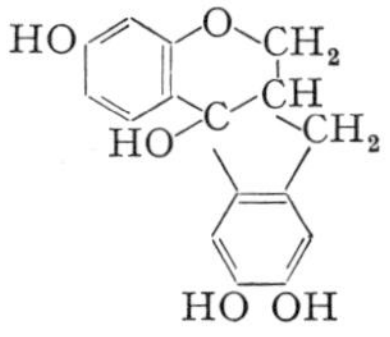

HO O CH_2 CH HO C CH_2 HO OH

(XV)

MeO O CH_2 CO CO CH_2 MeO OMe

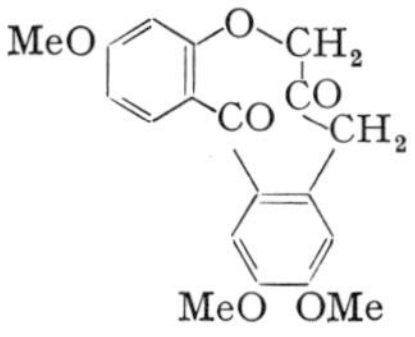

(XVI)
Trimethylbrazilone

Oxidation of O-trimethylbrazilin by means of chromic acid in acetic acid solution

The pale greenish-yellow product of this oxidation, termed **trimethylbrazilone,** $C_{19}H_{18}O_6$, was obtained at once in an almost pure condition. The composition change from trimethylbrazilin was + O and — 2H. The formula (XVI) was assigned to the substance in later years by Perkin and Robinson, consistent with the infra-red spectrum (*R. E. Richards* and *M. L. Tomlinson,* Nature, 1948, 162, 693) which shows a carbonyl band at 3,610 cm^{-1}. There is no band round 1,720 cm^{-1} attributable to a hydroxyl group, excluding an earlier ketol structure.

When heated with alcoholic potassium hydroxide, trimethylbrazilone loses the elements of water yielding the potassium salt of α-**anhydrotrimethylbrazilone** (XVII).

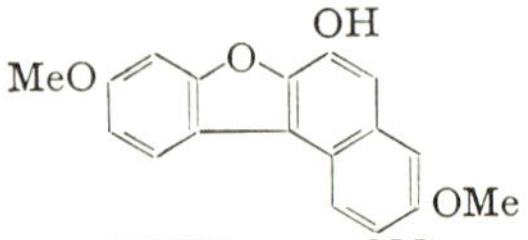

(XVII)

α-Anhydrotrimethylbrazilone

The acetyl-derivative of (XVII) is obtained by treatment of trimethylbrazilone with acetic anhydride. The reactions of α-anhydrotrimethylbrazilone indicated its character as a derivative of β-naphthol, azo-couplings, colour reaction with chloroform and aqueous alcoholic alkali, etc. Conversion to a β-naphthoquinone derivative, trimethoxy-α-brazanquinone, could be accomplished in stages as follows:

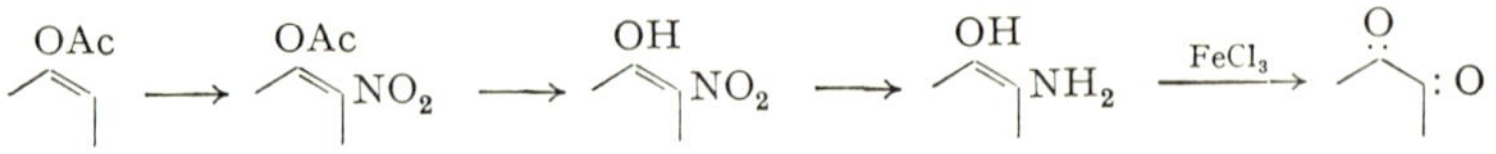

Trimethoxy-α-brazanquinone forms deep brown micro-crystals and condenses with *o*-tolylenediamine to a characteristic quinoxaline derivative. Furthermore, ferric chloride oxidizes α-anhydrotrimethylbrazilone to a very sparingly soluble dinaphthol derivative and from this the corresponding substituted dinaphthylene oxide can be obtained by the action of phosphoryl chloride.

An independent synthesis of α-anhydrotrimethylbrazilone was effected by *P. C. Johnson* and *A. Robertson* (J. chem. Soc., 1950, 2391).

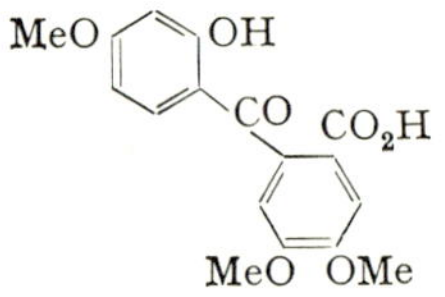

(XVIII) 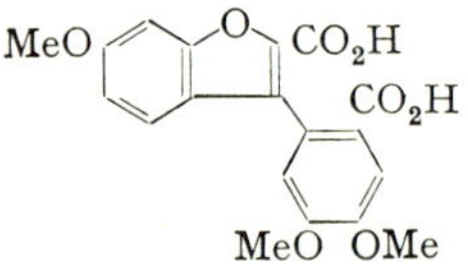 (XIX)

An improved synthesis of brazilinic acid was possible by reacting the known phenolic acid (XVIII) with ethyl bromacetate and potassium carbonate in an acetone medium. The ethyl brazilinate so produced was hydrolysed and the acid condensed with acetic anhydride and sodium acetate to the coumarone derivative (XIX). After eliminating the carboxyl in the coumarone ring, the Arndt-Eistert homologation was applied to the resulting monocarboxylic acid and the product (XX) furnished α-anhydrotrimethylbrazilone on gentle heating with concentrated sulphuric acid.

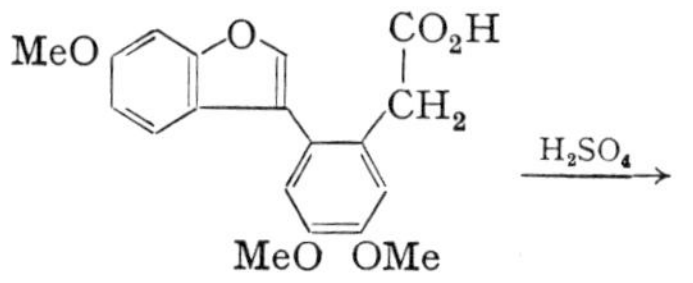

MeO CO$_2$H CH$_2$ MeO OMe (XX) $\xrightarrow{H_2SO_4}$ (XVII)

The action of sulphuric acid on trimethylbrazilone (XVI) was first studied by *Herzig, Pollak* and *Galitzenstein* (Ber., 1904, 37, 631) and the course of the reaction elucidated by *Perkin* and *Robinson* (J. chem., Soc., 1909, 95, 385; cf. *idem, ibid.*, 1908, 93, 501). The product, called **ψ-trimethylbrazilone,** is isomeric with the starting material. It is an acid and gives a good yield of 4:5-dimethoxyhomophthalic acid on oxidation by means of permanganate. When treated with dehydrating agents, ψ-trimethylbrazilone gives **β-anhydrotrimethylbrazilone.** A smooth conversion is that of the ψ-trimethylbrazilone to the acetyl-derivative of the anhydro-compound on heating with acetic anhydride. It transpires that the change may be represented as shown (XXI to XXII).

Trimethylbrazilone (XVI) $\xrightarrow{H_2SO_4}$ MeO HCO$_2$ OMe OMe CH$_2$ (XXI) ⟶

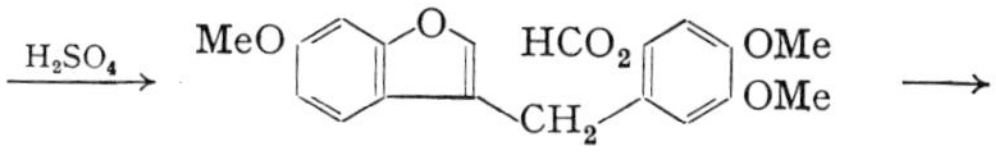

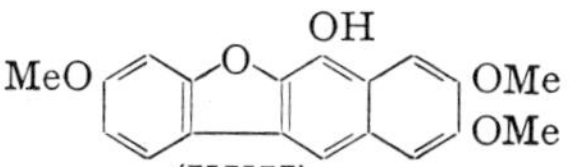

MeO OH OMe OMe (XXII) β-Anhydrotrimethylbrazilone

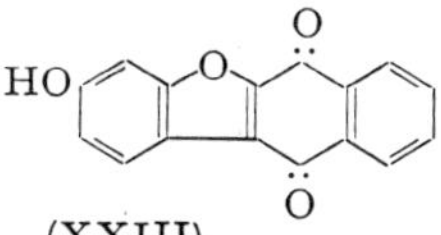

HO O O (XXIII)

Oxidation of β-anhydrotrimethylbrazilone by means of chromic acid furnishes a red quinone which is not a 1:2-quinone, but closely resembles the substance (XXIII) synthesised by reaction of 2:3-dichloronaphthoquinone and resorcinol in alcoholic sodium ethoxide (*Liebermann*, Ber., 1899, 32, 924).

An independent synthesis of β-anhydrotrimethylbrazilone was effected by *K. W. Bentley* and *Robinson* (J. chem. Soc., 1950, 1353). The processes are indicated in the following scheme:

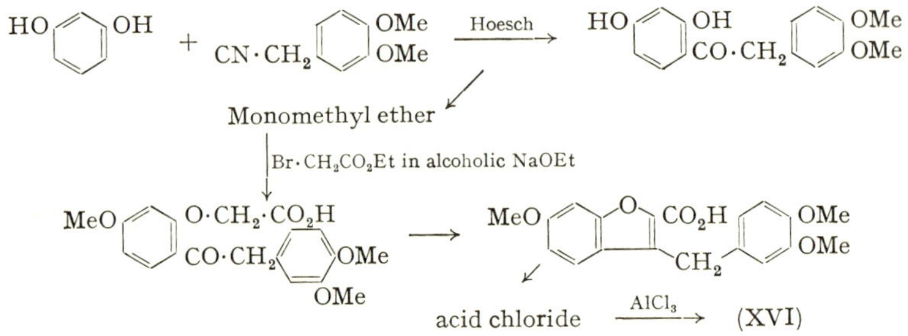

HO–C6H4–OH + CN·CH2–C6H3(OMe)2 —Hoesch→ HO(OH)C6H3–CO·CH2–C6H3(OMe)2

Monomethyl ether

Br·CH_2CO_2Et in alcoholic NaOEt

MeO–C6H3(O·CH2·CO2H)(CO·CH2–C6H3(OMe)2) ⟶ MeO–benzofuran–CO_2H, CH_2–C6H3(OMe)2

acid chloride —$AlCl_3$→ (XVI)

A noteworthy reaction occurs when trimethylbrazilone reacts with phenylhydrazine in hot alcoholic solution (cf. *J. Herzig* and *J. Pollak*, Ber., 1905, 38, 2166; 1906, 39, 265). Nitrogen is evolved and some benzene is produced whilst the diketone simply loses two atoms of oxygen, forming **deoxytrimethylbrazilone** (XXIV). The reaction is by no means smooth nor is the yield particularly favourable. Nevertheless it is substantial and the mechanism of the process is far from obvious. The constitution of deoxytrimethylbrazilone has been confirmed by synthesis (see below); the substance could also be termed anhydrotrimethylbrazilin. The reaction mechanism remains obscure and needs further investigation.

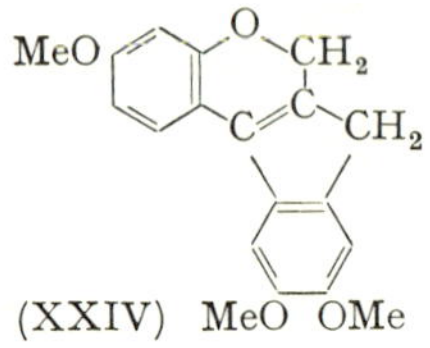

(XXIV)

Deoxytrimethylbrazilone (anhydrotrimethylbrazilin)

The action of concentrated nitric acid on trimethylbrazilone also led to remarkable results (*A. W. Gilbody* and *Perkin*, J. chem. Soc., 1902, 81, 1048; cf. *Perkin* and *Robinson*, *ibid.*, 1909, 95, 389; *Herzig*, *Pollak* and *Vonk*, Monatsh., 1904, 25, 871).

The elements of HNO_3 were added and the crystalline product obtained in nearly quantitative yield was termed nitrohydroxydihydrotrimethylbrazilone. It was recognised as a ketol-lactone of the structure (XXV). When

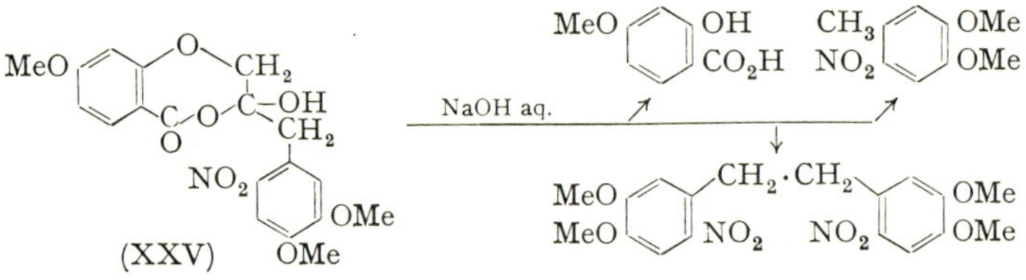

(XXV)

NaOH aq.

MeO–C6H3(OH)(CO2H) + CH3–C6H2(NO2)(OMe)2

MeO, MeO–C6H2(NO2)–CH2·CH2–C6H2(NO2)(OMe)2

treated with aqueous sodium hydroxide it dissolved to a purple solution which gradually became yellowish brown and deposited a precipitate. The latter consisted of 3:4-dimethoxy-6-nitrotoluene along with a small proportion of the related dinitrotetramethoxydiphenylethane. The solution contained 2-hydroxy-4-methoxybenzoic acid. Permanganate oxidation of (XXV) afforded the known 2-carboxy-6-methoxyphenoxyacetic acid (IV). Furthermore, the action of phenylhydrazine on (XXV) gave rise to the osazone (XXVI).

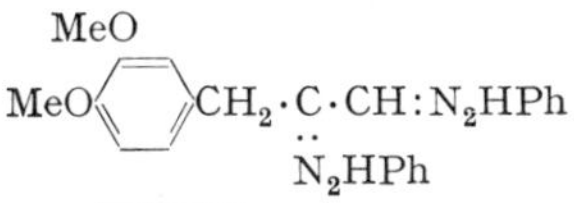

MeO
MeO⟨benzene⟩$CH_2 \cdot C \cdot CH{:}N_2HPh$
N_2HPh
(XXVI)

The formation of (XXV) from trimethylbrazilone is in full accord with the diketonic structure advocated, being simply a displacement of a carbonyl accompanying nitration and fully analogous to known processes such as the following (*V. J. Harding* and *C. Weizmann*, J. chem. Soc., 1910, 97, 1131; cf. *Harding, ibid.*, 1914, 105, 2790).

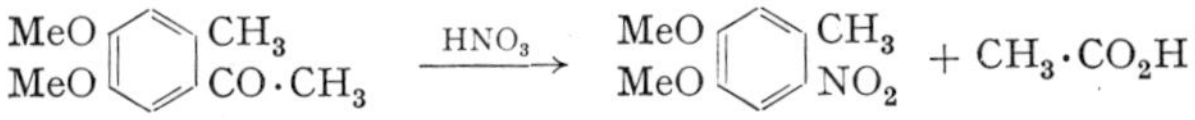

MeO, MeO ⟨benzene⟩ CH_3, $CO \cdot CH_3$ $\xrightarrow{HNO_3}$ MeO, MeO ⟨benzene⟩ CH_3, NO_2 + $CH_3 \cdot CO_2H$

The foregoing noteworthy series of transformations are due to different modes of attack by reagents, summarised in the following scheme:

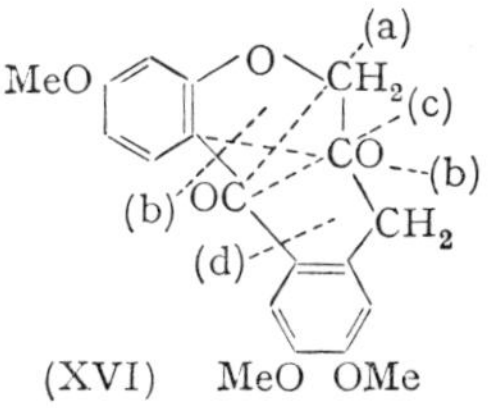

MeO; O; CH_2 (a); (c); CO (b); (b) OC; CH_2; (d); MeO OMe
(XVI)

Transformations of O-trimethylbrazilone (XVI)
(a) NaOH, Ac_2O etc → α-Anhydrotrimethylbrazilone (XVII)
(b) Fission by H_2SO_4 → ψ-trimethylbrazilone (XXI) leading to β-anhydrotrimethylbrazilone (XXII)
(c) Phenylhydrazine → deoxytrimethylbrazilone (XXIV)
(d) Fission by HNO_3 → nitrohydroxydihydrotrimethylbrazilone (XXV)

Brazilein, $C_{16}H_{12}O_5$, *and its derivatives*

Brazilin is a colourless substance readily convertible by oxidation into the colouring matter, brazilein. An alkaline solution of brazilin quickly acquires an intense crimson colour in contact with air and this method can be employed for the preparation of brazilein using ammoniacal solutions. A better method makes use of oxidation by means of iodine in alcoholic solution.

Brazilein separates in small crystals which are browish yellow by transmitted light under the microscope and in mass present a metallic glance resembling that of gun metal. It has only comparatively recently been observed that brazilein is optically active and that it can be reduced to brazilin by means of borohydride in aqueous alkaline solution (*F. Morsingh* and *Robinson*, Int. Congress Pure and Appl. Chem., Zürich, 1953).

Brazilin is obviously the leuco-compound of the quinonoid brazilein which could be formulated as (XXVII), (XXVIII) or (XXIX) on the basis of (XI) for brazilin.

We can proceed to eliminate the possibility (XXIX) and in the section on the synthesis of brazylium salts (below) it is shown that (XXVIII) is the correct structure.

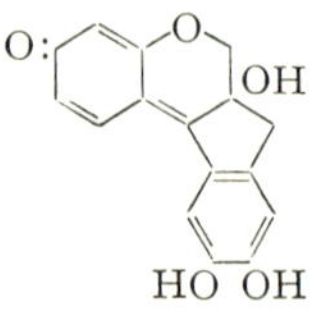

(XXVII) 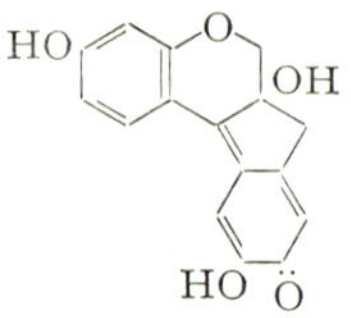 (XXVIII) Brazilein (XXIX) 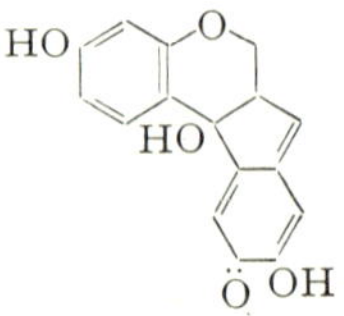 (XXIXa)

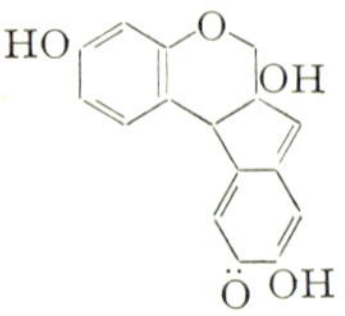

It should be noted that the quinoid pattern of (XXIX) as in (XXIXa) would be derived from the alternative structure (XV→XXIXa) for brazilin which, however, could not give rise to (XXVII) or (XXVIII). The exclusion of XXIX (XXIXa) accordingly also excludes this alternative and for the first time.

The methylation of brazilein by means of methyl sulphate and aqueous potassium hydroxide furnishes trimethylbrazilein and tetramethyldihydrobrazileinol as the main products (*Engels, Perkin* and *Robinson*; J. chem., Soc., 1908, 93, 1115). Less highly methylated products which are also obtained have been little investigated.

Trimethylbrazilein forms a magnificent scarlet formate and when recovered from this derivative may be crystallised in large orange-yellow prisms.

When treated with boiling sodium hydroxide it adds the elements of water affording the colourless trimethyldihydrobrazileinol which on the one hand is easily dehydrated to trimethylbrazilein (e.g. on warming with formic acid) and on the other hand can be methylated to the above mentioned tetramethyldihydrobrazileinol. The relation of these substances is clearly as shown in the following part structures:

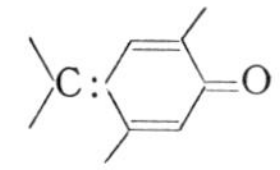

Trimethylbrazilein

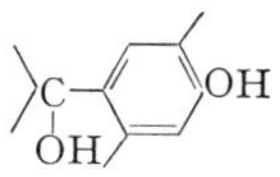

Dihydrotrimethyl-brazileinol

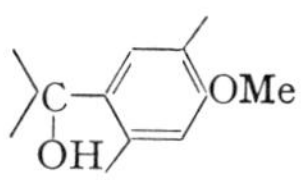

Tetramethyl-dihydrobrazileinol

The important result that excludes the structure (XXIX) and XXIXa) for brazilein is that oxidation of tetramethyldihydrobrazileinol with chromic acid in acetic acid solution affords trimethylbrazilone (XVI). Hence tetramethyldihydrobrazileinol must be (XXX) derived from (XXVII) or (XXVIII) for brazilein.

These considerations serve to determine the position of the alcoholic hydroxyl of brazilin (*Engels et al., loc. cit.*).

An interesting decomposition of brazilein and its derivatives occurs when these quinones are treated with hydrogen peroxide, best in acetic acid solution (*Engels et al.*). Oxidation takes place round the central carbon atom of the diphenylmethane structure and with formation of hydroxyquinol derivatives. An appropriate model was later found in the behaviour of aurine (XXXI) under similar circumstances.

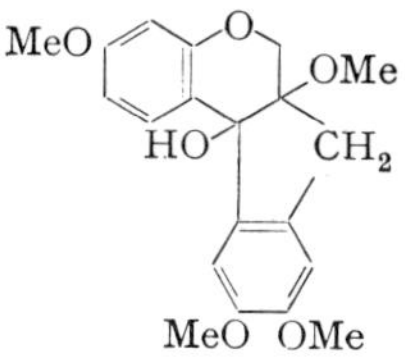

(XXX)

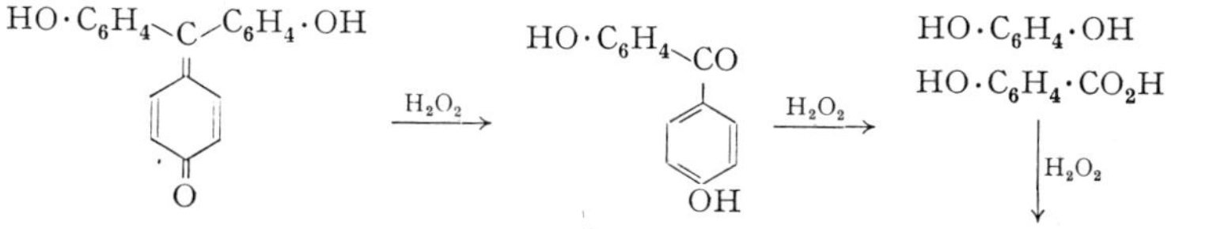

$HO\cdot C_6H_4\cdot C(=C_6H_4=O)\cdot C_6H_4\cdot OH$ (XXXI) $\xrightarrow{H_2O_2}$ $HO\cdot C_6H_4\cdot CO\cdot C_6H_4\cdot OH$ + $HO\cdot C_6H_4\cdot OH$ $\xrightarrow{H_2O_2}$ $HO\cdot C_6H_4\cdot OH$, $HO\cdot C_6H_4\cdot CO_2H$ $\xrightarrow{H_2O_2}$ $HO\cdot C_6H_4\cdot OH + CO_2$

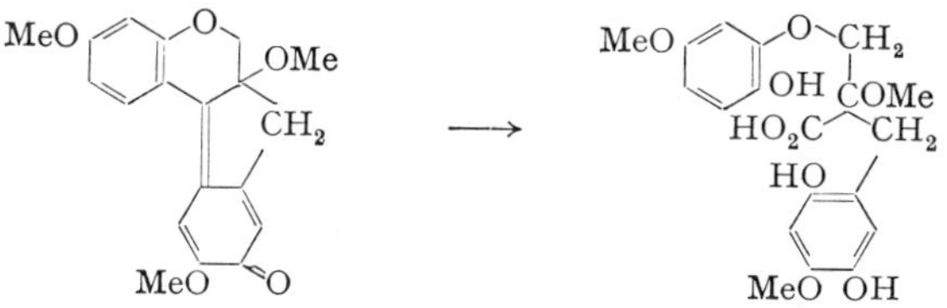

(XXXII) Trimethylbrazilein $\longrightarrow$ (XXXIII)

Thus trimethylbrazilein (XXXII anticipating a sequel) is readily converted into (XXXIII). A significant reaction of brazilein and its derivatives is mentioned immediately in the next section.

The iso*brazilein or brazylium salts*

So long ago as 1882 *J. J. Hummel* and *A. G. Perkin* (J. chem. Soc., 1882, 41, 367) studied the action of mineral acids on brazilein and discovered the so called *iso*brazilein salts. These are produced from their generators with loss of a molecule of water and are of the form B, HX. Thus the intense crimson solution of brazilein in concentrated sulphuric acid develops on gentle warming an orange brown and then exhibits a strong green fluorescence. On addition to water the orange *iso*brazilein hydrogen sulphate is precipitated.

$$C_{16}H_{12}O_5 + H_2SO_4 \longrightarrow C_{16}H_{11}O_4, HSO_4 + H_2O$$

This work was carried out before the theory of oxonium salts was established and long before the analogy of pyrylium with pyridinium was understood. Since 1907 the *iso*brazilein salts have been recognised (*Engels et al., loc. cit.*, cf. J. chem. Soc., 1908, 93, 490) as pyrylium salts and the skeleton of the cation may be termed "brazylium". Thus *iso*brazilein hydrogen sulphate can be formulated as (XXXIV). When tetramethyldihydrobrazileinol is warmed with sulphuric acid, the elements of methanol and of water are eliminated and O-trimethylbrazylium hydrogen sulphate is produced. This

HSO_4 HO HO OH (XXXIV)

$FeCl_4$ MeO MeO OMe (XXXV)

dissolves in water to a yellow solution exhibiting a striking green fluorescence. The corresponding *ferrichloride* (XXXV) can be readily recrystallized, possesses a definite melting point and is a suitable reference compound in the series.

A second method of formation of (XXXV) is from deoxytrimethylbrazilone (XXIV) by oxidation in acid solution. Thus treatment of deoxytrimethylbrazilone with bromine in acetone solution is an especially convenient process giving O-trimethylbrazylium bromide in quantitative yield. The ferrichloride (XXXV) is then produced by treatment with a solution of solid ferric chloride in concentrated hydrochloric acid and may be crystallized from acetic acid.

Synthesis of brazylium and other indenobenzopyrylium salts

In the course of synthetical experiments designed in relation to the older brazilin structure, *Perkin* and *Robinson* (*ibid.*, 1907, 91, 1073) had occasion to study the course of the condensation of salicylaldehyde and its derivatives with ketones such as indan-1-one and its dimethoxy-derivatives. In most cases basic catalysts were employed but these were unsatisfactory when β-resorcylaldehyde was used as the aldehyde component of the system. Turning to hydrogen chloride, for example in acetic acid solution, the formation of stable hydrochlorides was observed and these were equated with Bülow's pyranol salts (*C. Bülow* and *W. von Sicherer*, Ber., 1901, 34, 2368) and soon proved to be anhydro-pyranol salts.

An early preliminary communication (*Perkin* and *Robinson*, Proc. chem. Soc., 1907, 23, 149) described the identity of the salts made on the one hand from resorcinol and benzoylacetaldehyde and on the other from β-resorcylaldehyde and acetophenone.

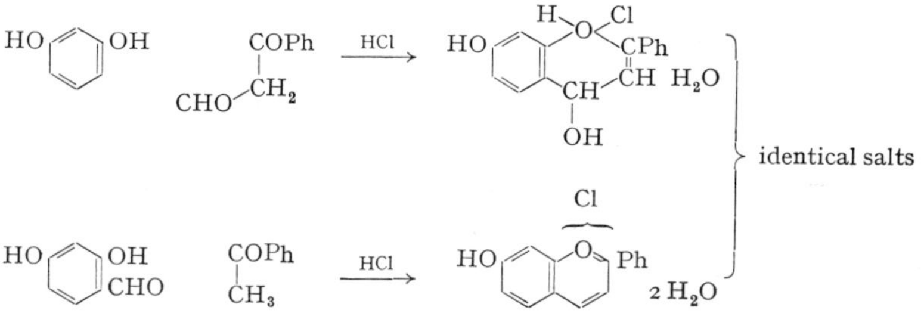

HO, OH; COPh, CH_2, CHO —HCl→ HO, H, O, Cl, CPh, CH, CH, OH, H_2O

identical salts

HO, OH, CHO; COPh, CH_3 —HCl→ HO, Cl, O, Ph, 2 H_2O

The salicylaldehyde-ketomethylene synthesis was generalised (*ibid.*, 1908, 93, 1086); parallel work by *H. Decker* and *T. von Fellenberg* was published at about the same time (Ber., 1907, 40, 3815). The method was later elaborated for the syntheses of anthocyanidins and anthocyanins (See this vol., 84).

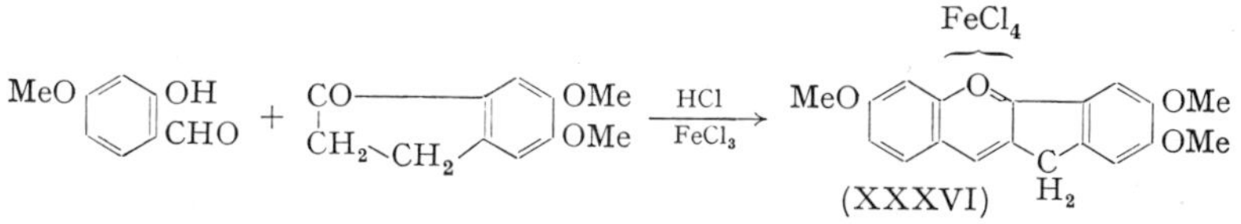

MeO, OH, CHO + CO, CH_2, CH_2, OMe, OMe —HCl, $FeCl_3$→ MeO, $FeCl_4$, O, OMe, OMe, C, H_2

(XXXVI)

MeO, OH, CO, CH_2, CH_2, MeO, OMe —HCl, HCO_2H→ MeO, Cl, O, CH_2, MeO, OMe —$FeCl_3$→ (XXXV)

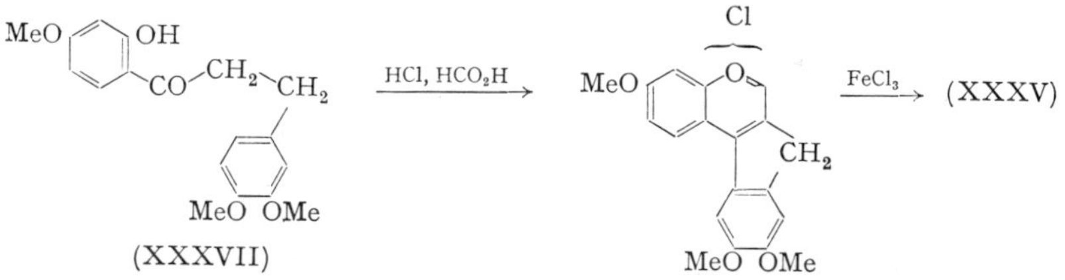

(XXXVII)

In 1907 it was of interest as helping to disprove the earlier brazilin formulations. The salt (XXXVI) prepared from *p*-methoxysalicylaldehyde and 5:6-dimethoxyindan-1-one with the help of hydrogen chloride, and conversion to ferrichloride (*Perkin* and *Robinson*, J. chem. Soc., 1908, 93, 1106), was quite different from O-trimethylbrazylium ferrichloride (XXXV).

The first synthesis of a substance containing the brazilin skeleton was that of O-trimethylbrazylium ferrichloride (XXXV) (*H. G. Crabtree* and *Robinson* J. chem. Soc., 1918, 113, 859). Veratrylidenepaeanol was catalytically hydrogenated to its dihydro-derivative (XXXVII). On treatment with boiling absolute formic acid and zinc chloride and subsequently with ferric

(XXXVIII) (XXXIX)

chloride dissolved in hydrochloric acid, the ferrichloride (XXXV) identical with O-trimethylbrazylium ferrichloride, was obtained. In other experiments it has not been found possible to convert the chromone (XXXVIII) into a brazylium salt and hence the mechanism probably involves ring-closure of (XXXVII) to an indene (XXXIX) followed by O-formylation and further ring-closure.

Synthesis of deoxytrimethylbrazilone and trimethylbrazilone

The synthesis of deoxytrimethylbrazilone (XXIV) was effected by Perkin and Robinson (J. chem. Soc., 1926, 941; 1927, 2094; 1928, 1504) and by Pfeiffer and his collaborators virtually simultaneously and by the same methods (cf. *P. Pfeiffer, O. Angern et al.*, Ber., 1928, 61, 839). The synthesis of veratrylidene-7-methoxychromanone (XL) was earlier described by *Perkin* and *Robinson* (Proc. chem. Soc., 1912, 28, 7). This substance was

(XL) → (XLI) $\xrightarrow{P_2O_5}$ (XXIV)

catalytically hydrogenated to a dihydro-derivative (XLI) which afforded deoxytrimethylbrazilone on cyclodehydration with phosphoric anhydride.

Catalytic hydrogenation of deoxytrimethylbrazilone affords its dihydro derivative (XLII) which is described as O-trimethylbrazilane-*a*. The pyran and cyclopentene rings of (XLII) can be fused in *cis*- or *trans*- configuration. Actually, an isomeric O-trimethylbrazilane-*b* which is possibly the *trans*-form is known (Chatterjea and Robinson, unpublished). The chief reason for this assumption is that the catalytic reduction of (XXIV) would be likely to yield the *cis* form.

Oxidation of (XLII) with chromic acid in acetic acid solution furnished trimethylbrazilone (XVI).

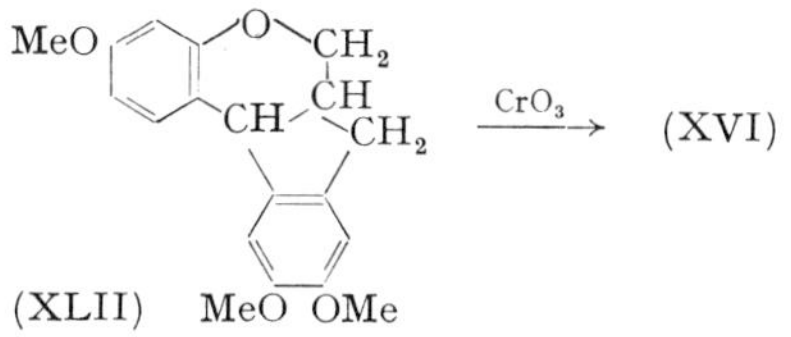

MeO O CH_2 CH CH CH_2 $\xrightarrow{CrO_3}$ (XVI)

(XLII) MeO OMe

The constitution of brazilein

It has been pointed out that addition of water to the quinonoid group of O-trimethylbrazilein produces a new phenolic hydroxyl. This marks the position of the quinone group and in order to label it this phenolic hydroxyl was ethylated. The O-ethyl-trimethyldihydrobrazileinol so obtained was converted by the method already described for the tetramethyl compound into an O-ethyldimethylbrazylium ferrichloride (XLIII or XLIV, dependent on the position of the quinonoid group of trimethylbrazilein).

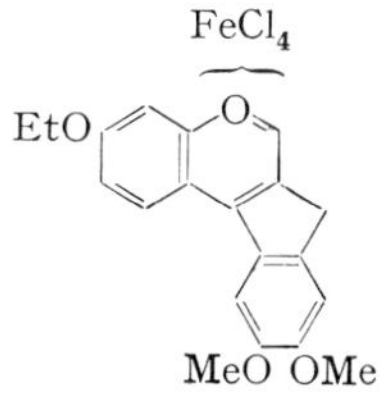

$FeCl_4$ EtO O MeO OMe (XLIII)

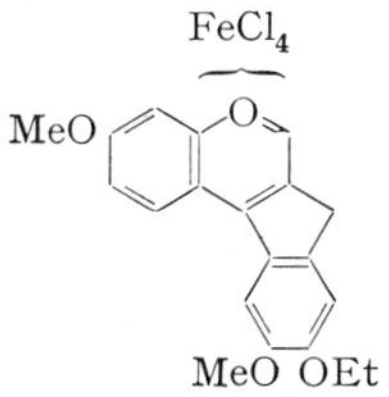

$FeCl_4$ MeO O MeO OEt (XLIV)

These two O-ethyldimethylbrazylium ferrichlorides were synthesized: (XLIII) was made by the Crabtree and Robinson method from veratrylidene-2-hydroxy-4-ethoxyacetophenone whilst (XLIV) was made from 3′-ethoxy-4′-methoxybenzylidene-7-methoxychromanone by conversion to the related deoxy-ethyldimethylbrazilone and thence to the brazylium salt. The synthesised salts proved to be sufficiently divergent in properties and a mixture showed a depressed melting point; (XLIV) was identical with the salt derived from trimethylbrazilein (*V. M. Mićović* and *Robinson,* J. chem. Soc., 1937, 43).

This proves that trimethylbrazilein is (XXXII) and hence brazilein is probably (XXVIII).

Synthesis of brazilin

The most obvious route to the synthesis of trimethylbrazilein was by way of deoxytrimethylbrazilone but all attempts to hydrate the double bond of this substance met with failure. In presence of strong acids oxidation to the brazylium salt occurred and even under oxygen-free conditions disproportionation was noted, the case being analogous to that of certain dihydroquinoline derivatives.

(XLV) $\xrightarrow{FeCl_3,\ HCl}$ (XXXV)

Several years before the synthesis was achieved it had been shown (*Perkin, W. N. Rây* and *Robinson,* J. chem. Soc., 1928, 1504) that reduction of trimethylbrazilone by means of zinc and acetic acid gave the pinacol (XLV) characterised by ready conversion into O-trimethylbrazylium ferrichloride.

(XLVI) Triacetylbrazilone $\xrightarrow[AcOH]{Zn}$ (XLVII) → $\xrightarrow{-H_2O}$ *dl*-brazilein (XXVIII) ↓ *dl*-brazilin ↓ *d*-brazilin

The synthesis depends on the preparation of a similar pinacol in the unmethylated series (*Morsingh* and *Robinson, loc. cit.*).

O-Triacetylbrazilone (XLVI) could be obtained by the chromic acid oxidation of O-tetra-acetylbrazilin and reduction of this substance by means of zinc and alcoholic acetic acid gave the pinacol (XLVII) in moderate yield. The conversion to *dl*-brazilein was realised when this pinacol was subjected to hydrolysis by means of alkalis and the solution acidified.

In order to synthesize triacetylbrazilone (XLVI) dihydrodeoxytrimethylbrazilone-*a* (XLII), already synthesized, was demethylated, the resulting trihydric phenol acetylated and the O-triacetylbrazilane oxidized by means of chromic acid. Triacetylbrazilone (XLVI), identical with the product from triacetylbrazilin, was obtained. This made *dl*-brazilein a synthetic compound. It was reduced by borohydride to *dl*-brazilin and the optical resolution effected by repeated crystallisation of the tetra-*d*-menthyloxyacetyl derivative. After hydrolysis of the resolved material, *d*-brazilin was obtained (not yet published). It is not known with certainty that naturally occurring brazilin is a member of the brazilane-*a* (*cis*-junction) series but this is probably the case in view of the fact that O-tetra-acetylbrazilin can be induced to lose the elements of acetic acid, in a catalysed pyrolysis, with formation of deoxy-O-triacetylbrazilone (XLVIII) (private communication, Ollis). The latter substance has been identified as the product (Herzig) of the acetylating zinc dust reduction of brazilein.

AcO O OAc CH_2 AcO OAc —AcOH→ AcO O CH_2 AcO OAc

(XLVIII)

2. The haematoxylin group

In general, progress in the haematoxylin group followed closely the brazilin developments but the various compounds are not quite so attractive physically, being usually more soluble in solvents, with less power to crystallise and the colours and fluorescences are not so bright as is the case with the brazilin derivatives.

Haematoxylin can be obtained from logwood extract in colourless crystals; the empirical formula is $C_{16}H_{14}O_6$. This contains one oxygen atom more than the formula of brazilin, due to the presence of an additional hydroxyl group, located in the 8 position of the benzopyran moiety of the structure by the synthesis of haematoxylinic acid from pyrogallol and by the formation of pyrogallol when haematoxylin is fused with potassium hydroxide. With four phenolic and an alcoholic hydroxyl group, haematoxylin forms a *pentacetate*, m.p. 165–166°; a *tetramethyl ether*, m.p. 139–140° (*acetate*, m.p. 178–180°); and a *pentamethyl ether*, m.p. 144–147°. Optical rotation, see p. 445.

Permanganate oxidation of d-*tetramethylhaematoxylin*

The analogue of brazilic acid was not isolated but haematoxylinic acid corresponded to brazilinic acid (*Perkin* and *J. Yates*, J. chem. Soc., 1902, 81, 235). On reduction this gave the lactone of dihydrohaematoxylinic acid (XLIX) which was synthesized (*Perkin* and *Robinson*, *ibid.*, 1908, 93, 515).

Condensation of metahemipinic anhydride and pyrogallol trimethyl ether with the help of aluminium chloride could be managed so as to hydrolyse only the *o*-methoxyl group. Reduction of the product and introduction of the $-CH_2\cdot CO_2H$ group furnished the required lactone (XLIX).

MeO MeO O·CH_2·CO_2H O CH CO MeO OMe (XLIX)

MeO MeO O CH_2 CO OC CH_2 MeO OMe (L)

Tetramethylhaematoxylone and *diethylenehaematoxylone*. Oxidation of tetramethylhaematoxylin by means of chromic acid in acetic solution gave tetramethylhaematoxylone (L) (*Perkin*, J. chem. Soc., 1902, 81, 1057) which is a substance entirely analogous to trimethylbrazilone. For example its behaviour with alkalis, acetic anhydride (α-anhydrotetramethylhaematoxylone and its acetyl derivative), nitric acid and phenylhydrazine corresponded to that of trimethylbrazilone. The ψ-tetramethylhaematoxylone series has been little investigated.

Deoxytetramethylhaematoxylone (*Pfeiffer*) and deoxydiethylenehaematoxylone (LI) were synthesized by applications of the method already described in connection with deoxytrimethylbrazilone (*Perkin, A. Pollard* and *Robinson*, J. chem. Soc., 1937, 49).

CH_2 H_2C O O O CH_2 O O (LI) H_2C–CH_2

The properties of these substances closely resembled those of deoxytrimethylbrazilone. Thus oxidation in acid solution gave the respective haematoxylium salts.

Haematein and its derivatives

The technically important point here is that haematein is a much better adjective dye than brazilein; otherwise the two series go in step together. The methylation of haematein gave a high yield of pentamethyldihydrohaemateinol (LII) which was exceptional in this series in that it had a considerable power of crystallisation (*Engels et al., ibid.*, 1908, 93, 1115). Tetramethylhaematoxylium ferrichloride (LIII) was synthesized by *H. G. Crabtree* and *Robinson* (*ibid.*, 1922, 121, 1033).

MeO MeO O OMe HO (LII) MeO OMe

$FeCl_4$ MeO MeO O (LIII) MeO OMe

Synthesis of d-*haematoxylin*

The stages and methods followed the brazilin model closely (*Morsingh* and *Robinson*, Internatl. Congress of Pure and Applied Chem., Zürich, 1953).

Possible natural occurrence of a haematoxylin methyl ether

Haematoxylon africanum, a South African member of the family *Caesalpiniaceae*, afforded an extract, the colour reactions and dyeing properties of which clearly indicated that it contained a substance similar to brazilin and to haematoxylin, but identical with neither. A. G. Perkin has suggested that this may be a methoxy-brazilin (O-methylhaematoxylin) (*Perkin* and *Everest*, The Natural Organic Colouring Matters, Longmans, Green and Co., London, 1918, p. 382; *A. G. Perkin*, J. Soc. Dy. Col., 1918, 34, 99).

If, on the other hand, the substance is brazilin, modified in properties by impurities, there is an interest of a different kind, namely the occurrence of brazilin in a botanical group in which only haematoxylin has previously been identified.

Optical rotatory powers of brazilin and haematoxylin and their derivatives

Brazilin (natural), $(\alpha)_{D}^{21.5°}$ + 121·5°; *brazilin* (reduction of brazilein), $(\alpha)_{D}^{21.5°}$ + 123·4°, both in methanol (*Morsingh* and *Robinson, loc cit.*). *O-tetra-acetylbrazilin*, $(\alpha)_{D}$ + 76·4°; *O-acetyltrimethylbrazilin*, $(\alpha)_{D}^{20°}$ + 128·2 (*Herzig et al.*, Monatsh., 1906, 27, 753). *Haematoxylin* (natural), $(\alpha)_{D}$ + 99·6° (*Hesse*, Ann. chim., 1859, 109, 532); *haematoxylin* (reduction of haematein), $(\alpha)_{D}^{20°}$ + 98·4° (M. and R.). *O-Acetyltetramethylhaematoxylin* $(\alpha)_{D}^{20°}$ + 151·5° (*Herzig et al., loc. cit.*).

Brazilinic acid and trimethylbrazilone are optically inactive, but brazilic acid, which should be optically active, was not examined for this property.

The stereochemical correspondence of brazilin and haematoxylin is probable but has not been directly demonstrated, for example, by X-ray crystallography.

Guide to the Index

This index is constructed in a similar manner to the volume indexes of the first edition of the Chemistry of Carbon Compounds. However, to make the index easier to use, more descriptive entries have been made for the commonly occurring individual, and groups of chemicals.

The indexes cover primarily the chemical compounds mentioned in the text, and also include reactions and techniques, where named, and some sources of chemical compounds such as plant and animal species, oils, etc.

Chemical compounds have been indexed alphabetically under the names used by authors, editing being restricted to ensuring uniformity of entries under the same heading. In view of the alternative nomenclature that can often be used, a limited amount of cross-referencing has been done where it is considered to be helpful, but attention is particularly drawn to Convention 2 below.

In this index, the page numbers are preceded by the letter indicating the volume part. The indexing conventions given below have been used.

1. Alphabetisation

(a) The following prefixes have not been counted for alphabetising:

n-	*o-*	*as-*	*meso-*	D	*C-*
sec-	*m-*	*sym-*	*cis-*	DL	*O-*
tert-	*p-*	*gem-*	*trans-*	L	*N-*
	vic-				*S-*
		lin-			*Bz-*
					Py-

Some prefixes and numbering have been omitted in the index, where they do not usefully contribute to the reference.

(b) The following prefixes have been alphabetised:

Allo	Epi	Neo
Anti	Hetero	Nor
Cyclo	Homo	Pseudo
	Iso	

(c) A letter by letter alphabetical sequence is followed for entries, firstly for the main entry, followed by the descriptive entry. The only exception to this sequence is the placing of plural entries in front of the corresponding individual entries to prevent these being overlooked by a strict alphabetical sequence which could lead to a considerable separation of plural from individual entries. Thus "butanes" will come before *n*-butane, "butenes" before 1-butene, and 2-butene, etc.

2. *Cross references*

In view of the many alternative trivial and systematic names for chemical compounds, the indexes should be searched under any alternative names which may be indicated in the main body of the text. Only a limited amount of cross-referencing has been carried out, where it is considered that it would be helpful to the user.

3. *Esters*

In the case of lower alcohols esters are indexed only under the acid, *e.g.* propionic methyl ester, not methyl propionate. Ethyl is normally omitted *e.g.* acetic ester.

4. *Derivatives*

Simple derivatives are not normally indexed if they follow in the same short section of the text.

5. *Collective and plural entries*

In place of "— derivatives" or "— compounds" the plural entry has normally been used. Plural entries have occasionally been used where compounds of the same name but differing numbering appear in the same section of the text.

6. *Main entries*

The main entry of the more common individual compounds is indicated by heavy type. Where entries relate to sections of three pages or more, the page number is followed by "ff".

Index

Acacetin, **181**, 182
Acacetinidin chloride, 80, 89
Acacetin 7-methyl ether, 182
Acacetin 7-rhamnoglucoside, 181, 182
Acacetin 7-β-rutinoside, 182
Acacia catechin, 245
Acacia catechu, 243
Acacia excelsa, 95
Acacia harpophylla, 95
Acacia intertexta, 95, 96
Acacia mearnsii, 95, 244
Acacia melanoxylon, 94, 95
Acacia mollissima, 95, 244, 245, 281
Acaciin, **181**
Acenaphthylo[5,6-*cd*]thiopyran, 401
3-[α-(Acetamidomethyl)benzyl]-4-hydroxy-coumarin, 116
3-Acetamidophenol, 317
3-Acetamidoxanthone, 317
Acetic anhydride, reaction with sulphuric acid, 30
Acetoacetic esters, 18
Acetoacetic ester, 11, 30, 98, 118, 135, 137, 373
–, reaction with *p*-cresol, 143
–, – with 1-naphthol, 338
–, – with phenols, 99, 100, 140, 142
–, – with phloroglucinol, 122
–, self-condensation, 35
3-Acetoacetyl-2-methylchromone, 146
Acetobromoglucose, 85, 210, 274
Acetonedicarboxylic acid, 338
Acetonedioxalic ester, 37, 39
3-(α-Acetonylbenzyl)-4-hydroxythio-coumarins, 372
4-Acetonyloxycoumarin, 134
Acetonyl phenacyl sulphide, 360
Acetophenone, 4, 5, 6, 70, 76, 80, 168, 179, 373, 439
2-Acetothienone, 70
3-Acetothionaphthenone, 70
Acetoveratrone, 188
ω-Acetoxyacetophenone, 84
7-Acetoxy-3-acetyl-2-methylchromone, 140
6-Acetoxy-7-alkyl-2,3-dihydrobenzofuran, 134
2-Acetoxybenzoic acid, 112
4α-Acetoxy-3-bromoflavan, 248
Acetoxychalcone dibromides, 271
4-Acetoxycoumarin, 112
8-Acetoxydaphnetin, 125
2-Acetoxy-3,4-dihydro-2-pyran, 2
5-Acetoxy-7,4′-dimethoxyflavanone, 280
7-Acetoxyeuxanthone, 325
3-Acetoxyflavanone, 272
7-Acetoxyflavone, 177
Acetoxyisochromanones, 300
8-(1-Acetoxyisopropyl)-9-angeloyloxy-2*H*-furo[2,3-*h*]chromen-2-one, 132
3-Acetoxymethyl-4,6-dimethoxy-2-(3,4-di-methoxyphenyl)coumaran, 242
7-Acetoxy-2-methylisoflavan-4-ols, 251
8-Acetoxymethyl-7-methoxyisoflavone, 213
2-Acetoxymethyl-2-methylthiochroman, 378
2-Acetoxy-6-methylthiochroman, 378
3-Acetoxynaringenin, 280
3-Acetoxy-1-oxaphenalene, 343
4-Acetoxy-ω-*O*-tetra-acetyl-β-glucosidoxy-acetophenone, 85
5-Acetoxytetramethoxyflavone, 192
3-Acetoxy-1-thiaphenalene, 399
2-Acetylaceto-6-methylphenol, 146
Acetylacetone, 2, 22, 31, 77, 356
–, reaction with resorcinol, 70
N-Acetyl-3-aminophenol, 312
Acetylated gossypetin hexamethyl ether, 198
1-Acetylbenzocyclobutene, 289
6-Acetylchroman, 223
2-Acetylcyclohexanone, 169

8-Acetyl-5,7-dihydroxy-3′,4′-dimethoxyflavone, 189
6-Acetyl-5,7-dihydroxyflavone, 180, 206
8-Acetyl-5,7-dihydroxyflavone, 169
Acetylene, reaction with thioxanthone, 391
Acetylenic esters, 12
Acetylenic ketones, 12, 357
3-Acetyl-6-ethoxycarbonyl-5-hydroxy-4,7-dimethylcoumarin, 100
3-Acetylflavone, 167
α-Acetylglutaric ester, 100
3-Acetyl-4-hydroxycoumarin, 112
6-Acetyl-5-hydroxycoumarin, 111
6-Acetyl-7-hydroxycoumarin, 118
7-Acetyl-8-hydroxycoumarin, 138
8-Acetyl-7-hydroxycoumarin, 120
6-Acetyl-5-hydroxy-7,4′-dimethoxyflavone, 176
6-Acetyl-7-hydroxy-5,4′-dimethoxyflavone, 177
2-Acetyl-7-hydroxy-4,5-dimethoxyindane-1,3-dione, 160
6-Acetyl-5-hydroxy-2,2-dimethylchromene, 58
3-Acetyl-2-hydroxyfuro[3,2-*l*]benzo-1,4-quinone, 156
6-Acetyl-7-hydroxy-5-methoxy-2,2-dimethyl chromene, 64
8-Acetyl-7-hydroxy-5-methoxy-2,2-dimethyl 2*H*-chromene, 64
3-Acetyl-7-hydroxy-5-methoxy-2-methylchromone, 153
3-Acetyl-6-hydroxy-2-methyl-4-pyrone, 31
3-Acetyl-6-hydroxy-2-phenyl-4-pyrone, 31
5-Acetyl-4-hydroxy-6-phenyl-2-pyrone, 11, 31
Acetylkaranjic acid, 205
Acetylmeconic acid, 38
4-Acetyl-7-methoxyisochroman-1,3-dione, 302
2-Acetyl-4-methylphloroglucinol, 65
3-Acetyl-6-methylpyrone-5-carboxylic ester, 12
6-Acetyl-8-methylthiochroman, 378
8-Acetyl-6-methylthiochroman, 378
1-Acetyl-2-naphthol, 339
4-Acetylresorcinol, 58
Acetylrotenolone, 261
Acetylsalicylaldehyde, 97, 110
Acetylsalicylic ester, 112
Acetylsalicyloyl chloride, 115
α-Acetylsuccinic ester, 100
O-Acetyltetramethylhaematoxylin, 445
2-Acetylthioxanthones, 394
5′-*O*-Acetylthymidine, 359
6-Acetyl-5,7,8-trimethoxy-2,2-dimethyl-2*H*-chromene, 64
8-Acetyl-5,6,7-trimethoxy-2,2-dimethyl-2*H*-chromene, 64
6-Acetyl-5,7,4′-trimethoxyflavone, 176
3-Acetyl-4,5,6-trimethoxyphthalide, 296
O-Acetyltrimethylbrazilin, 445
6-Acetylumbelliferone, 118
Achillea millefolium, 184
Achyrocline satureoides, 191
Acinos thymoides, 281
Acraldehyde, 41, 235
Acridines, 310
Acronycine, 58
Acrylic ester, reaction with naphthols, 339, 340
Acrylonitrile, 134, 256, 340
Acyl anhydrides, 4
Acylanthocyanins, 90, 93
6-Acyl-4-aralkyl-7-hydroxy-8-methylcoumarins, 138
6-Acyl-6-aryl-2*H*-thiopyran-2-thiones, 357
Acyl chlorides, α,β-unsaturated, 113
3-Acylchromones, 141
3-Acylcoumarins, 111
3-Acyl-3,4-dihydrocoumarin, 111
Acyldithioacetate esters, 357
Acyl halides, reaction with 4-pyrones, 28
O-Acyl-2-hydroxyacetophenones, 139, 140
3-Acyl-4-hydroxycoumarins, 112, 115
5-Acyl-4-hydroxycoumarins, 113
2-Acyloxyacetophenones, 141, 171
4-Acyloxycoumarins, 113
3-Acyl-2-phenylbenzofurans, 74
3-Acylthiopyran-2-thiones, 357
Adamantane, 398
Adonivernite, **184**
Aegle marmelos, 117
Aesculetin, **122**, 123
Aesculetin 7-benzyl ether, 122
Aesculetin dimethyl ether, 123
Aesculetin 6-β-glucoside, 122
Aesculin, **122**, 123

Aesculus hippocastanum, 122
Afromosia elata, 218
Afromosin, **218**
Afzelia sp., 243
Agathisflavone, **203**
Agathisflavone ethers, 203
Agathis palmerstoni, 203
Ageratochromene, 57, **63**
Ageratum sp., 63
Aglycones, 81, 82
Alcohols, α,β-unsaturated, 41
Aldehydes, reaction with acetic anhydride, 97
–, – with aldehydes, 69
–, – with ketones, 4, 5
–, – with malononitrile, 101
–, – with pyrylium salts, 80
–, α,β-unsaturated, 41
Aldol condensation, 71, 97
Alginetin, **152**
Alginic acid, 152
Alkaloids, 2
Alkenyl β-(dialkylamino) alkyl ketones, 365
5-Alkoxycarbinol-6-alkyl-3,4-dihydro-2*H*-thiopyrans, 359
3-Alkoxycarbonyl-2-phenylbenzofurans, 74
2-Alkoxycinnamic acids, 98
4-Alkoxy-4′-dialkylaminoflavylium salts, 73, 148
4-Alkoxy-2-(4-dialkylaminophenyl)benzo-[*b*]pyrylium salts, 148
2-Alkoxy-2,3-dihydro-4-pyrans, 41, 44
2-Alkoxy-5,6-dihydro-2-pyrans, 45
3-Alkoxyflavans, 60
2-Alkoxy-2*H*-flavenes, 62
3-Alkoxy-2*H*-flavenes, 60
α-Alkoxy-2′-hydroxychalcones, 60
2-Alkoxy-3-hydroxytetrahydridopyrans, 50
3-Alkoxy-2-phenyl-2*H*-chromenes, 60
2-Alkoxytetrahydropyrans, 50
2-Alkoxythiotetrahydrothiopyrans, 363
Alkylacetoacetic esters, 142
N-Alkyl-2-aminophenols, 314
Alkylangelicins, 127
4-Alkyl-3-arylcoumarins, 62
Alkyl aryl ketones, reaction with aromatic aldehydes, 5
2-Alkyl-2*H*-benzo[*b*]pyrans, 60
n-Alkylchlorosilanes, 415
Alkylchromenes, 52
2-Alkyl-2*H*-chromenes, 55, 60
Alkyl chromenol ethers, 77
Alkylchromones, 149
–, physical properties, 150
2-Alkylchromones, 140
Alkylcoumarins, 106, 122
Alkyl-3,5-dialkyl-2-furyl ketones, 7
2-Alkyl-2,3-dihydro-4-pyrans, 40
Alkyl-2′,4′-dihydroxychalcones, 269
4-Alkyl-5,7-dihydroxycoumarins, 122
2-Alkyl-1,3-dihydroxyxanthone, 322
3-Alkyl-6,7-dimethoxyisocoumarins, 295
10-Alkyl-9,9-dimethylthioxanthylium salts, 389
3-Alkylflavylium salts, **74**, 75
3-Alkyl-4-halogenotetrahydropyrans, 50
3-Alkyl-4-hydroxycoumarins, 115
Alkyl 2-hydroxyphenyl ketones, 145
2,2′-Alkylidenebiscyclohexanones, 395
4-Alkylidenechromans, 55
Alkylidenephthalides, 292
2-Alkylisoflav-3-enes, 211
2-Alkylisoflavones, 211
Alkyl-3-phenylangelicins, 127
Alkyl-4-phenylpsoralenes, 127
Alkylpsoralenes, 127
Alkylpyrylium salts, 8
Alkylresorcinols, 269
5-Alkyltetrahydrofurylmethanols, 40
2-Alkyltetrahydropyrans, 41
3-Alkyltetrahydropyrans, 363
3-Alkyltetrahydrothiopyrans, 363
6-(Alkylthio)pyran-2-ones, 354
2-Alkylthiotetrahydrothiopyrans, 363
Alkyltriphenylthiopyrans, 352
Alkylxanthenes, 304, 307, 308
Allobergapten, **130**
Alloevodione, **64**
Alloevodionol, **64**
Alloimperatorin, 128, 129
Allomaltol, **33**
Allopatulin, **46**
Allopteroxylin, **159**, 160
Allorottlerin, 66
Alloxanthoxyletin, **135**
Allyl aryl ethers, 235
O-Allylchrysin, 179
Allyldiols, 48

3-Allylflavanones, 269
Allylisoflavones, 213
2-Allylphenols, 104, 228
Allyl phenyl sulphide, 376
6-*C*-Allylquercetin tetra-allyl ether, 193
7-*O*-Allylquercetin tetramethyl ether, 193
8-*C*-Allylquercetin tetramethyl ether, 193
2-Allyltetramethoxyxanthone, 326
Allylthioacetyl chloride, 361
Allyl-δ-tocophenyl ether, 229
Allyl tocyl ether, 229
Alnus sieboldiana, 282
Alpinetin, **278**
Alpinia chinensis, 189, 278
Alpinia japonica, 189, 284
Alpinia officinarum, 189
Alpinone, **284**
Alpinumisoflavone, **220**
Altenuene, **335**
Alternaria solani, 19
Alternaria tenuis, 335
Alternaric acid, **19**
Alternariol, **335**
Alvaxanthone, **329**, 330
Amentoflavone, **202**, 203
Amentoflavone tetramethyl ether, 203
9-Aminobergapten, 130
2-Amino-(3-chloropropyl)benzene, 376
2-(β-Aminocinnamyl) phenols, 168
4-Aminocoumarin, **109**
2-Amino-3-cyano-4-acetamido-4*H*-benzo[*b*]-pyran, 107
3-Amino-4,7-dihydroxy-8-methylcoumarin, 120
5-(β-Aminoethyl)-1-phenylpyrazole, 25
4-Aminoflavans, 383
Aminoflavones, 164
3′-Amino-4′,4‴-flavonyl ether, 170
4-Amino-3-hydroxyflavylium chloride, 172
4-Amino-3-hydroxyflavylium salts, 277
4-Aminoisoflavans, 250
1-Amino-4-methylxanthone, 321
2-Aminophenols, 314
3-Aminophenols, 314
6-Amino-9-phenylfluorene, 312
6-Amino-9-phenylfluorone, 312
Aminothioacrylophenones, 354
4-Aminothiochroman, 381
9-Aminothioxanthene 10,10-dioxide, 391
Aminoxanthones, 320, 321, 394
3-Aminoxanthydrols, 311
Ammidin, **128**, 129
Ammi majus, 126, 132
Ammiol, 156, **157**
Ammi visnega, 137, 156
Ammoresinol, **120**
Amorphigenin, **161**
Amorphin, 161
Ampelopsin, **285**
Ampelopsis meliaefolia, 285
Amurensin, **285**
3-*n*-Amyl-1-hydroxy-6,6,9-trimethyldibenzo-[*b*, *d*]pyran, 336
5-*n*-Amylresorcinol, 336
Amyrolin, **135**
Andrographis serpyllifolia, 188
Andrographis wightiana, 187
Angelica, 118
Angelica anomala, 136
Angelica archangelica, 117, 127, 130, 133, 281
Angelic acid, 132, 133
Angelica decursiva, 136
Angelica edulis, 132
Angelica glabra, 121
Angelica japonica, 159
Angelica saxicola, 119
Angelicin, 126, **127**, 132
9-Angeloxyloxy-8-(1-angeloyloxy-1-methyl-ethyl)-8,9-dihydro-2*H*-furo[2,3-*h*]-chromen-2-one, 133
8-(1-Angeloyloxy-1-methylethyl)-8,9-dihydro-2*H*-furo[2,3-*h*]chromen-2-one, 133
Anhydro-bases, 78, 79
Anhydrobrazilic acid, 429, 430
Anhydrofulvic acid, 160
Anhydropyranol salts, 439
Anhydrosophorol, 287
α-Anhydrotetramethylhaematoxylone, 444
α-Anhydrotrimethylbrazilone, 432, 433
β-Anhydrotrimethylbrazilone, 433
Anhydrovisamminol, 159
4-Anilinocoumarin, 112
2-Anilinomalonylphenol, 112
1-Anilino-4-methylthioxanthone, 392

Anisaldehyde, 157, 190, 219
p-Anisylidenephthalide, 292
Anomalin, **136**
Anthamanta oreoselinum, 132
Anthemis fuscata, 297
Anthocyan, 81
Anthocyanidin β-glucosides, 85
Anthocyanidins, 52, 81–88, 94, 166, 173, 240, 247, 277, 439
–, benzoylated, 84
–, properties, 86
–, structural transformation, 89
–, synthesis, 84
Anthocyanins, 2, 52, 68, 81–88, 174, 439
–, acylated, 84
–, colour, 87
–, complex, 81
–, identification, 89
–, methylation, 83
–, precursors, 93
–, properties, 86
–, structural transformation, 89
–, sugar residues, 83
Anthocyanin 3-biosides, 84
– 3,5-diglucosides, 84
– 3-monogalactosides, 84
– 3-monoglucosides, 84
– 3-pentoseglucosides, 84
– 3-rhamnoglucosides, 84
Anthracene, 299, 402
4*H*-Anthra[1,2-*b*]pyran, 339
Anthra[2,1-*b*]pyrans, 341
Anticoagulants, 116
Antimycin, **290**
Antioxidants, tocopherols, 228
Antirrhinin chloride, **91**
Aphloia madagascariensis, 331
Aphloia theaeformis, 331
Aphloiol, **331**
Apigenidin chloride, **89**, 243
Apigenidin chloride gesneridin, **80**
Apigenin, 89, 170, **180**, 181, 183, 202, 204
Apigenin 7-apiosylglucoside, 181
Apigenin 7-(*p*-coumaroyl)-β-D-glucoside, 181
Apigenin-5,7-dimethyl ether, 182
Apigenin-7,4′-dimethyl ether, 182, 280
Apigenin 7-D-glucoside, 181
Apigenin 4′-methyl ether, 181
Apigenin 5-methyl ether, 181
Apigenin 7-methyl ether, 182
Apigenin 7-rhamnoglucoside, 181
Apigenin 5-rhamnosylglucoside, 181
Apigenin triacetate, 181
Apiin, **181**
Apiose, 181
Apotoxicarol, **266**
Apuleia leiocarpa, 199
Apuleidin, **199**
Apuleirin, **199**
Apuleisin, **199**
Apuleitrin, **199**
Arabinose, 161, 242
Aralkylidenemalononitriles, 354
Araucaria bidwillii, 203
Araucaria cookii, 203, 204
Araucaria cunninghamii, 204
Archangelenone, **281**
Archangelicin, 133
Archangelin, **130**
Arenediazonium salts, 113
–, reaction with malonic acid, 371, 372
Arenethiols, 372, 374
Aromadendrin, 88, **284**
3-Aroylflavones, 141
O-Aroyl-2-hydroxyacetophenones, 139, 140
5-Aroyl-4-hydroxycoumarins, 113
2-Aroyloxyacetophenones, 141
4-Aroyloxycoumarins, 113
Artemetin, **199**
Artemidin, **297**
Artemisetin, **199**
Artemisia absinthium, 199
Artemisia capillaris, 123
Artemisia dracunculus, 297
Artemisia scoparia, 123
Arthraxin, 189, **206**
Arthraxon hispidus, 206
Artocarpesin, **185**
Artocarpesin 7-methyl ether, 185
Artocarpetin, **185**
Artocarpus heterophyllus, 185, 206
Artocarpus integrifolia, 195, 237
Aryl acetoacetates, 99
N-Aryl-2-aminophenols, 314
Aryl β-arylthioethyl ketones, 370

2-Arylbenzofuran-3-carboxylic acids, 171
2-Aryl-4*H*-benzopyran-4-ones, 2
3-Arylbenzo[*c*]pyran-1-thiones, 385
Arylbenzoselenopyrylium cations, 407
2-Arylbenzo[*b*]thiopyrans, 369
3-Aryl-4*H*-benzo[*b*]thiopyran-4-ones, 375
2-Arylbenzothiopyrylium perchlorates, 369
2′-Arylchroman-3-spirocyclopropan-4-ones, 255
Arylchromenes, 52
Arylcoumaranones, 54
Arylcoumarins, 106, 107, 111
Aryl 1,1-dimethylpropargyl ethers, 57
10-Aryl-9,9-dimethylthioxanthylium salts, 389
4-Arylflavans, 247
O-Aryl-2-hydroxyacetophenones, 140
3-Aryl-4-hydroxycoumarins, 113, 115
2-Arylidenecoumaran-3-ones, 143
3-Arylisocoumarins, 294, 299
2-Aryloxybenzoic acids, 317
5-Aryloxy-4,6-diaryl-2-pyrones, 15
2-Aryloxytetrahydropyrans, 50
Aryl-2-pyrones, 12
β-Arylselenopropionic acids, 408
3-Arylthioacryloyl chlorides, 372
(Arylthio)fumaric acids, 374
3-Arylthioisocoumarins, 385
3-Arylthio-2-phenylpropanoic acids, 383
Arylxanthenes, 304, 307, 308
Asaronic acid, **261**
Ascochitine, **290**
Ascochyta fabae, 290
Ascochyta pisi, 290
Ascorbic acid, 167
Aspalathus acuminatus, 184
Aspergillus clavatus, 45
Aspergillus flavus, 33
Aspergillus malignus, 297
Aspergillus melleus, 303
Aspergillus niger, 154
Aspergillus ochraceus, 303
Aspergillus oryzae, 33
Aspergillus versicolor, 3
Asperuloside, **47**
Aster chinensis, 91
Astilbe odontophylla, 285
Astilbe thunbergii, 295
Astilbin, 284, **285**
Atanasin, **205**
Athamantin, **132**, 133
Atropic acid, 383
Atsinoside, **280**
Augustifolionol, 154
Aurantinidin, 86
Auraptenol, **119**
Auraptin, **129**
Auriculatin, **221**
Auriculatin 4′-methyl ether, 221
Auriculin, **221**
Aurine, 437
Aurones, 141, 142, 171, 272, 274, 275
Aurone epoxides, 115, 142
Autumnalin, **219**
Autumnariniol, **335**
Autumnariol, **335**
Avicularin, **193**
Axillarin, **199**, 201
Ayanin, **195**, 199
Ayapanin, 116, **117**
Ayapin, 116, **123**
Azaleatin 3-rhamnoside, 194
Azalein, **194**
Azaleitin, **194**
Azo compounds, 65

Backhousia angustifolia, 154
Badrakemin, **118**
Baeyer–Villiger type oxidation, 74
Baicalein 7-β-L-rhamnofuranoside, 180
Baicalin, 180
Baicalinase, 184
Baker–Venkataraman reaction, 139, 140, 141, 162, 187, 204
Baphia nitida, 288
ψ-Baptigenin, 207
Baptisia lecontei, 282
Baptisia tinctoria, 214
Batatifolin, **186**
Bayin, **182**
Beckmann rearrangement, 116, 381
Bellidifolin, **326**
2-Benzalcoumaran, 244
Benzaldehyde, 1, 4, 151, 269, 298, 311, 312
–, reaction with cyclohexanedione, 309
–, – with sodium polysulphide, 349
Benzaldehyde-2-carboxylic acid, 300
Benzaurins, 312
Benzeins, 312

Benzene, conversion to thioxanthone, 391
Benzeneselenol, 406, 407
Benzenethiol, 402, 406
4-Benzhydrylidenepyran, 26
Benzhydryl phenyl ether, 59
Benzo[*f*]chroman, 340
5,6-Benzochroman, 224, 340
7,8-Benzochroman, 224, 339
5,6-Benzochroman-2-ol, 106
Benzo[*f*]chroman-4-one, 340
5,6-Benzochroman-4-one, 340
7,8-Benzochroman-4-one, 338
7,8-Benzochrom-2-ene, 337
Benzochromenes, 336, 337
Benzo-2*H*-chromenes, **59**
Benzo-4*H*-chromenes, **55**
3,4-Benzochromenes, 334
5,6-Benzochromenes, 339
6,7-Benzochromenes, 341
Benzochromones, 337
5,6-Benzochromone, 339
7,8-Benzochromone, 337
6,7-Benzochromone-2-carboxylic acid, 341
3,4-Benzocoumarin, 334
5,6-Benzocoumarin, 106, 339
6,7-Benzocoumarin, 106, 342
7,8-Benzocoumarin, 106, 338
7,8-Benzocoumarin-4-acetic acid, 338
5*H*-2,3-Benzodiazepines, 294
Benzodifurans, 159
2,3-Benzo-1,1-diphenyl-1-silacyclohex-2-ene, 419
2,3-Benzo-1,1-diphenyl-1-silacyclohex-2-en-4-one, 420
5,6-Benzoflavone, 339
6,7-Benzoflavone, 342
7,8-Benzoflavone, 337
Benzofuranoxanthones, 346
Benzofurobenzofurans, 63
Benzofurobenzopyrans, 63
Benzofuro[3,2-*b*]benzopyran-6-one, 159
6*H*-Benzofuro[2,3-*c*]benzopyran-6-ones, 112
Benzoic acid, 168, 171, 173, 178, 274, 381
–, reaction with resorcinol, 312
Benzoic anhydride, 138, 169, 189
Benzopleinadene-8,13-quinone, 400
Benzo[*b*]pyrans, 51, 52
1*H*-Benzo[*c*]pyran, 288
2*H*-Benzo[*b*]pyran, **55**
4*H*-Benzo[*b*]pyran, **52**
3,4-Benzopyrans, 288
5,6-Benzopyrans, 51
Benzo[*b*]pyranols, 67
5,6-Benzopyranols, 67
1*H*-Benzo[*c*]pyran-1-one, 290
2*H*-Benzopyran-2-ones, 2, 10, 96
4*H*-Benzopyran-4-ones, 2
4*H*-Benzo[*b*]pyran-4-one, 112, 138
Benzopyrones, reaction with Grignard reagents, 71
Benzo-2-pyrones, 10
Benzo-α-pyrones, 10
Benzo[*b*]pyrones, 96
5,6-Benzo-2-pyrones, 96
5,6-Benzo-α-pyrones, 96
5,6-Benzo-4 (or γ) pyrone, 138
Benzopyrylium ferrichloride, 69, 75
Benzopyrylium perchlorates, 144
Benzopyrylium salts, 52, 67–80, 104
Benzoseleninium perchlorate, 406
Benzoselenophene-2-carbaldehyde, 406
Benzoselenopyrans, 405
2*H*-Benzoselenopyran-2-ones, 407
4*H*-Benzoselenopyran-4-one, 407
Benzoselenopyrylium perchlorate, 406, 407, 409
5,6-Benzothiapyrans, 367
Benzo[*b*]thiins, 367
2*H*-Benzo[*b*]thiin, 367
4*H*-Benzo[*b*]thiin, 368
2*H*-Benzothiin-2-ones, 370
4*H*-Benzo[*b*]thiin-4-ones, 372, 373
5,6-Benzothiochroman, 378
6,7-Benzothiochroman, 378
5,6-Benzothiochroman-4-one, 382
7,8-Benzothiochroman-4-one, 382
Benzo[*b*]thiophene carbaldehydes, 380
Benzo[*b*]thiophene-2-carboxylic acid, 371
Benzothiopyrans, 369, 399
Benzo[*b*]thiopyrans, 367, 368
1*H*-Benzo[*c*]thiopyran, 384
2*H*-Benzo[*b*]thiopyran, 367, 368
4*H*-Benzo[*b*]thiopyran, 368
3,4-Benzothiopyran, 384
1*H*-Benzo[*c*]thiopyran 2,2-dioxide, 390
2*H*-Benzo[*b*]thiopyran 1,1-dioxide, 367, 390
Benzo[*b*]thiopyranones, 367

1*H*-Benzo[*c*]thiopyran-1-ones, 384
2*H*-Benzothiopyran-2-ones, 370
4*H*-Benzo[*b*]thiopyran-4-ones, 372, 373
2*H*-Benzo[*b*]thiopyran-2-thione, 371, 377
4*H*-Benzo[*b*]thiopyran-4-thione, 377
Benzothiopyrylium perchlorates, 370, 380
Benzothiopyrylium salts, 368, 369, 383
Benzotrichloride, reaction with phenols, 312
1,2-Benzoxanthene, 345
3,4-Benzoxanthene, 346
1,2:7,8-Benzoxanthene, 346
7*H*-Benzo[*c*]xanthene, 346
12*H*-Benzo[*a*]xanthene, 345
7*H*-Benzo[*c*]xanthen-7-one, 346
12-Benzo[*b*]xanthen-12-one, 345
12*H*-Benzo[*a*]xanthen-12-one, 345
14*H*-Benzo[*a,j*]xanthen-14-one, 346
Benzoxanthones, 345
1,2-Benzoxanthone, 345
2,3-Benzoxanthone, 345
3,4-Benzoxanthone, 346
Benzoxazepinones, 255
Benzoylacetaldehyde, 439
Benzoylacetic acid, 100
Benzoylacetic ester, 70, 99, 163, 337
Benzoylacetone, 11, 31
Benzoylacetonitrile, 100, 115
2-Benzoylacetylcyclohexanone, 169
Benzoylanisoylmethane, 76
o-Benzoylbenzilic acid, 301
Benzoylbenzoic acids, 293, 314
3-Benzoylchromone, 151, 168
2-Benzoylcoumarin, 70
3-Benzoylcoumarin, 70
2-Benzoylcyclohexanone, 12
γ-Benzoyl-β,γ-diphenyl vinylacetic acid, 14
3-Benzoyl-3-hydroxyflavanone, 273
5-Benzoyl-2-hydroxy-3-methylacetophenone, 269
2-Benzoylisobutyraldehyde, 381
Benzoylmeconic acid, 38, 39
2-Benzoyl-4-methyl-4-butanethiolide, 359
6-Benzoyl-8-methylflavanones, 269
Benzoyl nitrate, as nitrating agent, 109
2-Benzoyloxyacetophenones, 140
2-Benzoyloxybenzyl ketones, 74
3-Benzoyloxybutan-2-ones, 292
2′-Benzoyloxychalcones, 250
7-Benzoyloxy-3-hydroxy-2,3,4′-trimethoxyflavanone, 250
Benzoyl peroxide, 122
2-Benzoylphenyl(phenyl)acetic acid, 290
3-Benzoyl-2-phenyl-4*H*,9*H*-pyran[3,2-*h*]-chromene-2,9-dione, 138
Benzoylphloroglucinol, 14
2-Benzoyltetrahydrothiophene, 363
α-Benzoyl-δ-valerothiollactone, 359
2-Benzoylxanthene, 306
2-Benzoylxanthone, 306
1-Benzyl-8-benzoylnaphthalene, 400
2-Benzylcoumaranones, 277
2-Benzylcoumarin-3-ones, 270
2-Benzyl-2,3-dihydrobenzofuran, 61
Benzyl 2,6-dihydroxy-3,4-dimethoxyphenyl ketone, 212
8-Benzyl-5,7-dihydroxy-6-methoxyflavone, 189
9-Benzyl-2,7-dimethylthioxanthene, 389
3-Benzyl-2,6-diphenyl-2*H*-thiopyran-5-carbaldehyde, 349
2-Benzyl-2-hydroxydihydrobenzofuran-3-ones, 171
Benzylhydroxyphenyl ketones, 207, 208, 209
Benzylideneacetone, 116
Benzylideneacetophenone, 4
3,3′-Benzylidenebis(4-hydroxycoumarins), 114
3-Benzylidenechroman-4-ones, 255
Benzylidenecoumaranones, 142
2-Benzylidenecoumarin-3-ones, 141
9-Benzylidene-2,7-dimethylthioxanthene, 389
Benzylidenephthalides, 292
3-Benzylideneselenochroman-4-one, 408
9-Benzylidenethioxanthene, 391
7-Benzyloroxylin A., 180
2′-Benzyloxychalcone epoxides, 270
7-Benzyloxy-5-hydroxytetramethoxyisoflavones, 212
ω-Benzyloxyphloroacetophenone, 189
o-Benzylphenylacetic acid, 299
Benzylpyrylium perchlorate, 102
S-Benzylthioacetyl chloride, 386
9-Benzylthioxanthydrol, 391

4-Benzyl-2,4,6-triphenyl-4-pyran, 3
3-Benzyl-2,4,6-triphenylpyrylium perchlorate, 9
2-Benzyl-2,4,6-triphenyl-2*H*-thiopyran, 349
4-Benzyl-2,4,6-triphenyl-4*H*-thiopyran, 251
7-Benzylwogonin, 180
9-Benzylxanthenes, 308
2-Benzylxanthene, 306
Berberine, 429
Bergamollin, **130**
Bergamot oil, 128, 130, 187
Bergapten, 126, **128**, 129, 130, 131
Bergaptenquinone, 129
Bergaptol, 128, 129, 130
Bergaptyl geranyl ether, 130
Bergenin, **295**
Betanin anthocyanins, 83
Betmidin, **196**
Betula ermanii, 196
Betula middendorfii, 193
Betula verrucosa, 193
Betuletol, **196**
Bhang, 336
8,8″-Biapigeninyl, 203
8,8″-Biapigeninyl hexamethyl ether, 182
4,4′-Bi(2,3-benzo-1,1-diphenyl-1-silacyclohex-2-ene), 420
Bichromonyls, **155**
Bicoumarinyls, 109
Bicyclo[2.2.1]heptane, 396
Bicyclo[3.3.1]nonane, 398
Bicyclo[3.3.1]nonane-2,6-dione, 398
Bicyclo[3.3.0]octane, 397
Bidens tripartita, 278
Biflavanones, 285
Biflavonyls, 201, 204
Biflavonyl ethers, 169, **170**
Bi(flavonyloxy)methanes, 174
Biflorin, **344**
Bignonia chika, 79
Bignonoside, **184**
Bileucofisetinidin, 95
Biochanin-A, 207, **215**
Bipigeninyl, 204
1,8-Bisbromomethylnaphthalene, 344
Bischalcones, 169
1,2-Bis(4-chlorobenzoyl)-ethylene, 319
Bis-3-chloro-2(2*H*-chromenyl) ether, 60
5,6-Bis(chloromethyl)acenaphthalene, 401
o,o′-Biscinnamaldehyde diselenide, 405
3,3′-Biscoumarins, 98
3,7-Bis(dimethylamino)-10*H*-dibenz[*b,e*]-iodinium iodide dihydriodide, 425
3,5-Bis(dimethylamino) methyltetrahydrothiopyran-4-one, 366
3,6-Bis(dimethylamino)selenoxanthene, 409
3,6-Bis(dimethylamino)selenoxanthone, 411
3,6-Bis(dimethylamino)selenoxanthylium chloride, 409
Bisdimethylaminoxanthene, 309, 313, 319
3,6-Bisdimethylaminoxanthone, 319
4,4′-Bis(1,1-diphenyl-1,2,3,4-tetrahydro-1-silanaphthyl), 419
3,3-Bisethoxycarbonyl-1,1-dimethylsilacyclohexane, 416
4,4′-Bis(flav-2-enes), 75
Bisflavenylidenes, 167
Bis(β-formylvinyl) sulphides, 348
5,7-Bis(β-D-glucopyranosyloxy)flavanones, 274
1,3-Bis(2-hydroxybenzoyl)propane, 116
Bis(4-hydroxycoumarinyl)methane, 113
1,8-Bis(hydroxymethyl)naphthalene, 400
3,6-Bis(hydroxymethyl)-4-oxo-1,4-dihydropyridazine, 34
2,2-Bis(2-hydroxy-4-methylphenyl)propane, 233
Bis(4-hydroxy-2-phenyl-4-chromanyl), 273
3,3′-Bis-(isochroman-1,4-dionyl)methane, 302
Bismabenzene, 424
2,2-Bis(2-methoxyphenyl)-5,6-benzo-2*H*-chromene, 59
2,2-Bis(4-methoxyphenyl)-5,6-benzo-2*H*-chromene, 59
2,4-Bis(2-methoxyphenyl)-5,6-benzo-4*H*-chromene, 55
2,4-Bis(3-methoxyphenyl)-5,6-benzo-4*H*-chromene, 55
2,4-Bis(4-methoxyphenyl)-5,6-benzo-4*H*-chrom-2-ene, 55
1,3-Bis(2-methoxyphenyl)-1*H*-naphtho-[2,1-*b*]pyran, 55
1,3-Bis(3-methoxyphenyl)-1*H*-naphtho-[2,1-*b*]pyran, 55
1.3-Bis(4-methoxyphenyl)-1*H*-naphtho-[2,1-*b*]pyran, 55

3,3-Bis(2-methoxyphenyl)-3*H*-naphtho-[2,1-*b*]pyran, 59
3,3-Bis(3-methoxyphenyl)-3*H*-naphtho-[2,1-*b*]pyran, 59
3,3-Bis(4-methoxyphenyl)-3*H*-naphtho-[2,1-*b*]pyran, 59
3,6-Bismethylaminoxanthone, 321
2,6-Bis(methylthio)-3,5-diphenyl-4*H*-thiopyran-4-thione, 358
1,3-Bis(4-pentenyldimethyl)disilacyclohexane, 418
1,8-Bis(phenylhydroxymethyl)naphthalene, 344
Bis(2-propynyl) sulphone, 348
Bis(selenopyrylium)platinum hexachloride, 403
1,3-Bistosyloxymethylcyclopentane, 396
3,3-Bis-(5,7,4′-trihydroxyflavone), 204
6,8″-Bis-(5,7,4′-trihydroxyflavone), 203
4,4′-Bithioflavylidene, 374
9,9′-Bithioxanthyl, 389
Bithioxanthylidene, 390, 391, 393
9,9′-Bixanthyl, **306**, 311
Bixanthylene, **306**, 307
Braylin, 57, **135**
Brazilane, 443
Brazilein, 428, 443
–, constitution, 441
Brazilic acid, **429**, 445
Brazilin, 151, **244**, 427, 428, 429, 430, 431, 435, 436, 443
–, optical rotary powers, 445
–, synthesis, 442
Brazilin acetates, 428
Brazilin ethers, 428
Brazilinic acid, **429**, 431, 432, 445
Brazylium salts, 438, 439
Bresil, 427
Brevifolinecarboxylic acid, 288
Brickelia pendula, 191
Brickelia squarrosa, 205
Bridged-ring sulphur compounds, 396
3-Broman-2,2-dimethylchroman-4-ol, 236
ω-Bromoacetophenone, 360
5-Bromo-*n*-amyl acetate, 48
3-Bromo-4-chlorochroman, 224
4-Bromo-3-chlorochroman, 224
Bromochromans, 223, 224
4-Bromochroman, 55
Bromochromanols, 236
3-Bromochroman-4-ones, 254
3-Bromo-4*H*-chromene, 53
6-Bromocomenic acid, 27
3-Bromocoumalic acid, 17
3-Bromocoumarin, **109**, 145
7-Bromocoumarin, **109**
Bromocyanidin tetramethyl ether, 240
4-Bromo-1,1-dichloro-1-silacyclohex-2-ene, 417
1-Bromo-2,3-dihydro-1*H*-naphtho[2,1-*b*]-thiopyran, 367
3,4-Bromo-3,4-dihydro-2*H*-naphtho[1,2-*b*]-thiopyran, 378
Bromodihydronaphthothiopyranols, 379
Bromodihydronaphthothiopyranones, 379
3-Bromo-5,6-dihydro-4-pyran, 41
5-Bromo-2,3-dihydro-4-pyran, 41
7-Bromo-2,2-dimethyl-2-silanorcarane, 418
3-Bromo-2,6-dimethylthiochromanone, 379
2-(2-Bromoethyl)benzaldehyde, 298
2-Bromoethylbenzo[*b*]thiophene, 379
2-(2-Bromoethyl)benzyl bromide, 386
1-Bromoethyl-2-β-bromoethylcyclohexane, 387
3-Bromoflavan, 53
4-Bromoflavan, 61
3-Bromoflavanone, 142, 271
8-Bromoflavanone, 275
Bromo-3-hydroxycoumarins, 111, 115
6-Bromo-4-hydroxycoumarin, 115
8-Bromo-5-hydroxyflavone, 166
5-Bromo-6-hydroxy-2-methylchromone, 152
2-Bromo-4-(2-hydroxyphenyl)butanoic acid, 225
2-Bromo-1-(2-hydroxyphenyl)-3-phenylpropan-3-ol, 53
1-Bromoisochroman, 298
Bromomalonic ester, 293
8-Bromo-5-methoxyflavone, 166
4-Bromo-7-methoxyindanone, 303
3-Bromo-4′-methoxy-6-methylflavan-4-ols, 250
6-Bromo-4-methylcoumarin, 109
7-Bromo-4-methylcoumarin, 109
6-Bromo-4-methyl-3-phenylcoumarin, 109

7-Bromo-4-methyl-3-phenylcoumarin, 109
2-Bromomethylphenylethyl bromide, 298
3-Bromo-6-methylthiochromanone, 379
2-Bromomethylthioxanthone, 394
4'-Bromo-3'-nitroflavone, 204
Bromonium ions, 426
1-Bromo-1-phenylgermacyclohexane, 420
3-(2-Bromophenyl)propyl bromide, 419
3-Bromopropyne, 347
3-Bromo-4-pyrone, 27
4-Bromotetrahydropyran, 44
2-Bromo-1-tetralone, 225
4-Bromo-6-thiabicyclo[3.2.1]octane, 397
3-Bromothiochroman-4-ol, 379
3-Bromothiochromanones, 372, 380
3-Bromothiochromones, 380
Bromothiocoumarins, 371
3-Bromothioflavanones, 374, 383
3-Bromothioxanthones, 394
Bromoxanthones, 319, 320
Buchanaxanthone, **329**
Buddleflavonoloside, 182
Budleia variabilis, 182
Bülow synthesis, 70
Butea frondosa, 93, 278
Butein, 271, 272, **278**
3-(But-1-enyl)isocoumarin, 297
Butin, 271, 272, **278**
Butin 7-*O*-D-glucopyranoside, 278
Butrin, **278**
5-Butylamino-5,6-dihydro-2*H*-thiopyran 1,1-dioxide, 361
tert-Butylhydroperoxide, 387, 396
4-*tert*-Butyltetrahydrothiopyran 1-oxide, 363
tert-Butyl-4-thiahexanoate, 365
6-*tert*-Butylthiochroman-4-one, 381
Byakangelicin, **130**
Byakangelicol, 129, **130**

Caesalpiniaceae, 445
Caesalpinia braziliensis, 427
Caesalpinia crista, 427
Caesalpinia digyna, 296
Calcicolin, **119**
Callistephin, 90
Callistephin chloride, 85, **90**
Callistephus chinensis, 90
Calophyllolide, 137
Calophyllum sp., 328
Calophyllum brasiliense, 328
Calophyllum canum, 329
Calophyllum fragrans, 329
Calophyllum inophyllum, 137
Calophyllum papuanum, 257
Calophyllum scriblitifolium, 329, 331
Calycopterin, **199**
Calycopteris floribunda, 199
Calycosin, **214**
Campherol, **190**
Canescin, **297**
Cannabichromene, 64, 231
Cannabicyclol, **231**
Cannabidiol, **336**
Cannabinol, **336**
Cannabis indica, 336
Cannabis sativa, 231, 336
Capraria biflora, 344
Carajurin, **79**
Carbohydrates, 2
–, conversion to kojic acid, 33
Carbon suboxide, 35
Carbonyl compounds, conversion to pyrans, 41
Carbonyl reactions, 146
2-(2-Carboxybenzoyl)-4,4'-ditolyl sulphide, 402
3-(2-Carboxybenzyl)isocoumarin, 291
2-Carboxybenzyl ketones, 292, 293
2-Carboxybenzyloxyacetic acid, 301
2-Carboxybenzylselenoacetic acid, 409
2-Carboxy-4,5-dimethoxyphenylacetaldehyde, 292
3-(2-Carboxy-4-methoxybenzyl)-7-methoxyisocoumarin, 291
2-Carboxy-5-methoxyphenoxyacetic acid, 429
2-Carboxymethylphenoxyacetic acids, 251
3-(2-Carboxy-4-nitrobenzyl)-7-nitroisocoumarin, 291
(2-Carboxyphenyl)acetic acids, 293
o-β-Carboxyphenylvinyl alcohol, 290
9-(2-Carboxyphenyl)xanthene, 313
9-(2-Carboxyphenyl)xanthydrol, 313
β-Carotene, 349
Carpenteles brefeldianum, 160
Carthamidin, **282**

Carthamin, **282**
Carthamus tinctorius, 282
Caryopteris clandonensis, 342
α-Caryopterone, **342**
Casimiroa edulis, 130, 185
Cassia marginata, 96
Cassia occidentalis, 331
Cassiaxanthone, **331**
Cassiollin, **331**
Castanospermum australe, 182, 214, 218
Casticin, **197**
Castor fiber, 232
Casuarin, **243**
Casuarina equisetifolia, 243
Catalpa bignoniodes, 184
Catechins, 88, 93, 96, 237, 238, 246, 284, 427
–, oxidation to anthocyanidins, 89
–, relation to tannins, 245
–, stereochemistry, 240, 241
–, structure, 238, 239, 240, 241
Catechin, 94, **242**, 243
Catechin-7-L-arabinoside, 242
Catechin tetramethyl ether, 95
Catechol, 112, 240
Catechu, 238
Catechu brown, 238
Caviunin, **217**
Caviunin ethers, 217
Ceanothus velutinus, 186
Cedrelopsis grevei, 160
Cedrus deodara, 284
Celebixanthone, **327**
Cellulose, dry distillation, 32
Centaurea cyanus, 90
Centaurea jacea, 197
Centaureidin, **197**
Centaureidin 7-glucoside, 197
Centaurein, **197**
Cephalotaxoside, **181**
Cephalotaxus drupacea, 181
Cercidiphyllum japonicum, 284
Cerdithiene, 402
Cerothiene, **401**
Chalcones, 69, 84, 172, 174, 269, 270, 274
–, conversion to isoflavones, 209
–, from flavanones, 271
–, oxidation, 272
Chalcone, 4, 76, 162, 163
Chalcone acetates, 280, 284
Chalcone dibromides, 162
Chalcone epoxides, 209
Chelidamic acid, 37
Chelidonic acid, 20, 22, 29, 36, **37**, 38
–, alkyl esters, 37
Chelidonic ester, 25, 26
Chelidonium majus, 20, 37
Chen-pi, 188
Chloro(chloromethyl)dimethylsilane, 418
Chlorochromans, 224
6-Chlorochroman, 222
Chlorochromanols, 236
4-Chlorochroman-3-ols, 251
3-Chlorochroman-4-ones, 254
4-Chlorocoumarin, **109**, 114, 151
α-Chloro-α-deoxykojic acid, 31
3-Chloro-6-dichloromethylflavone, 167
1-Chloro-2-(3-diethylchlorogermylpropyl)-cyclopropane, 421
8-Chloro-5,7-dihydroxy-2,6-dimethyl-chromone, 154
Chlorodimethylchromanols, 236
Chlorodiphenylchromanols, 236
2-Chloro-3,6-diphenylthiopyrylium chloride, 351
3-Chloroflavanones, 271
3-Chloroflavone, 109, 167, 271
3-Chloro-6-formylflavone, 167
3-Chloro-2-hydroxytetrahydropyran, 50
2-Chloro-6-mercaptobenzoic acid, 392
Chloromethoxydimethyl chromanols, 236
3-Chloro-7-methoxyflavone, 167
2-Chloromethyl-2*H*-benzothiopyran, 368
3-Chloro-3-methylbutyne, 118, 119
3-Chloromethylcoumarin, **110**
7-Chloro-4-methylcoumarin, 110
2-Chloromethyl-5,6-dihydro-2*H*-thiopyran, 360
4-Chloromethyl-3,4-dihydro-2*H*-thiopyran, 360
3-Chloromethyl-5,6-dimethyl-2-pyrone, 15
3-Chloro-6-methylflavone, 167
4′-Chloro-4-methylflavylium perchlorate, 80
2-Chloromethyl-5-hydroxy-4-pyrone, 31
2-Chloromethyl-5-mesyloxy-4-pyrone, 34
Chloromethyl 2-methoxydiphenylmethyl ketone, 251
8-Chloromethyl-7-methoxyisoflavone, 213

6-Chloromethyl-7-methoxy-4-methyl-
coumarin, 110
3-Chloromethyl-4-methylcoumarin, 110
3-Chloromethyl-6-methyl-5-propyl-2-pyrone,
15
1-Chloromethyl-1-methylsilacyclopentane,
414
7-Chloro-4-methyl-3-phenylcoumarin, 109
Chloromethyl β-phenylethyl ether, 297
1-Chloro-1-methylsilacyclohexane, 414
1-Chloro-4-methylthioxanthone, 392
1-Chloro-4-methylxanthone, 320
2-Chloronaphthalene, 368
Chloronitrobenzaldehydes, 114
5-Chloropentan-1-ol, 48
3-Chloroperbenzoic acid, 225, 390
2-Chlorophenylbenzoylacetylene, 163
6-(4-Chlorophenyl)-5-hydroxy-2-hydroxy-
methyl-4-pyrone, 34
2-Chloro-6-(phenylthio)benzoic acid, 392
Chlorophora tinctoria, 195
2-(γ-Chloropropyl)phenol, 222
γ-Chloropropyl phenyl ether, 222
2-Chloroselenoxanthone, 411
2-Chlorotetramethoxyisoflavan, 242
6-Chlorothiochroman-4-one, 382
Chlorothiochromones, 373
3-Chlorothiochromone, 374
Chlorothiocoumarins, 371
4′-Chlorothioflavanone, 383
6-Chlorothioisoflav-2-en-4-ol, 376
6-Chlorothioisoflavone 1,1-dioxide, 376
2-Chlorothioxanthene-9-thione, 393
1-Chlorothioxanthone, 392
2-Chlorothioxanthone, 393
1-Chlorothioxanthydrol, 391
5-Chloro-1-(trichlorosilyl)pent-1-ene, 417
Chlorotrimethylsilane, 418
β-Chlorovinylcarbaldehydes, 348, 357
β-Chlorovinyl ketones, 5, 12
1-Chloro-1-vinylstannacyclohexane, 422
Chloroxanthones, 320
4-Chloro-3,5-xylenol, 98
Cholestanol, 359
Chromans, 51, 52, 59, 64, 72, 75, 147, 222,
223
–, 3,4-disubstituted, 225
–, from chromenes, 56
–, naturally occurring, 226
Chromans, (*continued*)
–, rearrangement, 226
–, substituted, 224
Chroman, 52, 53, 56, 222, **223**, 225
Chroman-2-carboxylic acid, 225
Chroman-3,4-diol, 56
Chroman-2,3-dione, 111
Chromanochromanones, 257
Chromanols, 104, 105, 234, 236
Chroman-2-ol, 53, **234**
Chroman-3-ol, 108, **235**
Chroman-4-ols, 72, **236**, 254
–, oxidation, 237
–, 4-substituted, 55
Chroman-4-ol, 55
6-Chromanols, 229
Chromanones, 144, 147, 210, 223, 251
–, dehydrogenation, 72
Chroman-2-ones, 254
Chroman-2-one, **251**
Chroman-3-ones, **251**, 252
Chroman-4-ones, 52, 236, 251, **252**, 253,
254, 255
–, reaction with Grignard reagents, 71
Chroman-4-one, 64, **253**, 254
2-Chromanyl acetate, 53
Chromenes, 51, 52, 75, 78, 105, 145, 223,
250, 405
–, coloured photoproducts, 56
–, oxidation products, 52
α- or 2*H*-Chromenes, 55, 56, 71, 72, 234
2-Chromene, **52**
–, reactions, 56
3-Chromene, **55**, 236
4*H*- or β-Chromenes, 52, 53, 54
Chrom-2-ene mercurichlorides, 234
2*H*-Chromene-2-thione, **110**
Chromenols, 67, 71, 75
α- or 2*H*-Chromenols, 67
β- or 4*H*-Chromenols, 67, 71
Chromones, 2, 86, 96, 98, 99, 138, 139,
141, 142, 143, 254
–, carbonyl group reactivity, 166
–, hydrogenation, 147
–, oxonium salts, 73, 145, 148
–, properties, 144
–, reaction with Grignard reagents, 67, 71,
147
–, sulphonation, 148
–, synthesis, 140

Chromone, 20, 52, 144, 145, **149**, 170
–, dimerisation, 148
–, reaction with amines, 147
Chromone-2-carboxylic acid, 54, 148, 149, **151**
Chromone perbromides, 145
Chromonesulphonic acids, 148
Chromonoisochromans, 167
Chromonoisocoumarins, 161, 167
2-(3-Chromonyl)chromanone, 148
3-Chromonylphenylmethanol, 151
1-(3-Chromonyl)-2-salicyloylethylene, 148
Chromylium salts, 52, 67, 68, 69, 70
Chrysanthemin chloride, **91**
Chrysanthemum indicum, 91
Chrysergonic acid, **332**
Chrysin, 170, 175, **178**, 179, 180, 189
Chrysin 7-glucoside, 179
Chrysinidin chloride, **80**, 86
Chrysin 7-methyl ether, 179
Chrysosphenol, **199**
Chrysosplenetin, **199**
Chrysosplenetin 4′-glucoside, 199
Chrysosplenin, **199**
Chrysosplenium japonicum, 199
C.I. Acid Orange II, 316
C.I. Basic Red I, 314
Cicer arietinum, 215
Cichoriin, **123**
Cichotium intybus, 123
Cinenic acid, 50
Cinnamaldehyde, 62, 151
Cinnamic acid, 269, 274
Cinnamic anhydride, 208
Cinnamoyl chloride, 269
3-Cinnamoyl-4-hydroxy-2-pyrones, 36
2-Cinnamoylphenyl methyl selenide, 408
o-Cinnamylphenols, 62
Cirsium oleraceum, 185
Cistus ladanifera, 190
Citric acid, 34
Citrinin, **290**
Citromitin, **285**
Citromyces sp., 160
Citromycetin, **160**
Citromycin, 160
Citronetin, **281**
Citronin, **281**
Citropten, **120**
Citrus sp., 279, 281
Citrus aurantifolia, 130
Citrus aurantium, 181
Citrus bergamia, 130
Citrus decumana, 279
Citrus deliciosa, 188, 282
Citrus jambhiri, 188
Citrus limon, 197
Citrus mitis, 285
Citrus nobilis, 188
Citrus nobilis deliciosa, 188
Citrus oils, 120
Citrus paradisi, 281
Citrus poonensis, 282
Citrus reticulata, 188
Citrus sudachi, 187, 188
Cladinose, **51**
Cladrastin, **218**
Cladrastis lutea, 218
Cladrastis platycarpa, 218
Cladrin, **218**
Claisen rearrangement, 119, 120, 121, 128, 139, 140, 193, 229, 291, 333
Clausenidin, **136**
Clausenin, **136**
Clavacin, 45
Claviceps purpurea, 332
Clavitin, **45**
Clematis terniflora, 181
Clemmensen reduction, 211, 223, 232, 363, 364, 376, 385
Cocoa-red, 245
Coerulein, 313, **316**
Coladin, **118**
Coladonin, **118**
Colour bases, 78
Colours, of flowers, 81
Columbianadin, **133**
Columbianadin oxide, 133
Columbianetin, **133**
Columbianin, **133**
Comanic acid, 22, 29, **36**, 37, 51
Combretol, 191, **196**
Combretum quadrangulare, 196
Comenamic acid, 37
Comenic acid, 23, 26, 27, 33, **36**, 37, 38
–, methyl ether, 33

Complex anthocyanins, 81
Compound W13, **203**
Condensed tannins, 244, 245, 246
Cordyla africana, 218, 287
Corn oil, 226
Coronilla sp., 127
Corydalis ochotensis, 32
Corylopsis spicata, 295
Corymbiferin, **326**
Cosmetin, **181**
Cosmos bipinatus, 181
Cosmosiin, **181**
Cosmosine, 183
Cottonseed oil, 226
Coumalic acid, 10, 14, 17, **18**
Coumalic ester, 16
Coumalin, 17
Coumarans, 59, 228
Coumaric acids, 98, 102, 111
Coumaric acid, 132, 181
p-Coumarinic acid, 183
Coumarins, 2, 10, 57, 67, 71, 96, 141, 143, 145, 294
–, from malononitrile, 101
–, fused to 2,2-dimethylpyran rings, 97
–, hydrolysis, 102, 103
–, natural, 97
–, occurrence, 97
–, photodimerisation, 107
–, properties, 102
–, reaction with alkylmagnesium halides, 105
–, – with Grignard reagents, 104
–, synthesis, 97, 99, 100, 101, 102
Coumarin, 52, 54, 59, 75, 97, **106**, 222
–, arylation, 107
–, dimerisation, 107
–, hydrogenation, 251
–, hydrolysis, 102, 110
–, nitration, 108
–, oxidation, 235
–, photodimerisation, 107
–, protonation, 108
–, reactions, 107, 108, 109, 110
–, reaction with methyllithium, 57
–, reduction products, 104
–, synthesis, 102
Coumarin-3-carboxylic acid, 98
Coumarin-4-carboxylic acids, 101
Coumarin-8-carboxylic acid, **111**
Coumarin dibromide, 109
Coumarin-2,3-diones, 101
Coumarindisulphonic acids, 109
Coumarinic acids, 98, 102
Coumarin-6-sulphonic acid, **109**
Coumarones, 59
3-Coumaroxy-7-hydroxy-6-methoxycoumarin, 125
6-(7-Coumaroxymethyl)-7-methoxycoumarin, 124
Coumoxin, **115**
Coumurrayin, **121**
Cracca toxicaria, 264
Crataegus curvisepala, 183
Crataegus oxyacantha, 193
Crataegus phenophyrum, 281, 282
Cratenacin, **183**
Cratoxylon celebicum, 327
m-Cresol, reaction with acetone, 233
p-Cresol, reaction with acetoacetic ester, 99, 143
–, – with phthalic anhydride, 313
Crotonaldehyde, 44, 56
Crotonic esters, 11
–, phenols, 252
Crotonic ester, reaction with oxalic ester, 18
Crotonoyl chloride, 17
3-Crotonoyl-4-hydroxy-6-methyl-2-pyrone, 17
Croton sparsiflorus, 55
Crotyl phenyl sulphide, 377
Crotyl *m*-tolyl sulphide, 377
Cryptomeria japonica, 201
Cryptomium falcatum, 283
Cryptostrobin, **283**
Cube, 257, 264
Cube gambier, 238
Cunninghamia lanecolata, 203
Cupressuflavone, 203
Cupressuflavone ethers, 203
Cupressus sempervirens, 203
Cupressus torulosa, 203
Cyanamide, reaction with 4-pyrones, 24
Cyanidin, 82, 84, 89, 91, 243
Cyanidin bromide, 90
Cyanidin 3-cellobioside, 90
Cyanidin chloride, 82, 88, **90**, 93, 94, 96, 240

Cyanidin-3,5-di-β-glucoside, 90
Cyanidin 3-galactoside, 91
Cyanidin 3-gentiobioside, 90
Cyanidin 3-β-glucoside, 91
Cyanidin iodide, 240
Cyanidin pentamethyl ether, 88
Cyanidin 3-rhamnoglucoside, 91
Cyanin, 87, 91
Cyanin chloride, **90**
Cyanoacetamide, 104, 107
Cyanoacetic acid, 101
Cyanoacetic ester, 2, 115, 356
Cyanoalkynes, 349
3-Cyano-3,4-dihydro-4-acetamidocoumarin, 107
6-Cyano-5,6-dihydro-6-methylthio-2*H*-thiopyran, 361
Cyanodithioformic ester, 361
3-Cyanoflavone, 151, 168
α-Cyanoketones, 372
Cyanomaclurin, 237
3-Cyano-3-methylthiochroman-4-one, 383
1-Cyano-4-pyridones, 24
Cyanostegia angustifolia, 200
2-Cyanotetrahydropyrans, 41
3-Cyanothiochroman-4-one, 383
5-Cyano-2*H*-thiopyrans, 349
Cycas japonica, 201
Cycas revoluta, 201
Cyclamen persicum, 92
Cyclamin, 93
Cyclamin chloride, **92**
Cyclic pyrylium salts, 4
2,3-Cycloalkenothiochromones, 374
Cycloartocapesin, **206**
Cycloheptane, 396
1,3-Cyclohexadiene, 396
Cyclohexadienyl radicals, 334
Cyclohexane-1,3-dione, 309
6,7-Cyclohexanoflavone, 342
Cyclohexene-1,3-diones, 309
Cyclohexenocoumarins, 111
Cyclohexenones, 3
Cyclohexenonyl crotonates, 256
3-Cyclohexenylmethyl [4]bromobenzenesulphonate, 397
1,5-Cyclo-octadiene, 398
Cyclo-octene, 397
Cyclopentadiene, 356, 402
Cyclopentadienones, ring-enlargement, 13
Cyclopentamethylenediphenylstannane, 422
Cyclopentamethylenephenyltin iodide, 422
Cyclopentanol, 359
Cyclopentenes, 220
Cyclopropanes, 213
Cyclopropyl ketones, 255
Cyclopropyl 2-mercapto-4-methoxyphenyl ketone, 381
Cyclotelluropentane, 412
Cyclotelluropentane diiodide, 412
Cyrtominetin, **283**
Cyrtomium sp., 283
Cytisoside, **186**
Cytisus laburnum, 186, 215

Dactylin, **194**
Dahlia variabilis, 181
Daidzein, **214**, 234
Daidzin, **214**
Dakin reaction, 126, 180, 282
Dalbergia baroni, 124
Dalbergia nigra, 217
Dalbergia sissoo, 123, 216
Dalbergia variabilis, 234
Dalbergin, **123**, 124
Dalbergioidin, **287**
Daphne sp., 125
Daphne mezereum, 117
Daphnetin, **124**
Daphnetinic acid, 124
Daphnin, **124**
Daphniphyllum macropodum, 47
Daphnoretin, **125**
Datisca cannabina, 189, 190
Datiscetin, **189**
Datiscin, **189**
Decahydro-7,8-benzochroman, 339
Decahydrobenzothiopyran, 379
Decamethoxybisflavenylidene, 192
Decursin, **136**
Decussatin, **326**
Deguelin, 257, **264**, 265
Deguelinic acid, 265
Dehydracetic acid, 23, 30, 35, 36, 117
Dehydrocycloguanandin, **328**

Dehydrodeguelin, 265
Dehydrodicatechin A., **242**
Dehydrodihydrodeguelin, 265
Dehydrodihydromillettone, 268
Dehydrogeijerin, **119**
3,4-Dehydro-9-hydroxy-α-lapachone, 342
Dehydroisorotenone, 258, 260
Dehydromillettone, 268
Dehydroneotenone, **220**
Dehydronepseudin, **219**
Dehydropachyrrhizone, 267
Dehydrorotenone, 258, 259, 260, 261, 262
6-Dehydroxyjacareubin, 328
Delphin chloride, **91**
Delphinidin, 82, 89, 92
Delphinidin chloride, 82, **91**
Delphinidin chloride 3,5-diglucoside, 91
Delphinidin chloride 7,3′,5′-trimethyl ether, 93
Delphinidin 3-galactoside, 91
Delphinin chloride, **91**
Demethoxycentaureidin, 184
Demethoxykanugin, 196
Demethoxysudachitin, **187**
Demethylkhellin, 158
4′-Demethylmatteucinol, 283
5-*O*-Demethyltangeritin, 189
Demethylvisnagin, 158
Deodarin, **284**
Deoxybenzoins, 12, 208, 211, 298
Deoxydiethylenehaematoxylone, 444
5-Deoxyflavanonols, 275
6-Deoxyjacareubin, 328, 329
Deoxypatulin, 45
Deoxytetramethylhaematoxylone, 444
Deoxy-*O*-triacetylbrazilone, 443
Deoxytrimethylbrazilone, 438, 440, 441, 442, 444
5-Deoxyvitexin, 182
Derric acid, **260**, 265
Derris sp., 257, 267
Derris elliptica, 258, 264, 266
Derrisic acid, **260**
Derris malaccensis, 221, 266, 267
Derris scandens, 221
Derritol, **261**, 262
Derritol methyl ether, 261
5-*O*-Desmethylcitrometin, 285
5-*O*-Desmethyltangeritin, 188
Desoxybenzoins, 74
Desoxykarenin, **161**
β-Desylcinnamic acid, 14
Deuterioacetic acid, 64
3,6-Diacetamido-9-phenylxanthydrol, 312
5,7-Diacetoxy-4-(acetoxymethyl)coumarin, 122
5,7-Diacetoxy-4-(bromomethyl)coumarin, 122
5,7-Diacetoxycoumarin, 120
6,7-Diacetoxycoumarin, 123
Diacetoxy-2,3-flavans, 249
5,4′-Diacetoxyflavone, 181
5,7-Diacetoxyflavone, 170, 177
7,4′-Diacetoxy-5-hydroxyflavones, 182
5,7-Diacetoxyisoflavone, 212
5,7-Diacetoxy-4-methylcoumarin, 122
α,γ-Diacetylacetoacetic acid, 35
Diacetylacetone, 21, 28, 30
3,6-Diacetyl-4,7-dimethyl-5-hydroxycoumarin, 117
3,8-Diacetyl-4,7-dimethyl-5-hydroxycoumarin, 117
7,4′-Diacetylgenistein 5-methyl ether, 215
2,4-Diacetylglutaconic ester, 12
4-Diacetylmethylflav-2-ene, 77
4,6-Diacetylresorcinol, 169
5,8-Dialkoxychromones, 146
Dialkylanilines, 73
2,4-Dialkyl-3-aryl-2*H*-chromenes, 62
1,1-Dialkyl-3-arylisochromans, 299
2,2-Dialkylchromans, 223
Dialkylchromenes, 104, 105
Dialkyldichlorosilanes, 414
N,N-Dialkyl-α-ethoxycarbonylalkylacetamide, 340
1,1-Dialkylisochromans, 302
1,1-Dialkylisochroman-4-ones, 302
2,4-Dialkyl-2*H*-isoflavenes, 62
Dialkylnaphthopyrans, 337
2,5-Dialkylphenoxymagnesium bromide, 62
4,8-Dialkyl-2-thia-adamantane-4,8-diols, 399
Diallyldimethylsilane, 416
4,4′-Diaminobenzophenones, 314
4,6-Diamino-3,5-dicyano-2*H*-thiopyran-2-thione, 357
4,4′-Diamino-2,2′-dihydrodiphenylmethanes, 314

3,6-Diamino-9-phenylxanthydrol, 313
3,6-Diaminoxanthene dyes, 313
Diaminoxanthones, 314, 320, 394
3,6-Diaminoxanthydrols, 310, 313
Diamyl-2*H*-chromene, 58
Dianella revoluta, 233
Dianin's compound, 224
2,2-Diarylchromans, 223
Diarylchromenes, 59, 104
Diaryldichlorosilanes, 414
2,8-Diaryl-3,7-dihydroxy-4*H*,6*H*-benzo-[1,2-*b*:5,4-*b'*]dipyran-4,6-diones, 174
2,2-Diaryl-4-(2,2-diphenyl-vinyl)chromans, 59
1,1-Diarylethenes, 59
4,4-Diarylflavans, 247
Diarylmethane dyes, 312
Diarylnaphthopyrans, 337
1,4-Diaryloxybut-2-ynes, 63
2,6-Diaryl-4-pyrones, 24
3,6-Diarylthiopyran-2-ones, 357
1,5-Diazabicyclo[4.3.0]non-5-ene, 423
1,4-Diazabicyclo[2.2.2]octane, 404, 405
Diazoacetic ester, 300
Diazoaminobenzene, 65
Diazomethane, 16, 32, 116, 119, 122
Diazonium salts, 30, 33, 228, 242
9,9-Diazothioxanthene, 393
Dibenzodisalicylides, 346
10*H*-Dibenzo[*b,e*]iodininium salts, 425
10*H*-Dibenzo[*b,e*]iodinium chloride, 425
Dibenzo[*a,e*]pyran, 304
2,3,5,6-Dibenzopyran, 304
6*H*-Dibenzo[*b,d*]pyran, 334
6-Dibenzopyranone, 334
6*H*-Dibenzo[*b,d*]pyran-6-one, 334
Dibenzoselenopyran, 409, 411
Dibenzoselenopyrylium perchlorate, 411
Dibenzothiopyran, 388
Dibenzo[*bd*]thiopyran 5,5-dioxide, 390
Dibenzo[*bd*]thiopyrylium perchlorate, 395
7*H*-Dibenzo[*c,h*]xanthen-7-one, 346
14*H*-Dibenzo[*a,j*]xanthene, 309, 346
14*H*-Dibenzo[*a,h*]xanthen-14-one, 346
Dibenzoxanthones, 346
1,2:6,7-Dibenzoxanthone, 346
1,2:7,8-Dibenzoxanthone, 346
3,4:5,6-Dibenzoxanthone, 346
Dibenzoxanthylium chloride, 309
Dibenzylideneacetone, 365
3,5-Dibenzylidenetetrahydro-4-pyrones, 23
3,5-Dibenzyl-4-pyrones, 23
Diborane, 247, 388
2,3-Dibromochroman, 53
3,4-Dibromochroman, 56, 224
1,2-Dibromo-2,3-dihydro-1*H*-naphtho-[2,1-*b*]thiopyran, 378
3,4-Dibromo-3,4-dihydrothiocoumarin, 371
3,4-Dibromo-2,2-dimethylchroman, 224, 225
3,4-Dibromo-2,2-diphenylchroman, 224
4α,6-Dibromoflavan, 248
4,5-Dibromofluorescein, 316
1,1-Dibromogermacyclohexane, 420
Dibromo-3-hydroxycoumarins, 111
6,8-Dibromo-5-hydroxyflavone, 166
3,5-Dibromo-6-hydroxy-2-methylchromone, 152
6,8-Dibromo-5-methoxyflavone, 166
1,5-Dibromopentane, 40, 48, 404
3,5-Dibromo-4-pyrone, 27
1,1-Dibromostannacyclohexane, 422
2,3-Dibromotetrahydropyran, 41
3,5-Dibromotetraphenyl-4-pyran, 3
3,4-Dibromothiochroman, 378
3,3-Dibromothiochroman-4-one, 380
3,3-Dibromothioflavanones, 383
2,7-Dibromoxanthone, 319, 320
Dibutyl-2*H*-chromene, 58
1,1-Dibutyl-4,4-dimethylstannacyclohexane, 422
Di-*tert*-butyl peroxide, 49
Dibutylstannane, 423
Dibutyltin dibromide, 423
Dibutyltin dichloride, 423
2,6-Dicarboxythioxanthones, 394
Dichlorobenzaldehydes, 114
Dichlorocarbene, 53, 56, 60, 251, 368
3,4-Dichlorochroman, 56, 224
4,4-Dichloro-4*H*-chromene, 53, 54
4,4-Dichlorochromone-2-carbonyl chloride, 151
3,4-Dichlorocoumarins, 109, 110
Dichlorodicyanobenzoquinone, 62, 64, 323, 329
1,1-Dichloro-1a,7b-dihydro-2*H*-benzo[*b*]-cyclopropa[*d*]pyran, 56

1,1-Dichloro-1a,7a-dihydro-4*H*-benzo[*b*]-cyclopropa[*e*]pyran, 53
3,4-Dichloro-2,2-dimethylchroman, 224, 225
3,4-Dichloro-2,2-diphenylchroman, 224
Dichlorodiphenylsilane, 419
3,6-Dichlorofluoran, 314, 316
1,1-Dichlorogermacyclohexane, 420
(Dichloroiodo)benzene, 418
3,8-Dichloro-7-methoxyflavone, 167
4-Dichloromethyl-4*H*-benzothiopyran, 368
3-Dichloromethyl-1,1-diethyl-1-germa-cyclohexane, 421
3-Dichloromethyl-1,1-dimethylsilacyclo-hexane, 421
3-Dichloromethyl-1,1-dimethylstannacyclo-hexane, 423
Dichloromethyl methyl ether, 391
4,4-Dichloro-2-phenylthiochromene, 375
1,1-Dichloro-1-silacyclohexenes, 417
1,1-Dichlorosilacyclohexane, 413, 414, 415, 419
1,1-Dichloro-1-silacyclo-hexenes, 417
1,1-Dichlorotelluracyclohexane-3,5-dione, 413
7,7-Dichloro-2-thiabicyclo[4.1.0]heptane, 359
2,6-Dichloro-9-thiabicyclo[3.3.1]nonane, 398
9,9-Dichloroxanthene, 307
Dicoumarin, 116
Dicoumarol, 116, **154**
3,7-Dicyano-10*H*-dibenz[*b,e*]iodininium bi-sulphate, 425
3,3-Di(2-cyanomethyl)chroman-4-one, 253
4-(Dicyanomethylene)-2,6-dimethyl-4*H*-thiopyran, 356
4-(Dicyanomethylene)-4*H*-pyran, 24
Dicyclohexylcarbodiimide, 418
Di-*O*-dimethylcitromycin, 160
Didymine, **280**
Dieckmann condensation, 301, 364
Diels–Alder reaction, 13, 41, 104
Dienamines, reaction with carbon disulphide, 357
4,6-Diethoxy-2-(4-ethoxy-3-methoxybenzoyl-oxy)-5-methoxyacetophenone, 188
3,3-Diethoxypropionic ester, 123
Diethylaminoethylamine, 392
1-(2-Diethylaminoethylamino)-4-(hydroxy-methyl)thioxanthone, 393
1-(2-Diethylaminoethylamino)-4-methyl-thioxanthone, 392
1-(β-Diethylaminoethyl)amino-4-methyl-xanthone, 321
3-Diethylaminophenol, 314
Diethyl-5-bromopentyltin bromide, 422
Diethyl-3-bromo-4-pyrone-2,6-dicarboxylic acid, 39
2,2-Diethyl-2*H*-chromene, 58
Diethylenehaematoxylone, 444
1,1-Diethylgermacyclohexane, 420, 421
1,1-Diethyl-1-plumbacyclohexane, 424
1,1-Diethylstannacyclohexane, 422
Diethynylmethane, 423
Diflavone, **168**
Diflavonol, **174**
Diflavonyl ethers, 201, 204
Di(7-flavonyloxy)methane, 178
Diformylacetone, 28
Digicitrin, **200**
Digicitrin dimethyl ether, 200
Digitalis lanata, 185
Digitalis purpurea, 184, 200
Digitalis thapsi, 199
6,8-Di-*C*-β-D-glucopyranosylluteolin, 184
6,8-Di-*C*-β-D-glucopyranosyl-acacetin, 182
Dihalogenocarbenes, 60, 421
3,4-Dihalogenochromans, 225
Dihalogenodihydronaphthopyrans, 225
2,3-Dihalogenotetrahydropyrans, 41
Dihydroacenaphthylo[5,6-*cd*]thiopyran, 401
Dihydroanthracenes, 424
3,4-Dihydro-5,6-benzocoumarin, 340
[1a,7b]Dihydro-2*H*-benzo[*b*]cyclopropa[*d*]-pyran, 60
1,2-Dihydro-5*H*-benzo-2,3-diazepinones, 385
Dihydrobenzofuran-3-ones, 173
Dihydrobenzo[*b*]pyrans, 52
3,4-Dihydro-1*H*-benzo[*c*]pyran, 297
3,4-Dihydro-2*H*-benzo[*b*]pyran, 222
Dihydrobenzo[*b*]pyranones, 251
2,3-Dihydro-4*H*-benzo[*b*]selenopyran, 408
2,3-Dihydro-4*H*-benzo[*b*]selenopyran-4-one, 408
3,4-Dihydro-1*H*-benzoselenopyran-4-one, 409

2,3-Dihydro-4*H*-benzotelluropyran-4-one, 413
2,3-Dihydro-4*H*-benzo[*b*]thiin, 376
2,3-Dihydro-4*H*-benzo[*b*]thiopyran, 376
3,4-Dihydro-1*H*-benzo[*c*]thiopyran, 385
3,4-Dihydro-2*H*-benzo[*b*]thiopyrans, 368
2,3-Dihydro-4*H*-benzo[*b*]thiopyran-4-ones, 379, 381
Dihydrobenzoxepin, 57
2,3-Dihydro-1,3-bis-(methoxyphenyl)-1*H*-naphtho[2,1-*b*]pyran-3-ols, 235
Dihydrobrazilic acid, 429
Dihydrobrazilinic acid, 431
Dihydrochalcones, 62, 240, 280
1,4-Dihydro-1-chlorobismabenzene, 424
1,4-Dihydro-1-chlorostibabenzene, 423
5,8-Dihydrochromans, 225
Dihydrochromene, **222**
Dihydrocitrinin, 290
Dihydrocoumarins, 103, 223, 251, 340
Dihydrocoumarin, 104, **251**
3,4-Dihydrocoumarins, 100
–, 4,4-disubstituted, 106
3,4-Dihydrocoumarin-4-cyanoacetamide, 104
1,4-Dihydro-1,1-dibutylstannabenzene, 423
Dihydro-α,2-dihydroxychalcone, 244
8,9-Dihydro-5,6-dihydroxy-3,9,4′-trimethoxyfuro[2,3-*h*]flavone, 205
3,4-Dihydro-3,4-dimethylbenzo [*b*]thiopyran, 368
7,8-Dihydro-2,6-dimethylchromone, 151
10,11-Dihydro-10,10-dimethyl-5*H*-dibenzo-[*a,d*]cycloheptene, 299
2,3-Dihydro-2,2-dimethyl-6-methoxy-naphtho[1,2-b]pyran, 339
7,8-Dihydro-6,9-dimethyl-2 (4-methylpent-3-enyl)-1-oxaphenalene-7,8-dione, 344
2,3-Dihydro-3,3-dimethyl-1*H*-naphtho-[2,1-*b*]pyran-2-ol, 236
2,3-Dihydro-2,2-dimethyl-4*H*-naphtho-[1,2-*b*]pyran-4-one, 338
3,4-Dihydro-2,2-dimethylpyrans, 136
7,8-Dihydro-6,6-dimethyl-2*H*,6*H*-pyrano-[2,3-*f*]chromene-2,8-dione, 137
Dihydrodimethylpyranoisoflavone, 222
2,3-Dihydro-2,7-dimethylpyrano[4,3,b]-pyran-4,5-dione, 17
6,7-Dihydro-8,8-dimethyl-7-senecioyloxy-2*H*,8*H*-pyrano [3,2-*g*]chromen-2-one, 136
9,10-Dihydro-9,9-dimethyl-9-stannanthracene, 424
2,3-Dihydro-*N,N*-dimethyl-*p*-toluidine, 150
6,7-Dihydro-4,6-dimethylxanthyletin, 135
2,3-Dihydro-1,3-diphenyl-3-methoxy-1*H*-naphtho[2,1-*b*]pyran, 341
2,3-Dihydrodiphenyl-1*H*-naphtho[2,1-*b*]-pyrans, 340
2,3-Dihydrodiphenyl-4*H*-naphtho[1,2-*b*]-pyrans, 338
2,3-Dihydro-1,3-diphenyl-1*H*-naphtho-[2,1-*b*]pyran-3-ol, 235
2,3-Dihydro-2,4-diphenyl-4*H*-naphtho-[1,2-*b*]pyran-2-ol, 235
2,3-Dihydro-2,4-diphenyl-4*H*-naphtho-[2,3-*b*]pyran-2-ol, 235
2,3-Dihydro-3,3-diphenyl-1*H*-naphtho-[2,1-*b*]pyran-2-ol, 236
1,3-Dihydro-1,3-diphenylthiaphenalene, 400
Dihvdroevodione, 64
Dihydrofisetin, 281
3,4-Dihydroflavans, 94
Dihydroflavanols, 171
Dihydroflavonols, 209, 249, 250, 272
6,7-Dihydro-2*H*-furo[2,3-*f*]chromen-2-ones, 133
Dihydrofuscin, 231
Dihydrogalangin, 284
9,10-Dihydro-9-germanthracenes, 421
8,9-Dihydro-8-(1-glucosyloxy-1-methyl-ethyl)-2*H*-furo[2,3-*h*]chromen-2-one, 133
Dihydrohaematoxylinic acid, 443
Dihydrohydroxyflavonols, 88
8,9-Dihydro-9-hydroxy-8-(1-hydroxy-1-methylethyl)-2*H*-furo[2,3-*h*]chromen-2-one di-isovalerate, 132
3,4-Dihydro-8-hydroxy-6-methoxy-3-methylisocoumarin, 303
2,3-Dihydro-2-hydroxymethylbenzo[*b*]furan, 52
3,4-Dihydro-8-hydroxy-3-methylisocoumarin, 303
2,3-Dihydro-3-hydroxynaphtho[1.8-*b*]thiopyran, 399
2,3-Dihydro-3-hydroxy-2-phenyl-4*H*-benzo-[*b*]pyrans, 237

9,10-Dihydro-9-hydroxyseselin, 136
Dihydroisocoumarins, 293
3,4-Dihydroisocoumarin, 300
8,9-Dihydro-8-isopropenyl-2*H*-furo[2,3-*h*]-chromen-2-one, 132
9,10-Dihydro-9-isopropylanthracene, 299
6,7-Dihydro-7-isopropyl-2*H*-furo[3,2-*g*]-chromene-2,6-dione, 131
3,4-Dihydro-2-isopropyl-7-methyl-2*H*,5*H*-pyran[4,3-*b*]pyran-4,5-dione, 36
Dihydrojacareubin, 328
Dihydrokaempferide, 196
Dihydrokaempferol, 284
Dihydrokhellinone, 157
Dihydrolapachenole, 339
2,3-Dihydro-5-methoxy-1-oxaphehalene, 343
5,6-Dihydro-4-methoxy-6-styryl-2-pyrone, 19
2,3-Dihydro-2-methylbenzo[*b*]furan, 225
2,3-Dihydro-2-methylbenzo[*b*]selenophene, 408
1,4-Dihydro-1-methylbenzo[*c*]thiophene, 387
2,3-Dihydro-2-methyl-4*H*-benzo[*b*]thio-pyran-4-one, 382
2,6-Dihydro-3-methyl-5-oxo-5*H*-thiopyran 1,1-dioxide, 349
3,4-Dihydro-6-methyl-2*H*-thiopyran, 359
5,6-Dihydro-4-methyl-2*H*-thiopyran, 360
Dihydromilletone, 268
Dihydromorin, **285**
Dihydromuricetin, 285
2,3-Dihydro-1*H*-naphtho[2,1-*b*]pyran, 224, 340, 367, 378
3,4-Dihydro-2*H*-naphtho[1,2-*b*]pyran, 224, 339, 378
1,2-Dihydro-3*H*-naphtho[2,1-*b*]pyran-3-ones, 340, 341
2,3-Dihydro-1*H*-naphtho[2,1-*b*]pyran-1-ones, 340, 341
2,3-Dihydronaphtho[1,8-*bc*]pyran-3-one, 343
2,3-Dihydro-4*H*-naphtho[1,2-*b*]pyran-4-one, 338
3,4-Dihydro-2*H*-naphtho[1,2-*b*]pyran-2-one, 339
Dihydronaphthoselenopyrans, 412
2,3-Dihydronaphthoselenopyran-1-ols, 412
2,3-Dihydronaphthoselenopyran-1-ones, 411, 412
Dihydro-1*H*-naphtho[2,1-*b*]thiopyran, 378
3,4-Dihydro-2*H*-naphtho[2,3-*b*]thiopyran, 378
2,3-Dihydro-1*H*-naphtho[2,1-*b*]thiopyran-1-ol, 367, 382
3,4-Dihydro-2*H*-naphtho[1,2-*b*]thiopyran-4-ol, 368, 382
2,3-Dihydro-1*H*-naphtho[2,1-*b*]thiopyran-1-one, 382
3,4-Dihydro-2*H*-naphtho[1,2-*b*]thiopyran-4-one, 382
Dihydro-oreoselone, 131
Dihydro-oroxylin, **279**
1,2-Dihydro-1-oxoisoquinoline, 384
2,3-Dihydro-2-phenylbenzo[*b*]pyran, 231
3,4-Dihydrophenyl-2*H*-benzo[*b*]pyrans, 231
2,3-Dihydro-2-phenyl-4*H*-benzo[*b*]pyran-4-ones, 139, 142, 268
2,3-Dihydro-3-phenyl-4*H*-benzo[*b*]pyran-4-ones, 286
2,3-Dihydro-2-phenyl-4*H*-benzo[*b*]seleno-pyran-4-one, 408
2,3-Dihydro-2-phenyl-4*H*-benzo[*b*]thiopyran-4-one, 383
2,3-Dihydro-3-phenyl-4*H*-benzo[*b*]thiopyran-4-one, 383
5,6-Dihydro-3-phenyl-2*H*-thiopyran-5-ol 1,1-dioxide, 360
5,6-Dihydro-3-phenyl-2*H*-thiopyran-5-one 1,1-dioxide, 360, 361
1,2-Dihydrophthalic anhydride, 14
Dihydroprunetin, 287
Dihydropyrans, 39, 40, 359
–, reactions, 41
–, substituted, 40
Dihydropyran, 48
–, hydrolysis, 40
2,3-Dihydropyran, 50
–, oxidation, 42
2,3-Dihydro-4-pyrans, reaction with *N*-bromosuccin-imide, 41, 42
2,3-Dihydro-4-pyran, **39**, 40, 41
–, 5-alkoxycarbonyl derivatives, 43
–, halogenation, 42
–, hydroformylation, 43
–, ozonolysis, 42
–, reaction with diphenylacetylene, 44
–, – with hydroxy compounds, 40

2,3-Dihydro-4-pyran, reaction (*continued*)
–, – with Schiff bases, 43
2,3-Dihydro-γ-pyran, 39
2,3-Dihydro-4-pyran[2-^{14}C], 43
2,3-Dihydro-4-pyran[6-^{14}C], 43
5,6-Dihydro-2-pyran, **44**
5,6-Dihydro-2*H*-pyran, 44
5,6-Dihydro-α-pyran, 44
2,3-Dihydro-4-pyran-2-carbaldehyde, 41
4*H*-2,3-Dihydropyran-2,4-dione, 29
6,7-Dihydro-4*H*,8*H*-pyrano[3,2-*g*]benzopyran-4-one-2-carboxylic acid, 161
6,7-Dihydropyrano[3,2-*g*]chromone, 161
6,7-Dihydropyrano[3,2-*g*]chromone-2-carboxylic acid, 161
Dihydro-4-pyrones, 51
2,3-Dihydro-4-pyrone-6-carboxylic acid, 51
Dihydroquercetin, 90, 94, 172, 284
Dihydroquercetin 3-L-rhamnoside, 285
Dihydroquinoline, 405
Dihydroquinoxalines, 277
Dihydrorobinetin, 88, 276, **285**
Dihydrorotenone, 263, 264, 265
Dihydrosamidin, **137**
Dihydroseselin, 135
9,10-Dihydro-9-silanthracenes, 420
9,10-Dihydro-9-stannanthracenes, 424
2,3-Dihydro-1-thiaphenalen-3-ol, 399
Dihydrothiapyrans, 359
Dihydrothiins, 359
1,3-Dihydro-2-thiophenalene, 400
Dihydrothiopyrans, 359
3,4-Dihydro-2*H*-thiopyran, 359, 361
5,6-Dihydro-2*H*-thiopyran, 360, 361
2*H*-5,6-Dihydrothiopyran 1,1-dioxide, 364
3,4-Dihydro-2*H*-thiopyran-3-one, 361
5,6-Dihydro-2*H*-thiopyran-5-one, 361
2,3-Dihydro-4,6,7-trimethoxybenzo[*b*]-furan, 157
Dihydrotubaic acid, 259, 268
3,4-Dihydroumbelliferone, 118
Dihydrowogonin, 274, **279**
Dihydrowogonin 7-*O*-β-D-glucoside, 279
3,4-Dihydro-1(2*H*)-xanthenones, 309
Dihydroxanthyletin, 135
2,3-Dihydroxyacetophenone, 163
2,4-Dihydroxyacetophenone, 76, 312
2,5-Dihydroxyacetophenone, 269
3,4-Dihydroxyacetophenone, 183
2,4-Dihydroxyanisole, 123
2,4-Dihydroxybenzaldehyde, 117
2,5-Dihydroxybenzaldehyde, 117
2,6-Dihydroxybenzaldehyde, 116
Dihydroxybenzenes, reaction with cinnamic acids, 270
2,4-Dihydroxybenzene-1,5-dicarboxylic acid, 127
2,3-Dihydroxybenzoic acid, 322
2,4-Dihydroxybenzoic acid, 195
2,2′-Dihydroxybenzophenone, 317
2,3′-Dihydroxybenzophenone, 323
2,4-Dihydroxybenzophenone, 311
2,3-Dihydroxy-4*H*-benzo[*b*]thiopyran-4-one, 372
3,4-Dihydroxy-2*H*-benzo[*b*]thiopyran-2-one, 372
2,4-Dihydroxybenzoylbenzoic acid, 313, 315
2,4-Dihydroxybenzyl alcohol, 312
2,4-Dihydroxy-5-bromoacetophenone, 172
5,7-Dihydroxychroman-4-one, 219
3,7-Dihydroxychromone, **151**, 428
5,7-Dihydroxychromone, 147, **152**
7,8-Dihydroxychromone, **152**
3,4-Dihydroxycinnamic acid, 122
6,7-Dihydroxycoumaran-3-one, 128
Dihydroxycoumarins, 120
5,7-Dihydroxycoumarin, **120**
6,7-Dihydroxycoumarin, **122**
7,8-Dihydroxycoumarin, 124
2,3-Dihydroxy-6,7-dihydroxybenzofuran-3-one, 128
1,6-Dihydroxy-3,7-dimethoxy-2,8-diprenylxanthone, 330
3,4′-Dihydroxy-5,7-dimethoxyflavanone, 276
3,5-Dihydroxy-7,4′-dimethoxyflavanone, 280
5,6-Dihydroxy-7,4′-dimethoxyflavanone, 275
5,3′-Dihydroxy-7,4′-dimethoxyflavone, 186
5,3′-Dihydroxy-7,5′-dimethoxyflavone, 186
5,4′-Dihydroxy-3,7-dimethoxyflavone, 190
5,4′-Dihydroxy-7,3′-dimethoxyflavone, 186
5,7-Dihydroxy-3,8-dimethoxyflavone, 190
5,7-Dihydroxy-6,4′-dimethoxyflavone, 185
5,7-Dihydroxy-6,8-dimethoxyflavone, 186
5,3′-Dihydroxy-7,4′-dimethoxyflavonol, 201
5,7-Dihydroxy-6,4′-dimethoxyflavonol, 196

5,7-Dihydroxy-8,4′-dimethoxyflavonol, 196
7,3′-Dihydroxy-2′,4′-dimethoxyisoflavan, 234
5,7-Dihydroxy-2′,4′-dimethoxyisoflavanone, 286
5,3′-Dihydroxy-7,4′-dimethoxyisoflavone, 216
5,4′-Dihydroxy-6,7-dimethoxyisoflavone, 211, 216
5,4′-Dihydroxy-7,8-dimethoxyisoflavone, 211, 216
5,7-Dihydroxy-6,4′-dimethoxyisoflavone, 216, 217
6,4′-Dihydroxy-5,7-dimethoxyisoflavone, 216
7,4′-Dihydroxy-5,6-dimethoxyisoflavone, 211
7,4′-Dihydroxy-5,8-dimethoxyisoflavone, 211
1,2-Dihydroxy-6,8-dimethoxyxanthone, 326
1,3-Dihydroxy-4,5-dimethoxyxanthone, 327
1,3-Dihydroxy-7,8-dimethoxyxanthone, 326
1,7-Dihydroxy-3,8-dimethoxyxanthone, 327
1,8-Dihydroxy-3,5-dimethoxyxanthone, 326
1,8-Dihydroxy-3,7-dimethoxyxanthone, 327
3,8-Dihydroxy-1,7-dimethoxyxanthone, 327
5,7-Dihydroxy-2,2-dimethylchroman, 224
5,7-Dihydroxy-6,8-dimethylchroman-4-one, 283
7,8-Dihydroxy-2,2-dimethylchroman-4-one, 253
5,7-Dihydroxy-2,2-dimethyl-2*H*-chromene, 65
5,7-Dihydroxy-2,6-dimethylchromone, 153
5,7-Dihydroxy-2,8-dimethylchromone, 153
5,7-Dihydroxy-6,9-dimethylchromone, 155
5,7-Dihydroxy-2,6(and 2,8)dimethyliso-flavone, 207
5,4′-Dihydroxy-8,8-dimethyl-6-prenyl-8*H*-pyrano[2,3-*h*]isoflavone, 220
5,4′-Dihydroxy-8,8-dimethyl-10-prenyl-8*H*-pyrano[3,2-*g*]isoflavone, 220
5,4′-Dihydroxy-8,8-dimethyl-8*H*-pyrano-[3,2-*g*]isoflavone, 220
5,8-Dihydroxy-2,2-dimethyl-2*H*,6*H*-pyrano-[3,2-*b*]xanthen-6-one, 328
1,3-Dihydroxy-1,3-diphenylindan-2-one, 290
2,2′-Dihydroxydiphenylmethane, 304, 305
2′,4′-Dihydroxy-α-ethylchalcone, 269
1-(1,2-Dihydroxyethyl)-3-(hydroxymethyl)-4-phenylisochroman, 298
3,4-Dihydroxyflavans, 96
5,6-Dihydroxyflavanone, 269, 274, 275
5,7-Dihydroxyflavanones, 275, 277
5,8-Dihydroxyflavanone, 274
7,4′-Dihydroxyflavanone, 278
3,7-Dihydroxyflavone, 178
5,6-Dihydroxyflavone, 176
5,7-Dihydroxyflavone, 165, 170, 175, 178
5,8-Dihydroxyflavones, 175, 179
6,2′-Dihydroxyflavone, 176
5,7-Dihydroxyflavonol, 189
5,7-Dihydroxyflavylium chloride, 80
4′,7-Dihydroxyflavylium perchlorates, 77
3,6-Dihydroxyfluoran, 315
5,3′-Dihydroxyhexamethoxyflavone, 200
5,7-Dihydroxy-3-(4-hydroxybenzylidene)-6-methoxychroman-4-one, 219
5,7-Dihydroxy-3-(4-hydroxybenzylidene)-8-methoxychroman-4-one, 219
5,7-Dihydroxy-4-(hydroxymethyl)coumarin, 122
4,5-Dihydroxy-2-hydroxymethylpyridine, 33
2,4-Dihydroxy-3-isoamylbenzoic acid, 259
7,4′-Dihydroxyisoflavan, 234
5,6-Dihydroxyisoflavones, 210, 211
5,7-Dihydroxyisoflavones, 213
7,3′-Dihydroxyisoflavone, 214
7,4′-Dihydroxyisoflavone, 214
5,7-Dihydroxy-6-isopentenyl-8-isovaleroyl-4-propylcoumarin, 121
5,7-Dihydroxy-8-isopentenyl-6-isovaleroyl-4-propylcoumarin, 121
4,5-Dihydroxyisophthalic acid, 256
4,6-Dihydroxyisophthalic acid, 131
5,7-Dihydroxy-6-isovaleryl-8-isopentenyl-phenylcoumarin, 121
Dihydroxyketones, 74

2,6-Dihydroxy-4-methoxyacetophenone, 153
2,4-Dihydroxy-5-methoxybenzaldehyde, 123
2,6-Dihydroxy-4-methoxybenzoic ester, 15
5,7-Dihydroxy-3-(4-methoxybenzylidene)-chroman-4-one, 218
1,7-Dihydroxy-8-methoxycarbonyl-3-methylxanthone, 331
7,8-Dihydroxy-6-methoxycoumarin, 125
Dihydroxymethoxydeoxybenzoin, 211
4,6-Dihydroxy-2-methoxydihydrochalcone, 165
2,6-Dihydroxy-4-methoxy-3,5-dimethyl-benzoic acid, 154
5,7-Dihydroxy-4′-methoxy-6,8-dimethyl-flavanone, 283
2,5-Dihydroxy-4-methoxy-4,6-dimethyl-tetrahydro-pyran, 51
2,5-Dihydroxy-7-methoxyflavanone, 278
3,5-Dihydroxy-7-methoxyflavanone, 284
5,6-Dihydroxy-7-methoxyflavanone, 275
5,7-Dihydroxymethoxyflavanones, 274
5,7-Dihydroxy-2′-methoxyflavanone, 281
5,7-Dihydroxy-6-methoxyflavanone, 279
5,7-Dihydroxy-8-methoxyflavanone, 279
5,7′-Dihydroxy-4′-methoxyflavanone, 280
5,8-Dihydroxy-7-methoxyflavanone, 279
5,7-Dihydroxy-3-methoxyflavone, 178
5,7-Dihydroxy-4′-methoxyflavone, 181
5,7-Dihydroxy-6-methoxyflavone, 180
5,7-Dihydroxy-8-methoxyflavone, 180
5,8-Dihydroxy-4′-methoxyflavone, 175
5,7-Dihydroxy-2′-methoxyflavonol, 191
5,7-Dihydroxy-4′-methoxyflavylium chloride, 80
7,2′-Dihydroxy-4′-methoxyisoflavan, 234
5,4′-Dihydroxy-7-methoxyisoflavanone, 287
3′,4′-Dihydroxy-7-methoxyisoflavone, 213
5,4′-Dihydroxy-7-methoxyisoflavone, 215
5,7-Dihydroxy-4′-methoxyisoflavone, 215
3,7-Dihydroxy-4-methoxy-1-methyldi-benzo[*b,d*]pyran-6-one, 335
5,7-Dihydroxy-4-methoxy-8-methyliso-flavone-2-carboxylic ester, 207
5,6-Dihydroxy-8-methoxy-2-methyl-4*H*-naphtho[2,3-*b*]pyran-4-one, 341
5,8-Dihydroxy-6-methoxy-2-methyl-3-(2-oxopropyl)-1,4-naphthoquinone, 300
4,5-Dihydroxy-6-methoxy-3-(2,4,5-tri-methoxyphenyl)coumarin, 115
1,3-Dihydroxy-5-methoxyxanthone, 329
1,3-Dihydroxy-7-methoxyxanthone, 325
1,4-Dihydroxy-7-methoxyxanthone, 323
1,5-Dihydroxy-3-methoxyxanthone, 322
1,6-Dihydroxy-5-methoxyxanthone, 329
1,7-Dihydroxy-3-methoxyxanthone, 91, 325
3,4-Dihydroxy-2-methoxyxanthone, 329
5,6-Dihydroxy-2-methyl-7,8-benzo-chromone, 341
8-(2,3-Dihydroxy-3-methylbutoxy)psoralen, 131
2′,4′-Dihydroxy-α-methylchalcone, 269
3,8-Dihydroxy-2-methylchromone, 152
5,7-Dihydroxy-2-methylchromone, 153
5,7-Dihydroxy-8-methylchromone, 155
5,7-Dihydroxy-4-methylcoumarin, 120, 122
7,8-Dihydroxy-4-methylcoumarin, 124
3,7-Dihydroxy-1-methyldibenzo[*b,d*]pyran-6-one, 335
5,7-Dihydroxy-3′,4′-methylenedioxy-6,8-dimethoxyflavone, 186
7,2′-Dihydroxy-4′,5′-methylenedioxyiso-flavanone, 287
1,3-Dihydroxy-6,7-methylenedioxyxanthone, 327
5,7-Dihydroxy-6-methylflavanone, 282
5,7-Dihydroxy-8-methylflavanone, 283
5,7-Dihydroxy-6-methylflavone, 179
Dihydroxy-2-methyl-4*H*-naphtho[2,3-*b*]-pyran-4-ones, 341
4,5-Dihydroxy-2-methylpyridine, 33, 36
Dihydroxymethylxanthones, 321, 322
1,6-Dihydroxy-3-methylxanthone-8-carboxylic acid, 331
1,6-Dihydroxy-3-methylxanthone diacetate, 331
2,6-Dihydroxy-3-nitroacetophenone, 152
3,4-Dihydroxyoxane-2,6-dicarboxylic acid, 26
5,3-Dihydroxypentamethoxyflavone, 189
6,3′-Dihydroxypentamethoxyflavone, 199
2,4-Dihydroxyphenyl-4-hydroxybenzyl ketone, 214

2,7-Dihydroxy-9-phenylxanthydrol, 312
3,6-Dihydroxy-9-phenylxanthydrol, 313
3,4-Dihydroxyrottlerin, 66
3,4-Dihydroxytetrahydropyran, 44
4,5′-Dihydroxy-3,6,7,3′-tetramethoxyflavone, 197
5,3′-Dihydroxytetramethoxyflavone, 189, 197, 199
5,4′-Dihydroxytetramethoxyflavones, 199, 200
5,7-Dihydroxytetramethoxyflavones, 188
5,7-Dihydroxy-6,2′,4′,5′-tetramethoxyisoflavone, 217
2,3-Dihydroxythiochromone, 372
3,4-Dihydroxythiocoumarin, 372
1,4-Dihydroxythioxanthone, 393
5,3′-Dihydroxy-6,7,4′-trimethoxyflavonol, 187
5,3′-Dihydroxy-7,8,2′-trimethoxyflavone, 187
5,3′-Dihydroxy-7,8,4′-trimethoxyflavone, 187
5,4′-Dihydroxy-3,6,7-trimethoxyflavone, 191
5,4′-Dihydroxy-6,7,8-trimethoxyflavone, 187
5,7-Dihydroxy-6,3′,4′-trimethoxyflavone, 187
5,3′-Dihydroxy-6,7,4′-trimethoxyflavonol, 197
7,3′-Dihydroxy-8,2′,4′-trimethoxyisoflavan, 234
4,5-Dihydroxy-3-(2,4,5-trimethoxyphenyl)-coumarin, 115
7,8-Dihydroxy-2,2,5-trimethylchroman, 231
5,7-Dihydroxy-2,2,6-trimethylcinnamoylchromene, 66
7,4′-Dihydroxy-5,3′,5′-triprenylflavanone, 278
2,7-Dihydroxyxanthene, 312
3,6-Dihydroxyxanthene, 312
1,3-Dihydroxyxanthones, 317, 321, 322
1,5-Dihydroxyxanthone, 325, 329
1,7-Dihydroxyxanthone, 324
2,5-Dihydroxyxanthone, 322
3,4-Dihydroxyxanthone, 321
4,5-Dihydroxyxanthone, 323
1,1-Diiodogermacyclohexane, 420
1,5-Diiodopentane, 362, 425
Di-iron carbonyl complexes, 404
Di-isobutylaluminium hydride, 62
Di(isochroman-1-yl) ether, 299
Diketene, 30, 151
Diketones, cyclisation, 21
–, dimerisation, 35
1,3-Diketones, reaction with resorcinol, 70
1,5-Diketones, 2
–, conversion to thiopyrans, 251
–, cyclisation, 403
–, oxidative ring-closure, 4
β-Diketones, 12, 22, 69
–, conversion to 2-pyrones, 13
–, cyclisation, 139
–, reaction with ketones, 5
–, – with α-methylene-active ketones, 4
δ-Diketones, 3, 5
1,5-Dilithiopentane, 414, 422, 424
1,4-Dilithiotetraphenylbutadiene, 418
2,6-Dimercapto-3,5-dimethyl-4*H*-thiopyran-4-one, 366
Dimercaptothiopyran-4-ones, 356
5,6-Dimethoxyangelicin, 130
2,4-Dimethoxybenzene-1,3-dicarboxylic acid, 127
4,5-Dimethoxy-1,2-benzoquinone, 210, 255
Dimethoxybiflavonyls, 204
3,7-Dimethoxychromone, 146
2,5-Dimethoxycinnamic acid, 117
5,7-Dimethoxycoumarin, 120
6,7-Dimethoxycoumarin, 122
5,7-Dimethoxy-8-(2,3-dihydro-2,2-dimethyl-3-oxo-4-furyl)flavone, 179
5,7-Dimethoxy-6-(2,3-dihydroxyisopentyl)-coumarin, 120
5,7-Dimethoxy-8-(3,3-dimethylacryloyl)-coumarin, 121
5,7-Dimethoxy-8-(1,1-dimethylallyl)-coumarin, 121
6,7-Dimethoxy-2,2-dimethylchroman-4-one, 63
6,7-Dimethoxy-2,2-dimethyl-2*H*-chromene, 63
5,7-Dimethoxy-4,6-dimethylcoumarin, 122
3,5-Dimethoxy-6,7,3′,4′-dimethylenedioxyflavone, 200
7,4′-Dimethoxy-6,8-dimethyl-5-hydroxyflavone, 183

6,2′-Dimethoxy-8,8-dimethyl-4′,5′-methylenedioxy-8*H*-pyrano[2,3-*h*]isoflavone, 221
7,4′-Dimethoxyflavan-3,4-diol, 249
5,8-Dimethoxyflavanone, 274
5,7-Dimethoxy-4*H*-flavene, 62
2′,5′-Dimethoxyflavone, 176
3,7-Dimethoxyflavone, 167
5,8-Dimethoxyflavone, 176
2′,6′-Dimethoxyflavonol, 177
Di(3-methoxy-7-flavonyloxy)methane, 178
5,9-Dimethoxy-2*H*-furo[3,2-*g*]chromen-2-one, 130
5,9-Dimethoxyfuro[3,2-*g*]chromone-2-carbaldehyde, 157
5,9-Dimethoxyfuro[3,2-*g*]flavone, 204
Dimethoxyhomophthalic acid, 429, 430, 433
4,5-Dimethoxy-2-hydrobenzoic acid, 261
4,5-Dimethoxy-2-hydroxyacetophenone, 160
3,5-Dimethoxy-4-hydroxybenzoic acid, 83
5,7-Dimethoxy-6-hydroxycoumarin, 126
6,7-Dimethoxy-8-hydroxycoumarin, 126
6,8-Dimethoxy-7-hydroxycoumarin, 126
5,8-Dimethoxy-7-hydroxyflavone, 176
7,8-Dimethoxy-5-hydroxyisoflavone, 212
5,7-Dimethoxy-8-hydroxy-4-(4-methoxyphenyl)coumarin, 125
2,6-Dimethoxy-4-hydroxytoluene, 208
5,3′-Dimethoxy-2-hydroxy-4,6,4′-triethoxydibenzoylmethane, 188
5,7-Dimethoxy-4-hydroxy-3-(2,4,5-trimethoxyphenyl)coumarin, 115
1,2-Dimethoxy-8-hydroxyxanthone, 329
1,3-Dimethoxy-5-hydroxyxanthone, 322
5,6-Dimethoxyindan-1-one, 440
Dimethoxyisochromans, 300
Dimethoxyisochromanones, 300
6,7-Dimethoxyisochromene, 289
6,7-Dimethoxyisocoumarin, 292, 295
7,8-Dimethoxyisocoumarin, 295
5,7-Dimethoxyisoflavone, 213
5,8-Dimethoxyisoflavones, 211
7,4′-Dimethoxyisoflavone, 210, 213
5,7-Dimethoxy-6-(2-isopentenyl)coumarin, 120
5,7-Dimethoxy-8-(2-isopentenyl)coumarin, 121
5,7-Dimethoxy-8-methylchromone, 155
3,7-Dimethoxy-3′,4′-methylenedioxyflavone, 196
7,2′-Dimethoxy-4′,5′-methylenedioxyisoflavan, 234
6,7-Dimethoxy-3′,4′-methylenedioxyisoflavanone, 287
5,2′-Dimethoxy-6,7-methylenedioxyisoflavone, 215
6,7-Dimethoxy-3′,4′-methylenedioxyisoflavone, 217, 218
3′,4′-Dimethoxy-6-methylflavan-3,4-diol, 249
5,7-Dimethoxy-6-methylflavone, 179
4,9-Dimethoxy-7-methyl-5*H*-furo[3,2-*g*]-benzopyran-5-one, 156
5,9-Dimethoxy-2-methylfuro-[3,2-*g*]-chromone, 156
4,9-Dimethoxy-5-oxo-5*H*-furo-[3,2-*g*]-benzopyran-7-carbaldehyde, 157
2,3-Dimethoxy-2-phenyl-2*H*-chromene, 77
2,6-Di(4-methoxyphenyl)-4-(4-dimethylaminophenyl)pyrylium perchlorate, 9
2,6-Di(4-methoxyphenyl)-4-(3-nitrophenyl)-pyrylium perchlorate, 9
3-(3,4-Dimethoxyphenyl)pyrazoline, 241
Di(4-methoxyphenyl)thiocarbonyl, 375
2-(3,4-Dimethoxyphenyl)-3-(2,4,6-trimethoxyphenyl)propan-1-ol, 240
3,6-Dimethoxy-9-phenylxanthydrol-2′-carboxylic ester, 315
2,2-Dimethoxypropane, 250
3,4-Dimethoxypsoralene, 130
5,9-Dimethoxypsoralene, 130
3,8-Dimethoxy-5,7,3′,4′-tetrahydroxyflavone, 197
2,3-Dimethoxythiochromone, 372
3,4-Dimethoxythiocoumarin, 372
5,7-Dimethoxy-4-thionoflavone, 169
6,3′-Dimethoxy-5,7,4′-triethoxyflavone, 188
3′,5′-Dimethoxy-5,7,4′-trihydroxyflavone, 186
6,8-Dimethoxy-5,7,4′-trihydroxyflavone, 187
3′,5′-Dimethoxy-5,7,4′-trihydroxyflavylium, 89
7,3′-Dimethoxy-3,5,4′-trihydroxyflavylium, 92

2,3-Dimethoxyxanthone, 329
2,4-Dimethyl-6-acenaphthenylpyrylium perchlorate, 9
o(α,α-Dimethylacetonyl)benzoic acids, 300
β,β-Dimethylacrylic acid, 252, 338
γ,γ-Dimethylallyl bromide, 128
3-(1,1-Dimethylallyl)daphnetin dimethyl ether, 125
3,3-Dimethylallyl(diphenyl) phosphate, 223
6-Dimethylallyl-7-hydroxycoumarin, 127
3-(1,1-Dimethylallyl)-7-hydroxy-6-isopentenyl-8-methoxycoumarin, 125
3-(1,1-Dimethylallyl)-6-hydroxymethyl-7-methoxycoumarin, 119
Dimethylallylxanthones, 329
2-Dimethylaminomethylbenzyl-lithium, 298
3-[(Dimethylamino)methyl]-3-methyltetrahydrothiopyran-4-one, 365
3-(Dimethylamino)methyltetrahydrothiopyran-4-one, 366
m-Dimethylaminophenol, 309
4-(4-Dimethylaminophenyl)-2,6-diphenylpyrylium perchlorate, 9
9-(3-Dimethylaminopropyl)-selenoxanthone, 410
Di-*O*-methylanhydrofulvic ester, 160
4,7-Di-*O*-methylbellidin, 326
Dimethylbenzochromans, 56
2,2-Dimethyl-7,8-benzochroman-4-one, 338
2,2-Dimethyl-5,6-benzo-2*H*-chromene, 59
8,8-Dimethyl-2*H*,8*H*-benzo[1,2-*b*:5,4-*b*]-dipyran-2-one, 135
2,3-Dimethylbenzofurans, 159
3,4-Dimethyl-2*H*-benzo[*b*] thiopyran, 368
4,6-Dimethyl-2*H*-benzo[*b*]thiopyran, 367
9,9′-Dimethylbiselenoxanthyl, 410
Dimethyl-5-bromopentyltin bromide, 422
2,3-Dimethylbutadiene, 104
Dimethyl carbonate, 115
2,7-Dimethyl-9-(2-carboxyphenyl)thioxanthene, 402
2,2-Dimethylchroman, 65, 224
2,2-Dimethylchroman-3-ol, 236
4,7-Dimethylchroman-2-ol, 235
4,7-Dimethylchroman-3-ol, 235
2,2-Dimethylchromanones, 252, 253, 328
2,6-Dimethylchroman-4-one, 252
2,2-Dimethylchromenes, 57, 58, 64, 66
2,2-Dimethylchromene, 56, 57, 58, 105, 265
2,2-Dimethyl-2*H*-chromene, naturally occurring derivatives, 63
Dimethylchromones, physical properties, 150
2,3-Dimethylchromone, 142, 145, 148
2,8-Dimethylchromone, 146
Di-*O*-methylcitromycinol, 160
Di-*O*-methylcitromycinones, 160
Dimethylcoeroxene, 313
4,5-Di-*O*-methylcorymbin, 326
3,4-Dimethylcoumarin, **110**
4,6-Dimethylcoumarin, 143
4,7-Dimethylcoumarin, 103, 110, 235
4,8-Dimethylcoumarin, 110
5,8-Dimethylcoumarin, 110
6,8-Dimethylcoumarin, 110
4,4‴-Di-*O*-methylcupressuflavone, 204
1,3-Dimethylcyclopentane, 396
6,6-Dimethyldibenzo[*b,d*]pyran, 334
2,2-Dimethyl-7,7-dibromo-2-silanorcarane, 418
Dimethyldichlorosilane, 413
2,4-Dimethyl-2,3-dihydro-4-pyran-3-carbaldehyde, 45
2,6-Dimethyl-5,6-dihydro-2-pyran-3-carbaldehyde, 45
2,6-Dimethyldihydro-4-pyrone, 27
5,7-Dimethyl-4*a*,9*a*-dihydrothioxanthene, 395
3,3-Dimethyl-2,4-dioxochroman, 115
4,6-Dimethyl-2,2-diphenyl-2*H*-chromene, 59
2,6-Dimethyl-4,4-diphenyl-4-pyran, 26
3,5-Dimethyl-2,6-diphenyl-4*H*-thiopyran, 251
6,6′-Dimethyl-3,3′-dithiachromanol, 380
6,6′-Dimethyl-3,3′-dithiochromone, 380
6,7,3′,4′-Dimethylenedioxyisoflavone, 218
Dimethylenedioxy-1-methoxyxanthone, 327
2,2-Dimethyl-3,4-epoxychroman, 236
3,6-Dimethylflavone, 168
5,8-Dimethylflavone, 168
Di-*O*-methylflavonols, 178
4,6-Dimethylflavylium perchlorate, 80
2,7-Dimethylfluoran, 313

Dimethylformamide dimethyl acetal, 144
5,7-Dimethylhexahydrothioxanthene 10,10-dioxide, 395
2,8-Dimethyl-9-hydroxycerothiene, 402
2,3-Dimethyl-8-hydroxychromone, 152
3,4-Dimethyl-7-hydroxycoumarin, 118
3,4-Dimethyl-8-hydroxyisocoumarin, 297
2,6-Dimethyl-5-hydroxy-7-methoxychromone, 147
8,8-Dimethyl-4′-hydroxy-5-methoxy-6-prenyl-8*H*-pyrano-[2,3-*h*]isoflavone, 221
2,6-Dimethyl-4-hydroxytetrahydropyran, 49
Dimethyl-(5-iodo-3-methylpent-3-enyl)-sulphonium iodide, 360
Dimethylisochromanones, 300
3,4-Dimethylisocoumarins, 295
5,6-Dimethylisocoumarin, 295
Dimethylisothiochromans, 386
7,3′-Di-*O*-methylluteolin-4′-*O*-D-glucoapioside, 184
7,3′-Di-*O*-methylluteolin-4′-*O*-mono-D-glucoside, 184
Dimethylmaleic anhydride, 227
2,9-Dimethyl-5-methoxypsoralen, 129
2,6-Dimethyl-4-methoxypyridine, 6, 21
2,6-Dimethyl-4-methoxypyrylium cations, 8
Dimethylnaphthopyrans, 56
2,2-Dimethyl-2*H*-naphtho[1,2-*b*]pyran, 338, 340
3,3-Dimethyl-3*H*-naphtho[2,1-*b*]pyran, 59, 340
2,4-Dimethyl-6-(1-naphthyl)pyrylium perchlorate, 9
Dimethylnitrocoumarins, 109
6-(3,7-Dimethyl-2,6-octadienyl)-7-hydroxycoumarin, 118
2,2-Dimethyl-6-oxo-octanoic acid, 50
Dimethyl(4-pentenyl)silane, 414
o-(α,α-Dimethylphenacyl)benzoic acids, 300
4,4-Dimethyl-1-phenylisochroman, 299
6,8-Dimethyl-3-phenylisochroman, 300
4,10-Dimethyl-6-phenyl-2*H*,9*H*-pyrano-[3,2-*g*]chromene-2,9-dione, 138
2,4-Dimethyl-6-phenylpyrylium perchlorate, 9
2,2-Dimethyl-4-phenyltetrahydrothiopyran-4-ol, 365
2,5-Dimethyl-4-phenyltetrahydrothiopyran-4-ol, 365
2,2-Dimethyl-6-phenyltetrahydrothiopyran-4-one, 365
2,7-Dimethyl-9-phenylthioxanthene, 389
Dimethylpyranoisoflavones, 220
2,6-Dimethylpyran-4-one, 6, 7
2,6-Dimethyl-4-pyridone, 30
2,6-Dimethyl-4-pyridone-3-carboxylic acid, 35
4,6-Dimethyl-2-pyridone-5-carboxylic acid, 18
2,6-Dimethyl-4-pyrone, 6, 21, 23, 24, 25, 27, 28, 29, **30**
–, hydrogenation, 49
4,6-Dimethyl-2-pyrone, 14, 15, 17, 18
4,6-Dimethyl-2-pyrone-5-carboxylic acid, 18
5,7-Dimethylquercetin triacetate, 192
2,6-Dimethyl-1-selenocyclohexan-4-one-3,5-dicarboxylic ester, 404
2,6-Dimethylselenan-4-one-3,5-dicarboxylic ester, 404
Dimethylselenochromones, 407
Dimethylsilacyclohexadien-1-one, 417
1,1-Dimethylsilacyclohexane, 414
1,1-Dimethylsilacyclohexane-3-carboxylic acid, 416
4,4-Dimethyl-4-silacyclohexanol, 416, 417
4,4-Dimethyl-4-silacyclohexanone, 416, 417
1,1-Dimethylsilacyclohex-2-ene, 418
1,1-Dimethylsilacyclohex-3-ene, 418
4,4-Dimethyl-4-silacyclohex-2-enone, 417
1,1-Dimethylsilacyclopentane, 414, 415, 416
2,2-Dimethyl-2-silanorcarane, 418
Dimethylsilylenes, 415
5-(Dimethylsilyl)hex-1-ene, 414
1,1-Dimethylstannacyclohexane, 422, 423
7-Dimethylsuberosin, 135
Dimethylsulphoxonium methylide, 213, 220, 255, 264, 381
2,6-Dimethyltetrahydro-4-pyranol, 27
2,6-Dimethyltetrahydro-4-pyrone, 27, 51
4,6-Dimethyltetrahydro-2-pyrone, 17
2,5-Dimethyltetrahydrothiopyran-4-one, 365
1,1-Dimethyltetraphenyl-1-silacyclohexa-2,4-dien-1-yl, 418
Dimethylthiochromans, 378

4,6-Dimethylthiochroman-4-ol, 382
2,2-Dimethylthiochroman-4-one, 378
2,6-Dimethylthiochroman-4-one, 382
2,2-Dimethylthiochroman-1-one oxide, 378
2,2-Dimethylthiochroman-4-one 1-oxide, 380
3,3-Dimethylthiochroman-4-one 1-oxide, 381
Dimethylthiochromones, 373
2,3-Dimethylthiochromone, 373
2,6-Dimethylthio-3,5-diphenyl-4*H*-thiopyran-4-one, 356
Dimethylthioflavones, 375
2,7-Dimethylthiofluoran, 402
Dimethyl-4-thionochromones, 149
2,6-Dimethyl-4*H*-thiopyran-4-one, 355
Dimethylthioxanthones, 394
Dimethyltin diiodide, 422
5,7-Dimethyltocol, 230
5,8-Dimethyltocol, 230
7,8-Dimethyltocol, 230
N,N-Dimethyl-*p*-toluidine, 150
Dimethylxanthenes, 308
3,6-Dimethylxanthene, 307
Dimethylxanthones, 308
4,6-Dimethylxanthyletin, 135
2,6-Dimorpholinobicyclo[3.3.1]nona-2,7-diene, 398
3,5-Dimorpholinotetrahydrothiopyran 1,1-dioxide, 363
Dinaphthalene oxides, 432
2,6-Di(1-naphthyl)-4-(4-nitrophenyl)-pyrylium perchlorate, 9
Dinatin, **185**
1,1-Dineopentylstannacyclohexane, 422
1,5-Dinitroanthraquinone, 402
3,6-Dinitrocoumarin, 109
6,8-Dinitrocoumarin, 109
3,7-Dinitro-10*H*-dibenzo[*b,e*]iododininium bisulphate, 425
Dinitrodibromofluorescein, 316
4,4′-Dinitrodiphenylmethane, 425
4,5-Dinitrofluorescein, 316
Dinitro-3-hydroxycoumarins, 111
5,7-Dinitro-6-hydroxy-2-methyl-5,7-dinitrochromone, 152
2,7-Dinitro-9-iodonia-10*H*-anthracene bisulphate, 425
3,7-Dinitro-10-(4-nitrophenyl)-10*H*-dibenz-[*b,e*]iodininium bisulphate, 425
2,4-Dinitrophenylhydrazine, 2
Dinitrothioxanthones, 394
4,5-Dinitroveratrol, 94
Dinitroxanthones, 320, 321
2,7-Dinitroxanthone, 306, 320, 321
Diosmetin, **184**
Diosmin, 174, **184**
1,5-Dioxopimelic acid, 3
2,5-Dioxopyrano-[3,2-*c*]benzopyran-3-carboxylic ester, 115
1,5-Dioxo-1,2,4,5-tetraphenylpentane, 351
Dipentylplumbacyclohexane, 424
Diphenacyldiphenylmethane, 3
1,3-Diphenoxypropane, 222
1,2-Diphenylacenaphthalene, 345
Diphenylacetylene, 44
Diphenylbenzochroman-2-ol, 106
2,2-Diphenyl-5,6-benzo-2*H*-chromene, 59
2,2-Diphenyl-6,7-benzo-2*H*-chromene, 59
2,2-Diphenyl-7,8-benzo-2*H*-chromene, 59
2,4-Diphenyl-5,6-benzo-4*H*-chromene, 54
2,4-Diphenyl-6,7-benzo-4*H*-chrom-2-ene, 55
2,4-Diphenyl-7,8-benzo-4*H*-chrom-2-ene, 55
2,8-Diphenyl-4*H*,6*H*-benzo[1,2-*b*:5,4-*b*′]-dipyran-3,7-diol-4,6-dione, 174
2,8-Diphenyl-4*H*,6*H*-benzodipyran-4,6-dione, 168
1,4-Diphenylbenzo[*c*]pyran-3-one, 290
2,4-Diphenylbenzopyrylium ferrichloride, 67, 71
2,4-Diphenylbenzothiopyrylium perchlorate, 370
1,4-Diphenyl-1,3-butadiene, 351, 354
2,2-Diphenylchroman, 224
2,2-Diphenylchroman-3-ol, 236
2,4-Diphenylchroman-2-ol, 105
2,2-Diphenylchroman-4-one, 253
2,2-Diphenylchromene, 105, 106
2,2-Diphenyl-2*H*-chromene, 54, 59, 105
2,4-Diphenylchromenes, 106
2,4-Diphenyl-2*H*-chromene, 78
2,4-Diphenyl-4*H*-chromene, 54, 59, 71, 105
2,3-Diphenylchromen-2-ol, 68, 76
2,4-Diphenylchromen-2-ol, 67
Diphenyldiazomethane, 319, 358
Diphenyldihydronaphthopyran-2-ol, 106

Diphenyldihydronaphthopyran-3-ols, 106
4,6-Diphenyl-2-diphenyl-methylene-2*H*-thiopyran, 358
3,6-Diphenyl-4,5-disubstituted-2-pyrones, 17
Diphenylethene, 59
Diphenyl ether, 311
2,4-Diphenyl-2-ethoxychroman, 235
1,1-Diphenylgermacyclohexane, 420
4,6-Diphenyl-2-halogenothiopyrylium salts, 352
3,9-Diphenyl-6-hydroxy-1*H*,7*H*-benzo-[1,2-*b*:4,3-*b'*]dipyran-1,7-dione, 169
1,3-Diphenyl-3-(1-hydroxy-2-naphthyl)-propan-1-one, 235
1,3-Diphenyl-3-(3-hydroxy-2-naphthyl)-propan-1-one, 235
1,3-Diphenyl-3-(2-hydroxy-phenyl)propan-1-one, 235
Diphenyliodonium salts, 425
3,3-Diphenylisochroman, 300
1,4-Diphenylisochroman-3-one, 301
3,4-Diphenylisocoumarin, 295
Diphenylketene, reaction with 4-pyrone, 26
Diphenyllead dichloride, 424
Diphenyl malonate, 112
2,4-Diphenyl-2-methoxychroman, 71, 235
2,3-Diphenyl-6-methylchromone, 71
9-(Diphenylmethylene)xanthene, 319
2,3-Diphenyl-6-methyl-4-(4-tolyl)benzo-pyrylium salts, 71
1,3-Diphenylnaphthalene, 3
1,3-Diphenyl-1*H*-naphtho[2,1-*b*]pyran, 54
1,3-Diphenyl-1*H*,3*H*-naphtho[1,8-*cd*]pyran, 344
2,2-Diphenyl-2*H*-naphtho[1,2-*b*]pyran, 59
2,2-Diphenyl-2*H*-naphtho[2,3-*b*]pyran, 59
2,4-Diphenyl-4*H*-naphtho[1,2-*b*]pyran, 55
2,4-Diphenyl-4*H*-naphtho[2,3-*b*]pyran, 55
3,3-Diphenyl-3*H*-naphtho[2,1-*b*]pyran, 59, 339
1,3-Diphenylnaphtho[2,1-*b*]pyrylium ferri-chloride, 341
1,3-Diphenyl-1*H*,3*H*-naphtho[1,8-*cd*]thio-pyran, 400
7,8-Diphenyl-2-oxabicyclo[4.2.0]oct-7-ene, 44
1,1-Diphenyl-4-oxotetrahydro-1-silanaphthal-ene, 420
Diphenyl-β-picrylhydrazyl, 229
1,1-Diphenyl-1-plumbacyclohexane, 424
3,5-Diphenyl-4-pyrans, 2,6-disubstituted, 17
2,6-Diphenyl-4-pyrone, 24, 25
3,5-Diphenyl-4-pyrones, 2,6-disubstituted, 29
3,6-Diphenyl-2-pyrones, 4,5-disubstituted, 29
4,6-Diphenyl-2-pyrone, 13
2,6-Diphenyl-4-pyrone oxime, 24
2,6-Diphenylpyrylium tetrafluoroborate, 9
2,6-Diphenylpyrylium perchlorate, 5
2,6-Diphenylselenan-4-one, 405
9,10-Diphenyl-10-selenanthracene, 410
1,1-Diphenylsilacyclohexane, 415
1,1-Diphenylsilacyclohexan-2-ol, 417
Diphenylsilacyclohexanone, 417, 418
4,4-Diphenyl-4-silacyclohexanone, 417
1,1-Diphenylsilacyclohex-2-ene, 417
Diphenylsilane, 305, 388
1,1-Diphenyl-1-stannacyclohexane, 422
Diphenyl sulphide, 391
1,1-Diphenyltetrahydro-1-silanaphthalene, 419
2,6-Diphenyltetrahydrothiopyran, 351, 363
Diphenyltetrahydrothiopyran-4-one, 363, 365
1,2-Diphenyl-1,2-thianaphthalene, 385
3,5-Diphenylthiopyrans, 349, 350
2,6-Diphenyl-4*H*-thiopyran-4-one, 355
3,5-Diphenyl-4*H*-thiopyran-4-one, 355
3,6-Diphenyl-2*H*-thiopyran-2-one, 354
4,6-Diphenyl-2*H*-thiopyran-2-one, 352
2,6-Diphenyl-4*H*-thiopyran-4-thione, 358
3,6-Diphenyl-2*H*-thiopyran-2-thione, 357
4,6-Diphenyl-2*H*-thiopyran-2-thione, 358
2,6-Diphenylthiopyrylium perchlorate, 351
Diphenyltin dichloride, 422
Diphenylxanthenes, 308
Dipropyl-2*H*-chromene, 58
2,6-Dipyrrolidinobicyclo[3.3.1] nona-2,7-diene, 398
Disaccharides, conversion to kojic acid, 33
Dispiro(isothiochroman-1,4'-[1,3]dithiolane-5',1"-isothiochroman), 387
Distemonanthin, **161**
Distemonathus Benthamianus, 195
Distyryl ether, 405
Dithiachromanols, 380
3,9-Dithia[1,2:7,8]dibenzoperylene, 402

8,16-Dithiadibenzoperylene, 402
Dithiochromones, 377, 380
Dithiocoumarin, 377
1,4-Dithioflavone, 375
1,2-Dithiolylium acid sulphates, 357
2,6-Dithiono-3,5-dimethyltetrahydrothio-pyran-4-one, 366
Dithiophenylmalonic ester, 372
4,4'-Dithiopyrylium bistetrafluoroborate, 353
Dithiopyrylium dication, 353
Dithioxanthyl ether, 390
Dixanthyl ether, 310
Dixanthyl sulphides, 310
8-Dixanthylurea, 311
Dolichone, **268**
Dolineone, **268**
Dorema ammoniacum, 120
Dracaena draco, 79, 233
Dracorhodin, **79**
Duartin, **234**
Duff reaction, 155
Durmillone, **222**
Durohydroquinone, 227
Duroquinone, 100
Dypnone, 5

Edultin, **132**, 133
Eleutherin, 288
Eleutherine bulbosa, 154
Eleutherinol, **154**
Ellaborugin, **125**
Ellagic acid, **125**
Elliptone, 264, **266**
Empetrin chloride, **91**
Empetrum nigrum, 91
Emulsin, 181, 194
Enamines, reaction with carbon disulphide, 357
Eosin, 306
Eosine BN, 316
Eosin G., 316
Eosin S., 316
Epiafzelechin, 238, 242, **243**
Epicatechins, 88
–, stereochemistry, 240, 241
–, structure, 238, 239, 240, 241
Epicatechin, 88, 94, 237, 238, 242, **243**
Epicatechin ethers, 243
Epicatechin penta-acetate, 90
Epicatechin tetramethyl ether, 95
Epigallocatechin, **243**
Epigallocatechin gallate, 243
Epiquerbrachocatechin, **244**
Epoxyketones, 8
8-(2,3-Epoxy-3-methylbutoxy)psoralene, 131
5-(2,3-Epoxy-3-methylbutyl)-9-methoxy-psoralene, 131
3,α-Epoxy-3-methylthiochroman-4-ones, 380
Equisetum arvense, 184
Equol, **234**
Eremophila fraseri, 199
Ergochrysin A., **333**
Ergoflavin, **332**, 333
Ergot, 332, 333
Eriodictyol, 165, **282**, 284
Eriodictyol 5,3'-diglucoside, 282
Eriodictyol 7-glucoside, 274
Eriodictyon Californicum, 282
Erythrocentaurin, **303**
Erythromycin, 51
Erythrose, 239
Erythrosine, 316
Esculetin, **122**
Esculin, **122**
Ester-nitriles, 11
Ethoxalyl chloride, 207, 208
3-Ethoxalyltetrahydro-4-pyrone, 46
Ethoxyacetone, 38
4-Ethoxy-2*H*-benzopyran, 60
5-Ethoxycarbonyl-3,4-dihydro-6-phenyl-2*H*-thiopyran, 359
Ethoxycarbonyl-5,7-dihydroxy-4'-methoxy-isoflavone, 207
3-Ethoxycarbonyl-4-hydroxyisochroman-1-one, 300
2-Ethoxycarbonylisoflavones, 207
8-Ethoxycarbonylmethoxy-1-naphthoic ester, 343
3-Ethoxycarbonyl-2-methyl-4,6-diphenyl-pyrylium perchlorate, 9
O-Ethoxycarbonyloxycinnamoyl chloride, 270
3-Ethoxycarbonyl-2,4,6-triphenylpyrylium perchlorate, 9

2-Ethoxychroman, 53
4-Ethoxychroman, 55
4-Ethoxy-2*H*-chromene, 60
4-Ethoxy-4*H*-chromene, 76
2-Ethoxy-7-hydroxychroman, 235
Ethoxymagnesiomalonic ester, 12
3′-Ethoxy-7-methoxy-4′-methoxybenzylidene-chromanone, 441
4-Ethoxy-2-methyl-2*H*-chromene, 60
Ethoxymethyleneacetoacetic ester, 20
4-Ethoxy-2-pyrone-6-carboxylic acid, 13
2-Ethoxytetramethoxyisoflavan, 242
3-Ethylamino-4-cresol, 314
2-Ethylchroman, 147
6-Ethylchroman, 223
2-Ethylchroman-4-ol, 147
2-Ethylchroman-4-one, 147, 254
4-Ethyl-2*H*-chromene, 58
2-Ethylchrom-2-en-4-ol, 147
2-Ethylchromones, 148
2-Ethylchromone, 139
3-Ethylcoumarin, **110**
4-Ethylcoumarin, 110
3-Ethyl-5,7-dihydroxyflavanone, 269
O-Ethyldimethylbrazylium ferrichloride, 441
4-Ethyl-2,6-dimethylpyrylium salts, 5
6-Ethyl-2,2-dimethyltetrahydropyran-6-carboxylic acid, 50
9-Ethyl-2,7-dimethylthioxanthene, 389
9-Ethyl-2*H*-furo[3,2-*g*]chromen-2-ones, 134
α-Ethylglutaric acid, 16
6-Ethyl-7-hydroxy-3,4-dimethylcoumarin, 118
6-Ethyl-7-hydroxy-6-methylcoumarin, 118
3-Ethyl-7-hydroxyflavanone, 269, 270
Ethylidene-2,2′-bicyclohexanones, 395
3-Ethylisocoumarin-4-carboxylic acid, 295
3-Ethyl-7-methoxyisocoumarin, 295
4-Ethyl-7-methyl-2*H*-benzo[*b*]thiopyran, 367
2-Ethyl-6-methylchromone, 150
2-Ethylmethylcyclohexane, 387
6-Ethyl-4-methyl-7-hydroxycoumarin, 118
5-Ethyl-4-methylresorcinol, 290
6-Ethyl-8-methylthiochroman, 378
8-Ethyl-6-methylthiochroman, 378
4-Ethyl-6-methylthiochroman-4-ol, 382
5-Ethyl-2-pyrone, **17**
6-Ethylthiochroman, 378
O-Ethyltrimethyldihydrobrazileinol, 441
3-Ethyl-2,4,6-triphenylpyrylium perchlorate, 9
Ethyl vinyl sulphide, 347
α,β-Ethynyl ketones, reaction with phenols, 86
Ethynyl phenyl ketone, 341
Ethynylmethanols, 43
9-Ethynylthioxanthydrol, 392
Eucalyptin, **183**
Eucalyptus calophylla, 95
Eucalyptus cinerea, 183
Eucalyptus globulus, 183
Eucalyptus redunca, 245
Eucalyptus risdoni, 183
Eucalyptus sieberi, 278
Eucomin, **218**
Eucomin diacetate, 219
Eucomis autumnalis, 219, 335
Eucomis bicolor, 219
Eucomis punctata, 219
Eucomnalin, **219**
Eucomol, **218**
Eugenia caryophyllata, 153
Eugenin, **153**
Eugenitin, 147, **153**
Eugenitol, **153**
Eupafolin, **186**
Eupatilin, **187**
Eupatin, **197**
Eupatoretin, **197**
Eupatorin, **187**
Eupatoriochromene, 64
Eupatorium ayapana, 123
Eupatorium cuneifolium, 186
Eupatorium semiserratum, 187
Euxanthic acid, 304, **324**, 325
Euxanthogen, **326**
Euxanthone, 304, **324**, 325, 329, 330
Evodia elleryana, 64
Evodia glandulosum, 64
Evodia littoralis, 64
Evodia riparium, 64
Evodione, **64**
Evodionol, **64**
Evodionol methyl ether, 57, 58, 64

Exostemin, **125**
Exostemma caribaeum, 125
Exoticin, **200**

Fabaceae, 257
Fabiana imbricata, 123
Fagara macrophylla, 123
Fagara xanthoxyloides, 128
Farrerol, **283**
Fast Acid Violet A2R, 314
Ferreirea spectabilis, 279, 286
Ferreirin, **286**
Ferruol A, **122**
Ferula communis, 116
Ferula samarcandica, 118
Ferulenol, 111, **116**
Ficus carica, 127
Ficusin, **126**
Fisetin, 88, 174, 177, **191**, 428
Fisetinidin chloride, 75, 88
Fisetinidol, **244**
Fisetin 3-methyl ether, 191
Fisetol dimethyl ether, 151
Fish liver oils, 226
Fish poisons, 257
Flavans, 60, 62, 75, 231, 232
–, crystalline complexes, 233
Flavan, **231**, 232, 272
Flavan-2,3-diols, 96, 244
Flavan-3,4-diols, 94, 95, 96, 247, 248, 249, 250
Flavan-3,4-diol, 93, 172
Flavan-3,4-dione, 173
Flavanols, 61, 73, 88, 142, 232, 237, 276
Flavan-3-ol, 239
Flavan-4-ols, 247, 248, 249, 250, 272
Flavan-4-ol, 165
Flavanones, 52, 142, 152, 165, 170, 232, 248, 250, 257, 268, 269, 270
–, oxidation, 272
–, reduction, 272
–, ring fission, 271
Flavanone, 20, 77, 247, 269, **273**, 374
Flavanone ethylene thioacetals, 232
Flavanone glycosides, 174, 280
Flavanonols, 172, 250, 274, 276
3-Flavanyl methyl ether, 244
Flavasperone, **154**, 342
Flavenes, 52, 60, 75, 165
Flav-2-ene, 61, 78, 96, 165
Flav-3-ene, 61, 241
2*H*-Flavenes, 60, 61, 62
2*H*-Flavene, 53, **61**
4*H*-Flavene, **61**, 62
Flav-3-en-3-ol, 90
Flaven-4-ols, 165
4-Flavenylflavylium salts, 167
Flavones, 2, 82, 87, 89, 139, 141, 142, 143, 152, 162
–, carbonyl group reactivity, 166
–, from acetylenes, 163
–, hydrogenation, 270
–, oxidation, 175
–, physical properties, 166
–, reactions, 166
–, ring-fission, 164, 165
–, spectra, 178
–, structure, 164
–, synthesis, 140, 164
Flavone, 20, 52, 61, 76, 138, 140, 144, 164, **168**, 169, 170, 192, 247, 272
–, nitration, 166
–, oxidation, 167
Flavone aglycones, 177
Flavonebenzylimine, 169
Flavonecarboxylic acids, 164
Flavone-2′-carboxylic acid, 164
Flavone glycosides, 174, 178
Flavone-hydrazone, 169
Flavoneimines, 169
Flavone oxime, 169
Flavone pigments, 174
–, sugar residues, 175
Flavone semicarbazone, 169
Flavonoids, 93, 96, 162
Flavonols, 139, 143, 170, 171, 172, 173, 240, 241, 270
–, autoxidation, 200
–, *O*-methylated, 201
–, oxidation, 173
–, spectra, 178
Flavonol, 77, 139, **173**
Flavonol aglycones, 177
Flavonol 3-methyl ethers, 171
Flavonol pigments, 174
Flavonone, 139
Flavopinacols, 273
Flavoyadorinin A., **195**

Flavoyadorinin B., **184**
Flavylium cations, 60
Flavylium chloride, 76, 78, **80**
Flavylium ferrichloride, 80
Flavylium perchlorate, 70, 73, 77, 80
Flavylium salts, 68, 70, 73, 74, 76, 77, 82, 86, 89, 94, 168, 241
–, hydroxylated, 82
–, reduction, 75, 232
Fleminigen C, 64
Flindersia brayleyana, 136
Floroselin, **137**
Flowers, colouring, 87
Fluoran, **313**
Fluorene, 334, 335
9-Fluorenylsodium, 166
Fluorescein, 306, 313, 314, **315**
Fluorescein dichloride, 314, 316
Fluorescein dyes, 311
Fluorimes, 311
Fluorones, 311
Formic acid, 3, 28, 32, 33, 215
Formic esters, 149, 208, 287, 291
Formimidoyl chloride, 209
Formononetin, 211, **214**, 218
4-Formoxybutyraldehyde, 42
Formylacetic acid, 10, 14
Formylchromones, 155
o-Formylcinnamic acid, 384
Formylcrotonic acid, 17
ω-Formyl-1-hydroxyacetophenone, 151
ω-Formyl-2-hydroxyacetophenone, 148
3-Formyl-4-hydroxycoumarin, 115
3-Formyl-4-hydroxythiocoumarin, 372
2-Formylmethylene-2*H*-thiopyrans, 348
3-Formyl-2-naphthol, 342
4-Formyloxybutyric ester, 42
1-Formyl-2-phenylacetylene, 5
Fortunellin, **182**
Fragaria vesca, 90
Fragarin chloride, **90**
Franklinone, 64
Frasera albicaulis, 326
Frasera caroliniensis, 326, 327
Fraxetin, **125**
Fraxidin, **126**
Fraxin, **125**
Fraxinol, **126**
Fraxinus excelsior, 125, 126
Friedel–Crafts reaction, 28, 30
Friedel–Crafts reagents, 17
Fries rearrangement, 163
Fructose, 34
Fuchsone, 311
Fuchsonimine, 311
Fujikinetin, **218**
Fujikinin, **218**
Fukugetin, **202**
Fukugiside, 202
Fulvenes, 18
Fulvic acid, **160**
Fulvoplumierine, **18**
Furan-2-carboxylic acids, 5-substituted, 14
Furan-2-carboxylic acid, 14
Furan-2,3-dicarboxylic acid, 127, 156
Furfurylidenechalcones, 164
5*H*-Furo[3,2-*g*]benzopyran-5-ones, 156
2*H*-Furo[2,3-*h*]chromen-2-one, 127
2*H*-Furo[3,2-*g*]chromen-2-one, 126
Furochromones, 139, 155, 156, 159
Furocoumarins, 97, 126, 134
–, naturally occuring, 131, 132
Furocoumarinic acid, 127
Furodihydropyrans, naturally occurring, 45
Furoflavanones, 204
Furoflavones, 204
Furoisoflavones, 219
Furo[2,3-*h*]isoflavone, 220
5*H*-Furo[3,2-*b*]xanthen-5-one, 333
Furoxanthones, 329, 333
2-(2-Furyl)chromones, 164
Fusarium graminearum, 342
Fusarium solani, 300
Fusarubin, **300**
Fuscin, **231**
Fuscinic acid, 231
Fustic, 195
Fustin, 95, **281**
Fustin 3-*O*-β-D-glucoside, 282

Galactose, 36, 194
Galangin, 170, **189**, 276
Galanginidin chloride, 87
Galanginidin chloride 3-methyl ether, 86
Galangin 3-methyl ether, 189
Galangin 7-methyl ether, 189
Galega officinalis, 184, 191
Galein, **191**

Galetin, **191**
Galetin 3,6-dirhamnoside, 191
Galium aparine, 47
Gallein, **316**
Gallic acid, 82, 91, 196, 243, 296, 316
Gallic acid trimethyl ether, 197
Gallocatechins, 96, 238, 243
Galloflavin, **296**
3-Galloylepigallocatechin, 243
Galuteolin, 175, **183**
Gamatin, **206**
Garcinia buchananii, 285, 329
Garcinia eugeniifola, 285
Garcinia mangostana, 329, 330
Garcinia spicata, 202
Garcinia talboti, 202
Garcinia volkenii, 202
Garcinia xanthochymus, 286
Gardenia lucida, 189
Gardenin A., **189**
Gardenin B., **189**
Gardenin C., **189**
Gardenin D., **189**
Gattermann reaction, 152
Geijera parviflora, 118, 119
Geijera salicifolia, 118
Geijerin, **118**
Geiparvarin, **118**
Gelsemium sempervirens, 122
Genista tinctoria, 183, 214
Genistein, 207, 213, **214**, 215
Genistein 7-glucoside, 210
Genistein 4′-methyl ether, 215
Genistein 5-methyl ether, 215
Genistein 7-methyl ether, 215
Genistein triacetate, 215
Genistein trimethyl ether, 215
Genistin, 210, **214**
Genitisin, **325**
Genkwanin, **182**
Gentiakochianine, **326**
Gentiana acaulis, 91
Gentiana bellidifolia, 326
Gentiana kochiana, 326
Gentiana lutea, 47, 92, 325
Gentianin chloride, **91**
Gentiin, **325**
Gentiopicrin, **47**
Gentiopicroside, 47
Gentioside, **325**
Gentisein, **325**
Geranyl chloride, 130
5-Geranyloxy-2*H*-furo[3,2-*g*]chromen-2-one, 130
1-Germacyclohexa-2,4-dienes, 421
Germacyclohexanes, 420, 421
1-Germacyclopent-3-enes, 421
Germanium compounds, 420
6-Germaspiro[5,5]undecane, 420
Gesneria cardinalis, 89
Gesneria fulgens, 89
Gesneridin, 83
Gesneridin chloride, **89**
Gesnerin, 83, 89
Ginkgetin, **201**
Ginkgo biloba, 201
Glabralactone, **121**
Glaucophanic ester, **20**
Gleditschia monosperma, 196
Gleditsia tricanthos, 281
Globuxanthone, **330**
Glucoluteolin, **184**
Gluconoacetobacter roseum, 34
8-*C*-β-D-Glucopyranosylapigenin, 182
β-D-8-Glucopyranosyl-5,7-dihydroxy-4′-methoxyflavone, 186
6-*C*-β-D-Glucopyranosylgenkwanin, 182, 183
6-*C*-β-D-Glucopyranosylloteolin-7-methyl ether, 184
7-(β-D-Glucopyranosyloxy)-5-hydroxy-flavanones, 274
2-β-D-Glucopyranosyltetrahydroxyxanthone, 326
Glucose, 90, 91, 92, 93, 124, 133, 161
–, conversion to kojic acid, 33
D-Glucose, 181, 182, 183
α-Glucosidase, 194
4′-β-D-Glucosyl-7-β-neohesperidosylnaringenin, 218
2-Glucosyloxyaroylacetophenones, 174
4′-β-D-Glucosyl-7-rutinosylnaringenin, 281
Glutacondialdehyde, 7, 69
Glutaconic acids, cyclisation, 12
–, substituted, 18
Glutaconic acid, 18
Glutaraldehyde, 2, 350, 403
Glutaric acid, 42

Glucuronic acid, 184, 193
Glyceraldehyde, 113, 298
Glycosides, 81, 273
C-Glycosylflavones, 182
Glycyrrhiza glabra, 278
Gnaphaliin, **190**
Gnaphallium obtrusifolium, 191, 283
Golden balm, 90
Gossypetin, **197**, 198
Gossypetin 3,7-dimethyl ether, 201
Gossypetin ethers, 198
Gossypetin-7-β-D-glucopyranoside, 198
Gossypetin 3-glucoside, 198
Gossypetin hexamethyl ether, 197
Gossypetin 8-methyl ether, 198
Gossypitol tetramethyl ether, 197
Gossypitrin, **197**
Gossytrin, **198**
Grignard reaction, 59, 71, 254
Grignard reagents, 2, 4, 41, 62, 67, 105, 223, 247, 299, 318, 352, 380, 382, 390, 395, 410
–, reaction with benzopyrylium salts, 78
–, – with chromones, 71
–, – with coumarins, 104
–, – with isocoumarins, 294
–, – with 2-pyrones, 17
Grote reagent, 356
Guanandin, **328**
Guibourtacacidin, **95**
Guibourtia coleosperma, 95
Guibourtia demeusei, 95
Guibourtia tessmannii, 95

Haematein, 427, 444, 445
Haematoxylin, **244**, 427, **443**
–, optical rotatory powers, 445
Haematoxylinic acid, 443
Haematoxylin ethers, 443
Haematoxylin methyl ethers, 445
Haematoxylium salts, 444
Haematoxylon africanum, 445
Haematoxylon campechianum, 427
Halfordia kendack, 130
Halfordia scleroxyla, 130
Halfordin, **130**
Halkendin, **130**
Halogenochromanols, 236
3-Halogenochroman-4-ones, 254
3-Halogenocoumarins, 101
6- and 7-Halogenocoumarins, 109
3-Halogenodimethylchroman-4-ones, 254
4′-Halogenoflavones, 169
ω-Halogeno-2-hydroxyacetophenones, 143
Halogenomethoxybenzophenones, 319
3-Halogenomethoxychroman-4-ones, 254
3-Halogeno-4′-methoxyflavanones, 271
4-Halogeno-*N*-methylpyridinium iodides, 350
3-Halogeno-2-pyrones, 14
2-Halogenotetrahydropyrans, 41
3-Halogenothiochroman-4-ols, 379
3-Halogenothiochroman-4-ones, 379
Halogenoxanthones, 319
Hamamelis japonica, 193
Hamaudol, **159**
Hashish, 336
Helicin, 69
Helminthosporium ravenelii, 325
Helminthosporium turcicum, 325
Hemiacetals, 49
Hemiergoflavin, 332
Hemiergoflavin-2-carboxylic acid, 332
Hemp drugs, 336
Hepta-2,5-diyne-4-one, 30
Heptan-2,4-diol, 49
Heraclenin, **131**
Heraclenol, **131**
Heracleum candicans, 131
Heracleum sphondylium, 130, 131
Herbacetin, **196**
Herbacetin 3,7,8-trimethyl ether, 196
Herbacitrin, **196**
Herniaria hirsuta, 117
Herniarin, **117**
Hesperetin, 271, **279**
Hesperetin 7,3′-dimethyl ether, 279
Hesperidin, 174, **279**
Heterocyclic compounds with six-membered rings, 1
Hevea braseliensis, 203
Heveaflavone, **202**
Hexabromoxanthone, 319
Hexafluorobutyne, 424
Hexahydrocannabinol, 336
Hexahydroisochromans, 387
Hexahydroisochromanone, 304
Hexahydroisothiochromans, 387

Hexahydro-7-methyl-1-oxaphenalene, 343
Hexahydroxybiflavonyls, 202, 203, 204
Hexahydroxyflavan, 94
Hexahydroxyflavanones, 285
Hexahydroxyflavone, 199
Hexahydroxyflavylium chloride, 91
Hexahydroxyisoflavone, 217
Hexamethoxyflavones, 188, 198
Hexamethoxyisoflavone, 217
Hexamethoxyxanthone, 327
Hexamethylphosphoramide, 309
Hexamethyl-2,2′-spirobichroman, 233
Hexamine, 155, 210
Hexoses, 81
–, conversion to kojic acid, 33
Hibiscetin, **199**
Hibiscetin-3-glucoside, 199
Hibiscitrin, **199**
Hibiscus sabdariffa, 198, 199
Hinokiflavone, **201**
Hinokiflavone methyl ethers, 201
Hinokiflavone pentamethyl ether, 201
Hirsutidin chloride, 82, 87, **93**
Hirsutidin 3,5-diglucoside, 93
Hirsutin, 82, 87
Hirsutin chloride, **93**
Hispidin, **19**
Homoasaronic acid, **261**
Homoferreirin, **286**, 287
Homoflavoyadorinin, **184**
Homoisoflavones, 218
Homo-orientin, **184**
Homophthalic anhydride, 291, 301
Homophthalic esters, 107, 291
Homophthalyl alcohol, 297
Homopterocarpin, **288**
Hordeum vulgare, 184
Humiria balsamifera, 295
Humulus japonicus, 184
Hycanthone, **393**
Hydrangea arborescens, 117
Hydrangin, **117**
β-Hydrazino-β-(2-hydroxyphenyl) propionhydrazide, 108
Hydrocinnamic acid, 65
Hydrofluoranic acid, 313, 315
Hydrolysate tannins, 245, 246, 247
4-Hydroperoxy-4-methylbutyl formate, 42
Hydropyrans, 39
Hydroquinone, 227, 312
Hydroxyacetophenones, 275
2-Hydroxyacetophenones, 140, 269
2-Hydroxyacetophenone, 70, 112, 114, 139, 144, 145, 149, 168
4-Hydroxyacetophenone, 181
Hydroxyacetophenone glycosides, 274
4-Hydroxyamino-2,6-diphenylpyridine 1-oxide, 24
2-Hydroxyarenecarboxylic acids, 101
6-Hydroxyayanin, 195
2-Hydroxybenzaldehydes, 98
2-Hydroxybenzaldehyde, 70, 255, 256
Hydroxybenzoic acids, 82
4-Hydroxybenzoic acid, 81, 82, 84, 89, 91, 181, 190, 274, 278, 280, 286
2-Hydroxybenzoins, 207
Hydroxybenzophenones, 321
2′-Hydroxybenzophenone-2-sulphinic acids, 393
4-Hydroxybenzopyrylium perchlorates, 72
3-Hydroxybenzo[*b*] thiophene-2-carbaldehyde, 380
(2-Hydroxybenzoyl)acetic acid, 112
β-(2-Hydroxybenzoyl)acrylic acids, 151
7-Hydroxy-6-benzoyl-4,8-dimethylcoumarin, 138
2-Hydroxybenzoylmethanol, 173
2-Hydroxybenzoylpyruvic acid, 139
2-(2-Hydroxybenzoylvinyl)dimethylamine, 144
2-Hydrobenzyl alcohol, 309
4-(4-Hydroxybenzyl)chromans, 224
3-(2-Hydroxybenzylidene)chroman-2,4-dione, 255
2-Hydroxybenzyl ketones, benzoate esters, 73
2-(α-Hydroxybenzyl)-2-methoxycoumaran-3-ones, 172
Hydroxybiscoumarin, 255
α-Hydroxybutadiene-α,ω-dicarboxylic ester, 18
α-Hydroxycarboxylic acids, 10
9-Hydroxycerothiene, 401
Hydroxychalcones, 65, 272
2-Hydroxychalcones, 62, 69, 77, 141, 142, 171, 268
2-Hydroxychalcone, 76, 80

2′-Hydroxychalcones, 60
2′-Hydroxychalcone, 61, 247, 272, 273
4-Hydroxychalcones, 165
2-Hydroxychalcone dibromides, 270
4-Hydroxychromans, 71
6-Hydroxychromans, 226, 231
Hydroxychromenes, 78
Hydroxychromones, 151
–, naturally occurring, 153
2-Hydroxychromone, 112
5-Hydroxychromones, 147
7-Hydroxychromones, 152
Hydroxycinnamic acids, 84
2-Hydroxycinnamic acids, 99
4-Hydroxycinnamic acid, 81, 90, 91, 92, 274, 280
4-Hydroxycinnamic acid lactone, 97
2-Hydroxycinnamonitrile, 108
2-Hydroxycinnamyl alcohols, 55
3-Hydroxycoumaran, 239
Hydroxycoumarins, 97, 100
–, cyanoethylation, 119
–, dimethylallyl ethers, 120
–, spectra, 120
3-Hydroxycoumarin, **111**
4-Hydroxycoumarins, 137
–, 3-benzyl substituted, 113
–, reaction with acyl chlorides, 113
4-Hydroxycoumarin, 109, **112**, 116, 256
–, arylation, 113
–, reaction with formaldehyde, 116
5-Hydroxycoumarin, **116**
6-Hydroxycoumarin, **117**
7-Hydroxycoumarin, 111, **117**
8-Hydroxycoumarin, **119**
8-Hydroxycoumarin-7-carbaldehyde, 137
4-Hydroxycoumarin-3-carboxylic acids, 101
6-Hydroxycoumarin-3-carboxylic ester, 117
7-Hydroxycoumarin 1,1-dimethylallyl ether, 118
4-Hydroxycoumarin monohydrate, 115
3-Hydroxycoumarin-2-spiro-3′-isochroman, 168
7-(4-Hydroxy-3-coumarinyl)benzopyrano-[3,2-*c*]coumarins, 114
1-(4-Hydroxycoumarin-3-yl)-3-oxo-1-phenylbutane, 115
Hydroxydeguelins, 265, 266
6-Hydroxydeguelin, **266**
o-Hydroxydeoxybenzoins, 287
3-Hydroxy-6H-dibenzo[*b,d*]pyran-6-one, 334
2-Hydroxydibenzoylmethane, 168
2′-Hydroxydibenzyl-α-carboxylic acid, 110
6-Hydroxy-2,3-dihydrobenzofuran, 127, 219
6-Hydroxydihydrocoumarone, 127
6-Hydroxy-9-(2,4-dihydroxyphenyl)-fluorone, 312
2-Hydroxy-3,4-dimethoxybenzoic acid, 321
2-Hydroxy-4,5-dimethoxybenzoic acid, 265
4-Hydroxy-3,5-dimethoxybenzoic acid, 92
2-Hydroxy-4,5-dimethoxybenzophenone, 123
5-Hydroxy-7,3′-dimethoxy-4′-(3,6-dimethyl-hept-2-enyloxy)flavones, 184
3′-Hydroxy-7,4′-dimethoxyflavone, 183
5-Hydroxy-3,7-dimethoxyflavone, 191
7-Hydroxy-5,8-dimethoxyflavone, 180
4′-Hydroxy-5,7-dimethoxyflavonol, 276
2-Hydroxy-5,7-dimethoxy-8-isoflavanone, 288
5-Hydroxy-6,7-dimethoxyisoflavone, 212
7-Hydroxy-3′,4′-dimethoxyisoflavone, 218
7-Hydroxy-6,4′-dimethoxyisoflavone, 218
7-Hydroxy-5,8-dimethoxyisoflavone, 211
5-Hydroxy-7,4′-dimethoxy-6-methyl-flavanone, 275
5-Hydroxy-7,4′-dimethoxy-6-methyl-flavone, 183
5-Hydroxy-7,4′-dimethoxy-6-methyliso-flavone, 213
1-Hydroxy-6,8-dimethoxy-3-methyl-naphthalene, 154
2-Hydroxy-4,6-dimethoxy-5-methylphenyl-methoxymethyl ketone, 197
1-Hydroxy-3,6-dimethoxy-8-methyl-xanthone, 327
2-Hydroxy-1-(3,4-dimethoxyphenyl)-3-(2,4-dimethoxy-6-hydroxyphenyl)-1-oxopropane, 244
3-Hydroxy-4-(2,4-dimethoxyphenyl)flavan, 248
Hydroxy-2,3-dimethylbenzofurans, 134
7-Hydroxy-2,4-dimethylbenzopyrylium chloride, 70, 76

3-Hydroxy-2,2-dimethylchromans, 137
6-Hydroxy-2,2-dimethylchroman, 223
7-Hydroxy-2,2-dimethylchroman, 224
7-Hydroxy-2,2-dimethylchroman-4-one, 137, 253
7-Hydroxy-3,4-dimethylcoumarin, 118
7-Hydroxy-4,8-dimethylcoumarin, 118
7-Hydroxy-2,6-dimethylisoflavone, 210
3-Hydroxy-2,6-dimethyl-4-pyrone, 30
7-Hydroxy-6,8-dinitroflavone, 166
2-Hydroxy-5,5′-dinitro-1-piperidinobenzophenone, 321
Hydroxy-1,3-diphenylpropanetrione, 173
2′-Hydroxyepoxychalcone, 163
2-(β-Hydroxyethyl)benzoic acid, 300
Hydroxyflavans, 96, 165, 246
Hydroxyflavan-3,4-diols, 246
Hydroxyflavanones, 52, 247, 270, 273, 274, 275, 276
–, naturally occurring, 277
3-Hydroxyflavanones, 172, 232, 270, 271, 276, 277
5-Hydroxyflavanones, 279
6-Hydroxyflavanones, 269
7-Hydroxyflavanone, 163, 269, 280
Hydroxyflavanonols, 277
Hydroxyflavones, 167, 174, 178
–, absorption spectra, 170
–, reactions, 175
–, with a 3-hydroxyl group, 189
2′-Hydroxyflavones, 168, 176
3-Hydroxyflavone, 167, 170, **173**, 177, 178
3′-Hydroxyflavone, 168
4′-Hydroxyflavones, 168, 169, 204
5-Hydroxyflavones, 175, 177
5-Hydroxyflavone, 165, 166, **168**, 178
6-Hydroxyflavone, 168
7-Hydroxyflavone, 73, 165, 166, 168, 178, 256
8-Hydroxyflavones, 177
8-Hydroxyflavone, 162, 168
Hydroxyflavonols, 189
7-Hydroxyflavylio-4-carboxylate, 73
2′-Hydroxyflavylium chloride, 78
7-Hydroxyflavylium chloride, 69, 80
4′-Hydroxyflavylium perchlorates, 77
3-Hydroxyflavylium salts, 69, 74
3-Hydroxyfluoran, 313
2-Hydroxy-β-fluorenylidene-β-phenylpropiophenone, 166
6-Hydroxyfluorone, 312
3′-Hydroxyformononetin, 214
4-Hydroxy-2*H*-furo[2,3-*h*]chromen-2-one, 134
Hydroxyfurochromones, 158
6-Hydroxy-7-β-glucosidocoumarin, 123
1-Hydroxy-7-β-glucuronoxyxanthone, 325
6′-Hydroxyhesperetin, 279
5-Hydroxyhexamethoxyflavone, 189
1-Hydroxyhexan-5-one, 40
7-Hydroxy-8-(β-hydroxyisopropoxyethyl)coumarin, 119
6-Hydroxy-(3-hydroxy-4-methoxyphenyl)-7-methoxycoumarin, 124
4-Hydroxy-2-(hydroxymethylene)-3-methyl-2*H*-pyran, 3
5-Hydroxy-2-hydroxymethyl-4-pyrone, 33
β-(5-Hydroxy-2-hydroxymethyl-4-pyrone)propionic acid, 34
5-Hydroxy-7-(4-hydroxyphenyl)-8,8-dimethyl-6*H*-8*H*-benzodipyran-6-one, 220
4-Hydroxyiminoflavans, 269
3-Hydroxyiminoflavonones, 170
Hydroxyiminoindanones, 293
2-Hydroxyiminopyromeconic acid, 32
3-Hydroxyiminotrimethoxyflavanone, 190
Hydroxyindanones, 255
7-Hydroxyindanone, 255
3-Hydroxy-2-iodo-4-pyrone, 32
Hydroxyisobutyric acid, 265
4-Hydroxyisochroman-1-ones, 301
2-Hydroxyisoflavanones, 207, 208, 287
Hydroxyisoflavones, 210
2′-Hydroxyisoflavones, 264
5-Hydroxyisoflavone, 210, 211
7-Hydroxyisoflavones, 209, 210
2′-Hydroxyisoflavonoids, 264
7-Hydroxy-8-(2-isopentenyl)coumarin, 118
3-Hydroxyisovaleraldehyde dimethyl acetal, 57, 58
5-Hydroxy-6-isovaleryl-8,8-dimethyl-4-phenyl-2*H*,8*H*-pyrano[2,3-*h*]chromen-2-one, 136
6-Hydroxyluteolin, 186
Hydroxymercuricomenic anhydride, 27

2-Hydroxy-4-methoxyacetophenone, 183, 278
2-Hydroxy-4-methoxybenzaldehyde, 429
6-Hydroxy-4,7-methoxybenzofuran, 157
2-Hydroxy-4-methoxybenzoic acid, 146, 428
4-Hydroxy-3-methoxybenzoic acid, 194
Hydroxymethoxybenzophenones, 323
β-(6-Hydroxy-4-methoxybenzoyl) propionic acid, 430
5-Hydroxy-6-methoxycarbonylcoumarins, 111
2′-Hydroxy-α-methoxychalcones, 172
2-Hydroxy-3-methoxycinnamic acid, 98
3-Hydroxy-4-methoxycinnamic acid, 279
6-Hydroxy-7-methoxycoumarin, 122
7-Hydroxy-5-methoxycoumarin, 121
7-Hydroxy-6-methoxycoumarin, 123
7-Hydroxy-8-methoxycoumarin, 124, 128
5-Hydroxy-7-methoxy-2,6-dimethylchromone, 153
5-Hydroxy-7-methoxy-2,8-dimethylchromone, 153
4′-Hydroxy-7-methoxyflavan, 233
7-Hydroxy-5-methoxyflavan, 233
5-Hydroxymethoxyflavanones, 274, 275, 276
6-Hydroxy-5-methoxyflavanone, 269
3-Hydroxy-7-methoxyflavone, 178
3′-Hydroxy-7-methoxyflavone, 176
5-Hydroxy-4′-methoxyflavone, 175
5-Hydroxy-7-methoxyflavone, 170
7-Hydroxy-3-methoxyflavone, 178
7-Hydroxy-4″-methoxyflavone, 180
7-Hydroxy-5-methoxyflavone, 165
9-Hydroxy-5-methoxy-2*H*-furo[3,2-*g*]-chromen-2-one, 130
7-Hydroxy-4′-methoxyisoflavan, 211
4′-Hydroxy-7-methoxyisoflavone, 213, 214
5-Hydroxy-7-methoxyisoflavone, 213
7-Hydroxy-2′-methoxyisoflavone, 210
2-Hydroxy-4-methoxy-6-methylbenzoic acid, 16
5-Hydroxy-7-methoxy-2-methylchromone, 153
7-Hydroxy-5-methoxy-2-methylchromone, 153
7-Hydroxy-6-methoxy-3′,4′-methylenedioxyisoflavone, 218
4′-Hydroxy-7-methoxy-8-methylflavan, 233
7-Hydroxy-5-methoxy-6-methylflavan, 233
5-Hydroxy-7-methoxy-6-methylflavone, 175, 179
4′-Hydroxy-8-methoxy-3-methylflavylium chloride, 77
8-Hydroxy-6-methoxy-3-methylisocoumarin, 297, 303
5-Hydroxy-7-methoxy-6-methylisoflavone, 212
5-Hydroxy-7-methoxy-8-methylisoflavone, 212
7-Hydroxy-2′-methoxy-2-methylisoflavone-8-carbaldehyde, 210
4-Hydroxy-7-methoxy-2-methyl-4-vinylchroman, 144
1-Hydroxy-8-methoxy-3-methylxanthone, 325
5-Hydroxy-7-methoxy-4-phenylcoumarin, 137
6-Hydroxy-7-methoxy-4-phenylcoumarin, 123
9-Hydroxy-5-methoxypsoralene, 130
3-Hydroxy-2-methoxythiochromone, 372
4-Hydroxy-3-methoxythiocoumarin, 372
5-Hydroxy-7-methoxy-2,6,8-trimethylchromone, 154
1-Hydroxy-7-methoxyxanthone, 324
3-Hydroxy-1-methoxyxanthone, 322
7-Hydroxy-1-methoxyxanthone, 325
8-Hydroxy-2-methyl-6,7-benzochromone, 341
2-Hydroxymethylbenzo[*b*]thiophene, 379
1-Hydroxy-3-methyl-12*H*-benzo[*b*]xanthen-12-one, 346
1-Hydroxy-3-methyl-6,7-benzoxanthone, 346
2-Hydroxy-5-methylbiphenyl, 334
2-Hydroxy-3-(3-methyl-2-butenyl)-1,4-naphthoquinone, 342
8-Hydroxy-2-methylchroman-4-one-6-carboxylic acid, 256
Hydroxy-2-methylchromones, 148
6-Hydroxy-2-methylchromone, **152**
7-Hydroxy-2-methylchromone, 140, 256
2-Hydroxymethylcoumaran, 52
4-Hydroxy-3-methylcoumarin, 115
5-Hydroxy-4-methylcoumarin, 134
6-Hydroxy-4-methylcoumarin, 117, 134

7-Hydroxy-4-methylcoumarin, 118
6-Hydroxy-2-methyl-5,7-dinitrochromone, 152
6-Hydroxy-4-methyl-5,7-dinitrocoumarin, 117
7-Hydroxy-3′,4′-methylenedioxyisoflavone, 214
5-Hydroxy-3′,4′-methylenedioxy-3,6,7-trimethoxyflavone, 200
Hydroxymethylene ketones, 69, 209
Hydroxymethylflavanones, 270
7-Hydroxy-3-methylflavanone, 269
6-Hydroxy-9-methylfluorone, 312
8-Hydroxy-3-methylisochroman-1-one, 303
7-Hydroxy-2-methylisoflavone, 213, 256
7-Hydroxy-4′-methylisoflavone, 214
7-Hydroxy-6-methylisoflavone, 210
8-Hydroxymethyl-7-methoxyisoflavone, 213
2-Hydroxymethyl-5-methoxy-4-pyrone, 31
6-Hydroxymethyl-4-methoxy-2-pyrone, 20
10-Hydroxy-2-methyl-4*H*-naphtho[2,3-*b*]-pyran-4-one, 341
9-Hydroxy-3-methylnaphtho[2,3-*b*]pyran-2,5,10-trione, 342
2-Hydroxymethylnaphtho[1,2-*b*] thiophene, 379
2-Hydroxymethylnaphtho[2.1-*b*]thiophene, 379
5-Hydroxy-2-methyl-6-nitrochromone, 152
6-Hydroxy-2-methyl-5-nitrochromone, 152
7-Hydroxy-4-methyl-3-phenylcoumarin, 103
5-Hydroxy-2-methyl-4-pyridone, 37
3-Hydroxy-2-methyl-4-pyrone, 32
4-Hydroxy-6-methyl-2-pyrone, 17
5-Hydroxy-2-methyl-4-pyrone, 33
6-Hydroxy-3-methyl-2-pyrone, 12
2-Hydroxy-6-methyltetrahydropyran, 49
3-Hydroxymethylthiochroman-4-ones, 380
2′-Hydroxymethyl-7,4′,5′-trimethoxyflavonol, 200
Hydroxymethylxanthones, 394
2-Hydroxy-1-naphthaldehyde, 309, 339
Hydroxynaphthoic acids, 345, 346
2-Hydroxy-1-naphthoic acid, 345, 346
2-Hydroxy-3-naphthoic acid, 345, 346
β-(1-Hydroxy-2-naphthyl)acrylic acid, 338
β-(2-Hydroxy-1-naphthyl)acrylic acid, 340
3-Hydroxynaringenin, 280, 284
Hydroxynetoric acid, **266**
Hydroxynitroflavones, 166
7-Hydroxy-8-nitroflavone, 166
4-Hydroxy-2-oxodihydrofuro[3.4-*c*]pyran, 45
5-Hydroxy-4-oxo-4*H*-pyran-2-carboxylic acid, 36
3-Hydroxy-4-oxo-4*H*-pyran-2,6-dicarboxylic acid, 38
3-Hydroxy-6-oxo-7,8,9,10-tetrahydro-6*H*-dibenzo[*b,d*]pyran, 111
2′-Hydroxypentamethoxychalcone, 279
5-Hydroxypentamethoxyflavanone, 285
5-Hydroxypentamethoxyflavones, 188, 191, 199
7-Hydroxypentamethoxyflavone, 198
5-Hydroxypentamethoxyisoflavone, 217
Hydroxypentamethylchroman, 229
2′-Hydroxypentamethylflavan, 233
5-Hydroxypentanal, 49
4-Hydroxyperinaphthen-1-one, 341
9-Hydroxyperinaphthen-1-one, 340
4-Hydroxy-6-phenacyl-2-pyrone, 22
2-(2-Hydroxyphenoxy)benzophenone, 311
4-Hydroxy-3-phenoxycoumarin, 115
3-Hydroxy-2-phenoxytetrahydropyrans, 50
Hydroxyphenylacetic acids, 207
2-Hydroxyphenylacetic acid, benzoyl derivatives, 74
4-Hydroxyphenylacetic acid, 215
4-Hydroxy-6-phenylacyl-2-pyrone, 14
2-(2-Hydroxyphenyl)benzimidazole, 108
3-Hydroxy-2-phenyl-4*H*-benzo[*b*]pyran-4-one, 139, 170, 173
1-(2-Hydroxyphenyl)-2-benzoyloxyprop-1-enes, 74
3-Hydroxy-2-phenylchromans, 237
4-Hydroxy-2-phenylchromans, 247
2-Hydroxy-α-phenylcinnamic acid, 103
β-(4-Hydroxyphenyl)cinnamonitriles, 100
4-Hydroxy-3-phenylcoumarins, 115
3-(2-Hydroxyphenyl)-1,3-diarylpropan-1-ones, 104
3-(2-Hydroxyphenyl)-1,1-diarylprop-2-en-1-ols, 104
2-Hydroxyphenyl-β-diketones, 145
3-(2-Hydroxyphenyl)-1,3-diphenylpropan-1-one, 54, 105

3-(2-Hydroxyphenyl)-1,1-diphenylprop-2-en-1-ol, 54, 105
3-(2-Hydroxyphenyl)-1,5-diphenylpyrazoline, 272
6-Hydroxy-9-phenylfluorone, 311, 312
3-Hydroxy-3-phenylindan-1,2-dione, 173
7-Hydroxy-2-phenylisoflavone, 213
2-Hydroxyphenylisopropenyl compounds, 232
2-Hydroxyphenyl ketones, 145
1-(2-Hydroxyphenyl)-3-(4-methoxyphenyl)-propane, 232
3-(2-Hydroxyphenyl)-2-methoxy-1-phenyl-propan-1-ol, 244
4-(Hydroxyphenyl)-2-methylbut-3-en-2-ol, 57
5-(2-Hydroxyphenyl)-3-methyl-1-phenyl-pyrazole, 149
1-(2-Hydroxyphenyl)pentane-1,3-dione, 139
2-(Hydroxyphenyl)-1-phenylethane-1-carboxylic acid, 110
3-(2-Hydroxyphenyl)-5-phenylisooxazole, 168
1-(2-Hydroxyphenyl)-3-phenylpropane-1,3-dione, 140
1-(1-Hydroxyphenyl)-3-phenylpropan-1-ol, 61
1-(1-Hydroxyphenyl)-3-phenylpropenes, 61
3-(2-Hydroxyphenyl)propanol, 104
(2-Hydroxyphenyl)propiolamide, 109
β-(2-Hydroxyphenyl)propionic acids, 251
γ-(2-Hydroxyphenyl)propyl alcohols, 254
Hydroxyphenylpyrazoles, 149
(2-Hydroxyphenyl)pyrazole, 146
(2-Hydroxyphenyl)pyruvic acid, 111
4-(4′-Hydroxyphenyl)pyrylium salts, 4
(2-Hydroxyphenyl)succinic acid, 107
4-(4-Hydroxyphenyl)-2,2,4-trimethyl-chroman, 224
3-Hydroxy-9-phenylxanthene, 311
5-Hydroxypicolinic acid, 33
1-Hydroxy-3-prenyloxyxanthones, 333
2-Hydroxypropiophenones, 143, 145, 253
2-(3-Hydroxypropyl)-1-naphthol, 339
8-(2-Hydroxypropyl)-10-oxo-5,9,3′,4′-tetra-hydroxy-10*H*-pyrano[2,3-*h*]flavone, 206
2-(γ-Hydroxypropyl)phenol, 222
γ-Hydroxypropyl phenyl ether, 222
1-Hydroxy-2-pyridones, 15
1-Hydroxy-4-pyridones, 23, 24
Hydroxy-4-pyrones, 27, 30
–, reactions, 32
–, structures, 31
2-Hydroxy-4-pyrone, 11
3-Hydroxy-2-pyrones, 14
3-Hydroxy-4-pyrones, 30, 31, 32, 36
4-Hydroxy-2-pyrones, 13, 31
5-Hydroxy-4-pyrone-2-carboxylic acid, 36
3-Hydroxy-4-pyrone-2,6-dicarboxylic acid, 38
4-Hydroxyrottlerin, 66
Hydroxysalicylic acids, 321
4-Hydroxyselenocoumarins, 407
4-Hydroxy-6-styryl-2-pyrone, 19
2-Hydroxytetrahydropyrans, 48
2-Hydroxytetrahydropyran, 40, 42, 49
5-Hydroxytetrahydropyran-4-one-2-carboxylic acid, 26
2-Hydroxy-3-(2-tetrahydropyranyl)tetra-hydropyran, 49
4-Hydroxytetrahydrothiopyran 1,1-dioxide, 364
3-Hydroxytetramethoxyflavanone, 250
3′-Hydroxytetramethoxyflavone, 187
5-Hydroxytetramethoxyflavone, 177, 187, 188, 189, 192
3′-Hydroxy-5,6,7,4′-tetramethoxyflavonol, 197
5-Hydroxytetramethoxyisoflavones, 212, 217
1-Hydroxytetramethoxyxanthones, 327
2-Hydroxytetramethoxyxanthone, 328
6-Hydroxytetramethyl-2-(4,8,12-trimethyl-trideca-3,7,11-trienyl)chroman, 230
6-Hydroxy-9-thiabicyclo[3.3.1]nonane-2-one, 398
3-Hydroxythiochroman, 381
4-Hydroxythiochroman 1,1-dioxide, 367
4-Hydroxythiocoumarins, 371, 372
4-Hydroxythiocoumarin-3-carbaldehyde, 372
β-4-Hydroxythioflavan, 383
8-Hydroxythioflavone, 375
Hydroxythiopyran-4-ones, 356
2-Hydroxy-2′,3′,5′-trimethoxybenzo-phenone, 322

5-Hydroxy-2′,4′,5′-trimethoxy-8,8-dimethyl-8*H*-pyrano[2,3-*h*]isoflavone, 221
5-Hydroxy-7,3′,4′-trimethoxyflavanone, 279
4′-Hydroxy-3,5,7-trimethoxyflavone, 190
5-Hydroxy-3,7,8-trimethoxyflavone, 190
5-Hydroxy-6,2′,6′-trimethoxyflavone, 185
5-Hydroxy-6,7,4′-trimethoxyflavone, 182, 185, 216
5-Hydroxy-7,3′,4′-trimethoxyflavone, 176
7-Hydroxy-5,8,4′-trimethoxyflavone, 176, 185
5-Hydroxy-6,7,4′-trimethoxyflavonol, 191
5-Hydroxy-7,8,4′-trimethoxyisoflavone, 216
7-Hydroxy-6,3′,4′-trimethoxyisoflavone, 218
6-Hydroxy-5-(2,3,4-trimethoxyphenylacetyl)-2,3-dihydrobenzofuran, 219
1-Hydroxytrimethoxyxanthones, 326, 327
7-Hydroxy-2,2,5-trimethylchroman, 224
7-Hydroxy-2,2,5-trimethylchroman-4-one, 253
7-Hydroxy-2,2,4-trimethyl-2*H*-chromene, 135
6-Hydroxy-5,7,8-trimethylcoumarin-3-carboxylic ester, 100
6-Hydroxy-2,5,8-trimethyl-2-(4,8,12-trimethyltrideca-3,7,11-trienyl) chroman, 230
2-Hydroxy-2,4,6-triphenyl-2*H*-pyran, 4
δ-Hydroxyvaleraldehyde, 49
δ-Hydroxyvaleric acids, 49
Hydroxyvisnagin glucoside, 158
9-Hydroxyxanthene, 310
Hydroxyxanthones, 321, 324
–, naturally occurring, 324
1-Hydroxyxanthone, 323
2-Hydroxyxanthone, 323, 324
4-Hydroxyxanthone, 323, 324
8-Hydroxyxanthone-1,3-dicarboxylic acid, 331
Hydroxyxanthorrhone, 234
Hydroxyxanthydrols, 311, 312
Hymenoxin, **188**
Hymenoxys scaposa, 188, 189
Hypericum performatum, 193
Hyperin, **193**
Hypolaena fastigiata, 186
Hypolaetin, **186**
Hyssopus officinalis, 184

Ichthynone, **221**
Idaein chloride, **91**
2-Iminochrom-3-ene, 108
Iminocoumarins, 100
Iminodiethanol, 117
Imperatoria ostruthium, 118, 128, 129
Imperatorin, 126, **128**, 129, 131
Indanones, 293
Indan-1-one, 384, 439
Indan-2-one, 289, 301
Indene oxide, 289
Indenobenzopyrylium salts, 439
Indian Yellow, 325
Indigofera arrecta, 190
Iodocoumarins, 109
8-Iodoflavanones, 275, 276
Iodoform, 29
Iodoform reaction, 33
1-Iodogermacyclohexane, 420
3-(Iodomethyldimethylsilyl)propylmalonic ester, 416
8-Iodomethyl-7-methoxyisoflavone, 213
10-Iodomethylthioxanthylium perchlorate, 389
9-Iodonia-10*H*-anthracene chloride, 425
Iodonium ions, 426
1-Iodo-1-phenylstannacyclohexane, 422
8-Iodo-5,7,4′-trimethoxyflavone, 203
Iodoxanthones, 320
Ipoh, 257
Iresin celosioides, 215
Iretol, 274
Iridin, 217
Irigenin, **217**
Irigenin ethers, 217
Irigenol, **217**
Iris florentian, 217
Iris germanica, 217
Irisolidone, **217**
Irisolone, 217
Iris repalensis, 217
Iris tectorum, 216
2-Isoamylresorcinol, 259
Isoanhydroicaritin, **190**
Isoanthocyanidins, 234
Isoastilbin, 285
Isoauriculatin, **221**

Isobellidifolin, **326**
Isobenzopyran, 288
Isobenzopyrylium salts, 294
Isobergapten, **130**
Isobrazilein, 438
9-Isobutylthioxanthenes, 389
Isobutyraldehyde, 36
Isocarbostyrils, 294
Isocarthamidin, **282**
Isocarthamidin 5,7-dimethyl ether, 282
Isocarthamin, **282**
Isocaviunin 1-methyl ether, 217
Isocaviunin 7-methyl ether, 217
Isochroman, 289, **297**, 386
Isochroman-1,4-dione, 302
Isochroman-4-ol, 301
Isochromanone, 298
Isochroman-1-one, 299, **300**
Isochroman-3-one, **301**
Isochroman-4-ones, 301
Isochromenes, 168, 288, 294, 299
Isochromene, **289**, 301
Isochromenochromones, 299
Isoclavacin, **46**
Isocoumarins, 290, 301
–, reactions, 293
–, reaction with Grignard reagents, 294
–, 3-substituted, 292
Isocoumarin, 288, **295**
Isocoumarincarboxylic acids, **295**
Isocryptomerin, 201
Isocytisoside, 186
Isodehydracetic acid, **18**
Isoderritolisoflavone, 220
Isoelliptol, **220**
Isoergochrysin A., **333**
Isoeugenitin, **153**
Isoeugenitol, **153**, 154
Isoferulic acid, 279
Isoflavans, 231
Isoflavan-4-ols, 211, 250, 286
Isoflavanones, 210, 211, 286
Isoflavanone oximes, 250
Isoflavenes, 62, 241
Isoflavene, 250
Isoflavones, 66, 207, 270, 286, 430
–, biosynthesis, 209
–, colour tests, 213
Isoflavones, (*continued*)
–, from chalcones, 209
–, from isoflavanones, 210
–, hydrogenation, 211
–, isomerisation, 212
–, naturally occurring, 214
–, properties, 211
Isoflavone, 139, **210**, 272
Isoflavone-2-carboxylic acids, 208
Isoflavone glucosides, 210
Isoformononetin, 213, **214**
Isofraxidin, **126**
Isogalloflavin, **296**
Isogentisin, **325**
Isoginkgetin, **201**
Isognaphaliin, **191**
Isoguanandin, **328**
Isohyperoside, **193**
Isoimperatorin, **129**, 130
Isokojic acid, **34**
Isolapachenole, 65
Isolimocitril, **197**
Isomammein, **121**
Isomelacacidin, **95**
Iso-octyl phosphite, 397
Isooxypeucedanin, 129, **130**
Isopapuanic acid, **257**
8-Isopentenyl-5-isopentenyloxy-7-methoxycoumarin, 121
8-Isopentenylkaempferol, 190
8-Isopentenylrhamnocitrin, 190
6-Isopentenyltetrahydroxyflavone, 185
Isopimpinellin, 129, **130**
Isopolygalaxanthone, **327**
Isoprenylphenols, 64
8-Isopropenylangelicin, 132
8-Isopropenyl-2*H*-furo[2,3-*h*]chromen-2-one, 132
2-Isopropylbenzo[*b*]thiophene, 369
4-Isopropylidenetetrahydrothiophen-3-one, 361
3-Isopropylisochromene, 289
7-Isopropyl-6-methoxy-2*H*-furo[3,2-*g*]-chromen-2-one, 131
7-Isopropyl-6-methoxypsoralene, 131
Isopsoralene, 126, 127
Isopyrazolines, 348
Isoquercitrin, **193**, 194
Isoquinolones, 294

Isorhamnetin, **194**
Isorhamnetin 3,4′-diglucoside, 194
Isorhamnetin 3-*O*-β-glucofuranosyl-(1→4)-α-rhamnopyranoside, 194
Isorhamnetin 3-rutinoside, 194
Isorotenone, 258, 259, 261, 264
Isorottlerin, **66**
Isosakuranetin, **280**
Isosakuranetin 7-rhamnoglucoside, 280
Isosakuranetin 7-rutinoside, 280
Isosakuranin, **280**
Isoselenochroman-4-ol, 409
Isoselenochroman-4-one, 409
Isoselenochrom-3-ene, 409
Isoswertisin, **182**
Isotephrosin, **265**
Isotetracacidin, **95**
Isothiochroman, 385
Isothiochroman-4-ol, 384, 387
Isothiochroman-4-one, 386, 387
Isothiochroman-1-thione, 387
Isothiochromene, 384
Isothiocoumarins, 384
Isothiocoumarin-3-carboxylic acid, 384
Isotubaic acid, 259, 262
Isovaleric acid, 132, 231
Isovanillin, 123
Isovitexin, **182**
Isoxanthones, 311
Isoxazoles, 147
Izalpinin, **189**

Jacareubin, 58, **328**
Jack-wood, 195
Jamaicin, **221**
Jaranol, **190**
Jatamansin, **137**
Javanicin, 300
Juniperus macropoda, 186

Kaempferide, **190**
Kaempferitrin, **190**
Kaempferol, 88, **190**, 195, 196
Kaempferol 3-(*p*-coumaroylglucoside), 190
Kaempferol 3,7-dimethyl ether, 190
Kaempferol 5,4′-dimethyl ether, 190
Kaempferol 3,7-dirhamnoside, 190
Kaempferol 3-[2-(*O*-feruloyl)-*O*-β-D-glucopyranosyl-(1→2)-β-D-glucopyranoside], 190
Kaempferol 3-*O*-β-D-glucopyranosyl-(1→4)-α-arabofuranosyl-(1→2)-α-L-arabinopyranoside, 191
Kaempferol 7-glucoside, 190
Kaempferol 4′-methyl ether, 190
Kaempferol 7-methyl ether, 190
Kaempferol L-rhamnoside, 191
Kamala, 65, 66
Kampherol, 190
Kanjone, **205**
Kanugin, **196**
Karanjic acid, **205**
Karanjin, **205**
Karanjin ketone, 205
Karenin, **161**
Karsuranin, **284**
Kawaic acid, 19
Kawain, **19**, 47
Kawa resin, 19
Kayaflavone, **201**
Kellin, 156
Kellinin, 156, 158
Kellinol, 156
Kellinone, 204
Kellol glycoside, 156, 158
Keracyanin chloride, **91**
Keto-alcohols, 40
Ketones, α,β-ethynyl, 73
—, α-methylene active, 5
—, reaction with aldehydes, 4, 5
—, — with β-diketones, 5
—, — with ketones, 69
—, — with malonic esters, 12
—, — with malononitrile, 101
—, — with salicylaldehyde, 439
—, self-condensation, 4
—, α,β-unsaturated, 5, 73
—, —, reaction with phenols, 86
Ketonic esters, 70
β-Ketonic esters, 12, 137
—, reaction with phenols, 98, 100
Khellactone, 137
Khellin, **156**, 157, 158
Khellinin, 156, 158
Khellinol, 156, **158**
Khellinone, 156
Khellol, **158**

Kielmeyera, sp., 328, 329
Knoevenagel reaction, 354
Kojic acid, 20, 30, 31, **33**
–, reactions, 34
–, reaction with diazomethane, 34
–, – with vinyl cyanide, 34
Kojic acid methyl ethers, 33, 34
Kojic acid 7-triphenylmethyl ether, 34
Kolbe–Schmidt reaction, 112
Kostanecki–Robinson synthesis, 140
Kuhnia eupatorioides, 199
Kvanin, **132**

Laburnum alpinum, 220
Laburnum anagyroides, 215
Lactones, 31
δ-Lactones, 16
Lambertella sp., 342
Lambertellin, **342**
Lapachenole, 57, **64**
Lapachol, 338, 342
α-Lapachone, **342**
β-Lapachone, **338**
Laricin, **32**
Laricinic acid, **32**
Larix decidua, 284
Lasiocephalin, **124**
Lasiosiphon eriocephalus, 124
Lead compounds, 420
Lecanora rupicola, 154
Lecidea carpathica, 331
Lecontin, **282**
Leguminosae sp., 97, 190
Leonuris quinquelobatus, 183
Lepidophyllum quadrangulare, 199
Leptorumobra miqueliana, 283
Leptorumol, **155**
Lespedeza cyrtobotyra, 190
Lespedin, **190**
Lettadurone, **217**
Leucoanthocyans, 96
Leucoanthocyanidins, 88, 93, 94, 96, 246
Leucoanthocyanins, 93, 94, 250
Leucoanthodelphinidins, 96
Leucocyanidins, **96**
Leucocyanidin acetate, 90
Leucodelphinidins, **96**
Leucofisetinidin, **95**
Leucopelargonidin, **95**
Lichexanthone, **327**, 335
Lillium auratum, 194
Limettin, **120**
Limocitril, **197**
Limocitrin, **197**
Linaria vulgaris, 182, 185
Linarin, 181, **182**
Lindera lucida, 186
Lippia nodiflora, 186
Liquiritigenin, **278**
Liquiritin, **278**
Lisetin, **159**, 221
Lithium tetrahydridoaluminate, 50, 61, 75, 88, 94, 157, 165, 168
Litsea glauca, 285
Logwood, 443
Logwood extract, 427
Lomatin, **136**
Lomatium columbianum, 133
Lomatium nuttallii, 136
Lonchocarpic acid, 221
Lonchocarpin, 58
Lonchocarpus, 257
Lonchocarpus nicou, 264, 265
Lotoflavin, **195**
Lotus arabicus, 195
Lucanthone B.P, 392
Lucenin, **184**
Lucidin, **186**
Lumimethylquercetin, 192
Luteolin, 165, 174, **183**, 189, 214, 240
Luteolin 7-(6-*O*-*p*-cumaroyl)-D-glucoside, 184
Luteolin 7,3′-dimethyl ether, 184
Luteolin 7,3′-dimethyl ether 4′-D-gluco-apioside, 184
Luteolin *C*-glucosides, 184
Luteolin 5-glucoside, 183
Luteolin 7-glucoside, 184
Luteolinidin, **89**
Luteolin 4′-methyl ether, 184
Luteolin 7,3′,4′-trimethyl ether, 279
4-Lutidone, 30
Lutonaretin, 184
Luvanga scandens, 135
Luvangetin, 57, **135**
Luxtexin, 184

Machaerium sp., 254
Maclura pomifera, 220, 328, 329, 330
Macluraxanthone, **329**, 334
Majorana hortensis, 188
Majoranin, **188**
Majurin, **132**
Malaccol, 264, **267**
Maleic anhydride, adducts with 2-pyrones, 13
Malic acid, 10, 18, 34, 117, 124, 127, 137
Mallotus japonica, 295
Mallotus phillipinensis, 65
Malonic acids, reaction with arenethiols, 371, 372
Malonic acid, 84, 90, 98, 101
Malonic esters, reaction with ketones, 12
Malonic ester, 112, 117, 356
Malononitrile, 24, 107, 356, 357
–, reaction with aldehydes, 101
Malonyl chloride, 11, 31, 407
Maltol, 20, **32**, 33
Malva sylvestris, 92
Malvidin chloride, 82, 83, 84, **92**, 93
Malvidin chloride 3,5-di-β-glucoside, 92
Malvidin 3-β-glucoside, 93
Malvin, 82, 83
Malvin chloride, **92**
Malvone, 83
Mammea africana, 122
Mammea americana, 121, 133, 136, 325
Mammeigin, **136**
Mammein, **121**
Mammeisin, **121**, 136
Mandarin, **188**
Mangifera indica, 326
Mangiferin, **326**
Mangostin, **330**
β-Mangostin, **330**
Mannich bases, 117, 179
Mannich reaction, 33, 152, 166, 365, 366
Mannitol, 3
Marcoumar, **116**
Margicassidin, **96**
Marihuana, 336
Marmin, **117**
Marus lactea, 285
Matricaria chamomilla, 197
Matteucia orientalis, 283
Matteucinol, 283
Mecocyanin chloride, **90**
Meconic acid, 20, 23, 26, 32, 36, **38**, 39
–, acylation, 39
–, esters, 38
Medicosma cunninghamii, 64
Meerwein arylation reaction, 111, 113
Melacacidin, **94**, 95, 250
Melacacidin tetramethyl ether, 94
Melannein, **124**
Melibentin, **200**
Melicope broadbentiana, 200
Melicope sarcococca, 184
Melicope simplex, 200
Melicope ternata, **200**
Melilitus officinalis, 251
Melilotic acid, 103, 104, 251
Melisimplexin, **200**
Melisimplin, **200**
Meliternatin, **200**
Meliternin, 197, **200**
Mellein, 288, **303**
Mellein methyl ether, 303
Mercaptoacetic acid, 250
Mercaptoacetone, 360
2-Mercaptoacetophenone, 373
β-Mercaptoacroleines, 348
2-Mercaptobenzaldehyde, 370
2-Mercaptobenzoic acid, 373, 391, 394
2-Mercaptobenzoic ester, 383
4-Mercapto-1,3-butadiene-1-carbaldehyde, 353
2-Mercaptocinnamic acid, 370
2-Mercapto-5-methylacetophenone, 379
2-Mercapto-5-methylbenzoic acid, 379
2-Mercapto-ω-(methylsulphinyl)acetophenone, 375
Mesityl oxide, 114
–, reaction with phenol, 224
Mesoxalic ester, 45
Mesua ferrea, 122
Metahemipinic acid, 429
Metahemipinic anhydride, 431, 444
Methanesulphonic acid, 375
Methoxyacetone, 32, 33
2-Methoxyacetophenone, 253
4-Methoxyacetophenone, 76
7-Methoxy-8-(2-acetoxy-3,3-dimethylbutyl)-coumarin, 119
5-Methoxyangelicin, 130

6-Methoxyangelicin, 131
10-Methoxy-5*H*,7*H*-benzo[*b*]pyrano[4,3-*b*]-benzopyran-5,7-dione, 295
3-Methoxybenzo[*b*]thiophene, 374
3-(3-Methoxybenzoyloxy)butan-2-one, 292
3-(4-Methoxybenzyl)-4-hydroxycoumarin, 113
2-(4-Methoxybenzylidene)khellin, 157
2-(2-Methoxybenzylidene)-1-tetralone, 346
4-Methoxybenzyl phloroglucinyl ketone, 207
3-(4-Methoxybenzyl)-3,5,7-trihydroxy-chroman-4-one, 218
2′-Methoxybiphenyl-2-carboxylic acid, 334
Methoxybrazilin, 445
2-Methoxybutadiene, 225
3-(2-Methoxycarbonylbenzyl)coumarin, 294
5-*o*-Methoxycarbonylbenzylidenerhodanine, 384
3-Methoxycarbonyl-4-(2-methoxyphenyl)-pyrazoline, 108
4-Methoxycarbonyl-6,7-methylenedioxy-3-phenylisochromans, 299
3-Methoxycarbonyl-3-methylthiochroman-4-one, 383
Methoxychalcones, 275
Methoxychromans, 224
7-Methoxychromanone, 428
7-Methoxychromanone oximes, 255
2-Methoxychromone, 112
3-Methoxychromones, 143
2-Methoxycinnamic acid, 103
2-Methoxycinnamic ester, 99
Methoxycoumarins, 97
4-Methoxycoumarin, 112
5-Methoxycoumarin, 116
8-Methoxycoumarin, 98, 119
7-Methoxycoumarin-3-carbonyl chloride, 117
5-Methoxy-4,6-diaryl-2-pyrones, 15
2-Methoxydibenzoylmethanes, 139
5-Methoxy-7-*O*-(3,3-dimethylallyl)coumarin, 121
6-Methoxy-2,2-dimethyl-7,8-benzo-2*H*-chromene, 64
8-Methoxy-2,2-dimethyl-chroman-3-ol, 236
6-Methoxy-2,2-dimethylchroman-4-one, 253
8-Methoxy-2,2-dimethylchroman-4-one, 253
7-Methoxy-2,3-dimethylchromone, 148
7-Methoxy-4,8-dimethylcoumarin, 118
5-Methoxydimethylenedioxyprenyliso-flavones, 218
7-Methoxy-3,4-dimethylisocoumarin, 292
2′-Methoxy-8,8-dimethyl-4′,5′-methylene-dioxy-8*H*-pyrano[2,3-*h*]isoflavone, 221
6-Methoxy-8,8-dimethyl-4′,5′-methylene-dioxy-8*H*-pyrano[2,3-*h*]isoflavone, 222
6-Methoxy-2,2-dimethyl-2*H*-naphth[1,2-b]-pyran, 64
7-Methoxy-2,3-dimethyl-4-phenyl-chromylium chloride, 148
6-Methoxydiosmetin, 184
2′-Methoxydiphenyl-2-carboxylic acid, 334
7-Methoxy-2,4-diphenylchromen-4-ol, 71
3-Methoxyflavan, 244
4′-Methoxyflavan, 232
2-Methoxyflavan-3,4-diols, 250
2-Methoxy-3,4-flavandiones, 276
4′-Methoxyflavan-4β-ol, 248
Methoxyflavanones, 275
3-Methoxyflavanones, 271
7-Methoxyflavanone, 280
Methoxyflavones, demethylation, 176
3-Methoxyflavone, 167, 171, 176
7-Methoxyflavone, 165, 192
3-Methoxyflavylium chloride, 77
3-Methoxyflavylium salts, 74
5-Methoxy-2*H*-furo[3,2-*g*]chromen-2-one, 128
8-Methoxy-2*H*-furo[2,3-*f*]chromen-2-one, 130
9-Methoxy-2*H*-furo[3,2-*g*]chromen-2-one, 128
3-Methoxyfuro[2,3-*h*]flavone, 205
5-Methoxyfuro[3,2-*g*]flavone, 206
6-Methoxyfuro[2,3-*h*]flavone, 205
6-Methoxygenkwanin, 182
8-Methoxygravelliferone, 125
O-Methoxyhaematoxylin, 445
4-Methoxyhomophthalic acid, 302
4-Methoxyhomophthalic anhydride, 291
ω-Methoxy-2-hydroxyacetophenones, 171
3-Methoxy-5-hydroxyflavones, 167
3-Methoxy-7-hydroxyflavone, 178
7-Methoxy-8-(2-hydroxy-3-methyl-3-butenyl)coumarin, 119

Methoxyisochromanones, 300
Methoxyisocoumarins, 295
7-Methoxyisoflavan-4α-ol, 250
7-Methoxyisoflavanone, 210
Methoxyisoflavones, 211
7-Methoxyisoflavone, 209, 210, 213
8-Methoxyisoflavone, 210
7-Methoxy-8-isovalerylcoumarin, 118
6-Methoxykaempferol, 190
6-Methoxyluteolin, 184
2'-(Methoxymethoxy)chalcones, 250
2'-Methoxymethoxychalcone epoxides, 271
7-Methoxy-8-methoxymethylisoflavone, 213
2-Methoxymethylacetonitrile, 301
2-Methoxy-*N*-methylbenzamide, 303
2-Methoxy-2-methylchroman, 225, 235
7-Methoxy-2-methylchroman-4-one, 144, 254
Methoxy-2-methylchromones, 148
7-Methoxy-2-methylchromone, 144, 254
4-Methoxy-3-methylcoumarin, 115
7-Methoxy-4-methylcoumarin, 99
7-Methoxy-8-methylcoumarin, 118
5-Methoxy-7,8-methylenedioxycoumarin, 125
3-Methoxy-3',4'-methylenedioxyfuro[2,3-*h*]-flavone, 205
5-Methoxy-3',4'-methylenedioxyfuro[3,2-*g*]-flavone, 206
2'-Methoxy-4',5'-methylenedioxyfuro[3,2-*g*]-isoflavanone, 288
2'-Methoxy-5',4'-methylenedioxyfuro[3,2-*g*]-isoflavone, 220
4'-Methoxy-6-methylflavan, 232
4'-Methoxy-6-methylflavan-3,4-diol, 249, 250
4'-Methoxy-4-methylflavylium perchlorate, 80
4-Methoxy-7-methyl-5*H*-furo[3,2-*g*]benzo-pyran-5-one, 158
5-Methoxy-2-methylfuro[3,2-*g*]chromone, 158
5-Methoxy-7-methylisochromans, 300
7-Methoxy-3-methylisocoumarin, 295
7-Methoxy-3-methylisocoumarin-4-carboxylic acid, 295, 302
5-Methoxy-2-methyl-8-nitrochromone, 152
2-Methoxy-3-methyl-4-oxo-4*H*-chromene, 115
3-Methoxy-2-naphthoylacetone, 341
3-Methoxy-2-nitrobenzoic acid, 303
7-Methoxy-8-(1-oxo-3-methyl-3-butenyl)-coumarin, 119
4-Methoxyparacotoin, 19
7''-Methoxypentahydroxy-8,3'''-biflavonyl, 203
1-Methoxy-1-penten-3-yn-5-al diethyl acetal, 29
3-Methoxyphenoxyacetic acid, 429
3-Methoxyphenoxyacetic ester, 431
7-Methoxy-2-phenylchromone, 71
2-Methoxy-α-phenylcinnamic acid, 103
2-Methoxyphenyl 2,3-dimethoxybenzoate, 323
4-Methoxyphenyl 2,3-dimethoxybenzoate, 322
4-(4-Methoxyphenyl)-2,6-diphenyl-pyrylium tetrafluoroborate, 9
4-(4-Methoxyphenyl)flavan, 247
1-(2-Methoxyphenyl)indan-2-one, 252
3-(4-Methoxyphenyl)isocoumarin, 292
4-(2-Methoxyphenyl)-4-methylpentan-2-one, 235
1-(2-Methoxyphenyl)-1,3,3-tri(4-methoxy-phenyl)propane, 248
ω-Methoxyphloracetophenones, 171
ω-Methoxyphloroacetophenone tri-*p*-anisoate, 190
Methoxy-β-photomethylquercetin, 192
5-Methoxypsoralene, 128
9-Methoxypsoralene, 128
3-Methoxy-4-pyridone, 32
4-Methoxy-2-pyrone-6-acetic ester, 15
4-Methoxypyrylium salts, 28
p-Methoxysalicylaldehyde, 440
Methoxyseselin, 136
2-Methoxytetrahydropyran, 49
3-Methoxytetrahydrothiopyran 1,1-dioxide, 364
4-Methoxytetrahydrothiopyran 1,1-dioxide, 364
3'-Methoxytetrahydroxyflavone, 186
6-Methoxytetrahydroxyflavone, 186
7-Methoxytetrahydroxyflavone, 186
6-Methoxythiochroman-4-one, 382
7-Methoxythiochroman-4-one, 381
Methoxythiochromones, 373
Methoxythiocoumarins, 371

6-Methoxythioflavone, 375
4'-Methoxy-4-thionoflavone, 169
7-Methoxy-4-thionoflavone, 169
2-Methoxy-2*H*-thiopyran, 353
5-Methoxy-7,3',4'-trihydroxyflavonol, 194
6-Methoxy-5,7,4'-trihydroxyisoflavone, 216
6-Methoxy-1,3,5-trihydroxyxanthone, 321
3-Methoxyxanthone, 311
Methylacetoacetic ester, 70, 142
2-Methylacrylic ester, reaction with naphthols, 339, 340
β-Methylallyl phenyl sulphide, 377
Methylallylthioacetyl chlorides, 361
1-Methylamino-4-methylxanthone, 320
Methylangelicins, 127
6-Methylapigenin, 183
8-Methylapigenin, 183
4'-*O*-Methylapiin, 181
Methylbellidifolin, 326
2-Methyl-5,6-benzochroman, 340
2-Methyl-5,6-benzochrom-2-ene, 339
2-Methyl-6,7-benzochromone, 341
2-Methylbenzo[*b*]furan, 52
2-Methylbenzothiophene, 377
4-Methyl-2*H*-benzo[*b*]thiopyran, 367
6-Methyl-4*H*-benzothiopyran, 378
2-Methylbiphenyl, 334
2-Methylbutanoic acid, 331
2-Methylbut-2-ene, 5
3-Methylbut-2-en-1-ol, 128
3-Methyl-2-butenoyl chloride, 63
3-Methyl-2-butenylidenetriphenylphosphorane, 58
5-(3-Methyl-2-butenyloxy)-2*H*-furo[3,2-*g*]-chromen-2-one, 129
9-(3-Methyl-2-butenyloxy)-2*H*-furo[3,2-*g*]-chromen-2-one, 128
2-(3-Methylbut-2-enyloxy)naphthalene, 340
9-(3-Methyl-2-butenyloxy)psoralene, 128
1-Methyl-4-*tert*-butyl-1-silacyclohexanes, 415
α-Methylchalcones, 172, 269
Methylchromans, 223
2-Methylchroman, 225
6-Methylchroman, 222
2-Methylchroman-2-ol, 235
3-Methylchroman-4-ol, 108
Methylchroman-4-ones, 253
2-Methylchroman-4-one, 254
3-Methylchroman-4-one, 253, 255
2-Methylchrom-2-ene, 57, 235
Methylchromones, physical properties, 150
2-Methylchromones, 139, 140, 147, 148
2-Methylchromone, 146, 147
6-Methylchromone, dimerisation, 149
3-Methylchromone-2-carboxylic acid, 143
6-Methylchrysin, 179
8-Methylchrysin, 179
6-Methylcomenic acid, 33
6-Methylcoumalic acid, 16
6-Methylcoumalin-3,5-dicarboxylic ester, 11
3-Methylcoumarin, 108, **110**
4-Methylcoumarin, 100, 110
5-Methylcoumarin, 110
6-Methylcoumarin, 110
7-Methylcoumarin, 110
8-Methylcoumarin, 110
2-Methylcoumarone, 52
3-Methylcrotonaldehyde, 57
Methylcyclopropenedicarboxylic acid, 18
4-Methyldaphnetin, 124
2-Methyl-2,3-dihydrobenzothiophene, 377
3-Methyl-2,3-dihydro-1*H*-naphtho[2,1-*b*]-pyran, 340
3-Methyl-2,3-dihydro-6-phenyl-4-pyrone, 51
2-Methyl-5,6-dihydro-4-pyran, 40
Methyl-5,6-dihydro-2*H*-thiopyran-5-ones, 361
2-Methyl-5,7-dihydroxychromone, 147, 157
β-Methyl-2,4-dimethoxycinnamic ester, 99
6-Methyl-3,4'-dimethoxyflav-2-ene, 172
6-Methyl-3,4'-dimethoxyflavone, 172
6-Methyl-3',4'-dimethoxyflavonol, 172
Methyl-3,5-dimethoxyphenols, 322
Methoxy-2,2-dimethyl-2*H*-chromenes, 58
Methyl di-*O*-methylcitromycetin, 160
Methyl di-*O*-methylcitromycetinol, 160
Methyl-di-*O*-methylcitromycetinone, 160
3-Methyl-2,4-diphenyl-2*H*-chromene, 59
4-Methyl-2,2-diphenyl-2*H*-chromene, 59
4-Methyl-2,4-diphenyl-4*H*-chromene, 54
2-Methyl-4,6-diphenylpyrylium perchlorate, 5, 9
1-Methyl-3,5-diphenylthiabenzene, 350
1-Methyl-3,5-diphenylthiabenzene 1-oxide, 349

1-Methyl-3,5-diphenyl-2*H*-thiinium tetrafluoroborate, 350
4-Methyl-2,6-diphenylthiopyrylium perchlorate, 351
1-Methyl-3,5-diphenyl-2*H*-thiopyrylium tetrafluoroborate, 350
Methylene-2,2′-bicyclohexanones, 395
3,3′-Methylenebis-4-hydroxycoumarin, 10, **116**
3,3′-Methylenebis(4-hydroxythiocoumarin), 372
Methylenedicyclohexanones, 395
3,4-Methylenedioxybenzaldehyde, 301
6,7-Methylenedioxycoumarin, 123
7,8-Methylenedioxycoumarin, 125
3′,4′-Methylenedioxyfuro[2,3-*h*]flavone, 205
6,7-Methylenedioxyisochroman, 300
6,7-Methylenedioxyisochroman-1-one, 300
3′,4′-Methylenedioxykaranjin, 205
3-(3,4-Methylenedioxyphenyl)isochroman-1-one-4-carboxylic acid, 301
3′,4′-Methylenedioxy-7-prenylisoflavone, 218
3′,4′-Methylenedioxytetramethoxyflavone, 186, 200
6,7-Methylenedioxy-1,2,3-trimethoxyxanthone, 327
2-Methyl-1-ethylquinolinium salts, 148
Methyleupatoriochromene, 64
Methylevodionol, 64
3-Methylflavan-3-ols, 172
3-Methylflavanones, 269
C-Methylflavanones, 282
3-Methylflavone, **168**
5-Methylflavone, 168
6-Methylflavone, 99, 168
7-Methylflavone, 168
8-Methylflavone, 168
O-Methylflavonols, 178
4-Methylflavylium perchlorate, 70, 80
3-Methyl-2*H*-furo[3,2-*c*]chromen-2-one, 134
9-Methyl-2*H*-furo[3,2-*g*]chromen-2-ones, 134
2-Methylfuro[2,3-*h*]isoflavone, 220
2-Methylgenistein, 213
α-Methylglutaconic acid, 12
β-Methylglutaconic acids, 18
α-Methylglutaric acid, 251
4-Methyl-7-hydroxycoumarin, 118
4-Methylisochroman-4-ol, 301
3-Methylisochroman-1-one, 300
7-Methylisochroman-4-one, 301
3-Methylisochromene, 289
Methylisocoumarins, 295
3-Methylisocoumarin, 293
2-Methylisoflavan-4-ols, 251
2-Methylisoflavones, 208
6-Methylisoflavones, 208
3′-Methylisoirigenin, 188
7-Methylisotectorigenin, 216
1-Methylisothiochroman-4-ones, 386
4-Methylisothiocoumarin, 385
8-Methylkaranjin, 205
Methylketene dimer, 29
7-Methylmalvidin chloride, 93
Methyl matteucinol, 183
5-Methylmellein, 303
7-Methyl-5-methoxyisochroman-1-ones, 300
6-Methylmyricetin, 197
6-Methylmyricetin 3,3′,4′,5′-tetramethyl ether, 197
3-Methyl-1*H*-naphtho[2,1-*b*]pyran, 339
2-Methyl-4*H*-naphtho[1,2-*b*]pyran-4-one, 337
2-Methyl-4*H*-naphtho[2,3-*b*]pyran-4-one, 341
3-Methyl-1*H*-naphtho[2,1-*b*]pyran-1-one, 339
4-Methyl-2*H*-naphtho[1,2-*b*]pyran-2-one, 338
3-Methyl-1,4-naphthoquinon-2-ylmethyl anion, 305
6-*C*-Methylnaringenin, 281
Methylnitrocoumarins, 109
2-Methyl-6-nitrothioxanthone, 388
9-Methyloctahydrothioxanthene, 395
7-Methyloroxylin A, 180
8-Methylpentahydroflavanones, 284
5-Methylpentanolide-4-carboxylic ester, 16
7-Methylpeonidin, 92
9-Methylperhydrothioxanthene, 395
4-Methyl-2-phenylbenzoselenopyrylium perchlorate, 407
6-Methyl-4-phenyl-2*H*-benzo[*b*]thiopyran, 367
9-Methyl-10-phenyl-10-selenanthracene, 410

Methylphenylselenochrom-3-enes, 406
6-Methyl-4-phenylthiochroman-4-ol, 382
6-Methyl-3-phenyl-2*H*-thiopyran 1,1-dioxide, 350
6-Methyl-3-phenyl-4-(4-tolyl) coumarin, 71
C-Methylphloroglucinol, 65
2-Methylpropenylbenzene, 5
Methylpsoralenes, 127
2-Methyl-2-pyran, 3
4-Methyl-2*H*,5*H*-pyrano[3,2-*c*]chromene-2,5-dione, 137
2-Methylpyrones, 30
2-Methyl-4-pyrone, 22
3-Methyl-2-pyrone, 11
5-Methyl-2-pyrone, **17**
5-Methyl-2-pyrone-6-carboxylic acid, 11
6-Methyl-2-pyrone-3,5-dicarboxylic ester, 11
6-Methylquercetin, 195
8-Methylquercetin, 195
5-*O*-Methylquercetin tetra-acetate, 192
6-Methylquercetin 3,7,3′,4′-tetramethyl ether, 175
6-Methylquercetin 3,3′,4′-trimethyl ether, 195
Methylripariochromene A, 64
4-Methylsalicylic acid, 103
2-Methylselenane, 404
1-Methylselenanium iodide, 404
Methylselenochrom-3-enes, 406
2-(Methylseleno)cinnamic acids, 407
3-Methylselenocoumarin, 407
Methylselenoflavones, 408
4-Methylselenoflavylium perchlorate, 407
4-Methylselenopyran, 403
4-Methylselenopyrylium perchlorate, 403
9-Methylselenoxanthydrol, 410
9-Methylselenoxanthylium perchlorate, 410
β-Methylsorbic acid, 15
Methylsulphinyl ketones, 375
5-Methyltectorigenine, 211
7-Methyltectorigenin 4′-glucoside, 216
6-Methyltetrahydrobenzo[*e*]naphtho[1.8-*bc*]-thiopyran, 399
2-Methyltetrahydroselenopyran, 404
2-Methyltetrahydrothiopyran, 363
3-Methyltetrahydrothiopyran, 363
4-Methyltetrahydrothiopyran, 363
3-Methyltetrahydrothiopyran-4-one, 365
4-Methyltetrahydroxycoumarin, 126
2-Methyl-1-thiadecalin, 387
1-Methyl-2-thiaindane, 387
Methylthiochromans, 378
4-Methylthiochroman-4-ol, 382
Methylthiochroman-4-ones, 382
6-Methylthiochroman-4-one, 370, 379
Methylthiochromones, 373
6-Methylthiochromone, 370
2-Methylthiocinnamic acid, 371
Methylthiocoumarins, 371
6-Methylthioflavanone, 383
Methylthioflavones, 375
2-Methylthionochromone, **149**
3-Methyl-4-thionoflavone, 169
3-Methyl-4*H*-thiopyran, 251
9-Methylthioxanthene-10-oxide, 389
Methylthioxanthones, 394
9-Methylthioxanthydrol, 391
5-Methyltocol, 230, 231
7-Methyltocol, 230, 231
8-Methyltocol, 230, 231
Methyltrichlorosilane, 413
2-Methyl-5,7,8-trihydroxychromone, 152
3-Methyl-2,4,3′-trihydroxydiphenyl ether, 325
3-Methyl-2,4,6-trihydroxyphenyl 4-methoxybenzyl ketone, 207
6-Methyl-2,3,4-triphenylchromen-4-ol, 68
3-Methyl-2,4,6-triphenylpyrylium perchlorate, 9
4-Methylumbelliferone, 118
7-Methylwogonin, 180
Methylxanthenes, 308
3-Methyl-1*H*,4*H*-xanthene-1,9-dione, 146
Methylxanthones, 308, 321
Methysticine, 19, 47
Michael condensation, 100, 107, 116, 372
Michler's hydrol, 312
Micromelin, **119**
Micromelum minutum, 119
Micromelum pubescens, 119
Micromeria chamissonis, 187
Micropubescin, **119**
Mikania batataefolia, 186
Mikania cordata, 191
Mikanin, **191**
Milldurone, **217**

Millettia auriculata, 221
Millettia dura, 217, 222, 268
Millettone, **268**
Millettosin, **268**
Millon's test, 33
Miracil A., 321
Miracil-D, 392
Molecular orbital calculations, 4-pyrone, 21
Mollisacacidin, **95**
Monarda didyma, 90
Monardaein chloride, **90**
Monardin, **90**
3-Monoglucosides, 84
Monohydroxyxanthones, 324
Morelloflavone, **202**
Morelloflavone 7″-β-glucoside, 202
Morin, 189
Morindin chloride, 237
Morin ethers, 195
Morpholine, 163
β-Morpholino-2-chlorochalcone, 163
4-Morpholinoflavylium chloride, 163
1-Morphinolino-1-phenylethene, 144
1-Morpholino-1-phenylethylene, 164
Morus tinctoria, 195
Mucronulatol, **234**
Mundulea sericea, 66, 267
Mundulone, **66**
Munduserone, **267**
Muningin, **216**
Murraya exotica, 200
Murraya paniculata, 121
Myceluine, 335
Myricetin, **196**
Myricetin 3-α-L-arabinofuranoside, 196
Myricetin hexamethyl ether, 196
Myricetin methyl ether, 61, 196
Myricetin pentamethyl ether, 196
Myricitrin, **196**
Myrocarpus balsamum, 218
Myrocarpus fastigiatus, 218
Myrtillin-*b* chloride, **91**
Myrtle nagi, 196
Myrtus bullata, 51

Nacareubin, 329
Naphthacene-5,6,11,12-tetrone, 302
α-Naphthacoumarin, 338
β-Naphthacoumarin, 339
α-Naphthaflavone, 337
Naphthalenes, from pyrans, 3
Naphthalic anhydride, 345
Naphthoic acid, 346
1-Naphthol, 337, 338
2-Naphthol, 339, 340, 341
–, reaction with formaldehyde, 346
Naphtholcarboxylic acids, 345
Naphthol[3,2,1-*kl*]thioxanthene, 401
Naphthopyrans, 336, 337, 343, 344
Naphtho[2,1-*b*]pyrans, 339, 341
Naphtho[2,3-*b*]pyrans, 341
1*H*-Naphtho[2,1-*b*]pyran-1-one, 339
1*H*,3*H*-Naphtho[1,8-*cd*]pyran-1-one, 345
2*H*-Naphtho[1,2-*b*]pyran-2-one, 106, 338, 339
2*H*-Naphtho[2,3-*b*]pyran-2-one, 106, 342
3*H*-Naphtho[2,1-*b*]pyran-3-one, 106, 339
4*H*-Naphtho[1,2-*b*]pyran-4-one, 337, 338
Naphthopyrones, 154, 337
1,2-Naphthoquinone, 58, 340
Naphthoselenopyrans, 411, 412
Naphthothiopyrans, 367, 399
Naphtho[1,8-*bc*]thiopyran, 399, 400
2*H*-Naphtho[1,2-*b*]thiopyran, 368
3*H*-Naphtho[2,1-*b*]thiopyran, 367
4*H*-Naphtho[1,2-*b*]thiopyran, 368
β-(1-Naphthoxy)propionitrile, 338
2-(1-Naphthyl)coumarino-4-pyrones, 120
3-(2′-Naphthylseleno)propionic acid, 411
Narcissin, **194**
Narcussus tazetta, 194
Nardostachys jatamansi, 137
Naringenin, 202, 270, 271, 275, **279**, 280
Naringenin 5,7-diglucoside, 281
Naringenin 7,4′-diprenyl ether, 281
Naringenin ethers, 280
Naringenin 5-glucoside, 281
Naringenin 7-D-glucoside, 281
Naringenin 4′-methyl ether, 280
Naringenin 4′-prenyl ether, 281
Naringin, **280**
Neoastilbin, 285
Neocarthamin, **282**
Neocryptomerin, 201
Neoplathymenin, **282**
Neorautanenia pseudopachyrrhiza, 268, 288
Neotenone, 220, **288**

Nepseudin, 219, **288**
Netoric acid, **262**, 266
Nicouic acid, **265**
Nieshoutaol, **133**
Nilodin, 392
3′-Nitrobiflavonyl ether, 170
3-Nitrocoumarins, 109
5-Nitrocoumarin, 109
6-Nitrocoumarin, 108, 109
7-Nitrocoumarin, 108
8-Nitrocoumarin, 108, 109
Nitroflavones, 166
3-Nitroflavone, 167
4-Nitrohomophthalic anhydride, 291
Nitro-3-hydroxycoumarins, 111
Nitromesitylene, 7
Nitromethylenenephthalide, 292
3-Nitrophenol, 317
2-(3-Nitrophenoxy)benzoic acid, 317, 321
N-(4′-Nitrophenylamino)-4-pyridone, 25
4-(Nitrophenyl)-2,6-diphenylpyrylium perchlorate, 9
9-(3-Nitrophenyl)xanthylium perchlorate, 311
Nitrosalicylic acids, 109
3-Nitrosodimethylaniline, 386
4-Nitrosodimethylaniline, 380, 381
ω-Nitrostyrenes, 109
Nitrothiochromones, 373
Nitrothioxanthones, 394
4-Nitro-2-*p*-tolylthiobenzaldehyde, 388
Nitroxanthones, 320, 321
1-Nitroxanthone, 317
Nobiletin, **188**
Nodifloretin, **186**
Noranhydroicaritin, **190**
Norartocarpetin, **185**
Noreugenin, 147
Norkanugin, 196
5-Norkhellin, 158
Nor-rubrofusarin, 342
Norsantol, 217
Norwogin, 175
Norwogonin, **180**
Nothofagus dombeyi, 284
Novobiocin, 120

Obtusifolin, **283**
Obtusifoline, **20**
Ochracin, **303**
Octahydrorottlerone, 66
Octahydrothioxanthenes, 395
Octahydrothioxanthylium halides, 395
Octahydroxanthone, 318
Octahydroxanthylium perchlorate, 309
Octamethoxyflavone, 200
Oenin, 93
Oenin chloride, **92**
Oidiodendron fuscum, 231
Old fustic, 195
Olefins, diacylation, 5
–, reaction with hydroxytetrahydropyran, 49
–, – with paraformaldehyde and hydrogen halides, 50
Ombuin, 201
Ombuoside, **195**
Ononin, **214**
Ononis spinosa, 214
Oospoglycol, **297**
Oospolactone, **297**
Oosponol, **297**
Oospora astringenes, 297
Opium, 20, 38
Opuntia elatior, 20
Opuntiol, **20**, 47
Orange oil, 119
Orchidaceae sp., 97
Oreoselone, **131**
Organogermanium compounds, 420
Organolead compounds, 424
Organolithium reagents, 298, 389
Organomagnesium reagents, 389
Organoselenium compounds, 403
Organosilicon compounds, 413
Organotin compounds, 422
Orientin, **184**
Orientoside, 182
Orobol, **216**
Oroboside, **216**
Orobus tuberosus, 216
Oroselol, 132
Oroselol methyl ether, 132, 133
Oroselone, **132**, 133
Oroxylin, **180**
Oroxylum indicum, 180
Orthoformic ester, 114, 208, 219
Osajaxanthone, **328**, 329

Osajin, **220**
Osajin 4′-methyl ethers, 221
Osthenol, **118**, 119
Osthol, **118**
Ostruthin, **118**
Ougeinia dalbergioides, 287
Ougenin, **287**
Ovidia pillo-pillo, 186
Oxalic acid, 37, 38, 262, 346
Oxalic ester, 11, 22, 143
–, reaction with crotonic ester, 18
γ-Oxalylcrotonic esters, 11
Oxalyl halides, 352
Oxane, 48
2-Oxaperinaphthans, 344
1-Oxaphenalene, 343
1-Oxaphenalen-3-(2*H*)-one, 343
1-Oxaphenalen-3-yl acetate, 343
β-Oxoaldimines, 209
β-Oxocarbaldehydes, 209
β-Oxocarboxylic acid esters, 372
3-Oxo-2,3-dihydro-1-oxaphenalene-2-carboxylic ester, 343
β-Oxodithiocarboxylic acids, 357
3-Oxoflavans, 94
2-Oxoglutaraldehyde, 152
3-Oxoglutaric ester, 404
5-Oxohexan-1-ol, 40
2-Oxo-2*H*-naphtho[1,2-*b*]pyran-4-acetic acid, 338
4-Oxo-4*H*-naphtho[2,3-*b*]pyran-2-carboxylic acid, 341
Oxonium compounds, 87
Oxonium salts, 6, 69, 145, 156, 166, 196, 313
–, chromones, 73
6-Oxononanolide, 225
4-Oxo-3-phenyl-4*H*-benzo[*b*]pyrans, 207
γ-Oxopimelic acid, 26
2-Oxo-2*H*-pyran-5-carboxylic acid, 18
2-Oxo-2*H*-pyran-6-carboxylic acid, 18
4-Oxo-4*H*-pyran-2-carboxylic acid, 36
4-Oxo-4*H*-pyran-2,6-dicarboxylic acid, 37
4-Oxotetrahydrothiopyran-3-carboxylic ester, 364
4-Oxothiochroman-3-carboxylic acid, 383
Oxovinylenamines, 5
Oxyayanin A., **195**
Oxyayanin B., **195**, 199
Oxycoccicyanin chloride, **92**
Oxydihydroartocarpesin, **206**
Oxyjavanicin, 300
Oxypeucedanin, **130**

Pachyrrhisus erosus, 267
Pachyrrhizone, **267**
Padmakastein, 215, **286**
Padmakastin, 286
Paeonia arborea, 92
Paeonidin chloride, **92**
Palm oil, 226, 230
Papaveraceae, 20
Papaverine, 429
Papaver rhoeas, 90
Papilionaceae, 257
Papuanic acid, **257**
Paracotoin, **19**
Parascorbic acid, 19
Paratecoma alba, 64
Parmelia formosana, 327
Parmelia quercina, 327
Parthenium tomentosum, 199
Pasternoside, **194**
Patuletin, **198**
Patuletrin, **198**
Patulin, 20, **45**, 46
Patulin acetate, 46
Pechmann reaction, 34, 98, 117, 137, 142, 338, 339
Pectolinarigenin, **185**
Pectolinarin, **185**
Pedaliin, **187**
Pedalitin, **186**, 187
Pelargonidin, 82, 90
Pelargonidin chloride, 82, 84, 88, **89**, 90
Pelargonidin chloride 3-β-glucoside, 90
Pelargonidin β-glucosides, 85
Pelargonin, 85, 87
Pelargonin chloride, 85, **90**
Peltogyne porphyrocardia, 93, 94
Peltogyne pubescens, 93
Peltogynol, **93**, 94, 161
Peltogynol B, **94**
Peltogynol trimethyl ether, 94, 200
Peltogynone trimethyl ether, 200
Penduletin, **191**

Pendulin, **191**
Penicillium amarum, 331
Penicillium citrinum, 290
Penicillium claviforme, 45
Penicillium expansum, 290
Penicillium oxalicum, 332
Penicillium patulum, 45
Penicillium purpurogenum, 289
Penta-acetoxyflavone, 187
1,3-Pentadiene, 423
1,4-Pentadiene, 423
Pentahydroxychalcone, 282
Pentahydroxyflavan, 95, 240
Pentahydroxyflavan-3-ols, 243
Pentahydroxyflavanone, 284, 285
Pentahydroxyflavone, 178, 186, 187
Pentahydroxyflavonol, 196, 197, 198
Pentahydroxyflavylium chloride, 90, 94
Pentahydroxyflavylium, structure, 86
Pentamethoxyflavanone, 279, 282
Pentamethoxyflavone, 186, 187, 192, 198
Pentamethoxyisoflavone, 218
Pentamethoxy-3′,4′-methylenedioxyflavone, 200
Pentamethoxyxanthones, 327
Pentamethyldihydrohaemateinol, 445
Pentamethylene bromide, 48
Pentamethylene dibromide, 362
Pentamethylene diiodide, 412
Pentamethylene oxide, 48
Pentamethylene sulphide, **362**
Pentamethylepicatechin, 240
Pentane-1,5-dimagnesium dibromide, 413, 420, 422
Pentane-1,5-diols, 49
Pentane-1,5-diol diacetate, 48
Penthian-3-one, 364
Penthianone sulphone, 355
Pentoses, 81
–, conversion to kojic acid, 33
Peonidin chloride, 82, **92**
Peonidin chloride 3-β-glucoside, 92
Peonin, 82
Peonin chloride, **92**
Perbenzoic acid, 131, 225
Perfluoro-2-butyne, 421
Performic acid, 56
Perhydro-9*b*-thiaphenalenium halides, 398
Perhydrothioxanthene, 389, 395
Perinaphthopyrans, 343, 344
Perinaphthothiopyrans, 399
Perkin reaction, 116, 117, 119, 120, 122, 301, 342, 370
–, coumarins, 97, 98
Perphthalic acid, 225
Perunidin, 89
Petunia hybrida, 93, 190
Peonin chloride, **92**
Petunin, 82
Petunoside, **190**
Peucedanin, **131**, 132
Peucedanum formosanum, 136
Peucedanum morisoni, 119
Peucedanum officinale, 130, 131
Peucedanum ostruthium, 154
Peucedanum palustre, 133
Peucenin, **154**, 162
Peucenol, **119**
Peuformosin, **136**
Peumorisin, **119**
Phaseolus vulgaris, 193
Phellamurin, **285**
Phellodendron amurense, 285
Phellodendron japonicum, 285
Phellodendroside, **285**
Phellopterin, 129, **130**
9,10-Phenanthraquinone, 58
Phenols, 58, 247
–, polyhydric, 207
–, reaction with acetoacetic esters, 142
–, – with acetone, 232
–, – with acetylenecarboxylic acid, 163
–, – with cinnamoyl chloride, 270
–, – with ethynyl ketones, 86
–, – with hydroxymethyleneacetophenones, 85
–, – with β-ketonic esters, 98, 100
–, – with nitriles, 100
–, – with salicylic acids, 321
–, – with unsaturated acids, 252, 253
–, – with α,β-unsaturated carbonyl compounds, 235
–, – with α,β-unsaturated ketones, 73, 86
Phenol, 232
–, crotonic esters, 252
–, reaction with acetoacetic ester, 99
–, – with malonic acid, 112
–, – with mesityl oxide, 224

Phenol, reaction (*continued*)
–, – with 2-methoxybutadiene, 225
–, – with phthalic anhydride, 313
Phenolphthalein, 313, 314
ω-Phenoxyacetophenone cyanohydrins, 208
β-Phenoxyacrylic acid, 143
2-Phenoxybenzoic acid, 317
3-Phenoxycarbonyl-2-phenylbenzofurans, 74
β-Phenoxycinnamic acid, 143
β-Phenoxycrotonic esters, 99
Phenoxyfumaric acid, 143, 149
Phenoxymagnesium bromide, 62
2-Phenoxy-3-naphthoic acid, 345
1-Phenoxy-1-phenylpropane, 224
β-Phenoxypropionic acids, 252
Phenylacetic anhydride, 97
Phenylacetylenecarboxylic acid, 163
3-Phenylacyloxyxanthones, 334
Phenyl-2-aminobenzoate, 334
2-Phenyl-5,6-benzochromone, 339
2-Phenyl-7,8-benzochromone, 337
2-Phenyl-4*H*-benzo[*b*]pyran-4-ones, 138, 162, 168
3-Phenyl-4*H*-benzo[*b*]pyran-4-one, 139
2-Phenyl-5,6-benzopyrylium chloride, 341
2-Phenylbenzopyrylium perchlorate, 70
2-Phenylbenzopyrylium salts, 68, 81
2-Phenyl-4*H*-benzoselenopyran-4-ones, 407
2-Phenyl-4*H*-benzo[*b*]thiin-4-one, 374
3-Phenyl-4*H*-benzo[*b*]thiin-4-one, 375
2-Phenyl-4*H*-benzo[*b*]thiopyran-4-ones, 372, 374
2-Phenyl-4*H*-benzo[*b*]thiopyran-4-thione, 375
Phenylbenzothiopyrylium perchlorates, 369, 370
1-Phenylborabenzene, 423
Phenylboron dibromide, 423
Phenyl(bromodichloromethyl)mercury, 416, 421, 423
9-Phenylcerothiene, 401
Phenylchromans, 231
2-Phenylchroman, 231
3-Phenylchroman, 234
4-Phenylchroman, 223, 224
6-Phenylchroman, 224
8-Phenylchroman, 224
2-Phenylchroman-2-ol, 71
4-Phenylchroman-3-ol, 235
2-Phenylchroman-4-ones, 52, 257, 268
3-Phenylchroman-4-ones, 286
4-Phenylchroman-3-one, 102, 235
2-Phenyl-2*H*-chromenes, 60
4-Phenyl-2*H*-chromene, 59
2-Phenyl-2*H*-chromen-2-ol, 76
2-Phenyl-4*H*-chromen-4-ol, 76
Phenylchromones, 139
2-Phenylchromone, 52, 138, 162, **168**
3-Phenylchromones, 207
6-Phenylcoumalin, 19
3-Phenylcoumarin, 97, 103, **110**, 111
4-Phenylcoumarins, 100, 111, 123, 124
6-Phenylcoumarin, 111
8-Phenylcoumarin, 111
1-Phenyl-1,4-dihydroborabenzene, 423
4-Phenyldihydrocoumarin, 102, 273
o-Phenylenediamine, 108, 277
Phenyl epoxycinnamate, 163
β-Phenylethanol, 297
Phenylethynyllithium, 353
Phenylethynyl-2,4,6-triphenylthiopyrans, 353
3-Phenylflavylium perchlorate, 76
2-Phenylflavylium salts, 81
9-Phenylfluorime, 311
9-Phenylfluorone, 311
Phenyl-6*H*-furo[2,3-*c*]xanthen-6-ones, 334
1-Phenylgermacyclohexane, 420
3-Phenyl-4-hydroxycoumarin, 112
4-Phenyl-7-hydroxyflavylium chloride, 78
Phenyl-(2-hydroxyphenylethyl)methanol, 232
1-Phenylisobenzopyrylium perchlorate, 294
1-Phenylisochroman, 299
3-Phenylisochroman, 298, 300
4-Phenylisochroman-4-ol, 301
3-Phenylisochromene, 289
3-Phenylisocoumarin, 292, 294, 295
4-Phenylisocoumarin-3-carboxylic acids, 293
Phenyllithium, 8, 352, 365, 410, 417, 420
N-Phenylmaleimide, 400
1-Phenylmercaptoanthraquinone, 401
o-Phenylmercaptobenzaldehydes, 388
β-Phenylmercaptocinnamic acid, 383
2-(Phenylmercapto)propionic acids, 379

Phenylmercury chloride, 250
4-Phenyl-6-methylcoumarin, 99
2-Phenyl-4*H*-naphtho[1,2-*b*]pyran-4-one, 337
2-Phenyl-4*H*-naphtho[2,3-*b*]pyran-4-one, 342
3-Phenyl-1*H*-naphtho[2,1-*b*]pyran-1-one, 339
3-Phenylnaphtho[2,1-*b*]pyrylium chloride, 341
9-Phenyloctahydroxanthene-1,8-dione, 309
o-Phenylphenoxyacetic acid, 334
Phenyl(3-phenyl-1-propenyl)acetylene, 5
3-Phenylphthalides, 293
3-Phenyl-1-propanesulphonyl chloride, 377
Phenyl 2-propenylselenide, 408
Phenylpropiolic acid, 163
Phenylpropionic acids, 293
β-Phenylpropionic acid, 65
Phenylpropyl ether, 222
3-(1-Phenylpropyl)-4-hydroxycoumarin, 116
1-Phenyl-5-pyrazoleacetaldehyde, 25
2-Phenylpyridine, 19
6-Phenylpyrid-2-one, 19
Phenylpyronines, 214
6-Phenyl-4-pyronyl-2-acetic acid, 22
Phenyl salicylate, 316
2-(Phenylseleno)benzoic acid, 411
4-Phenylselenochroman-4-ol, 408
Phenyl-2*H*-selenochromenes, 407
2-Phenylselenochromones, 407, 408
3-Phenylselenochromone, 408
β-Phenylselenopropionic acid, 408
9-Phenylselenoxanthydrol, 410
9-Phenylselenoxanthyl, 410
2-Phenyltellurochroman-4-one, 413
2-Phenyltetrahydrothiopyran, 363
3-Phenyltetrahydrothiopyran 1,1-dioxide, 361
4-Phenyltetrahydrothiopyran-4-ol, 365
2-Phenylthiobenzoic acid, 391
3-Phenylthiochroman-3-ols, 369
2-Phenylthiochroman-4-one, 383
2-Phenyl-4*H*-thiochromene, 371
β-Phenylthiocinnamic acid, 374
3-Phenylthio-4-hydroxy-2-pyrones, 13
3-Phenylthiomalonic ester, 13
3-Phenyl-2*H*-thiopyran 1,1-dioxide, 348, 350, 360
9-Phenylthioxanthene, 391
9-Phenylthioxanthydrol, 390
9-Phenylthioxanthylium perchlorate, 389
Phenyl *o*-tolyl sulphide, 388
Phenyl(tribromomethyl)mercury, 418
δ-Phenylvaleric acid, 19
2-Phenyl-2-vinylcoumarin-3-one, 213
Phenylxanthenes, 308
9-Phenyl-3*H*-xanthen-3-one, 311
9-Phenylxanthydrol, 310, 311, 318
9-Phenylxanthydryl chloride, 310
9-Phenylxanthyl, 310
Phenylxanthylium perchlorate, 311
Phlobaphenes, 94, 96, 245, 246
Phlobatannins, 245
Phloroacetophenone, 58
Phloroacetophenone dimethyl ether, 190
Phloroglucinaldehyde, 85, 120
Phloroglucinaldehyde monobenzoate, 84
Phloroglucinol, 65, 77, 82, 83, 89, 90, 91, 92, 178, 181, 183, 189, 192, 194, 195, 215, 237, 238, 240, 242, 270, 274, 275, 279, 280, 286, 321
–, reaction with acetoacetic ester, 122
Phloroglucinol dimethyl ether, 238, 241, 322
Phoma terrestris, 332
Phosphoranes, 12
Photocatechuic acid, 82
Photo-Fries rearrangement, 322
Photomethylquercetin, 192
Phthalaldehydic ester, 384
Phthaldehydic acids, 292
Phthalic acid monoethyl ester chloride, 302
Phthalic anhydride, 214, 315, 316
–, reaction with 2-aminophenols, 314
–, – with *p*-cresol, 313
–, – with phenol, 313
Phytol, 228, 231
Phytolacca diocia, 195
Phytyl bromide, 228
Picea abies, 245
Pigments, in flowers, 81
–, naturally occurring, 174
Pilloin, **186**
Pimpinella sacifraga, 130

Pimpinellin, **130**
Pinnarin, **121**
Pinnaterin, **119**
Pinnatin, **206**
Pinobanksin, 276, **284**
Pinobanksin 7-methyl ether, 189, 284
Pinocembrin, **277**, 278
Pinocembrin ethers, 278
Pinomyricetin, **197**
Pinoquercetin, **195**
Pinostrobin, **278**
Pinselin, **331**
Pinus sp., 277
Pinus banksiana, 284
Pinus cembra, 277
Pinus clausa, 189
Pinus excelsa, 277, 278
Pinus nigra, 178
Pinus pyramidalis, 178
Pinus strobus, 179, 282, 283, 284
Pinus toringo, 179
Pinus virginiana, 277
Piperonal, 301
6-Piperonylcoumalin, 19
Piscerythrone, **217**
Piscidia erythrina, 159, 218, 221, 222, 268
Piscidone, **218**
Piuri, 325
Plantaginin, **185**
Plantago asiatica, 185
Plant pigments, 81
Plathymenia reticulata, 282
Plathymenin, **282**
Platonia insignis, 324
Pleiadenes, 400
Plumiera acutifolia, 18
Plumiera rubra varalba, 18
Podocarpus gracilior, 202
Podocarpus spicatus, 217
Podospicatin, **217**
Podospicatin trimethyl ether, 217
Polycladin, **199**
Polycyclic pyrylium salts, 4
Polydin, **242**
Polygala macradenia, 327
Polygala paenea, 327
Polygalaxanthone A., **327**
Polygalaxanthone B., **327**
Polygonum aviculare, 193
Polygonum orientale, 184
Polygonum polystachum, 193
Polygonum recumbens, 191
Polyhalogenoacroleins, 101
Polyhydroxyaryl acids, 101
Polyhydroxybenzoic acid, 207
Polyhydroxyflavans, 162
Polyhydroxyflavanones, 284
Polyhydroxyflavones, 162, 174, 175
Polyhydroxyisoflavones, 207, 212
Polyhydroxyxanthones, 322
Polymethoxyflavones, 176
Polypodium vulgare, 242
Polyporus hispidus, 19
Polystachoside, **193**
Pomiferin, **220**
Poncirin, **280**
Poncirus trifoliata, 280
Ponderosa, 195, 197
Pongaglabrone, **205**
Pongamia glabra, 196, 205
Pongamia pinnata, 205, 206
Pongapin, **205**
Ponkanetin, **188**, 282
Populnetin, 190
Populnin, **190**
Populus nigra, 278
Poriol, **281**
Potassium xanthate, 376
Pranferin, **119**
Prangenin, **131**
Prangos ferulaceae, 119
Prangos pabularia, 131
Prangos uloptera, 119
Pratol, 180, 214
2-Prenyltetrahydroxyxanthone, 329
2-Prenyl-1,3,5-trihydroxyxanthone, 328
2-Prenyl-1,3,7-trihydroxyxanthone, 329
Primetin, 175, **179**
Primetin 8-methyl ether, 179
Primflasine, **191**
Primula algida, 191
Primula hirsuta, 93
Primula modesta, 179
Primula rosea, 92
Primula viscosa, 92
Proban, **196**

Pro-oestrogens, 215
Propan-2-ol, 94
1-Propenyl 2-hydroxy-5-tolyl ketone, 252
Propionic esters, 12, 139
Propylene oxide, 303
2-Propylphenol, 222
2-*n*-Propyl-7,8,9-trimethoxy-5*H*-benzo[*c*]-pyrano[4,3-*b*]pyran-5-one, 296
2-Propynethiol, 349
Proteins, 244
Protocatechuic acid, 90, 191, 192, 197, 238, 278
Protofarrerol, **283**
Prudomestin, **196**
Prunetin, **215**
Prunetin dimethyl ether, 215
Prunetin 4′-glucoside, 215
Prunetol, **214**
Prunicyanin chloride, **91**
Prunin, **281**
Prunitrin, **215**
Prunus avium, 91, 190, 279
Prunus cerasus, 279
Prunus domestica, 196
Prunus donarium, 280
Prunus emarginata, 215
Prunus puddum, 182, 286
Prunus serotina, 215
Prunus verecunda, 278, 280
Prunus yedonensis, 280, 281
Pseudobaptigenin, **214**
Pseudobaptisin, **214**
Pseudocumen-6-ol, 227
Pseudocumoquinol, 230
Pseudoirigenin, **217**
Pseudotectorigenin, 216
Pseudotsuga menziesii, 281
Psoralea coryfolia, 126, 127
Psoralene, **126**, 127
Ptaeroxylan utile, 134
Ptaeroxylol, **191**
Ptaeroxylon obliquum, 122, 159, 161
Pterocarpin, **288**
Pterocarpus angolensis, 215, 216
Pterocarpus dalbergioides, 214
Pterocarpus indicus, 214
Pterocarpus santalinus, 216, 288, 306
Punctatin, **219**
Purpuranin B., **218**
Purpurogenone, **289**
Pyrans, 1, 2, 3, **4**
–, from carbonyl compounds, 41
2-Pyran, 2
2*H*-Pyrans, 2, 17
4-Pyrans, 3
–, conversion to naphthalene, 2
–, reactions, 2
–, 2,4,6-trisubstituted, 2, 8
4-Pyran, **2**, 3
4*H*-Pyran, 2
α-Pyran, 2
γ-Pyran, 2
Pyranhydrones, 3
Pyranochromene, 58
2*H*,5*H*-Pyrano[3,2-*c*]chromene-2,5-diones, 137
2*H*,6*H*-Pyrano[3,2-*g*]chromene-2,6-dione, 138
2*H*,9*H*-Pyrano[3,2-*h*]chromene-2,9-dione, 137
2*H*,9*H*-Pyrano[3,2-*g*]chromene-2,9-dione-3-carboxylic ester, 138
Pyranochromones, 159
Pyranocoumarins, 34, 134, 137, 221
Pyranodihydropyrans, naturally occurring, 45
Pyranoflavones, 206
Pyranoisoflavones, 219
Pyranols, 2, 3
Pyranol ethers, 78
2*H*-Pyran-2-ones, 9, 10
2*H*-Pyran-2-one, **17**
Pyran-4-one, chemical reactivities, 356
4*H*-Pyran-4-ones, 9, 10
4*H*-Pyran-4-one, 4, **29**
Pyranoxanthones, 329
4-Pyran-4-thiones, 25
4*H*-Pyran-4-thiones, 28
Pyrazoles, 25, 147, 149
Pyrazole-3-acetaldehyde, 25
Pyrazolines, 13, 108, 348
2*H*-Pyrazolothioxanthene, 392
Pyridine, 1, 7
(1-Pyridinio)sulphate, 7
Pyridinium perbromide, 142
Pyridones, 73, 146

2-Pyridones, 15
4-Pyridones, 31
–, from 4-pyrones, 20, 28
4-Pyridone, 30
4-Pyridone-2,6-dicarboxylic acid, 37
Pyridopiperazines, 38
4-(2-Pyridyl)coumarin, 106
α-Pyrocresol, 307
Pyrogallol, 91, 124, 315, 316, 321, 443
Pyrogallolcarbaldehyde, 124
Pyrogallol triethyl ether, 124
Pyrogallol trimethyl ether, 444
Pyrogallophthalein, **316**
Pyromecazonic acid, 32
Pyromeconic acid, **32**
Pyrones, 73, 146
–, conversion to pyrans, 2
–, reaction with acids, 6
α- or 2-Pyrones, 2, 9, 10, **17**
–, chloromethylation, 15
–, conversion to aromatic compounds, 15
–, 4,6-disubstituted, 12
–, from β-diketones, 13
–, from malonic esters, 12
–, from pyrazolines, 13
–, halogenation, 17
–, hydrogenation, 16
–, hydrolysis, 14
–, naturally occurring, 10
–, preparation, 10
–, reaction with amines, 15
–, – with ammonia and amines, 15
–, – with diazomethane, 16
–, – with Grignard reagents, 17
–, – with hydride ions, 14
–, 6-substituted, 12
–, 4,5,6-trisubstituted, 12
2-Pyrone, **17**
–, conversion to 2-pyridone, 15
γ- or 4-Pyrones, 4, 9, 10, 14, 20, 21, 22, 29, 31, 86, 144
–, aromaticity, 27
–, constitution, 20
–, fission, 28
–, from 4-pyronecarboxylic acids, 36
–, hydrates, 23
–, hydrogenated, 51
–, mercuration, 27
γ- or 4-Pyrones, (*continued*)
–, nitration, 27
–, polycyclic, 24
–, preparation, 21
–, properties, 23
–, reactions, 23
–, reaction with acyl halides, 28
–, – with ammonia and amines, 15, 28, 30
–, – with cyanamide, 24
–, – with diphenylketene, 26
–, – with Grignard reagents, 25
–, – with hydrazine, 25
–, – with phosphorus pentasulphide, 28
–, reduction, **26**, **27**
Pyronecarboxylic acids, decarboxylation, 22
2-Pyronecarboxylic acids, 11
α- (or 2-)Pyrone-5-carboxylic acid, 10, 18
2-Pyrone-6-carboxylic acid, 17, **18**
4-Pyronecarboxylic acids, 22, 36
4-Pyrone-2,6-dicarboxylic acid, 3, 22, 37
Pyronines, 309, 313
Pyronine dyes, 310
Pyronine G., 312, 313, 321
γ-Pyronochromanones, 256
Pyronones, 16
4-Pyronone, 29
Pyrylium cations, 1
Pyrylium compounds, 1
Pyrylium perchlorate, **7**
Pyrylium salts, 1, 2, 3, **4**, 5, 7, 8, 25, 28, 69, 73, 87, 93, 353, 438
–, aryl-substituted, 4
–, 2,6-disubstituted, 2, 8
–, from acetophenone, 6
–, from 1,5-diketones, 4
–, physical properties, 9
–, 2,4,6-trisubstituted, 2

Querbrachocatechin, **244**, 245
Querbracho colorado, 191, 244
Querbracho tannin, 244
Quercetagetin, **198**
Quercetagetin 3,3′-dimethyl ether, 199
Quercetagetin 3,6-dimethyl ether, 199
Quercetagetin 6,3′-dimethyl ether, 198
Quercetagetin 6-methyl ether, 198
Quercetagitrin, **198**
Quercetin, 88, 90, 170, 172, 174, 175, 177, 189, **191**, 192, 193, 194, 238, 240, 276

Quercetin 3-α-L-arabinoside, 193
Quercetin 3-β-L-arabinoside, 193
Quercetin 3,3′-dimethyl ether, 201
Quercetin 7,3′-dimethyl ether, 195
Quercetin 3-β-D-galactofuranoside, 193
Quercetin 3-D-galactoside, 193
Quercetin 3-β-D-glucoside, 193
Quercetin 4′-glucoside, 193
Quercetin 7-β-D-glucoside, 193
Quercetin methyl ethers, 194
Quercetin 3-methyl ether, 201
Quercetin penta-acetate, 94, 192
Quercetin pentamethyl ether, 61, 88, 177, 192, 240
Quercetin 3-L-rhamnoside, 193
Quercetin 3-rutinoside, 193
Quercetin tetra-acetate, 192
Quercetin 3,5,7,3′-tetramethyl ether, 195
Quercetin 3,7,4′-trimethyl ether, 195
Quercimeritrin, **193**, 194
Quercitrin, **193**
Quercituron, **193**
Quercus sp., 245
Quercus tinctoria, 192
Quinhydrone, 3
Quinol, 117, 223
Quinolines, 294
Quinoline, 79, 376
Quinones, 100
–, coenzyme Q group, 67
o-Quinones, 58
Quinone acetals, 229
o-Quinone allides, 56
Quinoxalines, 432
Quinqueloside, 183

Ranunculaceae sp., 190
Ranunculus repens, 198
Ranupin, **198**
Raphanus sativus, 93
Ravenelin, **325**
Resacetophenone, 140, 278
Reseda luteola, 183
Resocyanin, **118**
Reso-oxyayanin A., 201
Resorcinols, 14
Resorcinol, 69, 76, 103, 117, 118, 191, 232, 235, 248, 269, 278, 311, 312, 314, 315, 316, 428, 439
Resorcinol, (*continued*)
–, reaction with acetylacetone, 70
–, – with 1,3-diketones, 70
–, – with αβ-unsaturated acids, 252
Resorcinolbenzein, 312
Resorcinol dimethyl ether, 99, 431
Resorcinol monomethyl ether, 117
β-Resorcylaldehyde, 127, 439
β-Resorcylic acid, 237, 428
Reticulol, **296**
Rhamnazin, **195**
Rhamnazin-3-*O*-mono-D-glucoside, 195
Rhamnetin, **194**, 201
Rhamnetin tetra-acetate, 192
Rhamnetin 3,3′,4′-triacetate, 192
Rhamnocitrin, **190**, 276
Rhamnolutin, 190
Rhamnose, 91, 92, 194
L-Rhamnose, 182, 191
Rhamnus carthatica, 194
Rhamnus catharticus, 190
Rhamnus infectoria, 195
Rhamnus tintoria, 194
Rhizophoraceae, 245
Rhodamines, 313, 314, 316
Rhodamine B., **314**
Rhodamine 3B., 314
Rhodamine 6G., 314
Rhodamine dyes, 310, 311
Rhodanines, 356, 384
Rhododendron farrerae, 283
Rhododendron mucronatum, 194
Rhododendron simsii, 283
Rhoifolin, **181**
Rhus cotinus, 191, 281
Rhus succedanea, 181, 281
Ribonucleotides, 359
Rice germ oil, 226
Ricinocarpos muricatus, 197
Ricinocarpos stylosus, 199
Ripariachromenes A,B,C., 64
Rissic acid, **260**, 265
Robigenin, 190
Robinetin, **196**, 276
Robinetinidol, **243**
Robinia pseudacacia, 181, 191, 196, 244, 285
Robinin, **191**
Robinobiose, 191
Rosamines, 313, 314

Rose Bengal B., 316
Rose gallica, 90
Rosellinia necatrix, 256
Rosellinic acid, **256**
Rosinidin, **92**
Rotenic acid, 259
Rotenoids, 264
Rotenol, **261**
Rotenolone-1, **261**
Rotenolone-11, **261**
Rotenone, 257, **258**, 259, 261, 262, 264
Rotenonic acid, **263**
Rotenonone, **262**
β-Rotenonone, **262**
Rotenononic acid, **262**
Rotexen, 258
Rottlerin, **65**, 66
Rottlerone, 65, 66
Rubia tinctorum, 47
Rubrofusarin, **341**
Rutaceae sp., 97
Ruta graveolens, 125, 193
Ruta pinnata, 119, 121, 125
Rutin, **193**
Rutin 7,4′-dimethyl ether, 195

Sabandinin, **125**
Sabandinone, **125**
Sakuranetin, **280**, 284
Sakuranetin 5-D-glucoside, 280
Sakuranin, **280**
Salicylaldehydes, 109, 114
–, substituted, 69
Salicylaldehyde, 56, 59, 69, 70, 76, 80, 97, 98, 144, 164, 210
–, reaction with acetic anhydride, 97
–, – with ketones, 439
Salicylaldehyde glucoside, 69
Salicylic acids, 113, 322
Salicylic acid, 145, 168, 189, 273, 321, 345
Salicylic acid naphthyl ethers, 345, 346
Salicylic acid phenyl ether, 317
Salipurpin, **281**
Salipurposide, 281
Salix purpurea, 281
Salvianin chloride, **90**
Salvia patens, 91
Salvia splendens, 93
Salvia triloba, 182, 185
Salvigenin, 182, **185**
Salvinin, **90**
Samarcandin, **118**
Samarcandone, **118**
Samidin, **137**
Santal, 216
Santal ethers, 216
Santalin A., **309**
Santalin B., **309**
Santalin permethyl ether, 309
Sargentia areggii, 185
Saxifraga siberica, 295
Scandenin, 221
Scandenone, **220**
Scandenone 4′-methyl ethers, 221
Scandinone, **221**
Scaposin, **188**
Schiff bases, reaction with 2,3-dihydro-4-pyran, 43
Schinopsis balansae, 95
Schinopsis lorentzii, 245
Schinopsis quebracho-colorado, 95
Schmidt reaction, 116
Sciadopitysin, **201**, 203
Sciadopitys verticillata, 201
Scillaridin A., 10
Sclerin, **304**
Sclerotinia libertiana, 304
Sclerotinia sclerotiorum, 128
Scoparone, **123**
Scopoletin, **123**
Scopoletin 7-β-primveroside, 123
Scopolia sp., 123
Scopolin, **123**
Scriblitifolic acid, **331**
Scrophularia nodosa, 184
Scutellarein, 176, **184**
Scutellarein 7-glucoside, 185
Scutellarein 6-methyl ether, 185
Scutellaria baicalensis, 179, 180
Scutellaria galericulata, 180
Scutellarin, **184**
Secalonic acids, **332**
Selenane, **404**
Selenane-2,6-dicarboxylic acid, 404
Seleninium perchlorate, 403
Selenochroman, 408
Selenochroman-4-ol, 405, 408

Selenochroman-4-one, 407, 408
Selenochrom-3-enes, 405
Selenochromones, 406, 407
Selenochromylium perchlorate, 406
Selenocinnamic acid chlorides, 407
Selenocoumarins, 407
Selenoflavanone, 408
Selenoflavones, 407, 408
Selenoisoflavone, 408
Selenophenols, 407
Selenopyrans, 403
–, containing fused rings, 405
4*H*-Selenopyran, 403
Selenopyrylium perchlorate, 403
Selenopyrylium salts, 353, 403
Selenoxanthene, 409, 410
Selenoxanthen-9-one, 411
Selenoxanthone, 409, 410, 411
Selenoxanthone metal ketyls, 411
Selenoxanthydrols, 410
Selenoxanthylium perchlorate, 410, 411
Selinetin, 136
Selinidin, **136**
Selinone, **281**
Selinum vaginatum, 136, 281
Sequoiaflavone, **203**
Sequoia sempervirens, 203
Serpyllin, **188**
Sesamum indicum, 187
Seseli indicum, 135
Seselin, 57, **135**
Seseli sessiliflorum, 137
Seseli sibiricum, 121
Sesibiricin, **121**
Shimmetin, **117**
Shorea leprosula, 295
Silabenzene, 414
Silacyclohexadienes, 417
Silacyclohexadiene, 418
Silacyclohexanes, 413, 414, 415
Silacyclohexenes, 417
Silanes, cyclic, 414, 415, 416
–, polycyclic, 419
5-Silaspiro[4,5]decane, 419
10-Silaspirodicyclodecane, 419
11-Silaspirodicycloundecane, 419
6-Silaspiro[5,5]undecane, 419
Simonis reaction, 99, 143
Sinensetin, 186
Six-membered ring compounds, containing oxygen, 1
Skimmin, 117
Sodium bis(2-methoxyethoxy)aluminium hydride, 299
Sodium derritol, 261
Soja hispida, 214
Solanaceae sp., 123
Sophora angustifolia, 184, 190
Sophora japonica, 215, 287
Sophoranone, **278**
Sophora subprostrata, 278
Sophoricoside, **215**
Sophorol, **287**
Sorbifolin, **160**
Sotetsuflavone, **201**
Soya beans, acid hydrolysis, 32
Soya bean oil, 230
Spartium junceum, 184
Spathelia sorbifolia, 159, 160
Sphondin, **131**
Sphondylin, 131
Spinacetin, **198**
Spinacia oleracea, 198
Spiraea ulmaria, 193
Spiraeoside, **193**
Spirans, 168
Spiroacetal trimers, 229
Spirobarbiturates, 419
Spirocyclohexanes, 419
Spiro[dibenzosilole-5,1′-silacyclohexane], 419
Stachyflaside, **185**
Stachys annua, 185
6-Stannaspiro[5,5]undecane, 422
Starch, dry distillation, 32
Stereochemistry, catechins, 240
Steroids, with 2-pyrone side-chains, 10
Stibabenzene, 423
Stobochrysin, **179**
Stobochrysin 7-methyl ether, 179
Streptomyces mobaraensis, 297
Streptomyces rubrireticuli, 296
Streptomycin, 32
Strobobanksin, **283**
Strobopinin, **282**
Stypandra grandis, 233

Styphnic acid, 428
Styrylbis(chromon-3-yl)methane, 151
2-Styrylchromone, 319
2-Styrylchromone, physical properties, 150
2-Styrylchromylium salts, 148
Styrylcoumarins, 280
Styryl derivatives, 80
2-Styrylisoflavones, 208
Styryl ketones, 204
N-Styrylmorpholine, 210
Suberosin, **135**
Succinic acid, 48, 103
2-Succinimidyl-3-bromotetrahydropyran, 41
Sudachitin, **188**
Sugar residues, in anthocyanins, 83
Sulochrin, 331
Sulphonium salts, 379
Sumatrol, 264, 267
Swerchirin, **326**
Sweroside, **47**
Swertia chirata, 326
Swertia decussata, 326, 327
Swertia japonica, 47, 182, 183, 184, 326
Swertiajaponin, **184**
Swertiamarin, **47**
Swertiamorin, 303
Swertianol, **326**
Swertianolin, **326**
Swertia perennis, 327
Swertia tosaensis, 326
Swertinin, **326**
Swertisin, **182**
Swertisin dimethyl ether, 183
Symphonia globulifera, 330
Symphoxanthone, **330**
Syringic acid, 83
Syringidin chloride, **92**

Tachrosin, **179**
Tagetes erecta, 198
Tagetes patula, 198
Talbotaflavone, **202**
Tamarixin, **194**
Tamarix laxa, 194
Tamarix troupii, 194
Tambalin, **198**
Tambutetin, **198**
Tangeretin, **187**
Tangeritin, **187**
Tannins, 87, 96, 245, 246, 247
–, condensed, 237, 244
Tanning extracts, 245
Taxifolin, 88, 96, 193, 276, **284**
Taxus cuspidata, 201
Tectochrysin, 170, **179**
Tectoridin, **216**
Tectorigenin, **216**
Tectorigenin ethers, 216
Telluracyclohexane diiodide, 412
Telluracyclohexane-3,5-dione, 413
Tellurane, 412
Tellurane 1,1-diiodide, 412
Tellurane-3,5-dione, 413
Tellurane-3,5-dione 1,1-dichloride, 413
Telluroflavanone, 413
Tephrosia, 257
Tephrosia maxima, 218
Tephrosia polystachyoides, 179
Tephrosia toxicaria, 266
Tephrosia vogelii, 265
Tephrosin, 264, **265**, 267
Tephrosindicarboxylic acid, 265
Tephrosinmonocarboxylic acid, 265
Teracacidin, **96**
Ternatin, **200**
Terniflorin, **181**
O-Tetra-acetylbrazilin, 442, 443, 445
2-*O*-Tetra-acetyl-β-glucosidophloroglucinaldehyde, 85
Tetra-acetyl-α-glycosyl bromide, 174
Tetra-arylnitrobenzenes, 7
Tetra-arylphenols, 7
Tetrabromofluorescein, 316
Tetrabromotetrahydrothiopyran, 355
2,3,5,6-Tetrabromothiopyran 1,1-dioxide, 347
Tetrabromoxanthone, 319
Tetracacidin, **95**
Tetrachloro-1,2-benzoquinone, 369, 402
Tetrachlorobenzyne, 56
Tetrachlorobiphenyl, 33
5,6,7,8-Tetrachloro-2*H*-chromene, 56
Tetrachlorofluorescein, 316
Tetracyanoethylene 4-hydroxycoumarin adduct, 114
Tetrahydro-1,5-benzothiazepin-4-one, 381
Tetrahydro-1,5-benzoxazepinone, 255

Tetrahydrobicoumarinyls, 104
Tetrahydrocannabinols, 336
Tetrahydrochroman, 225
Tetrahydrochroman-2,5-diones, 256
Tetrahydrochroman-4,5-diones, 256
5,6,7,8-Tetrahydroflavone, **169**
Tetrahydrofurans, 44
Tetrahydrofurfuryl alcohols, 39, 40
5,6,7,8-Tetrahydroisochroman, 298
Tetrahydro-2-(3*H*)-naphthalenone, 225
Tetrahydropyrans, 27, 48, 49, 50, 51
Tetrahydropyran, 2, 40, **48**
Tetrahydro-4-pyrans, 51
Tetrahydropyran-3-carboxylic acid, 51
Tetrahydropyran-4-carboxylic acid, 51
Tetrahydropyran-2,6-dicarboxylic acid, 51
Tetrahydropyran-4,4-dicarboxylic acid, 51
Tetrahydropyran-2,3-diol, 42
Tetrahydropyran peroxides, 48
Tetrahydropyranols, 27
Tetrahydropyranotetrahydroquinolines, 43
Tetrahydropyranyl compounds, 43
Tetrahydropyrones, isomerisation, 23
Tetrahydro-4-pyrone, 27, 45, **51**
Tetrahydro-4-pyrone-3-acetic acid, 45
Tetrahydrorottlerin, 65
Tetrahydroselenopyran, 404
Tetrahydroselenopyran-2,6-dicarboxylic acid, 404
Tetrahydro-2*H*-tellurin, 412
Tetrahydro-2*H*-tellurin 1,1-diiodide, 412
Tetrahydrotelluropyran, 412
Tetrahydrotelluropyran derivatives, 412
Tetrahydrotelluropyran 1,1-diiodide, 412
Tetrahydrothiapyrans, 362
Tetrahydrothiopyrans, 352, 362
Tetrahydrothiopyran, 347, 356, 359, **362**
Tetrahydrothiopyran-4-acetic acid, 364
Tetrahydrothiopyran-4-carboxylic ester, 364
Tetrahydrothiopyran 1,1-dioxide, 362, 364
Tetrahydrothiopyranols, 365
Tetrahydrothiopyran-4-ol, 360, 364
Tetrahydrothiopyran-3-one, 366
Tetrahydrothiopyran-4-ones, 365, 366
Tetrahydrothiopyran-4-one, **364**
Tetrahydrothiopyran-4-one 1,1-dioxide, 355
Tetrahydrothiopyran-1-oxide, 359
Tetrahydrothiopyrones, 355
Tetrahydrothio-4-pyrone, 364
Tetrahydrotubaic acid, 259
Tetrahydrotubanol, 259
Tetrahydroxybenzophenone, 321
Tetrahydroxy-2-benzylcoumaran-3-one, 172
Tetrahydroxychalcone, 278
Tetrahydroxy-3′,5′-dimethoxyflavylium chloride, 92
5,7,3′,4′-Tetrahydroxy-6,8-dimethylflavanone, 283
Tetrahydroxydiphenylketimine, 317
Tetrahydroxyflavan, 75, 232
Tetrahydroxyflavan-3-ol, 191, 238, 239, 242, 243
3,5,7,4′-Tetrahydroxyflavanone, 284
3,7,3′,4′-Tetrahydroxyflavanone, 281
5,6,7,4′-Tetrahydroxyflavanone, 282
5,7,3′,4′-Tetrahydroxyflavanone, 282
5,7,8,4′-Tetrahydroxyflavanone, 282
6,7,3′,4′-Tetrahydroxyflavanone, 282
Tetrahydroxyflavone, 176, 183
5,6,7,4′-Tetrahydroxyflavone, 184
5,7,2′,4′-Tetrahydroxyflavone, 185
Tetrahydroxyflavone-7-*O*-β-D-glucopyranosyl-(2→1)-*O*-β-D-mannopyranoside, 185
5,6,7,4′-Tetrahydroxyflavonol, 191, 205
5,7,2′,4′-Tetrahydroxyflavonol, 195
5,7,8,4′-Tetrahydroxyflavonol, 196
7,3′,4′,5′-Tetrahydroxyflavonol, 196
3,5,7,4′-Tetrahydroxyflavylium chloride, 89
5,7,3′,4′-Tetrahydroxyflavylium chloride, 89
Tetrahydroxy-8-(γ-hydroxyisovaleryl)-flavanones, 285
5,7,2′,4′-Tetrahydroxyisoflavanone, 287
5,7,3′,4′-Tetrahydroxyisoflavone, 216
6,7,3′,4′-Tetrahydroxyisoflavone, 218
Tetrahydroxy-7-methoxyflavone, 187
5,7,3′,4′-Tetrahydroxy-8-methoxyflavonol, 198
Tetrahydroxy-3′-methoxyflavylium chloride, 92
Tetrahydroxy-5′-methoxy-3′-phenylisoflavone, 217
Tetrahydroxy-2-[2-(2-methyl-3-butenyl)]-8-prenylxanthone, 330

Tetrahydroxy-4-[2-(2-methyl-3-butenyl)]-xanthone, 330
Tetrahydroxy-2-methylchromone, 146
2,3,4,6-Tetrahydroxyoctane-1,8-dioic acid, 26
Tetrahydroxypimelic acid, 38
5,6,2′,3′-Tetrahydroxy-3,7,4′-trimethoxyflavone, 199
Tetrahydroxyxanthones, 321, 326, 330, 331
Tetramethoxybiflavonyl, 204
Tetramethoxyflavan-3,4-diol, 89
Tetramethoxy-2,3-flavan-3-ol, 95
5,7,3′,4′-Tetramethoxyflavanone, 275, 279
5,7,3′,4′-Tetramethoxyflav-2-ene, 244
Tetramethoxyflavone, 178
5,6,2′,6′-Tetramethoxyflavone, 185
5,6,7,2′-Tetramethoxyflavone, 185
5,6,7,8-Tetramethoxyflavone, 186
5,7,3′,4′-Tetramethoxyflavonol, 193, 194
5,7,3′,4′-Tetramethoxyisoflavan, 234
6,7,3′,4′-Tetramethoxyisoflavone, 218
3,5,6,7-Tetramethoxy-3′,4′-methylenedioxyflavone, 200
Tetramethoxy-3-phenyl-coumarin, 242
Tetramethoxy-5,7,5′-trihydroxyflavone, 188
Tetramethoxyxanthones, 327, 329
Tetramethoxyxanthone-2-carboxylic acid, 326
Tetramethylanhydroepicatechin, 239, 240, 241
Tetramethylcatechin, 240, 241, 242
Tetra-*O*-methylcupressuflavone, 204
5,7,3′,4′-Tetramethylcyanidin chloride, 89
Tetramethyldeoxyepicatechin, 240
Tetramethyldihydrobrazileinol, 436
Tetramethylepicatechin, 239, 240
Tetra-*O*-methylergoflavin, 333
Tetramethylflavonols, 95
Tetramethylhaematoxylin, 443, 444
Tetramethylhaematoxylium ferrichloride, 445
Tetramethylhaematoxylone, 444
Tetramethylisochromanone, 300
Tetramethyl luteolin, 240
2,4,4,6-Tetramethyl-4-pyran, 25
Tetramethylpyrylium salts, 5
Tetramethylsilacyclopentane, 414
Tetramethyl-6-stannaspiro[5,5]undecane, 422
Tetraphenyl-4,4′-di(thiopyranylidene), 358
2,4,4,6-Tetraphenyl-4-pyran, 3
2,3,4,6-Tetraphenylpyrylium perchlorate, 9
Tetraphenylselenopyran-4-one, 404
1,2,4,6-Tetraphenylthiabenzene, 352
2,3,5,6-Tetraphenyl-4*H*-thiopyran, 352
Tetroses, conversion to kojic acid, 33
Teucrium polium, 182
Thamnosin, **119**
Thamnosma montana, 119, 131
Thapsin, **199**
Thermochromism, bixanthylene, 306, 307
Thespesia populnea, 190
Thevetia peruviana, 181
Thia-adamantane, 347, 398, 399
2-Thia-adamantane-4,8-dione, 399
2-Thia-adamantane 2-oxide, 399
Thia-alkanes, 362
Thiabenzenes, 8
Thiabicycloalkanes, 396
Thiabicyclononanes, 397
9-Thiabicyclo[3.3.1]non-6-en-2-one, 398
Thiabicyclo[3.2.1]octanes, 396, 397
2-Thiabicyclo[2.2.2]octane, 396
8-Thiabicyclo[3.2.1]octan-3-one, 397
2-Thiabicyclo[2.2.2]oct-5-ene, 396
6-Thiabicyclo[3.2.1]oct-3-ene, 397
Thiabicyclotridecenes, 398
Thiacyanine dyes, 383
Thiacyclobutan-2-one, 365
Thiacyclohexanes, 347, 352, 356, 362
1-Thiadecalin, 379
2-Thiadecalins, 387
Thiadecalin dioxides, 387
2-Thianaphthalenes, 385
Thianaphthalenium salts, 368, 369
Thianes, 347, 362
Thiaphenalenes, 399
1-Thiaphenalene, 399
2-Thiaphenalene, 400
9-Thiaphenanthrenium perchlorate, 395
Thiapyrans, 347
1-Thiatetralin, 376
2-Thiatetralin, 385
Thiins, 347

2H-Thiin, 347
4H-Thiin, 350
2H-Thiin-2-one, 354
4H-Thiin-2-thione, 356
4H-Thiin-4-thione, 358
Thiinyl 1,1-dioxides, 250
Thioacetic acid, 393
α-Thioacyl-δ-thiolides, 359
Thiobenzopyrylium salts, 368
Thiochroman, 368, **376**, 377, 378
Thiochroman-3,4-dione, 381
Thiochroman 1,1-dioxide, **377**
Thiochroman-4-ols, 367, 382
Thiochromanones, 370, 373
–, oxo sulphones, 382
Thiochroman-3-ones, 386
Thiochroman-4-ones, 376, 379, 381
Thiochromanone-8-carboxylic ester, 382
Thiochroman-4-one 1,1-dioxide, 382
Thiochroman 1-oxide, 368
Thiochromenes, 367, 368, 369, 405
Thiochrom-2-ene, 368
Thiochrom-3-ene, 367
Thiochromones, 370, 372, 373
Thiochromone-2-carboxylic acids, 374
Thiochromono[3,2-*b*]benzofuran, 376
Thiocoumarin, **370**, 371
2-Thiocoumarin, **110**
β,β-Thiodipropionic ester, 364
Thioflavan-4-ols, 383
Thioflavanone, 383
Thioflavones, 372, 375, 383
Thioflavone, 371, 373, 374
Thiofluorenone, 402
Thioindigo, 375
Thioindoxyl, 375
Thioisoflavanones, 375, 383
Thioisoflavones, 375
Thionaphthalenium salts, 383
Thionochromones, 147, **149**
4-Thionochromone, 146
2-Thionocoumarin, 108
4-Thionoflavone, 149, **169**
Thionothiocoumarin, 371
4-Thionothioflavone, 374, 375
Thionothioxanthones, 389, 393
Thiopyrans, 1, 347
–, from 1,5-diketones, 351
2H-Thiopyran, **347**
4H-Thiopyran, 350, 351
α-Thiopyran, 347, 350
2H-Thiopyran 1,1-dioxide, 347, 348
Thiopyran-4-one, chemical reactivities, 356
2H-Thiopyran-2-ones, 354
4H-Thiopyran-4-ones, 355, 358, 364
Thiopyran-4-one 1,1-dioxides, 355
Thiopyran-2-thiones, 354, 357
2H-Thiopyran-2-thione, 354, 356, 357
4H-Thiopyran-4-thiones, 28, 355, 358
Thiopyranyl 1,1-dioxides, 350
Thiopyranylidenedihydropyridine *S,S*-dioxides, 350
Thiopyrones, 25, 347
Thiopyrylcyanines, 351
Thiopyrylium chloride, 355
Thiopyrylium salts, 347, 350, 351, 352, 353
Thiopyrylium trifluoroacetates, 352
Thiosalicylic ester, 375
Thiourea, 157, 158
Thioxanthene, 388, 389, 390, 392
Thioxanthene 10,10-dioxide, 389, 390
Thioxanthene oxides, 389
Thioxanthene-9-thione, 393
Thioxanthene-9-thione *S*-oxides, 393
Thioxanthen-9-ol, 390
Thioxanthen-9-ol 10-oxides, 390
Thioxanthen-9-one, 391
9-Thioxanthenyl esters, 309
Thioxanthones, substituted, 394
Thioxanthone, 388, 389, 390, **391**, 392, 393
–, complex with phosphorus pentachloride, 392
Thioxanthone-2-carbaldehyde, 393
Thioxanthone 10,10-dioxides, 391, 393
Thioxanthydrol, 390, 391, 392
Thioxanthydrol 10,10-dioxide, 390
Thioxanthydrol 10-oxides, 390
Thioxanth-9-yl ethers, 390
Thioxanthylium perbromides, 389
Thioxanthylium perchlorate, 389
Thioxanthylium salts, 388, 391
Thuringione, **331**
Thymelaecae sp., 124
Tiglic acid, 133, 136
Tilia argentea, 190
Tilianin, 181

Tiliroside, **190**
Timbo, 257
Tin compounds, 420
Tithonia tubaeformis, 183
Tithonine, **183**
Tlatlancuayin, **215**
Tocol, 67, 226, **231**
Tocopherols, 2, 226
–, properties, 228, 229
α-Tocopherol, 226, 227, 228, 229, 230
–, fatty acid esters, 230
–, oxidation, 229
β-Tocopherol, 226, 227, 228, 229, 230
γ-Tocopherol, 229, 230
δ-Tocopherol, 230
ε-Tocopherol, 230
ζ-Tocopherols, 230
η-Tocopherol, 230
Toddaculine, **120**
Toddalia aculeata, 120
Toddalolactone, **120**
p-Toluenesulphonic acid, 63
o-Toluidine, 314
Toluquinol, 231
p-Tolyl crotonate, 252
o-Tolylenediamine, 432
o-Tolylethanol, 298
Toringin, **179**
Torreya nucifera, 201
4-Tosyl-1,1-dimethylsilacyclohexane, 418
Toxicaric acid, **266**
Toxicarol, 257, 264, 267
α-Toxicarol, **266**
β-Toxicarol, **266**
Toxicarol-isoflavone, **221**
3,5,7-Triacetoxybenzofuran, 157
2,4,6-Triacetoxychalcone, 278
5,7,4′-Triacetoxyisoflavone, 212
5,7,4′-Triacetoxy-4-styrylcoumarin, 280
O-Triacetylbrazilane, 442
O-Triacetylbrazilone, 442
Triacetylfarrerol, 283
2,4,6-Trialkylpyrylium salts, 7
Triarylmethane dyes, 312
Triarylpropanes, 232
2,4,6-Triarylpyrylium salts, 7
Tribromofluorans, 315
Tricetin, **187**
Trichloroacetic ester, 374
3,4,4-Trichloro-4*H*-chromene, 53
3,4,6-Trichlorocoumarins, 110
2,4,7-Trichloro-1,6-dihydroxy-3-methoxy-8-methylxanthone, 331
2,3,3-Trichloroflavanone, 167, 271
Trichlorosilane, 349
2,9,9-Trichlorothioxanthene, 393
Tricin, **186**, 189
2,3,4-Triethoxybenzoic acid, 124
Trifluoroperoxyacetic acid, 69
Trigonella corniculata, 182
2,4,6-Trihydroxyanisole, 274
Trihydroxyanthocyanidins, 86
2,4,5-Trihydroxybenzaldehyde, 122
Trihydroxybenzenes, 315
Trihydroxybenzoic acid, 321
Trihydroxycoumarins, 125
5,7,4′-Trihydroxy-3,8-dimethoxyflavone, 200
5,7,4′-Trihydroxy-8,3-dimethoxyflavone, 187
5,7,4′-Trihydroxy-8,3′-dimethoxyflavonol, 197
5,7,2′-Trihydroxy-6,5′-dimethoxyisoflavone, 217
1,3,8-Trihydroxy-4,5-dimethoxyxanthone, 326
1,3,8-Trihydroxy-4,7-dimethoxyxanthone, 326
5,9,10-Trihydroxy-2,2-dimethyl-12-(1,1-dimethylallyl)-2*H*,6*H*-pyran[3,2-*b*]-xanthen-6-one, 329
5,7,4′-Trihydroxy-6,8-dimethylflavanone, 283
5,2′,4′-Trihydroxy-8,8-dimethyl-10-prenyl-8*H*-pyrano[3,2-*g*]isoflavone, 221
5,9,10-Trihydroxy-2,2-dimethyl-2*H*,6*H*-pyrano[3,2-*b*]xanthen-6-one, 328
7,3′,4′-Trihydroxyflavan, 232
Trihydroxy-2,3-flavan-3,4-diol, 96
5,7,4′-Trihydroxyflavan-3-ol, 243
7,3′,4′-Trihydroxyflavanol, 191, 244
3,5,7-Trihydroxyflavanone, 284
3,7,4′-Trihydroxyflavanone, 282
5,7,4′-Trihydroxyflavanone, 274, 279
7,3′,4′-Trihydroxyflavanone, 278
Trihydroxyflavone, 170
3,5,7-Trihydroxyflavone, 178
5,6,7-Trihydroxyflavones, 176, 179

5,7,4′-Trihydroxyflavone, 180
5,7,8-Trihydroxyflavone, 175, 176, 180
5,7,2′-Trihydroxyflavonol, 189
5,7,4′-Trihydroxyflavonol, 190
6,7,4′-Trihydroxyflavonol 4′-*O*-β-rhamnoside, 191
3,5,7-Trihydroxyflavylium system, 82
5,7,4′-Trihydroxyflavylium chloride, 80, 89
2,4,6-Trihydroxyheptane-1,7-dioic acid, 26
2′,4′,5′-Trihydroxyisoflavones, 210, 255
5,6,7-Trihydroxyisoflavone, 210
5,7,4′-Trihydroxyisoflavone, 214
1,3,6-Trihydroxy-7-methoxy-2,8-diphenylxanthone, 330
3,5,4′-Trihydroxy-7-methoxyflavanone, 276
5,7,3′-Trihydroxy-4′-methoxyflavanone, 279
5,6,8-Trihydroxy-4′-methoxyflavone, 175
5,7,3′-Trihydroxy-4′-methoxyflavone-7-rhamnoglucoside, 184
5,7,4′-Trihydroxy-6-methoxyflavonol, 190
6,7,4′-Trihydroxy-5-methoxyflavonol, 191
5,7,2′-Trihydroxy-4′-methoxyisoflavanone, 286
5,3′,4′-Trihydroxy-7-methoxyisoflavone, 216
5,7,4′-Trihydroxy-8-methoxyisoflavone, 216
5,7,8-Trihydroxy-3-methoxy-2-methylchromone, 146
5,3′,4′-Trihydroxy-3-methoxy-7,8-methylenedioxyflavone, 200
2,3,7-Trihydroxy-9-methoxy-4*a*-methyltetrahydrodibenzo[*b,d*]pyran-6-one, 335
3,4,8-Trihydroxy-2-methoxy-1-prenylxanthone, 327
1,3,7-Trihydroxy-5-methoxyxanthone, 326
1,3,8-Trihydroxy-5-methoxyxanthone, 326
1,5,8-Trihydroxy-3-methoxyxanthone, 326
1,7,8-Trihydroxy-3-methoxyxanthone, 326, 327
1,2,5-Trihydroxy-4-[2′-(2″-methyl-3″-butenyl)]xanthone, 330
3,7,9-Trihydroxy-1-methyldibenzo[*b,d*]-pyran-6-one, 335
3,5,7-Trihydroxy-6-methylflavanone, 283
5,7,4′-Trihydroxy-6-methylflavone, 183
5,7,4′-Trihydroxy-8-methylflavone, 183
5,6,7-Trihydroxy-2-methylisoflavone, 212
1,4,8-Trihydroxy-3-methylxanthone, 325
3,4,5-Trihydroxyphenylacetic acid, 231
2,3,4-Trihydroxypyridine, 32
1,2,3-Trihydroxy-4-pyridone, 32
5,7,4′-Trihydroxy-3-(5,7,3′,4′-tetrahydroxy-8-flavonyl)flavanone, 202
5,3′,5′-Trihydroxy-3,6,7,4′-tetramethoxyflavone, 199
5,6,3′-Trihydroxy-3,7,4′,5′-tetramethoxyflavone, 199
5,7,3′-Trihydroxy-3,8,4′,5′-tetramethoxyflavone, 201
5,7,4′-Trihydroxy-3-(5,7,4′-trihydroxy-8-flavonyl)flavanone, 202
Trihydroxytrimethoxybiflavonyls, 203
3,5,4′-Trihydroxy-6,7,3′-trimethoxyflavone, 199
5,2′,3′-Trihydroxy-3,7,4′-trimethoxyflavone, 199
5,2′,5′-Trihydroxy-3,7,4′-trimethoxyflavone, 195
5,3′,4′-Trihydroxy-3,7,8-trimethoxyflavone, 197
5,6,4′-Trihydroxy-3,7,3′-trimethoxyflavone, 199
5,7,3′-Trihydroxy-3,6,4′-trimethoxyflavone, 197
5,7,3′-Trihydroxy-3,8,3′-trimethoxyflavone, 201
5,7,3′-Trihydroxy-3,8,4′-trimethoxyflavone, 197
5,7,4′-Trihydroxy-6,8,3′-trimethoxyflavone, 188
5,7,3′-Trihydroxy-6,8,4′-trimethoxyflavonol, 197
5,7,4′-Trihydroxy-6,8,3′-trimethoxyflavonol, 197
3,5,4′-Trihydroxy-7,3′,5′-trimethoxyflavylium chloride, 93
5,7,3′-Trihydroxy-6,4′,5′-trimethoxyisoflavone, 217
5,7,3′-Trihydroxy-8,4′,5′-trimethoxyisoflavone, 217
4,7,8-Trihydroxy-3-(2,4,5-trimethoxyphenyl)coumarin, 115
Trihydroxyxanthones, 329

1,2,7-Trihydroxyxanthone, 323
1,3,7-Trihydroxyxanthone, 325
Triketones, 22, 28, 35
1,2,4-Trimethoxybenzene, 63
2,4,5-Trimethoxybenzoic acid, 261
3,4,5-Trimethoxybenzoylacetic ester, 163
2,3,4-Trimethoxybenzyl cyanide, 219
Trimethoxy-α-brazanquinone, 432
3,5,7-Trimethoxycoumarin, 84
5,7,4′-Trimethoxyflavan, 233
Trimethoxy-2,3-flavan-3-ol, 95
5,7,4′-Trimethoxyflavan-4-ols, 248
7,3′,4′-Trimethoxyflavan-4β-ols, 248
5,6,7-Trimethoxyflavanones, 180, 279
5,7,8-Trimethoxyflavanone, 279
4′,5,7-Trimethoxyflavone, 62
5,7,8-Trimethoxyflavone, 180
2′,3′,4′-Trimethoxyfuro[3,2-*g*]isoflavone, 219, 288
2′,4′,5′-Trimethoxyfuro[3,2-*g*]isoflavone, 220
2,4,6-Trimethoxyiodobenzene, 203
5,6,7-Trimethoxyisocoumarin-3-acetic acid, 296
5,6,7-Trimethoxyisocoumarin-3-carboxylic acid, 296
7,2′,4′-Trimethoxyisoflavan, 234
Trimethoxyisoflavones, 212
5,7,8-Trimethoxyisoflavone, 211, 212
7,3′,4′-Trimethoxyisoflavone, 213
3,7,5′-Trimethoxy-3′,4′-methylenedioxyflavone, 196
Trimethoxy-4′,5′-methylenedioxyisoflavone, 217, 218
1,2,4-Trimethoxy-6,7-methylenedioxyxanthone, 327
1,3,4-Trimethoxy-6,7-methylenedioxyxanthone, 327
1,2,3-Trimethoxy-6,7-methylenexanthone, 327
5,7,4′-Trimethoxy-6-methylflavone, 183
5,7,4′-Trimethoxy-8-methylflavone, 183
5,7,8-Trimethoxy-2-methylisoflavone, 212
2,4,5-Trimethoxyphenylacetic acid, 261
2,4,5-Trimethoxy-β-phenylcinnamic acid, 123
1-(2,4,6-Trimethoxyphenyl-3-(3,4-dimethoxyphenyl)propane, 238
1-(2,4,6-Trimethoxyphenyl)-3-(3,5-dimethoxyphenyl)propane, 243
1-(2,4,6-Trimethoxyphenyl)-3-(4-methoxyphenyl)propane, 233
2,4,6-Tri(4-methoxyphenyl)pyrylium tetrafluoroborate, 9
4,5,6-Trimethoxyphthalide-3-carboxylic acid, 296
3,5,9-Trimethoxypsoralene, 130
Trimethoxyxanthones, 329
Trimethyl-1,4-benzoquinone, 305
2,2,4-Trimethyl-2*H*-benzo[*b*]thiopyran, 369
O-Trimethylbrazilanes, 441
Trimethylbrazilein, 436, 437, 438, 441, 442
O-Trimethylbrazilin, 429, 431
Trimethylbrazilone, 431, 432, 433, 434, 440, 441, 442, 445
O-Trimethylbrazylium bromide, 438
O-Trimethylbrazylium ferrichloride, 438, 440, 442
O-Trimethylbrazylium hydrogen sulphate, 438
2,4,4-Trimethylchroman-2-ol, 54, 235
2,2,3-Trimethyl-2*H*-chromene, 58
2,4,4-Trimethylchromenes, 54, 235
Trimethylchromones, physical properties, 150
2,6,6-Trimethyldibenzo[*b,d*]pyran, 334
Trimethyldihydrobrazileinol, 436, 437, 438
Trimethyl-5,7-dioxo-5,6,7,8-tetrahydrocoumarin, 122
Trimethylhydroquinone, 228, 230
Trimethylindanyl hydroperoxide, 54, 235
2,3,9-Trimethylnaphtho[1.8.*bc*]thiopyran, 399
Trimethylnitrocoumarins, 109
2,4,6-Trimethyl-4*H*-pyranol, 25
1,2,6-Trimethyl-4-pyridone, 21
3,5,6-Trimethyl-2-pyrone, 17
2,4,6-Trimethylpyrylium perchlorate, 7, 9, 25
Trimethylselenochromones, 407, 408
Trimethylsulphonium iodide, 359
2,2,4-Trimethyltetrahydrothiopyran-4-ol, 365
4,6,8-Trimethylthiochroman-4-ol, 382
Trimethylthiochromones, 373
Trimethyl-4-thionochromones, 149

2,7,9-Trimethylthioxanthene, 389
5,7,8-Trimethyltocol, 229
5,7,8-Trimethyltocotrineol, 230
Tri-*O*-methylwedelolactone, 177
2,4,6-Trinitroresorcinol, 94
2,4,7-Trinitroxanthone, 320
2,3,4-Trioxochroman 3-arylhydrazones, 113
2,3,4-Trioxochroman 4-nitrophenyl-hydrazone, 116
Trioxymethylene, 113, 297, 301
2,3,4-Triphenylbenzopyrylium ferrichloride, 75
2,2,4-Triphenyl-2*H*-chromene, 59
2,3,4-Triphenylchromen-2-ol, 68, 76
5,3′,5′-Triphenylliquiritigenin, 278
Triphenylmethane, 351
Triphenylmethanol, 312
Triphenylmethyl chloride, 34, 370
Triphenylmethylmagnesium bromide, 318
Triphenylmethyl perchlorate, 5, 72, 144, 309, 339, 369, 373, 389, 403, 406
2,3,6-Triphenyl-4-methylpyrylium perchlorate, 9
Triphenylmethyl sulphate, 311
Triphenylmethyl tetrafluoroborate, 353
2,4,6-Triphenylphenol, 2
Triphenylphosphine, 393
2,4,6-Triphenyl-2-pyran, 3
2,4,6-Triphenyl-2*H*-pyranol, 3
3,5,6-Triphenyl-2*H*-pyran-2-one, 8
2,4,6-Triphenylpyridine, 4
4,5,6-Triphenyl-2-pyrone, 12, 14
2,4,6-Triphenylpyrylium chloride, 3
2,4,6-Triphenylpyrylium ferrichloride, 4
2,4,6-Triphenylpyrylium oxide, 8
2,4,6-Triphenylpyrylium perchlorate, 8, 9
2,4,6-Triphenyl-4*H*-thiopyran, 353
2,4,6-Triphenylthiopyrylium ions, 352
2,4,6-Triphenylthiopyrylium perchlorate, 8, 353
Tris(flavonyloxy)methanes, 174
Tris(nitrophenyl)pyrylium perchlorate, 8
Triticum diococcum, 187
Tropinone methiodide, 397
Tsuga canadensis, 245
Tuba, 257
Tubaic acid, **259**, 261
Tubatoxin, 257
Typha augustata, 194

Ubichromenol, **67**
Ubiquinone, 67
Ullmann reaction, 109, 202
Ulopterol, **119**
Umbelliferae sp., 97
Umbellifera resins, 117
Umbelliferone, 111, 116, **117**
–, naturally occurring ethers, 118
Umbelliferone-8-carbaldehyde, 132
Umbelliferone-6-carboxylic acid, 117
Umbelliferone-8-carboxylic acid, 127
Umbelliferone farnesyl ether, 117
Umbelliferono(7-hydroxycoumarin), 127
Umbelliprenin, **117**
Uncaria Gambier, 238, 245
Uranine A., 315
Urea, 158, 325

Vaccinium vitis idaea, 91
Vakerin, **296**
Valeric ester, 49
Vanillic acid, 92
o-Vanillin, 98, 119
Velutin, **186**
Veratric acid, 238, 246
Veratrylidene-4-ethoxy-2-hydroxyacetophenone, 441
Veratrylidene-7-methoxychromanone, 440
Veratrylidenepaeanol, 440
Verbenalin, **47**
Verbena officianalis, 47
Verecundin, **278**
Verreira spectabilis, 215
Versicolin, **3**
Vertiaflavone, **181**
Vestitol, **234**
Vicin, 91
Vicin chloride 1 and 11, **91**
Vilsmeier–Haack reaction, 340
o-Vinylbenzophenone, 299
Vinylcoumaranone, 264
Vinyl cyanide, 34, 253
Vinylmagnesium bromide, 144, 254
Violanin, 81, 87
Violanin chloride, **92**
Viola tricolor, 81, 92
Visammin, 156
Visamminol, 156, **158**
Viscum album, 184

Visnadin, **137**
Visnagidin, 156
Visnagin, 156, **158**, 180
Visnaginone, 158, 206
Vitamin A, 226
Vitamin D, 226
Vitamins E, 226
Vitamin K, 116
Vitex agnus-castus, 197
Vitexin, **182**
Vitex littoralis, 182
Vitis vinifera, 193
Vogeletin, **191**
Vogeletin-4′-*O*-β-rhamnoside, 191
Vogelin, **191**
Volkensiflavone, **202**

Wagner–Meerwein rearrangement, 242
Wanzlick oxidative coupling reaction, 112
Warangalone, **220**
Warfarin, **115**, 116
Wessely–Moser rearrangement, 146, 152, 155, 198, 202, 323
Wharangin, **200**
Wheat germ oil, 226
Wightin, **187**
Wittig reaction, 12, 58
Wogonin, **180**
Wolff–Kishner reduction, 304, 381, 385, 388, 399

Xanthanoic acid, 308
Xanthenes, 310
Xanthene, **304**, 305, 306, 307, 318, 320, 389
Xanthene-9-carboxylic acid, 308
Xanthene colouring matters, 310
Xanthene dyes, 312
Xanthene-9-thione, 389
3*H*-Xanthen-3-one, 311
9-Xanthenyl esters, 309
Xanthione, **318**
Xanthium pensylvanicum, 199
Xanthochymusside, **286**
Xanthomicrol, **187**
Xanthones, 307, 308, 316, 317, 393
Xanthone, 20, 304, 305, 306, 310, 311, 317, **318**, 319
Xanthonecarboxylic acids, 307
Xanthophanic ester, **20**
Xanthorhamin, **194**
Xanthorrhoea preissii, 233, 234, 343
Xanthorrhoea reflexa, 343
Xanthorrhoein, **343**
Xanthorrhone, **234**
Xanthotoxin, 126, **128**, 129
Xanthotoxol, **128**, 129, 131
Xanthoxyletin, **135**
Xanthoxylum americanum, 135
Xanthydrols, 307, 308, 310
Xanthydrol, 304, 305, 306, **310**, 311, 317, 318, 319
Xanthydryl chloride, 317
Xanthydryl ethers, 310
Xanthyletin, **134**, 135
Xanthylium perchlorate, 317
Xanthylium salts, 304, 306, 310
9-Xanthyllithium, 308
9-Xanthylmalonic acid, 310
9-Xanthylxanthydrol, 306
m-5-Xylenol, 7
p-Xylenol, 230
6-*O*-β-D-Xylopyranosylorientin, 184
Xylose, conversion to kojic acid, 33

Yangonin, **19**
Yerbabuena, 187
Yuen-hua, 182

Zanthoxylum sp., 279
Zanthoxylum acanthopodium, 198
Zanthoxylum nitidum, 184
Zanthoxylum setosum, 123
Zapotin, **185**
Zapotinin, **185**